Molecular Spectroscopy and Quantum Dynamics

Molecular Spectroscopy and Quantum Dynamics

Edited by

Roberto Marquardt
Laboratoire de Chimie Quantique
Institut de Chimie
Université de Strasbourg
Strasbourg, France

Martin Quack
ETH Zürich
Laboratorium für Physikalische Chemie
Zürich, Switzerland

ELSEVIER

Elsevier
3251 Riverport Lane
St. Louis, Missouri 63043

Molecular Spectroscopy and Quantum Dynamics ISBN: 978-0-12-817234-6

Publisher: Oliver Walter
Acquisitions Editor: Seans Simms
Editorial Project Manager: Kelsey Connors
Production Project Manager: Kiruthika Govindaraju
Designer: Alan Studholme

Contents

List of Contributors

Toshiaki Ando, Dr
Department of Chemistry, School of Science, The University of Tokyo, 7-3-1 Hongo, Bunkyo-ku, Tokyo 113-0033, Japan
ando@chem.s.u-tokyo.ac.jp

Thomas Baumert, Prof Dr
University of Kassel, Institute of Physics, 34132 Kassel, Germany
baumert@physik.uni-kassel.de

Tim Bayer, Dr
Carl von Ossietzky University of Oldenburg, Institute of Physics, 26129 Oldenburg, Germany
tim.bayer@uni-oldenburg.de

Denitsa Baykusheva, Dr
Laboratorium für Physikalische Chemie, ETH Zürich, 8093 Zürich, Switzerland
Stanford PULSE Institute, SLAC National Accelerator Laboratory, Menlo Park, CA 94025, United States
denitsab@stanford.edu

Tomislav Begušić, MSc
Laboratory of Theoretical Physical Chemistry, Institut des Sciences et Ingénierie Chimiques, EPF Lausanne, 1015 Lausanne, Switzerland
tomislav.begusic@epfl.ch

Hendrike Braun, Dr
University of Kassel, Institute of Physics, 34132 Kassel, Germany
braun@physik.uni-kassel.de

Lorenz S. Cederbaum, Prof Dr
Theoretische Chemie, Physikalisch-Chemisches Institut, Universität Heidelberg, 69120 Heidelberg, Germany
Lorenz.Cederbaum@pci.uni-heidelberg.de

Attila G. Császár, Prof Dr
MTA-ELTE Complex Chemical Systems Research Group and Laboratory of Molecular Structure and Dynamics, Institute of Chemistry, ELTE Eötvös Loránd University, 1117 Budapest, Hungary
csaszarag@caesar.elte.hu

Marko T. Cvitaš, Dr
Division of Physical Chemistry, Ruđer Bošković Institute, 10000 Zagreb, Croatia
Marko.Cvitas@irb.hr

Csaba Fábri, Sr Sci PhD
MTA-ELTE Complex Chemical Systems Research Group and Laboratory of Molecular Structure and Dynamics, Institute of Chemistry, ELTE Eötvös Loránd University, 1117 Budapest, Hungary
ficsaba@caesar.elte.hu

Kirill Gokhberg, Dr
Theoretische Chemie, Physikalisch-Chemisches Institut, Universität Heidelberg, 69120 Heidelberg, Germany
Kirill.Gokhberg@pci.uni-heidelberg.de

Atsushi Iwasaki, Dr
Department of Chemistry, School of Science, The University of Tokyo, 7-3-1 Hongo, Bunkyo-ku, Tokyo, Japan
aiwasaki@chem.s.u-tokyo.ac.jp

Alexander I. Kuleff, Dr
Theoretische Chemie, Physikalisch-Chemisches Institut, Universität Heidelberg, 69120 Heidelberg, Germany
Alexander.Kuleff@pci.uni-heidelberg.de

Roberto Marquardt, Prof Dr
Laboratoire de Chimie Quantique, Institut de Chimie, Université de Strasbourg, 67081 Strasbourg CEDEX, France
roberto.marquardt@unistra.fr

Martin Quack, Prof Dr Dr hc
Laboratorium für Physikalische Chemie, ETH Zürich, 8093 Zürich, Switzerland
martin@quack.ch

Jeremy O. Richardson, Prof Dr
Laboratorium für Physikalische Chemie, ETH Zürich, 8093 Zürich, Switzerland
jeremy.richardson@phys.chem.ethz.ch

Georg Seyfang, Dr
Laboratorium für Physikalische Chemie, ETH Zürich, 8093 Zürich, Switzerland
seyfang@phys.chem.ethz.ch

Tamás Szidarovszky, Sr Sci PhD
MTA-ELTE Complex Chemical Systems Research Group and Laboratory of Molecular Structure and Dynamics, Institute of Chemistry, ELTE Eötvös Loránd University, 1117 Budapest, Hungary
tamas821@caesar.elte.hu

Jiří Vaníček, Prof Dr
Laboratory of Theoretical Physical Chemistry, Institut des Sciences et Ingénierie Chimiques, EPF Lausanne, 1015 Lausanne, Switzerland
jiri.vanicek@epfl.ch

Matthias Wollenhaupt, Prof Dr
Carl von Ossietzky University of Oldenburg, Institute of Physics, 26129 Oldenburg, Germany
matthias.wollenhaupt@uni-oldenburg.de

Hans Jakob Wörner, Prof Dr
Laboratorium für Physikalische Chemie, ETH Zürich, 8093 Zürich, Switzerland
hansjakob.woerner@phys.chem.ethz.ch

Kaoru Yamanouchi, Prof Dr
Department of Chemistry, School of Science, The University of Tokyo, 7-3-1 Hongo, Bunkyo-ku, Tokyo 113-0033, Japan
kaoru@chem.s.u-tokyo.ac.jp

Preface

Molecular Spectroscopy and Quantum Dynamics: Molecules in Motion

That everything changes is an unescapable fact which from time immemorial has moved poets, exercised metaphysicians and excited the curiosity of natural philosophers.
(C.N. Hinshelwood)

Ever since Max Planck introduced "quanta" in the year 1900 in order to explain the spectral distribution of thermal black body radiation by a "quantum statistical" theory and Bohr's quantum theoretical interpretation of atomic line spectra in 1913, there has been a close relation between spectroscopy and quantum theory. This relation became even closer – one might speak of a fruitful marriage – with the advent of quantum mechanics in 1925. Indeed, quantum mechanics resulted in the discovery of the completely new world of microscopic dynamics, very different from the old world of classical mechanics describing so well the macroscopic dynamics of our daily life including celestial dynamics, which in the old world defined even our quantitative notion of time in terms of hours, days, months, and years. Today quantum dynamics provides an understanding of microscopic phenomena ranging from elementary particle physics to nuclei, atoms, and molecules.

As far as molecular spectroscopy and quantum mechanics are concerned it is probably fair to say that during much of the 20th century the analysis of spectra was dominated by the time-independent "structural" point of view in terms of stationary states, their energies, and wavefunctions. The three classic volumes on "Molecular Spectra and Molecular Structure" published by Gerhard Herzberg between 1939 and 1966 provide beautiful examples for this view with the quantum mechanical analysis of molecular spectra carrying an enormous information content. Further such examples can be found in many other books and the scientific journal literature during these decades. Also the three volumes of the "Handbook of High Resolution Spectroscopy" published in 2011, reporting many great and more recent advances both in theory and experiment, are dominated by a majority of chapters dealing with this stationary state point of view, although time-dependent molecular phenomena are dealt with as well to some extent.

On the other hand, the time-dependent phenomena of molecular kinetics during much of the 20th century were largely understood using the statistical point of view in terms of rate constants, for instance, in the framework of transition state theory, where quantum mechanics plays an important but subordinate, not truly "dynamical" role. This situation changed dramatically during the last decades of the 20th century, driven by the availability of strong, coherent laser radiation allowing for the observation of phenomena such as coherent infrared multiphoton excitation of polyatomic molecules, coherent control, and femtosecond kinetics. The now historical First Conference on "Femtosecond Chemistry" in Berlin 1993 is reflected by a special issue in the Journal of Physical Chemistry (Vol. 97, No. 48, pp. 12423–12649) and the two monumental volumes on "Femtosecond Chemistry" edited by Jörn Manz and Ludger Wöste (VCH, Weinheim 1995). While in Chapter 1 of these volumes the Lord George Porter as a pioneer in the field starting with microsecond to nanosecond kinetics claimed (on page 3) that with the "femtosecond timescale ... chemists are near the end of the race against time" the outlook of the last Chapter 27 by another author (on page 781) stated that "considering the possible time scales of molecular processes, it becomes clear that femtosecond (fs) resolution can only be a short time intermediate level in research and our goal must be to approach the yoctoseconds (10^{-9} fs) and beyond". Possibilities for this were outlined establishing the relation between molecular dynamics and the symmetries of high energy physics.

Indeed, it turned out that with the year 2000 vigorous developments in molecular spectroscopy and quantum dynamics on the attosecond time scale became a reality. Truly time-dependent dynamics with "Molecules in Motion" was the theme of a summarizing article in 2001 and recently also of the COST action MOLIM (2014–2019) combining efforts from numerous laboratories in many countries. It thus seemed timely to provide by 2020 a book summarizing some of these recent advances. No single author today can claim adequate expertise of the diverse fields related to these advances and it was therefore the strategy of the present book to collect contributions from leading authors in the field covering theory as well as experiment. Chapter 1 provides an introductory survey of the theoretical foundations by the editors of the book, starting from the basic concepts and dealing with some of the essential theoretical methods of treating time dependent quantum dynamics, including also a discussion of the important role of symmetries. Chapter 2 by Császár, Fábri, and Szidarovszky presents exact numerical methods for stationary state molecular quantum mechanics of polyatomic molecules. This provides the basis for an exact analysis of molecular spectra as also a starting point for many of the approaches towards time-dependent molecular quantum dynamics.

Chapter 3 by Braun, Bayer, Wollenhaupt, and Baumert reports on 2-Dimensional Strong Field Spectroscopy as applied to ultrafast phenomena in electronic dynamics and control schemes for molecules. Chapter 4 by Baykusheva and Wörner provides an overview of the state-of-the-art of experiments in attosecond molecular spectroscopy and dynamics and their theoretical description, including also photoionization and the dynamics of ions and further applications.

Chapter 5 by Gokhberg, Kuleff, and Cederbaum outlines the theoretical description of electronic decay cascades and interatomic Coulombic decay processes in chemical environments after excitation with high energy photons. Chapter 6 by Vaníček and Begušić provides, on the other hand, the theory of vibrationally resolved electronic spectra of polyatomic molecules by means of ab initio semiclassical methods with thawed Gaussians. Chapter 7 by Quack and Seyfang aims at an overview over Atomic and Molecular Tunnelling Processes in Chemistry, one of the central quantum effects in molecular dynamics, where the motion of "heavy" particles are involved (i.e., atoms or nuclei from protons, perhaps also muons or Muonium, to heavy atoms, but not electrons).

Chapter 8 by Ando, Iwasaki, and Yamanouchi demonstrates in beautiful experiments how Ultrafast Femtosecond Dynamics and the High Resolution Spectroscopy of Molecular Cations can be connected. Chapter 9 by Cvitaš and Richardson finally reports results on the quantum dynamics of water clusters as central systems in chemistry. This forms the basis for our spectroscopic and quantum dynamical understanding of the "liquid of life", for which many theoretical and experimental advances have been made in recent years.

To conclude we mention a further aspect of the timeliness of quantum dynamics today: The year 2019 has seen the introduction of important changes to the International System of Units, the SI (Système International). For the first time in the history of mankind, the units of measurements in science as in daily life are based on fundamental natural constants, including the quantum of action h. This concludes finally a development, which started with the atomic Cesium clock as standard (accepted at the 13ième Conférence Générale des Poids et Mesures, 1967) defining the second (as time unit s) through an atomic motion based on a hyperfine structure interval in the ground state of Cs, whereas formerly the second had been defined by an astronomical time interval with the planetary motion of the earth as an appropriate fraction of the tropical year 1900 (31 556 925.9747 s). Later the meter (m) as a unit of length was defined using a definition of the universal speed of light in vacuo, c, and the distance traveled in 1 s. Finally, according to the resolutions of the 26th Conference of Weights and Measures in Paris (2019) the unit for electric current was defined by fixing the value of the elementary charge (e) by definition, and the unit of mass, the kilogram kg, by fixing the Planck constant to a defined value. There is really a spectroscopic idea behind this. Given the relation of the meter and the second, and the definition of the latter through the Cs atom period or its inverse, the frequency ν, the mass m can be related to the frequency via the fundamental equations

$$\nu = \Delta m\, c^2 / h,$$

or

$$\Delta m = h\,\nu / c^2.$$

One can thereby obtain a definition of the kg, because one has for the unit of action $1\ \mathrm{J\,s} = 1\ \mathrm{m^2\,kg/s}$, and m and s are already defined. While the experiment to realize such a definition is more complicated, one might say that the relativistic mass of the photon emitted in the hyperfine transition of Cs, or equivalently the mass difference between the Cs atom in the lower state and in the upper state of the transition, is an appropriate fraction of the kg (through $\Delta m \approx 6.777265\ 10^{-41}$ kg), thereby defining the kilogram by "spectroscopy and quantum dynamics". In this sense our current century has become the true quantum century relating the microscopic and macroscopic quantities in terms of their units. These relations might perhaps remain in use for the millennium, in principle.

Thus to complement our preface, we add for the convenience of the readers a brief summary of the new "quantum dynamical" SI and a table of the new values for the fundamental constants, and further constants useful for molecular quantum dynamics and spectroscopy.

We should also conclude with our thanks to the authors contributing to this volume and many colleagues who gave us advice and support, too numerous to mention all of them individually, but they can be found cited in the references of individual chapters, and we give our particular thanks to Frédéric Merkt and Jürgen Stohner, and last but not least also to Regina and Roswitha.

Roberto Marquardt and Martin Quack

Strasbourg and Zurich
July 2020

Summary of the SI (excerpt from the SI Brochure, "The International System of Units (SI)", Bureau International des Poids et Mesures, 9th edition, 2019)

The SI defines all base units by means of fixed, defined values of certain natural constants.

TABLE 1
The seven defining constants and the corresponding units they define.

Defining constant	Symbol	Numerical value	Unit
hyperfine transition frequency of Cs	$\Delta\nu_{Cs}$	9 192 631 770	Hz
speed of light in vacuum	c	299 792 458	$m\,s^{-1}$
Planck constant	h	$6.626\,070\,15 \times 10^{-34}$	J s
elementary charge	e	$1.602\,176\,634 \times 10^{-19}$	C
Boltzmann constant	k	$1.380\,649 \times 10^{-23}$	$J\,K^{-1}$
Avogadro constant	N_A	$6.022\,140\,76 \times 10^{23}$	mol^{-1}
luminous efficacy	K_{cd}	683	$lm\,W^{-1}$

TABLE 2
The definition of the base units.

The second (s) $1\ s = 1\ Hz^{-1}$	$1\ Hz = \dfrac{\Delta\nu_{Cs}}{9\,192\,631\,770} \quad 1\ s = \dfrac{9\,192\,631\,770}{\Delta\nu_{Cs}}$
The meter (m)	$1\ m = \dfrac{c\ s}{299\,792\,458} = \dfrac{9\,192\,631\,770}{299\,792\,458}\dfrac{c}{\Delta\nu_{Cs}}$
The kilogram (kg)	$1\ kg = \dfrac{h\ s\,m^{-2}}{6.626\,070\,15 \times 10^{-34}} = \dfrac{(299\,792\,458)^2}{6.626\,070\,15 \times 10^{-34} \times 9\,192\,631\,770}\dfrac{h\,\Delta\nu_{Cs}}{c^2}$
The ampere (A)	$1\ A = \dfrac{e\ s^{-1}}{1.602\,176\,634 \times 10^{-19}} = \dfrac{1}{1.602\,176\,634 \times 10^{-19} \times 9\,192\,631\,770}\Delta\nu_{Cs}\,e$
The kelvin (K)	$1\ K = \dfrac{1.380\,649 \times 10^{-23}\ kg\,m^2\,s^{-2}}{k} = \dfrac{1.380\,649 \times 10^{-23}}{6.626\,070\,15 \times 10^{-34} \times 9\,192\,631\,770}\dfrac{h\,\Delta\nu_{Cs}}{k}$
The mole (mol)	$1\ mol = \dfrac{6.022\,140\,76 \times 10^{23}}{N_A}$
The candela (cd)[a]	$1\ cd = \dfrac{K_{cd}\ kg\,m^2\,s^{-3}\,sr^{-1}}{683} = \dfrac{1}{6.626\,070\,15 \times 10^{-34} \times (9\,192\,631\,770)^2 \times 683}h\,\Delta\nu_{Cs}^2\,K_{cd}\ sr^{-1}$

[a] The definition of the candela implies a definition of the numerical value of the luminous efficacy K_{cd}, see the SI Brochure for details.

TABLE 3
SI prefixes for decimal multiples and submultiples of SI units.

Factor	Name	Symbol	Factor	Name	Symbol
10^1	deca	da	10^{-1}	deci	d
10^2	hecto	h	10^{-2}	centi	c
10^3	kilo	k	10^{-3}	milli	m
10^6	mega	M	10^{-6}	micro	μ
10^9	giga	G	10^{-9}	nano	n
10^{12}	tera	T	10^{-12}	pico	p
10^{15}	peta	P	10^{-15}	femto	f
10^{18}	exa	E	10^{-18}	atto	a
10^{21}	zetta	Z	10^{-21}	zepto	z
10^{24}	yotta	Y	10^{-24}	yocto	y

TABLE 4
Useful physical constants[a].

Quantity	Symbol	Value
magnetic constant	μ_0	$1.256\ 637\ 062\ 12(19) \times 10^{-6}$ N A^{-2}
electric constant	$\varepsilon_0 = 1/\mu_0 c_0{}^2$	$8.854\ 187\ 8128(13) \times 10^{-12}$ F m^{-1}
electron mass	m_e	$9.109\ 383\ 7015(28) \times 10^{-31}$ kg
proton mass	m_p	$1.672\ 621\ 923\ 69(51) \times 10^{-27}$ kg
neutron mass	m_n	$1.674\ 927\ 498\ 04(95) \times 10^{-27}$ kg
atomic mass constant	$m_u = 1$ u $= 1$ Da	$1.660\ 539\ 066\ 60(50) \times 10^{-27}$ kg
molar mass constant	M_u	$0.999\ 999\ 999\ 65(30) \times 10^{-3}$ kg mol^{-1}
Faraday constant	$F = N_A e$	$9.648\ 533\ 212\ldots \times 10^4$ C mol^{-1} (exact)
molar gas constant	$R = N_A k$	$8.314\ 462\ 618\ldots$ J K^{-1} mol^{-1} (exact)
zero of the Celsius scale		273.15 K (defined)
molar volume of ideal gas, $p = 100$ kPa, $t = 0\,°$C	V_m	$22.710\ 954\ 64\ldots$ dm^3 mol^{-1} (exact)
fine-structure constant	$\alpha = \mu_0 e^2 c_0/2h$	$7.297\ 352\ 5693(11) \times 10^{-3}$
	α^{-1}	137.035 999 084(21)
Bohr radius	$a_0 = \varepsilon_0 h^2/\pi m_e e^2$	$5.291\ 772\ 109\ 03(80) \times 10^{-11}$ m
Hartree energy	$E_h = h^2/4\pi^2 m_e a_0{}^2$	$4.359\ 744\ 722\ 2071(85) \times 10^{-18}$ J
Rydberg constant	$R_\infty = E_h/2hc_0$	$1.097\ 373\ 156\ 8160(21) \times 10^7$ m^{-1}
electron volt	eV	$1.602\ 176\ 634 \times 10^{-19}$ J (exact)
atomic unit of time	$h/(2\pi\ E_h)$	$2.418\ 884\ 326\ 5857(47) \times 10^{-17}$ s
Bohr magneton	$\mu_B = eh/4\pi m_e$	$9.274\ 010\ 0783(28) \times 10^{-24}$ J T^{-1}
electron magnetic moment	μ_e	$-9.284\ 764\ 7043(28) \times 10^{-24}$ J T^{-1}
Landé g-factor for the free electron	$g_e = 2\mu_e/\mu_B$	$-2.002\ 319\ 304\ 362\ 56(35)$
nuclear magneton	$\mu_N = eh/4\pi m_p$	$5.050\ 783\ 7461(15) \times 10^{-27}$ J T^{-1}
Stefan–Boltzmann constant	$\sigma = 2\pi^5 k^4/15h^3 c_0{}^2$	$5.670\ 374\ 419\ldots \times 10^{-8}$ W m^{-2} K^{-4} (exact)
first radiation constant	$c_1 = 2\pi hc_0{}^2$	$3.741\ 771\ 852\ldots \times 10^{-16}$ W m^2 (exact)
second radiation constant	$c_2 = hc_0/k$	$1.438\ 776\ 877\ldots \times 10^{-2}$ m K (exact)
Newtonian constant of gravitation	G	$6.674\ 30(15) \times 10^{-11}$ m^3 kg^{-1} s^{-2}
standard acceleration of gravity	g_n	9.806 65 m s^{-2} (defined)
Josephson constant	$K_J = 2e/h$	$483\ 597.848\ 4\ldots\ 10^9$ Hz V^{-1} (exact)
von Klitzing constant	$R_K = h/e^2$	$25\ 812.807\ 45\ldots$ Ω (exact)
Fermi coupling constant	G_F	$1.435\ 8510(8) \times 10^{-62}$ J m^3
Weak mixing parameter	$\sin^2(\theta_W)$	0.2229(3)

[a] According to the view of https://physics.nist.gov/cuu/Constants/ as per July 31, 2020. The symbol c_0 is an alternative for the symbol of the speed of light in vacuum, c. The value of the weak mixing parameter $\sin^2(\theta_w)$ depends on the scheme used and upon momentum transfer. In the divisions all symbols to the right of the division sign are implied to be in the denominator, thus a/b c d corresponds to: a/(b c d). Defined constants are given in Table 1. Constants which can be calculated exactly from the defined constants are given with a finite number of digits followed by …, implying more digits than given here. Standard uncertainties are stated for the other constants in parentheses in terms of the last specified digits.

CHAPTER 1

Foundations of Time Dependent Quantum Dynamics of Molecules Under Isolation and in Coherent Electromagnetic Fields

ROBERTO MARQUARDT* • MARTIN QUACK†
*Laboratoire de Chimie Quantique, Institut de Chimie, Université de Strasbourg, Strasbourg, France
†Laboratorium für Physikalische Chemie, ETH Zürich, CH-8093 Zürich, Switzerland

Abstract

We discuss the foundations of molecules in motion as treated by time-dependent quantum dynamics from very short to long time scales. We consider molecules in isolation, as well as under the influence of coherent electromagnetic radiation, as relevant in many current time-dependent spectroscopic experiments.

1.1 INTRODUCTION

Starting with the analysis by Planck of thermal black body radiation using quantization (Planck, 1900a,b) and the photon concept introduced for the understanding of the photoelectric effect (Einstein, 1905), it was particularly Bohr's work on the analysis of atomic spectra, notably the hydrogen atom and the Balmer formula (Balmer, 1885a,b), which led to an early atomic and molecular "quantum" dynamics in relation to spectroscopy (Bohr, 1913a,b,c; Sommerfeld, 1919). In Bohr's picture of quantum dynamics, stationary atomic and molecular states corresponding to quantized energies, say, E_i and E_f, could undergo radiative transitions by emission of monochromatic radiation of frequency ν_{fi} satisfying the Bohr condition with Planck's constant h:[1]

$$|\Delta E_{fi}| = |E_f - E_i| = h\nu_{fi}. \tag{1.1}$$

The corresponding radiative "quantum jump" by emission or absorption of radiation was treated by Einstein quantitatively using statistical concepts (Einstein, 1916a,b, 1917). This was complemented by the more fundamental "quantum mechanics" (Heisenberg, 1925) and "wave mechanics" (Schrödinger, 1926a,b,c,d,e, see also the work of de Broglie, 1926 and Dirac, 1927, 1929), where also the stationary states and transitions between them were central concepts for understanding atomic and molecular spectra and structure (Herzberg, 1945, 1950, 1966). Indeed, high resolution spectroscopy has remained one of the most important tools in understanding atomic and molecular quantum dynamics until today (Merkt and Quack, 2011a,b). It is probably fair to say that during the first half of the 20th century the structural stationary state aspects of spectroscopy were dominant.

In the second half of the 20th century, much driven by the development of the MASER and LASER, the time-dependent aspects of molecular spectroscopy and quantum dynamics have become increasingly important. The dynamics on ever shorter time scales have become accessible experimentally, from microseconds to nanoseconds, to picoseconds and femtoseconds. Today the attosecond (10^{-18} s) time scale is the subject of intense investigations as exemplified by several chapters of the present book. And even the yoctosecond (10^{-24} s) from high energy physics can be shown to be of some relevance for molecular quantum dynamics (Quack, 1994, 1995a,b, 2001, 2006, 2011a,b), as we shall also briefly discuss here in Section 1.2. In parallel to the experimental developments, theoretical approaches were developed for treating explicitly time dependent molecular quantum dynamics, "molecules in motion" (Quack, 2001), which has been also the title

[1] According to Resolution 1 of the 26th Conference of Weights and Measures (International Bureau of Weights and Measures, 2019), as of 20 May 2019, the Planck constant has the fixed value of $6.626\,070\,15 \cdot 10^{-34}$ Js.

Molecular Spectroscopy and Quantum Dynamics. https://doi.org/10.1016/B978-0-12-817234-6.00006-4

TABLE 1.1
Summary of currently known elementary (pointlike) particles with their approximate masses and the charges Q in multiples of the elementary charge[a].

Standard model of particle physics: particles				
Leptons ($S = \hbar/2$, fermions)				
	ν_e	ν_μ	ν_τ	$Q = 0e$
Mass $m/(\text{GeV}\,c^{-2})$	$<2 \times 10^{-9}$	$<2 \times 10^{-4}$	$<2 \times 10^{-2}$	
	e^-	μ^-	τ^-	$Q = -1e$
Mass $m/(\text{GeV}\,c^{-2})$	5×10^{-4}	0.1	1.8	
Quarks ($S = \hbar/2$, fermions)				
	u_1, u_2, u_3	c_1, c_2, c_3	t_1, t_2, t_3	$Q = 2/3\,e$
Mass $m/(\text{GeV}\,c^{-2})$	5×10^{-3}	1.3	174	
	d_1, d_2, d_3	s_1, s_2, s_3	b_1, b_2, b_3	$Q = -1/3\,e$
Mass $m/(\text{GeV}\,c^{-2})$	10^{-2}	0.2	4.3	

[a] After Groom et al., 2000, Perkins, 2000.
The essence of experimental data from high-energy physics can be accounted for by these particles. To each particle, one has an antiparticle of opposite charge (not listed here Schopper, 1999, after Quack and Stohner, 2005, see also Quack, 2006, 2011a and CERN reference cited therein). The recent observations of neutrino oscillations indicate that also $m(\nu_\tau, \nu_\mu)\,c^2 < 2$ eV.

of a most recent transnational and transdisciplinary research effort (COST action "Molecules in Motion").

The goal of the present review is to provide a broad overview of various theoretical aspects and methods of time-dependent molecular quantum dynamics including also some of the foundations of the underlying physics. We shall take here the practical approach to time dependent quantum dynamics, where "time" is simply a parameter to be measured experimentally by some clock (say, an atomic clock) and "measurements" are considered to provide spectroscopically observed quantities. This approach circumvents some problems related to the foundations of time-dependent quantum mechanics. At this point, we thus take the theory as being used like a heuristic model describing and predicting experiments qualitatively and quantitatively (using in essence the "Copenhagen interpretation"). This is not to imply that there are no remaining basic conceptual problems, such as those considered by Bell (2004), Primas (1981), as well as by Fröhlich and Schnubel (2012). We shall return to some of the basic questions in conclusion, but they have no influence on the remainder of the review. We shall start out in Section 1.2 by a brief summary of the current theory of microscopic matter in terms of the standard model of particle physics (SMPP) and time dependent classical and quantum molecular dynamics, with a focus on the time evolution operator approach to time dependent quantum dynamics. In Section 1.3 we discuss in some detail various methods for solving the time-dependent Schrödinger equation (see also Tannor, 2007). Section 1.4 provides a brief discussion of relevant Hamiltonians, Section 1.5 deals with coordinates. In Section 1.6 we treat quite explicitly the time dependent quantum dynamics with excitation by coherent monochromatic radiation. In the concluding Section 1.7 we discuss the role of symmetries, constants of the motion and some related fundamental questions.

1.2 FOUNDATIONS OF MOLECULAR QUANTUM DYNAMICS BETWEEN HIGH ENERGY PHYSICS, CHEMISTRY AND MOLECULAR BIOLOGY

1.2.1 The Standard Model of Particle Physics (SMPP) as a Theory of Microscopic Matter Including the Low Energy Range of Atomic and Molecular Quantum Dynamics

The current theoretical understanding of microscopic matter is summarized in the so-called "Standard Model of Particle Physics". In spite of its modest name, "Model", it is really a fairly comprehensive theory of microscopic matter, particles, and fields, from high energy particle physics to atomic and molecular physics. Microscopic matter is built from elementary particles, which interact by four fundamental forces. These are summarized in Tables 1.1 and 1.2.

Fig. 1.1 summarizes the modern view of the origin of the fundamental interactions as publicized on the web-

TABLE 1.2
Summary of interactions and field particles[a].

Interactions and field particles	Strong	Electromagnetic	Weak	Gravitation
	SU(3)	SU(2) ⊗ U(1)		
Relative Strength	1	1/137	$\approx 10^{-5}$	$\approx 10^{-38}$
Range	0.1–1 fm	(∞)	<0.1 fm	(∞)
Gauge bosons ($S = 1\hbar$) (except graviton)	Gluons g_1–g_8	Photon γ	$W^{\pm}$, Z^0	(Still hypothetical), (graviton G, $S = 2\hbar$)
Mass m/(GeV c^{-2})	$<10^{-2}$	$<10^{-24}$	≈ 80, ≈ 91	
Acting on particles	Hadrons	Charged	Hadrons, leptons	All
Important in	Atomic nucleus	Atoms and molecules	Radioactive β-decay (neutrons), chiral molecules	Sun, planets, spacecraft, etc.

[a] After Quack and Stohner, 2005, see also Quack, 2006, 2011a.

The Forces in Nature			
Type	Intensity of Forces (Decreasing Order)	Binding Particle (Field Quantum)	Important in
Strong Nuclear Force	~ 1	Gluons (no mass)	Atomic Nucleus
Electro-Magnetic Force	$\sim 10^{-3}$	Photons (no mass)	Atoms and Molecules
Weak Nuclear force	$\sim 10^{-5}$	Bosons Z, W^+, W^-, (heavy)	Radioactive β-Decay, Chiral Molecules
Gravitation	$\sim 10^{-38}$	Gravitons (?)	Sun and Planets etc.

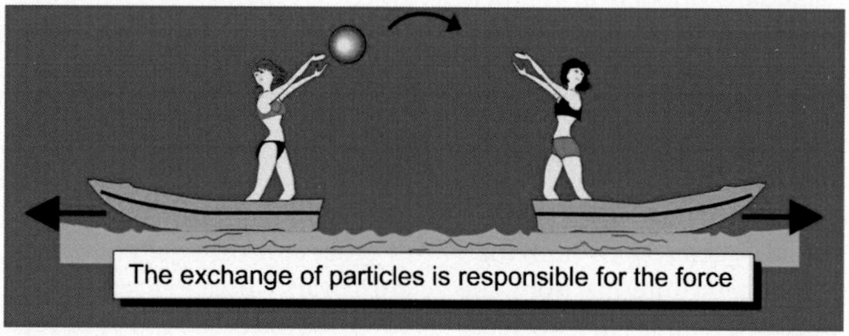

The exchange of particles is responsible for the force

FIG. 1.1 Forces in the standard model of particle physics (SMPP) and important effects. This is taken from the CERN website (CERN, 1992), but the importance of the weak interaction for chiral molecules has been added here from our work following Quack (2006) and by permission of CERN in public domain. We also note (while not mentioned by CERN) that the motif of lightly dressed ladies throwing a ball has been presented in a mosaic at Piazza Armerina, Sicily, 4th Century AD.

site of a large accelerator facility (CERN). According to this view, the electromagnetic force, which is included in the "Dirac-like" ordinary quantum chemistry, leads to the Coulomb repulsion, say, between two electrons in a molecule by means of photons as field particles. In the picture, the two electrons are compared to the ladies on two boats throwing a ball. If we do not see the exchange of the ball, we will observe only the motion of the boats resulting from the transfer of momentum in throwing the ball, and we could interpret this as resulting from a repulsive "force" between the two ladies on the boats. Similarly, we interpret the motion of the electrons resulting from "throwing photons as field particles" as arising from the Coulomb law, which forms the basis of ordinary quantum chemistry. The Coulomb force with the $1/r$ potential energy law is of long range. The other forces arise similarly. The strong force with very short range (0.1 to 1 fm) mediated by the gluons as field particles is important in nuclear physics but has only indirect influence in chemistry by providing the structures of the nuclei, which enter as parameters in chemistry, but there is otherwise usually no need to retain the strong force explicitly in chemistry. The weak force, on the other hand, is mediated by the $W^{\pm}$ and Z^0 bosons as field particles of very high mass (98 dalton for the Z^0 boson, with $m_0 c^2 = 91$ GeV) and short lifetime (0.26 yoctoseconds $= 0.26 \cdot 10^{-24}$ s). This force is thus very weak and of very short range (< 0.1 fm) and one might therefore think that similar to the even weaker gravitational force (mediated by the still hypothetical graviton of spin 2) it should not contribute significantly to the forces between the particles in molecules (nuclei and electrons). Indeed, the weak force, because of its short range, becomes effective in molecules, when the electrons penetrate the nucleus, and then it leads only to a very small perturbation on the molecular dynamics, which ordinarily might be neglected completely. It turns out, however, that because of the different symmetry groups of the electro-magnetic and the electroweak Hamiltonians there arises a fundamentally important, new aspect in the dynamics of chiral molecules, which we therefore have added in our Fig. 1.1 different from the figure from CERN, where this was not originally included.

When applying the standard model of particle physics (SMPP) to the low energy phenomena of atomic and molecular physics, one can do so at several levels of approximation. Firstly, the effects from the fundamental particles and the strong force generating the atomic nuclei are all incorporated in the properties of the specific nucleus, which are its mass, intrinsic angular momentum (usually called "nuclear spin" although it is not a pure spin but has contributions from the orbital motions of the nucleons within the nucleus) parity, magnetic dipole moment, nuclear quadrupole moment, etc. The nuclei as given by these parameters are thus the "effective elementary particles" of atomic and molecular physics, and neither the true elementary particles nor the strong nuclear force mediated by the gluons have to be considered explicitly in the usual approximations. The electrons are retained as elementary particles and interact with the nuclei through the electromagnetic force and the weak force. The gravitational force between electrons and nuclei is sufficiently weak to be neglected except for large assemblies of particles, with a large total mass. The weak nuclear force is frequently neglected, although it can be of importance under special circumstances, particularly in chiral molecules, to which we return in Section 1.7. Usually, quantum chemistry and quantum molecular dynamics retain only the electromagnetic force. One can then introduce further approximations in several steps.

Quantum chemistry in principle treats the quantum dynamics of atoms and molecules by solving the equations of motion for electrons and nuclei to obtain quantum states of atoms (see, for instance, Yamaguchi and Schaefer (2011) as well as Reiher and Wolf (2009)). In molecules one can introduce as a further step the Born–Oppenheimer approximation (or similar "adiabatic" approximations for the electronic structure), which provides effective potentials for the motions of nuclei or "atoms" as effective elementary particles, the dynamics of which is treated in a space of dimension $3N$, where N is the number of atoms, whereas the space of the complete atomic and molecular dynamics would be $3N + 3n$ where n is the often large number of electrons. Cederbaum (2004) presents a particularly clear account of the theory, which is also discussed in more detail in Section 1.4 below. The Born–Oppenheimer potential hypersurfaces defining the forces between atoms are given in a space of dimension $3N - 6(5)$ noting the 3 translational and 3(2) rotational degrees of freedom where the numbers in parentheses apply to linear diatomic molecules. Molecular quantum dynamics can often be treated with these approximations quite successfully in applications to molecular spectroscopy and kinetics (Carrington, 2011; Marquardt and Quack, 2011; Breidung and Thiel, 2011; Tennyson, 2011), see also Chapter 2 of the present book (Császár et al., 2020).

For a wide range of applications one introduces as a further approximation the use of the classical ("Newtonian") equations of motion for the atoms under the influence of the Born–Oppenheimer electronic potentials or other approximate potentials or force fields (Karplus, 2014; van Gunsteren et al., 2006; Car and Parrinello, 1985; Bunker, 1971, 1977; Hase, 1976, 1981, 1998) , see also Chapter 6 (Vaníček and Begušić, 2020).

We shall briefly summarize in the following subsections the foundations of the classical and quantum equations of motion.

1.2.2 Classical Mechanics and Quantum Mechanics

We follow here almost literally the presentation by Merkt and Quack (2011b). Many systems in both classical and quantum mechanics can be described by the motion of interacting point particles, where the physical "particles" are replaced by points of mass m_k with position at the center of mass of the particle. For planetary systems, the "particles" would be the sun and planets with their moons (plus planetoids and artificial satellites, etc.). For atomic and molecular systems the "point particles" can be taken to be the nuclei and electrons to within a very good approximation or the "atoms" within the less good Born–Oppenheimer approximation.

In classical dynamics one describes such an N particle system by a point in the mathematical phase space, which has dimension $6N$ with $3N$ coordinates (for instance, Cartesian coordinates x_k, y_k, z_k for each particle "k") and $3N$ momenta p_{x_k}, p_{y_k}, p_{z_k}. Such a point in phase space moving in time contains all mechanically relevant information of the dynamical system. In the 19th century Hamiltonian formulation of classical mechanics, one writes the Hamiltonian function H as a sum of the kinetic (T) and potential (V) energy,

$$H = T + V, \tag{1.2}$$

in terms of generalized coordinates q_k and their conjugate momenta p_k (Landau and Lifshitz, 1966; Goldstein, 1980; Iro, 2002). Following Hamilton, one obtains the canonical Hamiltonian differential equations of motion accordingly

$$\frac{\mathrm{d}q_k}{\mathrm{d}t} = \dot{q}_k = \left(\frac{\partial H}{\partial p_k}\right), \tag{1.3}$$

$$\frac{\mathrm{d}p_k}{\mathrm{d}t} = \dot{p}_k = -\left(\frac{\partial H}{\partial q_k}\right). \tag{1.4}$$

The dynamics of the classical system is thus obtained from the solution of $6N$ coupled differential equations. Provided that one knows some exact initial condition for one point in phase space, all future and past states of the system in terms of the set $\{q_k(t), p_k(t)\}$ can be calculated exactly. Further considerations arise if the initial state is not known exactly, but we shall not pursue this further.

One approach to quantum dynamics replaces the functions H, p_k, q_k by the corresponding quantum mechanical operators ($\hat{H}$, $\hat{p}_k$, $\hat{q}_k$) or their matrix representations ($\boldsymbol{H}$, $\boldsymbol{p}_k$, $\boldsymbol{q}_k$) resulting in the Heisenberg equations of motion (Heisenberg, 1925; Dirac, 1958):

$$\frac{\mathrm{d}\hat{q}_k}{\mathrm{d}t} = \frac{2\pi}{\mathrm{i}h}\left[\hat{q}_k, \hat{H}\right], \tag{1.5}$$

$$\frac{\mathrm{d}\hat{p}_k}{\mathrm{d}t} = \frac{2\pi}{\mathrm{i}h}\left[\hat{p}_k, \hat{H}\right], \tag{1.6}$$

which involve now Planck's quantum of action (or constant) h, and $\mathrm{i} = \sqrt{-1}$. Following Dirac (1958), these equations are the quantum-mechanical equivalent of the Poisson-bracket formulation of classical mechanics, and one can, in fact, derive the corresponding classical equations of motion from the Heisenberg equations of motion, if one uses quantum mechanics as the more fundamental starting point, as discussed by Sakurai (1985), for instance. Eqs. (1.5) and (1.6) contain the commutator of two operators $\hat{A}$ and $\hat{B}$ in general notation,

$$\left[\hat{A}, \hat{B}\right] = \hat{A}\hat{B} - \hat{B}\hat{A}. \tag{1.7}$$

As quantum mechanical operators and their matrix representations do not in general commute, this introduces a new element into quantum mechanics as compared to classical mechanics. For instance, in Cartesian coordinates the coordinate operator x_k is simply multiplicative, while the momentum operator $\hat{p}_{x_k}$ is given by the differential operator

$$\hat{p}_{x_k} = \frac{h}{2\pi\mathrm{i}}\frac{\partial}{\partial x_k}, \tag{1.8}$$

leading to the commutator

$$\left[\hat{x}_k, \hat{p}_{x_k}\right] = \mathrm{i}h/(2\pi) \tag{1.9}$$

and the corresponding Heisenberg uncertainty relation (Messiah, 1961)

$$\Delta x_k \Delta p_{x_k} \geq h/(4\pi) \tag{1.10}$$

where Δx_k and Δp_{x_k} are defined as the root mean square deviations of the corresponding ideal measurement results for the coordinates x_k and momenta p_{x_k}. Similar equations apply to y_k, z_k with p_{y_k}, p_{z_k}, etc., for all particles labeled by their index k. It is thus impossible in quantum mechanical systems to know experimentally the position of the "point in phase space" to better than allowed by the Heisenberg uncertainty relation in a quantum mechanical state. In classical mechanics, on the other hand, x_k and p_{x_k}, etc., commute, and the point in phase space can be defined and measured with arbitrary accuracy, in principle.

A somewhat more complex reasoning leads to a similar "fourth" uncertainty relation for energy E and time t,

$$\Delta E \Delta t \geq h/(4\pi). \tag{1.11}$$

We note that Eqs. (1.10) and (1.11) are strictly *inequalities*, not equations in the proper sense. Depending on the system considered, the uncertainty can be *larger* than what would be given by the strict equation. If the equality sign in Eqs. (1.10), (1.11) applies, one speaks of a "minimum uncertainty state or wavepacket[2]" (see below). The commutators in Eqs. (1.5), (1.6) are readily obtained from the form of the kinetic energy operator in Cartesian coordinates:

$$\hat{T} = \frac{1}{2}\sum_{k=1}^{N}\left(\frac{\hat{p}_{xk}^2}{m_k} + \frac{\hat{p}_{yk}^2}{m_k} + \frac{\hat{p}_{zk}^2}{m_k}\right) \tag{1.12}$$

and

$$\hat{H} = \hat{T} + \hat{V} \tag{1.13}$$

if the potential energy $\hat{V}$ is a multiplicative function of the coordinates of the particles (for instance, with the Coulomb potential for charged particles).

While this so-called Heisenberg representation of quantum mechanics is of use for some formal aspects and also certain calculations, frequently the "Schrödinger representation" turns out to be useful in spectroscopy and quantum dynamics.

1.2.3 Time Evolution Operator Formulation of Quantum Dynamics

The time dependence of the operators $\hat{p}_k$ and $\hat{q}_k$ in the Heisenberg equations of motion and, indeed, the time dependence of every operator $\hat{Q}$ in the Heisenberg representation is given by Eq. (1.14),

$$\hat{Q}(t) = \hat{U}^{\dagger}(t, t_0)\hat{Q}(t_0)\hat{U}(t, t_0). \tag{1.14}$$

Here t_0 is the initial time and t the time after some evolution. The operator $\hat{U}$ satisfies the differential equation

$$\mathrm{i}\frac{h}{2\pi}\frac{\partial \hat{U}(t, t_0)}{\partial t} = \hat{H}\hat{U}(t, t_0). \tag{1.15}$$

Thus, in general one has to solve this differential equation in order to obtain $\hat{U}(t, t_0)$. If, however, $\hat{H}$ does not depend upon time, $\hat{U}(t, t_0)$ is given by the equation

$$\hat{U}(t, t_0) = \exp\left[-\frac{2\pi\mathrm{i}}{h}\hat{H}\cdot(t - t_0)\right]. \tag{1.16}$$

The exponential function of an operator $\hat{Q}$, as well as that of a matrix representation of this operator, is given by Eq. (1.17),

$$\exp(\hat{Q}) = \sum_{n=0}^{\infty}\frac{\hat{Q}^n}{n!}, \tag{1.17}$$

$\hat{U}(t, t_0)$ thus "propagates" the operators $\hat{p}_k$, $\hat{q}_k$, etc., from time t_0 to time t and is often called "propagator". Also $\hat{U}$ provides the solution for the time-dependent Schrödinger equation for the wave function[2] Ψ,

$$\mathrm{i}\frac{h}{2\pi}\frac{\partial\Psi(x_1, y_1, z_1, \ldots, x_n, y_n, z_n, t)}{\partial t} = \hat{H}\,\Psi(x_1, y_1, z_1, \ldots, x_n, y_n, z_n, t). \tag{1.18}$$

In the Schrödinger formulation of quantum mechanics ("wave mechanics"), one introduces the "wave function" $\Psi(x_1, y_1, z_1, \ldots x_n, y_n, z_n, t)$ depending on the particle coordinates and time, and satisfying the differential equation (time-dependent Schrödinger equation, Eq. (1.18)).

The physical significance of the wave function Ψ (also called state function) can be visualized by the probability density

$$\begin{aligned} P(x_1, y_1, z_1, \ldots, x_n, y_n, z_n, t) &= \Psi(x_1, \ldots, z_n, t)\Psi^*(x_1, \ldots, z_n, t) \\ &= |\Psi(x_1, \ldots, z_n, t)|^2 \end{aligned} \tag{1.19}$$

where P is real, positive or zero, whereas Ψ is, in general, a complex-valued function. Moreover, $P(x_1, y_1, z_1, \ldots, z_n\, t)\,\mathrm{d}x_1\mathrm{d}y_1\mathrm{d}z_1\cdots\mathrm{d}z_n$ gives the probability of finding the quantum mechanical system of point particles in the volume element $(\mathrm{d}x_1\cdots\mathrm{d}z_n)$ at position $(x_1, \ldots, z_n)$ at time t.

The differential operator in Eq. (1.18) is sometimes called energy operator $\hat{E}$,

$$\hat{E} = \mathrm{i}\frac{h}{2\pi}\frac{\partial}{\partial t}, \tag{1.20}$$

thus one can write

$$\hat{E}\,\Psi(r, t) = \hat{H}\,\Psi(r, t), \tag{1.21}$$

where we introduce the convention that r represents in general a complete set of space (and spin) coordinates and includes the special case of systems depending only on one coordinate which then can be called r.

The solution of Eq. (1.18) has the form

$$\Psi(r, t) = \hat{U}(t, t_0)\Psi(r, t_0). \tag{1.22}$$

One of the most important properties of Ψ is that it satisfies the principle of linear superposition. If $\Psi_1(r, t)$

[2] In this book, these terms are written in a single word 'wavepacket', 'wavefunction' or in separate words 'wave packet', 'wave function'; while the separate form is more frequently found, and is also sometimes separated by a hyphen, all forms are currently being used and are accepted.

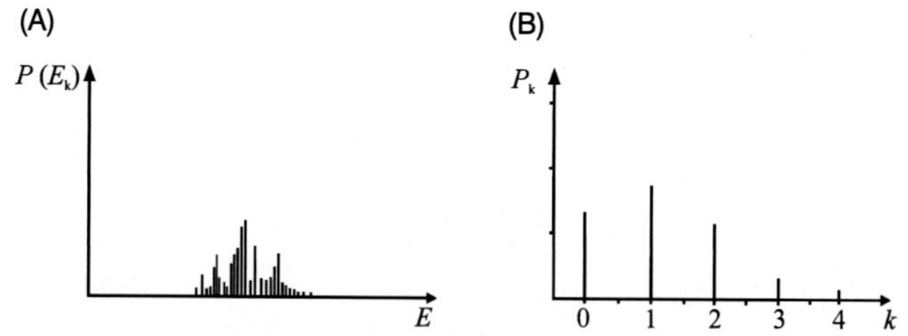

FIG. 1.2 Spectral decomposition schemes: Illustration of spectral decomposition of a time-dependent state where $p_k(E_k) = |c_k|^2$ is the probability of measuring the eigenvalue E_k in the time-dependent state given by $\Psi(r,t)$: (A) irregular spectrum and distribution; (B) harmonic oscillator with a Poisson distribution (after Merkt and Quack, 2011b).

and $\Psi_2(r,t)$ satisfy Eq. (1.18) as possible representations of the dynamical state of the system, then the linear superposition

$$\Psi(r,t) = c_1\Psi_1(r,t) + c_2\Psi_2(r,t) \tag{1.23}$$

is also a possible dynamical state satisfying Eq. (1.18), as is readily shown, given that $\hat{H}$ is a linear operator and c_1, c_2 are complex coefficients. However, $\Psi(r,t)$, in general, is not an eigenstate of $\hat{H}$. In the special case of stationary states leading to the time-independent Schrödinger equation, we assume that $\hat{H}$ does not depend on time. We consider the special case where $\Psi_k(r,t)$ is an eigenfunction of $\hat{H}$ with eigenvalue E_k. Thus

$$\hat{H}\,\Psi_k(r,t) = \hat{E}\Psi_k(r,t) = E_k\Psi_k(r,t). \tag{1.24}$$

The solution for this special case is given by Eq. (1.25),

$$\mathrm{i}\frac{h}{2\pi}\frac{\partial\Psi_k(r,t)}{\partial t} = E_k\Psi_k(r,t) = E_k\psi_k(r)\exp(-2\pi\mathrm{i}\frac{E_k t}{h}), \tag{1.25}$$

where, $\hat{H}$ being independent of time, one can divide both sides in Eq. (1.24) by $\exp(-2\pi\mathrm{i}E_k t/h)$ and obtain

$$\hat{H}\,\psi_k(r) = E_k\psi_k(r). \tag{1.26}$$

The eigenfunctions of $\hat{H}$ are called *stationary states,*

$$\Psi_k(r,t) = \psi_k(r)\exp(-2\pi\mathrm{i}\frac{E_k t}{h}). \tag{1.27}$$

The name for stationary states is related to the time independence of the corresponding probability density

$$P(r,t) = \Psi_k(r,t)\,\Psi_k^*(r,t) = |\Psi_k(r,t)|^2 = |\psi_k(r)|^2. \tag{1.28}$$

The time-independent Schrödinger equation (1.26) is thus derived as a special case from the time-dependent Schrödinger equation.

Making use of the superposition principle (Eq. (1.23)), the general solution of the Schrödinger equation results as follows:

$$\Psi(r,t) = \sum_k c_k\psi_k(r)\exp(-2\pi\mathrm{i}\frac{E_k t}{h}) = \sum_k c_k\Psi_k(r,t). \tag{1.29}$$

If $\hat{H}$ does not depend on time, such as in the case of isolated atomic and molecular systems, the coefficients c_k are time independent, generally complex coefficients. According to the principle of spectral decomposition, the probability of measuring energy E_k in the time-dependent state given by Eq. (1.29) is

$$p_k(E_k) = |c_k|^2 = c_k c_k^*. \tag{1.30}$$

Thus, with time-independent $\hat{H}$, the p_k are independent of time, as is also the expectation value of the energy

$$\langle E(t)\rangle = \sum |c_k|^2 E_k. \tag{1.31}$$

Fig. 1.2 illustrates the spectral decomposition for two types of spectra. The energy in a time-dependent state is therefore not a well-defined quantity but is characterized by a statistical distribution given by p_k in Eq. (1.30). This distribution satisfies the uncertainty relation given by Eq. (1.11). For further discussion and the numerical approaches to realize solutions of the Schrödinger equation, we refer to Section 1.3.

We conclude this section by mentioning the special limiting case of scattering theory and S-matrix theory used therein. Formally, the S-matrix in a collision between two (or more) collision partners can be considered to be a limiting case of the matrix representation of the time evolution operator in the basis of the scattering channels related to the quantum states of the scattering partners at infinite distance "i" before and "f" after the collision), i.e.,

$$S_{fi} = U_{fi}(t = +\infty, t = -\infty). \tag{1.32}$$

For a more detailed introduction of collision and S-matrix theory, we refer to the books of Newton (1966), Clary (1986), and Schatz and Ratner (1993).

1.2.4 Further Approaches to Quantum Mechanics and Molecular Dynamics

The Schrödinger and Heisenberg approaches are certainly the most widely used approaches towards time dependent and time independent quantum dynamics (often introduced as the Schrödinger and Heisenberg "pictures" of quantum mechanics). We shall briefly mention here a few further approaches to molecular quantum dynamics which have found wider use. Apart from the entirely classical molecular dynamics approaches, which we have already mentioned, there are also the so-called "semiclassical" methods of quantum dynamics, which have their historical roots in the "old quantum theory" of Bohr (1913a,b,c). One of these is the Wentzel (1926), Kramers (1926), and Brillouin (1926) (abbreviated WKB) approximation to quantum mechanics, which has found wide use, for instance, also for quantum mechanical tunneling problems, as discussed in Chapters 7 (Quack and Seyfang, 2020) and 9 of this book (Cvitaš and Richardson, 2020). A more recent development is the semiclassical limit quantum mechanics by Miller (1974, 1975b). Another, in principle rigorous approach is the so-called path integral quantum mechanics, which is commonly attributed to Feynman (1948), but has its historical origin in the early work of Gregor Wentzel (1924) (the successor of Schrödinger in Zürich in 1928, see also Freund et al., 2009; Antoci and Liebscher, 1996). Path integral quantum mechanics with its important numerical implementations has been extremely fruitful in recent times as an alternative approach to quantum dynamics, and substantial books have been written on this approach (Feynman and Hibbs, 1965; Kleinert, 2009). Marx and Parrinello (1996), Tuckerman et al. (1996) as well as Chapters 6 (Vaníček and Begušić, 2020) and 9 (Cvitaš and Richardson, 2020) in the present book refer also to path integral methods. Numerical implementations of path integral methods were published in computer code packages (Ceriotti et al., 2010, 2014; Kapil et al., 2019).

Finally, Diffusion Quantum Monte Carlo (DQMC) methods have found much recent application as a rigorous approach to numerically solve the time-independent Schrödinger equation as a first step towards solving then also the time-dependent Schrödinger equation. DQMC follows an idea originally attributed to Fermi (Metropolis and Ulam, 1949) and introduced into quantum chemistry as a numerically practical approach in the algorithmic implementation by Anderson (1975, 1976). DQMC makes use of the interesting isomorphism between the N-body time-dependent Schrödinger equation propagated in an imaginary time equivalent $\tau = 2\pi i t/h$ with dimensions of a reciprocal energy and a $3N$ dimensional transport equation (with diffusion and source/sink terms) in Cartesian coordinate space,

$$\frac{\partial \Psi}{\partial \tau} = -\hat{H}\Psi, \tag{1.33}$$

$$\hat{H} = \hat{T} + \hat{V} = -\sum_{k-1}^{N} \frac{\nabla_k^2}{2m_k} + V. \tag{1.34}$$

Here $\hat{H}$ is a time-independent Hamiltonian with kinetic energy $\hat{T}$ and potential energy $\hat{V}$, having eigenvalues $E_0 \leq E_1 \leq \cdots \leq E_k$ and eigenfunctions Ψ_0, Ψ_1, etc. By numerically simulating a diffusion process as a quasi-statistical process, one can converge towards obtaining the ground state energy and wavefunction, as well as, with appropriate techniques making use of symmetry and nodal properties, also excited state results. The approach is conceptually and numerically interesting as it provides statistical upper *and* lower bounds on the energies E_0 (possibly E_1, etc). It has been used for both electronic structure and vibrational–rotational dynamics (Anderson, 1975, 1976; Reynolds et al., 1982; Ceperley and Alder, 1986; Coker and Watts, 1986; Garmer and Anderson, 1988; Bernu et al., 1990; Quack and Suhm, 1991; Lewerenz and Watts, 1994; Quack and Suhm, 1998; Tanaka et al., 2012). The possibility of simulating the quantum mechanics of a relatively large number of particles, as well as the upper and lower bound property of the solutions, is of interest. Limitations arise in obtaining excited state energies and wavefunctions, although this is possible as well, as discussed in Chapter 7 of this book (Quack and Seyfang, 2020) in applications to tunneling.

1.2.5 Time-Dependent Quantum Statistical Dynamics

When one wishes to consider the time evolution of a physical system, the initial state of which might be characterized by a statistical distribution of a mixture of different "pure quantum states", it is useful to define a density operator given by Eq. (1.35) (Messiah, 1961;

Sakurai, 1985):

$$\hat{\rho}(t) = \sum_n p_n |\Psi_n\rangle \langle\Psi_n| \quad (1.35)$$

satisfying the Liouville–von Neumann equation

$$\mathrm{i}\frac{h}{2\pi}\frac{\mathrm{d}\hat{\rho}(t)}{\mathrm{d}t} = \left[\hat{H}, \hat{\rho}(t)\right] \quad (1.36)$$

with the solution

$$\hat{\rho}(t) = \hat{U}(t, t_0)\hat{\rho}(t_0)\hat{U}^{\dagger}(t, t_0). \quad (1.37)$$

This equation is of particular importance for statistical mechanics.

In many applications one can usefully introduce reduced density matrices, which describe only a subsystem of the total quantum statistical mechanical system, with the understanding that only the knowledge of the time-dependent behavior of this small subsystem is of interest in an experiment. Often one uses then simple kinetic models for the matrix representation of the reduced operator. For instance, for the simple reduced density matrix with just two states, the diagonal elements (P_{11}, P_{22}) describe the time-dependent populations of the two levels and the off-diagonal elements their "coherences" (P_{12}, P_{21}). The time-dependent relaxation of the populations towards equilibrium might be assumed to be exponential with a relaxation time τ_1, whereas the coherences (P_{12}, P_{21}) decay to zero with a relaxation time τ_2. Of course, there is no guarantee that such a simple model will be a good approximation, and there is no need to restrict to just two states. In any case the idea of the reduced density matrix description is to treat a problem of small size (perhaps matrices of the order of 1000), whereas the complete quantum statistical system might have to be described by matrices easily exceeding 10^{1000}. These reduced density matrix approaches are widely used in magnetic resonance (Ernst et al., 1987; Schweiger and Jeschke, 2001), but also more generally (Blum, 1981). In principle, one can also simulate statistical behavior by random ensembles of solutions of the Schrödinger equation (Marquardt and Quack, 1994).

Another approach to simplify the quantum dynamical treatment of large microscopic or "mesoscopic", or even macroscopic systems, by statistical methods goes back to Pauli (1928). Here one starts from the time-dependent Schrödinger equation of the complete system involving a very large number of quantum states but considers only coarse-grained sums (or averages) of individual state populations p_k to derive coarse-grained level populations

$$p_K = \sum_k{}' p_{k(K)} = \sum_{k=x+1}^{x+N_K} b_{k(K)} b^*_{k(K)}. \quad (1.38)$$

By a nontrivial reasoning, which considers the emergence of simple structures for such average (or summed coarse grained) quantities, one obtains Master Equations of low dimension (Quack, 1981, 2014a,b):

$$\frac{\mathrm{d}\boldsymbol{p}(t)}{\mathrm{d}t} = \boldsymbol{K}\,\boldsymbol{p}(t), \quad (1.39)$$

$$\boldsymbol{p}(t) = \boldsymbol{Y}(t, t_0)\,\boldsymbol{p}(t_0), \quad (1.40)$$

$$\boldsymbol{Y}(t, t_0) = \exp[\boldsymbol{K}(t - t_0)]. \quad (1.41)$$

The sums $\sum{}'$ in Eq. (1.38) with $p_{k(K)}$ and $b_{k(K)}$ are implied to be restricted to quantum "states" k, ranging between some counting index $x+1$ and $x+N_k$ and belonging to the "level" K, with N_k being the number of such states, which may possibly be very large. The rate coefficient matrix $\boldsymbol{K}$ has matrix elements K_{MN} which can be derived from quantum-mechanical perturbation theory or by other methods. In recent times classical trajectory calculations (i.e., classical molecular dynamics) have been proposed and used to calculate the "rate coefficient matrix" K_{MN} (Nüske et al., 2014), following otherwise similar lines of thinking as in Quack (1978, 1979, 1981), where a calculation of K_{MN} by means of quantum mechanical perturbation theory or other quantum approaches was implied.

Using a theorem originally due to Frobenius (von Mises, 1931), one can further simplify the mathematics by considering the rapid convergence of the solutions of Eq. (1.39) with only a small number of eigenvalues $(\lambda_1, \lambda_2, \lambda_3, \ldots)$ of $\boldsymbol{K}$. For some of the earlier discussion of these approaches, we refer to Quack (1979, 1981, 1982). We also note that the differential equation (1.39) can be of the "Pauli Master equation" type (Pauli, 1928, case B in Quack, 1978) or of a more general nature (cases A, B, C, D in Quack, 1978), where the "case A" is the well known "Fermi Golden Rule" and is a very special long known case (Wentzel, 1927, 1928) (some historical aspects are discussed by Merkt and Quack, 2011a), which has been rigorously derived for a model of electronic relaxation in large molecules by Bixon and Jortner (1968), and Jortner et al. (1969).

This "Fermi Golden Rule" (case A) can be considered as a "statistical" case because the product state populations are summed following Eq. (1.38). We can note here also that classical molecular dynamics by classical trajectories can be considered to be a statistical approximation to quantum dynamics when averaging over the initial conditions corresponding to a pure quantum state, when the latter is simulated by a statistical distribution in phase space (Quack and Troe, 1981). Although one might assume that statistical averages in classical dynamics might be a better approximation to quantum dynamics than just a straight phase space trajectory, there are, of course, quantum phenomena such as tunneling, which are not "averaged out" by statistical averaging (Quack and Seyfang, 2020, Chapter 7 of this book). Sometimes in classical molecular dynamics simulations of biomolecular systems such as proteins (Karplus, 2014; van Gunsteren et al., 2006), it is argued that, while the motion of the light H-atoms in the protein may well be quantum-like, the motion of the "heavy atom" framework (C, N, O, etc.) of the protein behaves classically. However, simulations of processes involving essential motion of even heavier atoms such as fluorine in the dissociation of the dimer $(HF)_2$ indicate large differences between quantum and classical results (Manca et al., 2008). Sometimes one might consider a combination of classical trajectory calculations for a part of the problem with a quantum statistical theory such as the statistical adiabatic channel model (Quack and Troe, 1981, 1998; Troe et al., 2005; Troe, 2006). Of course, the ultimate quantum statistical limit widely used in reaction kinetics is transition state theory for which various quantum dynamical versions have been formulated, such as the statistical adiabatic channel model (SACM, Quack and Troe, 1974, 1998) or semiclassical and quantum transition state theory (Miller, 1975a, 2014) beyond the original theory for thermal rate constants both in the classical mechanical and quantum mechanical versions (for the historical references see Chapter 7 in the present book, Quack and Seyfang, 2020).

In the debate on the validity of classical dynamics for describing the atomic motions on quantum Born–Oppenheimer potential hypersurfaces, it is often argued that the high degree of averaging in thermal situations justifies the use of classical mechanics. This point of view can be rejected with an argument given by Quack and Troe (1981): If we calculate the forward and backward rates of a thermal reaction by classical dynamics, the ratio of the rate constants results in the classical statistical thermodynamic limit for the equilibrium constant, which is known to be highly inaccurate by comparison with the easily accessible quantum statistical equilibrium constants. Thus the individual rate constants cannot be accurate.

1.3 METHODS FOR SOLVING THE TIME-DEPENDENT SCHRÖDINGER EQUATION

In this section we write the time-dependent Schrödinger equation in the form

$$\mathrm{i}\frac{h}{2\pi}\frac{\partial|\Psi(t)\rangle}{\partial t} = \hat{H}\,|\Psi(t)\rangle \tag{1.42}$$

where $|\Psi(t)\rangle$ represents the time dependent state of the system under investigation. Following the mathematical foundations of quantum mechanics (von Neumann, 2018), states are vectors, for which Dirac's notation is used here and essentially throughout the following section, where methods to solve Eq. (1.42) will be reviewed. The specific form of the molecular Hamiltonian is addressed in Section 1.4, and appropriate choices of coordinates used to describe the position of the particles composing a molecule are discussed in Section 1.5.

To solve Eq. (1.42) in practice, in particular to obtain numerical solutions for it, states and operators are represented in a finite set of states $\mathbb{L}_N = \{|\chi_1\rangle, \ldots, |\chi_N\rangle\}$ that is well defined in advance, and for which a scalar product $\langle\chi_n|\chi_m\rangle$ can also be defined; these states can always be defined to be orthonormal, i.e., $\langle\chi_n|\chi_m\rangle = \delta_{nm}$, where δ_{nm} is the Kronecker symbol (Cohen et al., 2007).

The state $|\Psi(t)\rangle$ is then represented by a time-dependent vector $\boldsymbol{b}(t)$, the so-called *state vector*, the components of which are the projections $b_n(t) = \langle\chi_n|\Psi(t)\rangle$:

$$|\Psi(t)\rangle = \sum_{n=1}^{N} b_n(t)\,|\chi_n\rangle. \tag{1.43}$$

Operators are represented by matrices, i.e., $\hat{H}$ is represented by a matrix $\boldsymbol{H}$; the element H_{nm} is given by the scalar product of $\langle\chi_n|$ with $\hat{H}|\chi_m\rangle$, $H_{nm} = \langle\chi_n|\hat{H}|\chi_m\rangle$. In this representation, Eq. (1.42) becomes

$$\mathrm{i}\frac{h}{2\pi}\frac{\mathrm{d}}{\mathrm{d}t}\boldsymbol{b}(t) = \boldsymbol{H}\,\boldsymbol{b}(t). \tag{1.44}$$

The set $\mathbb{L}_N$ is a subspace of the entire linear space in which the quantum mechanical states exist. Because of the finiteness of N, the representation given by Eqs. (1.43) and (1.44) is normally an approximation of the true physical situation, which can be improved systematically, the larger N is made. The symbols $|\Psi(t)\rangle$ and $\boldsymbol{b}(t)$ denote two different types of vectors: the former is defined in the actual space of quantum states, the latter is defined in the dual space of the linear subspace $\mathbb{L}_N$; mathematically the former is a *covariant* vector, the latter a *contravariant* vector. It is important to note that, while a quantum mechanical state $|\Psi\rangle$ is basis set independent, the state vector is a contravariant vector, and as such dependent on the specific choice made for the basis states used to set up $\mathbb{L}_N$. A more detailed notation for it would therefore be $\boldsymbol{b}^{(\chi)}$; if a different basis of orthonormal states is used, say $\mathbb{L}_N = \{|\eta_1\rangle, \ldots, |\eta_N\rangle\}$, the corresponding state vector would be $\boldsymbol{b}^{(\eta)}$. However, for the sake of simplicity, and when any ambiguity can be discarded, we drop the specific indication to the chosen basis in the notation.

If the system under investigation is isolated, i.e., the Hamiltonian does not depend on time, $\hat{H}(t') = \hat{H}(t'')$, for all t' and t'', its total energy is conserved, and Eq. (1.42) has the special solutions

$$|\Psi_n\rangle(t) = |\psi_n\rangle \exp\left(-2\pi\mathrm{i}\frac{E_n}{h}t\right) \tag{1.45}$$

where the states $|\psi_n\rangle$ are the solutions of the time-independent Schrödinger equation

$$\hat{H}\,|\psi_n\rangle = E_n\,|\psi_n\rangle \tag{1.46}$$

at the specific energies E_n. Mathematically they are the eigenstates of $\hat{H}$ and the E_n are their energies; they describe the *spectroscopic states* of the isolated system, and energy differences $E_n - E_m$ correspond to potentially observable spectral lines $\nu_{nm} = (E_n - E_m)/h$. Because of their simple time dependence, as noted from Eq. (1.45), spectroscopic states are also called stationary states. Eq. (1.45) is indeed equivalent to Eq. (1.27), because $\Psi_k(r,t) = \langle r|\Psi_k(t)\rangle$ and $\psi_k(r) = \langle r|\psi_k\rangle$. And so are Eqs. (1.46) and (1.42) equivalent to Eqs. (1.21) and (1.18), respectively.

For isolated systems, Eq. (1.44) may be solved formally, by setting $\boldsymbol{b}(t) = \boldsymbol{U}(t,t_0)\,\boldsymbol{b}(t_0)$, where

$$\boldsymbol{U}(t,t_0) = \exp\left(-\mathrm{i}\frac{2\pi}{h}\,\boldsymbol{H}\,(t-t_0)\right) \tag{1.47}$$

is the matrix representation of the time evolution operator (see also Eq. (1.16)). For time-dependent Hamiltonians, the formal integration is more complex. For instance, in the so-called Magnus expansion (Magnus, 1954), as reviewed by Quack (1978, 1982) and Blanes et al. (2009), the integration involves nested commutators of the Hamiltonian at different times, see also Eq. (1.100), Section 1.6.3.

1.3.1 Spectral Decomposition Method

When the Hamiltonian is independent of time, the natural method suggested by Schrödinger (1926a,b,c,d,e) to solve Eq. (1.42) is to determine prior solutions of Eq. (1.46) and use an appropriate finite subset of solutions $\mathbb{L}_N = \{|\psi_1\rangle, \ldots, |\psi_N\rangle\}$. The time-dependent wave function may then be given such as in Eq. (1.43), see also Eq. (1.29),

$$|\Psi(t)\rangle = \sum_{n=1}^{N} b_n(t)\,|\psi_n\rangle \tag{1.48}$$

with

$$b_n(t) = b_n(0)\,\exp(-2\pi\mathrm{i}\,E_n\,t/h) \tag{1.49}$$

and $b_n(0) = \langle\phi_n|\Psi(0)\rangle$

However, it is difficult to know the stationary states in advance (see also Chapter 2 of this book (Császár et al., 2020)). It is reasonable to conjecture, that an approximate knowledge of these states could help simplify the calculation and interpretation of molecular quantum dynamics.

In practice, the direct way in this case is to solve Eq. (1.46) by first representing $\hat{H}$ in a given basis set $\mathbb{L}_N = \{\chi_1, \ldots, \chi_N\}$ and diagonalizing the thus obtained matrix $\boldsymbol{H}$. The time evolution operator is then given as

$$\boldsymbol{U}(t-t_0) = \boldsymbol{Z}\,\boldsymbol{U}^{(\mathrm{d})}(t-t_0)\,\boldsymbol{Z}^{\dagger} \tag{1.50}$$

where $\boldsymbol{U}^{(\mathrm{d})}(t-t_0)$ is a diagonal matrix,

$$U_{nm}^{(\mathrm{d})}(t-t_0) = \begin{cases} \exp(-2\pi\mathrm{i}E_n(t-t_0)/h), & n=m, \\ 0, & n\neq m, \end{cases} \tag{1.51}$$

E_n are the eigenvalues of $\boldsymbol{H}$, and the matrix $\boldsymbol{Z}$ is composed of N column vectors $z_1, \ldots, z_N$, which are the representations of the eigenvectors of $\boldsymbol{H}$ in $\mathbb{L}_N$. Eq. (1.52) is the corresponding representation of Eq. (1.46),

$$\boldsymbol{H}\,z_n = E_n\,z_n. \tag{1.52}$$

The number of floating point operations involved in the diagonalization with full determination of all eigenstates increases with N^3. Currently, diagonalization algorithms in standard linear algebra program libraries straightforwardly handle matrices with ranks up to $N = 50\,000$, although computation time and storage space of vectors and matrices increase rapidly with the rank.

In multidimensional spaces, the linear space of wave functions can be represented by a simple tensor product of one-dimensional spaces. Let d be the number of dimensions to be considered and let M be the average number of one-dimensional basis functions; then $N = M^d$, and the number of floating point operations is M^{3d}. For $d > 5$, typically, the spectral decomposition method becomes essentially impractical, unless some special measures are taken to optimize the size of the original representation basis, e.g., by suitably compressing basis vectors, or by applying collocation methods (Avila and Carrington, 2015). This issue is also discussed in Chapter 2 of this book (Császár et al., 2020).

These technical drawbacks are the only serious disadvantages of the spectral decomposition method. Whenever possible, this method should be given preference to other methods discussed below for three main reasons: Firstly, one can compare the calculated energy values with those derived from high-resolution spectroscopy, which are frequently available with very high accuracy, and thus test some of the underlying approximations, for instance, the potential energy surface (PES) used for the nuclear dynamics; other observables, such as the transition dipole moments, can also be directly compared. Secondly, it is very easy to vary the initial condition of the dynamical calculation with almost no additional computational effort. Finally, with the spectral decomposition method, one can design suitable approximations, such as the quasiresonant approximation for coherent excitation (see Section 1.6 below), which allows for accurate long-time propagation that is not easily accessible with the direct approaches to be discussed in the following sections.

1.3.2 Linearization

The operational simplest method to obtain the time evolution operator is to consider the Taylor expansion of the exponential function in Eq. (1.47). For "small" displacements along the time axis, $\Delta t = t - t_0$, the operator in Eq. (1.47) may be approximated by the linearized operator

$$\boldsymbol{U}_{\text{lin}}(\Delta t) = \boldsymbol{I} - \mathrm{i}\frac{2\pi}{h}\,\boldsymbol{H}\,\Delta t. \tag{1.53}$$

Here, $\boldsymbol{I}$ is the identity matrix. Hence, the state vector at time $t_k = k\,\Delta t + t_0$ $(k > 0)$ is given as a linear function of the state vector at time t_{k-1},

$$\boldsymbol{b}(t_k) = \boldsymbol{b}(t_{k-1}) - \mathrm{i}\frac{2\pi}{h}\Delta t\,\boldsymbol{H}\,\boldsymbol{b}(t_{k-1}). \tag{1.54}$$

There is an error of the order Δt^2 at each time step, and the integrated error increases with the total evolution time by error propagation. Following the uncertainty relation Eq. (1.11), appropriate time step sizes should satisfy the condition $\Delta t \ll h/\Delta E$, where ΔE is a typical transition or coupling energy.

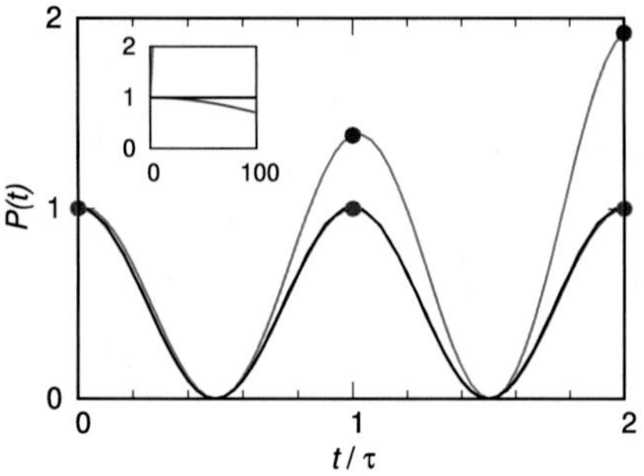

FIG. 1.3 Population evolution $P(t) = \cos^2(\pi t/\tau)$ of an initially 100% populated level in a simple scheme of two isoenergetic levels coupled by an interaction energy $V = h/(2\tau)$ (black line). The blue line yields the result from the simple linearized formula in Eq. (1.54) with a propagation step $\Delta t = 0.033\,\tau$. The red line is from the second order formula in Eq. (1.55) – it essentially overlaps with the black line in the main figure. The dots are the results at times $t_n = n \times \tau$ $(n = 0, 1, 2, \ldots)$ and their collection is displayed in the insert by lines in the corresponding color – the blue line is hardly seen because of the fast increase of the propagation error, the black line is the exact solution.

What is typically a sufficiently small Δt? This question seems odd, in particular in studies of ultrafast processes, as one might naively think that the process will be over before error has accumulated significantly. Fig. 1.3 shows, as an example, a simple two-level dynamics, where a quantum state is coupled resonantly via a coupling constant V to a second, isoenergetic quantum state – this is the simplest model of a quantum mechanical tunneling motion and can also be used to discuss the monochromatic excitation between two quantum states, see Section 1.6.6 below. The exact evolution $P(t) = \cos^2(\pi t/\tau)$ of the population of the initial state is shown there as a black line; $\tau = h/2V$ is the typical evolution time. The blue line shows the same evolution calculated via Eq. (1.54). Quite impressively, the accrued propagation error becomes as large as 30% after just one period of the evolution, if Δt is only 3% of the typical evolution time τ. There is also error propagation in terms of the phases of the time evolving state vector, which is not shown here.

As can be seen from the figure, the error increases rapidly and is already 100% after just two periods because of error propagation. The strong increase of the population goes inline with a severe nonconservation of the norm, which is an unacceptable nonphysical behavior. The error can be much reduced via the use of a second order formula which relies on the evaluation of the wave function at two earlier time steps t_k and t_{k-1} (Kosloff and Kosloff, 1983a) – this formula is related to the Crank–Nicholson method (Bachau et al., 2001):

$$\boldsymbol{b}(t_{k+1}) = \boldsymbol{b}(t_{n-1}) - \mathrm{i}\frac{4\pi}{h}\Delta t\, \boldsymbol{H}\, \boldsymbol{b}(t_n). \tag{1.55}$$

The corresponding evolution is given in Fig. 1.3 by the red line. Even if the error can be much reduced by the second-order linear formula in the case of the simple example discussed here, it can still be quite important at a longer time evolution, as shown by the insert of Fig. 1.3 or in more complex situations where many states with different sizes of couplings are involved. The sole remedy for this simple algorithm is to reduce drastically the linear evolution time step Δt, at the cost of having a huge number of sequential matrix vector multiplications to perform if a longer time evolution is needed.

A very detailed discussion on error accumulation was given by Marquardt and Quack (1989) on the basis of an exactly solvable model for femtosecond multiphoton excitation in the infrared (see also Section 1.6 below).

Despite the important error propagation inherent to the linearized propagator, the simplicity of the method is appealing, in particular when the system is very complex. In such cases even a simple matrix–vector multiplication as that of Eq. (1.54) can become highly time consuming because the dimension of the linear space becomes very large. Therefore the simpler the algorithm, the easier its implementation. More advanced methods exist, however. Sophisticated predictor–corrector algorithms are typically employed to solve nonlinear ordinary differential equations (Gear, 1971; Shampine and Gordon, 1975; Beck et al., 2000) and may be implemented to optimize dynamically the time step and in this way contribute to reduce error accumulation. Such algorithms are used in the MCTDH program package discussed below.

1.3.3 The "Chebychev" Method

Some methods make explicit use of higher order expansions of the propagator. As an alternative to expanding the exponential function in Eq. (1.47) in terms of powers of the argument, it may be expanded in terms of polynomials.

Tal-Ezer and Kosloff (1984) used an expansion in terms of Chebychev polynomials $T_k(z)$ (Courant and Hilbert, 1968), which are defined for $-1 \le z \le 1$. Light and Carrington (2000) and Beck et al. (2000) discuss several characteristic properties of these polynomials, relevant for the application in quantum dynamics. This method has since then been used in varied forms mainly in scattering quantum dynamics.

This method works as follows: First, upper ($E_{\max}$) and lower ($E_{\min}$) bounds for the largest and lowest eigenvalues of $\boldsymbol{H}$, respectively, need to be known, at least approximately. These may be estimated straightforwardly for finite matrix representations (Carrington, 2011). Then, the time evolution operator may be written as (Tal-Ezer and Kosloff, 1984)

$$\begin{aligned}\boldsymbol{U}_{\text{cheb}}(t+\Delta t, t) &= \exp\left(\mathrm{i}\frac{2\pi}{h} E_{\text{S}}\,\Delta t\right)\\ &\times \sum_{k=0}^{k_{\max}} (2-\delta_{k0})\, J_k\left(\frac{2\pi}{h} E_{\text{D}}\,\Delta t\right) T_k\left(\frac{2\pi}{h}\boldsymbol{H}\,\Delta t\right)\end{aligned} \tag{1.56}$$

where $E_{\text{S}} = 1/2(E_{\max} + E_{\min})$, $E_{\text{D}} = 1/2(E_{\max} - E_{\min})$, and J_k are Bessel functions of the first kind of order k (Courant and Hilbert, 1968). To evaluate the Chebychev polynomial of degree k on the state vector, the recursion formula

$$T_k(\boldsymbol{M}) = 2\boldsymbol{M}\, T_{k-1}(\boldsymbol{M}) + T_{k-2}(\boldsymbol{M}) \tag{1.57}$$

is used, with $T_0(\boldsymbol{M}) = \boldsymbol{I}$ and $T_1(\boldsymbol{M}) = \boldsymbol{M}$. The Bessel functions of the first kind have the property that $J_k(x) \approx 0$ for $k \ge x$. If the maximal order considered in Eq. (1.56) is chosen such that $k_{\max} \ge 2\pi E_{\text{D}}\,\Delta t/h$, the propagation error can be made as small as desired.

The Chebychev method has been recently applied in the scattering quantum dynamics of triatomic (Zhou and Xie, 2015) and tetraatomic systems (Song and Guo, 2015), in full-dimensional rovibrational quantum dynamics of ammonia isotopomers (Fábri et al., 2019), but also in spin noise calculations of semiconductor quantum dots (Hackmann and Anders, 2014), in the hydrodynamical formulation of quantum mechanics (Cruz-Rodriguez et al., 2016), and in continuous time quantum walks (Izaac and Wang, 2015). The method was also extensively reviewed by Kosloff (1994).

1.3.4 "Short-Iterative" Lanczos Method

Quite often, only a rather small number of spectroscopic states is effectively involved in the representation of the time evolution operator during a given time interval Δt. This means that, if $|\Psi(t_j)\rangle$ is the solution of Eq. (1.42) at time t_j, it can be decomposed in a limited set of states $\{|\tilde{\psi}_1\rangle, \ldots, |\tilde{\psi}_{N(t_j)}\rangle\}$,

$$|\Psi(t_j)\rangle = \sum_{n=1}^{N(t_j)} b_n(t_j)\,|\tilde{\psi}_n\rangle \tag{1.58}$$

with

$$b_n(t_j) = \langle \tilde{\psi}_n | \Psi(t_j) \rangle \tag{1.59}$$

such that

$$\sum_{n=1}^{N(t_j)} |b_n(t_j)|^2 \approx 1 \tag{1.60}$$

with a hopefully rather small value of $N(t_j)$. Since the method described here may also apply to potentially time-dependent Hamiltonians $\hat{H}(t)$ and corresponding pseudoeigenvalues $E_n(t)$, the notation is here such that $|\tilde{\psi}_n\rangle$ are pseudoeigenstates, for which

$$\hat{H}(t_j)\,|\tilde{\psi}_n\rangle = E_n(t_j)\,|\tilde{\psi}_n\rangle \tag{1.61}$$

holds at the given time step t_j, and $N(t_j)$ is the number of pseudoeigenstates contributing importantly to the dynamics at time t_j. This number can vary smoothly as a function of time.

Following an idea formulated by Park and Light (1986), we consider here the integration step from a time t_j to a time $t_{j+1} = t_j + \Delta t$. The Hamiltonian at time t_j is diagonalized within the Lanczos method (Cullum and Willoughby, 1985), such as to yield a limited number of pseudoeigenstates $|\tilde{\psi}_n\rangle$ and corresponding energies $E_n(t_j)$. The Lanczos iteration is started from the state $|\Psi(t_j)\rangle$, which helps ensure that the set of calculated pseudoeigenfunctions includes all eigenfunctions $\tilde{\psi}_n$ that contribute significantly to the dynamics at this particular time step, i.e., for which the overlap integrals $b_n(t_j)$ fulfill the condition of Eq. (1.60). With this method, $N(t_j)$ varies typically between 20 and 30.

The state at time t_{j+1} is then calculated from a restricted expansion of the type of Eq. (1.48):

$$|\Psi(t_{j+1})\rangle = \sum_{n}^{N(t_j)} b_n(t_j)\,\exp\left(-\mathrm{i}\frac{2\pi}{h}\,E_n(t_j)\,\Delta t\right)\,|\tilde{\psi}_n\rangle. \tag{1.62}$$

The sum in Eq. (1.62) is restricted to the subset of pseudoeigenstates obtained within the Lanczos algorithm that satisfy Eq. (1.60) within a given accuracy threshold. With this recipe, the solution of Eq. (1.42) can then be obtained by iteration. The accrued propagation error depends clearly on the iteration time step Δt used, but it can be largely reduced depending on the threshold used to verify Eq. (1.60).

The Lanczos method to solve Eq. (1.42) has been reviewed more recently from an applied mathematical (Lubich, 2015) and physical prospective (Bader et al., 2018), and also applied in the context of qubit dynamics (Cangemi et al., 2018).

1.3.5 "Split-Operator" Technique

Another method that makes use of higher order expansions of the propagator is the *split-operator* method (Feit et al., 1982). Let $\boldsymbol{H} = \boldsymbol{H}_0 + \boldsymbol{H}_1$ be such that the norms of these Hamiltonians satisfy $||\boldsymbol{H}_1|| \ll ||\boldsymbol{H}_0||$. The quantity E_D defined in Section 1.3.3, is one possible norm of a matrix. The time evolution operator

$$\boldsymbol{U}_{\text{split}}(t+\Delta t, t) = \exp\left(-\mathrm{i}\frac{\pi}{h}\,\boldsymbol{H}_0\,\Delta t\right) \times \exp\left(-\mathrm{i}\frac{2\pi}{h}\,\boldsymbol{H}_1\,\Delta t\right)\,\exp\left(-\mathrm{i}\frac{\pi}{h}\,\boldsymbol{H}_0\,\Delta t\right) \tag{1.63}$$

is an approximation to $\boldsymbol{U}(t+\Delta t, t)$ and, compared to the latter, has quadratic convergence in Δt (Bandrauk and Shen, 1993).

The Hamiltonian used in Eq. (1.63) can often be considered to be split into the kinetic and potential energy part $\boldsymbol{H} = \boldsymbol{T} + \boldsymbol{V}$, with either $\boldsymbol{H}_0 = \boldsymbol{T}$ and $\boldsymbol{H}_1 = \boldsymbol{V}$, or vice versa, depending on which operator satisfies the aforementioned norm condition better. It is then possible and even appropriate to evaluate $\boldsymbol{U}_{\text{split}}$ in two steps involving the momentum and space representations of the time evolving quantum state. Let $\boldsymbol{H}_0 = \boldsymbol{T}$.

The state at time t is first represented in the momentum space. When the kinetic energy is evaluated in Cartesian coordinates, the action of the first factor on the right-hand side of Eq. (1.63) is to add the angle $-(2\pi t/h)\sum_k p_k^2/2m_k$ to the phase of the state's wave function in momentum space.

Then the representation is switched to a space representation by a Fourier transformation and the central factor in Eq. (1.63) adds a phase $-(2\pi t/h)\,V(\boldsymbol{x})$ to the state's wave function.

Finally, this function is back-transformed into momentum space and its phase is incremented by the same angle as in the first step.

The split operator technique takes advantage of available routines that perform fast Fourier transformations (Kosloff and Kosloff, 1983a,b) and has since then been much used to solve Eq. (1.42) (see, for instance, Kolba et al., 1992 and, for recent applications and developments, Sun et al., 2015, Greene and Batista, 2017, Blanes et al., 2017 for nuclear dynamics and Sabzyan and Jenabi, 2016 for electron transfer processes). This method can be applied also when $\boldsymbol{H}_1$ is an explicit function of time, supposing that the variation of $\boldsymbol{H}_1$ with time t is itself of order Δt^3 in the time interval between t and

$t + \Delta t$. In this context, "commutator-free" propagators have been discussed recently, by which the performance can be significantly improved (Bader et al., 2018).

One interesting aspect of the split operator technique is to chose a representation basis for the dynamics that diagonalizes the kinetic energy. The advantage of the fast Fourier transform is then lost, however. The numerical effort of such a procedure has been tested (Quack and Stohner, 1993).

1.3.6 The "Multiconfigurational Time-Dependent Hartree" Method

Quantum dynamics are in general multidimensional. The time dependent quantum state $|\Psi(t)\rangle$ contains simultaneously information on all particles involved in a system, each particle being essentially characterized by its position in the three-dimensional space. In molecules, one considers as particles the nuclei of the atoms composing these systems and the electrons surrounding them.

Accordingly, the quantum state has to be studied with respect to its projections on the spaces of the different particles. If the total dimension to be considered is d, the quantity to be determined is the d-dimensional wave function $\Psi(x_1, \ldots, x_d; t) = \langle x_1, \ldots, x_d|\psi(t)\rangle$.

Similarly to the idea underlying the short iterative Lanczos method explained above, one exploits the fact that the number of states that strongly participate at the dynamics in a given time interval Δt is normally small. In the time-dependent Hartree method (Jungwirth and Gerber, 1999), the time dependent, d-dimensional wave function is set up as a product of d one-dimensional, time dependent *single particle functions* $\varphi_j(x_j, t)$,

$$\Psi(x_1, \ldots, x_d; t) = \varphi_1(x_1, t) \cdots \varphi_d(x_d, t).$$

The single particle functions can be obtained as solutions of nonlinear coupled partial differential equations, which can be rigorously derived from the Dirac–Frenkel variational principle (see Beck et al., 2000, and references therein).

Alike the configuration interaction expansion in electronic structure theory, it is possible to improve the accuracy of the description via a Hartree product by considering a "multiconfigurational" expansion ("MCTDH", Meyer et al., 1990)

$$\Psi(x_1, \ldots, x_d; t) = \sum_{I=1}^{N_{\mathrm{MCTDH}}} b_I(t)\, \Phi_I(x_1, \ldots, x_d; t). \tag{1.64}$$

The number I is a superindex that counts set of indexes $i_1, \ldots, i_d$. It defines a particular configuration of single particle functions $\Phi_I(x_1, \ldots, x_d, t) = \varphi_{i_1}^{(1)}(x_1, t) \cdots \varphi_{i_d}^{(d)}(x_d, t)$. The quantities $b_I(t)$ are time-dependent coefficients. For the numerical evaluation, single particle functions are expanded in terms of time-independent *primitive functions* $\xi_1^{(k)}, \ldots, \xi_{n_k}^{(k)}$:

$$\varphi_i^{(k)}(x_k, t) = \sum_{j=1}^{n_k} c_{ji}^{(k)}(t)\, \xi_j^{(k)}(x_k). \tag{1.65}$$

The maximal number of single particle functions of coordinate x_k is n_k. The equations of motion for the expansion coefficients $c_{ji}^{(k)}(t)$ and $b_I(t)$ can again be rigorously obtained from the Dirac–Frenkel variational principle. The number n_k of primitive functions can quite often be chosen sufficiently small for converged calculations. The number $N_{\mathrm{MCTDH}} = n_1 \cdots n_d$ of configurations participating in the expansion Eq. (1.64) can hence be made considerably smaller than the number N of stationary states needed such as in Eq. (1.43).

While the idea underlying the MCTDH method allows us to solve high-dimensional problems that would otherwise not be solvable, in practice it has a serious conceptual drawback: the linearity originally embedded in Eq. (1.44) (or Eq. (1.42), or Eq. (1.18)) has to be given up. The equations of motion are nonlinear, ordinary differential equations that can be solved by any integration method, in principle, with variable efficiency and accuracy, which depends strongly on the precise predictor–corrector algorithm used (see Beck et al., 2000 and references given therein; a specific algorithm with step size control was also devised for MCTDH). Also, as for all initial value problems, the integration has to be restarted from time t_0 for each new initial condition. In contrast to this, the spectral decomposition scheme described in the previous subsections yields a universal time evolution operator $\boldsymbol{U}(t, t_0)$ which holds for all initial conditions $\boldsymbol{b}(t_0)$.

The great advantage of the MCTDH approach in high dimensional problems is, however, that it simply renders these problems treatable, as $N_{\mathrm{MCTDH}} \ll N$. A second, more technically oriented advantage of the method is related to the implicit product form of the wave function. As a consequence, multidimensional integrals related to practical evaluation of scalar products may be carried out as products of one dimensional integrals, if the operators can also be written in the form of sums of products of one dimensional operators. While this condition can be met quite generally by specially devised forms of the potential energy operators, a more judicious choice of the coordinates used is necessary to ensure that the kinetic energy operator is of this form. Such coordinates are, for instance, polyspherical coordinates. The MCTDH method has meanwhile been applied to a plethora of systems. Both method and applications were reviewed by Gatti (2014) and Gatti et al. (2017).

Marquardt et al. (2010) give a detailed comparison between results obtained with the MCTDH and URIMIR codes of the time evolution of populations and relative phases of a state vector defined in a four-dimensional vibrational subspace in ammonia. The URIMIR code discussed in Section 1.6 below allows us to also obtain numerically exact solutions of Eq. (1.42) under the influence of an external radiation field.

1.3.7 Specific Methods for the Electronic Motion

Electronic motion can in principle be assessed by all methods discussed in the previous sections. Several problems hamper the straightforward implementation of the equations, however. One problem is connected with the representation of the Hamiltonian: Basis sets need to be very large, as the long range Coulomb potential leads to interactions in very much extended regions of space. A second problem arises in the correct time dependent treatment of the fermionic character of the electrons. Some approaches are related to the MCTDH method, described above, where the total, time-dependent electronic function is described as a multiconfigurational expansion of completely antisymmetric configuration functions, very much like in the configuration interaction method of static electronic structure calculations (Kato and Kono, 2004; Caillat et al., 2005; Nest, 2006); contrary to the static case, the orbitals used to compose the time dependent electronic wave function are time-dependent, however. These "MCTDHF" called methods allow us also to calculate excited state properties (Nest et al., 2005). However, they still have a limited domain of applications.

In most cases treated so far, the solutions of Eq. (1.42) are given for the electronic motion alone in the so-called "sudden" approximation, in which the nuclei are essentially frozen at their positions during the actual interaction time of a few tens or hundreds of attoseconds. Additionally, methods used so far are generally based on the *single active electron* (SAE) approximation. This approximation is used to simulate processes involved in high harmonic generation (Farrell et al., 2011) and attosecond molecular dynamics (Wörner and Corkum, 2011), see also Chapter 4 (Baykusheva and Wörner, 2020) of this book.

Related, highly specific methods have been used for instance to discuss $He^{2+} + H_2$collisions (Sisourat et al., 2011), or the interaction of atoms and molecules with high intensity radiation, as reviewed by Maquet and Grobe (2002) and Salières et al. (2012), where the *strong field approximation* (SFA) is made. The latter consists essentially in describing the wave function of the "active" electron in terms of analytic Gordon–Volkov solutions for Eq. (1.42), to describe the quantum dynamics of an electron in the presence of a strong laser field (Reiss, 1992). Lorin et al. (2007) solved the Maxwell–Schrödinger equation numerically for H_2^+ in a four-dimensional space; here, the time propagation was evaluated via a modified version of Eq. (1.55).

In Chapter 5 of this book (Gokhberg et al., 2020), methods are described for the theoretical treatment of several fast electronic processes involving molecular ionization and charge migration in molecular aggregates. They rely on the calculation of a time dependent hole density using the Heisenberg representation (Eq. (1.14)) and the short iterative Lanczos technique explained in Section 1.3.4 above. For the calculation of the hole density matrix the algebraic diagrammatic construction scheme was used, a Green's function formalism. For more details on this technique, we refer to Kuleff et al. (2005), Sansone et al. (2012), Kuleff (2019), and the references cited therein, as well as to Chapter 5.

Theoretical methods used in attosecond molecular dynamics were reviewed in the book edited by Vrakking and Lépine (2019). We also refer to the work by Ambrosek et al. (2004), Barth and Manz (2007), Bredtmann et al. (2015) and Jia et al. (2019) on femtosecond and attosecond electron and nuclear quantum fluxes and currents in molecular compounds. The sub-fs electronic time scale for charge migration in small peptides was discussed by Remacle and Levine (2006).

1.4 HAMILTONIANS

In nonrelativistic quantum mechanics the molecular Hamiltonian is indeed set up as the sum of the Coulomb potential energy terms between the charged nuclei and the electrons as point masses, as well as the kinetic energy of these particles expressed in the usual way in terms of the Cartesian coordinates of these points and their masses (see Eqs. (1.12) and (1.13)). Solving for the complete molecular dynamics within this general scheme is very difficult and limited to small systems, such as the H_2^+ cation, for which the full three-body problem was solved numerically by Chelkowski et al. (1995) and Lorin et al. (2007) (see also the references cited therein).

The focus of current theoretical developments is still based on the Born–Huang expansion of the molecular state (Born and Huang, 1954),

$$|\Psi(t)\rangle = \sum_k |\Psi_k^{(n)}(t)\rangle \otimes |\psi_k^{(e)}\rangle. \tag{1.66}$$

The symbol $|\Psi^{(n)}(t)\rangle \otimes |\psi^{(e)}\rangle$ means a tensor product of "nuclear" and "electronic states".

The "electronic state" $|\psi^{(e)}\rangle$ is supposed to be time independent, while the "nuclear state" $|\Psi^{(n)}(t)\rangle$ contains the information on the time dependence of the molecular state. Variations of Eq. (1.66) have been suggested (Cederbaum, 2008; Abedi et al., 2010), in which the electronic state is of the form $|\Psi_k^{(e)}(t)\rangle$, i.e., it also contains an explicit time dependence. While this ansatz is legitimate and interesting from a fundamental point of view, so far it yields no practical method for the solution of Eq. (1.42) for the combined nuclear and electronic motions in molecules. In practice, all electronic states $|\psi_k^{(e)}\rangle$ are set to depend parametrically on the nuclear coordinates, as will be discussed below, and the product ansatz does not correspond to a true separation of the dynamics.

The expansion in Eq. (1.66) is usually interpreted as corresponding to a representation of the molecular state in a specific, countable basis of electronic states $|\psi_1^{(e)}\rangle, |\psi_2^{(e)}\rangle, \ldots$ Clearly, this interpretation provides only an approximate description of ionization processes, where the state of the dissociated electron belongs to a continuum of states.

The time-dependent Schrödinger equation, Eq. (1.42), then reads

$$\mathrm{i}\frac{h}{2\pi}\frac{\partial}{\partial t}\begin{pmatrix} |\Psi_1^{(n)}(t)\rangle \\ |\Psi_2^{(n)}(t)\rangle \\ \vdots \end{pmatrix} = \begin{pmatrix} \hat{H}_{11}^{(n)} & \hat{H}_{12}^{(n)} & \cdots \\ \hat{H}_{21}^{(n)} & \hat{H}_{22}^{(n)} & \cdots \\ \vdots & \vdots & \ddots \end{pmatrix} \begin{pmatrix} |\Psi_1^{(n)}(t)\rangle \\ |\Psi_2^{(n)}(t)\rangle \\ \vdots \end{pmatrix}. \tag{1.67}$$

The matrix in Eq. (1.67) is a representation of the total Hamiltonian in the chosen basis of electronic states $|\psi^{(e)}\rangle$. Its matrix elements are Hamiltonian operators $\hat{H}_{ik}^{(n)}$ that act exclusively on the nuclear degrees of freedom:

$$\hat{H}_{ik}^{(n)} = \langle\psi_i^{(e)}||\hat{H}||\psi_k^{(e)}\rangle = \begin{cases} \hat{T} + V_k & (i = k), \\ \hat{H}_{ik}^{(\text{non-a})} & (i \neq k). \end{cases} \tag{1.68}$$

In this equation, the symbol $\langle\cdot||\cdot||\cdot\rangle$ means a scalar product involving the electronic states only; V_k is normally a function of the relative positions of the nuclei in the molecular complex and is interpreted as the "potential energy surface" for the motion of the nuclei in the electronic state k; $\hat{T}$ is an adequate representation of the operator for the kinetic energy in the coordinate space of the nuclei and does generally not depend on any specific electronic state. Such representations will be addressed in the following section. It will be shown that $\hat{T}$ is essentially a derivative operator, which might depend on the position of the nuclei, however.

Potential energy surfaces can be obtained from electronic structure calculations. In these calculations the electronic wave functions $\psi_k^{(e)}(\boldsymbol{x}^{(e)}) = \langle\boldsymbol{x}^{(e)}||\psi_k^{(e)}\rangle$ are obtained as eigenstates of the time independent molecular Schrödinger equation in the so-called *clamped nuclei approximation* (see Cederbaum, 2004), i.e., by fixing the position of the nuclei and neglecting their kinetic energy. Here $\boldsymbol{x}^{(e)}$ is a generalized electronic position vector. The functions obtained in this way depend parametrically on the positions of the nuclei (see above). The eigenvalue corresponding to a state $|\psi_k^{(e)}\rangle$ yields the multidimensional potential energy (hyper-)surface $V_k(\boldsymbol{x}^{(n)})$, where $\boldsymbol{x}^{(n)}$ is a generalized nuclear position vector. The states $|\psi_k^{(e)}\rangle$ obtained in the clamped approximation are called *adiabatic* electronic states, as the electrons are supposed to follow adiabatically the motion of the nuclei.

Many methods exist today in electronic structure theory with a variety of applications. Some benchmark calculations have been discussed by Schreiber et al. (2008). The status of methods used in electronic structure calculations to obtain potential energy surfaces for electronically excited states has been reviewed by González et al. (2012) and, in particular in reference to transition metal complexes by Daniel (2002, 2015a,b) and Penfold et al. (2018).

The clamped nuclei technique enforces a point-wise determination of the functions. In molecular dynamics and spectroscopy, the knowledge of potential energy surfaces is essential and often it is desirable to have analytical representations of these surfaces that are global, i.e., that are physically well defined in the entire space of nuclear coordinates. Methods for deriving global analytical representations of potential energy surfaces from electronic structure calculations and from spectroscopic data are discussed by Marquardt and Quack (2011). It is worth mentioning that the very first analytical representations of molecular potential energy surfaces were derived from spectroscopic experiments prior to modern quantum mechanics (Bjerrum, 1914). The use of permutation symmetry of identical nuclei in the formulation of analytical potential hypersurfaces was discussed by Marquardt and Quack (1998) and Qu et al. (2018) (see also Fábri et al., 2017).

The operators $\hat{H}_{ik}^{(\mathrm{non-a})}$ occurring in Eq. (1.68) represent the *nonadiabatic* couplings between the adiabatic electronic states. They account for the coupling between the adiabatic states and therefore play an important role in the description of ultrafast processes that may occur mainly in electronically excited molecules (González et al., 2012; Daniel, 2002, 2015a; Vrakking and Lépine, 2019), see also Chapter 6 of this book (Vaníček and Begušić, 2020).

Eq. (1.67) is in principle an exact representation of Eq. (1.42), whenever the expansion in Eq. (1.66) can be considered exact. It cannot be considered exact, as explained above, in processes involving the ionization of molecules. To solve this equation in a general case is difficult, however, and one has normally to resort to several approximations, as discussed in Section 1.3.7 above.

In the *Born–Oppenheimer approximation*, the action of the nonadiabatic couplings is neglected with respect to that of the diagonal operators. This approximation is adequate, even excellent in many cases, as long as the energies of two electronic states are not too close one to the other, i.e., as long as $|V_k(\boldsymbol{x}^{(\mathrm{n})}) - V_i(\boldsymbol{x}^{(\mathrm{n})})| \gg 0$. These terms can hardly be neglected, when the potential energy surfaces of two electronic states cross, such as at conical intersections (Domcke et al., 2004; Domcke and Yarkony, 2012). These are the regions of the nuclear configuration space where the multidimensional potential energy surfaces of two or more electronic states intersect, even if they belong to the same irreducible representation space of the molecular symmetry group (Teller, 1937; Herzberg and Longuet-Higgins, 1963). The mathematical foundations of conical intersections are clearly explained by Longuet-Higgins (1975).

The exact evaluation of the nonadiabatic couplings is difficult for several reasons (Cederbaum, 2004): as for potential energy surfaces, these terms need to be evaluated in pointwise electronic structure calculations in the space of nuclear configurations, there are generally no global analytical representations of these operators; furthermore, they may become singular, in particular at conical intersections.

Singularities may theoretically be removed by performing an appropriate unitary basis transformation (Cederbaum, 2004):

$$|\tilde{\Psi}_k^{(\mathrm{n})}(t)\rangle = \sum_l Y_{lk}^* \,|\Psi_l^{(\mathrm{n})}(t)\rangle, \tag{1.69}$$

$$|\tilde{\psi}_k^{(\mathrm{e})}\rangle = \sum_l Y_{kl} \,|\psi_l^{(\mathrm{e})}\rangle, \tag{1.70}$$

where $\sum_k Y_{lk}^* Y_{kl'} = \delta_{ll'}$. The unitarity of the matrix $\boldsymbol{Y}$ ensures the invariance of the Born–Huang expansion in Eq. (1.66). Because the basis of electronic states depends parametrically on the relative nuclear position vectors $\boldsymbol{x}^{(\mathrm{n})}$, it is expected that the matrix $\boldsymbol{Y}$ will also depend on $\boldsymbol{x}^{(\mathrm{n})}$.

It can be shown that, under appropriate conditions, the singularities expected to occur in the derivative coupling terms $\hat{H}_{ik}^{(\mathrm{non-a})}$ in the adiabatic basis representations can be made to vanish and be replaced by multiplicative, generally nonsingular coupling operators (Cederbaum, 2004). The states $|\tilde{\psi}_k^{(\mathrm{e})}\rangle$ introduced by this procedure are called *diabatic*; in practice this technique leads only to an approximate treatment of the coupling, since the transformation runs over a finite number of countable states. A quite valuable technique to obtain nonadiabatic couplings uses a block diagonalization of the electronic Hamiltonian (Pacher et al., 1988). These operators may then be represented by an analytical Taylor expansion limited to the vicinity of the conical intersection, at which they were derived. This method leads to the *vibronic coupling Hamiltonian*, which may be used for the study of the ultrafast dynamics in this limited region of configuration space (González et al., 2012; Daniel, 2015a), also in competition with the spin–orbit coupling prominent in transition metal complexes (Penfold et al., 2018).

Many theoretically fundamental or methodological articles on nonadiabatic couplings and conical intersections, as well as the dynamics resulting from them and examples of applications, can be found in the book by Domcke et al. (2004) or in the reviews by Chu et al. (2006), Matsika and Krause (2011) or Domcke and Yarkony (2012).

Vibrational wave packet dynamics associated with ultrafast electron transfer reactions in a condensed phase environment has been studied by Thoss et al. (2004). A special issue of the *Journal of Chemical Physics* has been devoted to nonadiabatic molecular dynamics (see the keynote article by Tully (2012)).

For systems involving more than 5 atoms, typically, finding accurate solutions of Eq. (1.67) outside the vibronic coupling Hamiltonian approach becomes truly difficult, and classical or semiclassical approaches for the motion of the nuclei are considered instead, as discussed in Chapter 6 of this book (Vaníček and Begušić, 2020). In order to describe the transfer of population from one adiabatic state to the other, many of these approaches use a *surface hopping* idea, such as that proposed by Tully (1990), or variations thereof (Tully, 2012, and the references given therein). However, one must not consider these classical or semiclassical treatments as actual approximations of the quantum dynamics. Rather, they yield a classical picture of a process that can be quite misleading, indeed, due to the wave mechanical nature of the actual dynamics that they cannot fully recover.

One way to simplify the great complexity of the Hamiltonian for multidimensional molecular motion is a quasiadiabatic separation of vibrational degrees of freedom. For instance, one may treat the Hamiltonian quite explicitly in one coordinate (the "reaction path") or perhaps a small subset of coordinates, whereas all other coordinates are treated implicitly by their effect on this selected subset. This is the idea of the "Reaction Path Hamiltonian" (RPH, Miller et al., 1980) and the extension to the quasiadiabatic channel RPH approach (Quack and Suhm, 1991; Fehrensen et al., 1999, 2007). These approaches are discussed in more detail in Chapter 7 of this book (Quack and Seyfang, 2020).

1.5 COORDINATES

Cartesian coordinates defined in a space fixed reference frame give the simplest spatial representation of the quantum mechanical kinetic energy operator. As in classical mechanics, the Cartesian coordinates of all particles can be used to describe the internal molecular dynamics. However, in addition to the latter, they do also describe the center of mass motion and the overall rotation of the molecular system, which are constant quantities in the absence of external forces or torques. In order to separate external from internal motions, two types of coordinate transformations can be performed. In the first, linear type of transformation, the center of mass is separated by the definition of space fixed relative position vectors of the atoms. The second, nonlinear transformation defines a reference frame that rotates with the molecule.

Despite linearity, the kinetic energy operator becomes more complicated and nonseparable when expressed in the space fixed relative position vectors which are obtained, when the center of mass is separated. There is a class of *orthogonal relative position vectors*, the definition of which depends on the masses of the particles, and in which the kinetic energy operator becomes maximally separable. These are three dimensional vectors represented by the symbol $\vec{r}_n$. The issue is discussed very pedagogically by Mladenović (2000). For N particles, there are $N-1$ orthogonal relative position vectors $\vec{r}_n$. Examples of such vectors are Jacobi and Radau vectors. When each of the three-dimensional position vectors is described in terms of its polar coordinates r_n, ϑ_n and φ_n, one obtains a representation of the molecular system in polyspherical coordinates (Nauts and Chapuisat, 1987; Gatti et al., 1998). For a system of N particles, the kinetic energy operator then gains the following general, very compact form,

$$^{(\mathrm{SF})}\hat{T} = \sum_{n=1}^{N-1} \left\{ -\frac{h^2}{8\pi^2 \mu_n} \left(\frac{\partial^2}{\partial {r_n}^2} + \frac{2}{r_n}\frac{\partial}{\partial r_n} \right) + \frac{h^2}{8\pi^2 \mu_n r_n^2}\, ^{(\mathrm{SF})}\hat{\Lambda}_n \right\}, \tag{1.71}$$

where

$$^{(\mathrm{SF})}\hat{\Lambda} = -\left(\frac{1}{\sin(\vartheta_n)}\frac{\partial}{\partial \vartheta_n}\sin(\vartheta_n)\frac{\partial}{\partial \vartheta_n} + \frac{1}{\sin^2(\vartheta_n)}\frac{\partial^2}{\partial {\varphi_n}^2} \right). \tag{1.72}$$

The quantities μ_n ($n = 1, \ldots, N-1$) are appropriately defined reduced masses (Mladenović, 2000; Gatti et al., 1998; Gatti and Iung, 2009). In principle, N can be the total number of particles in the molecule, i.e., electrons and nuclei. In practice, electronic motion has usually been separated adiabatically from the nuclear motion, as discussed in the previous section, before the actual nuclear dynamics is treated, and only the nuclei are considered for the definition of orthogonal relative position vectors; $N = N_{\mathrm{a}}$ is then the number of atoms. Note that, in this case, an error is introduced, as the electrons are generally not considered in the definition of the center of mass. This error, which is typically smaller than the error usually made by applying the Born–Oppenheimer approximation with well separated electronic states, was discussed by Kutzelnigg (2007).

The kinetic operator in Eq. (1.71) has singularities at $r_n = 0$ and $\vartheta_n = 0$ or π. The singularity at $r_n = 0$ can easily be removed. Those in terms of ϑ_n can be removed in the evaluation of matrix elements using nonseparable bases such as spherical harmonics in ϑ_n and φ_n, or using special *discrete variable representations* (DVR) (Light and Carrington, 2000; Beck et al., 2000) derived from these functions. Using nonseparable basis functions may render calculations longer.

In order to transform the space fixed coordinates into molecule fixed coordinates that rotate with the molecule, a direction-cosine matrix is used in connection with the three Euler angles α, β and γ (Zare, 1988). In order to define these angles, the body fixed axes system ${}^{(\mathrm{BF})}z$, ${}^{(\mathrm{BF})}y$ and ${}^{(\mathrm{BF})}x$ has to be defined. Formerly (see, e.g., Wilson et al., 1955), the system of principal axes of the instantaneous inertia tensor was usually chosen. In more recent work, an *embedding idea* following Tennyson and Sutcliffe (1982) has been adopted, which consists of using two of the orthogonal relative position vectors to define the body fixed reference axis system (see also Tennyson, 2011). The highest indexed vector $\vec{r}_{N_\mathrm{a}-1}$ may be chosen to define the body fixed z-axis. Hence, we identify the Euler angles $\alpha = \varphi_{N_\mathrm{a}-1}$ and $\beta = \vartheta_{N_\mathrm{a}-1}$. Then, $\vec{r}_{N_\mathrm{a}-2}$ is chosen to define the body fixed xz-plane. The rotation angle γ about the body fixed axis ${}^{(\mathrm{bf})}z$ as well as all remaining angles can be determined analytically (Mladenović, 2000) from the space fixed angles. The corresponding transformation equations, which are nonlinear and transcendental, however, lead to a transformation among angles alone:

$$\underbrace{\vartheta_{N_\mathrm{a}-1}, \varphi_{N_\mathrm{a}-1}, \ldots, \vartheta_1, \varphi_1}_{2\times(N_\mathrm{a}-1)\ \text{angles}}$$

$$\longrightarrow \underbrace{\alpha, \beta, \gamma, \theta_{N_\mathrm{a}-2}, \theta_{N_\mathrm{a}-3}, \phi_{N_\mathrm{a}-3}, \ldots, \theta_1, \phi_1}_{2\times(N_\mathrm{a}-1)\ \text{angles}}.$$

In the body fixed system, since $\vec{r}_{N_\mathrm{a}-2}$ lies in the body fixed xz-plane, the azimuthal angle $\phi_{N_\mathrm{a}-2}$ is fixed (usually to π), and does not appear as a dynamical variable. Note also that this transformation does not alter the lengths of the orthogonal relative position vectors.

The set of $(3 \times N_\mathrm{a} - 6)$ internal coordinates composed of $(N_\mathrm{a} - 1)$ radii and $(2 \times (N_\mathrm{a} - 1) - 3)$ angles,

$$r_{N_\mathrm{a}-1}, \ldots, r_1, \theta_{N_\mathrm{a}-2}, \theta_{N_\mathrm{a}-3}, \phi_{N_\mathrm{a}-3}, \ldots, \theta_1, \phi_1,$$

defines a set of *polyspherical coordinates*.

Polyspherical coordinates are a good choice for multidimensional dynamics, as the expression of the kinetic operator can be rendered quite simple. Additionally, the differential volume element, which is given as the Jacobian determinant of the nonlinear transformation, will always be separable. This has two important aspects: First, multidimensional integrals can always be evaluated as products of one-dimensional integrals; secondly, reduced dimensionality treatments can easily be defined, as the integration domains have no mutual dependencies. We may mention here also the use of "polar normal coordinates" in treating anharmonic Fermi resonances (Lewerenz and Quack, 1988; Luckhaus and Quack, 1992) who used this efficiently in contraction schemes in conjunction with DVR techniques (Bačić et al., 1988).

Such coordinates were used to describe a multidimensional tunneling motion in the femtosecond time domain in vibrationally highly excited ammonia (Marquardt et al., 2010) and methane vibrational states (Zhao et al., 2018), in the proton transfer dynamics in malonaldehyde (Joubert-Doriol et al., 2012), and even in the complex ring-opening dynamics of indolinobenzospiropyran (Joubert-Doriol et al., 2014).

Because of the nonlinear transformation between the space fixed and the body fixed reference system, the body-fixed form of the kinetic energy operator will in general be more complicated than the expression in Eq. (1.71). However, it is possible to show that it can always be written in terms of a sum of products of one dimensional operators (see the review by Gatti and Iung, 2009), which facilitates its use in particular in connection with the MCTDH method for solving Eq. (1.42).

Polyspherical coordinates will generate technical problems when chemical reaction are to be described, during which the set of orthogonal relative position vectors enabling the separation of reactant spaces in the entrance channel is not the same as that enabling the separation of product spaces in the exit channel. In order to evaluate expectation values such as scattering matrix elements, for instance, complicated coordinate transformations need to be applied, in such cases. Elegant solutions to this problem have been proposed by Xiao et al. (2011). An alternative would be to use hyperspherical coordinates (Nauts and Chapuisat, 1987; Kuppermann, 1996; Aquilanti and Cavalli, 1997; Aquilanti et al., 1999). The latter would allow us to obtain a more balanced treatment of entrance and exit channels in a collision, and the calculation of integrals become more easy. One disadvantage of using hyperspherical coordinates is the increased number of angular variables. As can be followed from inspection of Eq. (1.72), angular variables are accompanied by singularities in the kinetic energy operator, the number of which will hence increase. The necessity to use a larger number of nonseparable bases to avoid these singularities is a drawback.

1.6 QUANTUM DYNAMICS UNDER EXCITATION WITH COHERENT MONOCHROMATIC RADIATION

1.6.1 Introductory Remarks

The excitation with coherent monochromatic (or nearly monochromatic) radiation is an important special case of molecular spectroscopy and quantum dynamics to be discussed in more detail in this section. While, in principle, any of the direct integration schemes for the time dependent Schrödinger equation discussed in Section 1.3 might be used, we concentrate here on the approach where the electromagnetic field is treated classically and the stationary states of the field free molecule are obtained in a first step by solving the time independent Schrödinger equation for the isolated molecule, as described in Chapter 2 of this book (Császár et al., 2020).

In a second step one then solves the time dependent Schrödinger equation in the basis of these molecular eigenstates. The advantages of this two-step approach have been pointed out for the case of coherent infrared excitation by Quack (1978, 1989b), and in Fábri et al. (2019) (see also Section 1.3.1). One important advantage of this approach is the possibility to check the theoretical results of the first step against very accurate experimental results from high resolution spectroscopy (Merkt and Quack, 2011a). The division of the approach in two steps has been called a divide-and-rule (divide et impera, DEI)-Ansatz by Fábri et al. (2019).

We follow here closely the treatment given in Quack (1998) and Merkt and Quack (2011a), in part even literally. One should note, however, that some of the advantages of this DEI approach are lost when one considers very short excitation times, very strong fields and very high radiation frequencies (VUV, Röntgen, etc.) where some of the approximations in the starting point of the approach are lost.

1.6.2 General Aspects of Atomic and Molecular Systems in Electromagnetic Field

High-power coherent laser light sources allow for a variety of phenomena ranging from coherent single-photon transitions to multiphoton transitions of different types. Fig. 1.4 provides a summary of mechanisms for such transitions.

While excitation with incoherent light can be based on a statistical treatment (Einstein, 1916a,b, 1917; Merkt and Quack, 2011b), excitation with coherent light can be handled by means of quantum dynamics. Intense, coherent laser radiation as also electromagnetic radiation in the radiofrequency domain used in nuclear magnetic resonance (NMR) spectroscopy (Ernst et al., 1987) can be treated as a classical electromagnetic wave satisfying the general wave equations (1.73) and (1.74) resulting from Maxwell's theory:

$$\nabla^2 \vec{E} = \mu\mu_0\varepsilon\varepsilon_0 \frac{\partial^2 \vec{E}}{\partial t^2}, \tag{1.73}$$

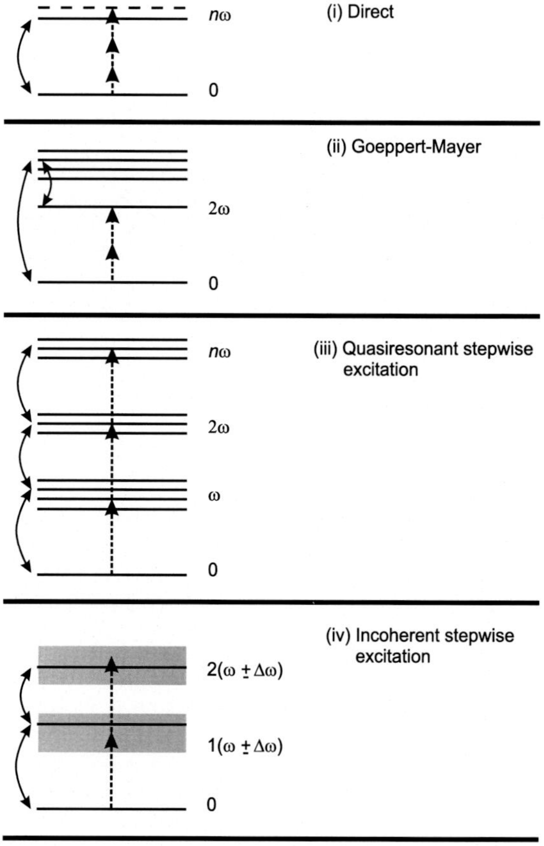

FIG. 1.4 Mechanisms for radiative excitation [after Quack, 1998]. Dotted lines give the transitions, curved full lines the dipole coupling.

$$\nabla^2 \vec{B} = \mu\mu_0\varepsilon\varepsilon_0 \frac{\partial^2 \vec{B}}{\partial t^2}, \tag{1.74}$$

where $\vec{E}$ is the electric field vector and $\vec{B}$ the vector of the magnetic induction, μ, μ_0, ε, ε_0 are the usual field constants, see (Cohen et al., 2007), with $\varepsilon = \mu = 1$ in vacuo. In Eqs. (1.73) and (1.74), $\nabla^2 = \vec{\nabla}^2$, and the nabla operator $\vec{\nabla}$ is defined in Eq. (1.75):

$$\vec{\nabla} = \vec{e}_x \frac{\partial}{\partial x} + \vec{e}_y \frac{\partial}{\partial y} + \vec{e}_z \frac{\partial}{\partial z} \tag{1.75}$$

where $\vec{e}_x$, $\vec{e}_y$, $\vec{e}_z$ are the unit vectors in a (right-handed) Cartesian coordinate system. The classical electromagnetic wave can be understood as the coherent state description of the quantum field in the limit of very large average number $\langle n \rangle$ of quanta per field mode (Glauber, 1963a,b; Perelomov, 1986). Coherent laser radiation and also radiofrequency radiation are frequently charac-

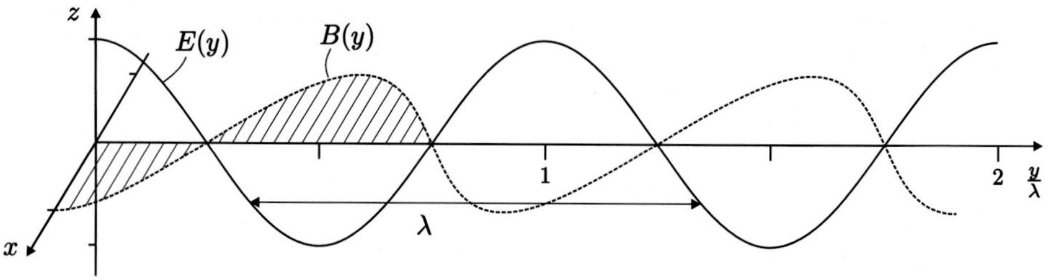

FIG. 1.5 Schematic representation of a z-polarized monochromatic wave (after Merkt and Quack, 2011b).

terized by $\langle n \rangle > 10^{10}$. Thus the classical approximation to the electromagnetic field is excellent. The situation of weak thermal light sources in the optical domain is very different ($\langle n \rangle < 1$), requiring a quantum statistical treatment.

We consider here, for simplicity, the special case of a classical z-polarized electromagnetic wave propagating in vacuo in the y-direction with slowly varying (or constant) field amplitudes $E_0(t)$ and $B_0(t)$ (see Fig. 1.5):

$$E_z(y,t) = |E_0(t)| \cos(\omega t + \eta' - k_\omega y), \tag{1.76}$$

$$B_x(y,t) = |B_0(t)| \cos(\omega t + \eta' - k_\omega y), \tag{1.77}$$

$\omega = 2\pi\nu$ is the angular frequency, $k_\omega = 2\pi/\lambda$ the angular wavenumber, $\nu = c/\lambda$ the ordinary frequency, and λ the wavelength. At a given position y, the phase η' can be combined with the phase $-k_\omega y$ to an overall phase ($\eta = k_\omega y - \eta'$).

The extension to more general cases is straightforward (see also Quack, 1998). The intensity of the radiation is, in general,

$$I(y,t) = |E_z(y,t)|^2 \sqrt{\frac{\varepsilon\varepsilon_0}{\mu\mu_0}}, \tag{1.78}$$

and averaging over time with $\left\langle \cos^2 x \right\rangle = 1/2$, one has from Eqs. (1.76) and (1.78)

$$I(t) = \frac{1}{2} |E_0(t)|^2 \sqrt{\frac{\varepsilon\varepsilon_0}{\mu\mu_0}}. \tag{1.79}$$

For the speed of light, one has in some medium with refractive index n_m

$$c_m = (\mu\mu_0\varepsilon\varepsilon_0)^{-1/2} = c/n_m \tag{1.80}$$

and in vacuo ($\mu = \varepsilon = 1$)

$$c = (\mu_0\varepsilon_0)^{-1/2}. \tag{1.81}$$

We can mention here some practical equations for calculating electric and magnetic field strengths when irradiating with monochromatic radiation of given intensity I:

$$\left|\frac{E_0}{\mathrm{V\,cm^{-1}}}\right| \simeq 27.44924 \sqrt{\frac{I}{\mathrm{W\,cm^{-2}}}}, \tag{1.82}$$

$$\left|\frac{B_0}{\mathrm{T}}\right| \simeq 9.156 \cdot 10^{-6} \sqrt{\frac{I}{\mathrm{W\,cm^{-2}}}}. \tag{1.83}$$

A further quantity characterizing the irradiation over some period of time t is the fluence $F(t)$ defined by Eq. (1.84) as

$$F(t) = \int_0^t I(t')\mathrm{d}t'. \tag{1.84}$$

For wavelengths $\lambda > 100$ nm, one can assume $\vec{E}$ and $\vec{B}$ to vary little over the extension of the atomic or molecular system ($\Delta y < 1$ nm) at any given time, which leads to the dipole approximation for the interaction energy between molecule and field,

$$\hat{V}_{\text{el.dipole}} = -\hat{\vec{\mu}}_{\text{el}} \cdot \vec{E}, \tag{1.85}$$

where $\hat{\vec{\mu}}_{\text{el}}$ is the electric dipole operator vector given by Eq. (1.86), with charges q_i for the particles with position operator vector $\hat{\vec{r}}_i$,

$$\hat{\vec{\mu}}_{\text{el}} = \sum_i q_i \hat{\vec{r}}_i. \tag{1.86}$$

Similarly, one has the interaction energy with a magnetic dipole operator vector $\hat{\vec{\mu}}_{\text{magn}}$,

$$\hat{V}_{\text{magn.dipole}} = -\hat{\vec{\mu}}_{\text{magn}} \cdot \vec{B}. \tag{1.87}$$

For the present quantum dynamical treatment of coherent excitation, we restrict our attention to electric dipole transitions in a field given by Eq. (1.76), and therefore we can write, with the z-component $\hat{\mu}_z$ of the electric dipole operator vector (and abbreviating $\eta = k_\omega y - \eta'$), as follows:

$$\hat{V}_{\text{el.dipole}} = -\hat{\mu}_z E_z(y,t) = -\hat{\mu}_z \left|E_0(t)\right| \cos(\omega t - \eta). \tag{1.88}$$

The extension to magnetic dipole transitions is straightforward. We give here only a brief summary and refer to Quack (1978, 1982, 1998) for more details.

1.6.3 Time-Dependent Quantum Dynamics in an Oscillatory Electromagnetic Field

Consider now the time-dependent Schrödinger equation (1.18) (or Eq. (1.42)) with a time-dependent Hamiltonian

$$\hat{H}(t) = \hat{H}_{\text{Mol}} - \hat{\mu}_z \left|E_0(t)\right| \cos(\omega t - \eta), \tag{1.89}$$

with $\hat{H}_{\text{Mol}}$ being the time-independent Hamiltonian for the isolated molecule in the absence of fields and the interaction Hamiltonian being a time-dependent, oscillatory function. We assume the solution of the time-independent Schrödinger equation for the isolated molecule to be given by Eq. (1.90) with $\hbar = h/2\pi$,

$$\hat{H}_{\text{Mol}}\varphi_k = E_k\varphi_k = \hbar\omega_k\varphi_k, \tag{1.90}$$

and write the solution of the time-dependent Schrödinger equation in the basis φ_k of molecular eigenstates with time-dependent coefficients as

$$\Psi(r,t) = \sum_k b_k(t)\varphi_k(r). \tag{1.91}$$

Here r stands again for a set of molecular coordinates, as in Eq. (1.21). Inserting this into Eq. (1.18), we obtain a set of coupled differential equations:

$$i\hbar\frac{\mathrm{d}b_j}{\mathrm{d}t} = \sum_k H_{jk}b_k(t), \tag{1.92}$$

or in matrix notation,

$$i\hbar\frac{\mathrm{d}\boldsymbol{b}(t)}{\mathrm{d}t} = \boldsymbol{H}(t)\boldsymbol{b}(t). \tag{1.93}$$

This is again, in essence, a matrix representation of the original Schrödinger equation (see Section 1.3 and Eq. (1.44)). Assuming molecular states of well-defined parity, the diagonal electric dipole matrix elements vanish and we have the diagonal elements of $\boldsymbol{H}(t)$,

$$H_{ii} = E_i = \left\langle \varphi_i \middle| \hat{H}_{\text{Mol}} \middle| \varphi_i \right\rangle \equiv \hbar\omega_i. \tag{1.94}$$

For other situations such as for chiral molecules or if parity violation were important (see Quack, 2011a), one would have also a diagonal contribution from the electric dipole interaction energy. Disregarding such cases here, the electric dipole interaction energy leads to time-dependent off-diagonal matrix elements

$$H_{kj}(t) = \left\langle \varphi_k \middle| \hat{V}_{\text{el.dipole}}(t) \middle| \varphi_j \right\rangle. \tag{1.95}$$

Dividing H_{kj} by $\hbar\cos(\omega t - \eta)$, we obtain a matrix element V_{kj}, which is independent of time, if we can assume $|E_0(t)|$ to be sufficiently slowly varying in time that it can be taken to be constant for the time period under consideration, as we shall do, replacing $E_0(t)$ by E_0 leads to

$$V_{kj} = \frac{H_{kj}}{\left[\hbar\cos(\omega t - \eta)\right]} = -\left\langle \varphi_k \middle| \hat{\mu}_z \middle| \varphi_j \right\rangle \frac{|E_0|}{\hbar} = V_{jk}^{*}. \tag{1.96}$$

We then obtain a set of coupled differential equations in matrix notation:

$$\mathrm{i}\frac{\mathrm{d}}{\mathrm{d}t}\boldsymbol{b}(t) = \{\boldsymbol{W} + \boldsymbol{V}\cos(\omega t - \eta)\}\,\boldsymbol{b}(t) \tag{1.97}$$

where we have defined the diagonal matrix $\boldsymbol{W}$ by the matrix elements $W_{kk} \equiv \omega_k$.

This is still a practically exact representation of the original time-dependent Schrödinger equation for the physical situation considered here. Because of the essential time dependence in $\boldsymbol{V}\cos(\omega t - \eta)$, in general there is no simple closed expression in the form of the exponential function analogous to Eqs. (1.16), (1.22) or (1.47) and (1.50), with a few exceptions, such as the harmonically driven harmonic oscillator (see Marquardt and Quack, 1989, and the references cited therein, as discussed at the end of this section). Apart from numerical, stepwise solutions discussed by Quack (1998), one can make use of series expansions such as the Magnus expansion. This solves Eq. (1.93) by means of the following series for $\boldsymbol{U}(t,t_0)$, with the unit matrix $\boldsymbol{I}$:

$$\boldsymbol{b}(t) = \boldsymbol{U}(t,t_0)\,\boldsymbol{b}(t_0), \tag{1.98}$$

$$\boldsymbol{U}(t_0,t_0) = \boldsymbol{I}, \tag{1.99}$$

$$\boldsymbol{U}(t,t_0) = \exp\left(\sum_{n=0}^{\infty} \boldsymbol{C}_n\right). \tag{1.100}$$

The first two terms are given by the following expressions:

$$\mathrm{i}\hbar \boldsymbol{C}_0 = \int_{t_0}^{t} \boldsymbol{H}(t')\mathrm{d}t', \tag{1.101}$$

$$\mathrm{i}\hbar \boldsymbol{C}_1 = -\frac{1}{2}\int_{t_0}^{t}\left\{\int_{t_0}^{t''}\left[\boldsymbol{H}(t'),\boldsymbol{H}(t'')\right]\mathrm{d}t'\right\}\mathrm{d}t''. \tag{1.102}$$

Higher terms contain more complex combinations of commutators of the type $[\boldsymbol{H}(t'), \boldsymbol{H}(t'')]$. From this, one recognizes that the series terminates after the first term given by Eq. (1.78), if $\boldsymbol{H}(t')$ and $\boldsymbol{H}(t'')$ commute at all t', t'', which is true if $\boldsymbol{H}$ does not depend on time, resulting in the exponential solutions already discussed. There are other (rare) cases of time-dependent $\boldsymbol{H}(t)$, but with $[\boldsymbol{H}(t'), \boldsymbol{H}(t'')] = 0$. One can, however, also make use of the periodicity of the field using Floquet's theorem (Quack, 1978, 1998).

1.6.4 Floquet Solution for Hamiltonians With Strict Periodicity

With $\boldsymbol{H} = \hbar\{\boldsymbol{W} + \boldsymbol{V}\cos(\omega t - \eta)\}$ from Eq. (1.74), one has obviously

$$\boldsymbol{H}(t+\tau) = \boldsymbol{H}(t) \tag{1.103}$$

with period $\tau = 2\pi/\omega$.

Making use of the Floquet theorem (or Floquet–Liapunoff theorem, see Quack, 1978, 1998 and Quack and Sutcliffe, 1985 for the historical references), one has the following form for the time-evolution matrix (with some integer n):

$$\boldsymbol{U}(t,t_0) = \boldsymbol{F}(t,t_0)\exp\left(\boldsymbol{A}(t-t_0)\right), \tag{1.104}$$

$$\boldsymbol{F}(t_0,t_0) = \boldsymbol{I}, \tag{1.105}$$

$$\boldsymbol{F}(t+n\tau) = \boldsymbol{F}(t), \tag{1.106}$$

$$\boldsymbol{A}(t') = \boldsymbol{A}(t'') \qquad (\text{all } t', t''). \tag{1.107}$$

It is then sufficient to integrate Eq. (1.93) numerically over one period τ using methods discussed in Section 1.3, and then obtain the evolution for all times by matrix multiplications according to Eqs. (1.104)–(1.107). In particular, at multiples of the period τ, one finds (with $t_0 = 0$):

$$\boldsymbol{U}(\tau) = \exp(\boldsymbol{A}\tau), \tag{1.108}$$

$$\boldsymbol{U}(n\tau) = [\boldsymbol{U}(\tau)]^n. \tag{1.109}$$

There has been considerable literature making use of Floquet's theorem for the treatment of coherent excitation and there also exist computer program packages (Quack and Sutcliffe, 1986; Quack, 1998; Marquardt et al., 2019). We shall discuss here a further useful approximation.

1.6.5 Weak-Field Quasiresonant Approximation (WF-QRA) for Coherent Monochromatic Excitation

We consider a level scheme for coherent excitation with levels near the resonance as shown in Fig. 1.6. One can then associate with each molecular level of energy $E_k = \hbar\omega_k$ an integer photon number n_k for near-resonant excitation such that

$$\omega_k = n_k\omega + x_k \tag{1.110}$$

where x_k is a frequency mismatch for exact resonance at the best choice of n_k.

Under the conditions that (i) there is a sequential near-resonant excitation path, (ii) only levels with a general resonance mismatch satisfying $|x_k| \ll \omega$ contribute effectively to excitation (quasiresonant condition), and (iii) the coupling matrix elements satisfy $|V_{kj}| \ll \omega$ (weak-field condition), one can approximately derive a set of coupled equations with an effective Hamiltonian that does not depend on time. For this purpose, one makes the simple substitution (Quack, 1978, 1998), namely

$$a_k = \exp(\mathrm{i}n_k\omega t)b_k, \tag{1.111}$$

resulting in the set of differential equations

$$\mathrm{i}\frac{\mathrm{d}a_k}{\mathrm{d}t} = x_k a_k + \frac{1}{2}\sum_{j\neq k} V_{kj}a_j, \tag{1.112}$$

or in matrix form (with the diagonal matrix $\boldsymbol{X}$ and $X_{kk} = x_k$)

$$\mathrm{i}\frac{\mathrm{d}\boldsymbol{a}}{\mathrm{d}t} = \left\{\boldsymbol{X} + \frac{1}{2}\boldsymbol{V}'\right\}\boldsymbol{a}. \tag{1.113}$$

where $\boldsymbol{V}'$ differs from $\boldsymbol{V}$ by explicitly setting all V_{kj} equal to zero if the levels E_k and E_j are far off resonance, e.g. if they are not adjacent quasiresonant levels.

One can interpret this equation by means of an "effective Hamiltonian"

$$\boldsymbol{H}_{\mathrm{eff}}^{(a)} = \hbar\left\{\boldsymbol{X} + \frac{1}{2}\boldsymbol{V}'\right\} \tag{1.114}$$

and the corresponding effective time-evolution matrix

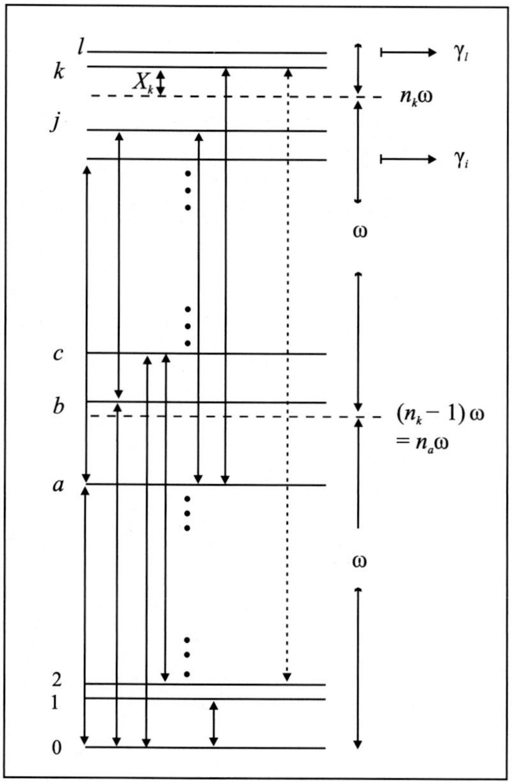

FIG. 1.6 Energy-level scheme [after Quack, 1982]. The molecular energy levels are marked as horizontal full lines. The horizontal dashed lines correspond to the energies $E_0 + n_k \hbar\omega$ of the ground state (E_0) plus the energy of n_k photons.

$$U_{\text{eff}}^{(a)}(t, t_0) = \exp\left[-2\pi \mathrm{i} \frac{H_{\text{eff}}^{(a)}(t - t_0)}{h}\right]$$

$$= \exp\left[-\mathrm{i}\left(X + \frac{1}{2}V'\right)(t - t_0)\right], \quad (1.115)$$

$$a(t) = U_{\text{eff}}^{(a)}(t, t_0)\, a(t_0). \quad (1.116)$$

It is sometimes useful to describe decay phenomena by adding an imaginary energy contribution, for instance, $E_l = \Re(E_l) - \mathrm{i}\gamma_l/2$ as indicated, where $\Re(z)$ designates the real part of the complex number z (see Quack, 1982 and Quack and Sutcliffe, 1984). The introduction of an imaginary part to the energy of a spectroscopic state introduces into the field of non-Hermitian quantum mechanics.

Non-Hermitian quantum mechanics is covered in great detail in the book of Moiseyev (2011) (see also Hehenberger et al., 1974). In principle exponential decay can be derived also with the aid of Hermitean quantum mechanics, as exemplified by the Bixon-Jortner model (Bixon and Jortner, 1968). Related to exponential decay are quasi-Lorentzian lineshapes with a lower energy bound which were shown to arise from Hermitean quantum mechanics in an analytical exact solution of the time dependent Schrödinger equation (Marquardt, 2019). Various aspects of exponential decay and the related spectroscopic lineshapes were discussed by Merkt and Quack (2011b), were one can also find a critical discussion of the incorrect but frequently used "textbook" derivation of exponential decay lineshapes from the uncertainty relation Eq. (1.11), which often is presented as "didactically useful" but in fact is quite fallacious.

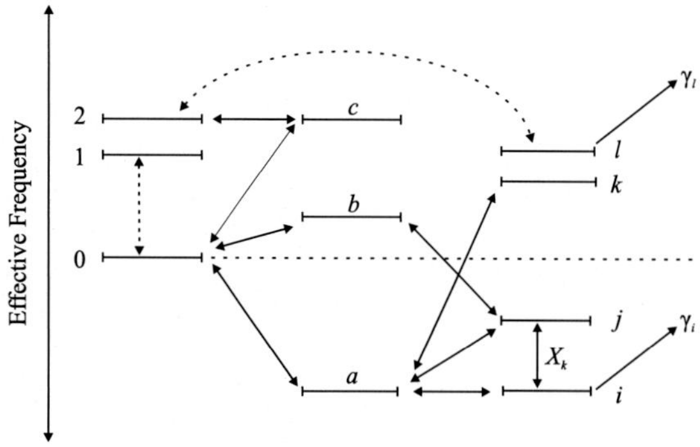

FIG. 1.7 Effective-frequency scheme corresponding to the energy-level scheme of Fig. 1.6 [after Quack, 1982].

The result in Eqs. (1.113)–(1.116) is quite remarkable as it corresponds to replacing the molecular energies E_k by new effective energies $\hbar X_{kk}$ and the couplings V_{kj} by new effective couplings $(V_{kj}/2)$ for near-resonant levels (and implicitly by zero for far off-resonant levels). We can therefore use the time independent $V'/2$ rather than $V\cos(\omega t - \eta)$ for the general coupling matrix in Eq. (1.97). This is graphically shown in Fig. 1.7 for the same level scheme as in Fig. 1.6, but with effective energies that are "on the same energy shell" and thus effective couplings between levels of similar effective energy. We note the close analogy to the dressed atom (dressed molecule) picture by Cohen-Tannoudji et al. (1992) and Haroche (2012), which uses, however, a different derivation. We note that the quasiresonant transformation as given in Quack (1978, 1998) can be written in matrix notation:

$$a = S\, b, \quad (1.117)$$

with the diagonal matrix

$$S_{kk} = \exp(\mathrm{i} n_k \omega t). \quad (1.118)$$

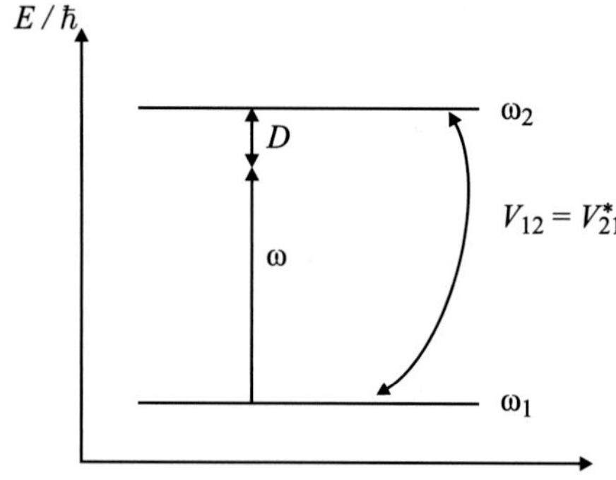

FIG. 1.8 Two-level scheme for coherent radiative excitation with frequency ω [after Merkt and Quack, 2011b].

Similarly, a transformation for the density matrix $\boldsymbol{P}^{(a)}$ from $\boldsymbol{P}^{(b)}$ can be derived in this approximation, resulting in the solution of the Liouville–von Neumann equation for $\boldsymbol{P}(t)$ by

$$\boldsymbol{P}^{(a)}(t) = \boldsymbol{S}\,\boldsymbol{P}^{(b)}\,\boldsymbol{S}^{\dagger}, \tag{1.119}$$

$$\boldsymbol{P}^{(a)}(t) = \boldsymbol{U}_{\text{eff}}^{(a)}(t,t_0)\,\boldsymbol{P}^{(a)}(t_0)\,\boldsymbol{U}_{\text{eff}}^{(a)\dagger}(t,t_0). \tag{1.120}$$

For details, we refer to Quack (1978, 1982, 1998) and the discussions in Donley et al. (2001), Marquardt et al. (2019), Quack and Sutcliffe (1984, 1985) as well as in Whaley and Light (1984). We turn now to a simple application to the special case of coherent radiative excitation connecting just two quantum states.

1.6.6 Coherent Monochromatic Excitation Between Two Quantum States

If only two quantum states are considered, one obtains a scheme for the coherent monochromatic radiative excitation as shown in Fig. 1.8.

Eq. (1.93) simplifies to the set of just two coupled differential equations:

$$\mathrm{i}\frac{\mathrm{d}b_1}{\mathrm{d}t} = \omega_1 b_1 + V_{12}\cos(\omega t - \eta)\; b_2, \tag{1.121}$$

$$\mathrm{i}\frac{\mathrm{d}b_2}{\mathrm{d}t} = V_{21}\cos(\omega t - \eta) b_1 + \omega_2 b_2. \tag{1.122}$$

One might think that this rather simple set of coupled differential equations has a simple solution, but, in fact, it seems that until today no simple general analytical solution in the form of a closed expression is known (see the discussion in Quack (1978, 1998), also for special cases). Of course, one can write down series expansions (Sections 1.3.2 and 1.3.3) or quite easily solve the equations numerically, by other methods discussed in Section 1.3.

However, following the discussion of Section 1.6.5, an analytical solution in the form of a closed expression can be derived in the weak-field quasiresonant approximation (WF-QRA). The most general solution in terms of the time-evolution matrix $\boldsymbol{U}_{\text{eff}}^{(a)}$ is given by the matrix elements, omitting the index "eff" to simplify the notation, and taking a real $V_{12} = V_{21}^{*} = V$ as parameter,

$$U_{11}^{(a)} = \exp(-\mathrm{i}\lambda_1 t)\left[x^2 + y^2\exp(\mathrm{i}\omega_R t)\right], \tag{1.123}$$

$$U_{22}^{(a)} = \exp(-\mathrm{i}\lambda_1 t)\left[y^2 + x^2\exp(\mathrm{i}\omega_R t)\right], \tag{1.124}$$

$$U_{12}^{(a)} = U_{21}^{(a)} = \exp(-\mathrm{i}\lambda_1 t)xy\left[1 - \exp(\mathrm{i}\omega_R t)\right], \tag{1.125}$$

where we have used the following parameters as abbreviations:

$$D = \omega_2 - \omega_1 - \omega, \tag{1.126}$$

$$\omega_R = (\lambda_1 - \lambda_2) = \sqrt{V^2 + D^2} = \frac{2\pi}{\tau_R}, \tag{1.127}$$

$$\lambda_1 = \frac{1}{2}\left(D + \sqrt{V^2 + D^2}\right), \tag{1.128}$$

$$\lambda_2 = \frac{1}{2}\left(D - \sqrt{V^2 + D^2}\right), \tag{1.129}$$

$$x = \left[\frac{1}{2} - \frac{D}{2\omega_R}\right]^{1/2}, \tag{1.130}$$

$$y = \left[\frac{1}{2} + \frac{D}{2\omega_R}\right]^{1/2}. \tag{1.131}$$

We note that in Quack (1998) some of the expressions where misprinted, and these are corrected here (see also Merkt and Quack, 2011b). We also give the explicit form of the effective Hamiltonian corresponding to Eq. (1.114):

$$\frac{\boldsymbol{H}_{\text{eff}}^{(a)}}{\hbar} = \begin{pmatrix} 0 & 0 \\ 0 & D \end{pmatrix} + \frac{1}{2}\begin{pmatrix} 0 & V \\ V & 0 \end{pmatrix} \tag{1.132}$$

$$= \boldsymbol{X} + \frac{1}{2}\boldsymbol{V}. \tag{1.133}$$

In the two-level case, the distinction between V and V' is not necessary.

Now $\boldsymbol{U}_{\text{eff}}^{(a)}$ is explicitly derived by means of the eigenvalues and eigenstates of $\boldsymbol{H}_{\text{eff}}^{(a)}$ (see Section 1.3.1 and Quack, 1998) as

$$\boldsymbol{Z}^{-1}\left(\boldsymbol{X} + \frac{1}{2}\boldsymbol{V}\right)\boldsymbol{Z} = \boldsymbol{\Lambda} = \boldsymbol{Diag}(\lambda_1, \lambda_2), \tag{1.134}$$

$$\boldsymbol{U}_{\text{eff}}^{(a)}(t - t_0) = \boldsymbol{Z}\exp\left[-\mathrm{i}\boldsymbol{\Lambda}(t - t_0)\right]\boldsymbol{Z}^{-1}. \tag{1.135}$$

In Eq. (1.134) $\boldsymbol{\Lambda}$ is a diagonal matrix with eigenvalues λ_1 and λ_2.

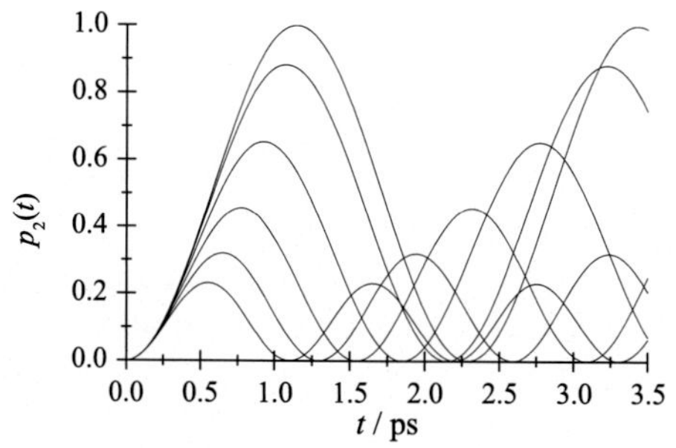

FIG. 1.9 Time-dependent level populations from Eq. (1.136). The population p_2 of the upper level in the scheme of Fig. 1.8 is shown with different resonant defects $D = (0, 1, 2, 3, 4, 5) \times 10^{12}$ s^{-1} for an electric dipole transition with the laser wave number $\tilde{\nu} = 1000$ cm^{-1}, intensity $I = 1$ GW cm^{-2} and transition moment $|\mu_{21}| = \langle\varphi_2|\hat{\mu}_z|\varphi_1\rangle =$ 1 Debye [after Merkt and Quack, 2011b].

While these general equations can be used to derive numerous properties of the coherent monochromatic excitation in the two-level problem, we conclude here with the result for the time-dependent population of the excited level $p_2(t)$, if, initially, at time zero, only the ground state is populated ($p_1(t = 0) = 1$).

One obtains Eq. (1.136) for the population of the upper level:

$$p_2(t) = |b_2(t)|^2 = \frac{V^2}{V^2 + D^2}\left[\sin\left(\frac{t}{2}\sqrt{V^2 + D^2}\right)\right]^2$$
$$= 1 - p_1(t). \tag{1.136}$$

This is a periodic exchange of population between ground and excited states with a period

$$\tau_R = \frac{2\pi}{\sqrt{V^2 + D^2}}. \tag{1.137}$$

This period is called the *Rabi period*, with ω_R being the *Rabi frequency*, as these equations were derived on the basis of the so-called rotating wave approximation by Rabi (1937) (see also Rabi et al., 1938) in the context of early NMR experiments (in beams, with a magnetic dipole transition matrix element V, of course).

The rotating wave approximation becomes identical with the quasiresonant approximation for the special case of a two-level problem. Eq. (1.136) is frequently called the *Rabi formula*.

Fig. 1.9 shows the time-dependent level populations for a fairly typical case. With increasing resonant defect D, the amplitude of the oscillation decreases, but the frequency of oscillation increases. The initial time evolution is independent of the resonant defect as is readily seen from the series expansion of the $\sin(x)$ function for small arguments x ($\sin(x) = x + \cdots$) giving at sufficiently small times

$$p_2(t) \simeq V^2 t^2/4 \quad \text{(small } t\text{)}. \tag{1.138}$$

One can also consider the time-averaged population $\langle p_2(\omega)\rangle_t$ as a function of the exciting laser frequency ω at fixed resonance frequency $\omega_{12} = \omega_2 - \omega_1$. Because $\left\langle \sin^2(x)\right\rangle_x = 1/2$, one has

$$\langle p_2(\omega)\rangle_t = \frac{1}{2}\frac{V^2}{V^2 + (\omega - \omega_{12})^2} \tag{1.139}$$

so that $\langle p_2(\omega)\rangle$ is proportional to the average absorbed energy as a function of frequency ω, and one can interpret this expression as the effective absorption lineshape under intense coherent excitation. Indeed, Eq. (1.139) corresponds to a Lorentzian lineshape with full width at half-maximum $\Gamma_{\text{FWHM}} = 2V$. This effect is called *power broadening* because $V \propto \sqrt{I}$ (cf. Eqs. (1.82) and (1.96), sometimes the term *intensity broadening* is used as well). We have neglected here effects from spontaneous emission or collisions, as is obviously appropriate for the timescales applicable to Fig. 1.9, but not necessarily always so.

We may finally conclude with an estimate of errors arising when the conditions of the WF-QRA are not fulfilled. The special case of the degenerate two-level problem has been solved exactly (Quack, 1978). In this case, one has $\omega_1 = \omega_2$ and therefore $D^2 = \omega^2$. The Rabi formula Eq. (1.136) thus would give

$$p_2^{\text{Rabi}}(t) = \frac{V^2}{V^2 + \omega^2}\left[\sin\left(\frac{t}{2}\sqrt{V^2 + \omega^2}\right)\right]^2. \tag{1.140}$$

The exact solution (Quack, 1978) is

$$p_2^{\text{ex}}(t) = \left\{\sin\left[\left(\frac{V}{\omega}\right)\sin(\omega t)\right]\right\}^2. \tag{1.141}$$

One can consider the limit $|V| \ll \omega$ because this must be assumed for the validity of the Rabi formula, and considering this limit, one obtains

$$p_2^{\text{Rabi}}(t) = \frac{V^2}{\omega^2}\left[\sin\left(\frac{\omega t}{2}\right)\right]^2, \tag{1.142}$$

$$p_2^{\text{ex}}(t) = \frac{V^2}{\omega^2}[\sin(\omega t)]^2. \tag{1.143}$$

Thus the exact solution gives the same amplitude as the Rabi formula, but the period differs by a factor of 2 and the short time limit for the population p_2 differs by a factor of 4. Of course, the second condition for the validity of the WF-QRA is not satisfied, as one cannot have $|D| \ll \omega$ (rather by definition of the special case, one has $|D| = \omega$). If $|D| \ll \omega$, the Rabi formula gives a very good approximation, as one can show numerically. We may note here that closed analytical expressions for the solutions are available for the excitation of the harmonic oscillator both with the exact treatment (Eq. (1.93)) and within the WF-QRA (Eq. (1.113)), taking an infinite number of levels into account, which may, perhaps, seem surprising. The derivation has been given by Marquardt and Quack (1989) and leads to a further estimate of the ranges of validity and uncertainties introduced by the WF-QRA, both for populations and phases, in this case for a many-level system. A very interesting simple situation for a two-level population transfer by a three-level mechanism which can be treated with the QRA or Floquet approximation is STIRAP (Stimulated Raman Adiabatic Passage) reviewed recently by Bergmann et al. (2019).

1.7 CONCLUDING REMARKS

We measure time by motion and motion by time (Aristotle, as cited and translated in Quack (2004a,b)).

Time is what you read from your clock ("Zeit ist das, was man an der Uhr abliest", Einstein, 1922, as cited and translated in Quack (1999, 2004a,b)).

We shall conclude here with some general considerations on time as defined by atomic and molecular clocks and the important role of symmetries and approximate or exact constants of the motion in molecular quantum dynamics. We follow here Quack (2011a), see also Quack and Hacker (2016) for the role of symmetries in a broader context.

1.7.1 Time-Dependent Quantum Motion, Spectroscopy and Atomic and Molecular Clocks

Simple periodic quantum motions can be considered to be the basis of atomic and molecular clocks as special cases of general intramolecular dynamics. Indeed, the general quantum dynamics of isolated molecules are described by Eqs. (1.18) and (1.26) (or the formally equivalent Eqs. (1.42) and (1.46)). The quantities ψ_k and E_k are obtained from the solution of the stationary Schrödinger equation (1.26). The eigenfunctions ψ_k depend only on space (and spin) coordinates,

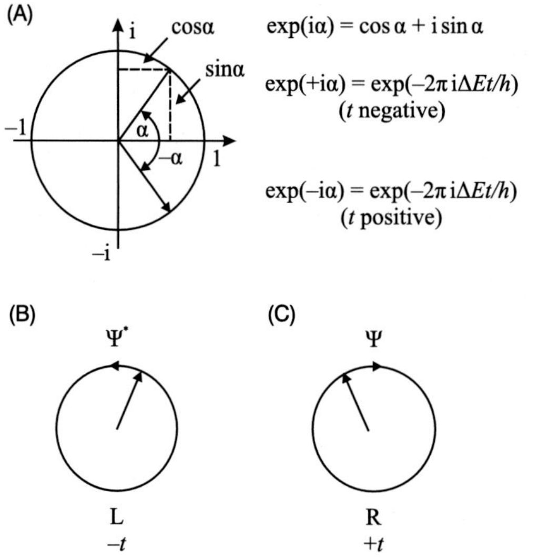

FIG. 1.10 Explanation of the phase factor of the complex, time-dependent wavefunction for the atomic and molecular clock. (A) The phase factor is graphically depicted as an arrow in the Gaussian plane (α is taken to be positive). For positive times t, one obtains $\exp(-i\alpha)$, the hand of the clocklike phase factor turns in the clockwise direction, in the sense of increasingly negative α. (B) and (C) These illustrate the wavefunction Ψ and its complex conjugate Ψ^*, which both describe a symmetrically equivalent solution of the Schrödinger equation, given time-reversal symmetry. The picture also describes the corresponding orbits of planets around the Sun, symmetric under time reversal. These can move in a clockwise as well as counterclockwise direction, providing acceptable solutions to the classical equations of motion [after Quack, 1999 by permission].

and the energy eigenvalues E_k can be subject to spectroscopic observation by means of the Bohr condition from Eq. (1.1), $|\Delta E_{jk}| = |E_k - E_j| = h\nu_{jk}$, with the transition frequency ν_{jk}. If we consider just two levels as equally populated, we obtain a periodic motion simplified to some time-dependent probability function:

$$p(t) = |a + b\exp(-2\pi i\,|\Delta E_{jk}|\,t/h)|^2. \tag{1.144}$$

The time dependence of such an atomic or a molecular clock is fully described by a time-dependent, complex periodic phase factor $\exp(i\alpha)$ represented graphically in Fig. 1.10 in terms of the complex Gaussian plane. The hand of the clock, figuratively speaking, is given by the vector describing $\exp(i\alpha)$ in the plane that moves clockwise for positive time and counterclockwise for negative time. The period of the clock

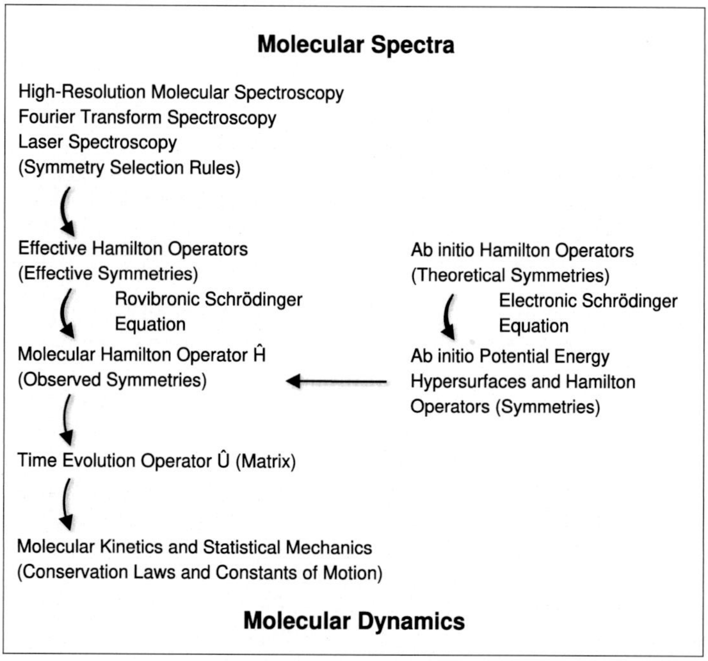

FIG. 1.11 Spectroscopic approach to intramolecular kinetics (see also Quack, 1997, 2003).

is given by one energy interval $|\Delta E_{jk}|$. The modern definition of the second is given by the separation ΔE_{jk} of the two hyperfine levels in ^{133}Cs (total angular momenta $F = 4$ and $F = 3$ resulting from combining nuclear spin $I = 7/2$ and electron spin $s = 1/2$) ($\Delta E_{jk} \approx 0.306\,633\,1899$ hc cm^{-1}) such that 1 s = $9\,192\,631\,770\,\tau$ exactly, with $\tau = h/\Delta E_{jk}$ (see Stohner and Quack, 2011).

Combining Eq. (1.144) for the atomic clock, and the analysis of molecular spectra in terms of solutions of the time-independent Schrödinger equation for energies and wave functions and the solution of the time-dependent Schrödinger equation in terms of these wavefunctions and energies as well, this can be made the basis of a spectroscopic approach to intramolecular kinetics following the scheme in Fig. 1.11.

We have discussed elsewhere how this spectroscopic approach can be used to obtain deep insights into intramolecular processes with time resolution starting from about 200 attoseconds (Quack, 1990, 2003, 2004a) (see also Albert et al., 2011 and Hippler et al., 2011). Here, we shall furthermore address the relation to the breaking and violation of symmetries and various time scales for atomic and molecular primary processes.

1.7.2 Hierarchy of Interactions and Hierarchy of Timescales for the Successive Breaking of Approximate Dynamical Symmetries in Intramolecular Primary Processes

The hierarchy of time scales for symmetry breaking can be related to the size of contributions in the molecular Hamiltonian. For instance, one might write the Hamiltonian in the following practical order of contributions to a sum, which decrease (roughly) in the following order (Quack, 1983):

$$\begin{aligned}\hat{H} = \hat{T}_{\mathrm{e}} + \hat{V}_{\mathrm{nn}} + \hat{V}_{\mathrm{ne}} + \hat{V}_{\mathrm{ee}} \\ + \hat{T}_{\mathrm{n}} \\ + \hat{H}_{\mathrm{SO}} + \hat{H}_{\mathrm{SS}} + \hat{H}_{\mathrm{rel}} \\ + \hat{H}_{\mathrm{hfs}} \\ + \hat{H}_{\mathrm{mol,rad}} + \hat{H}_{\mathrm{mol,environment}} \\ + \hat{H}_{\mathrm{weak}} + \hat{H}_{\mathrm{gravitational}} + \cdots\end{aligned} \tag{1.145}$$

where $\hat{T}_{\mathrm{e}}$ and $\hat{T}_{\mathrm{n}}$ are the kinetic energy operators for electrons and nuclei and $\hat{V}_{\mathrm{nn}}$, $\hat{V}_{\mathrm{ne}}$ and $\hat{V}_{\mathrm{ee}}$ are nucleus–nucleus, nucleus–electron, and electron–electron Coulomb potentials. The contributions to the Hamiltonian can be characterized by their underlying symmetry groups. For instance, the initial terms in Eq. (1.145) all show "parity symmetry". $\hat{H}_{\text{weak}}$ would be, in the example of parity violation, the relevant term contributing to the dynamical symmetry breaking in an intramolecular process as discussed by Quack (1986, 2011a), on the timescale of about 1 s; $\hat{H}_{\text{so}}$, $\hat{H}_{\text{ss}}$ are spin–orbit and spin–spin, $\hat{H}_{\text{hfs}}$ is the hyperfine coupling and $\hat{H}_{\text{rel}}$ contains other relativistic contributions to the Hamiltonian, with obvious notation for the other terms. However, one could also separate the molecular Hamiltonian into other types of contributions, according to what one considers a practical separation of the Hamiltonian, in some abstract manner, say

$$\hat{H} = \hat{H}_0 + \hat{H}_1 + \hat{H}_2 + \hat{H}_3 + \hat{H}_4 + \cdots. \tag{1.146}$$

In such an abstract description, a large $\hat{H}_0$ may show some (very high) symmetry, corresponding to a symmetry group of high order; $\hat{H}_1$ may be smaller (in terms of contributions to total energy), but might have a lower symmetry (smaller subgroup of the symmetry group of $\hat{H}_0$). Thus, considering $\hat{H}_1$, some of the symmetries of $\hat{H}_0$ will be violated or broken, and so forth with $\hat{H}_2$, $\hat{H}_3$, etc. The formal aspects of the time evolution with more or less symmetry can be summarized by just a few equations.

A dynamical variable represented by the operator $\hat{Q}$ satisfies the Heisenberg equation of motion, Eq. (1.14)).

If for a certain dynamical variable corresponding to the operator $\hat{C}$, one has the commutation relation (for an isolated system, $\hat{H}$ being time-independent)

$$\hat{C}\hat{H} = \hat{H}\hat{C}. \tag{1.147}$$

It follows with Eq. (1.16) that $\hat{U}$, being a function of $\hat{H}$, commutes with $\hat{C}$ as well,

$$\hat{U}\hat{C} = \hat{C}\hat{U}, \tag{1.148}$$

and therefore,

$$\begin{aligned}\hat{C}(t) &= \hat{U}^{\dagger}(t,t_0)\,\hat{C}(t_0)\,\hat{U}(t,t_0) = \hat{U}^{\dagger}(t,t_0)\,\hat{U}(t,t_0)\,\hat{C}(t_0)\\ &= \hat{C}(t_0).\end{aligned} \tag{1.149}$$

Thus, in the Heisenberg representation, $\hat{C}$ does not depend on time and is a "constant of the motion" or constant of evolution. Considered as transformations acting on $\hat{H}$ the $\hat{C}$ form a group, the symmetry group of $\hat{H}$ (and $\hat{U}$). One can readily see (Quack, 1983) that with the Liouville–von Neumann equation (1.37) for the density operator $\hat{P}$ or the corresponding density matrix $\boldsymbol{P}$ the expectation value of $\hat{C}$ is constant,

$$\langle \hat{C}(t) \rangle = \mathrm{Tr}(\hat{P}\hat{C}) = \langle \hat{C}(t_0) \rangle, \tag{1.150}$$

and also if $\Psi(t)$ is an eigenfunction $\zeta_n(t)$ of $\hat{C}$ with eigenvalue C_n, one has

$$\langle \hat{C}(t) \rangle = \langle \zeta_n(t) | \hat{C} | \zeta_n(t) \rangle = C_n. \tag{1.151}$$

The C_n are "good quantum numbers", which do not change in time with the evolution.

These equations can be applied in an analogous fashion to each part of the Hamiltonian in Eq. (1.146), i.e., successively to $\hat{H}_0$, $\hat{H}_1$, $\hat{H}_2$, ..., etc. A constant of the motion $\hat{C}_0$ of $\hat{H}_0$ will be strictly independent of time at the level of description corresponding to $\hat{H}_0$, which may be the largest contribution to the total energy. If then a smaller contribution $\hat{H}_1$ has a lower symmetry, some of the constants of motion $\hat{C}_0$ may show a time dependence due to this symmetry breaking term. If $\hat{H}_1$ contributes only a small part to the total energy, the time evolution of $\hat{C}_0$ will happen on correspondingly longer time scales and so forth for even smaller contributions due to $\hat{H}_2$, $\hat{H}_3$, $\hat{H}_4$, ...

Indeed, we have made this abstract approach a basis for understanding the separation of timescales in intramolecular processes as derived from spectroscopy, and Table 1.3 provides a summary of such results for intramolecular kinetics.

For instance, if $\hat{H}_0$ is taken to correspond to a description of an N-atomic nonlinear molecule by $3N - 6$ uncoupled harmonic oscillators and separable rotation, all quantum numbers v_k for the individual harmonic oscillators correspond to conserved "good" quantum numbers or constants of the motion.

As this is not a very good approximation, the anharmonic coupling terms (corresponding then to $\hat{H}_1$) are quite large and will lead to symmetry breaking on a very short timescale (depending on the case ranging from 10 fs to 10 ps as shown in Table 1.3 in the top group of processes). One can then introduce a number of successive coupling terms $\hat{H}_k$ that lead to further symmetry breaking and specific timescales for primary processes. The case of parity violation appears in group 4 counted from the top of Table 1.3 and the time evolution of parity as a kinetic primary process is one of the current frontiers in molecular quantum dynamics – interestingly on long timescales (Quack, 1986, 1994, 2001, 2011a,b). This example is also useful for illustrating another related concept arising for tunneling stereomutation in chiral molecules, as discussed in Chapter 7 of this book

TABLE 1.3
Timescales for intramolecular primary processes as successive symmetry breaking[a].

Symmetric state	Coupling and symmetry breaking	Timescale
Conservation of separable vibrational quantum numbers (for harmonic oscillators of normal modes)	Selective vibrational CH-stretch-bend-Fermi-Resonance in R_3CH (Marquardt et al., 1986; Marquardt and Quack, 1991; Quack, 1990; Quack and Stohner, 1993; Quack, 1995b; Beil et al., 1996a,b, 2000; Pochert et al., 2000)	10–200 fs
	Ordinary nonselective anharmonic couplings in CF_3R, Δl coupling in asymmetric $R_1R_2R_3CH$ (Quack, 1990; Pochert et al., 2000; He et al., 2002; Albert et al., 2017)	500 fs to 10 ps
Uncoupled oscillators (nearly adiabatically separable)	Adiabatically decoupled dynamics R–C≡C–H (Quack and Stohner, 1993; Quack, 1995b; von Puttkamer et al., 1983; Lehmann et al., 1994); $(HF)_2$ (Quack and Suhm, 1998; von Puttkamer and Quack, 1989; Quack, 2001, 2003); Δl coupling in C_{3v}-symmetric R_3CH (Luckhaus and Quack, 1993; Kushnarenko et al., 2018)	10 ps to 1 ns
Separable rotation-vibration-nuclear spin states (conservation of nuclear spin symmetry)	Violation of nuclear spin symmetry (nuclear spin–rotation–vibration coupling) (Quack, 1977, 1983; Chapovsky and Hermans, 1999)	1 ns to 1 s
Space inversion symmetry Parity conservation P	Parity violation (Quack, 1986, 1989a; Bakasov et al., 1996, 1998; Bakasov and Quack, 1999; Berger and Quack, 2000; Quack, 2002, 2006; Quack and Stohner, 2005; Prentner et al., 2015; Dietiker et al., 2015)	1 ms to 1 ks
Time-reversal symmetry T	T-violation in chiral and achiral molecules (Quack, 1997; Luckhaus and Quack, 1993; Luckhaus et al., 1993)	Molecular timescale not known
CPT symmetry	Hypothetical CPT violation (Quack, 1994, 1995a, 2003, 2008)	∞ (?)

[a] After Quack et al., 2008, see also Quack, 1990, Marquardt and Quack, 2001, Quack, 1995a, 2001, 2003, 2007.

(Quack and Seyfang, 2020). In the case of a very high barrier (tunneling splitting $\Delta E_{\pm} \approx 0$), the ground state of a chiral molecule is twofold degenerate because of space inversion symmetry, but this degeneracy is lifted by parity violation, resulting in a parity-violating energy difference $\Delta_{\text{pv}}E$, which can be interpreted as the "splitting" of the degenerate levels by a symmetry-violating contribution $\hat{H}_{\text{pv}}$ arising from H_{weak} to the Hamiltonian. The smallness of the symmetry-violating splitting of the degeneracy is directly related to the long timescale (1 ms to 1 ks, depending on the molecular example) for the dynamical, time-dependent symmetry breaking, simply by the equation for the general period of motion $\Delta\tau$:

$$\Delta\tau = h/\Delta E \tag{1.152}$$

which can be called the *elementary timescale for the symmetry breaking*. The ordering of the Hamiltonian in Eq. (1.146) by contributions from terms of different magnitude and different symmetry has one further aspect: It allows one to determine small contributions separately from large contributions, like weighing a captain directly and not as a difference of weights of ship with captain and ship without the captain. This is important in experimental and theoretical approaches.

The examples of Table 1.3 are largely drawn from the work of the Zurich group with emphasis on the connection between symmetries, approximate constants of the motion and time scales for intramolecular processes arising from various symmetry breakings. Of course, if one just considers time scales in kinetics, there is an enormous body of work from worldwide efforts. Particularly noteworthy for the remainder of the book is the "short time frontier", which would continue the table at the top end toward the attosecond range and shorter. There is considerable early history (Manz and Wöste, 1995; Quack, 2003, 2014c) and a nice personal account of the history of short time quantum dynamical processes from the early days of electronic relaxation in molecules to cluster Coulomb explosion and table top nuclear fusion ending with the question "How fast is ultrafast?" can be found in Jortner (2003). The various chapters in the present book provide beautiful examples of current work by these authors in the field addressing and in part also answering such questions.

A brief note is also useful concerning the role of relativistic effects. As is well known, these are important for the dynamics of electron motion, whenever the heavier elements are involved in the molecules considered. These effects can be calculated by relativistic quantum chemistry (Reiher and Wolf, 2009; Mastalerz and Reiher, 2011) and can lead, indeed, to dramatic changes in the effective Born–Oppenheimer potential hypersurfaces for molecular quantum dynamics, for example. Once these effects are included, the motion of atoms and molecules can be computed and understood in very much the same way using the Schrödinger equation as discussed for nonrelativistic potentials. If the molecules move at relativistic speeds, one has to consider the changes in the definition of time which is then to be measured by an atomic clock moving at relativistic speed. Indeed, one then has the well understood (and, in fact, experimentally observed) relativistic effects, such as an atomic and molecular "twin paradox" due to Einstein (1922). Further considerations arise when considering violations of time reversal symmetry and possibly a hypothetical violation of *CPT* symmetry, and we refer to Quack (2011a,b) where one can also find a discussion of the "42 open problems", some of which are related to the measurement and definition of time, time reversal symmetry, *CPT* symmetry and time-dependent entropy in statistical mechanics and molecular quantum dynamics. The investigation of possible violations of *CPT* symmetry is clearly one of the frontiers of fundamental physics (Quack, 1994, 2003; Gabrielse, 2016; Ahmadi et al., 2020) and cosmology (Boyle et al., 2018).

ACKNOWLEDGMENTS

We gratefully acknowledge support, help from and discussions with Frédéric Merkt, Georg Seyfang, Jürgen Stohner and Gunther Wichmann, as well as financial support from Université de Strasbourg, CNRS, ETH Zürich, the laboratory of Physical Chemistry, and an Advanced Grant of the European Research Council ERC. We are much indebted to Sandra Jörimann for substantial help in the preparation of the manuscript.

REFERENCES

Abedi, A., Maitra, N.T., Gross, E.K.U., 2010. Exact factorization of the time-dependent electron-nuclear wave function. Phys. Rev. Lett. 105, 123002.

Ahmadi, M., Alves, B.X.R., Baker, C.J., Bertsche, W., Capra, A., Carruth, C., Cesar, C.L., Charlton, M., Cohen, S., Collister, R., Eriksson, S., Evans, A., Evetts, N., Fajans, J., Friesen, T., Fujiwara, M.C., Gill, D.R., Granum, P., Hangst, J.S., Hardy, W.N., Hayden, M.E., Hunter, E.D., Isaac, C.A., Johnson, M.A., Jones, J.M., Jones, S.A., Jonsell, S., Khramov, A., Knapp, P., Kurchaninov, L., Madsen, N., Maxwell, D., McKenna, J.T.K., Menary, S., Michan, J.M., Momose, T., Munich, J.J., Olchanski, K., Olin, A., Pusa, P., Rasmussen, C.Ø., Robicheaux, F., Sacramento, R.L., Sameed, M., Sarid, E., Silveira, D.M., So, C., Starko, D.M., Stutter, G., Tharp, T.D., Thompson, R.I., van der Werf, D.P., Wurtele, J.S., Collaboration, T.A., 2020. Investigation of the fine structure of antihydrogen. Nature 578, 375–380.

Albert, S., Albert, K.K., Hollenstein, H., Manca Tanner, C., Quack, M., 2011. Fundamentals of rotation-vibration spectra. In: Quack, M., Merkt, F. (Eds.), Handbook of High-Resolution Spectroscopy. Wiley, Chichester, pp. 117–173. Chapter 3.

Albert, S., Bekhtereva, E., Bolotova, I., Chen, Z., Fábri, C., Hollenstein, H., Quack, M., Ulenikov, O., 2017. Isotope effects on the resonance interactions and vibrational quantum dynamics of fluoroform $^{12,13}CHF_3$. Phys. Chem. Chem. Phys. 19, 26527–26534.

Ambrosek, D., Chatzidimitriou-Dreismann, C., Krause, P., Manz, J., Naumann, H., van Wüllen, C., 2004. Attosecond dynamics of nuclear wavepackets induced by neutron Compton scattering. Chem. Phys. 302, 229–241.

Anderson, J.B., 1975. A random-walk simulation of the Schrödinger equation: H_3^+. J. Chem. Phys. 63, 1499.

Anderson, J.B., 1976. Quantum chemistry by random walk: H ^2P, H_3^+ D_{3h} $^1A_1'$, H_2 $^3\sigma_u^+$, H_4 $^1\sigma_g^+$, Be ^1S. J. Chem. Phys. 65, 4121.

Antoci, S., Liebscher, D.-E., 1996. The third way to quantum mechanics is the forgotten first. Ann. Fond. Broglie 21, 349–366.

Aquilanti, V., Capecchi, G., Cavalli, S., 1999. Hyperspherical coordinates for chemical reaction dynamics. Adv. Quantum Chem. 36, 341–363.

Aquilanti, V., Cavalli, S., 1997. The quantum-mechanical Hamiltonian for tetraatomic systems in symmetric hyperspherical coordinates. J. Chem. Soc. Faraday Trans. 93, 801–809.

Avila, G., Carrington Jr., T., 2015. A multi-dimensional Smolyak collocation method in curvilinear coordinates for computing vibrational spectra. J. Chem. Phys. 143, 214108.

Bachau, H., Cormier, E., Decleva, P., Hansen, J.E., Martín, F., 2001. Applications of B-splines in atomic and molecular physics. Rep. Prog. Phys. 64, 1815.

Bačić, Z., Whitnell, R., Brown, D., Light, J., 1988. Localized representations for large amplitude molecular vibrations. Comput. Phys. Commun. 51, 35–47.

Bader, P., Blanes, S., Kopylov, N., 2018. Exponential propagators for the Schrödinger equation with a time-dependent potential. J. Chem. Phys. 148, 244109.

Bakasov, A., Ha, T.K., Quack, M., 1996. Ab initio calculation of molecular energies including parity violating interactions.

In: Chela-Flores, J., Raulin, F. (Eds.), Chemical Evolution, Physics of the Origin and Evolution of Life, Proceedings of the 4th Trieste Conference. 1995. Kluwer Academic Publishers, Dordrecht, pp. 287–296.

Bakasov, A., Ha, T.K., Quack, M., 1998. Ab initio calculation of molecular energies including parity violating interactions. J. Chem. Phys. 109, 7263–7285.

Bakasov, A., Quack, M., 1999. Representation of parity violating potentials in molecular main chiral axes. Chem. Phys. Lett. 303, 547–557.

Balmer, J.J., 1885a. Notiz über die Spektrallinien des Wasserstoffs. Verh. Nat.forsch. Ges. Basel 7, 548–560.

Balmer, J.J., 1885b. Zweite Notiz über die Spectrallinien des Wasserstoffs. Verh. Nat.forsch. Ges. Basel 7, 750–752.

Bandrauk, A.D., Shen, H., 1993. Exponential split operator methods for solving coupled time-dependent Schroedinger equations. J. Chem. Phys. 99, 1185.

Barth, I., Manz, J., 2007. Electric ring currents in atomic orbitals and magnetic fields induced by short intense circularly polarized π laser pulses. Phys. Rev. A 75, 012510.

Baykusheva, D., Wörner, H.J., 2020. Attosecond molecular dynamics and spectroscopy. In: Marquardt, R., Quack, M. (Eds.), Molecular Spectroscopy and Quantum Dynamics. Elsevier, Amsterdam. Chapter 4 (this book).

Beck, M.H., Jäckle, A., Worth, G.A., Meyer, H.D., 2000. The multiconfiguration time-dependent Hartree (MCTDH) method: a highly efficient algorithm for propagating wavepackets. Phys. Rep. 324, 1–105.

Beil, A., Hollenstein, H., Monti, O.L.A., Quack, M., Stohner, J., 2000. Vibrational spectra and intramolecular vibrational redistribution in highly excited deuterobromochlorofluoromethane CDBrClF: experiment and theory. J. Chem. Phys. 113, 2701–2718.

Beil, A., Luckhaus, D., Quack, M., 1996a. Fermi resonance structure and femtosecond quantum dynamics of a chiral molecule from the analysis of vibrational overtone spectra of CHBrClF. Ber. Bunsenges. Phys. Chem. 100, 1853–1875.

Beil, A., Luckhaus, D., Quack, M., Stohner, J., 1996b. Intramolecular vibrational redistribution and unimolecular reaction: concepts and new results on the femtosecond dynamics and statistics in CHBrClF. Ber. Bunsenges. Phys. Chem. 101, 311–328.

Bell, J.S., 2004. Speakable and Unspeakable in Quantum Mechanics, second edition. Cambridge University Press, Cambridge. With foreword by Alain Aspect.

Berger, R., Quack, M., 2000. Multiconfiguration linear response approach to the calculation of parity violating potentials in polyatomic molecules. J. Chem. Phys. 112, 3148–3158.

Bergmann, K., Nägerl, H.C., Panda, C., Gabrielse, G., Miloglyadov, E., Quack, M., Seyfang, G., Wichmann, G., Ospelkaus, S., Kuhn, A., Longhi, S., Szameit, A., Pirro, P., Hillebrands, B., Zhu, X.F., Zhu, J., Drewsen, M., Hensinger, W.K., Weidt, S., Halfmann, T., Wang, H.L., Paraoanu, G.S., Vitanov, N.V., Mompart, J., Busch, T., Barnum, T.J., Grimes, D.D., Field, R.W., Raizen, M.G., Narevicius, E., Auzinsh, M., Budker, D., Pálffy, A., Keitel, C.H., 2019. Roadmap on STIRAP applications. J. Phys. B, At. Mol. Opt. Phys. 52, 202001.

Bernu, B., Ceperley, D.M., Lester, W.A., 1990. The calculation of excited states with quantum Monte Carlo. II. Vibrational excited states. J. Chem. Phys. 93, 552.

Bixon, M., Jortner, J., 1968. Intramolecular radiationless transitions. J. Chem. Phys. 48, 715–726.

Bjerrum, N., 1914. Über die ultraroten Spektren der Gase. III. Die Konfiguration des Kohlenstoffdioxydmoleküls und die Gesetze der intramolekularen Kräfte. Verh. Dtsch. Phys. Ges. 16, 737–753.

Blanes, S., Casas, F., Murua, A., 2017. Symplectic time-average propagators for the Schrodinger equation with a time-dependent Hamiltonian. J. Chem. Phys. 146, 114109.

Blanes, S., Casas, F., Oteo, J., Ros, J., 2009. The Magnus expansion and some of its applications. Phys. Rep. 470, 151–238.

Blum, K., 1981. Density Matrix Theory and Applications. Plenum Press, New York.

Bohr, N., 1913a. On the constitution of atoms and molecules, Part I. Philos. Mag. 26, 1–25.

Bohr, N., 1913b. On the constitution of atoms and molecules, Part II, systems containing only a single nucleus. Philos. Mag. 26, 476–502.

Bohr, N., 1913c. On the constitution of atoms and molecules, Part III, systems containing several nuclei. Philos. Mag. 26, 857–875.

Born, M., Huang, K., 1954. Dynamical Theory of Crystal Lattices. Oxford University Press, London.

Bredtmann, T., Diestler, D.J., Li, S.-D., Manz, J., Pérez-Torres, J.F., Tian, W.J., Wu, Y.B., Yang, Y., Zhai, H.J., 2015. Quantum theory of concerted electronic and nuclear fluxes associated with adiabatic intramolecular processes. Phys. Chem. Chem. Phys. 17, 29421–29464.

Boyle, L., Finn, K., Turok, N., 2018. CPT-symmetric universe. Phys. Rev. Lett. 121, 251301.

Breidung, J., Thiel, W., 2011. Prediction of vibrational spectra from ab initio theory. In: Quack, M., Merkt, F. (Eds.), Handbook of High-Resolution Spectroscopy. Wiley, Chichester, pp. 389–404.

Brillouin, M.L., 1926. Remarques sur la mécanique ondulatoire. J. Phys. Radium 7, 353–368.

Bunker, D.L., 1971. Classical trajectory methods. Methods Comput. Phys. 10, 287.

Bunker, D.L., 1977. Some aspects of the reactions of highly excited molecules. Ber. Bunsenges. Phys. Chem. 81, 155.

Caillat, J., Zanghellini, J., Kitzler, M., Koch, O., Kreuzer, W., Scrinzi, A., 2005. Correlated multielectron systems in strong laser fields: a multiconfiguration time-dependent Hartree–Fock approach. Phys. Rev. A 71, 012712.

Cangemi, L.M., Passarelli, G., Cataudella, V., Lucignano, P., De Filippis, G., 2018. Beyond the Born–Markov approximation: dissipative dynamics of a single qubit. Phys. Rev. B 98, 184306.

Car, R., Parrinello, M., 1985. Unified approach for molecular dynamics and density-functional theory. Phys. Rev. Lett. 55, 2471–2474.

Carrington Jr., T., 2011. Using iterative methods to compute vibrational spectra. In: Quack, M., Merkt, F. (Eds.), Handbook of High-Resolution Spectroscopy. Wiley, Chichester, pp. 572–586.

Cederbaum, L.S., 2004. Born-Oppenheimer approximation and beyond. In: Domcke, W., Yarkony, D.R., Köppel, H. (Eds.), Conical Intersections. In: Advanced Series in Physical Chemistry, vol. 15. World Scientific, London, pp. 3–40.

Cederbaum, L.S., 2008. Born–Oppenheimer approximation and beyond for time-dependent electronic processes. J. Chem. Phys. 128, 124101.

Ceperley, D., Alder, B., 1986. Quantum Monte Carlo. Science 231, 555–560.

CERN, 1992. CERN-AC Z04. http://cds.cern.ch/record/39722?ln=en.

Ceriotti, M., More, J., Manolopoulos, D.E., 2014. i-pi: a python interface for ab initio path integral molecular dynamics simulations. Comput. Phys. Commun. 185, 1019–1026.

Ceriotti, M., Parrinello, M., Markland, T.E., Manolopoulos, D.E., 2010. Efficient stochastic thermostatting of path integral molecular dynamics. J. Chem. Phys. 133, 124104.

Chapovsky, P.L., Hermans, L.J.F., 1999. Nuclear spin conversion in polyatomic molecules. Annu. Rev. Phys. Chem. 50, 315–345.

Chelkowski, S., Foisy, C., Bandrauk, A.D., 1995. Electron-nuclear dynamics of multiphoton H_2^+ dissociative ionization in intense laser fields. Phys. Rev. A 57, 1176–1185.

Chu, T.S., Zhang, Y., Han, K.L., 2006. The time-dependent quantum wave packet approach to the electronically nonadiabatic processes in chemical reactions. Int. Rev. Phys. Chem. 25, 201–235.

Clary, D.C., 1986. The Theory of Chemical Reaction Dynamics. Reidel Publ., Boston.

Cohen, E.R., Cvitaš, T., Frey, J.G., Holmström, B., Kuchitsu, K., Marquardt, R., Mills, I., Pavese, F., Quack, M., Stohner, J., Strauss, H.L., Takami, M., Thor, A.J., 2007. Quantities, Units and Symbols in Physical Chemistry ('The Green Book'), 3rd edition. RSC Publishing, Cambridge, UK.

Cohen-Tannoudji, C., Dupont-Roc, J., Grynberg, G., 1992. Atom–Photon Interaction. John Wiley and Sons, New York.

Coker, D.F., Watts, R.O., 1986. Quantum simulation of systems with nodal surfaces. Mol. Phys. 58, 1113–1123.

Courant, R., Hilbert, D., 1968. Methoden der Mathematischen Physik. Heidelberger Taschenbücher, Band I, 3. Auflage, Band II, 2. Auflage. Springer, Berlin.

Cruz-Rodriguez, L., Tremblay, J.C., Martinez-Mesa, A., Uranga-Pina, L., 2016. A Chebyshev expansion of hydrodynamical fields for ultrafast vibrational wave packet dynamics. Comput. Theor. Chem. 1078, 104–112.

Császár, A., Fábri, C., Szidarovszky, T., 2020. Stationary state quantum dynamics. In: Marquardt, R., Quack, M. (Eds.), Molecular Spectroscopy and Quantum Dynamics. Elsevier, Amsterdam. Chapter 2 (this book).

Cullum, J.K., Willoughby, R.A., 1985. Lanczos Algorithm for Large Symmetric Eigenvalue Computations, vol. 2. Birkhäuser, Basel.

Cvitaš, M., Richardson, J., 2020. Quantum dynamics of water clusters. In: Marquardt, R., Quack, M. (Eds.), Molecular Spectroscopy and Quantum Dynamics. Elsevier, Amsterdam. Chapter 9 (this book).

Daniel, C., 2002. Electronic spectroscopy and photoreactivity in transition metal complexes. Coord. Chem. Rev. 238–239, 143–146.

Daniel, C., 2015a. Photochemistry and photophysics of transition metal complexes. In: Quantum Chemistry. Coordination Chemistry Reviews 282–283, 19–32. Proceedings From the 20th International Symposium on the Photophysics and Photochemistry of Coordination Compounds.

Daniel, C., 2015b. Spectroscopy and photophysics involving transition metal complexes, theoretical perspectives. In: Reference Module in Chemistry, Molecular Sciences and Chemical Engineering. Elsevier, Amsterdam.

Dietiker, P., Miloglyadov, E., Quack, M., Schneider, A., Seyfang, G., 2015. Infrared laser induced population transfer and parity selection in $^{14}NH_3$: a proof of principle experiment towards detecting parity violation in chiral molecules. J. Chem. Phys. 143, 244305.

de Broglie, L., 1926. Ondes et Mouvement. Gauthier-Villars, Paris.

Dirac, P.A.M., 1927. On quantum algebra. Proc. Camb. Philos. Soc. 23, 412–418.

Dirac, P.A.M., 1929. Quantum mechanics of many-electron systems. Proc. R. Soc. Lond. Ser. A 123, 714–733.

Dirac, P.A.M., 1958. The Principles of Quantum Mechanics, 4th edition. Clarendon Press, Oxford.

Domcke, W., Yarkony, D.R., 2012. Role of conical intersections in molecular spectroscopy and photoinduced chemical dynamics. Annu. Rev. Phys. Chem. 63, 325–352.

Domcke, W., Yarkony, D.R., Köppel, H. (Eds.), 2004. Conical Intersections. Advanced Series in Physical Chemistry, vol. 15. World Scientific, London.

Donley, E., Marquardt, R., Quack, M., Stohner, J., Thanopulos, I., Wallenborn, E.U., 2001. Some simple mechanisms of multiphoton excitation in many level systems. Mol. Phys. 99, 1275–1287.

Einstein, A., 1905. Über einen die Erzeugung und Verwandlung des Lichtes betreffenden heuristischen Gesichtspunkt. Ann. Phys. 17, 132–148.

Einstein, A., 1916a. Strahlungs-Emission und -Absorption nach der Quantentheorie. Verhand. der Deut. Phys. Ges. 13/14, 318–324.

Einstein, A., 1916b. Zur Quantentheorie der Strahlung. Mitt. Phys. Ges. Zür. 18, 47–62.

Einstein, A., 1917. Zur Quantentheorie der Strahlung. Phys. Z. 18, 121–128.

Einstein, A., 1922. Grundzüge der Relativitätstheorie. Vieweg, Wiesbaden.

Ernst, R., Bodenhausen, G., Wokaun, A., 1987. Principles of Nuclear Magnetic Resonance in One and Two Dimensions. Clarendon Press, Oxford.

Fábri, C., Quack, M., Császár, A.G., 2017. On the use of nonrigid-molecular symmetry in nuclear motion computations employing a discrete variable representation: a case study of the bending energy levels of CH_5^+. J. Chem. Phys. 147, 134101.

Fábri, C., Marquardt, R., Császár, A.G., Quack, M., 2019. Controlling tunneling in ammonia isotopomers. J. Chem. Phys. 150, 014102.

Farrell, J.P., Petretti, S., Förster, J., McFarland, B.K., Spector, L.S., Vanne, Y.V., Decleva, P., Bucksbaum, P.H., Saenz, A., Gühr, M., 2011. Strong field ionization to multiple electronic states in water. Phys. Rev. Lett. 107, 083001.

Fehrensen, B., Luckhaus, D., Quack, M., 1999. Mode selective stereomutation tunneling in hydrogen peroxide isotopomers. Chem. Phys. Lett. 300, 312–320.

Fehrensen, B., Luckhaus, D., Quack, M., 2007. Stereomutation dynamics in hydrogen peroxide. Chem. Phys. 338, 90–105.

Feit, M.D., Fleck Jr., J.A., Steiger, A., 1982. Solution of the Schroedinger equation by spectral method. J. Comput. Phys. 47, 412.

Feynman, R.P., 1948. Space-time approach to non-relativistic quantum mechanics. Rev. Mod. Phys. 20, 367–387.

Feynman, R.P., Hibbs, A.R., 1965. Quantum Mechanics and Path Integrals. McGraw-Hill, New York.

Freund, P.G.O., Goebel, C.J., Nambu, V., Oehme, R., 2009. Gregor Wentzel 1898–1978, A Biographical Memoir. National Academy of Sciences Biographical Memoir, Washington D.C..

Fröhlich, J., Schnubel, B., 2012. Do we understand quantum mechanics – finally? arXiv:1203.3678. Proceedings from the Conference in Memory of Erwin Schrödinger, Vienna, 2012.

Gabrielse, G., 2016. Probing nature's fundamental symmetries. In: Quack, M., Hacker, J. (Eds.), Symmetrie und Asymmetrie in Wissenschaft und Kunst. Nova Acta Leopoldina, NF Band 127, Nr. 412, pp. 91–98.

Garmer, D.R., Anderson, J.B., 1988. Potential energies for the reaction $F+H_2 \rightarrow HF+H$ by the random walk method. J. Chem. Phys. 89, 3050.

Gatti, F. (Ed.), 2014. Molecular Quantum Dynamics. Springer Verlag, Berlin–Heidelberg.

Gatti, F., Iung, C., 2009. Exact and constrained kinetic energy operators for polyatomic molecules: the polyspherical approach. Phys. Rep. 484, 1–69.

Gatti, F., Iung, C., Menou, M., Justum, Y., Nauts, A., Chapuisat, X., 1998. Vector parametrization of the N-atom problem in quantum mechanics. I. Jacobi vectors. J. Chem. Phys. 108, 8804–8820.

Gatti, F., Lasorne, B., Meyer, H.D., Nauts, A. (Eds.), 2017. Applications of Quantum Dynamics in Chemistry. Springer Verlag, Berlin–Heidelberg.

Gear, C.W., 1971. Numerical Initial Value Problems in Ordinary Differential Equation. Prentice-Hall, Englewood Cliffs, NJ.

Glauber, R.J., 1963a. Coherent and incoherent states of radiation field. Phys. Rev. 131, 2766–2788.

Glauber, R.J., 1963b. Quantum theory of optical coherence. Phys. Rev. 130, 2529–2539.

Gokhberg, K., Kuleff, A., Cederbaum, L.S., 2020. Electronic decay cascades in chemical environment. In: Marquardt, R., Quack, M. (Eds.), Molecular Spectroscopy and Quantum Dynamics. Elsevier, Amsterdam. Chapter 5 (this book).

Goldstein, H., 1980. Classical Mechanics, second edition. Addison Wesley, Menlo Park.

González, L., Escudero, D., Serrano-Andrés, L., 2012. Progress and challenges in the calculation of electronic excited states. ChemPhysChem 13, 28–51.

Greene, S.M., Batista, V.S., 2017. Tensor-train split-operator Fourier transform (TT-SOFT) method: multidimensional nonadiabatic quantum dynamics. J. Chem. Theory Comput. 13, 4034–4042.

Groom, D., Aguilar-Benitez, M., Amsler, C., Barnett, R., Burchat, P., Carone, C., Caso, C., Conforto, G., Dahl, O., Doser, M., Eidelman, S., Feng, J., Gibbons, L., Goodman, M., Grab, C., Gurtu, A., Hagiwara, K., Hayes, K., Hernández, J., Hikasa, K., Honscheid, K., Kolda, C., Mangano, M., Manohar, A., Masoni, A., Mónig, K., Murayama, H., Nakamura, K., Navas, S., Olive, K., Pape, L., Piepke, A., Roos, M., Tanabashi, M., Törnqvist, N., Trippe, T., Vogel, P., Wohl, C., Workman, R., Yao, W., Armstrong, B., Casas Serradilla, J., Filimonov, B., Gee, P., Lugovsky, S., Nicholson, F., Babu, K., Besson, D., Biebel, O., Bloch, P., Cahn, R., Cattai, A., Chivukula, R., Cousins, R., Damour, T., Desler, K., Donahue, R., Edwards, D., Erler, J., Ezhela, V., Fassò, A., Fetscher, W., Froidevaux, D., Fukugita, M., Gaisser, T., Garren, L., Geer, S., Gerber, H., Gilman, F., Haber, H., Hagmann, C., Hinchliffe, I., Hogan, C., Höhler, G., Igo-Kemenes, P., Jackson, J., Johnson, K., Karlen, D., Kayser, B., Klein, S., Kleinknecht, K., Knowles, I., Kolb, E., Kreitz, P., Landua, R., Langacker, P., Littenberg, L., Manley, D., March-Russell, J., Nakada, T., Quinn, H., Raffelt, G., Renk, B., Rolandi, L., Ronan, M., Rosenberg, L., Sadrozinski, H., Sanda, A., Schmitt, M., 2000. Review of particle physics. Eur. Phys. J. C 15, 1–878.

Hackmann, J., Anders, F.B., 2014. Spin noise in the anisotropic central spin model. Phys. Rev. B 89.

Haroche, S., 2012. Controlling Atoms in a Box and Exploring the Quantum to Classical Boundary (Nobel Lecture). The Nobel Foundation, Stockholm.

Hase, W.L., 1976. Dynamics of unimolecular reactions. In: Miller, W.H. (Ed.), Modern Theoretical Chemistry: Dynamics of Molecular Collisions Part B. Plenum Press, New York, pp. 121–170.

Hase, W.L., 1981. Overview of unimolecular dynamics. In: Truhlar, D.G. (Ed.), Potential, Energy Surfaces and Dynamics Calculations. Plenum Press, New York, p. 1.

Hase, W.L., 1998. Classical trajectory simulations. In: von Ragué Schleyer, P., Allinger, N.L., Clark, T., Gasteiger, J., Kollman, P.A., Schaefer III, H.F., Schreiner, P.R. (Eds.), Encyclopedia of Computational Chemistry. John Wiley and Sons, New York, p. 399.

He, Y.B., Hollenstein, H., Quack, M., Richard, E., Snels, M., Bürger, H., 2002. High resolution analysis of the complex symmetric CF_3 stretching chromophore absorption in CF_3I. J. Chem. Phys. 116, 974–983.

Hehenberger, M., McIntosh, H.V., Brändas, E., 1974. Weyl's theory applied to the Stark effect in the hydrogen atom. Phys. Rev. A 10, 1494–1506.

Heisenberg, W., 1925. Über quantentheoretische Umdeutung kinematischer und mechanischer Beziehungen. Z. Phys. 33, 879–893.

Herzberg, G., 1945. Molecular Spectra and Molecular Structure, Vol. II: Infrared and Raman Spectra. Van Nostrand Reinhold Company Inc., New York, Cincinnati, Toronto, London and Melbourne.

Herzberg, G., 1950. Molecular Spectra and Molecular Structure, Vol. I, Spectra of Diatomic Molecules. van Nostrand Reinhold Company, New York.

Herzberg, G., 1966. Molecular Spectra and Molecular Structure, Vol. III; Electronic Spectra and Electronic Structure of Polyatomic Molecules. van Nostrand Reinhold Company, New York.

Herzberg, G., Longuet-Higgins, H.C., 1963. Intersection of potential energy surfaces in polyatomic molecules. Discuss. Faraday Soc. 35, 77–82.

Hippler, M., Miloglyadov, E., Quack, M., Seyfang, G., 2011. Mass and isotope-selective infrared spectroscopy. In: Quack, M., Merkt, F. (Eds.), Handbook of High-Resolution Spectroscopy. Wiley, Chichester, pp. 1069–1118.

International Bureau of Weights and Measures, 2019. Comptes Rendus des séances de la vingt-sixième Conférence Générale des Poids et Mesures. Résolution 1, Annexe 3. BIPM, Paris.

Iro, H., 2002. Classical Mechanics. World Scientific, Singapore.

Izaac, J.A., Wang, J.B., 2015. pyCTQW: a continuous-time quantum walk simulator on distributed memory computers. Comput. Phys. Commun. 186, 81–92.

Jia, D., Manz, J., Yang, Y., 2019. Timing the recoherences of attosecond electronic charge migration by quantum control of femtosecond nuclear dynamics: a case study for hcci$^+$. J. Chem. Phys. 151, 244306.

Jortner, J., 2003. Unended quest in science. Isr. J. Chem. 43, 169–217.

Jortner, J., Rice, S.A., Hochstrasser, R.M., 1969. Radiationless transitions in photochemistry. Adv. Photochem. 7, 149.

Joubert-Doriol, L., Lasorne, B., Gatti, F., Schröder, M., Vendrell, O., Meyer, H.D., 2012. Suitable coordinates for quantum dynamics: applications using the multiconfiguration time-dependent Hartree (MCTDH) algorithm. Comput. Theor. Chem. 990, 75–89.

Joubert-Doriol, L., Lauvergnat, D., Meyer, H.D., Gatti, F., 2014. A generalized vibronic-coupling Hamiltonian model for benzopyran. J. Chem. Phys. 140, 044301.

Jungwirth, P., Gerber, R.B., 1999. Quantum molecular dynamics of ultrafast processes in large polyatomic systems. Chem. Rev. 99, 1583–1606.

Kapil, V., Rossi, M., Marsalek, O., Petraglia, R., Litman, Y., Spura, T., Cheng, B., Cuzzocrea, A., Meissner, R.H., Wilkins, D.M., Helfrecht, B.A., Juda, P., Bienvenue, S.P., Fang, W., Kessler, J., Poltavsky, I., Vandenbrande, S., Wieme, J., Corminboeuf, C., Kühne, T.D., Manolopoulos, D.E., Markland, T.E., Richardson, J.O., Tkatchenko, A., Tribello, G.A., Van Speybroeck, V., Ceriotti, M., 2019. i-pi 2.0: a universal force engine for advanced molecular simulations. Comput. Phys. Commun. 236, 214–223.

Karplus, M., 2014. Development of multiscale models for complex chemical systems: from H+H_2 to biomolecules (Nobel lecture). Angew. Chem., Int. Ed. 53, 9992–10005.

Kato, T., Kono, H., 2004. Time-dependent multiconfiguration theory for electronic dynamics of molecules in an intense laser field. Chem. Phys. Lett. 392 (4–6), 533–540.

Kleinert, H., 2009. Path Integrals in Quantum Mechanics, Statistics, Polymer Physics and Financial Markets, 5th edition. World Scientific Publisher, Singapore.

Kolba, E., Manz, J., Schreier, H.J., Trisca, I., 1992. A reflection principle in continuum resonance Raman scattering. Chem. Phys. Lett. 189, 505–512.

Kosloff, R., 1994. Propagation methods for quantum molecular dynamics. Annu. Rev. Phys. Chem. 45, 145–178.

Kosloff, D., Kosloff, R., 1983a. A Fourier method solution for the time dependent Schrödinger equation as a tool in molecular dynamics. J. Comput. Phys. 52, 35–53.

Kosloff, R., Kosloff, D., 1983b. A Fourier method solution for the time dependent Schrödinger equation: a study of the reaction $H^+ + H_2$, $D^+ + HD$, and $D^+ + H_2$. J. Chem. Phys. 79, 1823–1833.

Kramers, H.A., 1926. Wellenmechanik und halbzahlige Quantisierung. Z. Phys. 39, 828–840.

Kuleff, A.I., 2019. Ultrafast electron dynamics as a route to explore chemical processes. In: Vrakking, M.J.J., Lépine, F. (Eds.), Attosecond Molecular Dynamics. The Royal Society of Chemistry, pp. 103–138.

Kuleff, A.I., Breidbach, J., Cederbaum, L.S., 2005. Multielectron wave-packet propagation: general theory and application. J. Chem. Phys. 123, 044111.

Kuppermann, A., 1996. Reactive scattering with row-orthonormal hyperspherical coordinates. 1. Transformation properties and Hamiltonian for triatomic systems. J. Phys. Chem. 100, 2621–2636.

Kushnarenko, A., Miloglyadov, E., Quack, M., Seyfang, G., 2018. Intramolecular vibrational energy redistribution in $HCCCH_2X$ (X = Cl, Br, I) measured by femtosecond pump-probe experiments in a hollow waveguide. Phys. Chem. Chem. Phys. 20, 10949–10959.

Kutzelnigg, W., 2007. Which masses are vibrating or rotating in a molecule? Mol. Phys. 105, 2627–2647.

Landau, L.D., Lifshitz, E.M., 1966. Mécanique. Editions Mir, Moscow.

Lehmann, K.K., Scoles, G., Pate, B.H., 1994. Intramolecular dynamics from eigenstate-resolved infrared-spectra. Annu. Rev. Phys. Chem. 45, 241–274.

Lewerenz, M., Quack, M., 1988. Vibrational Spectrum and Potential Energy Surface of the CH Chromophore in CHD_3. J. Chem. Phys. 88, 5408–5432.

Lewerenz, M., Watts, R.O., 1994. Quantum Monte Carlo simulation of molecular vibrations. Mol. Phys. 81, 1075–1091.

Light, J.C., Carrington Jr., T., 2000. Discrete-variable representations and their utilization. Adv. Chem. Phys. 114, 263–310.

Longuet-Higgins, H.C., 1975. The intersection of potential energy surfaces in polyatomic molecules. Proc. R. Soc. Lond. Ser. A, Math. Phys. Sci. 344, 147–156.

Lorin, E., Chelkowski, S., Bandrauk, A., 2007. A numerical Maxwell–Schrödinger model for intense laser–matter interaction and propagation. Comput. Phys. Commun. 177, 908–932.

Lubich, C., 2015. Time integration in the multiconfiguration time-dependent Hartree method of molecular quantum dynamics. Appl. Math. Res. Express 2015, 311–328.

Luckhaus, D., Quack, M., 1992. Spectrum and dynamics of the CH chromophore in CD_2HF. I. Vibrational Hamiltonian and analysis of rovibrational spectra. Chem. Phys. Lett. 190, 581–589.

Luckhaus, D., Quack, M., 1993. The role of potential anisotropy in the dynamics of the CH-chromosphere in CHX_3 (C_{3v}) symmetrical tops. Chem. Phys. Lett. 205, 277–284.

Luckhaus, D., Quack, M., Stohner, J., 1993. Femtosecond quantum structure, equilibration and time reversal for the CH-chromophore dynamics in CHD_2F. Chem. Phys. Lett. 212, 434–443.

Magnus, W., 1954. On the exponential solution of differential equations for a linear operator. Commun. Pure Appl. Math. 7, 649–673.

Manca, C., Quack, M., Willeke, M., 2008. Vibrational predissociation in hydrogen bonded dimers: the case of $(HF)_2$ and its isotopomers. Chimia 62 (4), 235–239.

Manz, J., Wöste, L. (Eds.), 1995. Femtosecond Chemistry. Proceedings of Berlin Conference Femtosecond Chemistry. Berlin, March 1993, vols. 1 and 2. Verlag Chemie, Weinheim.

Maquet, A., Grobe, R., 2002. Atoms in strong laser fields: challenges in relativistic quantum mechanics. J. Mod. Opt. 49, 2001–2018.

Marquardt, R., 2019. A formula for the contribution of a resonance to the canonical partition function. Mol. Phys. 117, 1964–1970.

Marquardt, R., Quack, M., 1989. Infrared-multiphoton excitation and wave packet motion of the harmonic and anharmonic oscillators – exact-solutions and quasiresonant approximation. J. Chem. Phys. 90 (11), 6320–6327.

Marquardt, R., Quack, M., 1991. The wave packet motion and intramolecular vibrational redistribution in CHX_3 molecules under infrared multiphoton excitation. J. Chem. Phys. 95, 4854–4867.

Marquardt, R., Quack, M., 1994. Statistical aspects of the radiative excitation of the harmonic oscillator. J. Phys. Chem. 98, 3486–3491.

Marquardt, R., Quack, M., 1998. Global analytical potential hypersurfaces for large amplitude motion and reactions in methane. I. Formulation of the potentials and adjustment of the parameters to *ab initio* data and experimental contraints. J. Chem. Phys. 109, 10628–10643.

Marquardt, R., Quack, M., 2001. Energy redistribution in reacting systems. In: Moore, J., Spencer, N. (Eds.), Encyclopedia of Chemical Physics and Physical Chemistry, vol. I. IOP Publishing, Bristol, pp. 897–936. Chapter A3.13.

Marquardt, R., Quack, M., 2011. Global analytical potential energy surfaces for high-resolution molecular spectroscopy and reaction dynamics. In: Quack, M., Merkt, F. (Eds.), Handbook of High-Resolution Spectroscopy. Wiley, Chichester, pp. 511–550.

Marquardt, R., Quack, M., Stohner, J., Sutcliffe, E., 1986. Quantum-mechanical wavepacket dynamics of the CH group in the symmetric top X_3CH compounds using effective Hamiltonians from high-resolution spectroscopy. J. Chem. Soc. Faraday Trans. 2 (82), 1173–1187.

Marquardt, R., Quack, M., Stohner, J., Thanopulos, I., 2019. Quantum dynamics and spectra of the iodine atom in a strong laser field as calculated with the URIMIR package. Mol. Phys. 117, 3132–3147.

Marquardt, R., Sanrey, M., Gatti, F., Le Quéré, F., 2010. Full-dimensional quantum dynamics of vibrationally highly excited NHD_2. J. Chem. Phys. 133, 174302.

Marx, D., Parrinello, M., 1996. Ab initio path integral molecular dynamics: basic ideas. J. Chem. Phys. 104, 4077–4082.

Mastalerz, R., Reiher, M., 2011. Relativistic electronic structure theory for molecular spectroscopy. In: Quack, M., Merkt, F. (Eds.), Handbook of High-Resolution Spectroscopy. Wiley, Chichester, pp. 405–442.

Matsika, S., Krause, P., 2011. Nonadiabatic events and conical intersections. Annu. Rev. Phys. Chem. 62, 621–643.

Merkt, F., Quack, M., 2011a. Preface by the Editors Frédéric Merkt and Martin Quack. Handbook of High-Resolution Spectroscopy. Wiley, Chichester.

Merkt, F., Quack, M., 2011b. Molecular quantum mechanics and molecular spectra, molecular symmetry, and interaction of matter with radiation. In: Quack, M., Merkt, F. (Eds.), Handbook of High-Resolution Spectroscopy. Wiley, Chichester, pp. 1–55.

Messiah, A., 1961. Quantum Mechanics. North-Holland, Amsterdam.

Metropolis, N., Ulam, S., 1949. The Monte Carlo method. J. Am. Stat. Assoc. 44, 335.

Meyer, H.D., Manthe, U., Cederbaum, L.S., 1990. The multi-configurational time-dependent Hartree approach. Chem. Phys. Lett. 165, 73.

Miller, W.H., 1974. Classical limit quantum mechanics and the theory of molecular collision. Adv. Chem. Phys. 25, 69–190.

Miller, W.H., 1975a. Semiclassical limit of quantum mechanical transition state theory for nonseparable systems. J. Chem. Phys. 62, 1899–1906.

Miller, W.H., 1975b. The classical S-matrix in molecular collisions. Adv. Chem. Phys. 30, 77–148.

Miller, W.H., 2014. A journey through chemical dynamics. Annu. Rev. Phys. Chem. 65, 1–19.

Miller, W.H., Handy, N.C., Adams, J.E., 1980. Reaction path hamiltonian for polyatomic molecules. J. Chem. Phys. 72, 99–112.

Mladenović, M., 2000. Rovibrational Hamiltonians for general polyatomic molecules in spherical polar parametrization. I. Orthogonal representations. J. Chem. Phys. 112, 1070–1081.

Moiseyev, N., 2011. NHQM - Non Hermitian Quantum Mechanics. Cambridge University Press.

Nauts, A., Chapuisat, X., 1987. Hamiltonians for constrained N-particle systems. Chem. Phys. Lett. 136, 164–170.

Nest, M., 2006. Quantum carpets and correlated dynamics of several fermions. Phys. Rev. A 73, 023613.

Nest, M., Klamroth, T., Saalfrank, P., 2005. The multiconfiguration time-dependent Hartree–Fock method for quantum chemical calculations. J. Chem. Phys. 122, 124102.

Newton, R.G., 1966. Scattering Theory of Waves and Particles. McGraw Hill, New York.

Nüske, F., Keller, B.G., Pérez-Hernández, G., Mey, A.S.J.S., Noé, F., 2014. Variational approach to molecular kinetics. J. Chem. Theory Comput. 10, 1739–1752.

Pacher, T., Cederbaum, L.S., Köppel, H., 1988. Approximately diabatic states from block diagonalization of the electronic Hamiltonian. J. Chem. Phys. 89, 7367–7381.

Park, T.J., Light, J.C., 1986. Unitary quantum time evolution by iterative Lanczos reduction. J. Chem. Phys. 85, 5870.

Pauli, W., 1928. Über das H-theorem vom Anwachsen der Entropie vom Standpunkt der neuen Quantenmechanik. Probleme der Modernen Physik, Arnold Sommerfeld zum 60. Geburtstage gewidmet von seinen Schülern. Hirzel, Leipzig.

Penfold, T.J., Gindensperger, E., Daniel, C., Marian, C.M., 2018. Spin-vibronic mechanism for intersystem crossing. Chem. Rev. 118, 6975–7025.

Perelomov, A., 1986. Generalized Coherent States and Their Applications. Springer, Berlin.

Perkins, D.H., 2000. Introduction to High Energy Physics. Cambridge University Press, Cambridge.

Planck, M., 1900a. Über das Gesetz der Energieverteilung im Normalspectrum. Verh. Dtsch. Phys. Ges. 2, 202–236.

Planck, M., 1900b. Zur Theorie des Gesetzes der Energieverteilung im Normalspektrum. Verh. Dtsch. Phys. Ges. 2, 237–245.

Pochert, J., Quack, M., Stohner, J., Willeke, M., 2000. Ab initio calculation and spectroscopic analysis of the intramolecular vibrational redistribution in 1, 1,1, 2-tetrafluoroiodoethane CF_3CHFI. J. Chem. Phys. 113, 2719–2735.

Prentner, R., Quack, M., Stohner, J., Willeke, M., 2015. Wavepacket dynamics of the axially chiral molecule Cl-O-O-Cl under coherent radiative excitation and including electroweak parity violation. J. Phys. Chem. A 119, 12805–12822.

Primas, H., 1981. Chemistry, Quantum Mechanics and Reductionism. Springer, Berlin.

Qu, C., Yu, Q., Bowman, J.M., 2018. Permutationally invariant potential energy surfaces. Annu. Rev. Phys. Chem. 69, 151–175.

Quack, M., 1977. Detailed symmetry selection rules for reactive collisions. Mol. Phys. 34, 477–504.

Quack, M., 1978. Theory of unimolecular reactions induced by monochromatic infrared radiation. J. Chem. Phys. 69, 1282–1307.

Quack, M., 1979. Master equations for photochemistry with intense infrared light. Ber. Bunsenges. Phys. Chem. 83, 757–775.

Quack, M., 1981. Statistical mechanics and dynamics of molecular fragmentation. Il Nuovo Cimento 63B, 358–377.

Quack, M., 1982. Reaction dynamics and statistical mechanics of the preparation of highly excited states by intense infrared radiation. Adv. Chem. Phys. 50, 395–473.

Quack, M., 1983. Detailed symmetry selection rules for chemical reactions. In: Maruani, J., Serre, J. (Eds.), Symmetries and Properties of Non-rigid Molecules: a Comprehensive Survey, vol. 23. Elsevier, Amsterdam, pp. 355–378.

Quack, M., 1986. On the measurement of the parity violating energy difference between enantiomers. Chem. Phys. Lett. 132, 147–153.

Quack, M., 1989a. Structure and dynamics of chiral molecules. Angew. Chem. 101, 588–604. See also Angew. Chem., Int. Ed. 28, 571–586.

Quack, M., 1989b. Infrared laser chemistry and the dynamics of molecular multiphoton excitation. Infrared Phys. 29, 441–466.

Quack, M., 1990. Spectra and dynamics of coupled vibrations in polyatomic molecules. Annu. Rev. Phys. Chem. 41, 839–874.

Quack, M., 1994. On the measurement of CP-violating energy differences in matter–antimatter enantiomers. Chem. Phys. Lett. 231, 421–428.

Quack, M., 1995a. Molecular femtosecond quantum dynamics between less than yoctoseconds and more than days: experiment and theory. In: Manz, J., Wöste, L. (Eds.), Femtosecond Chemistry, Proceedings of Berlin Conference Femtosecond Chemistry. Berlin, March 1993. Verlag Chemie, Weinheim, pp. 781–818. Chapter 27.

Quack, M., 1995b. Molecular infrared-spectra and molecular-motion. J. Mol. Struct. 347, 245–266.

Quack, M., 1997. Comments on intramolecular dynamics and femtosecond kinetics. In: Proc. 20th Solvay Conference "Chemical Reactions and Their Control on the Femtosecond Time Scale". Adv. Chem. Phys. 101, 377–381.

Quack, M., 1998. Multiphoton excitation. In: von Ragué Schleyer, P., Allinger, N.L., Clark, T., Gasteiger, J., Kollman, P.A., Schaefer III, H.F., Schreiner, P.R. (Eds.), Encyclopedia of Computational Chemistry, vol. 3. John Wiley and Sons, New York, pp. 1775–1791.

Quack, M., 1999. Intramolekulare Dynamik: Irreversibilität, Zeitumkehrsymmetrie und eine absolute Moleküluhr. Nova Acta Leopoldina 81, 137–173.

Quack, M., 2001. Molecules in motion. Chimia 55, 753–758.

Quack, M., 2002. How important is parity violation for molecular and biomolecular chirality? Angew. Chem., Int. Ed. Engl. 114, 4618–4630.

Quack, M., 2003. Molecular spectra, reaction dynamics, symmetries and life. Chimia 57, 147–160.

Quack, M., 2004a. Time and time reversal symmetry in quantum chemical kinetics. In: Brändas, E.J., Kryachko, E.S. (Eds.), Fundamental World of Quantum Chemistry. A Tribute to the Memory of per-Olov Löwdin, vol. 3. Kluwer Academic Publishers, Dordrecht, pp. 423–474.

Quack, M., 2004b. Zeit und Zeitumkehrsymmetrie in der molekularen Kinetik. Schriftliche Fassung des Vortrages am 7. Symposium der deutschen Akademien der Wissenschaften. In: Simon, D. (Ed.), Berlin-Brandenburgische Akademie der Wissenschaften Berlin, Zeithorizonte in den Wissenschaften 31.10 und 01.11.2002. De Gruyter, Berlin, pp. 125–180.

Quack, M., 2006. Electroweak quantum chemistry and the dynamics of parity violation in chiral molecules. In: Naidoo, K.J., Brady, J., Field, M.J., Gao, J., Hann, M. (Eds.), Modelling Molecular Structure and Reactivity in Biological Systems, Proc. 7th WATOC Congress. Cape Town, January 2005. Royal Society of Chemistry, Cambridge, pp. 3–38.

Quack, M., 2007. Recent results in quantum chemical kinetics from high resolution spectroscopy. In: Simos, T.E., Maroulis, G. (Eds.), Computation in Modern Science and Engineering: Proceedings of the International Conference on Computational Methods in Science and Engineering, vol. 2, Part A. ICCMSE, 2007, Corfu, Greece, 25–30 September 2007. In: AIP Conference Proceedings, vol. 963. American Institute of Physics, pp. 245–248.

Quack, M., 2008. Tunneling dynamics and symmetry in simple hydrogen compounds. In: Abstract Brijuni (Brioni) Conference on "Hydrogen, a Universal Saga". Croatia, pp. 25.08.2008–29.08.2008.

Quack, M., 2011a. Fundamental symmetries and symmetry violations from high resolution spectroscopy. In: Quack, M., Merkt, F. (Eds.), Handbook of High-Resolution Spectroscopy. Wiley, Chichester, pp. 659–722.

Quack, M., 2011b. Frontiers in spectroscopy (concluding paper to Faraday discussion 150, 2011). Faraday Discuss. 150, 533–565.

Quack, M., 2014a. On the emergence of simple structures in complex phenomena: concepts and some numerical examples. Adv. Chem. Phys. 157, 97–116.

Quack, M., 2014b. On biomolecular homochirality as a quasi-fossil of the evolution of life. Adv. Chem. Phys. 157, 249–290.

Quack, M., 2014c. The concept of law and models in chemistry. Eur. Rev. 22, S50–S86.

Quack, M., Hacker, J. (Eds.), 2016. Symmetrie und Asymmetrie in Wissenschaft und Kunst. Nova Acta Leopoldina, NF Band 127, Nr. 412. Book with contributions in English and German by several authors.

Quack, M., Seyfang, G., 2020. Atomic and molecular tunneling processes in chemistry. In: Marquardt, R., Quack, M. (Eds.), Molecular Spectroscopy and Quantum Dynamics. Elsevier, Amsterdam. Chapter 7 (this book).

Quack, M., Stohner, J., 1993. Femtosecond quantum dynamics of functional-groups under coherent infrared multiphoton excitation as derived from the analysis of high-resolution spectra. J. Phys. Chem. 97 (48), 12574–12590.

Quack, M., Stohner, J., 2005. Parity violation in chiral molecules. Chimia 59 (7–8), 530–538.

Quack, M., Stohner, J., Willeke, M., 2008. High-resolution spectroscopic studies and theory of parity violation in chiral molecules. Annu. Rev. Phys. Chem. 59, 741–769.

Quack, M., Suhm, M.A., 1991. Potential energy surfaces, quasiadiabatic channels, rovibrational spectra, and intramolecular dynamics of $(HF)_2$ and its isotopomers from quantum Monte Carlo calculations. J. Chem. Phys. 95, 28–59.

Quack, M., Suhm, M.A., 1998. Spectroscopy and quantum dynamics of hydrogen fluoride clusters. In: Bačić, Z., Bowman, J. (Eds.), Advances in Molecular Vibrations and Collision Dynamics, vol. 3. JAI Press, Stamford, Conn. and London, England, pp. 205–228.

Quack, M., Sutcliffe, E., 1984. On the possibility of mode-selective IR-multiphoton excitation of ozone. Chem. Phys. Lett. 105, 147–152.

Quack, M., Sutcliffe, E., 1985. On the validity of the quasiresonant approximation for molecular infrared-multiphoton excitation. J. Chem. Phys. 83 (8), 3805–3812.

Quack, M., Sutcliffe, E., 1986. Program 515. URIMIR: unimolecular reactions induced by monochromatic infrared radiation. QCPE Bull. 6 (3), 98.

Quack, M., Troe, J., 1974. Specific rate constants of unimolecular processes II. Adiabatic channel model. Ber. Bunsenges. Phys. Chem. 78, 240–252.

Quack, M., Troe, J., 1981. Statistical methods in scattering. In: Henderson, D. (Ed.), Theoretical Chemistry: Advances and Perspectives, vol. 6B. Academic Press, New York, pp. 199–276.

Quack, M., Troe, J., 1998. Statistical adiabatic channel model. In: von Ragué Schleyer, P., Allinger, N., Clark, T., Gasteiger, J., Kollman, P.A., Schaefer III, H.F., Schreiner, P.R. (Eds.), Encyclopedia of Computational Chemistry, vol. 4. John Wiley and Sons, pp. 2708–2726.

Rabi, I.I., 1937. Space quantization in a gyrating magnetic field. Phys. Rev. 51, 652–654.

Rabi, I.I., Zacharias, J.R., Millman, S., Kusch, P., 1938. A new method of measuring nuclear magnetic moment. Phys. Rev. 53, 318.

Reiher, M., Wolf, A., 2009. A Relativistic Quantum Chemistry: The Fundamental Theory of Molecular Science, 1st edition. Wiley-VCH, Weinheim.

Reiss, H.R., 1992. Theoretical methods in quantum optics: S-matrix and Keldysh techniques for strong-field processes. Prog. Quantum Electron. 16, 1–71.

Remacle, F., Levine, R.D., 2006. An electronic time scale in chemistry. Proc. Natl. Acad. Sci. 103, 6793–6798.

Reynolds, P.J., Ceperley, D.M., Alder, B.J., Lester, W.A., 1982. Fixed-node quantum Monte Carlo for molecules. J. Chem. Phys. 77, 5593.

Sabzyan, H., Jenabi, M.J., 2016. Electron quantum dynamics in atom-ion interaction. J. Chem. Phys. 144.

Sakurai, J.J., 1985. Modern Quantum Mechanics. Benjamin/Cummings Publishing, Menlo Park.

Salières, P., Maquet, A., Haessler, S., Caillat, J., Taïeb, R., 2012. Imaging orbitals with attosecond and Ångström resolutions: toward attochemistry? Rep. Prog. Phys. 75, 062401.

Sansone, G., Pfeifer, T., Simeonidis, K., Kuleff, A.I., 2012. Electron correlation in real time. ChemPhysChem 13, 661–680.

Schatz, G.C., Ratner, M.A., 1993. Quantum Mechanics in Chemistry. Prentice Hall, Englewood Cliffs.

Schopper, H., 1999. Lebenszeiten im Mikrokosmos – von ultrakurzen bis zu unendlichen und oszillierenden. Nova Acta Leopoldina 81 (314), 109–134.

Schreiber, M., Silva-Junior, M.R., Sauer, S.P.A., Thiel, W., 2008. Benchmarks for electronically excited states: CASPT2, CC2, CCSD, and CC3. J. Chem. Phys. 128, 134110.

Schrödinger, E., 1926a. Quantisierung als Eigenwertproblem I. Ann. Phys. 79, 361–376.

Schrödinger, E., 1926b. Quantisierung als Eigenwertproblem II. Ann. Phys. 79, 489–527.

Schrödinger, E., 1926c. Quantisierung als Eigenwertproblem III. Ann. Phys. 80, 437–490.

Schrödinger, E., 1926d. Quantisierung als Eigenwertproblem IV. Ann. Phys. 81, 109–139.

Schrödinger, E., 1926e. Der stetige Übergang von der Mikro- zur Makromechanik. Naturwissenschaften 14, 664–666.

Schweiger, A., Jeschke, G., 2001. Principles of Pulse Electron Paramagnetic Resonance. Oxford University Press, Oxford.

Shampine, L.F., Gordon, M.K., 1975. Computer Solution of Ordinary Differential Equations: The Initial Value Problem. W. H. Freeman, San Francisco.

Sisourat, N., Pilskog, I., Dubois, A., 2011. Nonperturbative treatment of multielectron processes in ion-molecule scattering: application to He^{2+}-H_2 collisions. Phys. Rev. A 84, 052722.

Sommerfeld, A., 1919. Atombau und Spektrallinien. Vieweg, Braunschweig.

Song, H., Guo, H., 2015. Vibrational and rotational mode specificity in the Cl + H_2O → HCl plus OH reaction: a quantum dynamical study. J. Phys. Chem. A 119, 6188–6194.

Stohner, J., Quack, M., 2011. Conventions, symbols, quantities, units and constants for high-resolution molecular spectroscopy. In: Quack, M., Merkt, F. (Eds.), Handbook of High-Resolution Spectroscopy. Wiley, Chichester, pp. 263–324.

Sun, Z., Yu, D., Xie, W., Hou, J., Dawes, R., Guo, H., 2015. Kinetic isotope effect of the $^{16}O + ^{36}O_2$ and $^{18}O + ^{32}O_2$ isotope exchange reactions: dominant role of reactive resonances revealed by an accurate time-dependent quantum wavepacket study. J. Chem. Phys. 142, 174312.

Tal-Ezer, H., Kosloff, R., 1984. An accurate and efficient scheme for propagating the time dependent Schrödinger equation. J. Chem. Phys. 81, 3967–3971.

Tanaka, S., Rothstein, S.M., Lester, W.A. (Eds.), 2012. Advances in Quantum Monte Carlo. ACS Symposium Series, vol. 1094. Oxford University Press.

Tannor, D.J., 2007. Introduction to Quantum Mechanics: A Time-Dependent Perspective. University Science Books, Sausalito.

Teller, E., 1937. The crossing of potential surfaces. J. Phys. Chem. 41, 109–116.

Tennyson, J., 2011. High accuracy rotation-vibration calculations on small molecules. In: Quack, M., Merkt, F. (Eds.), Handbook of High-Resolution Spectroscopy. Wiley, Chichester, pp. 853–896.

Tennyson, J., Sutcliffe, B.T., 1982. The ab initio calculation of the vibrational–rotational spectrum of triatomic systems in the close-coupling approach, with KCN and H_2Ne as examples. J. Chem. Phys. 77, 4061–4072.

Thoss, M., Domcke, W., Wang, H., 2004. Theoretical study of vibrational wave-packet dynamics in electron-transfer systems. Chem. Phys. 296, 217–229.

Troe, J., 2006. Reaction kinetics: an addiction. J. Phys. Chem. A 110, 2831–2834.

Troe, J., Ushakov, V.G., Viggiano, A.A., 2005. Classical trajectory and statistical adiabatic channel study of the dynamics of capture and unimolecular bond fission. VII. Thermal capture and specific rate constants $k(E, J)$ for the dissociation of molecular ions. Z. Phys. Chem. 219, 715–741.

Tuckerman, M.E., Marx, D., Klein, M.L., Parrinello, M., 1996. Efficient and general algorithms for path integral Car-Parrinello molecular dynamics. J. Chem. Phys. 104, 5579–5588.

Tully, J.C., 1990. Molecular dynamics with electronic transitions. J. Chem. Phys. 93, 1061–1071.

Tully, J.C., 2012. Perspective: nonadiabatic dynamics theory. J. Chem. Phys. 137, 22A301.

van Gunsteren, W.F., Bakowies, D., Baron, R., Chandrasekhar, I., Christen, M., Daura, X., Gee, P., Geerke, D.P., Glättli, A., Hünenberger, P.H., Kastenholz, M.A., Oostenbrink, C., Schenk, M., Trzesniak, D., van der Vegt, N.F.A., Yu, H.B., 2006. Biomolecular modeling: goals, problems, perspectives. Angew. Chem., Int. Ed. 45, 4064–4092.

Vaníček, J., Begušić, T., 2020. Ab initio semiclassical evaluation of vibrationally resolved electronic spectra with thawed gaussians. In: Marquardt, R., Quack, M. (Eds.), Molecular Spectroscopy and Quantum Dynamics. Elsevier, Amsterdam. Chapter 6 (this book).

von Mises, R., 1931. Wahrscheinlichkeitsrechnung und ihre Anwendung in der Statistik und Theoretischen Physik. Leipzig.

von Neumann, J., 2018. Mathematical Foundations of Quantum Mechanics, new edition. Princeton University Press, New Jersey, New. Transl. Robert Beyer; original German edition "Mathematische Grundlagen der Quantenmechanik", Springer Heidelberg, 1932, 1996.

von Puttkamer, K., Dübal, H.R., Quack, M., 1983. Temperature-dependent infrared band-structure and dynamics of the CH chromophore in C_4F_9CC-H. Chem. Phys. Lett. 95, 358–362.

von Puttkamer, K., Quack, M., 1989. Vibrational-spectra of $(HF)_2$, $(HF)_n$ and their D-isotopomers – mode selective rearrangements and nonstatistical unimolecular decay. Chem. Phys. 139, 31–53.

Vrakking, M.J.J., Lépine, F. (Eds.), 2019. Attosecond Molecular Dynamics. Theoretical and Computational Chemistry Series. The Royal Society of Chemistry.

Wentzel, G., 1924. Zur Quantenoptik. Z. Phys. 22, 193–199.

Wentzel, G., 1926. Eine Verallgemeinerung der Quantenbedingungen für die Zwecke der Wellenmechanik. Z. Phys. 38, 518–529.

Wentzel, G., 1927. Über strahlungslose Quantensprünge. Z. Phys. 43, 524–530.

Wentzel, G., 1928. Die unperiodischen Vorgänge in der Wellenmechanik. Phys. Z. 29, 321–337.

Whaley, K.B., Light, J.C., 1984. Rotating-frame transformations: a new approximation for multiphoton absorption and dissociation in laser fields. Phys. Rev. A 29, 1188.

Wilson Jr., E.B., Decius, J.C., Cross, P.C., 1955. Molecular Vibrations. The Theory of Infrared and Raman Vibrational Spectra. McGraw-Hill Book Company, Inc.. Dover edition, New York, 1980.

Wörner, H.J., Corkum, P.B., 2011. Attosecond spectroscopy. In: Quack, M., Merkt, F. (Eds.), Handbook of High-resolution Spectroscopy. John Wiley & Sons, Ltd., pp. 1781–1803. Chapter 85.

Xiao, C., Xu, X., Liu, S., Wang, T., Dong, W., Yang, T., Sun, Z., Dai, D., Xu, X., Zhang, D.H., Yang, X., 2011. Experimental and theoretical differential cross sections for a four-atom reaction: $HD + OH \rightarrow H_2O + D$. Science 333, 440–442.

Yamaguchi, Y., Schaefer III, H.F., 2011. Analytic derivative methods in molecular electronic structure theory: a new dimension to quantum chemistry and its applications to spectroscopy. In: Quack, M., Merkt, F. (Eds.), Handbook of High-Resolution Spectroscopy. Wiley, Chichester, pp. 325–362.

Zare, R.N., 1988. Angular Momentum. Baker Lecture Series. Wiley, New York.

Zhao, Z., Chen, J., Zhang, Z., Zhang, D.H., Wang, X.G., Carrington Jr, T., 2018. Computing energy levels of CH_4, CHD_3, CH_3D, and CH_3F with a direct product basis and coordinates based on the methyl subsystem. J. Chem. Phys. 148, 074113.

Zhou, L., Xie, D., 2015. Full-dimensional quantum dynamics of vibrational mediated photodissociation of HOD in its B band. J. Phys. Chem. A 119, 12062–12072.

CHAPTER 2

Exact Numerical Methods for Stationary-State-Based Quantum Dynamics of Complex Polyatomic Molecules

ATTILA G. CSÁSZÁR • CSABA FÁBRI • TAMÁS SZIDAROVSZKY
MTA-ELTE Complex Chemical Systems Research Group and Laboratory of Molecular Structure and Dynamics, Institute of Chemistry, ELTE Eötvös Loránd University, Budapest, Hungary

Abstract

The fourth age of quantum chemistry offers fully flexible, black-box-type protocols for the accurate and detailed study of nuclear motions, applicable equally well to semirigid, floppy, flexible, polytopic, fluxional, and quasistructural polyatomic molecular systems, including complexes and clusters. Several codes, based on advanced fourth-age protocols, have been developed for the variational (or variational-like) solution of the time-independent nuclear-motion (rotational-vibrational) Schrödinger equation. These codes yield accurate rovibrational energy levels, wavefunctions, and to some extent quantum-number assignments for bound, resonance, and scattering states, revealing important spectroscopic and dynamical characteristics about the systems studied. When no approximations are introduced to the kinetic energy part of the rovibrational Hamiltonian, the accuracy of the computed results, assuming the validity of the Born–Oppenheimer approximation, depends solely on the accuracy of the representation of the potential energy surface utilized during these computations. From the point of view of potential applications it is important to emphasize that the most general codes can be employed both in full and any number of reduced dimensions. Several a posteriori analysis tools are available to improve the understanding of the extreme amount of numerical results produced by the stationary-state nuclear-motion computations. As shown through a few examples, these stationary-state solutions can straightforwardly be utilized for detailed quantum-dynamics studies. The applications briefly detailed at the end of this chapter help appreciate the power of the fourth-age quantum-chemical techniques developed and available to the spectroscopic and dynamics communities.

2.1 INTRODUCTION

Over the last few decades there has been a steady increase in the number of gas-phase species, including molecules, molecular complexes, and clusters, in all feasible neutral and charged (cationic and anionic) forms, whose internal motions (rotations and vibrations) have been studied at high resolution by advanced techniques of molecular spectroscopy (Merkt and Quack, 2011). For systems confined to a finite region of configuration space or adsorbed on a surface, translational degrees of freedom (dof) may also need to be considered explicitly in a quantum treatment of nuclear motions (Dosch, 2001; Xu et al., 2008b; Xu et al., 2009b; Xu et al., 2008a; Xu et al., 2009a; McAfee and Poirier, 2009; Matanović et al., 2009; McAfee and Poirier, 2011; Matanović et al., 2012; Xu et al., 2013; Firmino et al., 2014; Xu et al., 2015; Poirier, 2015; Felker and Bačić, 2017; Zanuttini et al., 2018); nevertheless, such systems and situations are *not* considered in what follows.

It is highly desirable, almost mandatory in certain cases, to accompany high-resolution experimental spectroscopic studies with ever more sophisticated quantum-chemical [electronic structure (Helgaker et al., 2000; Császár et al., 2000) and nuclear motion (Merkt and Quack, 2011; Bowman et al., 2008; Császár et al., 2012; Tennyson, 2016; Carrington, 2017)] computations (note that the emphasis here is on the joint sophisticated treatment of the two distinct but equally relevant fields of quantum chemistry for a proper dynamical description of simple as well as complex molecular systems). This is especially important when the experimental absorption, emission, and action spectra become highly complex, in the simplest sense dense, which happens due to (a) the sheer size of the molecule under investigation, (b) the complexity of the internal motions when the dynamics leaves the so-called "semi-

Molecular Spectroscopy and Quantum Dynamics. https://doi.org/10.1016/B978-0-12-817234-6.00007-6

rigid" region [in cases when the results based on the *rigid rotor* (RR) (Kroto, 1992) and *harmonic oscillator* (HO) (Wilson et al., 1955) approximations, perhaps after a slight extension based on second-order vibrational perturbation theory (VPT2) (Papoušek and Aliev, 1982; Nielsen, 1951; Clabo et al., 1988; Allen et al., 1990; Barone et al., 2014), provide an outstanding qualitative and even a semiquantitative understanding of spectral regularities, the molecule is said to be *semirigid*], (c) high level of rovibronic excitation (at higher excitation energies all molecules become "floppy"), and (d) interaction of several electronic states [in what follows we mostly neglect these cases, focusing our attention on a single, well-separated electronic state, by default the ground electronic state (Császár et al., 2000)]. In the 21st century sophisticated quantum-chemical computations have been used to aid (or even make possible) the deduction of information encoded in exceedingly complex observed spectra.

There is growing evidence that the rovibrational energy-level structure of certain molecular species, due to their complex rovibrational dynamics, defies the conceptually simple, traditional interpretation attempts, building upon the RRHO approximation. This means that certain molecular parameters [e.g., the geometric structure (Demaison et al., 2011) and the rotational constants (Kroto, 1992) corresponding to a minimum on the potential energy surface (PES)] as well as the rovibrational energy level structure of these species strongly disagree with the corresponding data computed by (even the most sophisticated) electronic-structure techniques (Helgaker et al., 2000) within the RRHO approximation. It is thus important to note that the structure and spectra of "flexible", "polytopic" (Clementi et al., 1973; Nielsen et al., 1997), "fluxional" (McKee, 2011), and "quasistructural" (Fábri et al., 2014c; Fábri et al., 2017; Fábri and Császár, 2018; Császár et al., 2020) molecules, the course of reactions and collisions of molecular systems (Schinke, 1993; Wyatt and Zhang, 1996; Zhang, 1999), and the extended dynamics characterizing rovibrational resonances (Papp et al., 2017a,b; Papp et al., 2018), playing a special role, for example, in bimolecular reactions, can all be computed and interpreted via sophisticated fourth-age (Császár et al., 2012) *variational* quantum-chemical techniques. In this chapter all time-independent nuclear-motion techniques which employ basis sets, represent the rovibrational Hamiltonian, $\hat{H}_{VR}$, in this basis as a matrix, $\boldsymbol{H}$, and solve the resulting matrix eigenvalue problem via a "diagonalization" technique will be referred to as "variational", irrespective whether the matrix elements are computed accurately or not (Szalay et al., 2012). Detailed description of the best of these variational techniques utilizing exact kinetic energy operators expressed in curvilinear internal (shape) coordinates is the main aim of the first part of this chapter. This restriction on the algorithms addressed is dictated partly by the fact that only these techniques will generally be applicable for complex stationary-state-based dynamical computations. This choice also means that certain popular nuclear-motion techniques, especially those built upon the *Eckart–Watson Hamiltonians* (Eckart, 1935; Watson, 1968, 1970), are not considered here in detail. The reader is referred to review articles (Bowman et al., 2008; Carrington, 2017; Carney et al., 1978; Beck et al., 2000; Christiansen, 2007; Neff and Rauhut, 2009; Seidler et al., 2010) if interested in other extremely useful nuclear-motion techniques, whose applicability, however, is basically constrained to low excitations of semirigid molecular systems. It is also not the purpose of this chapter to discuss certain highly useful simplification techniques (Sadri et al., 2014) built upon certain choices of the form of the PES or computational techniques (Fábri et al., 2014a; Pavlyuchko et al., 2015) not designed to treat systems exhibiting arbitrarily large-amplitude motions (LAM).

At this point it is instructive to return briefly to what is considered to be the standard model in the field of stationary rovibrational states of free molecules. First, without external forces the three translational dofs of molecular systems can be separated exactly from the rest of the motions. Second, the approximate separation of the three rotational dofs from the $3N-6$ vibrational dofs of an N-atomic non-linear molecule is part of the RRHO-based treatment of molecular vibrations and rotations (Wilson et al., 1955; Kroto, 1992). Third, using normal coordinates within the HO approximation to the PES leads to perfect separation of the vibrational modes, i.e., to one-dimensional (1D) model vibrations [the adiabatic separation of certain molecular vibrations remains a highly useful concept much beyond the HO approximation (Hougen et al., 1970; Miller et al., 1980; Carrington and Miller, 1984; Lauvergnat, 2001; Fehrensen et al., 2003)]. Finally, note that in linear molecules, including diatomics, rotation around the internuclear axis needs special treatment. Linear molecules have only two rotational dofs and one extra vibrational dof.

As the next point, it is noted that observable rovibrational energy-level splittings, assuming a molecular Hamiltonian without hyperfine interaction terms, are considered by many to be due to (multidimensional) quantum tunneling (Hund, 1927; Nordheim, 1928; Frenkel, 1932; Bell, 1980) (think of ammonia and its "umbrella" motion (Al Derzi et al., 2015) as a principally 1D tunneling case). Clearly, if tunneling occurs in a molecule then all of its vibrational and rotational dofs are affected (Császár and Furtenbacher, 2016). This simple picture is complicated by the fact that splitting, perhaps with a rather complex splitting pattern (Fábri et al., 2017; Fábri and Császár, 2018), also occurs if LAMs among symmetrically equivalent versions (Longuet-Higgins, 1963; Bunker and Jensen, 1998) of a molecule take place. A LAM coordinate can be associated with chemical isomerization [e.g., HNC → HCN for the [H,C,N] system (van Mourik et al., 2001; Mellau, 2011a,b)], but here the picture of tunneling splitting is less useful due partly to the asymmetry of the wells. Energy-level splitting in just slightly asymmetric potentials introduces interesting dynamical phenomena, like tunneling switching (Quack, 1986; Albert et al., 2013, 2016; Šmydke et al., 2019). Deep tunneling (Bell, 1980; Schreiner et al., 2008) along a single path can be treated as a simple, well understood and utilized concept similar in spirit to the RRHO approximation. However, there are complex motions where treatment of tunneling and energy-level splittings requires sophisticated variational modeling efforts, way beyond the simple 1D picture. It must be emphasized that often it is hard to separate tunneling and other large-amplitude motions. For the variational nuclear-motion techniques described in this chapter this separation causes no particular difficulties.

A molecule is considered semirigid if its (ground) electronic state contains a single, well-defined, and conveniently deep minimum. For the $3N-3$ internal degrees of freedom of a semirigid molecule the rovibrational eigenstates can be labeled by the irreducible representations (irreps) of the point group [isomorphic to a molecular symmetry (MS) group (Longuet-Higgins, 1963; Bunker and Jensen, 1998) for semirigid molecules] characterizing the unique equilibrium structure of the molecule, the vibrational spacing decreases, almost without exception, with the increase in vibrational excitation, and the rotational states can be assigned to a certain vibrational state, as implied by the RRHO approximation (from now on these vibrational states will be called the *vibrational parents* (Mátyus et al., 2010; Wang et al., 2011; Szidarovszky et al., 2012) of certain rovibrational states). For molecules considered to be semirigid, (a) the vibrational and rotational motions can be treated separately (they have sufficiently different energy and time scales); (b) the rotational spectrum provides important and accurate information about the (temperature-dependent, effective) structure (Demaison et al., 2011) of the molecule; and (c) the simple RR picture is perfectly adequate to explain not only the characteristics of the observed microwave (MW) or millimeterwave (MMW) spectra, though hyperfine splittings may also need to be considered, but it is also sufficient to derive the (equilibrium) structure (Demaison et al., 2011) of the molecule via the determination of rotational constants [the lowest-order terms in an effective rotational Hamiltonian (Kroto, 1992)]. For semirigid molecules there are traditional quantum-chemical treatments, such as vibrational perturbation theory based on the Eckart–Watson (EW) Hamiltonians (Eckart, 1935; Watson, 1968, 1970) carried out to second order (Nielsen, 1951; Mills, 1972; Papoušek and Aliev, 1982) and beyond (Sibert, 1988; Aarset et al., 2000; Iung et al., 2006) or even variational treatments utilizing the exact EW Hamiltonians. Note that extensions offered to the variational EW Hamiltonian treatment [like the Reaction Path Hamiltonian (RPH) (Miller et al., 1980) extension of the widely utilized protocol and code Multimode (Carter et al., 1998; Bowman et al., 2003)] would still allow the treatment of a LAM only along a single coordinate.

As noted above, there is a large number of molecules for which RRHO-based treatments are not adequate. The extent how far one needs to go beyond the RRHO, the VPT2, and the EW-based variational treatments depends greatly upon the complexity of the nuclear motions. Nuclear motions of molecules exhibiting LAM, as well as polytopic, fluxional, and quasistructural molecules require quantum treatments of differing sophistication. Furthermore, treating dynamical processes (e.g., reactions and collisions) also calls for sophisticated variational procedures. Dealing with motions of such molecules and molecular systems leads us closer to the techniques required in complex quantum-dynamical treatments. Systems where understanding complex internal motions, as well as their spectral manifestations, required the development of fourth-age variational quantum-chemical techniques insensitive to the number of minima and the corresponding equilibrium structures, r_e^{BO}, the PES possesses, are as follows: fluxional (McKee, 2011), polytopic (Clementi et al., 1973; Nielsen et al., 1997; Bogey et al., 1991), and quasistructural (Fábri et al., 2014c; Fábri et al., 2017; Fábri and Császár, 2018; Mills and Thompson, 1954; Bunker and Longuet-Higgins, 1964; Sarka et al., 2015; Sarka and Császár, 2016; Császár et al., 2020) molecules, as well as van der Waals (vdW) complexes (Nesbitt and Naaman, 1989; Sarka et al., 2016, 2017).

In the fourth age of quantum chemistry (Császár et al., 2012) the nuclear motions of rigid, semirigid, floppy, flexible, polytopic, fluxional, and quasistructural molecular systems, including molecular clusters, van der Waals complexes, and molecules exhibiting highly unusual bonds, like three-center–two-electron (3c–2e) bonds [e.g., in protonated H_2, H_3^+ (Polyansky et al., 2012), and protonated methane, CH_5^+ (Fábri et al., 2017)], can all be treated with sophisticated variational techniques solving the time-independent nuclear Schrödinger equation in a numerically exact fashion. Time-independent variational techniques employed in computational molecular spectroscopy, a field of nuclear-motion theory, have traditionally been built around the use of rectilinear normal coordinates, the Eckart embedding, and the resulting EW Hamiltonians (Eckart, 1935; Watson, 1968, 1970). Nevertheless, for large-amplitude motions one must use curvilinear internal coordinates as only they provide a suitable and efficient, physically motivated description of the dynamics and allow to move away arbitrarily far from a single reference configuration and allow the use of arbitrary body-fixed frame embeddings. Furthermore, since in many cases there is only a small number of (but more than one) internal coordinates that dominate the large-amplitude motions (the other vibrational dofs can be considered as "spectator" modes) even in relatively complex and large molecules, development of reduced-dimensional (variational) nuclear-motion treatments is highly desirable. It is hard to expect that "regular" stretching dofs will have a significant role in complex nuclear dynamics, unless dissociation is part of the motions considered. On the other end of vibrational softness, torsional vibrations and other LAMs appear to be the best candidates of the vibrational dofs which could couple extremely strongly with the rotational dofs. During the last two decades efficient and black-box-type protocols relying on the use of a Hamiltonian expressed in arbitrary internal coordinates and body-fixed frames have been developed (Fehrensen et al., 1999a,b; Luckhaus, 2000, 2003; Lauvergnat and Nauts, 2002; Yurchenko et al., 2007; Mátyus et al., 2009; Changala, 2019). The most desirable features of these protocols are as follows: (a) they are completely general in the sense that a single code can treat *all* molecular systems of feasible size, irrespective of the fact whether their PESs contain a single minimum or easily accessible multiple minima and what the choices of internal coordinates and coordinate system embeddings are; (b) the always very complex form of the exact kinetic energy operator in internal coordinates, see, e.g., Handy (1987), Császár and Handy (1995a), and Császár and Handy (1995b), may not need to be known explicitly; and (c) they allow the use of arbitrary reduced-dimensional nuclear motion treatments and PESs of arbitrary functional form within the same code. All these features of a time-independent variational nuclear-motion protocol become especially important when the aim is to treat large-amplitude motions of larger fluxional and quasistructural molecular systems or highly excited rovibrational states of molecules [the extreme case is the determination of the complete set of rovibrational (bound) states (Szidarovszky et al., 2010; Császár et al., 2010)]. Furthermore, these are the computations which can provide all the necessary rovibrational states for dynamical studies. Several codes have been developed which can perform these numerically very demanding quantum-chemical computations: ElVibRot (Lauvergnat and Nauts, 2002; Lauvergnat, 2019), GENIUSH (Mátyus et al., 2009; Fábri et al., 2011a), MCTDH (Meyer et al., 2009), NITROGEN (Changala, 2019), TROVE (Yurchenko et al., 2007), codes due to Carrington (Manzhos et al., 2009, 2015; Manzhos and Carrington, 2016), Poirier (Poirier, 2003b; Chen and Poirier, 2006, 2010; Petty and Poirier, 2014), Yu (Yu, 2002, 2004b,a, 2006), and perhaps there are others, as well. Note that for triatomic systems, for which a completely general internal-coordinate Hamiltonian, often referred to as the Sutcliffe–Tennyson Hamiltonian (Sutcliffe and Tennyson, 1991), could be developed, there are lot more codes available which are similar in spirit to those mentioned above, e.g., DVR3D (Tennyson et al., 2004), ScalIT (Petty and Poirier, 2014), D^2FOPI (Szidarovszky et al., 2010), and codes due to Guo (Ma et al., 1999), Schwenke et al. (Schwenke, 1992; Klepeis et al., 1993) and Carter and Handy (1986). For four-atomic molecules it is not possible to develop a Hamiltonian which would cover all possible internal coordinate systems. Nevertheless, there are excellent variational nuclear-motion codes which can handle a number of internal coordinate systems, e.g., WAVR4 (Kozin et al., 2004, 2005) and codes developed by Carter and Handy (1988), Schwenke (1996) and Mladenović (2002a,b).

It is appropriate to add at the end that there exist advanced variational codes which do not make the BO separation of nuclear and electronic dofs and allow the "full" treatment of quantum systems, at present up to five bodies (Armour et al., 2005; Stanke et al., 2006; Pavanello et al., 2010; Pachucki and Komasa, 2010; Mátyus et al., 2011; Mátyus and Reiher, 2012; Mátyus, 2013). These extremely accurate computations are not yet applicable for polyelectronic and polyatomic systems and the eigenstates may not be amenable to a straightforward dynamical interpretation.

2.2 MOLECULAR HAMILTONIANS

Let our isolated molecular system contain N nuclei with masses m_i, $i = 1, \dots, N$, and let $\boldsymbol{X}_i$ be the position vectors of the nuclei in the *space-fixed* (SF) Cartesian coordinate frame (X, Y, Z). The set of nuclear positions $\{\boldsymbol{X}_1, \boldsymbol{X}_2, \dots, \boldsymbol{X}_N\}$ is called the *configuration* of the molecular system. Then,

$$\boldsymbol{X}_i = \boldsymbol{X}^{\mathrm{COM}} + \boldsymbol{R}\boldsymbol{x}_i, \quad i = 1, \dots, N, \tag{2.1}$$

where $\boldsymbol{R}$ is an orthogonal rotation matrix depending on the three rotational coordinates ρ_1, ρ_2, and ρ_3, and $\boldsymbol{x}_i = \boldsymbol{x}_i(t)$ are the *body-fixed* position vectors of each nucleus, a function of time t, in the *body-fixed* (BF) reference frame (x, y, z), and COM stands for center-of-mass. It is customary to describe the motions of the molecular system in terms of some set of scalar variables $q_i(t)$ $(i = 1, 2, \dots, A_{\mathrm{s}})$. The $\boldsymbol{x}_i = \boldsymbol{x}_i(q_1, q_2, \dots, q_{A_{\mathrm{s}}}; t)$ BF atomic position vectors are functions of the q_i internal (shape) coordinates, of which A_{s} are active ($A_{\mathrm{s}} < 3N - 6$ holds for reduced-dimensional, while $A_{\mathrm{s}} = 3N - 6$ holds for full-dimensional rovibrational treatment of non-linear molecules). Furthermore, let V denote the PES of the system, depending on a given set of shape coordinates (in this chapter we are dealing solely with conservative systems).

2.2.1 Coordinate Systems

When choosing the $3N$ coordinates, sufficient to describe the complete nuclear dynamics of an N-atomic molecular system, it is desirable to choose coordinates which simultaneously fulfill the criteria of *simplicity*, *separability*, and *factorizability*. Since there is no single coordinate system which exhibits all these important characteristics at the same time, several proposals have been made (Stefanski and Taylor, 1985; Bačić and Light, 1989; Colbert and Sibert, 1989; Bramley et al., 1991; Mayrhofer and Sibert, 1995; Rauhut, 2007; Yagi et al., 2012; Thomsen et al., 2014; Klinting et al., 2015) how to generate optimal/optimized coordinate systems for treating nuclear dynamics.

Simplicity in this context means that the form of the kinetic energy operator (KEO), based on the chosen set of coordinates, is as simple as possible. The simplest way to achieve simplicity goes through the selection of orthogonal shape coordinates, including the Jacobi (body-fixed scattering) (Jacobi, 1843; Smith, 1959; Tennyson and Sutcliffe, 1982) and Radau (Radau, 1868; Smith, 1980) coordinates.

Separability means that the complete nuclear-motion (rotational-vibrational) Hamiltonian, $\hat{H}_{\mathrm{VR}}$, can be written as the finite sum of independent operators:

$$\hat{H}_{\mathrm{VR}} = \hat{H}_{1,i}(q_1, \dots, q_i) + \hat{H}_{i+1,j}(q_{i+1}, \dots, q_j) + \dots + \hat{H}_{k,A_{\mathrm{s}}}(q_k, \dots, q_{A_{\mathrm{s}}}) + \hat{H}'(q_1, \dots, q_{A_{\mathrm{s}}}), \tag{2.2}$$

and $\hat{H}'$ remains small. In an ideal case each term is one-dimensional and the product(s) of the one-dimensional eigenfunctions approximates well the exact eigenfunction(s) of $\hat{H}_{\mathrm{VR}}$. We talk about exact separation when $\hat{H}' = 0$, but this is rarely the case. Note that in a field-free case the translational motion can be separated exactly from the other nuclear motions.

Factorizability means that the complete Hamiltonian can be written as the sum of products (SOP) (Carrington, 2017) of one-dimensional operators,

$$\hat{H}_{\mathrm{VR}} = \sum_i \prod_{j=1}^{A_{\mathrm{s}}+3} \hat{H}_{ij}(q_j). \tag{2.3}$$

This form of the Hamiltonian is advantageous from a computational point of view, as well, as the matrix elements of a representation of $\hat{H}_{\mathrm{VR}}$ of this form can be computed through simple one-dimensional integrations.

Next, let us briefly discuss the different choices of shape coordinates that can be employed during nuclear-motion computations.

Let us start with a set of rectilinear *Cartesian coordinates*. Although Cartesian coordinates are simple and their use is simple, as well, as in Cartesian coordinates both the classical and the quantum-mechanical expressions for the kinetic energy have by far the simplest form, Cartesian coordinates (Cartesians) are considered to be not well suited for rovibrational computations. Nevertheless, some aspects of the use of SF Cartesians in variational nuclear motion computations have been explored (Blanco and Heller, 1983; Broeckhove and Lathouwers, 1993; Suarez et al., 2009; Manzhos and Carrington, 2016). Manzhos and Carrington (2016) demonstrated that if the kinetic energy operator is expressed in space-fixed Cartesians but the basis functions depend only on shape coordinates then one obtains only vibrational energy levels. The use of KEOs in SF Cartesians is still uncommon, partly because these days derivation (Handy, 1987; Lukka, 1995; Császár and Handy, 1995a,b; Pesonen, 2013, 2014) or numerical handling (Meyer, 1979; Laane et al., 1982; Harthcock and Laane, 1982; McCoy et al., 1991; Luckhaus, 2000; Lauvergnat and Nauts, 2002; Mátyus et al., 2009; Fábri et al., 2011a) of kinetic energy operators in curvilinear coordinates is relatively straightforward. In nuclear-motion theory there are only a few methods, like the quantum Monte Carlo technique (Hammond et al., 1994), which work in Cartesian space, mostly out of algorithmic convenience.

Normal coordinates are defined in quantum chemistry analogously to their classical mechanical definition (Hestenes, 1999; Goldstein et al., 2014). They are defined with respect to a single minimum and they are excellent choice for the description of small-amplitude vibrations of semirigid molecules. Nevertheless, normal coordinates are unsuitable to describe LAMs as they are strongly attached to a single deep minimum and its immediate surrounding. Note that not only rectilinear but also curvilinear normal coordinates have been defined (Quade, 1976).

Curvilinear *internal (shape) coordinates* appear to be the best choice for performing variational nuclear-motion computations. Even in harmonic vibrational analysis (Wilson et al., 1955; Califano, 1970) they have been employed extensively by spectroscopists and quantum chemists (Pulay et al., 1979). Properly selected internal coordinates can describe straightforwardly the full configuration space. For variational nuclear-motion computations the use of orthogonal variants (e.g., generalized Jacobi (Jacobi, 1843; Hirschfelder and Dahler, 1956; Jepsen and Hirschfelder, 1959; Smith, 1959; Pack, 1984) and Radau (Radau, 1868; Smith, 1980) coordinates) is advantageous, so they should be selected whenever feasible. For traditional reasons, bond length–bond angle internal coordinates are often referred to as valence coordinates. A number of KEOs have been obtained analytically for valence coordinates (Handy, 1987; Sutcliffe and Tennyson, 1991; Chapuisat and Iung, 1992; Császár and Handy, 1995a,b; Lukka, 1995; Mladenović, 2000b; Pesonen and Halonen, 2003). Curvilinear internal coordinates tend to reduce the coupling in the potential, though usually at the expense of enhanced coupling in the KEO. *Polyspherical coordinates* (Chapuisat and Iung, 1992; Gatti and Iung, 2009; Mladenović, 2000a,b,c), providing a spherical polar parametrization, have also been used to derive analytic KEOs for nuclear-motion computations. For an N-atomic molecule polyspherical coordinates are comprised of $N-1$ radial coordinates ("stretches") and $2N-5$ angular coordinates, usually further classified to $N-2$ bends and $N-3$ torsional (dihedral) angles in a spherical polar parametrization. Many valence coordinate systems can be considered as polyspherical coordinate systems. As to *hyperspherical coordinates*, they are defined by one hyperradius, and all the other coordinate elements are hyperangles (Delves, 1959; Smith, 1959; Louck and Galbraith, 1972; Johnson, 1980, 1983; Pack, 1984). Hyperspherical coordinates have special symmetry properties, these can be exploited during nuclear-motion computations (Louck and Galbraith, 1972).

A difficulty in nuclear-motion computations of flexible systems is that if one uses normal coordinates or certain internal coordinates not describing well the actual complex motions of the system, the basis functions associated with these coordinates are often strongly coupled. A judicious choice of internal coordinates and basis functions is thus fundamental to converge the usually very large number of computed (ro)vibrational eigenstates (De Leon and Heller, 1984; Colbert and Sibert, 1989) and allow for the improved interpretation of the wavefunctions, important to understand the dynamics of the system.

From what is noted above it follows that the choice of curvilinear (generalized, orthogonal) internal coordinate systems is by far the best choice for variational nuclear-motion computations (Watson, 2004). Redundancy conditions (Wilson et al., 1955; Califano, 1970; Hoy et al., 1972) should always be checked when selecting the shape coordinates.

2.2.2 Formulation of the Classical Hamiltonian in Generalized Internal Coordinates

We follow the generally applied approach to derive the quantum Hamiltonian by first deriving the classical Lagrangian and Hamiltonian forms and then performing the quantization. The non-relativistic Lagrangian L of the molecule is of the form (Sørensen, 1979; Hestenes, 1999; Pesonen, 2013, 2014)

$$L = \frac{1}{2}\sum_{k=1}^{A_s+6}\sum_{l=1}^{A_s+6} g_{q_k q_l}\dot{q}_k\dot{q}_l - V, \tag{2.4}$$

where

$$g_{kl} \equiv g_{q_k q_l} = \sum_{i=1}^{N} m_i \frac{\partial \boldsymbol{X}_i^{\mathrm{T}}}{\partial q_k}\frac{\partial \boldsymbol{X}_i}{\partial q_l} = \sum_{i=1}^{N} m_i \boldsymbol{t}_{ik}^{\mathrm{T}}\boldsymbol{t}_{il}, \quad k,l = 1,\dots,A_s+6 \tag{2.5}$$

is the *covariant metric tensor* of the coordinate transformation from the space-fixed mass-weighted Cartesian coordinates to the generalized coordinates q_k, based on the *covariant measuring vectors* $e_k^{(i)} = \frac{\partial \boldsymbol{X}_i}{\partial q_k}$ (Hestenes, 1999). As in Eq. (2.5), the covariant measuring vectors are often referred to as the $\boldsymbol{t}$ vectors (Sørensen,

1979), corresponding to the q_k generalized coordinate on atom i.

The classical Hamiltonian can then be expressed as

$$H = \frac{1}{2} \sum_{k=1}^{A_s+6} \sum_{l=1}^{A_s+6} G^{kl} p_k p_l + V, \tag{2.6}$$

where $\boldsymbol{G} = \boldsymbol{g}^{-1}$ is the *contravariant metric tensor* and $p_k = \frac{\partial L}{\partial \dot{q}_k}$ $(k = 1, \ldots, A_s + 6)$ is the canonical momentum conjugate to q_k. There is a well-known reciprocality relation (Hestenes, 1999) connecting the covariant and contravariant measuring vectors:

$$\sum_{i=1}^{N} e_j^{(i)} \cdot e_{(i)}^k = \delta_{jk}. \tag{2.7}$$

The contravariant measuring vectors are known in the theory of the rovibrational motion of molecules as $\boldsymbol{s}$ vectors (Wilson et al., 1955; Pesonen, 2014).

To construct the covariant $\boldsymbol{g}$ and the contravariant $\boldsymbol{G}$ metric tensors, let us describe the configuration of the system by the $q_k = q_k$ active $(k = 1, \ldots, A_s)$ and constrained $(k = A_s + 1, \ldots, 3N - 6)$ internal coordinates, the three rotational $(\rho_1, \rho_2$, and $\rho_3)$ and the three center-of-mass $(X_1^{\mathrm{COM}}, X_2^{\mathrm{COM}}$, and $X_3^{\mathrm{COM}})$ coordinates describing the translation of the system. Derivation of the g_{kl} matrix elements is equivalent, see Eq. (2.5), to constructing the $\boldsymbol{t}_{ik}$ vectors (Sørensen, 1979) in terms of the generalized coordinates.

The *translational* $\boldsymbol{t}_{i,k+A_s+3}$ $(k = 1, 2, 3)$ vectors are simply

$$t_{i,k+A_s+3}^a = \frac{\partial X_{ia}}{\partial X_k^{\mathrm{COM}}} = \delta_{ak}, \tag{2.8}$$

where $a = 1, 2, 3$ refers to the three components of the vector $\boldsymbol{t}$. Thus, the translational $\boldsymbol{g}$ matrix elements are obtained as

$$g_{k+A_s+3,l+A_s+3} = M \cdot \delta_{kl}, \quad k, l = 1, 2, 3, \tag{2.9}$$

where $M = \sum_{i=1}^{N} m_i$ is the total mass of the system.

The *rotational-translational* and the *vibrational-translational* coupling matrix elements of $\boldsymbol{g}$ are all equal to zero. Therefore, the COM motion can be separated exactly from the rest of the coordinates. This allows the introduction of the

$$H^{\mathrm{int}} = T^{\mathrm{int}} + V = \frac{1}{2} \sum_{k=1}^{A_s+3} \sum_{l=1}^{A_s+3} G^{kl} p_k p_l + V \tag{2.10}$$

internal-motion (rovibrational) Hamiltonian.

The *rotational* $\boldsymbol{t}_{i,k+A_s}$ $(k = 1, 2, 3)$ vectors take the form

$$t_{i,k+A_s}^a = \frac{\partial X_{ia}}{\partial \rho_k} = \sum_{b=1}^{3} \frac{\partial R_{ab}}{\partial \rho_k} x_{ib} \tag{2.11}$$

and the *rotational* $\boldsymbol{g}$ matrix elements are equal to

$$g_{k+A_s,l+A_s} = \sum_{i=1}^{N} m_i (\boldsymbol{u}_k \times \boldsymbol{x}_i)^{\mathrm{T}} (\boldsymbol{u}_l \times \boldsymbol{x}_i), \tag{2.12}$$

where the direction of the unit vector $\boldsymbol{u}_k$ coincides with the axis of rotation corresponding to the rotational coordinate ρ_k.

The *vibrational* $\boldsymbol{t}_{ik}$ $(k = 1, \ldots, A_s)$ vectors are

$$t_{ik}^a = \frac{\partial X_{ia}}{\partial q_k} = \sum_{b=1}^{3} R_{ab} \frac{\partial x_{ib}}{\partial q_k}. \tag{2.13}$$

Thus, the *vibrational* $\boldsymbol{g}$ matrix elements are given as

$$g_{kl} = \sum_{i=1}^{N} m_i \frac{\partial \boldsymbol{x}_i^{\mathrm{T}}}{\partial q_k} \frac{\partial \boldsymbol{x}_i}{\partial q_l}, \tag{2.14}$$

where $k, l = 1, \ldots, A_s$.

Finally, the *rotational-vibrational coupling* (often called Coriolis coupling) $\boldsymbol{g}$ matrix elements have the form

$$g_{k,l+A_s} = \sum_{i=1}^{N} m_i \frac{\partial \boldsymbol{x}_i^{\mathrm{T}}}{\partial q_k} (\boldsymbol{u}_l \times \boldsymbol{x}_i), \tag{2.15}$$

where $k = 1, \ldots, A_s$ and $l = 1, 2, 3$. To determine $\boldsymbol{g}$, the body-fixed frame embedding has to be defined, which gives the dependence of the $\boldsymbol{x}_i$ body-fixed nuclear position vectors on the q_k shape coordinates (see also Section 2.2.4).

If larger molecules are examined, it is often necessary and at the same time a good approximation to introduce reduced-dimensional rovibrational models. The two possible ways of reducing the dimensionality of the problem are to delete rows and the corresponding columns of $\boldsymbol{g}$ or $\boldsymbol{G}$. The first case, when the ith row and column $(i = 1, \ldots, 3N-6)$ of $\boldsymbol{g}$ are deleted, implies the $\dot{q}_i = 0$ constraint, while the second approach, when the ith row and column of $\boldsymbol{G}$ are discarded, is equivalent to $p_i = 0$. If orthogonal coordinates are employed to describe the rovibrational dynamics of molecules, the two approaches are equivalent. However, in general the two strategies provide different reduced-dimensional models and numerical results (Fábri et al., 2009; Mátyus et al., 2009).

2.2.3 Formulation of the Quantum-Mechanical Hamiltonian in Generalized Internal Coordinates

In this subsection the internal-motion (rovibrational) quantum-mechanical Hamiltonian $\hat{H}^{\text{int}}$ is introduced in analogy to the internal-motion (rovibrational) classical Hamiltonian H^{int} (Nauts and Chapuisat, 1985; Littlejohn and Reinsch, 1997), see Eq. (2.10). Within the Born–Oppenheimer (BO) separation of nuclear and electronic dofs (Born and Oppenheimer, 1927, see also Marquardt and Quack, 2020, Chapter 1 of this book), the potential energy acting on the nuclei, $\hat{V}$, can be obtained by electronic-structure computations (Murrell et al., 1984; Mezey, 1987; Császár et al., 2000, 2001) (see Section 2.2.5). Here, we focus on constructing the rovibrational kinetic energy operator in the set of q_k ($k = 1, \ldots, A_\text{s}$) vibrational and ρ_k ($k = 1, 2, 3$) rotational coordinates. According to differential geometry (Podolsky, 1928; Schutz, 1980), $\hat{T}^{\text{int}}$ becomes

$$\hat{T}^{\text{int}} = \frac{1}{2} \sum_{k=1}^{A_\text{s}+3} \sum_{l=1}^{A_\text{s}+3} \tilde{g}^{-1/4} \hat{p}_k^\dagger G^{kl} \tilde{g}^{1/2} \hat{p}_l \tilde{g}^{-1/4}, \quad (2.16)$$

where $\tilde{g} = \det(\boldsymbol{g})$, the integration volume element contains no extra factors, and in units of $\hbar$, for the vibrational coordinates $\hat{p}_k = -\mathrm{i}\frac{\partial}{\partial q_k}$ ($k = 1, \ldots, A_\text{s}$), while for the rotational coordinates $\hat{p}_{k+A_\text{s}} = -\mathrm{i}\frac{\partial}{\partial \rho_k}$ ($k = 1, 2, 3$), where $\mathrm{i}^2 = -1$.

Next, let us utilize that the angular momentum $\hat{\boldsymbol{J}}$ is the infinitesimal generator of rotations (Zare, 1988),

$$\boldsymbol{n}\hat{\boldsymbol{J}} = -\mathrm{i}\frac{\partial}{\partial \phi}, \quad (2.17)$$

where $\boldsymbol{n}$ is a unit vector specifying the rotational axis and ϕ is an angle associated with the rotation. According to Eq. (2.17), it is straightforward to identify the rotational $\hat{p}_{k+A_\text{s}}$ operators as the projections of the total angular momentum. After specifying three unique rotational axes, three successive rotations can be performed, defining the transformation between the SF and BF frames. As the ρ_k rotational coordinate describes a rotation around the kth of the three rotational axes,

$$\hat{p}_{k+A_\text{s}} = -\mathrm{i}\frac{\partial}{\partial \rho_k} = \hat{J}_k, \quad (2.18)$$

where $\hat{J}_k$ is the component of the total angular momentum vector along the kth rotational axis. If the three rotational axes are chosen to coincide with the three axes of the body-fixed frame, the $\hat{J}_k$ operators correspond to the angular momentum components expressed in the body-fixed frame, and the ρ_k rotational angles define three successive rotations around the three orthogonal axes of the body-fixed system. It is advantageous to employ these infinitesimal rotational coordinates (Lukka, 1995; Colwell and Handy, 1997) instead of the widely-used Eulerian angles for at least two reasons: (a) one can directly insert the body-fixed components of the total angular momentum into the rovibrational Hamiltonian by utilizing Eq. (2.18); and (b) the matrix elements g_{kl} can be computed trivially, as according to Eqs. (2.12) and (2.15) one only needs to evaluate the $\boldsymbol{u}_k \times \boldsymbol{x}_i$ cross-products.

2.2.4 Body-Fixed Frame Embeddings

While the SF form of the nuclear-motion Hamiltonian expressed in internal coordinates is available (Mladenović, 2000a), it is much more common to work with BF frame embeddings. Defining a convenient set of rotating BF axes has become an important issue in modern nuclear-motion theory (Born and Heisenberg, 1924; Louck and Galbraith, 1976; Bunker and Jensen, 1998; Mladenović, 2000a,b,c). The orientation of the BF axis system with respect to the SF frame is specified by the three rotational coordinates. Although all embeddings provide the same rovibrational energy levels and eigenstates during numerically exact variational rovibrational computations, it is of great importance to find embeddings that provide an optimal separation between rotations and vibrations. Many conceptually simple embeddings based on geometric definitions do not meet this requirement. For example, in the case of XY_2 molecules the bisector embedding, especially if it is coupled with the choice of orthogonal Radau shape coordinates, provides a much more perfect separation of rotations and vibrations than the so-called r_1 or r_2 embeddings (Sarka et al., 2020).

It was Eckart (Eckart, 1935) who formulated equations that lead to a good separation of rotations and vibrations, yielding exactly zero rotational-vibrational coupling at a chosen reference structure (actually this is how the Eckart embedding is defined). The optimal separation of the rotational and vibrational dofs is often crucial for obtaining converged rovibrational energy levels and eigenstates in practical nuclear-motion computations. Other possible motivations for using the Eckart frame include the assignment of rovibrational eigenstates (Mátyus et al., 2010; Szidarovszky et al., 2012), the computation of vibrational band intensities (Le Sueur et al., 1992), and minimizing the root-mean-square deviation between two molecular conformations (Kudin and Dymarsky, 2005).

The translational Eckart condition,

$$\sum_{i=1}^{N} m_i \boldsymbol{x}_i = \boldsymbol{0}, \tag{2.19}$$

where the vectors $\boldsymbol{x}_i$ denote the instantaneous position vectors of the N nuclei in the Eckart frame, is supplemented with the rotational Eckart condition,

$$\sum_{i=1}^{N} m_i (\boldsymbol{x}_i \times \boldsymbol{a}_i) = \boldsymbol{0}, \tag{2.20}$$

where the vectors $\boldsymbol{a}_i$ specify the reference configuration, whereby the rovibrational coupling vanishes. The translational Eckart condition can be satisfied by placing the origin of the body-fixed frame at the center of mass, while fulfilling the rotational Eckart condition is equivalent to finding a $\boldsymbol{T}$ (pseudo)rotation matrix which transforms the $\boldsymbol{x}'_i$ initial position vectors into the $\boldsymbol{x}_i = \boldsymbol{T}\boldsymbol{x}'_i$ position vectors corresponding to the Eckart frame. Following Eckart (1935), Strauss and Pickett (1970) devised a procedure for finding the $\boldsymbol{T}$ transformation matrix. An important shortcoming of the methods of Eckart (1935) and of Strauss and Pickett (1970) is the need for computing the inverse of a matrix which can be singular for certain nuclear configurations. Two new methods, achieving the same task but free of the singularity problem, have been developed recently (Dymarsky and Kudin, 2005; Krasnoshchekov et al., 2014).

As the Eckart frame is crucial to obtain good separation of vibrations and rotations, there has been a considerable effort to derive Eckart-embedded KEOs. For rectilinear vibrational coordinates, including normal coordinates, the theory has been worked out by Watson (1968, 1970). For curvilinear internal coordinates, analytic Eckart-embedded KEOs have been derived for the limiting class of triatomic (Wei and Carrington, 1997a,b, 1998) as well as the more general class of planar molecules (Wei, 2003b,a). Unfortunately, the Eckart-embedded KEOs expressed in curvilinear coordinates have rather bulky forms preventing their widespread use. It is worth noting at this point that flexible reference configurations, as opposed to the choice of a rigid, e.g., equilibrium structure, have been introduced in several spectroscopic models, including the Hougen–Bunker–Johns approach (Hougen et al., 1970) and several of its extensions and variants (Špirko, 1983; Szalay, 1988; Yurchenko et al., 2005).

The challenge to construct general analytic Eckart-embedded KEOs expressed in arbitrary curvilinear coordinates for arbitrary molecules can be circumvented if the KEO is treated numerically (Luckhaus, 2000; Lauvergnat and Nauts, 2002; Yurchenko et al., 2007; Mátyus et al., 2009; Strobusch and Scheurer, 2011b,a; Fábri et al., 2011a). This can be achieved in grid-based codes (Fábri et al., 2014b) by transforming the body-fixed position vectors $\boldsymbol{x}_i$ to the Eckart frame using the quaternion-based transformation method outlined in (Krasnoshchekov et al., 2014) and then evaluating the $\frac{\partial \boldsymbol{x}_i}{\partial q_k}$ vectors [see Eq. (2.14)] by numerical differentiation of the Eckart-embedded $\boldsymbol{x}_i$ vectors (Szidarovszky et al., 2012; Fábri et al., 2014b), while the computation of the $\boldsymbol{u}_k \times \boldsymbol{x}_i$ vectors used in Eqs. (2.12) and (2.15) simply require the Eckart-embedded $\boldsymbol{x}_i$ vectors (Szidarovszky et al., 2012; Fábri et al., 2014b). This allows the user to employ the Eckart frame for arbitrary vibrational coordinates with either fixed or flexible reference structures. Other approaches for the application of the Eckart frame have also been reported (McCoy et al., 1991; Wang and Carrington, 2013; Sadri et al., 2014; Pesonen, 2014; Szalay, 2014, 2015b,a; Yachmenev and Yurchenko, 2015; Szalay, 2017).

2.2.5 Potential Energy Hypersurfaces

The concept of a PES (Murrell et al., 1984; Mezey, 1987; Császár et al., 2001; Marquardt and Quack, 2011) is fundamental to the understanding of most modern fields of chemistry, PESs are especially relevant for studies in high-resolution spectroscopy, detailed chemical kinetics, and nuclear dynamics. For an arbitrary chemical system, PESs describe the variation of the total electronic energy as a function of the (relative) nuclear positions. In computational molecular spectroscopy it is still usual to focus on cases where a single BO-PES is sufficiently uncoupled from the other surfaces (electronic states) that their interaction may safely be ignored. This is usually true for the PES characterizing the lowest (ground) electronic state, as the error thus introduced is smaller than the intrinsic error of the PES. In dynamical studies it is more usual to couple a few selected PESs, perhaps just two, and define adiabatic and diabatic representations (Smith, 1969; Song and Gao, 2008).

Although PESs developed for rovibrational spectroscopy, chemical reactions, and intermolecular interactions representing systems of non-covalent chemistry have a lot in common, there are many important differences in their construction. These differences originate from the fact that these surfaces need to concentrate on different configuration regions where different electronic-structure techniques are suitable, and that they aim to interpret different types of experimental information of widely different accuracy. The four distinct types of (global) PESs available in the literature are empirical, semiempirical, semitheoretical, and theoretical. Empirical potentials are rather inaccurate even for triatomic systems and they are basically unavailable for larger ones. As to theoretical PESs, due to well recognized inadequacies of most electronic structure computations (Császár et al., 2001), it is still not possible to even approach the accuracy required by high-resolution spectroscopic measurements, except perhaps for the smallest systems (*vide infra*). Even today, the highest quality PESs are obtained by a combination of theoretical and experimental approaches.

Excellent books (Murrell et al., 1984; Mezey, 1987) and reviews (Császár et al., 2001, 2000; Majumder et al., 2016) are available discussing the generation and utilization of local as well as global PESs; thus, they are not discussed in detail here. We are only mentioning a couple of recent developments worth emphasizing.

The n-mode (or multimode) representation of the PES (Carter et al., 1997a,b; Bowman et al., 2003; Rauhut, 2004; Sibaev and Crittenden, 2016; Ziegler and Rauhut, 2016) is highly useful for several types of nuclear-motion applications, especially those employing the EW Hamiltonians. These PESs may prove less useful when large configuration regions need to be sampled. The use of permutationally invariant PESs (Braams and Bowman, 2009) is particularly advantageous for systems containing more than 4–5 atoms and this concept leads to an almost automatic generation of a "global" PES. Almost automatic generation of the PES is highly desirable; thus, use of neural networks and artificial intelligence have been explored (Handley and Popelier, 2010; Majumder et al., 2015; Jiang et al., 2016; Petty et al., 2018) for this purpose. For chemical reactions proper representation of the surface close to an intrinsic reaction path (IRP) (Fukui, 1981), a minimum-energy path (MEP) (Fukui, 1970), or some other path (Müller, 1980) is often what is used to determine the location of the points underlying the PES (Klepeis et al., 1993). Stationary points (beyond minima and first-order transition states) characterizing the given PES form are rarely reported, not even for molecular complexes, though they may serve important qualitative and quantitative purposes.

2.2.6 Basis Sets and Representations

Basis sets and the representation of $\hat{H}^{\text{int}}$ in the chosen basis play an extremely important role in the effectiveness of the variational approaches characterizing nuclear-motion theory. Thus, it is not surprising that a considerable number of efficient numerical techniques have been developed, with a concerted effort to tailor the basis sets and the representations to the usual need of computing an exceedingly large number of rovibrational eigenstates.

Today it appears that the solution of the time-independent rovibrational Schrödinger equation is achieved perhaps most efficiently by grid-based techniques. The utility and the basic aspects of grid-based techniques were realized by different groups working in different fields independently and thus several somewhat distinct methods were developed, including the discrete variable representation (DVR) (Harris et al., 1965; Dickinson and Certain, 1968; Lill et al., 1982; Light et al., 1985; Light and Carrington, 2000), the Lagrange-mesh (Baye and Heenen, 1986), the quadrature discretization (Shizgal and Blackmore, 1984; Blackmore and Shizgal, 1985), and the Fourier grid (Kosloff and Kosloff, 1983; Marston and Balint-Kurti, 1989; Layton, 1993) methods. Of these we will concentrate here only on the DVR technique.

When the Hamiltonian is represented by a DVR, one basically employs both a set of basis functions and a set of grid points. Within the DVR and finite basis representation (FBR) methods the operators of the physical quantities considered are represented by matrices with indices referring either to grid points or spectral basis functions, respectively. The DVR of the potential energy operator is extremely simple, a diagonal matrix with the (nn)th diagonal element equal to the value of the PES function taken at the nth grid point. The FBR of the potential energy operator corresponds to evaluating integrals in the spectral basis by numerical quadrature, whereby the quadrature grid is identical to the one defining the DVR. The DVR and FBR of an operator of a physical quantity are related by a unitary transformation.

At the beginning (Harris et al., 1965; Dickinson and Certain, 1968), the DVR developed was based on standard orthogonal polynomial bases and the associated Gaussian quadratures. Within this simplest DVR scheme the same number of basis functions and quadrature points are employed. Later, generalizations of the DVR method to more general basis functions depending on a single variable and to multidimensional non-direct product basis sets have been developed (Light et al., 1985; Kanfer and Shapiro, 1984; Corey and Tromp, 1995; Szalay, 1996; Littlejohn et al., 2002; Szalay et al., 2003; Dawes and Carrington, 2004; Yu, 2005). A family of generalized DVR (GDVR) techniques, including the optimal GDVR, was introduced by Szalay (Szalay, 1996). Similar to the relation between DVR and FBR schemes, GDVR techniques have a strong connection with generalized FBR (GFBR) techniques (Szalay, 1996; Czakó et al., 2006). In a GFBR one is allowed to use more grid points than basis functions, leading to an improved accuracy of the results. Consideration of boundary conditions and singularities (Szalay et al., 2012) is extremely important for the application of all flavors of FBR and DVR techniques. This becomes especially important when one does not know the actual form of the kinetic energy operator as it is used just numerically on a grid. A particularly important characteristic of DVR bases is that they provide numerically exact solution of the time-independent rovibrational Schrödinger equation without computing the Hamiltonian matrix elements exactly (Szalay et al., 2012). For this reason, techniques based on DVR should not be considered strictly variational.

There are different strategies toward selecting the most useful and practical form of the rovibrational Hamiltonian. One possible strategy is to define lower-dimensional but non-sparse and non-direct-product matrices. Contraction of the vibrational basis (Carter and Handy, 1988; Bačić and Light, 1989) offers considerable numerical advantages, while other strategies, based on direct-product bases, produce often enormous-dimensional but sparse matrices with a special structure. There is not much choice how rotational basis functions can be chosen. A very practical and efficient choice of a rotational basis involves Wang functions (Zare, 1988), as discussed in Section 2.3.1. Naturally, different "diagonalization" techniques are required for the different Hamiltonians (see Section 2.2.7). Some further details, relevant to this chapter, will be given in Section 2.3.

2.2.7 Determination of Eigenstates

Each variational technique, at least in principle, requires the diagonalization of the Hamiltonian yielding rovibrational eigenvalues and eigenvectors. In order to compute the required many (often tens of thousands of) eigenpairs for medium-sized molecules, an iterative eigensolver [the usual choice is the Lanczos technique (Lanczos, 1950; Cullum and Willoughby, 1985)] must be implemented, adapted to the features and requirements of nuclear-motion theory (Wang and Carrington, 2001; Mátyus et al., 2009).

The conventional iterative Lanczos algorithm (Lanczos, 1950) converges to the dominant eigenvalues of the matrix. The convergence rate of the Lanczos iteration is largely determined by the relative separation of the eigenvalues. To compute the lowest or interior eigenvalues and speed up convergence, in variational nuclear-motion theory one needs to introduce spectral transformation techniques (Ericsson and Ruhe, 1980; Kono, 1993; Wyatt, 1995). Thus, families of polynomial transformation techniques (Sorensen, 1992; Wang and Zunger, 1994), exponential filters using a Chebyshev expansion (Yu and Nyman, 1998, 1999), and shift-invert (Ericsson and Ruhe, 1980; Mátyus et al., 2009) filtering techniques using the conjugate gradient method (CGM) have been introduced. Nevertheless, construction of an efficient black-box Lanczos variant, for instance, a method that is efficient for any spectral range of a rotation-vibration Hamiltonian, remains a challenge. In each Lanczos step the original matrix is transformed, requiring a certain number of matrix–vector multiplications. These multiplications form the most CPU-intensive part of the computation; thus, the number of matrix–vector multiplications required determines the "cost" of a specific spectral transformation. The spectral properties of matrices corresponding to different spectral transformation methods can be very different. This is an important consideration as the relative separation of the eigenvalues of the matrix introduced in the Lanczos iteration influences strongly the convergence rate of the iteration.

Under exact arithmetics, the Lanczos vectors are orthogonal by construction (Saad, 2003). In the presence of computational round-off errors, this orthogonality is not maintained. The loss of orthogonality manifests itself in the appearance of (a) spurious eigenvalues, and (b) copies of correct ones. To remove the extra and spurious eigenvalues a posteriori, Cullum and Willoughby (Cullum and Willoughby, 1985) as well as Wang and Carrington (Wang and Carrington, 2002) suggested useful and relatively inexpensive algorithms. A somewhat expensive alternative is to maintain (semi)orthogonality among the Lanczos vectors throughout the iterations. It has also been demonstrated (Simon, 1984; Wu and Simon, 1998; Mátyus et al., 2009) that maintaining full orthogonality is not necessary and semiorthogonality of Lanczos vectors, provided by partial reorthogonalization, is sufficient, accurate eigenpairs without extra and spurious levels entering the spectrum can be computed this way. To keep storage requirements under control, the Lanczos algorithm must be occasionally restarted (Wu and Simon, 1998; Wu et al., 1999; Mátyus et al., 2009).

2.3 COMPUTATION OF BOUND ROVIBRATIONAL STATES

There are three distinct classes of methods for the variational solution of the time-independent nuclear

Schrödinger equation,

$$\hat{H}^{\text{int}}\Psi_i = E_i\Psi_i, \tag{2.21}$$

based on Hamiltonians detailed in Section 2.2: (a) the first one is based on tailor-made Hamiltonians, with explicitly derived kinetic energy operators expressed in terms of orthogonal or non-orthogonal internal coordinates (Shipsey, 1998; Bramley et al., 1991; Chapuisat and Iung, 1992; Klepeis et al., 1993; Henderson and Tennyson, 1993; Schwenke, 1996; Mladenović, 2000a,b,c; Czakó et al., 2004; Tennyson et al., 2004; Kozin et al., 2004; Huang et al., 2008; Szidarovszky et al., 2010); (b) the second one utilizes the EW Hamiltonians, relying on a single reference structure, rectilinear coordinates, and the Eckart embedding (Carter et al., 1998; Tew et al., 2001; Bowman et al., 2003; Mátyus et al., 2007; Mátyus et al., 2009); and (c) the third one uses arbitrary vibrational (shape) coordinates preferably supplemented with the numerical construction of the kinetic-energy operator (Lauvergnat and Nauts, 2002; Yurchenko et al., 2007; Mátyus et al., 2009; Sadri et al., 2014; Changala, 2019), so that the exact form of the kinetic-energy operator does not need to be known a priori.

Tailor-made Hamiltonians have been employed extensively (Sutcliffe and Tennyson, 1991; Klepeis et al., 1993; Tennyson et al., 2004; Kozin et al., 2004; Szidarovszky et al., 2010). As noted more than three decades ago, the "disadvantage with [tailor-made Hamiltonians] is that a separate program has to be developed for each molecule and for each chosen coordinate system" (Handy, 1987). Therefore, codes belonging to this class cannot be considered as "black-box" approaches to variational nuclear-motion computations for polyatomic molecules. For this reason this class of approaches is not discussed here further.

There are a number of codes available which are based on the use of the EW Hamiltonians (Bowman et al., 2003; Mátyus et al., 2007; Christiansen, 2007; Neff and Rauhut, 2009). These codes are completely general, can be employed for rather large systems, especially after the introduction of special sophisticated algorithms. Nevertheless, they cannot be applied to complex nuclear motions and for molecules exhibiting several LAMs. Therefore, their discussion is terminated here, the interested reader is referred to the extensive literature (Bowman et al., 2003; Mátyus et al., 2007) on their form and use. From here on only approach (c) from above is considered.

2.3.1 On the Variational Solution

To compute bound rovibrational states variationally, the matrix representation of $\hat{H}^{\text{int}}$ needs to be considered. For this, it is advantageous to split $\hat{T}^{\text{int}}$ into three terms:

$$\hat{H}^{\text{int}} = \hat{T}^{\text{int}} + \hat{V} = \hat{T}^{\text{v}} + \hat{T}^{\text{r}} + \hat{T}^{\text{rv}} + \hat{V}, \tag{2.22}$$

where (see Eq. (2.16))

$$\hat{T}^{\text{v}} = \frac{1}{2}\sum_{k=1}^{A_s}\sum_{l=1}^{A_s} \tilde{g}^{-1/4}\hat{p}_k^{\dagger} G^{kl}\tilde{g}^{1/2}\hat{p}_l\tilde{g}^{-1/4}, \tag{2.23}$$

$$\hat{T}^{\text{r}} = \frac{1}{2}\sum_{k=1}^{3} G^{k+A_s,k+A_s}\hat{J}_k^2 + \frac{1}{2}\sum_{k=1}^{3}\sum_{l>k}^{3} G^{k+A_s,l+A_s}[\hat{J}_k,\hat{J}_l]_+, \tag{2.24}$$

and

$$\hat{T}^{\text{rv}} = \frac{1}{2}\sum_{l=1}^{3}\sum_{k=1}^{A_s}\left(\hat{p}_k^{\dagger}G^{k,l+A_s} + G^{k,l+A_s}\hat{p}_k\right)\hat{J}_l. \tag{2.25}$$

In Eqs. (2.23)–(2.25), $\hat{T}^{\text{v}}$ and $\hat{T}^{\text{r}}$ are the vibrational and rotational kinetic energy operator parts, respectively, and $\hat{T}^{\text{rv}}$ gives the coupling between vibrations and rotations (often referred to as the Coriolis term). In Eq. (2.24), $\hat{J}_k$ is the kth body-fixed component of $\hat{\boldsymbol{J}}$ and $[\hat{J}_k,\hat{J}_l]_+$ refers to the anticommutator of $\hat{J}_k$ and $\hat{J}_l$. As the $\hat{J}_k$ angular momentum components correspond to the body-fixed frame, they satisfy the anomalous commutation relations (Zare, 1988)

$$[\hat{J}_k,\hat{J}_l] = -\mathrm{i}\epsilon_{klm}\hat{J}_m, \tag{2.26}$$

where ϵ_{klm} is the Levi-Cività permutation symbol, and the Einstein summation convention is applied.

The multidimensional vibrational basis functions can be constructed as the direct product of one-dimensional primitive or potential-optimized (PO) (Echave and Clary, 1992; Wei and Carrington, 1992; Szalay et al., 2003) DVR functions,

$$F_i(\boldsymbol{q}) = \prod_{\alpha=1}^{A_s} f_{i_\alpha}(q_\alpha), \tag{2.27}$$

where $f_{i_\alpha}(q_\alpha)$ denote a 1D DVR function associated with vibrational dof α and 1D DVR grid point $q_{i_\alpha}^{(\alpha)}$. This choice assures that the coordinate-dependent G^{kl},

$\tilde{g}$, and $\hat{V}$ operators are represented by diagonal matrices in the direct-product DVR basis,

$$\langle F_i|\hat{O}|F_j\rangle = O(q_{i_1}^{(1)},\ldots,q_{i_{A_s}}^{(A_s)})\prod_{\alpha=1}^{A_s}\delta_{i_\alpha,j_\alpha}. \quad (2.28)$$

The diagonal matrix elements are equal to the values of the given coordinate-dependent operator $\hat{O}$ evaluated at the DVR grid points. Therefore, the $\boldsymbol{t}$ vectors needed to construct $\hat{T}^{\text{int}}$ have to be evaluated at the DVR grid points, which can be done either analytically or numerically for a specific choice of internal coordinates and body-fixed frame embeddings. This is what makes the numerical construction of $\hat{T}^{\text{int}}$ possible and obviates the need for tailor-made kinetic-energy operators. Matrix elements of the vibrational $\hat{p}_k$ operators can be expressed as

$$\langle F_i|\hat{p}_k|F_j\rangle = \langle f_{i_k}(q_k)|\hat{p}_k|f_{j_k}(q_k)\rangle \prod_{\alpha=1,\alpha\neq k}^{A_s}\delta_{i_\alpha,j_\alpha}, \quad (2.29)$$

where $\langle f_{i_k}(q_k)|\hat{p}_k|f_{j_k}(q_k)\rangle$ is proportional to the corresponding matrix element of the first derivative operator $\frac{\partial}{\partial q_k}$ in the DVR basis applied (Szalay, 1993).

For a given rotational angular momentum quantum number J (note that the molecular system is isolated and no external fields are present), the set of $2J+1$ orthonormal $|JKM\rangle$ symmetric rigid-rotor eigenfunctions serve as a suitable basis to set up the matrix representation of $\hat{H}^{\text{int}}$. According to Eqs. (2.24) and (2.25), the matrix representation of $\hat{J}_k$, $\hat{J}_k^2$ and $[\hat{J}_k,\hat{J}_l]_+$ is required to solve the rovibrational problem. The complete set of non-zero $\hat{J}_k$ matrix elements (Zare, 1988) is given by

$$\langle JKM|\hat{J}_x|J(K\pm 1)M\rangle = \frac{1}{2}\sqrt{J(J+1)-K(K\pm 1)},$$
$$\langle JKM|\hat{J}_y|J(K\pm 1)M\rangle = \mp\frac{\mathrm{i}}{2}\sqrt{J(J+1)-K(K\pm 1)},$$
$$\langle JKM|\hat{J}_z|JKM\rangle = K, \quad (2.30)$$

where $K=-J,\ldots,J$ corresponds to the body-fixed z, while $M=-J,\ldots,J$ to the space-fixed Z components of the overall angular momentum. The $\hat{J}_k^2$ and $[\hat{J}_k,\hat{J}_l]_+$ matrices can be computed exactly by inserting the resolution of identity between $\hat{J}_k$ and $\hat{J}_l$, and thus

$$\langle JKM|\hat{J}_k\hat{J}_l|JK'M\rangle = \sum_{K''=-J}^{J}\langle JKM|\hat{J}_k|JK''M\rangle\langle JK''M|\hat{J}_l|JK'M\rangle. \quad (2.31)$$

In practice, a good choice for the rotational basis involves the Wang functions that yield real Hamiltonian matrix elements and transform according to the irreducible representations of the D_2 rotational group (Zare, 1988).

Construction of $\boldsymbol{H}^{\text{int}}$ requires the introduction of a rovibrational basis. For this purpose one can employ the direct product of the multidimensional direct-product DVR vibrational basis functions and the $2J+1$ Wang rotational basis functions. Using Eqs. (2.22)–(2.25), $\boldsymbol{H}^{\text{int}}$ takes the form

$$\boldsymbol{H}^{\text{int}} = \boldsymbol{T}^{\text{int}} + \boldsymbol{V} = \boldsymbol{T}^{\text{v}} + \boldsymbol{T}^{\text{r}} + \boldsymbol{T}^{\text{rv}} + \boldsymbol{V}, \quad (2.32)$$

where

$$\boldsymbol{T}^{\text{v}} = \frac{1}{2}\boldsymbol{I}_{2J+1}\otimes\sum_{k=1}^{A_s}\sum_{l=1}^{A_s}\tilde{\boldsymbol{g}}^{-1/4}\boldsymbol{p}_k^{\dagger}\boldsymbol{G}^{kl}\tilde{\boldsymbol{g}}^{1/2}\boldsymbol{p}_l\tilde{\boldsymbol{g}}^{-1/4}, \quad (2.33)$$

$$\boldsymbol{T}^{\text{r}} = \frac{1}{2}\sum_{k=1}^{3}\boldsymbol{J}_k^2\otimes\boldsymbol{G}^{k+A_s,k+A_s} + \frac{1}{2}\sum_{k=1}^{3}\sum_{l>k}^{3}[\boldsymbol{J}_k,\boldsymbol{J}_l]_+\otimes\boldsymbol{G}^{k+A_s,l+A_s}, \quad (2.34)$$

$$\boldsymbol{T}^{\text{rv}} = \frac{1}{2}\sum_{l=1}^{3}\boldsymbol{J}_l\otimes\sum_{k=1}^{A_s}\left(\boldsymbol{p}_k^{\dagger}\boldsymbol{G}^{k,l+A_s}+\boldsymbol{G}^{k,l+A_s}\boldsymbol{p}_k\right), \quad (2.35)$$

and

$$\boldsymbol{V} = \boldsymbol{I}_{2J+1}\otimes\boldsymbol{V}_{J=0}, \quad (2.36)$$

where $\boldsymbol{I}_{2J+1}$ is the identity matrix of dimension $2J+1$, $\otimes$ refers to the direct-product operation, $\boldsymbol{V}_{J=0}$ denotes the diagonal potential-energy matrix in the direct-product DVR vibrational basis, and multiple resolutions of identity have been inserted in between some of the neighboring operators.

The iterative Lanczos algorithm (Lanczos, 1950; Cullum and Willoughby, 1985) can be utilized to compute the required eigenvalues and eigenvectors of $\boldsymbol{H}^{\text{int}}$. Then one must evaluate the effect of $\boldsymbol{H}^{\text{int}}$ on an arbitrary vector of the same dimension. As $\boldsymbol{H}^{\text{int}}$ has a special and very sparse structure, the explicit construction and storage of $\boldsymbol{H}^{\text{int}}$ can be avoided during the computation. Instead, one has to implement the $\boldsymbol{H}^{\text{int}}\boldsymbol{x}$ matrix–vector multiplication, which can be done efficiently due to the special sparse structure of $\boldsymbol{H}^{\text{int}}$ (Mátyus et al., 2009; Fábri et al., 2011a).

An alternative solution of the rovibrational problem is provided by the *vibrational subspace* (VS) method (Fábri et al., 2011b; Fábri et al., 2019) relying on the

$$\hat{H}^{\text{int}} = \hat{H}^{\text{v}} + \hat{T}^{\text{r}} + \hat{T}^{\text{rv}} \tag{2.37}$$

partition of $\hat{H}^{\text{int}}$, where $\hat{H}^{\text{v}} = \hat{T}^{\text{v}} + \hat{V}$ refers to the vibrational Hamiltonian. After solving the

$$\hat{H}^{\text{v}}\Phi_j = E_j\Phi_j \tag{2.38}$$

vibrational Schrödinger equation and obtaining the E_j vibrational energy levels and Φ_j vibrational eigenstates, the rovibrational eigenstates can be expanded in the form

$$\Psi_i = \sum_{j=1}^{n}\sum_{k=1}^{2J+1} c_{jk}^{(i)}\Phi_j R_k, \tag{2.39}$$

where R_k denotes rotational basis functions (symmetric top eigenfunctions, Wang functions, or rigid rotor eigenfunctions of the molecule under investigation). The VS ansatz of Eq. (2.39) offers a compact representation of the rovibrational eigenstates as the size of the VS rovibrational Hamiltonian remains moderate compared to the size of the DVR Hamiltonian even for high J values. We found (Fábri et al., 2011b) that it is inevitable to apply the Eckart frame in conjunction with the VS method to achieve satisfactory convergence of the computed rovibrational energy levels and eigenstates.

Finally, note that it is possible to simplify $\hat{T}^{\text{v}}$ of Eq. (2.23) at the cost of adding the so-called extrapotential term

$$\hat{U} = \frac{1}{32}\sum_{k=1}^{A_s}\sum_{l=1}^{A_s}\left[\frac{G^{kl}}{\tilde{g}^2}\frac{\partial\tilde{g}}{\partial q_k}\frac{\partial\tilde{g}}{\partial q_l} + 4\frac{\partial}{\partial q_k}\left(\frac{G^{kl}}{\tilde{g}}\frac{\partial\tilde{g}}{\partial q_l}\right)\right] \tag{2.40}$$

to the potential energy. The so-called rearranged vibrational Hamiltonian becomes

$$\hat{H}^{\text{v,rearr}} = \frac{1}{2}\sum_{k=1}^{A_s}\sum_{l=1}^{A_s}\hat{p}_k^{\dagger}G^{kl}\hat{p}_l + \hat{U} + \hat{V}, \tag{2.41}$$

while the operators $\hat{T}^{\text{r}}$ and $\hat{T}^{\text{rv}}$ do not change. Although the vibrational kinetic energy operator is simplified with the introduction of $\hat{U}$, the rearranged form is more prone to numerical instabilities than the so-called Podolsky form defined in Eq. (2.16); therefore, the Podolsky form is preferred over the rearranged form in practical applications (Mátyus et al., 2009).

2.3.2 Symmetry in Nuclear-Motion Computations

The symmetry group of $\hat{H}^{\text{int}}$ contains all symmetry operations that commute with $\hat{H}^{\text{int}}$. These symmetry operations are the permutations of identical nuclei and the operation of space inversion, and they form the complete nuclear permutation inversion (CNPI) group (Bunker and Jensen, 1998), while the symmetry group constituted by the so-called feasible permutation-inversion symmetry elements is called the molecular symmetry (MS) group, $G = \{g_i | i = 1, \ldots, |G|\}$, where $|G|$ denotes the order of G (Longuet-Higgins, 1963; Quack, 1977; Bunker and Jensen, 1998; Mills and Quack, 2002; Quack, 2011; Schnell, 2011; Oka, 2011).

As $[g_i, \hat{H}^{\text{int}}] = 0$, the n-fold degenerate ψ_{ij} eigenfunctions corresponding to the energy eigenvalue E_i span an n-dimensional irreducible representation of G. According to the grand orthogonality theorem (GOT) (Wigner, 1959), basis functions spanning the Γ_α irreducible representation can be constructed by the projector

$$P_{\alpha j} = \frac{n_\alpha}{|G|}\sum_{i=1}^{|G|} D_{jj}^{\alpha *}(g_i)g_i, \tag{2.42}$$

where the dimension of the irreducible representation Γ_α is n_α, $j = 1, \ldots, n_\alpha$, and $\boldsymbol{D}^\alpha(g_i)$ is the matrix representation of g_i in Γ_α. Although $P_{\alpha j}$ commutes with $\hat{H}^{\text{int}}$, in practical computations the matrix representations of certain symmetry operations may not commute with the Hamiltonian matrix due to numerical errors (Wang and Carrington, 2003; Poirier, 2003a; Wang and Carrington, 2003b, 2005). In this section we provide a brief account of the implementation of molecular symmetry in a DVR-based code (Fábri et al., 2017). For alternative approaches we refer the reader to the literature (Wang and Carrington, 2001, 2003; Poirier, 2003a; Wang and Carrington, 2003b,a, 2004, 2005; Yurchenko et al., 2017).

In order to obtain symmetry-adapted linear combinations of the grid-based vibrational basis functions, $P_{\alpha j}$ must act on direct-product DVR functions denoted by F_k. We require that each direct-product DVR function (or equivalently the associated multidimensional grid point) is mapped onto another DVR function (or onto itself in special cases) by any of the g_i symmetry operations. The symmetry-adapted vibrational basis functions

$$\chi_i^\alpha = \sum_k A_{ik}^\alpha F_k \tag{2.43}$$

can be expressed as the linear combination of direct-product DVR functions.

As the iterative Lanczos algorithm is used usually to diagonalize the vibrational Hamiltonian, we need to devise a procedure to evaluate matrix–vector products

$$\begin{aligned} y_i^\alpha &= \sum_j \langle \chi_i^\alpha | \hat{H}^{\text{int}} | \chi_j^\alpha \rangle x_j^\alpha \\ &= \sum_k A_{ik}^{\alpha *} \sum_l \langle F_k | \hat{H}^{\text{int}} | F_l \rangle \sum_j A_{jl}^\alpha x_j^\alpha \end{aligned} \quad (2.44)$$

in the symmetry-adapted vibrational basis. First, a symmetrized input vector $\boldsymbol{x}^\alpha$ from Γ_α is transformed to the original unsymmetrized vibrational basis (sum over j), then the resulting full-dimensional vector is multiplied by the unsymmetrized Hamiltonian (sum over l), and, finally, the full-dimensional result vector is transformed back to the symmetry-adapted basis (sum over k), yielding the symmetry-adapted output vector $\boldsymbol{y}^\alpha$. Thus, while the dimension of the input $\boldsymbol{x}$ and output $\boldsymbol{y}$ vectors is reduced due to symmetry, the second step involves a multiplication with the Hamiltonian represented in the unsymmetrized vibrational basis. Although it is possible to evaluate matrix–vector products directly in the symmetry-adapted basis, we have found it more convenient to apply the matrix–vector product algorithm outlined in Eq. (2.44), also related to the symmetry-adapted Lanczos method (Wang and Carrington, 2001, 2003; Poirier, 2003a; Wang and Carrington, 2003b,a, 2004, 2005). The symmetry-adapted computation provides degenerate energy levels only once, since projectors $P_{\alpha j}$ are applied with a single j index. This implies that we get only one eigenstate (ψ_{ij}) for each degenerate manifold of dimension n. The missing $n-1$ eigenstates can be obtained by applying symmetry operations on ψ_{ij} and generating an orthonormal basis from the $g_k\psi_{ij}$ functions, similar to Wang and Carrington (2008).

It is important to emphasize that the symmetrization algorithm presented is completely general and has the following advantages over the unsymmetrized treatment: (a) the dimension of the Lanczos vectors spanning the so-called Krylov subspace (Lanczos, 1950; Cullum and Willoughby, 1985) is reduced, therefore the memory requirement is reduced; (b) the computed energy levels and eigenstates are automatically labeled with Γ_α; (c) as the convergence rate of the Lanczos algorithm deteriorates with the increasing density of the energy levels, it is advantageous to decrease the spectral density of the Hamiltonian by separating the eigenstates according to symmetry. The only limitation of the vibrational symmetrization approach outlined is that the image of each direct-product grid point generated by any kind of symmetry operation has to be an element of the multidimensional DVR grid. Unfortunately, this requirement may limit the practically applicable symmetry group to a subgroup of the full MS group. One such case is that of the CH_5^+ molecule, described in Section 2.5.

2.3.3 Nuclear Spin Statistics

Knowledge of the nuclear spin (ns) statistical weights (Landau and Lifshitz, 1977; Bunker and Jensen, 1998) of rovibronic (in the simplest case rovibrational) eigenstates of a molecule is necessary for a detailed understanding of molecular high-resolution spectra for at least two important reasons: (a) only levels with a nonzero ns statistical weight will be involved in measured transitions, and (b) the intensities of measured lines is directly related to the ns statistical weights. The interaction of nuclear-spin angular momenta with other angular momenta characterizing the molecule is extremely weak; thus, it is an excellent approximation to separate the total molecular wavefunction into nuclear spin and rotation–vibration–electronic–electron-spin parts:

$$|\Psi^{\text{total}}\rangle = |\Psi^{\text{ns}}\rangle \otimes |\Psi^{\text{r,v,e,es}}\rangle. \quad (2.45)$$

Knowing the symmetry of $|\Psi^{\text{total}}\rangle$ and $|\Psi^{\text{ns}}\rangle$ restricts the symmetry of $|\Psi^{\text{r,v,e,es}}\rangle$. The nuclear spin statistics, and consequently the degeneracy factors also depend on the nuclear spin I of the non-permuting nuclei. The ns statistical weights can be determined *via* different methods (Landau and Lifshitz, 1977; Bunker and Jensen, 1998).

Molecules containing identical nuclei with non-zero nuclear spins, I, can exist as more than one nuclear-spin isomers. An old example is the H_2 molecule (Bonhoeffer and Harteck, 1929), with two spin-1/2 H nuclei, it can exist in *ortho* ($I=1$) and *para* ($I=0$) forms. *para*-H_2, once prepared (Bonhoeffer and Harteck, 1929), can exist in an appropriate container at room temperature for months. Polyatomic molecules may have more nuclear-spin isomers than H_2. CH_4, for example, has *ortho* ($I=1$), *meta* ($I=2$), and *para* ($I=0$) forms. Larger molecules may have even more nuclear-spin isomers.

Related to nuclear-spin isomers and the related states one must note an important peculiarity of time-independent variational nuclear-motion computations: they tend to yield eigenstates which may not exist in nature due to the Pauli exclusion principle (Herzberg, 1945). This is the result of not considering nuclear spin explicitly during the standard variational rovibrational computations (this is advantageous as this way the same code, the same BO-PES, and the same variational parameters can be used for nuclear-motion computations involving different isotopologues and isotopomers). Thus, computed rovibrational eigenstates must be checked a posteriori whether they exist or not.

As a straightforward example, let us consider a triatomic molecule where two identical spin-1/2 nuclei (fermions) can be exchanged. In the case of the three $H_2{}^nO$, $n = 16, 17$, and 18, isotopologues of water, there are two separate rovibrational energy level sets corresponding to the two distinct molecules. The two sets are traditionally called *ortho* and *para*, with the convention that the *ortho* form has the higher spin-statistical weight. In the case of water, these distinct principal components of the spectroscopic network (SN) (Császár and Furtenbacher, 2011; Furtenbacher and Császár, 2012) of the $H_2{}^nO$ molecules are not connected by measured transitions (Tóbiás et al., 2020). Note, however, that due to the allowed coupling of nuclear spin with rovibronic motions, nuclear spin symmetry-breaking mixings and *ortho* to *para* transitions have been observed (Ozier et al., 1970; Bordé et al., 1980; Pique et al., 1984; Quack, 2011).

2.3.4 Wavefunction Analysis Tools Via Projection Techniques

Besides the exact J and Γ_α labels introduced in the previous sections it is useful to assign approximate quantum numbers to the computed rovibrational eigenstates, such as the zeroth-order HO and RR quantum numbers, as these labels are widely used in experimental high-resolution spectroscopy and they also help capture the dynamics embodied in the eigenstates. This task can be completed by decomposing the numerically exact variational eigenstates in the basis of appropriately chosen zeroth-order states.

The normal mode decomposition (NMD) technique (Mátyus et al., 2010) has been developed to facilitate the assignment of zeroth-order HO quantum numbers to variational vibrational eigenstates Φ_i by calculating the

$$C_{ij}^{(\mathrm{NMD})} = |\langle \Phi_i | \Phi_j^{\mathrm{HO}} \rangle|^2 \tag{2.46}$$

coefficients, where Φ_j^{HO} denotes a HO eigenstate. Labeling Φ_i with HO quantum numbers can be accomplished by finding the dominant $C_{ia}^{(\mathrm{NMD})}$ coefficient (if it exists) and assign Φ_i with the HO quantum numbers of Φ_a^{HO}.

The rigid rotor decomposition (RRD) method (Mátyus et al., 2010; Szidarovszky et al., 2012) is a useful tool for the assignment of variationally-computed rovibrational energy levels and eigenstates. The initial step is the evaluation of the

$$C_{i,jk}^{(\mathrm{RRD})} = |\langle \Psi_i | \Phi_j R_k \rangle|^2 \tag{2.47}$$

coefficients, where Ψ_i is the rovibrational eigenstate to be assigned with approximate vibrational and rotational quantum numbers, while Φ_j and R_k denote variational vibrational and rigid-rotor eigenstates, respectively. After finding the dominant $\Phi_a R_b$ contribution in Ψ_i, it is straightforward to assign Ψ_i with the vibrational labels of Φ_a and the rigid-rotor quantum numbers [(J, K_a, K_c) for asymmetric tops, and (J, K) for symmetric tops] of R_b.

Projection techniques are also useful to attach rovibrational labels to the intermolecular motions of clusters, in particular dimers, highlighting their dynamical behavior. These techniques can be employed both for bound and resonance states, as demonstrated for the $CH_4{\cdot}H_2O$ dimer (Sarka et al., 2016, 2017) and the HeH_2^+ complex (Papp et al., 2018), respectively.

The so-called coupled-rotor decomposition (CRD) technique (Sarka et al., 2017) is relevant to weakly-bound dimers, where the intermonomer bonding is very weak compared to the bonding within the monomer units, denoted as A and B. In these dimers, an example is $CH_4{\cdot}H_2O$, considering the two monomer units rigid is an excellent model. Then the description of the low-energy vibrational motions of the dimer requires only six dofs. To obtain a description of the 6D eigenstates of the dimer based on the picture of coupled monomers with hindered rotation, one needs to do the following: (a) fix the $A - B$ separation at a given value; (b) set the PES to zero and compute the rovibrational states in this 5D model using exactly the same coordinate and grid representation as employed for the 6D model; (c) calculate the eigenenergies of the CR model, characterized by the J total angular momentum quantum number, with respect to the energies of states of the rigid monomers characterized by j_A and j_B:

$$E^{\mathrm{CR}}(j_A, j_B, j, J) = E_A^{\mathrm{RR}}(j_A) + E_B^{\mathrm{RR}}(j_B) + E_c(j, J), \tag{2.48}$$

where j is the internal angular momentum quantum number of the AB diatomic, responsible for coupling the angular momenta of A and B, and $E_c(j, J)$ is the related coupling energy, whose computation is detailed in Brocks et al. (1983); (d) at "infinite" separation, $E^{\mathrm{CR}}(j_A, j_B, j, J)$ is simply the sum of the rigid-rotor energies of the monomers, allowing straightforward labeling of the CR states; (e) in a DVR representation it is also straightforward to compute the overlaps of the 5D and 6D eigenstates (Sarka et al., 2017), computation and analysis of the CRD overlap coefficients completes the CRD assignment of the 6D eigenstates. A scheme very similar to the CRD one can be employed for the characterization of computed rovibrational resonance states, as described by Papp et al. (2018).

Note in the end that sometimes it seems more effective to base the assignment of dynamical labels to the computed eigenstates not on wavefunctions but on one- and two-mode reduced density matrices (Šmydke and Császár, 2019). Overlaps computed between densities of states already assigned and the list of new, assignable states seemingly provides an efficient route toward developing a large set of semiautomatically assigned states.

2.4 COMPUTATION OF ROVIBRATIONAL RESONANCES

Rovibrational resonances, also known as metastable or quasibound rovibrational states, are not mere curiosities of nuclear-motion theory arising within the BO approximation but they are at the heart of chemistry, especially when bimolecular chemical reactions, including reactive collisions, are considered. Rovibrational resonances are also met in predissociation and photodissociation studies as well as in unimolecular decays of polyatomics. The efficient computation and characterization of rovibrational resonances forms an essential and extremely important part of modern quantum chemistry.

While bound rovibrational states are assumed to have infinite lifetimes, resonances are usually characterized by two parameters: resonance positions and resonance lifetimes. These parameters can be computed within time-dependent and time-independent frameworks (Moiseyev, 2011; Klaiman and Gilary, 2012).

Hereby we restrict ourselves to time-independent approaches, in which resonances are identified as eigenstates of the rovibrational Hamiltonian having complex eigenvalues. These eigenvalues are usually written, in atomic units, as

$$E_n^{\mathrm{res}} = \epsilon_n - \frac{\mathrm{i}}{2}\Gamma_n, \tag{2.49}$$

where $\epsilon_n = \mathrm{Re}(E_n^{\mathrm{res}})$ is the resonance position, while Γ_n is the resonance width, related to the inverse lifetime,

$$\rho(\boldsymbol{Q}, t) = |\Psi_n^{\mathrm{res}}|^2 \propto e^{-\Gamma_n t}, \tag{2.50}$$

where $\boldsymbol{Q}$ represents a point in coordinate space. Due to the complex nature of the eigenvalues, the corresponding Ψ_n^{res} wavefunctions are not normalizable, i.e., they diverge exponentially as the dissociation coordinate goes to infinity. This makes it necessary to apply special computational tools, some of which are detailed below, for the determination of resonance eigenstates. Similar to bound states, analysis of resonance wavefunctions gives insight into the physical properties of the metastable eigenstates, for example into the mechanisms responsible for their dissociation (or other dynamical behavior).

We briefly review in this section three techniques for computing rovibrational resonances: the stabilization method (Hazi and Taylor, 1970; Lefebvre, 1985; Riera, 1993; Mandelshtam et al., 1993; Moiseyev, 2011), the technique of complex coordinate scaling (CCS) (Moiseyev et al., 1981; Moiseyev, 2011), and the use of complex absorbing potentials (CAP) (Vibók and Balint-Kurti, 1992; Riss and Meyer, 1993; Skokov et al., 1999; Halász and Vibók, 2000; Mussa and Tennyson, 2002; Poirier and Carrington, 2003a,b; Muga et al., 2004).

2.4.1 The Stabilization Method

The simplest approach to compute long-lived rovibrational resonances is offered by the so-called stabilization method (Hazi and Taylor, 1970; Lefebvre, 1985; Riera, 1993; Mandelshtam et al., 1993; Moiseyev, 2011). The attractiveness of the stabilization method lies in its simplicity, since it allows the use of bound-state variational nuclear-motion codes for computations yielding rovibrational resonance positions and even resonance widths (Simons, 1981; Macías and Riera, 1984, 1989).

In the stabilization method the eigenvalues above the first dissociation asymptote are monitored, while several (on the order of 10–20) standard, Hermitian, variational nuclear-motion computations are carried out with slightly different computational parameters chosen along the coordinate describing dissociation. These parameters can be quite different: a scaling factor in a basis (Macías and Riera, 1984), the size of the box containing the system (Lefebvre, 1985), and/or the number of DVR basis functions. The "continuum" energies obtained vary significantly with changes in the parameters. Resonance energies are obtained by identifying eigenvalues which are converged well and are insensitive to the (small) changes in the parameters chosen.

The results obtained from the stabilization method can most easily be visualized in the form of a histogram. In the stabilization histogram, the eigenvalues computed at different ranges of the dissociation coordinate that fall in the respective bins are counted. Resonance energies, which are located above the dissociation threshold and are converged tightly appear in the histogram as clear peaks (Hazi and Taylor, 1970; Mandelshtam et al., 1993; Papp et al., 2017a). Resonance widths can also be obtained using the stabilization method; see, for example, Simons (1981), Macías and Riera (1989), Mandelshtam et al. (1994), and Haritan and Moiseyev (2017).

2.4.2 The Technique of Complex Coordinate Scaling (CCS)

An alternative approach for computing resonance eigenstates using bound-state nuclear-motion algorithms is to transform the time-independent nuclear Schrödinger equation, $\hat{H}\Psi^{\text{res}} = E^{\text{res}}\Psi^{\text{res}}$, to

$$\hat{S}\hat{H}\hat{S}^{-1}\hat{S}\Psi^{\text{res}} = E^{\text{res}}\hat{S}\Psi^{\text{res}} = (\hat{S}\hat{H}\hat{S}^{-1})\Phi = E^{\text{res}}\Phi, \tag{2.51}$$

such that $\Phi = \hat{S}\Psi^{\text{res}}$ becomes square integrable and can be expanded using standard bound-state basis sets. The probably best known transformation is the so-called complex coordinate scaling (CCS) (Moiseyev et al., 1981; Moiseyev, 2011). The simplest implementation of CCS involves a parameter-dependent operator $\hat{S}(\theta)$, which rotates the dissociation coordinate R in the complex plane by the angle θ, $\hat{S}(\theta)f(R) = f(Re^{i\theta})$ (Moiseyev, 2011). In practice, the eigenvalues of the non-Hermitian $\hat{S}(\theta)\hat{H}\hat{S}^{-1}(\theta)$ operator are computed for many different values of θ, forming eigenvalue trajectories on the complex plane, and resonance eigenvalues are identified as cusps or similar features (Moiseyev et al., 1981) in the trajectories.

The authors' implementation of the CCS method, called D^{2}FOPI-CCS (Szidarovszky et al., 2010; Szidarovszky and Császár, 2013), was obtained by extending the tailor-made variational nuclear-motion code D^{2}FOPI (Szidarovszky et al., 2010) with the conventional CCS as well as the standard exterior CCS methods, detailed to some extent in Szidarovszky and Császár (2013) and Papp et al. (2018), respectively. In the D^{2}FOPI-CCS protocol all the bound rovibrational eigenstates and hundreds of eigenpairs above the first dissociation threshold are computed as a first step, then these eigenstates are used as basis functions to construct the matrix representation of the complex-coordinate-scaled rovibrational Hamiltonian.

2.4.3 Complex Absorbing Potentials (CAP)

A third possibility to compute rovibrational resonances is offered by the technique of complex absorbing potentials (CAP) (Riss and Meyer, 1993; Muga et al., 2004; Mussa and Tennyson, 2002; Skokov et al., 1999), which involves perturbing the rovibrational Hamiltonian by a CAP function that is non-zero only in the asymptotic region of the dissociative coordinate. The CAP is chosen such that the perturbed non-Hermitian Hamiltonian has square-integrable wavefunctions with corresponding complex eigenvalues that are good approximations to the resonance eigenvalues. In practice, the CAP function is associated with a strength parameter η, so that the perturbation of the Hamiltonian can be written as

$$\hat{H} \to \hat{H}(\eta) = \hat{H} - \mathrm{i}\eta\hat{W}, \tag{2.52}$$

where $\hat{W}$ is the CAP function, its argument(s) are the dissociation coordinate(s); $\hat{W}$ is typically chosen to be non-zero only in the asymptotic regions of the PES, and the optimal functional form of $\hat{W}$ has been the topic of many studies, see, for example, Vibók and Balint-Kurti (1992), Halász and Vibók (2000), Poirier and Carrington (2003a), and Poirier and Carrington (2003b). The eigenvalues of the non-Hermitian operator $\hat{H}(\eta)$ are computed for many different values of η, forming eigenvalue trajectories on the complex plane, and resonance eigenvalues are identified as cusps in the trajectories.

The authors' implementation of the CAP method, called GENIUSH-CAP (Papp et al., 2017b; Simkó et al., 2019), is an extension of the GENIUSH code (Mátyus et al., 2009; Fábri et al., 2011a) in which a CAP is added to the standard rovibrational Hamiltonian, and the matrix representation of the CAP-perturbed Hamiltonian is constructed in the basis of the eigenvectors of the unperturbed Hamiltonian. The GENIUSH-CAP code inherits all the advantages of the GENIUSH code; thus, it allows full- and reduced-dimensional computations of rovibrational resonances in a black-box-type fashion. Efficient algorithms moving the computation of rovibrational resonances toward a black-box-type technique have appeared (Tremblay and Carrington, 2005; Simkó et al., 2019).

2.4.4 Wavefunction Analysis Tools

One of the advantages of the GENIUSH-CAP code (Papp et al., 2017b; Simkó et al., 2019) is that beyond full- and reduced-dimensional resonance computations, reduced-dimensional model computations can be carried out with the GENIUSH code (Mátyus et al., 2009; Fábri et al., 2011a) for the system whose resonances are investigated. This allows for a straightforward computation of overlaps between the resonance wavefunctions determined by GENIUSH-CAP and model wavefunctions computed by GENIUSH. Based on such overlaps, quantum numbers, dissociation branching ratios, etc., can be determined in a rigorous and efficient way, see, for example, Papp et al. (2017a) and Papp et al.

(2018). This type of overlap-based analysis is very similar in spirit to the RRD (Mátyus et al., 2010) and the CRD (Sarka et al., 2017) schemes described above.

2.5 APPLICATIONS

Up to now we have focused on the most important technical details characterizing some of the fourth-age quantum-chemical nuclear-motion protocols, allowing the efficient variational computation of rovibrational bound states and resonances. Next, we present a couple of representative results of computations based on these protocols, obtained mostly in our laboratory. The results, i.e., energy levels, eigenstates, assignments, and time-dependent quantum-dynamical results, have applications in many different areas of science and engineering, which, in order to save space, we are going to cover only occasionally.

The computations whose results are reported in this section utilized the following in-house codes: D^2FOPI (Czakó et al., 2004; Furtenbacher et al., 2006; Szidarovszky et al., 2010), D^2FOPI-CCS (Szidarovszky et al., 2010; Szidarovszky and Császár, 2013), GENIUSH (Mátyus et al., 2009; Fábri et al., 2011a), and GENIUSH-CAP (Papp et al., 2017b). D^2FOPI is a tailor-made DVR-based code designed for computing bound rovibrational states of triatomic molecules. D^2FOPI-CCS is an extension of D^2FOPI, utilizing the CCS (Moiseyev, 2011) method, and allows the computation of rovibrational resonance states. In D^2FOPI-CCS the CCS Hamiltonian is represented in the eigenstate basis of the untransformed rovibrational Hamiltonian. GENIUSH, which stands for a general (GE) rovibrational code with a numerical (N), internal-coordinate (I), user-specified (US) Hamiltonian (H), computes rovibrational bound states by numerically representing not only the potential (through a DVR) but also the kinetic energy operator of a molecular system. Any number of reduced-dimensional models of vibrating-rotating molecules can straightforwardly be defined within GENIUSH. GENIUSH-CAP is an extension of GENIUSH, in which a CAP (Riss and Meyer, 1993; Skokov et al., 1999; Mussa and Tennyson, 2002; Muga et al., 2004) is added to the standard rovibrational Hamiltonian, allowing for the computation of resonance states. In GENIUSH-CAP the matrix representation of the CAP-perturbed Hamiltonian is constructed using the basis of the eigenvectors of the unperturbed Hamiltonian. In all the codes mentioned different versions of iterative Lanczos eigensolvers are used to determine the desired eigenvalues and eigenvectors.

2.5.1 Computation of All the Bound (Ro)Vibrational Eigenstates

There are only a few strongly-bound polyatomic molecules for which all the bound (ro)vibrational eigenstates have been computed. This is not surprising when one considers the fact that even for a triatomic molecule this may involve the computation of a couple of million eigenstates (Simkó et al., 2017). The formidable task of computing these eigenstates is helped tremendously by the fact that the quantum number corresponding to overall rotation, J, is one of the good quantum numbers in a field-free case; thus, the Hamiltonian $\boldsymbol{H}$ is block-diagonal in J, with block sizes proportional to $2J+1$. Due to dissociation limit(s) the largest block is at a J value considerably smaller than the $J_{\max}$ value characterizing the highest bound rovibrational state.

2.5.1.1 $H_2{}^{16}O$ and Its Isotopologues

Several BO and adiabatic PESs are available for water and its isotopologues (Partridge and Schwenke, 1997; Polyansky et al., 2003; Barletta et al., 2006; Shirin et al., 2008; Polyansky et al., 2018), most facilitate the determination of all bound states of these molecules (Maksyutenko et al., 2007). The first-principles (*ab initio*) PESs of water, especially after including several small correction terms [scalar relativistic (Császár et al., 1998b; Quiney et al., 2001), diagonal Born–Oppenheimer (Zobov et al., 1996), and quantum electrodynamics (QED) (Pyykkö et al., 2001) corrections], all become fairly accurate. In fact, water is the first polyatomic, polyelectronic molecule for which a purely *ab initio* PES could provide sub-cm^{-1} accuracy in a variational computation of all measured transitions (Polyansky et al., 2003). When *ab initio* PESs are adjusted to the large number of "measured" rovibrational energy levels (Tennyson et al., 2013), the accuracy of the thus derived semiempirical PESs increases by somewhat more than an order of magnitude. Nevertheless, the accuracy of rovibrational states computed with the best semiempirical PES and an exact KEO is still orders of magnitude lower than those of high-resolution spectroscopic measurements (Tennyson et al., 2009, 2010, 2013; Tennyson et al., 2014a,b).

$H_2{}^{16}O$ has about 1150 bound vibrational ($J=0$) states (Császár et al., 2010). The corresponding numerical results, the wavefunction plots (Császár et al., 2010), and the assignments reveal interesting characteristics about the energy-level structure and the dynamics of the $H_2{}^{16}O$ molecule. The plots and the assignments show, for example, how the convenient normal-mode picture of vibrations characterizing states of $H_2{}^{16}O$ with low excitation changes to a local-mode picture, the effect the low barrier to linearity (Császár et al., 1998a; Valeev et al., 2001) has on the characteristics of the states, and the onset of quantum monodromy (Zobov et al., 2005).

As water is the most prevalent polyatomic molecule in the universe, generation of a detailed line list of the rovibrational transitions is of utmost importance for many scientific and engineering applications. Thus, it is not surprising that several line lists have been developed, the most important ones are probably Partridge and Schwenke (1997), Barber et al. (2006), Gordon et al. (2017), and Polyansky et al. (2018). Note that while Partridge and Schwenke (1997) and Barber et al. (2006) contain 300 and 500 million transitions, respectively, Polyansky et al. (2018) lists 5 billion rovibrational transitions. These transitions have been computed variationally based on a highly accurate semiexperimental PES (Furtenbacher et al., 2016; Polyansky et al., 2018).

If all the bound rovibrational energies are available from detailed nuclear-motion computations, they can be used to determine ideal-gas partition functions of molecules (Mayer and Mayer, 1940; Herzberg, 1945; Lewis et al., 1961; McQuarrie, 2000; Furtenbacher et al., 2016; Simkó et al., 2017). For this purpose, the total partition function is assumed to be the product of the internal and the translational partition functions. The latter can be computed exactly, only the internal partition function will be approximate, even for an ideal gas. As also well known, the internal partition function, Q_{int}, of a free molecule can be calculated as

$$Q_{\text{int}} = g_s \sum_i g_i (2J_i + 1) \exp\left(\frac{-c_2 E_i}{T}\right), \tag{2.53}$$

called direct summation, where $c_2 = hc/k_B$ is the second radiation constant, J_i is the rotational quantum number, E_i is the rovibrational energy level given in cm^{-1} (the zero is taken as the lowest existing (ro)vibrational state), T is the thermodynamic temperature in K, g_s is a state-independent nuclear-spin degeneracy factor for atoms not exchanged under rotation (Herzberg, 1945), g_i is the nuclear-spin degeneracy factor for identical atoms interchanged under rotation, and the index i runs over all possible rovibronic energies considered. The availability of about 20 000 "measured" (better say "empirical") rovibrational energy levels (with an almost complete coverage up to 9200 cm^{-1}) for $H_2{}^{16}O$ (Tennyson et al., 2013), augmented with all the computed bound energy levels, means that Q_{int} of $H_2{}^{16}O$ (Furtenbacher et al., 2016), as well as Q_{int}s of $D_2{}^{16}O$ (Simkó et al., 2017), $D_2{}^{17}O$ (Simkó et al., 2017), and $D_2{}^{18}O$ (Simkó et al., 2017), and thus that of heavy water (Simkó et al., 2017), have the ultimate accuracy at lower temperatures and an accuracy better than 1% even at 6000 K. Note that due to their lower ZPVE values and lower fundamentals, the D_2O species have almost three million bound rovibrational states (Simkó et al., 2017), while $H_2{}^{16}O$ has only less than one million.

2.5.1.2 H_3^+ and Its Deuterated Isotopologues

H_3^+ is an extremely important molecule in interstellar space as it governs basically the whole gas-phase chemistry of the interstellar medium (ISM) (Herbst, 2000).

In 2006, based on a PES (Munro et al., 2006) with correct asymptotic behavior, Tennyson et al. (Tennyson et al., 2006; Munro et al., 2006) computed almost all of the vibrational energy levels corresponding to the electronic ground state of H_3^+. Szidarovszky et al. (2010), using the same PES but a different protocol, confirmed the results of the earlier study and determined 1287 (counting the E-symmetry states twice) bound vibrational states. It is believed that a convergence of at least 1 cm^{-1} was achieved even for states very close to the first dissociation asymptote, placed at $D_0 = 34\,911.6$ cm^{-1}.

H_3^+ is the polyatomic molecule for which the most accurate *ab initio* adiabatic PES has been computed (Pavanello et al., 2012b,a; Polyansky et al., 2012). To achieve this accuracy required the use of special techniques of electronic structure theory and the consideration of (scalar) relativistic (Cowan and Griffin, 1976; Tarczay et al., 2001), quantum electrodynamics (Pyykkö et al., 2001), and diagonal Born–Oppenheimer (Handy et al., 1986) corrections obtained based on highly-accurate electronic wavefunctions. The accuracy achieved allowed the straightforward assignment of near-IR measured spectra (Pavanello et al., 2012b). The accuracy of the adiabatic PES is in fact so high that comparison of the computed and measured rovibrational energies clearly revealed the role of non-adiabatic effects on the computed levels (Furtenbacher et al., 2013b). Modeling of non-adiabatic effects is one of the last frontiers of computational molecular spectroscopy and it is likely that the advances will employ H_3^+ as the test molecule. These tests will likely utilize the empirical rovibrational energy level sets of H_3^+ (Furtenbacher et al., 2013b), H_2D^+ (Furtenbacher et al., 2013a), and HD_2^+ (Furtenbacher et al., 2013a).

One of the great remaining challenges of H_3^+ spectroscopy is the understanding of the Carrington–Kennedy experiments (Carrington et al., 1982; Carrington and Kennedy, 1984), revealing an extraordinary number of rovibrational transitions around the first dissociation asymptote. This requires an extremely accurate

PES, new methodological developments for the computation of bound and resonance states, and even perhaps new experiments at low temperatures producing much less cluttered spectra.

2.5.2 Rovibrational Computations on Quasistructural Molecules

2.5.2.1 H_5^+

The H_5^+ molecular ion lacks the usual (heavier) central atom(s) that may form multiple strong (covalent, often 2c–2e) bonds, typical for virtually all small molecular species. Thus, unlike in the case of semirigid molecules (like H_2O and H_3^+), the lack of a well-defined (effective) structure makes the low- and especially the high-resolution spectra of H_5^+ particularly challenging to anticipate and interpret (Fábri et al., 2014c; Sarka et al., 2015; Sarka and Császár, 2016). It has also become clear that it is not trivial to set up meaningful low-dimensional model Hamiltonians to describe the complex internal dynamics of this ion, governed by the following three large-amplitude motions: hopping of the central proton (PH), torsion (T), and scrambling of the hydrogens of the formal H_3^+ subunit (SC). Nevertheless, even for this cation (formally H_3^+ solvated by a H_2 molecule at one of the vertices of H_3^+) (a) free motion (no hindering potential) works well for describing the 1D torsional motion of the two H_2 moieties in the sides of the central proton; (b) coupling the torsional motion with one of the rotational dofs provides a meaningful model to interpret the structure of the lowest rovibrational energy levels; and (c) the unfeasible scrambling motion, hindered by a relatively substantial barrier, lowers the high permutational symmetry of the system.

The very unusual nuclear dynamics of H_5^+ holds true for all D-substituted isotopomers and isotopologues, $H_{5-n}D_n$, $n = 0 - 5$ (Sarka and Császár, 2016). The computations strongly suggest that the unusual dynamical behavior is connected to the shape of the PES and not to the five-fold permutation symmetry of the molecule. Briefly, the detailed nuclear-motion computations (Fábri et al., 2014c; Sarka et al., 2015; Sarka and Császár, 2016) suggest that (a) the six possible D-substituted isotopologues form 12 isotopomers; (b) to interpret the rovibrational energy levels, one needs to explicitly consider that the effective value of the *A* rotational constant is twice as large as its equilibrium value and the nearly-free torsion is strongly coupled to the rotational dof corresponding to the torsional axis; (c) though some isotopomers, namely $[DH–H–DH]^+$ and $[DH–D–DH]^+$, show more chaotic behavior than others, labeling of a number of states involving coordinates corresponding to the T, PH, and SC motions is possible, yielding some understanding of the quantum dynamics of these ions; (d) the splittings shown by certain torsion-rotation energy level pairs can be attributed to the perturbation of free internal rotation by the weak torsional potential instead of tunneling between the two equivalent torsional potential wells; and (e) the hopping mode is sensitive to the identity of the middle atom but more or less independent of the sides. These characteristics of the internal dynamics make H_5^+ and its deuterated isotopologues members of the family of quasistructural molecules (Császár et al., 2020).

2.5.2.2 CH_5^+

Based on a considerable number of experimental (White et al., 1999; Asvany et al., 2005; Ivanov et al., 2010; Oka, 2015; Asvany et al., 2015; Brackertz et al., 2017), computational (Schreiner et al., 1993; Padma Kumar and Marx, 2006; Huang et al., 2006; Wang and Carrington, 2008; Wodraszka and Manthe, 2015; Wang and Carrington, 2016; Fábri et al., 2017), and modeling (Bunker, 1996; Kolbuszewski and Bunker, 1996; East and Bunker, 1997; East et al., 1997; Bunker et al., 2004; Schmiedt et al., 2016, 2017b,a; Fábri and Császár, 2018) studies, it is transparent that the rovibrational energy level pattern of the protonated methane molecule, CH_5^+, defies any description attempts based on the RRHO model. The computations reported in Fábri et al. (2017), Wang and Carrington (2016), and Wang and Carrington (2008) yielded vibrational and rovibrational energy levels for CH_5^+ for 7D bend and 12D models. These results prove numerically that the commonly applied separation of rotations and vibrations completely breaks down for CH_5^+, as predicted by group-theoretical arguments (Wodraszka and Manthe, 2015; Schmiedt et al., 2015).

Given the extraordinary difficulties related to the dynamics of CH_5^+, it is of considerable interest to note that a simple and intuitive model, the vibrational quantum-graph model (Fábri and Császár, 2018), is able to provide a qualitative explanation for the unusual low-energy vibrational quantum dynamics of CH_5^+. The vertices of the quantum graph represent versions of the equilibrium structure with distinct atom numbering (120 possibilities for CH_5^+), while the edges refer to collective nuclear motions (torsional and flip motions for CH_5^+) transforming the versions of the equilibrium structure into one another. Note that it was (Nefedova et al., 1993) who introduced to chemistry the graphical representation "of the symmetries of [the] PES of non-rigid molecule[s], whose nuclear motion wavefunctions are delocalized over several equal-energy global minima" (the molecules considered were ArH_3^+, $C_2H_3^+$, CH_4^+, and $LiBH_4^+$). The novel vibrational quantum-graph model allows the mapping of the complex vibrational quantum dynamics of CH_5^+ onto the motion of a particle confined in a quantum graph. The vibrational energy levels are obtained by solving the one-dimensional time-independent nuclear Schrödinger equation, subject to appropriate boundary conditions. Furthermore, due to the low barriers hindering the torsional and flip motions in CH_5^+, the motion along the edges of the quantum graph can safely be assumed to be free. The quantum-graph model is able to reproduce the lowest-lying vibrational energy levels of CH_5^+ (and CD_5^+) with remarkable accuracy (Fábri and Császár, 2018). Note that a five-dimensional rigid-rotor model, based on the combination of the three rotational dofs with the two soft vibrational modes (torsion and flip), is also able to explain the unusual rovibrational energy structure of CH_5^+ (Schmiedt et al., 2016, 2017b,a).

In summary, while we are close to understanding the highly unusual nuclear dynamics, and even perhaps the related spectra, of CH_5^+ via variational nuclear-motion computations, it is safe to say that this cation remains the premier representative of quasistructural molecules (Császár et al., 2020), providing a very challenging playground for both experimental and theoretical spectroscopy.

2.5.3 Computation of Rovibrational Resonances

2.5.3.1 H_2O

Szidarovszky and Császár (2013) computed the low-lying rovibrational resonances of the strongly-bound H_2O molecule using the CCS method. The calculations aided the proper assignment of some observed rovibrational transitions beyond the first dissociation limit of $H_2{}^{16}O$ (Zobov et al., 2011). Furthermore, by inspecting the vibrational probability density plots from the stationary resonance calculations, several types of (dynamical) dissociation behavior, varying among the states, could be identified.

2.5.3.2 $Ar{\cdot}NO^+$

Vibrational resonances of the vdW complex $Ar{\cdot}NO^+$ were studied in Papp et al. (2017a) using the CAP technique and the stabilization method. It was found that the results obtained are in good agreement with those derived from close-coupling scattering computations (Stoecklin et al., 2002), and that the close-coupling scattering and rovibrational resonance approaches complement each other remarkably well. Comparison with the limited experimental information confirmed the experimental results obtained, and shed light to more subtle dynamics than initially predicted based on the observations.

A particularly interesting feature of the computed rovibrational resonances of $Ar{\cdot}NO^+$ is the repetition of the basically same set of levels for each vibrational excitation of the NO^+ stretch. The dissociation energy of $Ar{\cdot}NO^+$ is about 800 cm^{-1} but due to the almost perfect adiabatic separation of the NO^+ stretch from the other two vibrational dofs, basically the same energy-level structure is computed even at about 8000 cm^{-1}.

2.5.3.3 H_2He^+

Using the CCS, the CAP, and the stabilization methods, a large number of rovibrational resonances have been computed and characterized in Papp et al. (2018) for H_2He^+. It is hoped that these accurate computations facilitate the first experimental observation of rovibrational transitions of this fundamental molecule, made up of the two most abundant elements of the universe. Note that HHe^+ has just been observed in the interstellar medium (Güsten et al., 2019).

Alongside the spectroscopic data valuable for future high-resolution experiments, the stabilization mechanism of the long-lived resonances, possible dissociation pathways, and dissociation branching ratios could also be determined from the quantum-chemical computations. These results provide insight into the complex physics and the rich dynamics characterizing this system. Because the H–H stretching fundamental lies above the first dissociation threshold of $D_0 = 1775.4$ cm^{-1}, resonances are expected to play a crucial role in the collision and association reaction involving H_2He^+, including radiative association and radiative charge-transfer reactions.

2.5.3.4 $H_2{\cdot}CO$

Papp et al. (2017b) computed the vibrational resonances of the weakly-bound complex $H_2{\cdot}CO$ using the GENIUSH-CAP approach and a four-dimensional model PES. This was one of the first examples where vibrational resonances could be determined for a system containing more than three atoms.

The computations allowed to identify and characterize resonances of both *para*- and *ortho*-$H_2{\cdot}CO$. Quantum number assignments for the resonances were achieved by inspecting the vibrational probability density plots and by computing wavefunction overlaps with eigenstates of reduced-dimensional models, as discussed briefly in Section 2.4.4.

2.5.4 Stationary-State Computations Serving Dynamical Studies

In this penultimate section we mention briefly a few enlightening examples where variational stationary-state computations have been utilized to obtain detailed information on certain dynamical processes of molecular systems.

Szidarovszky and Yamanouchi (2017) simulated the laser-induced alignment dynamics of the weakly-bound H_2He^+ complex to investigate the effects of non-rigidity and rovibrational couplings on the rotational dynamics of the system. For this, variationally-computed rovibrational states of the H_2He^+ complex were used as basis functions to expand the laser-induced rovibrational wave packet. It was shown that centrifugal distortion and rovibrational couplings indeed influence strongly the laser-induced rotational dynamics of the system; thus, the standard protocol of applying the RR approximation for simulating such processes needs to be used with considerable care.

In the case of H_2O, a molecule much more rigid than H_2He^+, deviations from the RR approximation in its laser-induced alignment dynamics can also arise, mostly when one-photon transitions inducing vibrational excitations occur (Szidarovszky and Yamanouchi, 2018). In this case the origin of the deviation is due less to centrifugal distortion, it stems mostly from the inadequacy of the RR model to describe optical selection rules of rotational transitions associated with vibrational excitation.

In the experiments described in Larimian et al. (2016), the $C_2H_2^{2+}$ and $C_2H_4^{2+}$ dications were generated by doubly ionizing their neutral parents using few-cycle intense laser pulses. For both $C_2H_2^{2+}$ and $C_2H_4^{2+}$, in the deprotonation pathway of their Coulomb explosion, an ultraslow, microsecond timescale exponential decay channel was observed. Reduced-dimensional resonance-state computations revealed that in both cases the slow decay channel is due to quasibound states along the C–H vibrational mode, where tunneling through a barrier is responsible for the exponential decay.

Due to the rapidly increasing amount of high-quality data placed into molecular spectroscopy databases, it is worth mentioning here that laser-induced dynamics can also be simulated for molecules for which spectroscopic databases, containing rovibrational energy levels as well as transition amplitudes between the corresponding eigenstates, are available. This interesting approach was pursued, for example, in Owens et al. (2017) and Schuh et al. (2017).

Fábri et al. (2019) reported the extension of the time-independent nuclear-motion code GENIUSH with time-dependent quantum-dynamical features and results are presented there for the coherent inhibition and enhancement of tunneling in NH_3 isotopomers. The time-dependent Schrödinger equation was solved in the basis of the rovibrational eigenstates of the isolated molecule and the interaction of the molecule with a classical electromagnetic field was described within the framework of the electric dipole approximation. The rovibrational eigenstates were computed using contracted vibrational basis functions and the vibrational subspace method introduced in Section 2.3.1. The quantum-dynamical schemes utilized in Fábri et al. (2019) rely heavily on the concept of light-dressed states (Grossmann et al., 1991; Holthaus, 1992) and achieve the coherent inhibition and enhancement of tunneling by non-resonant laser fields. An important feature of the work of Fábri et al. (2019) is that, in contrast to previous vibration-only studies (Marquardt et al., 2003, 2010; Gatti and Marquardt, 2012; Sala et al., 2012, 2014), all rotational and vibrational degrees of freedom are treated in a numerically exact way and neither the alignment nor the orientation of the molecule are assumed.

2.6 SUMMARY AND OUTLOOK

Molecules are assemblies of electrons and nuclei. "Effects" due to the rotational and/or the vibrational motions of the nuclei are omnipresent whether one takes them into account explicitly or not. Nevertheless, the Born–Oppenheimer separation of the motion of the electrons and the nuclei is a remarkably good approximation for most of chemistry. As a result, in the majority of quantum-dynamics computations one first solves the electronic motion problem adiabatically decoupled from the motions of the nuclei. This can be done today in an almost black-box-like fashion with outstanding accuracy for quite large systems. Then the motion of the nuclei is considered on a potential energy hypersurface (PES) provided by electronic-structure theory.

One of the simplest dynamical descriptions of molecules in motion is given by considering the classical movement of the nuclei on a PES. This classical description is useful in a semi-quantitative understanding of many dynamical processes. Nevertheless, the zero-point energy of molecules and the tunneling of nuclei, most importantly that of protons, are not readily incorporated in this model and must be considered as special "effects", showing limitations of this approach.

Thus, one needs to move toward a quantum treatment of all the particles.

Spectroscopy, dealing with the results of the interaction between matter and electromagnetic radiation, has been one of the greatest inventions of humankind. Its often extreme accuracy and precision facilitates the understanding of the world around us from the extremely small (atoms and molecules) to the extremely large (astronomical) objects. The remarkable successes of experimental and observational spectroscopy could not have happened without the help from theory, mostly quantum chemistry and molecular physics. Spectroscopic experiments provide a very large amount of data that need to be interpreted and explained and the aim of a large part of nuclear-motion computations is exactly this.

In the fourth age of quantum chemistry the algorithms and the codes developed became sophisticated enough that they allow the interpretation of most high-resolution spectroscopic measurements. Nevertheless, black-box-type techniques, like those of electronic structure theory, still do not exist in nuclear-motion theory though the field is moving toward that direction with considerable pace. Attempts to develop black-box-type quantum dynamics techniques will remain with us for the foreseeable future.

As in all fields under intense development, there are various trends within nuclear-motion theory. These include time-dependent and time-independent descriptions, perturbational and variational algorithms, various forms of rovibrational Hamiltonians, as well as different numerical representations of the Hamiltonians. In this chapter only time-independent variational techniques have been treated in detail. One may wonder which is the best combination of the numerous possible choices or whether a unified protocol would soon emerge.

At this point we have to leave these questions open and say that at present the application in question determines the most appropriate choice. What can be said with confidence is that the extensive use of the different algorithms and protocols indicates that the current quantum-chemical adaptation of quantum mechanics to molecular systems is highly successful all the way from the semirigid to the quasistructural regime. As a result, the outcome of quantum-chemical simulations based on the technology developed can be used in several applied fields of science and engineering, an example is the improvement of the outcome of atmospheric radiative models.

In the most modern times we are witnessing the merging of time-independent and time-dependent approaches. This is due to the fact that time-dependent nuclear motion computations may follow two basic approaches, a direct or an indirect one. The direct approach implies the numerical solution of the time-dependent Schrödinger equation by an appropriate time-propagation method, not requiring the usually expensive computation of eigenstates. In the indirect approach the time-dependent problem is solved in two consecutive steps. First, the time-independent Schrödinger equation is solved via techniques described in this chapter, resulting in rovibrational energy levels and eigenstates. The results of this step can be compared with experimental transitions and transition moments. In the second step the time-dependent wavefunction is represented in the basis of eigenstates and the resulting set of coupled first-order differential equations is solved, yielding the time-dependent wavefunction and other time-dependent molecular quantities. This step can be executed either for the isolated molecule or for the system interacting with an electromagnetic field. In the second step one can design suitable approximations, such as the quasiresonant or the Floquet–Lyapunov approximation for coherent excitation, which facilitate long-time propagation, not easily amenable to direct approaches.

Without discounting the considerable achievements of nuclear-motion theory, it must also be stressed that there are several outstanding challenges which should be met in the near future and require further developments of the existing methodologies. The so-called Carrington bands close to the first dissociation limit of the molecular ion H_3^+ still remain unassigned. In particular, efficient and black-box-type computation of resonance states requires further concerted method developments. Treatment of highly-excited states of semirigid molecules, of almost all rovibronic states of molecules "with no structure" (called quasistructural molecules, like CH_5^+), as well as of weakly-bound molecular systems, like those held together by dispersive interactions, is far from being solved, especially not in full dimension. Computation of the hyperfine structure of measured spectra as well as the efficient joint treatment of several PESs await further extensive studies. Extending the applicability of nearly exact nuclear-motion treatments to much larger, flexible molecules requires further significant efforts. Quantum-chemical computation of rate constants and advancing quantum control are two areas where a lot of developments are expected in the near future. In summary, it is clear that in the fourth age of quantum chemistry method development and the application of the new techniques to outstanding problems of chemistry continue and continue with a hopefully increased speed.

REFERENCES

Aarset, K., Császár, A.G., Sibert III, E.L., Allen, W.D., Schaefer III, H.F., Klopper, W., Noga, J., 2000. Anharmonic force field, vibrational energy levels, and barrier to inversion of SiH_3^-. J. Chem. Phys. 112, 4053–4063.

Al Derzi, A.R., Furtenbacher, T., Tennyson, J., Yurchenko, S.N., Császár, A.G., 2015. Marvel analysis of the measured high-resolution spectra of $^{14}NH_3$. J. Quant. Spectrosc. Radiat. Transf. 161, 117–130.

Albert, S., Chen, Z., Fábri, C., Lerch, P., Prentner, R., Quack, M., 2016. A combined Gigahertz and Terahertz (FTIR) spectroscopic investigation of meta-D-phenol: observation of tunneling switching. Mol. Phys. 114, 2751–2768.

Albert, S., Lerch, P., Prentner, R., Quack, M., 2013. Tunneling and tunneling switching dynamics in phenol and its isotopomers from high-resolution FTIR spectroscopy with synchrotron radiation. Angew. Chem., Int. Ed. 52, 346–349.

Allen, W.D., Yamaguchi, Y., Császár, A.G., Clabo Jr., D.A., Remington, R.B., Schaefer III, H.F., 1990. A systematic study of molecular vibrational anharmonicity and vibration-rotation interaction by self-consistent-field higher derivative methods. Linear polyatomic molecules. Chem. Phys. 145, 427–466.

Armour, E.A.G., Richard, J.-M., Varga, K., 2005. Stability of few-charge systems in quantum mechanics. Phys. Rep. 413, 1–90.

Asvany, O., Padma Kumar, P., Redlich, B., Hegemann, I., Schlemmer, S., Marx, D., 2005. Understanding the infrared spectrum of bare CH_5^+. Science 309, 1219–1222.

Asvany, O., Yamada, K.M.T., Brünken, S., Potapov, A., Schlemmer, S., Marx, D., 2015. Experimental ground-state combination differences of CH_5^+. Science 347, 1346–1349.

Bačić, Z., Light, J.C., 1989. Theoretical methods for rovibrational states of floppy molecules. Annu. Rev. Phys. Chem. 40, 469–498.

Barber, R.J., Tennyson, J., Harris, G.J., Tolchenov, R.N., 2006. A high-accuracy computed water line list. Mon. Not. R. Astron. Soc. 368, 1087–1094.

Barletta, P., Shirin, S.V., Zobov, N.F., Polyansky, O.L., Tennyson, J., Valeev, E.F., Császár, A.G., 2006. The CVRQD ab initio ground-state adiabatic potential energy surfaces for the water molecule. J. Chem. Phys. 125, 204307.

Barone, V., Biczysko, M., Bloino, J., 2014. Fully anharmonic IR and Raman spectra of medium-size molecular systems: accuracy and interpretation. Phys. Chem. Chem. Phys. 16, 1759–1787.

Baye, D., Heenen, P.H., 1986. Generalised meshes for quantum mechanical problems. J. Phys. A 19, 2041.

Beck, M.H., Jäckle, A., Worth, G.A., Meyer, H.-D., 2000. The multiconfiguration time-dependent Hartree method: a highly efficient algorithm for propagating wavepackets. Phys. Rep. 324, 1–105.

Bell, R.P., 1980. The Tunnel Effect in Chemistry. Chapman and Hall, New York.

Blackmore, R., Shizgal, B., 1985. Discrete-ordinate method of solution of Fokker–Planck equations with nonlinear coefficients. Phys. Rev. A 31, 1855–1868.

Blanco, M., Heller, E.J., 1983. Angular momentum projection operators and molecular bound states. J. Chem. Phys. 78, 2504–2517.

Bogey, M., Demuynck, C., Destombes, J.L., Walters, A.D., 1991. Laboratory rotational spectrum of SiC_2 in the ν_3 vibrationally excited state. Astron. Astrophys. 247, L13–L16.

Bonhoeffer, K.F., Harteck, P., 1929. Über Para- und Orthowasserstoff. Z. Phys. Chem. 4B, 113.

Bordé, J., Bordé, C.J., Salomon, C., Van Lerberghe, A., Ouhayoun, M., Cantrell, C.D., 1980. Breakdown of the point-group symmetry of vibration-rotation states and optical observation of ground-state octahedral splittings of $^{32}SF_6$ using saturation spectroscopy. Phys. Rev. Lett. 45, 14–17.

Born, M., Heisenberg, W., 1924. Zur Quantentheorie der Molekeln. Ann. Phys. 379, 1.

Born, M., Oppenheimer, J.R., 1927. Zur Quantentheorie der Molekeln. Ann. Phys. 389, 457.

Bowman, J.M., Carrington Jr., T., Meyer, H.-D., 2008. Variational quantum approaches for computing vibrational energies of polyatomic molecules. Mol. Phys. 106, 2145–2182.

Bowman, J.M., Carter, S., Huang, X., 2003. MULTIMODE: a code to calculate rovibrational energies of polyatomic molecules. Int. Rev. Phys. Chem. 22, 533–549.

Braams, B., Bowman, J., 2009. Permutationally invariant potential energy surfaces in high dimensionality. Int. Rev. Phys. Chem. 28, 577–606.

Brackertz, S., Schlemmer, S., Asvany, O., 2017. Searching for new symmetry species of CH_5^+ – from lines to states without a model. J. Mol. Spectrosc. 342, 73–82.

Bramley, M.J., Green, W.J., Handy, N.C., 1991. Vibration-rotation coordinates and kinetic energy operator for polyatomic molecules. Mol. Phys. 73, 1183.

Brocks, G., van der Avoird, A., Sutcliffe, B.T., Tennyson, J., 1983. Quantum dynamics of non-rigid systems comprising two polyatomic fragments. Mol. Phys. 50, 1025–1043.

Broeckhove, J., Lathouwers, L., 1993. Quantum molecular dynamics and angular momentum projection. In: Cerjan, C. (Ed.), Numerical Grid Methods and Their Application to Schrödinger's Equation. Springer Netherlands, Dordrecht, pp. 49–56.

Bunker, P.R., 1996. A preliminary study of the proton rearrangement energy levels and spectrum of CH_5^+. J. Mol. Spectrosc. 176, 297–304.

Bunker, P.R., Jensen, P., 1998. Molecular Symmetry and Spectroscopy. NRC Research Press, Ottawa.

Bunker, P.R., Longuet-Higgins, H.C., 1964. The infra-red spectrum of dimethylacetylene and the torsional barrier. Proc. R. Soc. Lond. Ser. A 280, 340.

Bunker, P.R., Ostojić, B., Yurchenko, S., 2004. A theoretical study of the millimeterwave spectrum of CH_5^+. J. Mol. Struct. 695–696, 253–261.

Califano, S., 1970. Vibrational States. Wiley, New York.

Carney, G.D., Sprandel, L.I., Kern, C.W., 1978. Variational approaches to vibration-rotation spectroscopy for polyatomic molecules. Adv. Chem. Phys. 37, 305.

Carrington, A., Buttenshaw, J., Kennedy, R.A., 1982. Observation of the infrared spectrum of H_3^+ ion at its near dissociation limit. Mol. Phys. 45, 753–758.

Carrington, A., Kennedy, R.A., 1984. Infrared predissociation limit of the H_3^+ ion. J. Chem. Phys. 81, 91–112.

Carrington Jr., T., Miller, W.H., 1984. Reaction surface Hamiltonian for the dynamics of reactions in polyatomic systems. J. Chem. Phys. 81, 3942–3950.

Carrington, T., 2017. Perspective: computing (ro-)vibrational spectra of molecules with more than four atoms. J. Chem. Phys. 146, 120902.

Carter, S., Bowman, J.M., Handy, N.C., 1998. Extensions and tests of MULTIMODE: a code to obtain accurate vibration/rotation energies of many-mode molecules. Theor. Chem. Acc. 100, 191–198.

Carter, S., Bowman, J.M., Harding, L.B., 1997a. Ab initio calculations of force fields for H_2CN and ClHCN and vibrational energies of H_2CN. Spectrochim. Acta 53A, 1179–1188.

Carter, S., Culik, S.J., Bowman, J.M., 1997b. Vibrational self-consistent field method for many-mode systems: a new approach and application to the vibrations of CO adsorbed on Cu(100). J. Chem. Phys. 107, 10458–10469.

Carter, S., Handy, N.C., 1986. The variational method for the calculation of rovibrational energy levels. Comput. Phys. Rep. 5, 117–171.

Carter, S., Handy, N., 1988. A variational method for the determination of the vibrational ($J = 0$) energy levels of acetylene, using a Hamiltonian in internal coordinates. Comput. Phys. Commun. 51, 49–58.

Changala, P.B., 2019. NITROGEN, numerical and iterative techniques for Rovibronic energies with general internal co-ordinates, a program by P.B. Changala. http://www.colorado.edu/nitrogen.

Chapuisat, X., Iung, C., 1992. Vector parametrization of the N-body problem in quantum mechanics: polyspherical coordinates. Phys. Rev. A 45, 6217–6235.

Chen, W., Poirier, B., 2006. Parallel implementation of efficient preconditioned linear solver for grid-based applications in chemical physics. II: QMR linear solver. J. Comput. Phys. 219, 198–209.

Chen, W., Poirier, B., 2010. Quantum dynamics on massively parallel computers: efficient numerical implementation for preconditioned linear solvers and eigensolvers. J. Theor. Comput. Chem. 9, 825–846.

Christiansen, O., 2007. Vibrational structure theory: new vibrational wave function methods for calculation of anharmonic vibrational energies and vibrational contributions to molecular properties. Phys. Chem. Chem. Phys. 9, 2942–2953.

Clabo Jr., D.A., Allen, W.D., Remington, R.B., Yamaguchi, Y., Schaefer III, H.F., 1988. A systematic study of molecular vibrational anharmonicity and vibration-rotation interaction by self-consistent-field higher derivative methods. Asymmetric top molecules. J. Chem. Phys. 123, 187.

Clementi, E., Kistenmacher, H., Popkie, H., 1973. Study of the electronic structure of molecules. XVIII. Interaction between a lithium atom and a cyano group as an example of a polytopic bond. J. Chem. Phys. 58, 2460–2466.

Colbert, D.T., Sibert III, E.L., 1989. Variable curvature coordinates for molecular vibrations. J. Chem. Phys. 91, 350–363.

Colwell, S.M., Handy, N.C., 1997. The derivation of vibration-rotation kinetic energy operators in internal coordinates. II. Mol. Phys. 92, 317–330.

Corey, G.C., Tromp, J.W., 1995. Variational discrete variable representation. J. Chem. Phys. 103, 1812–1820.

Cowan, R.D., Griffin, D.C., 1976. Approximate relativistic corrections to atomic radial wave functions. J. Opt. Soc. Am. 66, 1010.

Császár, A.G., Allen, W.D., Schaefer III, H.F., 1998a. In pursuit of the ab initio limit for conformational energy prototypes. J. Chem. Phys. 108, 9751–9764.

Császár, A.G., Allen, W.D., Yamaguchi, Y., Schaefer, H.F., 2000. Ab initio determination of accurate ground electronic state potential energy hypersurfaces for small molecules. In: Computational Molecular Spectroscopy. Wiley, New York, pp. 15–68.

Császár, A.G., Fábri, C., Sarka, J., 2020. Quasistructural molecules. WIREs Comput. Mol. Sci. 10, e1432.

Császár, A.G., Fábri, C., Szidarovszky, T., Mátyus, E., Furtenbacher, T., Czakó, G., 2012. Fourth age of quantum chemistry: molecules in motion. Phys. Chem. Chem. Phys. 13, 1085–1106.

Császár, A.G., Furtenbacher, T., 2011. Spectroscopic networks. J. Mol. Spectrosc. 266, 99–103.

Császár, A.G., Furtenbacher, T., 2016. Promoting and inhibiting tunneling via nuclear motions. Phys. Chem. Chem. Phys. 18, 1092–1104.

Császár, A.G., Handy, N.C., 1995a. The exact quantum mechanical vibrational kinetic energy operator of sequentially-bonded molecules in valence internal coordinates. J. Chem. Phys. 102, 3962–3967.

Császár, A.G., Handy, N.C., 1995b. On the form of the exact quantum mechanical vibrational kinetic energy operator for penta-atomic molecules in internal coordinates. Mol. Phys. 86, 959–979.

Császár, A.G., Kain, J.S., Polyansky, O.L., Zobov, N.F., Tennyson, J., 1998b. Relativistic correction to the potential energy surface and vibration-rotation levels of water. Chem. Phys. Lett. 293, 317–323. Erratum: Chem. Phys. Lett. 312 (1999) 613.

Császár, A.G., Mátyus, E., Szidarovszky, T., Lodi, L., Zobov, N.F., Shirin, S.V., Polyansky, O.L., Tennyson, J., 2010. First-principles prediction and partial characterization of the vibrational states of water up to dissociation. J. Quant. Spectrosc. Radiat. Transf. 111, 1043–1064.

Császár, A.G., Tarczay, G., Leininger, M.L., Polyansky, O.L., Tennyson, J., Allen, W.D., 2001. Dream or reality: complete basis set full configuration interaction potential energy hypersurfaces. In: Spectroscopy from Space. Kluwer, Dordrecht, pp. 317–339.

Cullum, J.K., Willoughby, R.A., 1985. Lanczos Algorithms for Large Symmetric Eigenvalue Computations. Birkhäuser, Boston.

Czakó, G., Furtenbacher, T., Császár, A.G., Szalay, V., 2004. Variational vibrational calculations using high-order anharmonic force fields. Mol. Phys. 102, 2411–2423.

Czakó, G., Szalay, V., Császár, A.G., 2006. Finite basis representations with nondirect product basis functions having structure similar to that of spherical harmonics. J. Chem. Phys. 124, 014110.

Dawes, R., Carrington Jr., T., 2004. A multidimensional discrete variable representation basis obtained by simultaneous diagonalization. J. Chem. Phys. 121, 726–736.

De Leon, N., Heller, E.J., 1984. Vector fields, line integrals, and the Hamilton–Jacobi equation: semiclassical quantization of bound states. Phys. Rev. A 30, 5–18.

Delves, L.M., 1959. Tertiary and general-order collisions. Nucl. Phys. 9, 391–399.

Demaison, J., Boggs, J.E., Császár, A.G. (Eds.), 2011. Equilibrium Molecular Structures. CRC Press, Boca Raton.

Dickinson, A.S., Certain, P.R., 1968. Calculation of matrix elements for one-dimensional quantum-mechanical problems. J. Chem. Phys. 49, 4209–4211.

Dosch, H., 2001. Some general aspects of confinement in nanomaterials. Appl. Surf. Sci. 182, 192–195.

Dymarsky, A.Y., Kudin, K.N., 2005. Computation of the pseudorotation matrix to satisfy the Eckart axis conditions. J. Chem. Phys. 122, 124103.

East, A.L.L., Bunker, P.R., 1997. A general rotation-contortion Hamiltonian with structure relaxation: application to the precessing internal rotor model. J. Mol. Spectrosc. 183 (1), 157–162.

East, A.L.L., Kolbuszewski, M., Bunker, P.R., 1997. *Ab initio* calculation of the rotational spectrum of CH_5^+ and CD_5^+. J. Phys. Chem. A 101 (36), 6746–6752.

Echave, J., Clary, D.C., 1992. Potential optimized discrete variable representation. Chem. Phys. Lett. 190, 225–230.

Eckart, C., 1935. Some studies concerning rotating axes and polyatomic molecules. Phys. Rev. 47, 552–558.

Ericsson, T., Ruhe, A., 1980. The spectral transformation Lánczos method for the numerical solution of large sparse generalized symmetric eigenvalue problems. Math. Comput. 35, 1251–1268.

Fábri, C., Császár, A.G., 2018. Vibrational quantum graphs and their application to the quantum dynamics of CH_5^+. Phys. Chem. Chem. Phys. 20, 16913–16917.

Fábri, C., Czakó, G., Tasi, G., Császár, A.G., 2009. Adiabatic Jacobi corrections on the vibrational energy levels of H_2^+ isotopologues. J. Chem. Phys. 130, 134314.

Fábri, C., Furtenbacher, T., Császár, A.G., 2014a. A hybrid variational-perturbational nuclear motion algorithm. Mol. Phys. 112, 2462–2467.

Fábri, C., Marquardt, R., Császár, A.G., Quack, M., 2019. Controlling tunneling in ammonia isotopomers. J. Chem. Phys. 150, 014102.

Fábri, C., Mátyus, E., Császár, A.G., 2011a. Rotating full- and reduced-dimensional quantum chemical models of molecules. J. Chem. Phys. 134, 074105.

Fábri, C., Mátyus, E., Császár, A.G., 2014b. Numerically constructed internal-coordinate Hamiltonian with Eckart embedding and its application for the inversion tunnelling of ammonia. Spectrochim. Acta A 119, 84–89.

Fábri, C., Mátyus, E., Furtenbacher, T., Nemes, L., Mihály, B., Zoltáni, T., Császár, A.G., 2011b. Variational quantum mechanical and active database approaches to the rotational-vibrational spectroscopy of ketene. J. Chem. Phys. 135, 094307.

Fábri, C., Quack, M., Császár, A.G., 2017. On the use of nonrigid-molecular symmetry in nuclear-motion computations employing a discrete variable representation: a case study of the bending energy levels of CH_5^+. J. Chem. Phys. 147, 134101.

Fábri, C., Sarka, J., Császár, A.G., 2014c. Communication: rigidity of the molecular ion H_5^+. J. Chem. Phys. 140, 051101.

Fehrensen, B., Luckhaus, D., Quack, M., 1999a. Inversion tunneling in aniline from high resolution infrared spectroscopy and an adiabatic reaction path Hamiltonian approach. Z. Phys. Chem. (Münich) 209, 1.

Fehrensen, B., Luckhaus, D., Quack, M., 1999b. Mode selective stereomutation tunnelling in hydrogen peroxide isotopomers. Chem. Phys. Lett. 300, 312–320.

Fehrensen, B., Luckhaus, D., Quack, M., Willeke, M., Rizzo, T.R., 2003. *Ab initio* calculations of mode selective tunneling dynamics in $^{12}CH_3OH$ and $^{13}CH_3OH$. J. Chem. Phys. 119, 5534.

Felker, P.M., Bačić, Z., 2017. Electric-dipole-coupled $H_2O{\cdot}C_{60}$ dimer: translation-rotation eigenstates from twelve-dimensional quantum calculations. J. Chem. Phys. 146, 084303.

Firmino, T., Marquardt, R., Gatti, F., Dong, W., 2014. Diffusion rates for hydrogen on Pd(111) from molecular quantum dynamics calculations. J. Phys. Chem. Lett. 5, 4270–4274.

Frenkel, J., 1932. Wave Mechanics, Elementary Theory. Clarendon Press, Oxford, UK.

Fukui, K., 1970. Formulation of the reaction coordinate. J. Phys. Chem. 74, 4161–4163.

Fukui, K., 1981. The path of chemical reactions – the IRC approach. Acc. Chem. Res. 14, 363–368.

Furtenbacher, T., Császár, A.G., 2012. The role of intensities in determining characteristics of spectroscopic networks. J. Mol. Struct. 1009, 123–129.

Furtenbacher, T., Czakó, G., Sutcliffe, B.T., Császár, A.G., Szalay, V., 2006. The methylene saga continues: stretching fundamentals and zero-point energy of $\tilde{X}^3B_1$ CH_2. J. Mol. Struct. 780–781, 283.

Furtenbacher, T., Szidarovszky, T., Fábri, C., Császár, A.G., 2013a. MARVEL analysis of the rotational-vibrational states of the molecular ions H_2D^+ and D_2H^+. Phys. Chem. Chem. Phys. 15, 10181–10193.

Furtenbacher, T., Szidarovszky, T., Hrubý, J., Kyuberis, A.A., Zobov, N.F., Polyansky, O.L., Tennyson, J., Császár, A.G., 2016. Definitive high-temperature ideal-gas thermochemical functions of the $H_2{}^{16}O$ molecule. J. Phys. Chem. Ref. Data 45, 043104.

Furtenbacher, T., Szidarovszky, T., Mátyus, E., Fábri, C., Császár, A.G., 2013b. Analysis of the rotational-vibrational states of the molecular ion H_3^+. J. Chem. Theory Comput. 9, 5471–5478.

Gatti, F., Iung, C., 2009. Exact and constrained kinetic energy operators for polyatomic molecules: the polyspherical approach. Phys. Rep. 484, 1–69.

Gatti, F., Marquardt, R., 2012. Stereomutation in vibrationally excited NHD_2. Theor. Comput. Chem. 990, 90–93.

Goldstein, H., Poole Jr., C.P., Safko, J.L., 2014. Classical Mechanics, 3rd edn. Pearson, Harlow.

Gordon, I., Rothman, L., Hill, C., Kochanov, R.V., Tan, Y., Bernath, P., Birk, M., Boudon, V., Campargue, A., Chance, K.V., Drouin, B.J., Flaud, J.-M., Gamache, R.R., Jacquemart, D., Perevalov, V.I., Perrin, A., Smith, M.A.H., Tennyson, J., Tran, H., Tyuterev, V.G., Toon, G.C., Hodges, J.T., Shine, K.P., Barbe, A., Császár, A.G., Devi, M.V., Furtenbacher, T., Harrison, J.J., Jolly, A., Johnson, T., Karman, T., Kleiner, I., Kyuberis, A., Loos, J., Lyulin, O.M., Mikhailenko, S.N., Moazzen-Ahmadi, N., Müller, H.S.P., Naumenko, O.V., Nikitin, A.V., Polyansky, O.L., Rey, M., Rotger, M., Sharpe, S., Starikova, E., Tashkun, S.A., Vander Auwera, J., Wagner, G., Wilzewski, J., Wcislo, P., Yu, S., Zak, E., 2017. The HITRAN2016 molecular spectroscopic database. J. Quant. Spectrosc. Radiat. Transf. 203, 3–69.

Grossmann, F., Dittrich, T., Jung, P., Hänggi, P., 1991. Coherent destruction of tunneling. Phys. Rev. Lett. 67, 516–519.

Güsten, R., Wiesemeyer, H., Neufeld, D., Menten, K.M., Graf, U.U., Jacobs, K., Klein, B., Ricken, O., Risacher, C., Stutzki, J., 2019. Astrophysical detection of the helium hydride ion HeH^+. Nature 568, 357–359.

Halász, G., Vibók, A., 2000. Using a multi-step potential as an exact solution of the absorbing potential problem on the grid. Chem. Phys. Lett. 323, 287–292.

Hammond, B.L., Lester Jr., W.A., Reynolds, P.J., 1994. Monte Carlo Methods in ab initio Quantum Chemistry. World Scientific, Singapore.

Handley, C.M., Popelier, P.L.A., 2010. Potential energy surfaces fitted by artificial neural networks. J. Phys. Chem. A 114, 3371–3383.

Handy, N.C., 1987. The derivation of vibration-rotation kinetic energy operators, in internal coordinates. Mol. Phys. 61, 207–223.

Handy, N.C., Yamaguchi, Y., Schaefer III, H.F., 1986. The diagonal correction to the Born–Oppenheimer approximation: its effect on the singlet–triplet splitting of CH_2 and other molecular effects. J. Chem. Phys. 84, 4481–4484.

Haritan, I., Moiseyev, N., 2017. On the calculation of resonances by analytic continuation of eigenvalues from the stabilization graph. J. Chem. Phys. 147, 014101.

Harris, D.O., Engerholm, G.G., Gwinn, W.D., 1965. Calculation of matrix elements for one-dimensional quantum-mechanical problems and the application to anharmonic oscillators. J. Chem. Phys. 43, 1515–1517.

Harthcock, M.A., Laane, J., 1982. Calculation of kinetic energy terms for the vibrational Hamiltonian: application to large-amplitude vibrations using one-, two-, and three-dimensional models. J. Mol. Spectrosc. 91, 300–324.

Hazi, A.U., Taylor, H.S., 1970. Stabilization method of calculating resonance energies: model problem. Phys. Rev. A 1, 1109–1120.

Helgaker, T., Jørgensen, P., Olsen, J., 2000. Molecular Electronic-Structure Theory. Wiley, Chichester.

Henderson, J.R., Tennyson, J., 1993. DVR1D: programs for mixed pointwise/basis set calculation of ro-vibrational spectra. Comput. Phys. Commun. 75, 365–378.

Herbst, E., 2000. The astrochemistry of H_3^+. Philos. Trans. R. Soc. Lond. Ser. A 358, 2523–2534.

Herzberg, G., 1945. Molecular Spectra and Molecular Structure, Vols. 1-3. Van Nostrand Reinhold, Melbourne.

Hestenes, D., 1999. New Foundations for Classical Mechanics, 2nd edn. Kluwer Academic Publishers, Dordrecht.

Hirschfelder, J.O., Dahler, J.S., 1956. The kinetic energy of relative motion. Proc. Natl. Acad. Sci. 42, 363–365.

Holthaus, M., 1992. Pulse-shape-controlled tunneling in a laser field. Phys. Rev. Lett. 69, 1596–1599.

Hougen, J.T., Bunker, P.R., Johns, J.W.C., 1970. The vibration-rotation problem in triatomic molecules allowing for a large-amplitude bending vibration. J. Mol. Spectrosc. 34, 136–172.

Hoy, A.R., Mills, I.M., Strey, G., 1972. Anharmonic force constant calculations. Mol. Phys. 24, 1265–1290.

Huang, X., McCoy, A.B., Bowman, J.M., Johnson, L.M., Savage, C., Dong, F., Nesbitt, D.J., 2006. Quantum deconstruction of the infrared spectrum of CH_5^+. Science 311, 60–63.

Huang, X., Schwenke, D.W., Lee, T.J., 2008. An accurate global potential energy surface, dipole moment surface, and rovibrational frequencies for NH_3. J. Chem. Phys. 129, 214304.

Hund, F., 1927. Zur Deutung der Molekelspektren. I. Z. Phys. 40, 742–764.

Iung, C., Ribeiro, F., Sibert III, E.L., 2006. Comparison of perturbative and variational treatments of molecular vibrations: application to the vibrational spectrum of HFCO up to 8000 cm^{-1}. J. Phys. Chem. A 110, 5420–5429.

Ivanov, S.D., Asvany, O., Witt, A., Hugo, E., Mathias, G., Redlich, B., Marx, D., Schlemmer, S., 2010. Quantum-induced symmetry breaking explains infrared spectra of CH_5^+ isotopologues. Nature Chem. 2, 298–302.

Jacobi, C.G.J., 1843. Sur l'élimination des noeuds dans le problème des trois corps. C. R. Acad. Sci. Paris 15, 236–255.

Jepsen, D.W., Hirschfelder, J.O., 1959. Set of co-ordinate systems which diagonalize the kinetic energy of relative motion. Proc. Natl. Acad. Sci. 45, 249–256.

Jiang, B., Li, J., Guo, H., 2016. Potential energy surfaces from high fidelity fitting of *ab initio* points: the permutation invariant polynomial – neural network approach. Int. Rev. Phys. Chem. 35, 479–506.

Johnson, B.R., 1980. On hyperspherical coordinates and mapping the internal configurations of a three body system. J. Chem. Phys. 73, 5051–5058.

Johnson, B.R., 1983. The quantum dynamics of three particles in hyperspherical coordinates. J. Chem. Phys. 79, 1916–1925.

Kanfer, S., Shapiro, M., 1984. Inversion of the $CH_3I(\tilde{X}\,^1A_1)$ potential by the discrete position operator method. J. Phys. Chem. 88, 3964–3968.

Klaiman, S., Gilary, I., 2012. On resonance: a first glance into the behavior of unstable states. Adv. Quantum Chem. 63, 1–31.

Klepeis, N.E., East, A.L.L., Császár, A.G., Allen, W.D., Lee, T.J., Schwenke, D.W., 1993. The $[FHCl]^-$ molecular anion: structural aspects, global surface, and vibrational eigenspectrum. J. Chem. Phys. 99, 3865–3897.

Klinting, E.L., König, C., Christiansen, O., 2015. Hybrid optimized and localized vibrational coordinates. J. Phys. Chem. A 119, 11007–11021.

Kolbuszewski, M., Bunker, P.R., 1996. Potential barriers, tunneling splittings, and the predicted $J = 1 \leftarrow 0$ spectrum of CH_5^+. J. Chem. Phys. 105, 3649–3653.

Kono, H., 1993. Extraction of eigenstates from an optically prepared state by a time-dependent quantum-mechanical method. Toward simulation of "intermediate case" radiationless transitions. Chem. Phys. Lett. 214, 137–143.

Kosloff, D., Kosloff, R., 1983. A Fourier method solution for the time dependent Schrödinger equation as a tool in molecular dynamics. J. Comput. Phys. 52, 35–53.

Kozin, I.N., Law, M.M., Tennyson, J., Hutson, J.M., 2004. New vibration–rotation code for tetraatomic molecules exhibiting wide-amplitude motion: WAVR4. Comput. Phys. Commun. 163, 117–131.

Kozin, I.N., Law, M.M., Tennyson, J., Hutson, J.M., 2005. Calculating energy levels of isomerizing tetra-atomic molecules. II. The vibrational states of acetylene and vinylidene. J. Chem. Phys. 122, 064309.

Krasnoshchekov, S.V., Isayeva, E.V., Stepanov, N.F., 2014. Determination of the Eckart molecule-fixed frame by use of the apparatus of quaternion algebra. J. Chem. Phys. 140, 154104.

Kroto, H.W., 1992. Molecular Rotation Spectra. Dover, New York.

Kudin, K.N., Dymarsky, A.Y., 2005. Eckart axis conditions and the minimization of the root-mean-square deviation: two closely related problems. J. Chem. Phys. 122, 224105.

Laane, J., Harthcock, M.A., Killough, P.M., Bauman, L.E., Cooke, J.M., 1982. Vector representation of large-amplitude vibrations for the determination of kinetic energy functions. J. Mol. Spectrosc. 91, 286–299.

Lanczos, C., 1950. An iteration method for the solution of the eigenvalue problem of linear differential and integral operators. J. Res. Natl. Bur. Stand. 45, 255.

Landau, L.D., Lifshitz, E.M., 1977. Quantum Mechanics, 3rd ed. Pergamon, Oxford.

Larimian, S., Erattupuzha, S., Lötstedt, E., Szidarovszky, T., Maurer, R., Roither, S., Schöffler, M., Kartashov, D., Baltuška, A., Yamanouchi, K., Kitzler, M., Xie, X., 2016. Fragmentation of long-lived hydrocarbons after strong field ionization. Phys. Rev. A 93, 053405.

Lauvergnat, D., 2001. A harmonic adiabatic approximation to calculate highly excited vibrational levels of "floppy molecules". J. Chem. Phys. 114, 6592–6604.

Lauvergnat, D., 2019. ElVibRot quantum dynamics code. http://www.lcp.u-psud.fr/Pageperso/lauvergnat/ElVibRot.html.

Lauvergnat, D., Nauts, A., 2002. Exact numerical computation of a kinetic energy operator in curvilinear coordinates. J. Chem. Phys. 116, 8560–8570.

Layton, E.G., 1993. The Fourier-grid formalism: philosophy and application to scattering problems using R-matrix theory. J. Phys. B 26, 2501–2522.

Le Sueur, C.R., Miller, S., Tennyson, J., Sutcliffe, B.T., 1992. On the use of variational wavefunctions in calculating vibrational band intensities. Mol. Phys. 76, 1147–1156.

Lefebvre, R., 1985. Theory of resonances in molecular systems. EuroPhys. News 16, 4–7.

Lewis, G.N., Randall, M., Pitzer, K.S., Brewer, L., 1961. Thermodynamics. McGraw-Hill, New York.

Light, J.C., Carrington Jr., T., 2000. Discrete variable representations and their utilization. Adv. Chem. Phys. 114, 263–310.

Light, J.C., Hamilton, I.P., Lill, J.V., 1985. Generalized discrete variable approximation in quantum mechanics. J. Chem. Phys. 82, 1400–1409.

Lill, J.V., Parker, G.A., Light, J.C., 1982. Discrete variable representations and sudden models in quantum scattering theory. Chem. Phys. Lett. 89, 483–489.

Littlejohn, R.G., Cargo, M., Mitchell, K., Carrington Jr., T., Poirier, B., 2002. A general framework for discrete variable representation basis sets. J. Chem. Phys. 116, 8691–8703.

Littlejohn, R.G., Reinsch, M., 1997. Gauge fields in the separation of rotations and internal motions in the n-body problem. Rev. Modern Phys. 69, 213–275.

Longuet-Higgins, H.C., 1963. The symmetry groups of nonrigid molecules. Mol. Phys. 6, 445–460.

Louck, J.D., Galbraith, H.W., 1972. Application of orthogonal and unitary group methods to the N-body problem. Rev. Modern Phys. 44, 540–601.

Louck, J.D., Galbraith, H.W., 1976. Eckart vectors, Eckart frames, and polyatomic molecules. Rev. Modern Phys. 48, 69.

Luckhaus, D., 2000. 6D vibrational quantum dynamics: generalized coordinate discrete variable representation and (a)diabatic contraction. J. Chem. Phys. 113, 1329–1347.

Luckhaus, D., 2003. The vibrational spectrum of HONO: fully coupled 6D direct dynamics. J. Chem. Phys. 118, 8797–8806.

Lukka, T.J., 1995. A simple method for the derivation of exact quantum-mechanical vibration-rotation Hamiltonians in terms of internal coordinates. J. Chem. Phys. 102, 3945–3955.

Ma, G., Chen, R., Guo, H., 1999. Quantum calculations of highly excited vibrational spectrum of sulfur dioxide. I. Eigenenergies and assignments up to 15 000 cm^{-1}. J. Chem. Phys. 110, 8408–8416.

Macías, A., Riera, A., 1984. Constrained variational procedure for the calculation of autoionization lifetimes. Phys. Lett. A 103, 377–380.

Macías, A., Riera, A., 1989. Accurate resonance lifetimes from real stabilization graphs. Chem. Phys. Lett. 164, 359–363.

Majumder, M., Hegger, S.E., Dawes, R., Manzhos, S., Wang, X.-G., Carrington Jr., T., Li, J., Guo, H., 2015. Explicitly-correlated MRCI-F12 potential energy surfaces for methane fit with several permutation invariant schemes and full-dimensional vibrational calculations. Mol. Phys. 113, 1823–1833.

Majumder, M., Ndengue, S.A., Dawes, R., 2016. Automated construction of potential energy surfaces. Mol. Phys. 114, 1–18.

Maksyutenko, P., Zobov, N.F., Shirin, S.V., Polyansky, O.L., Muenter, J.S., Rizzo, T.R., Boyarkin, O.V., 2007. Approaching the full set of energy levels of water. J. Chem. Phys. 126, 241101.

Mandelshtam, V.A., Ravuri, T.R., Taylor, H.S., 1993. Calculation of the density of resonance states using the stabilization method. Phys. Rev. Lett. 70, 1932–1935.

Mandelshtam, V.A., Taylor, H.S., Ryaboy, V., Moiseyev, N., 1994. Stabilization theory for computing energies and widths of resonances. Phys. Rev. A 50, 2764–2766.

Manzhos, S., Carrington Jr., T., 2016. Using an internal coordinate gaussian basis and a space-fixed cartesian coordinate kinetic energy operator to compute a vibrational spectrum with rectangular collocation. J. Chem. Phys. 145, 224110.

Manzhos, S., Carrington, T., Laverdure, L., Mosey, N., 2015. Computing the anharmonic vibrational spectrum of UF_6 in 15 dimensions with an optimized basis set and rectangular collocation. J. Phys. Chem. A 119, 9557–9567.

Manzhos, S., Yamashita, K., Carrington, T., 2009. Using a neural network based method to solve the vibrational Schrödinger equation for H_2O. Chem. Phys. Lett. 474, 217–221.

Marquardt, R., Quack, M., 2011. Global analytical potential energy surfaces for high-resolution molecular spectroscopy and reaction dynamics. In: Quack, M., Merkt, F. (Eds.), Handbook of High-Resolution Spectroscopy, Vol. 1. John Wiley & Sons, Ltd, pp. 511–549.

Marquardt, R., Quack, M., 2020. Foundations of time dependent quantum dynamics of molecules under isolation and in coherent electromagnetic fields. In: Marquardt, R., Quack, M. (Eds.), Molecular Spectroscopy and Quantum Dynamics. Elsevier, Amsterdam. Chapter 1 (this book).

Marquardt, R., Quack, M., Thanopulos, I., Luckhaus, D., 2003. Tunneling dynamics of the NH chromophore in NHD_2 during and after coherent infrared excitation. J. Chem. Phys. 118, 643–658.

Marquardt, R., Sanrey, M., Gatti, F., Le Quéré, F., 2010. Full-dimensional quantum dynamics of vibrationally highly excited NHD_2. J. Chem. Phys. 133, 174302.

Marston, C.C., Balint-Kurti, G.G., 1989. The Fourier grid Hamiltonian method for bound state eigenvalues and eigenfunctions. J. Chem. Phys. 91, 3571–3576.

Matanović, I., Belof, J.L., Space, B., Sillar, K., Sauer, J., Eckert, J., Bačić, Z., 2012. Hydrogen adsorbed in a metal organic framework-5: coupled translation-rotation eigenstates from quantum five-dimensional calculations. J. Chem. Phys. 137, 014701.

Matanović, I., Xu, M., Moskowitz, J.W., Eckert, J., Bačić, Z., 2009. Methane molecules confined in the small and large cages of structure I clathrate hydrate: quantum six-dimensional calculations of the coupled translation-rotation eigenstates. J. Chem. Phys. 131, 224308.

Mátyus, E., 2013. On the calculation of resonances in pre-Born–Oppenheimer molecular structure theory. J. Phys. Chem. A 117, 7195–7206.

Mátyus, E., Czakó, G., Császár, A.G., 2009. Toward black-box-type full- and reduced-dimensional variational (ro)vibrational computations. J. Chem. Phys. 130, 134112.

Mátyus, E., Czakó, G., Sutcliffe, B.T., Császár, A.G., 2007. Variational vibrational calculations with arbitrary potentials using the Eckart–Watson Hamiltonians and the discrete variable representation. J. Chem. Phys. 127, 084102.

Mátyus, E., Fábri, C., Szidarovszky, T., Czakó, G., Allen, W.D., Császár, A.G., 2010. Assigning quantum labels to variationally computed rotational-vibrational eigenstates of polyatomic molecules. J. Chem. Phys. 133, 034113.

Mátyus, E., Hutter, J., Müller-Herold, U., Reiher, M., 2011. On the emergence of molecular structure. Phys. Rev. A 83, 052512.

Mátyus, E., Reiher, M., 2012. Molecular structure calculations: a unified quantum mechanical description of electrons and nuclei using explicitly correlated Gaussian functions and the global vector representation. J. Chem. Phys. 137, 024104.

Mátyus, E., Šimunek, J., Császár, A.G., 2009. On variational computation of a large number of vibrational energy levels and wave functions for medium-sized molecules. J. Chem. Phys. 131, 074106.

Mayer, J.E., Mayer, M.G., 1940. Statistical Mechanics. Wiley, New York.

Mayrhofer, R.C., Sibert III, E.L., 1995. Investigating optimal coordinates for describing vibrational motion. Theor. Chem. Acc. 92, 107–122.

McAfee, J.L., Poirier, B., 2009. Quantum dynamics of hydrogen interacting with single-walled carbon nanotubes. J. Chem. Phys. 130, 064701.

McAfee, J.L., Poirier, B., 2011. Quantum dynamics of hydrogen interacting with single-walled carbon nanotubes: multiple H-atom adsorbates. J. Chem. Phys. 134, 074308.

McCoy, A.B., Burleigh, D.C., Sibert III, E.L., 1991. Rotation-vibration interactions in highly excited states of SO_2 and H_2CO. J. Chem. Phys. 95, 7449–7465.

McKee, M.L., 2011. Fluctional molecules. WIREs Comput. Mol. Sci. 1, 943–951.

McQuarrie, D.A., 2000. Statistical Mechanics. University Science Books, Sausalito.

Mellau, G.C., 2011a. Complete experimental rovibrational eigenenergies of HCN up to 6880 cm^{-1} above the ground state. J. Chem. Phys. 134, 234303.

Mellau, G.C., 2011b. Highly excited rovibrational states of HNC. J. Mol. Spectrosc. 269, 77–85.

Merkt, F., Quack, M., 2011. Molecular quantum mechanics and molecular spectra, molecular symmetry, and interaction of matter with radiation. In: Quack, M., Merkt, F. (Eds.), Handbook of High-Resolution Spectroscopy, Vol. 1. John Wiley & Sons, Ltd, pp. 1–55.

Meyer, H.D., Gatti, F., Worth, G.A., 2009. Multidimensional Quantum Dynamics: MCTDH Theory and Applications. John Wiley & Sons.

Meyer, R., 1979. Flexible models for intramolecular motion, a versatile treatment and its application to glyoxal. J. Mol. Spectrosc. 76, 266–300.

Mezey, P.G., 1987. Potential Energy Hypersurfaces. Elsevier, New York.

Miller, W.H., Handy, N.C., Adams, J.E., 1980. Reaction path Hamiltonian for polyatomic molecules. J. Chem. Phys. 72, 99–112.

Mills, I.M., 1972. Vibration-rotation structure in asymmetric- and symmetric-top molecules. In: Molecular Spectroscopy: Modern Research. Academic Press, New York, pp. 115–140.

Mills, I.M., Quack, M., 2002. Comment. Mol. Phys. 100, 9–10.

Mills, I.M., Thompson, H.W., 1954. Internal rotation in dimethyl acetylene. Proc. R. Soc. Lond. Ser. A 226, 306.

Mladenović, M., 2000a. Rovibrational Hamiltonians for general polyatomic molecules in spherical polar parametrization. I. Orthogonal representations. J. Chem. Phys. 112, 1070–1081.

Mladenović, M., 2000b. Rovibrational Hamiltonians for general polyatomic molecules in spherical polar parametrization. II. Nonorthogonal descriptions of internal molecular geometry. J. Chem. Phys. 112, 1082–1095.

Mladenović, M., 2000c. Rovibrational Hamiltonians for general polyatomic molecules in spherical polar parametrization. III. Global vs local axis system and angular coordinates. J. Chem. Phys. 113, 10524–10534.

Mladenović, M., 2002a. Discrete variable approaches to tetratomic molecules: part I: DVR(6) and DVR(3)+DGB methods. Spectrochim. Acta, Part A: Mol. Biomol. Spectrosc. 58, 795–807.

Mladenović, M., 2002b. Discrete variable approaches to tetratomic molecules: Part II: Application to H_2O_2 and H_2CO. Spectrochim. Acta, Part A: Mol. Biomol. Spectrosc. 58, 809–824.

Moiseyev, N., 2011. Non-hermitian Quantum Mechanics. Non-Hermitian Quantum Mechanics. Cambridge University Press.

Moiseyev, N., Friedland, S., Certain, P.R., 1981. Cusps, θ trajectories, and the complex virial theorem. J. Chem. Phys. 74, 4739–4740.

Muga, J.G., Palao, J.P., Navarro, B., Egusquiza, I.L., 2004. Complex absorbing potentials. Phys. Rep. 395, 357–426.

Müller, K., 1980. Reaction paths on multidimensional energy hypersurfaces. Angew. Chem., Int. Ed. 19, 1–13.

Munro, J.J., Ramanlal, J., Tennyson, J., Mussa, H.Y., 2006. Properties of high-lying vibrational states of the H_3^+ molecular ion. Mol. Phys. 104, 115–125.

Murrell, J.N., Carter, S., Farantos, S.C., Huxley, P., Varandas, A.J.C., 1984. Molecular Potential Energy Surfaces. Wiley, New York.

Mussa, H.Y., Tennyson, J., 2002. Calculating quasi-bound rotation-vibrational states of HOCl using massively parallel computers. Chem. Phys. Lett. 366 (5–6), 449–457.

Nauts, A., Chapuisat, X., 1985. Momentum, quasi-momentum, and Hamiltonian operators in terms of arbitrary curvilinear coordinates, with special emphasis on molecular Hamiltonians. Mol. Phys. 55, 1287–1318.

Nefedova, V.V., Boldyrev, A.I., Simons, J., 1993. Graphical representation of the symmetries of potential energy surfaces. J. Chem. Phys. 98, 8801–8809.

Neff, M., Rauhut, G., 2009. Toward large scale vibrational configuration interaction calculations. J. Chem. Phys. 131, 124129.

Nesbitt, D.J., Naaman, R., 1989. On apparent spectroscopic rigidity of floppy molecular systems. J. Chem. Phys. 91, 3801–3809.

Nielsen, H.H., 1951. The vibration-rotation energies of molecules. Rev. Modern Phys. 23, 90–136.

Nielsen, I.M.B., Allen, W.D., Császár, A.G., Schaefer, H.F., 1997. Toward resolution of the silicon dicarbide (SiC_2) saga: *ab initio* excursions in the web of polytopism. J. Chem. Phys. 107, 1195–1211.

Nordheim, L., 1928. Zur Theorie der thermischen Emission und der Reflexion von Elektronen an Metallen. Z. Phys. 46, 833–855.

Oka, T., 2011. Orders of magnitude and symmetry in molecular spectroscopy. In: Quack, M., Merkt, F. (Eds.), Handbook of High-Resolution Spectroscopy, Vol. 1. John Wiley & Sons, Ltd, pp. 633–658.

Oka, T., 2015. Taming CH_5^+, the "enfant terrible" of chemical structures. Science 347, 1313–1314.

Owens, A., Zak, E.J., Chubb, K.L., Yurchenko, S.N., Tennyson, J., Yachmenev, A., 2017. Simulating electric field interactions with polar molecules using spectroscopic databases. Sci. Rep. 7, 45068.

Ozier, I., Yi, P., Khosla, A., Ramsey, N.F., 1970. Direct observation of ortho-para transitions in methane. Phys. Rev. Lett. 24, 642–645.

Pachucki, K., Komasa, J., 2010. Rovibrational levels of HD. Phys. Chem. Chem. Phys. 12, 9188–9196.

Pack, R.T., 1984. Coordinates for an optimum CS approximation in reactive scattering. Chem. Phys. Lett. 108, 333–338.

Padma Kumar, P., Marx, D., 2006. Understanding hydrogen scrambling and infrared spectrum of bare CH_5^+ based on ab initio simulations. Phys. Chem. Chem. Phys. 8, 573–586.

Papoušek, D., Aliev, M.R., 1982. Molecular Vibrational-Rotational Spectra. Elsevier Scientific Publishing, Amsterdam.

Papp, D., Császár, A.G., Yamanouchi, K., Szidarovszky, T., 2018. Rovibrational resonances in H_2He^+. J. Chem. Theory Comput. 14, 1523–1533.

Papp, D., Sarka, J., Szidarovszky, T., Császár, A.G., Mátyus, E., Hochlaf, M., Stoecklin, T., 2017a. Complex rovibrational dynamics of the Ar·NO^+ complex. Phys. Chem. Chem. Phys. 19, 8152–8160.

Papp, D., Szidarovszky, T., Császár, A.G., 2017b. A general variational approach for computing rovibrational resonances of polyatomic molecules. Application to the weakly bound H_2He^+ and H_2·CO systems. J. Chem. Phys. 147, 094106.

Partridge, H., Schwenke, D.W., 1997. The determination of an accurate isotope dependent potential energy surface for water from extensive *ab initio* calculations and experimental data. J. Chem. Phys. 106, 4618–4639.

Pavanello, M., Adamowicz, L., Alijah, A., Zobov, N.F., Mizus, I.I., Polyansky, O.L., Tennyson, J., Szidarovszky, T., Császár, A.G., 2012a. Calibration-quality adiabatic potential energy surfaces for H_3^+ and its isotopologues. J. Chem. Phys. 136, 184303.

Pavanello, M., Adamowicz, L., Alijah, A., Zobov, N.F., Mizus, I.I., Polyansky, O.L., Tennyson, J., Szidarovszky, T., Császár, A.G., Berg, M., Petrignani, A., Wolf, A., 2012b. Precision measurements and computations of transition energies in rotationally cold triatomic hydrogen ions up to the mid-visible spectral range. Phys. Rev. Lett. 108, 023002.

Pavanello, M., Tung, W.C., Adamowicz, L., 2010. Determination of deuteron quadrupole moment from calculations of the electric field gradient in D_2 and HD. Phys. Rev. A 81, 042526.

Pavlyuchko, A.I., Yurchenko, S.N., Tennyson, J., 2015. A hybrid variational-perturbational calculation of the ro-vibrational spectrum of nitric acid. J. Chem. Phys. 142, 094309.

Pesonen, J., 2013. Constrained molecular vibration-rotation Hamiltonians: contravariant metric tensor. J. Chem. Phys. 139, 144310.

Pesonen, J., 2014. Eckart frame vibration-rotation Hamiltonians: contravariant metric tensor. J. Chem. Phys. 140, 074101.

Pesonen, J., Halonen, L., 2003. Recent advances in the theory of vibration-rotation Hamiltonians. Adv. Chem. Phys. 125, 269–349.

Petty, C., Poirier, B., 2014. Using ScalIT for performing accurate rovibrational spectroscopy calculations for triatomic molecules: a practical guide. Appl. Math. 5, 2756–2763.

Petty, C., Spada, R.F.K., Machado, F.B.C., Poirier, B., 2018. Accurate rovibrational energies of ozone isotopologues up to $J = 10$ utilizing artificial neural networks. J. Chem. Phys. 149, 024307.

Pique, J.P., Hartmann, F., Bacis, R., Churassy, S., Koffend, J.B., 1984. Hyperfine-induced ungerade-gerade symmetry breaking in a homonuclear diatomic molecule near a dissociation limit: $^{127}I_2$ at the $^2P_{\frac{3}{2}} - ^2P_{\frac{1}{2}}$ limit. Phys. Rev. Lett. 52, 267–270.

Podolsky, B., 1928. Quantum-mechanically correct form of Hamiltonian function for conservative systems. Phys. Rev. 32, 812–816.

Poirier, B., 2003a. Exploiting both C_{3v} symmetry and sparsity in vibrational calculations for methanelike molecules. J. Chem. Phys. 119, 90–93.

Poirier, B., 2003b. Using wavelets to extend quantum dynamics calculations to ten or more degrees of freedom. J. Theor. Comput. Chem. 2, 65–72.

Poirier, B., 2015. Communication: the $H_2{\cdot}C_{60}$ inelastic neutron scattering selection rule: expanded and explained. J. Chem. Phys. 143, 101104.

Poirier, B., Carrington Jr., T., 2003a. Semiclassically optimized complex absorbing potentials of polynomial form I. Pure imaginary case. J. Chem. Phys. 118, 17–28.

Poirier, B., Carrington Jr., T., 2003b. Semiclassically optimized complex absorbing potentials of polynomial form. II. Complex case. J. Chem. Phys. 119, 77–89.

Polyansky, O.L., Alijah, A., Zobov, N.F., Mizus, I.I., Ovsyannikov, R.I., Tennyson, J., Lodi, L., Szidarovszky, T., Császár, A.G., 2012. Spectroscopy of H_3^+ based on a new high-accuracy global potential energy surface. Philos. Trans. R. Soc. A 370, 5014–5027.

Polyansky, O.L., Császár, A.G., Shirin, S.V., Zobov, N.F., Barletta, P., Tennyson, J., Schwenke, D.W., Knowles, P.J., 2003. High-accuracy *ab initio* rotation-vibration transitions for water. Science 299, 539–542.

Polyansky, O.L., Kyuberis, A.A., Zobov, N.F., Tennyson, J., Yurchenko, S.N., Lodi, L., 2018. ExoMol molecular line lists XXX: a complete high-accuracy line list for water. Mon. Not. R. Astron. Soc. 480, 2597–2608.

Pulay, P., Fogarasi, G., Pang, F., Boggs, J.E., 1979. Systematic ab initio gradient calculation of molecular geometries, force constants, and dipole moment derivatives. J. Am. Chem. Soc. 101, 2550–2560.

Pyykkö, P., Dyall, K.G., Császár, A.G., Tarczay, G., Polyansky, O.L., Tennyson, J., 2001. Lamb shift effects in rotation-vibration spectra of water. Phys. Rev. A 63, 024502.

Quack, M., 1977. Detailed symmetry selection rules for reactive collisions. Mol. Phys. 34, 477–504.

Quack, M., 1986. On the measurement of the parity violating energy difference between enantiomers. Chem. Phys. Lett. 132, 147–153.

Quack, M., 2011. Fundamental symmetries and symmetry violations from high-resolution spectroscopy. In: Quack, M., Merkt, F. (Eds.), Handbook of High-Resolution Spectroscopy, Vol. 1. John Wiley & Sons, Ltd, pp. 659–722.

Quade, C.R., 1976. Internal coordinate formulation for the vibration-rotation energies of polyatomic molecules. J. Chem. Phys. 64, 2783–2795.

Quiney, H.M., Barletta, P., Tarczay, G., Császár, A.G., Polyansky, O.L., Tennyson, J., 2001. Two-electron relativistic corrections to the potential energy surface and vibration-rotation levels of water. Chem. Phys. Lett. 344, 413–420.

Radau, R., 1868. Sur une transformation des équations différentielles de la dynamique. Ann. Sci. Éc. Norm. Supér., 5–311.

Rauhut, G., 2004. Efficient calculation of potential energy surfaces for the generation of vibrational wave functions. J. Chem. Phys. 121, 9313–9322.

Rauhut, G., 2007. Configuration selection as a route towards efficient vibrational configuration interaction calculations. J. Chem. Phys. 127, 184109.

Riera, A., 1993. Some properties and pitfalls of the stabilization method. A unified approach. J. Phys. Chem. 97, 1558–1565.

Riss, U.V., Meyer, H.D., 1993. Calculation of resonance energies and widths using the complex absorbing potential method. J. Phys. B 26, 4503–4536.

Saad, Y., 2003. Iterative Methods for Sparse Linear Systems. Society for Industrial and Applied Mathematics, Philadelphia, PA.

Sadri, K., Lauvergnat, D., Gatti, F., Meyer, H.-D., 2014. Rovibrational spectroscopy using a kinetic energy operator in Eckart frame and the multi-configuration time-dependent Hartree (MCTDH) approach. J. Chem. Phys. 141, 114101.

Sala, M., Gatti, F., Guérin, S., 2014. Coherent destruction of tunneling in a six-dimensional model of NHD_2: a computational study using the multi-configuration time-dependent Hartree method. J. Chem. Phys. 141, 164326.

Sala, M., Guérin, S., Gatti, F., Marquardt, R., Meyer, H.-D., 2012. Laser-induced enhancement of tunneling in NHD_2. J. Chem. Phys. 136, 194308.
Sarka, J., Császár, A.G., 2016. Interpretation of the vibrational energy level structure of the astructural molecular ion H_5^+ and all of its deuterated isotopomers. J. Chem. Phys. 144, 154309.
Sarka, J., Császár, A.G., Althorpe, S.C., Wales, D.J., Mátyus, E., 2016. Rovibrational transitions of the methane-water dimer from intermolecular quantum dynamical computations. Phys. Chem. Chem. Phys. 18, 22816–22826.
Sarka, J., Császár, A.G., Mátyus, E., 2017. Rovibrational quantum dynamical computations for deuterated isotopologues of the methane-water dimer. Phys. Chem. Chem. Phys. 19, 15335–15345.
Sarka, J., Fábri, C., Szidarovszky, T., Császár, A.G., Lin, Z., McCoy, A.B., 2015. Modelling rotations, vibrations, and rovibrational couplings in astructural molecules – a case study based on the H_5^+ molecular ion. Mol. Phys. 113, 1873–1883.
Sarka, J., Poirier, B., Szalay, V., Császár, A.G., 2020. On neglecting Coriolis and related couplings in first-principles rovibrational spectroscopy: considerations of symmetry, accuracy and simplicity. Sci. Rep. 10, 4872.
Schinke, R., 1993. Photodissociation Dynamics. Cambridge University Press, Cambridge.
Schmiedt, H., Jensen, P., Schlemmer, S., 2016. Collective molecular superrotation: a model for extremely flexible molecules applied to protonated methane. Phys. Rev. Lett. 117, 223002.
Schmiedt, H., Jensen, P., Schlemmer, S., 2017a. The role of angular momentum in the superrotor theory for rovibrational motion of extremely flexible molecules. J. Mol. Spectrosc. 342, 132–137.
Schmiedt, H., Jensen, P., Schlemmer, S., 2017b. Rotation-vibration motion of extremely flexible molecules – the molecular superrotor. Chem. Phys. Lett. 672, 34–46.
Schmiedt, H., Schlemmer, S., Jensen, P., 2015. Symmetry of extremely floppy molecules: molecular states beyond rotation-vibration separation. J. Chem. Phys. 143, 154302.
Schnell, M., 2011. Group theory for high-resolution spectroscopy of nonrigid molecules. In: Quack, M., Merkt, F. (Eds.), Handbook of High-Resolution Spectroscopy, Vol. 1. John Wiley & Sons, Ltd, pp. 607–632.
Schreiner, P.R., Kim, S.-J., Schaefer III, H.F., von Ragué Schleyer, P., 1993. CH_5^+: the never-ending story or the final word? J. Chem. Phys. 99, 3716–3720.
Schreiner, P.R., Reisenauer, H.P., Pickard, F.C., Simmonett, A.C., Allen, W.D., Mátyus, E., Császár, A.G., 2008. Capture of hydroxymethylene and its fast disappearance through tunnelling. Nature 453, 906–909.
Schuh, K., Rosenow, P., Kolesik, M., Wright, E.M., Koch, S.W., Moloney, J.V., 2017. Nonlinear rovibrational polarization response of water vapor to ultrashort long-wave infrared pulses. Phys. Rev. A 96, 043818.
Schutz, B.F., 1980. Geometrical Methods of Mathematical Physics. Cambridge University Press.
Schwenke, D.W., 1992. On the computation of ro-vibrational energy levels of triatomic molecules. Comput. Phys. Commun. 70, 1–14.
Schwenke, D.W., 1996. Variational calculations of rovibrational energy levels and transition intensities for tetratomic molecules. J. Phys. Chem. 100, 2867–2884.
Seidler, P., Hansen, M.B., Gyorffy, W., Toffoli, D., Christiansen, O., 2010. Vibrational absorption spectra calculated from vibrational configuration interaction response theory using the Lanczos method. J. Chem. Phys. 132, 164105.
Shipsey, E.J., 1998. Rotationless molecular vibrations. J. Chem. Phys. 89, 5473–5485.
Shirin, S.V., Zobov, N.F., Polyansky, O.L., Tennyson, J., 2008. High accuracy potential energy surfaces for the $H_2{}^{16}O$, $H_2{}^{17}O$ and $H_2{}^{18}O$ molecules. J. Chem. Phys. 128, 224306.
Shizgal, B., Blackmore, R., 1984. A discrete ordinate method of solution of linear boundary value and eigenvalue problems. J. Comput. Phys. 55, 313–327.
Sibaev, M., Crittenden, D.L., 2016. Balancing accuracy and efficiency in selecting vibrational configuration interaction basis states using vibrational perturbation theory. J. Chem. Phys. 145, 064106.
Sibert III, E.L., 1988. Theoretical studies of vibrationally excited polyatomic molecules using canonical Van Vleck perturbation theory. J. Chem. Phys. 88, 4378–4390.
Simkó, I., Furtenbacher, T., Dénes, N., Szidarovszky, T., Hrubý, J., Zobov, N.F., Polyansky, O.L., Tennyson, J., Császár, A.G., 2017. Recommended ideal-gas thermochemical functions for heavy water and its substituent isotopologues. J. Phys. Chem. Ref. Data 46, 023104.
Simkó, I., Szidarovszky, T., Császár, A.G., 2019. Toward automated variational computation of rovibrational resonances. A case study of the H_2 dimer. J. Chem. Theory Comput. 15 (7), 4156–4169.
Simon, H.D., 1984. The Lanczos algorithm with partial reorthogonalization. Math. Comp. 42, 115–142.
Simons, J., 1981. Resonance state lifetimes from stabilization graphs. J. Chem. Phys. 75, 2465–2466.
Skokov, S., Bowman, J., Mandelshtam, V., 1999. Calculation of resonance states of non-rotating HOCl using an accurate ab initio potential. Phys. Chem. Chem. Phys. 1 (6), 1279–1282.
Smith, F.T., 1959. Participation of vibration in exchange reactions. J. Chem. Phys. 31, 1352–1358.
Smith, F.T., 1969. Diabatic and adiabatic representations for atomic collision problems. Phys. Rev. 179, 111–123.
Smith, F.T., 1980. Modified heliocentric coordinates for particle dynamics. Phys. Rev. Lett. 45, 1157–1160.
Šmydke, J., Császár, A.G., 2019. On the use of reduced density matrices for the semi-automatic assignment of vibrational states. Mol. Phys. 117 (13), 1682–1693.
Šmydke, J., Fábri, C., Sarka, J., Császár, A.G., 2019. Rovibrational quantum dynamics of the vinyl radical and its deuterated isotopologues. Phys. Chem. Chem. Phys. 21, 3453–3472.
Song, L., Gao, J., 2008. On the construction of diabatic and adiabatic potential energy surfaces based on ab initio valence bond theory. J. Phys. Chem. A 112, 12925–12935.

Sorensen, D.S., 1992. Implicit application of polynomial filters in a K-step Arnoldi method. SIAM J. Matrix Anal. Appl. 13, 357.

Sørensen, G.O., 1979. A new approach to the Hamiltonian of nonrigid molecules. Top. Curr. Chem. 82, 99.

Špirko, V., 1983. Vibrational anharmonicity and the inversion potential function of NH_3. J. Mol. Spectrosc. 101, 30–47.

Stanke, M., Kedziera, D., Molski, M., Bubin, S., Barysz, M., Adamowicz, L., 2006. Convergence of experiment and theory on the pure vibrational spectrum of HeH^+. Phys. Rev. Lett. 96, 233002.

Stefanski, K., Taylor, H.S., 1985. New approach to understanding quasiperiodicity in nonintegrable Hamiltonian systems. Phys. Rev. A 31, 2810–2820.

Stoecklin, T., Voronin, A., Rayez, J.C., 2002. Vibrational quenching of $N_2(v = 1, j_{rot} = j)$ by ^3He: surface and close-coupling calculations at very low energy. Phys. Rev. 66, 042703.

Strauss, H.L., Pickett, H.M., 1970. Conformational structure, energy, and inversion rates of cyclohexane and some related oxanes. J. Am. Chem. Soc. 92, 7281–7290.

Strobusch, D., Scheurer, C., 2011a. The hierarchical expansion of the kinetic energy operator in curvilinear coordinates extended to the vibrational configuration interaction method. J. Chem. Phys. 135, 144101.

Strobusch, D., Scheurer, C., 2011b. Hierarchical expansion of the kinetic energy operator in curvilinear coordinates for the vibrational self-consistent field method. J. Chem. Phys. 135, 124102.

Suarez, J., Farantos, S., Stamatiadis, S., Lathouwers, L., 2009. A method for solving the molecular Schrödinger equation in Cartesian coordinates via angular momentum projection operators. Comput. Phys. Commun. 180, 2025–2033.

Sutcliffe, B.T., Tennyson, J., 1991. A general treatment of vibration-rotation coordinates for triatomic molecules. Int. J. Quant. Chem. 39, 183–196.

Szalay, V., 1988. Derivation of the nonrigid rotation–large-amplitude internal motion Hamiltonian of the general molecule. J. Mol. Spectrosc. 128, 24–61.

Szalay, V., 1993. Discrete variable representations of differential operators. J. Chem. Phys. 99, 1978–1984.

Szalay, V., 1996. The generalized discrete variable representation. An optimal design. J. Chem. Phys. 105, 6940–6956.

Szalay, V., 2014. Eckart–Sayvetz conditions revisited. J. Chem. Phys. 140, 234107.

Szalay, V., 2015a. Aspects of the Eckart frame ro-vibrational kinetic energy operator. J. Chem. Phys. 143, 064104.

Szalay, V., 2015b. Understanding nuclear motions in molecules: derivation of Eckart frame ro-vibrational Hamiltonian operators via a gateway Hamiltonian operator. J. Chem. Phys. 142, 174107.

Szalay, V., 2017. Eckart ro-vibrational Hamiltonians via the gateway Hamilton operator: theory and practice. J. Chem. Phys. 146, 124107.

Szalay, V., Czakó, G., Nagy, A., Furtenbacher, T., Császár, A.G., 2003. On one-dimensional discrete variable representations with general basis functions. J. Chem. Phys. 119, 10512–10518.

Szalay, V., Szidarovszky, T., Czakó, G., Császár, A.G., 2012. A paradox of grid-based representation techniques: accurate eigenvalues from inaccurate matrix elements. J. Math. Chem. 50, 636–651.

Szidarovszky, T., Császár, A.G., 2013. Low-lying quasibound rovibrational states of $H_2{}^{16}O$. Mol. Phys. 111, 2131–2146.

Szidarovszky, T., Császár, A.G., Czakó, G., 2010. On the efficiency of treating singularities in triatomic variational vibrational computations. The vibrational states of H_3^+ up to dissociation. Phys. Chem. Chem. Phys. 12, 8373–8386.

Szidarovszky, T., Fábri, C., Császár, A.G., 2012. The role of axis embedding on rigid rotor decomposition (RRD) analysis of variational rovibrational wave functions. J. Chem. Phys. 136, 174112.

Szidarovszky, T., Yamanouchi, K., 2017. Full-dimensional simulation of the laser-induced alignment dynamics of H_2He^+. Mol. Phys. 115, 1916–1926.

Szidarovszky, T., Yamanouchi, K., 2018. Laser-Induced Alignment and Orientation Dynamics Beyond the Rigid-Rotor Approximation. In: Progress in Ultrafast and Intense Laser Science XIV. Springer Nature Switzerland AG, pp. 17–36.

Tarczay, G., Császár, A.G., Klopper, W., Quiney, H.M., 2001. Anatomy of relativistic energy corrections in light molecular systems. Mol. Phys. 99, 1769–1794.

Tennyson, J., 2016. Perspective: accurate ro-vibrational calculations on small molecules. J. Chem. Phys. 145, 120901.

Tennyson, J., Barletta, P., Munro, J.J., Silva, B.C., 2006. The role of asymptotic vibrational states in H_3^+. Philos. Trans. R. Soc. Lond. Ser. A 364, 2903–2916.

Tennyson, J., Bernath, P.F., Brown, L.R., Campargue, A., Carleer, M.R., Császár, A.G., Gamache, R.R., Hodges, J.T., Jenouvrier, A., Naumenko, O.V., Polyansky, O.L., Rothman, L.S., Toth, R.A., Vandaele, A.C., Zobov, N.F., Daumont, L., Fazliev, A.Z., Furtenbacher, T., Gordon, I.F., Mikhailenko, S.N., Shirin, S.V., 2009. Critical evaluation of the rotational-vibrational spectra of water vapor. Part I. Energy levels and transition wavenumbers for $H_2{}^{17}O$ and $H_2{}^{18}O$. J. Quant. Spectrosc. Radiat. Transf. 110, 573–596.

Tennyson, J., Bernath, P.F., Brown, L.R., Campargue, A., Carleer, M.R., Császár, A.G., Gamache, R.R., Hodges, J.T., Jenouvrier, A., Naumenko, O.V., Polyansky, O.L., Rothman, L.S., Toth, R.A., Vandaele, A.C., Zobov, N.F., Fazliev, A.Z., Furtenbacher, T., Gordon, I.F., Hu, S.-M., Mikhailenko, S.N., Voronin, B., 2010. Critical evaluation of the rotational-vibrational spectra of water vapor. Part II. Energy levels and transition wavenumbers for $HD^{16}O$, $HD^{17}O$, and $HD^{18}O$. J. Quant. Spectrosc. Radiat. Transf. 110, 2160–2184.

Tennyson, J., Bernath, P.F., Brown, L.R., Campargue, A., Császár, A.G., Daumont, L., Gamache, R.R., Hodges, J.T., Naumenko, O.V., Polyansky, O.L., Rothman, L.S., Vandaele, A.C., Zobov, N.F., Dénes, N., Fazliev, A.Z., Furtenbacher, T., Gordon, I.E., Hu, S.-M., Szidarovszky, T., Vasilenko, I.A., 2014a. IUPAC critical evaluation of the rotational-vibrational spectra of water vapor. Part IV. Energy levels and transition wavenumbers for $D_2{}^{16}O$, $D_2{}^{17}O$ and $D_2{}^{18}O$. J. Quant. Spectrosc. Radiat. Transf. 142, 93–108.

Tennyson, J., Bernath, P.F., Brown, L.R., Campargue, A., Császár, A.G., Daumont, L., Gamache, R.R., Hodges, J.T., Naumenko, O.V., Polyansky, O.L., Rothman, L.S., Vandaele, A.C., Zobov, N.F., 2014b. A database of water transitions from experiment and theory (IUPAC technical report). Pure Appl. Chem. 86, 71–83.

Tennyson, J., Bernath, P.F., Brown, L.R., Campargue, A., Császár, A.G., Daumont, L., Gamache, R.R., Hodges, J.T., Naumenko, O.V., Polyansky, O.L., Rothman, L.S., Vandaele, A.C., Zobov, N.F., Al Derzi, A.R., Fábri, C., Fazliev, A.Z., Furtenbacher, T., Gordon, I.E., Lodi, L., Mizus, I.I., 2013. IUPAC critical evaluation of the rotational-vibrational spectra of water vapor. Part III: Energy levels and transition wavenumbers for $H_2^{16}O$. J. Quant. Spectrosc. Radiat. Transf. 117, 29–58.

Tennyson, J., Kostin, M.A., Barletta, P., Harris, G.J., Polyansky, O.L., Ramanlal, J., Zobov, N.F., 2004. DVR3D: a program suite for the calculation of rotation-vibration spectra of triatomic molecules. Comput. Phys. Commun. 163, 85–116.

Tennyson, J., Sutcliffe, B.T., 1982. The *ab initio* calculation of the vibrational-rotational spectrum of triatomic systems in the close-coupling approach, with KCN and H_2Ne as examples. J. Chem. Phys. 77 (8), 4061–4072.

Tew, D.P., Handy, N.C., Carter, S., 2001. Glyoxal studied with multimode, explicit large amplitude motion and anharmonicity. Phys. Chem. Chem. Phys. 3, 1958–1964.

Thomsen, B., Yagi, K., Christiansen, O., 2014. Optimized coordinates in vibrational coupled cluster calculations. J. Chem. Phys. 140, 154102.

Tóbiás, R., Furtenbacher, T., Simkó, I., Császár, A.G., Diouf, M.L., Cozijn, F.M.J., Staa, J.M.A., Salimbides, E.J., Ubachs, W., 2020. Spectroscopic-network-assisted precision spectroscopy and its application to water. Nat. Commun. 11, 1708.

Tremblay, J.C., Carrington, T., 2005. Computing resonance energies, widths, and wave functions using a Lanczos method in real arithmetic. J. Chem. Phys. 122, 24.

Valeev, E.F., Allen, W.D., Schaefer III, H.F., Császár, A.G., 2001. The second-order Møller–Plesset limit for the barrier to linearity of water. J. Chem. Phys. 114, 2875–2878.

van Mourik, T., Harris, G.J., Polyansky, O.L., Tennyson, J., Császár, A.G., Knowles, P.J., 2001. *Ab initio* global potential, dipole, adiabatic and relativistic correction surfaces for the HCN/HNC system. J. Chem. Phys. 115, 3706–3718.

Vibók, A., Balint-Kurti, G.G., 1992. Reflection and transmission of waves by a complex potential — a semiclassical Jeffreys–Wentzel–Kramers–Brillouin treatment. J. Chem. Phys. 96, 7615–7622.

Wang, L.-W., Zunger, A., 1994. Solving Schrödinger's equation around a desired energy: application to silicon quantum dots. J. Chem. Phys. 100, 2394–2397.

Wang, X.-G., Carrington Jr., T., 2001. A symmetry-adapted Lanczos method for calculating energy levels with different symmetries from a single set of iterations. J. Chem. Phys. 114, 1473–1477.

Wang, X.-G., Carrington Jr., T., 2002. New ideas for using contracted basis functions with a Lanczos eigensolver for computing vibrational spectra of molecules with four or more atoms. J. Chem. Phys. 117, 6923–6934.

Wang, X.-G., Carrington Jr., T., 2003a. A contracted basis-Lanczos calculation of vibrational levels of methane: solving the Schrödinger equation in nine dimensions. J. Chem. Phys. 119, 101–117.

Wang, X.-G., Carrington Jr., T., 2003b. Using C_{3v} symmetry with polyspherical coordinates for methane. J. Chem. Phys. 119, 94–100.

Wang, X.-G., Carrington Jr., T., 2004. Contracted basis Lanczos methods for computing numerically exact rovibrational levels of methane. J. Chem. Phys. 121, 2937–2954.

Wang, X.-G., Carrington Jr., T., 2005. Improving the calculation of rovibrational spectra of five-atom molecules with three identical atoms by using a $C_{3v}(G_6)$ symmetry-adapted grid: applied to CH_3D and CHD_3. J. Chem. Phys. 123, 154303.

Wang, X.-G., Carrington Jr., T., 2008. Vibrational energy levels of CH_5^+. J. Chem. Phys. 129, 234102.

Wang, X.-G., Carrington Jr., T., 2013. Computing rovibrational levels of methane with curvilinear internal vibrational coordinates and an Eckart frame. J. Chem. Phys. 138, 104106.

Wang, X.-G., Carrington Jr., T., 2016. Calculated rotation-bending energy levels of CH_5^+ and a comparison with experiment. J. Chem. Phys. 144, 204304.

Wang, X.-G., Carrington Jr., T., Dawes, R., Jasper, A.W., 2011. The vibration-rotation-tunneling spectrum of the polar and T-shaped-N-in isomers of $(NNO)_2$. J. Mol. Spectrosc. 268, 53–65.

Wang, X.-G., Carrington, T., 2003. A finite basis representation Lanczos calculation of the bend energy levels of methane. J. Chem. Phys. 118, 6946–6956.

Watson, J.K.G., 1968. Simplification of the molecular vibration-rotation Hamiltonian. Mol. Phys. 15, 479–490.

Watson, J.K.G., 1970. The vibration-rotation Hamiltonian of linear molecules. Mol. Phys. 19, 465–487.

Watson, J.K.G., 2004. The molecular vibration-rotation kinetic-energy operator for general internal coordinates. J. Mol. Spectrosc. 228, 645–658.

Wei, H., 2003a. An Eckart-frame kinetic energy operator for tetra-atomic planar molecules. J. Chem. Phys. 118, 7208–7214.

Wei, H., 2003b. Eckart frames for planar molecules. J. Chem. Phys. 118, 7202–7207.

Wei, H., Carrington Jr., T., 1992. The discrete variable representation of a triatomic Hamiltonian in bond length–bond angle coordinates. J. Chem. Phys. 97, 3029–3037.

Wei, H., Carrington Jr., T., 1997a. Explicit expressions for triatomic Eckart frames in Jacobi, Radau, and bond coordinates. J. Chem. Phys. 107, 2813–2818.

Wei, H., Carrington Jr., T., 1998. An exact Eckart-embedded kinetic energy operator in Radau coordinates for triatomic molecules. Chem. Phys. Lett. 287, 289–300.

Wei, H., Carrington, T., 1997b. The triatomic Eckart-frame kinetic energy operator in bond coordinates. J. Chem. Phys. 107, 9493–9501.

White, E.T., Tang, J., Oka, T., 1999. CH_5^+: the infrared spectrum observed. Science 284 (5411), 135–137.

Wigner, E.P., 1959. Group Theory and Its Application to the Quantum Mechanics of Atomic Spectra. Academic Press, New York.

Wilson Jr., E.B., Decius, J.C., Cross, P.C., 1955. Molecular Vibrations. McGraw-Hill, New York.

Wodraszka, R., Manthe, U., 2015. CH_5^+: symmetry and the entangled rovibrational quantum states of a fluxional molecule. J. Phys. Chem. Lett. 6, 4229–4232.

Wu, K., Canning, A., Simon, H.D., Wang, L.-W., 1999. Thick-restart Lanczos method for electronic structure calculations. J. Comput. Phys. 154, 156–173.

Wu, K., Simon, H.D., 1998. Thick-restart Lanczos method for the symmetric eigenvalue problems. Lawrence Berkeley National Laboratory Report No. 41412.

Wyatt, R.E., 1995. Matrix spectroscopy: computation of interior eigenstates of large matrices using layered iteration. Phys. Rev. E 51, 3643–3658.

Wyatt, R.E., Zhang, J.Z.H., 1996. Dynamics of Molecules and Chemical Reactions. Marcel Dekker, New York.

Xu, M., Sebastianelli, F., Bačić, Z., 2009a. Coupled translation-rotation eigenstates of H_2, HD, and D_2 in the large cage of structure II clathrate hydrate: comparison with the small cage and rotational Raman spectroscopy. J. Phys. Chem. A 113, 7601–7609.

Xu, M., Sebastianelli, F., Bačić, Z., Lawler, R., Turro, N.J., 2008a. H_2, HD, and D_2 inside C_{60}: coupled translation-rotation eigenstates of the endohedral molecules from quantum five-dimensional calculations. J. Chem. Phys. 129, 064313.

Xu, M., Sebastianelli, F., Bačić, Z., Lawler, R., Turro, N.J., 2008b. Quantum dynamics of coupled translational and rotational motions of H_2 inside C_{60}. J. Chem. Phys. 128, 011101.

Xu, M., Sebastianelli, F., Gibbons, B.R., Bačić, Z., Lawler, R., Turro, N.J., 2009b. Coupled translation-rotation eigenstates of H_2 in C_{60} and C_{70} on the spectroscopically optimized interaction potential: effects of cage anisotropy on the energy level structure and assignments. J. Chem. Phys. 130, 224306.

Xu, M., Ye, S., Bačić, Z., 2015. General selection rule in the inelastic neutron scattering spectroscopy of a diatomic molecule confined inside a near-spherical nanocavity. J. Phys. Chem. Lett. 6, 3721–3725.

Xu, M., Ye, S., Powers, A., Lawler, R., Turro, N.J., Bačić, Z., 2013. Inelastic neutron scattering spectrum of $H_2{\cdot}C_{60}$ and its temperature dependence decoded using rigorous quantum calculations and a new selection rule. J. Chem. Phys. 139, 064309.

Yachmenev, A., Yurchenko, S.N., 2015. Automatic differentiation method for numerical construction of the rotational-vibrational Hamiltonian as a power series in the curvilinear internal coordinates using the Eckart frame. J. Chem. Phys. 143, 014105.

Yagi, K., Keceli, M., Hirata, S., 2012. Optimized coordinates for anharmonic vibrational structure theories. J. Chem. Phys. 137, 204118.

Yu, H., 2002. Two-layer Lanczos iteration approach to molecular spectroscopic calculation. J. Chem. Phys. 117, 8190–8196.

Yu, H., 2004a. Converged quantum dynamics calculations of vibrational energies of CH_4 and CH_3D using an ab initio potential. J. Chem. Phys. 121, 6334–6340.

Yu, H., 2004b. Full-dimensional quantum calculations of vibrational spectra of six-atom molecules. I. Theory and numerical results. J. Chem. Phys. 120, 2270–2284.

Yu, H., 2006. A rigorous full-dimensional quantum dynamics calculation of the vibrational energies of $H_3O_2^-$. J. Chem. Phys. 125, 204306.

Yu, H.-G., 2005. A coherent discrete variable representation method for multidimensional systems in physics. J. Chem. Phys. 122, 164107.

Yu, H.-G., Nyman, G., 1998. A spectral transform Krylov subspace iteration approach to quantum scattering. Chem. Phys. Lett. 298, 27–35.

Yu, H.-G., Nyman, G., 1999. A four dimensional quantum scattering study of the Cl + CH_4 → HCl + CH_3 reaction via spectral transform iteration. J. Chem. Phys. 110, 7233–7244.

Yurchenko, S.N., Carvajal, M., Jensen, P., Lin, H., Zheng, J., Thiel, W., 2005. Rotation-vibration motion of pyramidal XY_3 molecules described in the Eckart frame: theory and application to NH_3. Mol. Phys. 103, 359–378.

Yurchenko, S.N., Thiel, W., Jensen, P., 2007. Theoretical Rovibrational energies (TROVE): a robust numerical approach to the calculation of rovibrational energies for polyatomic molecules. J. Mol. Spectrosc. 245, 126–140.

Yurchenko, S.N., Yachmenev, A., Ovsyannikov, R.I., 2017. Symmetry-adapted ro-vibrational basis functions for variational nuclear motion calculations: TROVE approach. J. Chem. Theory Comput. 13, 4368–4381.

Zanuttini, D., Gatti, F., Marquardt, R., 2018. CO quantum dynamics diffusion on Cu(100). Chem. Phys. 509, 3–12.

Zare, R.N., 1988. Angular Momentum: Understanding Spatial Aspects in Chemistry and Physics. Wiley-Interscience, New York.

Zhang, J.Z.H., 1999. Theory and Application of Quantum Molecular Dynamics. World Scientific, Singapore.

Ziegler, B., Rauhut, G., 2016. Efficient generation of sum-of-products representations of high-dimensional potential energy surfaces based on multimode expansions. J. Chem. Phys. 144, 114114.

Zobov, N.F., Polyansky, O.L., Le Sueur, C.R., Tennyson, J., 1996. Vibration-rotation levels of water beyond the Born–Oppenheimer approximation. Chem. Phys. Lett. 260, 381–387.

Zobov, N.F., Shirin, S.V., Lodi, L., Silva, B.C., Tennyson, J., Császár, A.G., Polyansky, O.L., 2011. First-principles rotation-vibration spectrum of water above dissociation. Chem. Phys. Lett. 507, 48–51.

Zobov, N.F., Shiron, S.V., Polyansky, O.L., Tennyson, J., Coheur, P.-F., Bernath, P.F., Carleer, M., Colin, R., 2005. Monodromy in the water molecules. Chem. Phys. Lett. 414, 193–197.

CHAPTER 3

2D Strong-Field Spectroscopy to Elucidate Impulsive and Adiabatic Ultrafast Electronic Control Schemes in Molecules

HENDRIKE BRAUN* • TIM BAYER† • MATTHIAS WOLLENHAUPT† • THOMAS BAUMERT*
*University of Kassel, Institute of Physics, Kassel, Germany
†Carl von Ossietzky University of Oldenburg, Institute of Physics, Oldenburg, Germany

Abstract

We present a multidimensional spectroscopy approach based on strong-field coherent electronic excitation to elucidate intramolecular dynamics and unveil control mechanisms. Traditional 2D spectroscopy generally relies on weak-field interactions and the well-known control mechanism of spectral interference (SI) with its characteristic I^n-scaling of transition probabilities (I, laser intensity; n, photonic order). For non-perturbative strong-field interactions, completely different scenarios become relevant. Especially, electronic coherences play a significant role to control the population dynamics and the bidirectional Stark shifts. To that end, we study the potassium dimer prototype both experimentally and with the help of quantum dynamics simulations. We employ two distinct types of shaped femtosecond laser pulses that induce either impulsive or adiabatic excitation of atoms and molecules – both directly related to coherent strong-field interaction.

3.1 INTRODUCTION

Exploiting coherence properties of laser light together with quantum mechanical matter interferences in order to steer a chemical reaction into a predefined target channel is the basis of coherent control (see Shapiro and Brumer, 2011; Faraday Discussion vol. 153, 2011 and references therein). The strong-field interaction regime is of particular interest, as it allows for efficient and robust population transfer. "No field no yield" is the simplified phrase for this statement. This is the regime beyond perturbative descriptions of light–matter interactions and below the regime where the ionization probability reaches unity. It is the regime of resonant and not too far off-resonant excitations, i.e. the regime where Rabi cycling is important (Shore, 2011), see Chapter 1 of this book (Marquardt and Quack, 2020).

Using laser pulses short in comparison to nuclear time scales in molecules, the coupling between electronic and nuclear degrees of freedom comes into play. The control of this coupled electron-nuclear dynamics is a fascinating perspective. Without nuclear dynamics pure coherent control of electronic excitation is best understood with the help of a resonantly coupled two-state system. Two extreme scenarios are attractive. The first scenario is the analogue to the NMR spin locking pulse sequence (Hartmann and Hahn, 1962). In the optical regime, a first laser pulse creates a state of maximum electronic coherence, i.e. a charge oscillation with maximized induced electronic dipole moment. During this interaction, the oscillating electric field of the driving laser pulse is out of phase by 90° with respect to the induced dipole. In this configuration, no interaction energy arises as the interaction energy of an electric dipole in an external electric field is given by the (negative) scalar product of both quantities (cf. Eq. (3.1)). This preparation step is followed by a control pulse whose electric field is in-phase or in anti-phase with the induced dipole. As a result, the interaction energy is decreased or increased, respectively. In terms of a dressed state picture, the first laser pulse prepares two dressed states with equal population, which can also be interpreted as the first order AC Stark splitting. The second laser pulse selects one of these dressed states (one AC Stark component) and – depending on the electric field strength of the second laser pulse – shifts this Stark component unidirectionally. This control scenario was coined selective population of dressed states

Molecular Spectroscopy and Quantum Dynamics. https://doi.org/10.1016/B978-0-12-817234-6.00008-8

(SPODS) and is reviewed in Bayer et al. (2014). The second scenario is the well-known rapid adiabatic passage (RAP) where a laser pulse with sufficient pulse area is frequency swept over a resonance resulting in complete and adiabatic population transfer being robust to typical experimental imperfections such as laser intensity fluctuations and focal volume effects. The question that arises is: What would be an appropriate approach to elucidate the contribution of these fundamental mechanisms in molecules – taking into account the much richer electronic structure, compared to the simple two-level atom model, and in addition taking into account the coupled electronic and nuclear motion? Here, multidimensional spectroscopy comes into play which was described in a recent review by T. Brixner and coworkers (Nuernberger et al., 2015) in general terms as follows: "Spectroscopic approaches that map out a characteristic signal not only as a function of one external parameter but rather of several ones generally provide more insight into the underlying physics. As a main benefit, one can often separate contributions that would be overlapping in a one-dimensional measurement. The power of multidimensional techniques lies in the disclosure of a connection between an initially excited and a successively probed behavior." This statement holds true for a plethora of multidimensional spectroscopy techniques, where usually higher-order perturbation theory is behind the experimental implementations. It is noteworthy that very recently 2D spectroscopy was extended to electronic mass spectrometry that allows 2D measurements on effusive molecular beams and thus on quantum systems with minimum system–bath interaction (Roeding and Brixner, 2018). In the spirit of this general concept, we extend multidimensional spectroscopy to strong-field interactions and the next question to be addressed is: What would be a reasonable choice for the external parameters? The SPODS approach is based on pulse sequences with adjustable relative intensities, temporal separations and most importantly precisely adjustable temporal phase discontinuities. To allow for impulsive excitation, the pulses should be short compared to the molecular dynamics which are typically in the few 10 fs to few 100 fs range. The induced charge oscillation is determined by the Bohr frequency. For typical valence bond excitations in the visible (VIS) to ultraviolet (UV) spectral range, the Bohr oscillation period is in the several 100 as to few fs regime. Thus the temporal phase precision has to be in the 10 as regime, where experimentally a phase precision down to the zeptosecond regime has been demonstrated (Köhler et al., 2011). Using optical Fourier transform pulse shaping techniques (Brixner and Gerber, 2001; Weiner, 2000; Wollenhaupt et al., 2012), the above-mentioned parameters are conveniently obtained with the help of sinusoidal phase modulation in the frequency domain (Herek et al., 2002; Meshulach and Silberberg, 1998; Wollenhaupt et al., 2006b). Important natural pulse parameters to study RAP-type strong-field effects in order to induce adiabatic population transfer are frequency sweeps and time varying intensity profiles. Again, with the help of optical Fourier transform pulse shaping, these attributes are obtained by implementing group delay dispersion (GDD) and third order dispersion (TOD) in the frequency domain (Wollenhaupt et al., 2012) and the resulting temporal pulse shapes are chirped Airy pulses (Ruge et al., 2013; Schneider et al., 2011). These approaches have already been tested on different photophysical systems (Bayer et al., 2013; Schneider et al., 2011). So far, however, a conclusive test on a prototype system that is accessible to time-dependent quantum mechanical simulations and at the same time to experiments was missing. This gap is filled by this contribution.

The chapter is structured as follows. In Section 3.2, we discuss control of coupled electron-nuclear dynamics in the potassium dimer prototype, i.e. the SPODS mechanism extended to molecular excitations. In Section 3.3, adiabatic control scenarios in molecules based on chirped Airy pulses are presented. In both sections the experimental results are compared to quantum dynamics simulations to analyze the underlying physical mechanisms. We use the unit ångström 1 Å = 100 pm for lengths.

3.2 CONTROL OF COUPLED ELECTRON-NUCLEAR DYNAMICS IN THE POTASSIUM MOLECULE

The steering of photochemical processes especially in large, biologically relevant molecules is one of the big perspectives of coherent control (Prokhorenko et al., 2006; Zewail, 2000a). To that end, tailored laser fields are used to steer the system from an initial state via a laser-induced process into a predefined target state. Processes such as the fragmentation or the isomerization of molecules are based on the motion of the nuclei in the molecule and take place on a time scale of few to several hundred femtoseconds (Remacle and Levine, 2006; Zewail, 2000b). The result is the formation, rearrangement or cleavage of chemical bonds, which ultimately are constituted by the valence electrons. Due to the coupling of electronic and nuclear degrees of freedom, coherent control of valence electron excitation by specifically tailored laser fields can be used to initiate these processes and, at the same time, guide the ensuing nuclear dynamics (von den Hoff et al., 2012; Thallmair et al., 2014). Both femtosecond and attosecond laser pulses are routinely used to address valence electrons. The latter are characterized by their extremely short duration and allow for the direct observation of electron dynamics (Corkum and Krausz, 2007; Kling and Vrakking, 2008; Krausz and Ivanov, 2009; Remacle and Levine, 2006). However, due to the large photon energies of extreme UV or X-ray pulses, the cross-sections for valence bond excitation, either direct or via Raman-type processes, are comparatively small. Theoretical investigations found a population transfer of up to 13% for resonance supported excitation pathways via the continuum (Greenman et al., 2015). Attosecond pulses are more likely to trigger direct ionization of the molecule. Femtosecond laser pulses in the near infrared or visible spectral range are better suited for the steering of neutral systems and benefit from large transition moments. In combination with high pulse intensities the latter are a prerequisite for efficient population transfer. In addition, the intense, resonant laser fields allow the use of the dynamical Stark effect to manipulate the potential energy landscape by bidirectional energy shifts (Autler and Townes, 1955; Meier and Engel, 1994; Palacios et al., 2007; Wollenhaupt et al., 2006a; Wollenhaupt et al., 2003). This opens up new target channels that ensue specific nuclear dynamics. In the recent years, several mechanisms have been devised to control ultrafast electron dynamics in atoms. Here, we highlight the selective population of dressed states (SPODS), which relies on the initial preparation of the quantum system in a state of maximum electronic coherence and the subsequent tailored manipulation of the interaction energy between the driving laser field and the induced charge dynamics.

These concepts are to be transferred to molecular systems, to control the coupled electron-nuclear dynamics and achieve efficient and selective population of neutral target states. The coupling of the target states to the electronic coherence is non-perturbative.

The basic principle of SPODS is elucidated on a two-level system driven by an intense resonant ultrashort laser pulse $E_{\mathrm{mod}}(t)$. The expectation value of the interaction energy $\langle V \rangle (t)$ between the laser-induced dipole moment $\langle \mu \rangle (t)$ and the driving electric field depends on the relative phase and amplitude of the two oscillating quantities (Bayer et al., 2016):

$$\langle V \rangle (t) = - \langle \mu \rangle (t) \cdot E_{\mathrm{mod}}(t). \tag{3.1}$$

For maximum amplitude of the dipole oscillation, i.e. maximum coherence in the two-level system, the upper (lower) light-induced, i.e. dressed state is selectively populated when the interaction energy is maximized (minimized). In these cases, $\langle \mu \rangle (t)$ and $E_{\mathrm{mod}}(t)$ oscillate either out of phase (maximization of $\langle V \rangle (t)$) or in phase (minimization of $\langle V \rangle (t)$). Spectral phase modulation allows precisely tailoring the shape of the femtosecond laser pulse and, thereby, efficiently controlling the interplay between field and dipole in order to manipulate the populations of the light-induced states. Quantum dynamics simulations in multi-level systems show that the selective population of dressed states in a resonantly driven two-level system enables the selective excitation of neutral target states which are inaccessible in the weak-field regime (Bayer et al., 2016). Moreover, these simulations indicate selective switching between target states that are non-perturbatively coupled to the two-level system.

The coupling of electronic and nuclear degrees of freedom (see Marquardt and Quack, 2020, Chapter 1 of this book) is the basis for the control of nuclear dynamics. At the same time, they render the extension of the SPODS mechanism from atoms to molecules a challenge. Electronic excitations not only induce nuclear dynamics but are interdependently influenced by the motion of the nuclei (Thallmair et al., 2014). The amplitude of the induced electronic dipole – and therefore also the interaction energy – is determined by the overlap of the nuclear wave packets in the participating electronic states (see Eq. (3.3)). To achieve efficient control over the induced dynamics, a large overlap is desirable, however, it can change due to the propagation of the nuclear wave packets. Moreover, the dynamics lead to a change of the electronic resonance and thereby to a time-dependent phase of the induced dipole moment. The electric field of the laser pulse must adapt to the additional dynamics in order to maintain a predefined phase relation to the charge oscillation. Hence, it is not to be expected that a generic double pulse sequence, where the first pulse induces the oscillating dipole moment and the second pulse couples to the dipole with an appropriate phase relation, will achieve optimal control results. Instead, we use a sinusoidal spectral phase function in the experiments for coherent control of coupled electron-nuclear dynamics in molecules. Its parameters allow for extensive flexibility of the temporal phase and amplitude of the resulting laser pulse. The temporal phase influences the instantaneous frequency of the laser pulse as well as the relative phase of the electric field and the induced dipole moment. In the experiment, we chose the potassium dimer (K_2) as the molecular prototype system. This diatomic molecule is readily available in the measurements and can be de-

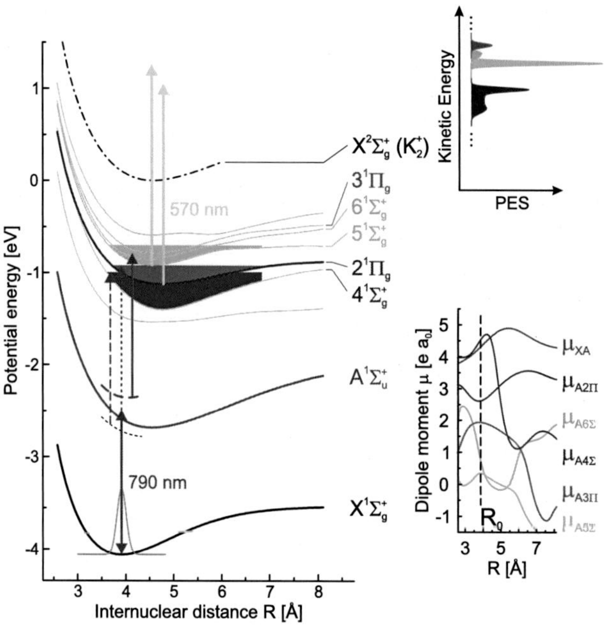

FIG. 3.1 Potential energy functions of the potassium dimer and the relevant dipole transition moments between the electronic states. When excited by a weak laser pulse the $2^1\Pi_g$-state gets populated via two-photon absorption from the ground state. For an intense resonant laser field the strong coupling between the $X^1\Sigma_g^+$- and the $A\Sigma_u^+$-state results in a dynamic Stark splitting. This energetic splitting is indicated by dashed lines above and below the $A^1\Sigma_u^+$ energy curve around the equilibrium distance R_0. Higher-lying states of the molecule, which cannot be populated in the perturbative weak field regime, are accessible from these light-induced potentials. In the experiment the detection of the population in the target channels after excitation by the control pulse is achieved by subsequent ionization of the molecule by a second laser pulse of 570 nm wavelength. A simulation of the emerging relevant signals is shown in the upper right inset.

scribed accurately by theory (von den Hoff et al., 2011; Petersen and Mitric, 2012; Wollenhaupt and Baumert, 2006). The experimental results discussed in this section are partially published in Bayer et al. (2016), Bayer et al. (2013), Braun et al. (2014), Braun et al. (2013).

3.2.1 The Model System K_2

The potassium dimer serves as a prototype molecule for both the experiments and the quantum dynamical simulations. The relevant potential energy curves and the corresponding electronic dipole couplings as a function of the internuclear distance R are shown in Fig. 3.1. The molecular data was provided by the group of Prof. Ch. Meier at the Université Paul Sabatier in Toulouse. The wave packet in the $X^1\Sigma_g^+$-state represents the molecule in its ground state, centered at the equilibrium distance $R_0 = 3.91$ Å. The first electronically excited state is the $A\Sigma_u^+$-state (red). The transition $X^1\Sigma_g^+ \rightarrow A\Sigma_u^+$ is resonant at about 845 nm. By interaction with a weak femtosecond laser pulse of central wavelength $\lambda_0 = 790$ nm the molecule is excited via the $A\Sigma_u^+$-state to the $2^1\Pi_g$-state (blue) by two-photon absorption. Higher lying states in the neutral molecule, such as the $5^1\Sigma_g^+$-, the $6^1\Sigma_g^+$- and the $3^1\Pi_g$-state (light green, green, dark green), as well as lower lying states such as the $4^1\Sigma_g^+$-state (dark blue), are energetically inaccessible in the perturbative limit. The dipole coupling between these states and the $A\Sigma_u^+$-state at R_0 are comparable to $\mu_{X\text{-}A}(R_0)$. Therefore, efficient population of these states is in principle possible.

In contrast to weak field excitation, interaction with an intense, resonant laser pulse (indicated by red arrows) creates an electronic coherence, i.e. wave packet in the X–A-subsystem. In a spatiotemporal picture, this electronic wave packet represents a coherent charge oscillation. In analogy to the atomic case, the light-induced potentials (LIPs) split up energetically due to the strong coupling of states $X^1\Sigma_g^+$ and $A\Sigma_u^+$. In Fig. 3.1

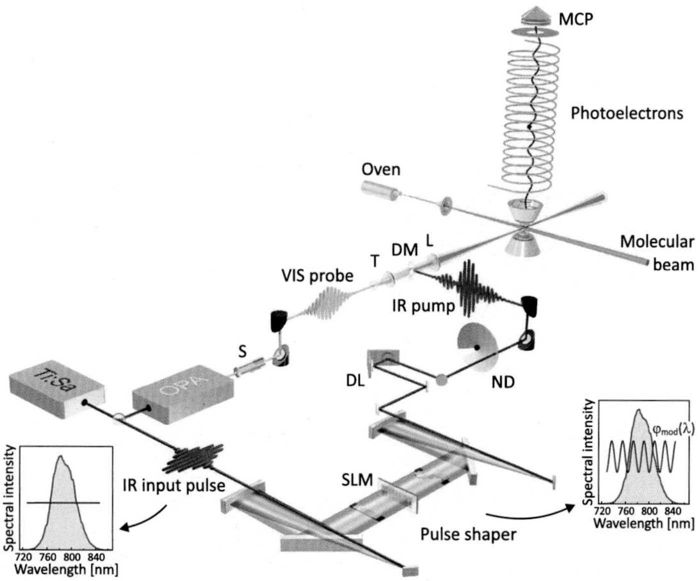

FIG. 3.2 Experimental two-color setup featuring an IR beamline, to generate intense shaped IR pump pulses, and a VIS probe beamline, to provide time-delayed probe pulses of a different color. Both beams are focussed collinearly into a supersonic beam to interact with isolated K atoms and molecules. Photoelectrons released during the interaction are measured by an energy-calibrated TOF-spectrometer. The following abbreviations are used: SLM, Spatial light modulator; DL, Delay line; ND, Continuous neutral density filter; L, Lens; S, Stretcher; T, Telescope; DM, Dichroic mirror; MCP, Multi-channel plate detector. Adapted from Bayer et al. (2014).

the upper (thick, dashed, red) and the lower (thin, dashed, black) LIP are represented by line segments around R_0. If the splitting is large enough, i.e. the driving laser pulse is sufficiently intense, additional neutral target states become accessible. For example, by absorption of a second photon, the molecule can be excited from the lower LIP to the $4^1\Sigma_g^+$-state. From the upper LIP the states $5^1\Sigma_g^+$, $6^1\Sigma_g^+$ and $3^1\Pi_g$ are accessible. If the molecule is excited by a resonant bandwidth-limited pulse (real pulse), the two LIPs are populated equally. As a consequence, the population transfer to the high-lying target states is also symmetric and no selectivity among the target states is attainable.

The aim of the control experiments described here is the selective and efficient population of neutral target channels in the potassium molecule. Because the bandwidth of the 790 nm, 25 fs laser pulse (IR pulse) is too large to address only a single state out of the dense-lying manifold of target states, we combine these states into two target channels: The lower target channel comprises the $4^1\Sigma_g^+$- and $2^1\Pi_g$-state (blue curves in Fig. 3.1), while the states $5^1\Sigma_g^+$, $6^1\Sigma_g^+$ and $3^1\Pi_g$ (green curves) constitute the upper target channel. Fig. 3.1 also indicates the photoionization of the molecule from the excited target states into the ionic ground state $X^2\Sigma_g^+(K_2^+)$ (yellow arrows). Experimentally, this ionization probe step is realized by a time-delayed second laser pulse with a central wavelength of 570 nm (VIS pulse). The generated photoelectrons map the vibrational wave packets excited by the IR pulse into the energy-resolved photoelectron spectrum. A simulated photoelectron spectrum is shown in the upper right inset to Fig. 3.1. The integrated photoelectron yield of these signals provides information on the final population of the target channels.

3.2.2 Experimental Two-Color Setup

Here we give a brief description of the apparatus and strategy used in the experiments presented below. Fig. 3.2 shows an overview of the experimental two-color setup. An amplified 1 kHz Ti:Sapphire laser system provides intense 790 nm, 25 fs FWHM laser pulses which are referred to as infrared (IR) input pulses. The IR input pulses are spectrally phase-modulated by a home-built Fourier-transform pulse shaper based on a computer-controlled Liquid Crystal Spatial Light Modulator (LC SLM) in the Fourier plane of a $4f$ zero dispersion compressor (Köhler et al., 2011; Präkelt et al., 2003; Wollenhaupt et al., 2009). Application of the desired phase mask $\varphi_{\text{mod}}(\omega)$ to the SLM yields shaped IR pulses at the shaper output. The shaped IR pulses are attenuated to pulse energies of about 0.1 to 2.0 μJ and focussed into the interaction region of a time-of-flight (TOF) photoelectron spectrometer. Here the laser beam intersects a supersonic potassium beam from an adjacent oven chamber. To promote the dimer formation, the particle beam is seeded with argon gas. Photoelectrons released during the strong-field interaction of shaped IR pulses with K_2 molecules (and K atoms) are collected by a magnetic bottle and detected using a multi-channel plate (MCP) detector. After calibration of the spectrometer using different narrowband laser sources and buffer gases we obtain energy-resolved photoelectron spectra which are studied as a function of the IR pulse shape, i.e. the applied phase mask $\varphi_{\text{mod}}(\omega)$.

In order to probe the population of neutral target states in K_2, excited by the IR pulse, we employ an Optical Parametric Amplifier (OPA) pumped by a fraction of the IR input pulse. The OPA is adjusted to produce visible (VIS) pulses with a central wavelength of 570 nm and a duration of 20 fs. This wavelength is specifically chosen to enable background-free detection of the relevant two-color photoelectron signals. The VIS probe pulses are stretched in time to about 2 ps, in order to avoid direct multi-photon ionization as a background signal. In addition, due to the temporal stretching of the VIS pulse, nuclear wave packet dynamics launched in the target states by the IR pulse are averaged out, which smoothes the resulting photoelectron signals. Subsequently, the probe beam is magnified by a telescope (magnification factor 2.7) and focussed tightly into the center of the IR laser focus. By this means we reduce averaging over the focal intensity distribution of the IR laser, which typically washes out strong-field effects. A delay line in the IR one-color beamline serves to adjust the time-delay between the IR pump and the VIS probe pulse. Both beams are combined by a dichroic mirror and focussed collinearly into the interaction region of the photoelectron spectrometer.

Prior to the experiments, the residual spectral phase of the IR pulse, introduced by dispersive elements in the IR beamline, is compensated by *in situ* adaptive optimization of the total photoelectron yield from multiphoton ionization of ground state xenon (Xe) atoms. Xe is supplied by an effusive gas inlet leading to the interaction region. The trial phase is parametrized by a polynomial function up to 5th order. In the experiments, the optimal result $\varphi_{\text{opt}}(\omega)$ is applied to the SLM in addition to the desired phase mask $\varphi_{\text{mod}}(\omega)$, providing compressed, i.e. bandwidth-limited, IR pulses for zero phase $\varphi_{\text{mod}}(\omega) \equiv 0$.

3.2.3 Molecular Dynamics Simulations

The discussion of the physical mechanism in Section 3.2.4.2 is based on numerical simulations of the molecular dynamics during the interaction with the shaped laser pulse. In these simulations the time-dependent Schrödinger-equation (TDSE) is solved numerically with the help of a grid-based split operator method, described in Chapter 1 of this book (Marquardt and Quack, 2020). The results are the time-dependent amplitudes and phases of the nuclear wave functions in the electronic states. Furthermore, energy-resolved photoelectron spectra and the evolution of expectation values of several operators, like the internuclear distance and the dipole moment, can be calculated. As a figure of merit, we assign each pulse with a scalar contrast value. In the simulations this value is calculated as the normalized difference of the final populations $P^{\infty}_{\text{up/low}}$ of the respective target channels:

$$C = \frac{P^{\infty}_{\text{up}} - P^{\infty}_{\text{low}}}{P^{\infty}_{\text{up}} + P^{\infty}_{\text{low}}}. \tag{3.2}$$

The contrast ranges between $[-1, 1]$, where $C = -1$ and $+1$ correspond to the selective population of the lower and upper target channel, respectively. In the experiments, the contrast is derived from the photoelectron yield measured in the target channel signals.

In order to reproduce the experimental conditions as accurately as possible, two types of averaging are taken into account in our numerical model. By calculating the photoelectron spectra for different IR pulse amplitudes, we account for the focal intensity distribution in the IR laser focus. By considering different angles between the internuclear axis and the IR laser polarization direction, we account for the interaction of the laser pulse with a randomly aligned molecular ensemble.

3.2.3.1 Calculation of the Induced Dipole Moment

The SPODS control scenario depends on the relative phase between the electric field of the driving laser pulse and the induced dipole moment in the X–A-subsystem. The time-dependent induced dipole moment is calculated according to

$$\begin{aligned} &\langle \mu_{\text{X-A}} \rangle (t) \\ &= \int_0^{\infty} \mu_{\text{XA}}(R) \cdot \left\{ \psi_{\text{A}}^{*}(R,t)\, \psi_{\text{X}}(R,t) + c.c. \right\} \mathrm{d}R. \end{aligned} \tag{3.3}$$

Here $\psi_X(R,t)$ and $\psi_A(R,t)$ are the nuclear wave functions of the XΣ- and the AΣ-state. The corresponding vibrational wave packets propagate during the interaction with the laser field in the potentials. This propagation leads to a time-dependent expectation value of the internuclear separation $\langle R\rangle(t) = \int_0^\infty |\psi(R,t)|^2 R dR$. Similar to the dipole moment $\langle \mu_{X-A}\rangle(t)$, $\langle R\rangle(t)$ is calculated for the X–A-subsystem. Because the wave packets remain well-localized, the analysis based on this single parameter is sufficient.

3.2.3.2 Intensity and Orientation Averaging

To extract a clear physical picture of the strong-field interaction of intense ultrashort pulses with molecules from the experimental data, the averaging over different excitation conditions needs to be taken into account. Because the laser focus typically exhibits a broad intensity distribution, the volume average over an ensemble of molecules exposed to the different intensities needs to be calculated. Especially for strong-field interaction scenarios that are sensitive to the laser intensity, such as Rabi cycling or photon locking, this inherent average has a strong influence on the measured spectra. In the simulations we assume the intensity distribution of a Gaussian beam propagating in z-direction (Siegmann, 1986)

$$I(r,z) = I_0 \frac{1}{1+(z/z_R)^2} \exp\left[-\frac{2r^2}{w_0^2\left\{1+(z/z_R)^2\right\}}\right], \tag{3.4}$$

with the peak intensity I_0 at $(r,z) = (0,0)$, the Rayleigh range z_R and the beam waist w_0. In the simulations, no significant excitation of the target channels was obtained for intensities smaller than $I = 0.05\, I_0$, where $I_0 \leq 1 \times 10^{12}$ W/cm^2 corresponds to the experimental conditions. Therefore, no lower intensities were included in the calculations.

Besides the average over the focal intensity distribution, the average over an ensemble of randomly aligned molecules distributed homogeneously over the laser focus needs to be considered. Due to the vectorial character of the laser–molecule interaction, i.e. the transition dipole moments, the interaction with the linearly polarized laser field depends on the relative orientation of field and molecules. While the Σ–Σ-transitions are excited by the field component parallel to the internuclear axis, Σ–Π-transitions are driven by the perpendicular field component. Hence, the coupling strength of the electronic transitions depends on the projection of the electric field vector onto the internuclear axis and, therefore, on the angle between both. For randomly oriented molecules the excitation will take place with quite different efficiencies. In the simulations the possible orientations of the dimers are taken into account by a corresponding average. In supersonic beams of alkali metals that are seeded with lighter carrier-gases the generated molecules are partially aligned in an edge-on orientation (Sinha et al., 1974). Here, the total angular momenta are perpendicular to the molecular beam axis (Brixner et al., 2004). However, in the simulations the assumption of an isotropic distribution of molecular orientations leads to the best agreement with the experimental results. When both the average over the focal intensity distribution and the molecular orientation average are taken into account, good accordance with the measurements is achieved.

3.2.4 Coherent Control of Coupled Electron-Nuclear Dynamics

The experiment was performed in two steps. First we investigated the intensity dependence of the molecular excitation using bandwidth-limited IR pulses in order to identify the onset of the strong-field regime. Subsequently, we employed intense shaped pulses from sinusoidal phase modulation to exert control on the coupled electron-nuclear dynamics and steer the molecule selectively into different target channels. The experimental results are presented in Section 3.2.4.1. The neutral electron dynamics induced by the phase-shaped pulses are analyzed in Section 3.2.4.2 employing molecular dynamics simulations. In particular, the effects of the electron-nuclear coupling on the induced charge oscillation and the energetic accessibility of the target channels is discussed. Finally, the nuclear dynamics induced in the target channels by the phase-shaped pulses is discussed in Section 3.2.4.3.

3.2.4.1 Experimental Results

To characterize the experiment, Fig. 3.3A shows different photoelectron spectra, recorded with either the IR pulse or the VIS pulse or with both pulses combined, in the relevant region of photoelectron kinetic excess energies between 0.9 and 1.8 eV. The one-color spectra generated by the IR and VIS pulse are depicted as red and yellow lines, respectively. The two-color spectrum arising from the combined interaction of both pulses with potassium molecules (and atoms) is depicted as a black line. By design of the experiment, the VIS probe pulse is too weak to ionize ground state molecules. Direct ionization of potassium atoms and molecules by the IR pulse via a REMPI (resonance-enhanced multiphoton ionization) process leads to pronounced photoelectron

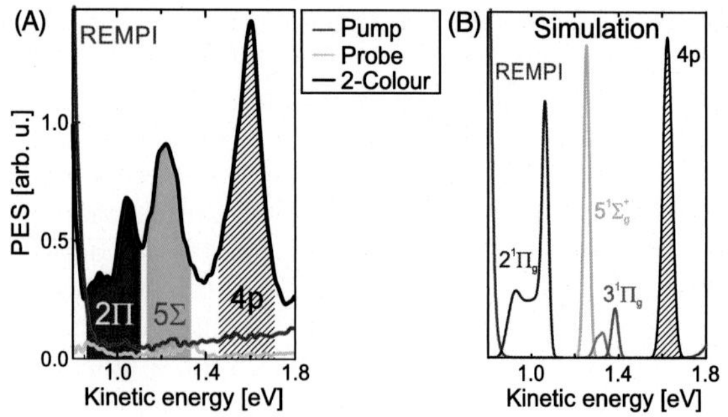

FIG. 3.3 (A) Measured photoelectron spectra generated by the interaction of potassium dimers (and atoms) with the bandwidth limited IR laser pulse and the VIS probe pulse. Shown are ionization signals created by the IR control pulse (red line), the VIS probe pulse (yellow line) and the combination of both pulses. In the energy window between 0.9 and 1.4 eV, two-color photoelectron signals from the molecular target states are detected without interfering single-color background. (B) Simulated photoelectron spectrum for the assignment of the various molecular and atomic signals. Adapted from Braun et al. (2014).

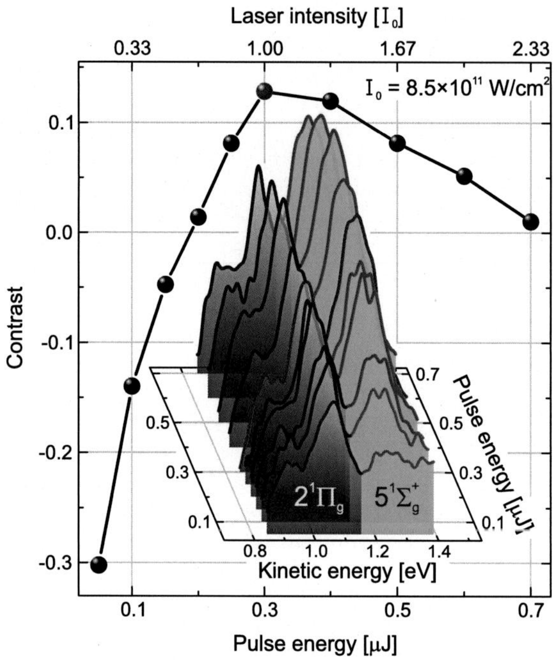

FIG. 3.4 Intensity dependence of the population transfer to the electronic target states. Photoelectron signals from the target channels excited by the bandwidth-limited IR pulse are recorded as a function of the laser intensity. Adapted from Braun et al. (2014).

signals located below 0.9 eV and above 1.8 eV (above threshold ionization). The intermediate energy window is open for background-free detection of two-color signals from the molecular target channels. Signals from the lower target channel, mainly the 2Π-state (blue-shaded area) are centered around 1.0 eV. The double-hump structure of the signal results from the preferential ionization of the molecule at the classical turning points of the nuclear vibration. Due to a steep difference potential between the 2Π- and the ionic ground state in the relevant R-window, the two turning points are mapped onto different energies in the photoelectron spectrum (see also discussion in Section 3.2.4.3). Photoelectrons from the 5Σ-state (green-shaded area) centered around 1.25 eV dominate the signal from the upper target channel. In addition, two-color ionization of residual potassium atoms, excited to the 4p-state by the IR pulse, generates photoelectrons with a kinetic energy of about 1.6 eV. Fig. 3.3B shows simulated photoelectron spectra. The position and shape of the calculated molecular and atomic signals agrees well with the measured contributions confirming the signal assignment. From the measured two-color spectra, we calculate the contrast C (cf. Eq. (3.2)) as

$$C = \frac{S_{\mathrm{up}} - S_{\mathrm{low}}}{S_{\mathrm{up}} + S_{\mathrm{low}}} \tag{3.5}$$

where S_{up} and S_{low} denote the integrated photoelectron yields from the upper and lower target channel, respectively. Characterizing each photoelectron spectrum by the scalar fitness parameter C allows us to map out two-dimensional parameter spaces with respect to the control objective, i.e. the selective excitation of a single target channel, and visualize the result in terms of 2D contrast landscapes. The photoelectron yields S_{up} and S_{low} are extracted from the measured spectra by integration over the respective energy windows (cf. Fig. 3.3A). Comparison with a more sophisticated method based on a multi-Gaussian Levenberg–Marquardt fit delivered only minor differences.

3.2.4.1.1 Intensity Dependence. To study the intensity dependence of the electronic excitation of the target channels, we varied the pulse energy of the bandwidth-limited IR pulse between 0.05 and 0.70 μJ, covering roughly one order of magnitude. By analyzing the photoelectron yield from the target channels we examine the transition from weak-field excitation (pulse energy < 0.1 μJ) to non-perturbative laser–molecule interaction. The recorded two-color spectra are shown in Fig. 3.4. For the lowest pulse energy, only the signal from the 2Π-state is observed. According to the discussion in Section 3.2.1, the upper target channel is not accessible under perturbative conditions. With increasing intensity, however, the upper target channel is populated efficiently due to the AC Stark splitting in the resonant X–A-subsystem. This is indicated by the rapid increase of the 5Σ-signal in the photoelectron spectrum and the steep rise of the derived contrast curve. At an IR intensity of $I_0 = 8.5 \times 10^{11}$ W/cm^2, highlighted in red in Fig. 3.4, the contrast reaches a maximum of $C = 0.12$. The positive value is explained by the blue shift of the IR pulse with respect to the X–A-resonance. This detuning already leads to a preferential population of the upper LIP, thus favoring the upper target channel once this one is accessible energetically. For larger IR intensities, the contrast decreases again. Our simulations suggest that in this intensity regime multi-state Rabi-oscillations start to play a role in the distribution of population among the target channels, making the interpretation of the physical mechanism non-transparent (Baumert et al., 1992).

The manipulation of the branching ratio between the target channels observed in Fig. 3.4 cannot be explained by the weak-field control mechanisms of spectral interference, since the neutral excitation underlying the photoelectron spectra is driven by IR pulses of the same shape, only scaled in intensity. In particular, the efficient population of the upper target channel with increasing intensity is a genuine strong-field effect based on the non-perturbative modification of the potential energy landscape by the AC Stark effect.

In the coherent control experiments described in the following section, the pulse energy was set to 0.3 μJ (red spectrum in Fig. 3.4). At the corresponding intensity, both target channels are efficiently populated and one-color background signals from direct ionization are yet negligible.

3.2.4.1.2 Phase Control. Next we fixed the pulse energy to 0.3 μJ, to ensure non-perturbative excitation conditions (see Section 3.2.4.1.1), and applied sinusoidal spectral phase modulation to exert phase control on the electron-nuclear excitation dynamics. Sinusoidal spectral phase modulation has been described in detail in Herek et al. (2002), Meshulach and Silberberg (1998), Wollenhaupt et al. (2006b), Wollenhaupt et al. (2005). In general, by application of phase modulation functions of the form

$$\varphi_{\mathrm{mod}}(\omega) = A \cdot \sin\left[(\omega - \omega_0) \cdot T + \phi\right], \tag{3.6}$$

the input pulse splits up into a train of subpulses temporally separated by the sine frequency T. The nth subpulse is scaled by $J_n(A)$, i.e. the Bessel function of first kind and order n evaluated at the sine amplitude. The sine phase ϕ determines the relative phase of adjacent subpulses. Compared to double pulse sequences, sinusoidal phase modulation provides a multitude of pulse shapes ranging from multi-pulse sequences with precisely adjustable relative phases for $T \gg \Delta t$ to complex-shaped single pulses with continuously varying (chirped) temporal phases for $T \leq \Delta t$. Due to this versatility, shaped pulses from sine- modulation are well-suited to adapt to the complex phase dynamics of the induced molecular charge oscillation.

In view of the molecular SPODS scenario, the canonical control parameter of the sine mask in Eq. (3.6) are the sine frequency T and the sine phase ϕ. The former allows the shaped pulse to adapt to the timing of vibrational wave packets, whereas the latter serves to adjust the instantaneous phase of the field to the phase of the molecular dipole. A coarse-grained scan of the parameter space revealed that for a sine amplitude of $A = 0.8$ rad a high degree of control could be exerted on the population of both the upper and lower target channel. Therefore, we fixed the sine amplitude to this value and subsequently mapped the 2D control space $\{T, \phi\}$. Fig. 3.5 displays the result in terms of a 2D contrast landscape. For $T = 0$, the sine-function in Eq. (3.6) reduces to a constant phase which does not alter the shape of the IR input pulse effectively. Consequently, the corresponding ϕ-section through the landscape is constant with the contrast $C = 0.12$ obtained for the bandwidth-limited IR pulse (cf. Section 3.2.4.1.1). For $T > 0$, we observe a rich control topology with a distinct maximum of $C = 0.21$ at $T = 50$ fs and $\phi = 1.8$ rad. Taking into account that the measured landscape is subject to focal intensity and molecular orientation averaging (see Section 3.2.3.2), the contrast achieved under ideal conditions is presumed to be significantly larger. The corresponding photoelectron spectrum, shown on the left-hand side of Fig. 3.5, is dominated by the 5Σ-signal which clearly exceeds the signal from the 2Π-state. This indicates the selective population of the upper target channel by the shaped IR pulse, outperforming the bandwidth-limited pulse. The dashed curve shows the calculated photoelectron spectrum from the molecular dynamics simulations performed for the experimental conditions. The complete simulated control landscape displayed in Fig. 3.6 is in reasonable agreement with the experimental result.

According to earlier SPODS studies on atoms (Wollenhaupt et al., 2006b), a π-shift of the optical phase should invert the phase relation between driving field and induced dipole and, thus, switch the population transfer to the lower target channel. In fact, also

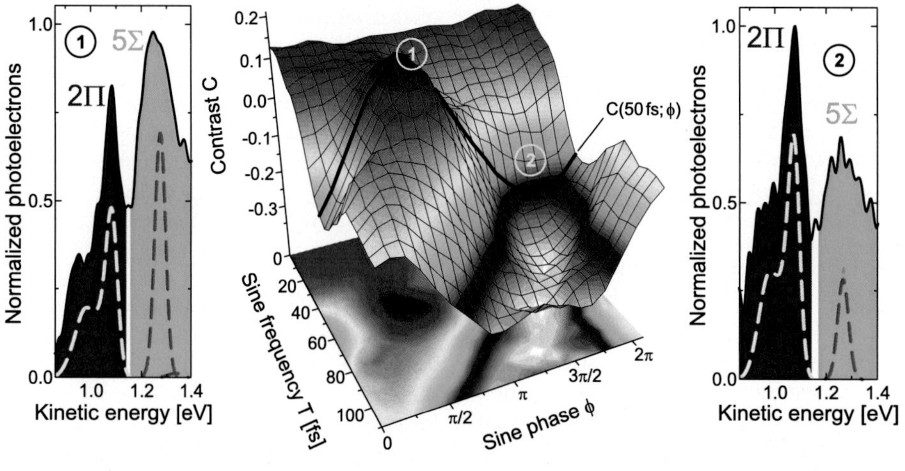

FIG. 3.5 (Center) Measured 2D contrast landscape as a function of the pulse separation T and the optical phase ϕ. (Left and right) Photoelectron spectra for the highlighted points (1) and (2) in the landscape. At the maximum (1), the spectrum is dominated by photoelectrons from the upper target channel. Variation of the optical phase by π, switches the population transfer to the lower target channel, indicated by the increase of increased photoelectron yield from the $2^1\Pi_g$-state at the expense of $5^1\Sigma_g^+$-electrons. Calculated photoelectron spectra are depicted as dashed lines. From Braun et al. (2014).

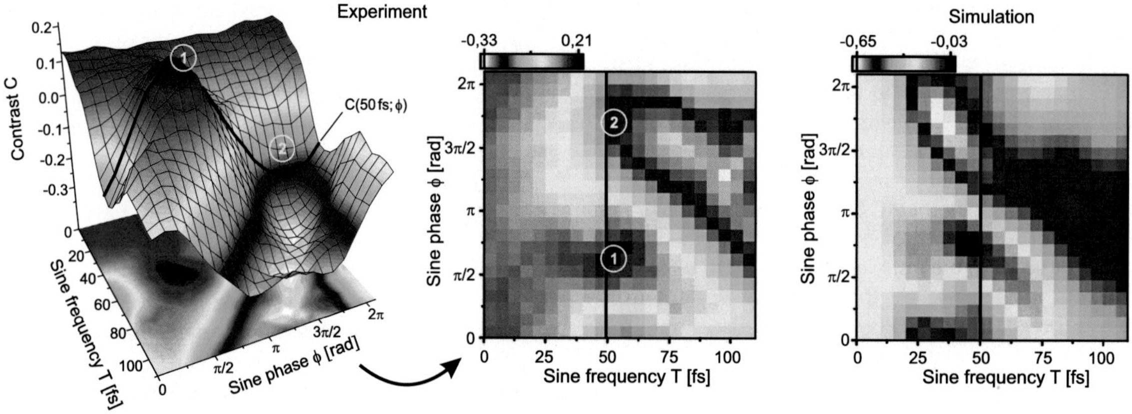

FIG. 3.6 Reproduction of the experimental results by molecular dynamics simulations. The measured landscape is shown in the left and middle frame. The right frame displays the result from the numerical simulations. In the simulations, intensity and orientation averages were taken into account to match the experimental conditions. Adapted from Braun et al. (2014).

in the molecular case discussed here, the contrast is inverted upon variation of the sine phase by π. The corresponding point in the landscape, with $T = 50$ fs and $\phi = 4.9$ rad, exhibits a contrast of $C = -0.20$, indicating the selective population of the lower target channel. The related photoelectron spectrum, shown on the right-hand side of Fig. 3.5, displays a promoted 2Π-signal, while the 5Σ-signal is largely suppressed. We conclude, that basic SPODS concepts devised for atomic systems are valid in molecular scenarios as well.

The global minimum of the contrast landscape is situated around $T = 100$ fs and $\phi = 4.0$ rad. Here, due to the large subpulse separation, the wave packet in the $A\Sigma$-state propagates out of the transition window in between adjacent subpulses. This impedes the SPODS mechanism and results in accumulative population of the lower target channel. In the following section, we provide a more detailed discussion of the physical mechanisms underlying the prominent landscape points highlighted above.

3.2.4.2 Physical Mechanism for Ultrafast Electronic Switching

The reproduction of the experimental results by the molecular dynamics simulations (see Fig. 3.6) provides access to the underlying neutral wave packet dynamics and allows us to analyze the physical mechanism behind the selective population of the electronic target channels. For a maximum transparent discussion, we consider ideal excitation conditions of a fixed laser intensity (corresponding to the peak intensity in the focal center) and an orientation of the molecule of 45° with respect to the laser polarization. The latter implies, that Σ–Σ- and Σ–Π-transitions are addressed with the same efficiency.

Selective population of the upper target channel. The shaped IR pulse associated with the selective population of the upper target channel – marked by (1) in Figs. 3.5 and 3.6 – is presented in Fig. 3.7B along with the induced electronic population dynamics in (A). Initially, around $t = -50$ fs, the pulse creates a coherence in the X–A-subsystem (black dashed and red dashed–dotted line), i.e. excites an electronic wave packet oscillating at the molecular Bohr frequency. Some population is transferred directly to the two-photon resonant 2Π-state (blue solid line). However, around $t = -10$ fs, the most intense part of the pulse transfers population efficiently from these states to the upper target channel (green solid line). By the end of the pulse, the upper target state is populated by 75%. For a detailed analysis of the underlying physical mechanism, we consider the nuclear motion $\langle R \rangle (t)$, the temporal dynamics of the induced dipole moment $\langle \mu_{X\text{–}A} \rangle (t)$ and the LIPs related to the X–A-subsystem in the following.

The time evolution of the induced dipole is shown in Fig. 3.7C (blue line) and compared to the electric field of the shaped laser pulse (red line). Both oscillations are strictly out of phase, i.e. phase-shifted by π, in accordance with the blue-detuned excitation of the electronic wave packet. After the initial excitation, the phase of the electric dipole is influenced by vibrational wave packet dynamics in the X–A-subsystem. Due to the increasing internuclear distance $\langle R \rangle (t)$ shown in Fig. 3.7E, the eigenfrequency of the electronic wave packet, determined by the difference potential between states $X\Sigma$ and $A\Sigma$, gradually decreases, leading to a phase slip of the dipole oscillation. The shaped pulse counteracts this phase slip by a continuously decreasing temporal phase in the rising edge. By this means, the field maintains its initial anti-phase relation to the dipole which leads to the selective population of the upper LIP in the X–A-subsystem (cf. Eq. (3.1)). During the time window of maximum field intensity (grey-shaded background), the AC Stark splitting of the LIPs is sufficient to shift the upper LIP into resonance with the upper target channel (black circle in Fig. 3.7D). Therefore, once the high-lying target states are energetically accessible, the population flows rapidly into the upper target channel. As the field amplitude decreases, the upper target channel is closed and the population transfer is interrupted leaving 75% of the population stored in the high-lying target states. During the entire interaction with the pulse, the internuclear distance increases by 9%. This results in a total decrease of the Bohr frequency of 100 meV (0.167 rad/fs) and a red-shift of the molecular X–A-resonance to 926 nm.

To verify the strong-field character of the scenario discussed above, we performed the same simulations for a reduced IR peak intensity of $0.08 \cdot I_0$. The results are shown in the right column of Fig. 3.9. Again we obtain an anti-phase relation between the laser field and the induced dipole. Nevertheless, no population is transferred to the upper target channel in this case. Only the 2Π-state is populated by about 12%, since this state is resonant also in the perturbative regime. The upper target channel, however, is not accessible due to the insufficient AC Stark splitting in the X–A-subsystem. This underscores that the population of the 5Σ-state observed in the experiment is a genuine strong-field effect.

Selective population of the lower target channel. Next we analyze the selective population of the lower target channel at the landscape point (2) in Figs. 3.5 and 3.6. The corresponding simulation results are depicted in Fig. 3.8. In this case, the shaped IR pulse starts with a sequence of two weak prepulses, the first of which builds up the electronic coherence in the X–A-subsystem. Around $t = -45$ fs, the second prepulse already triggers an efficient population transfer to the low-lying target states. Due to the constant energetic accessibility of states 2Π and 4Σ, Rabi-type population oscillations are observed between the lower target channel and (mainly) the X–A-subsystem. After approximately 2.5 Rabi cycles, 60% of the population is transferred to the lower target channel.

To elucidate the physical mechanism, we consider again the induced dipole oscillation (Fig. 3.8C) and the temporal dynamics of the LIPs (Fig. 3.8D). During its initial (blue-detuned) excitation, the dipole oscillation starts again in anti-phase to the laser field. However, the intensity of the prepulses is not sufficient to open up the upper target channel for efficient population. Only during the intense main part of the pulse, around $t = 0$ fs,

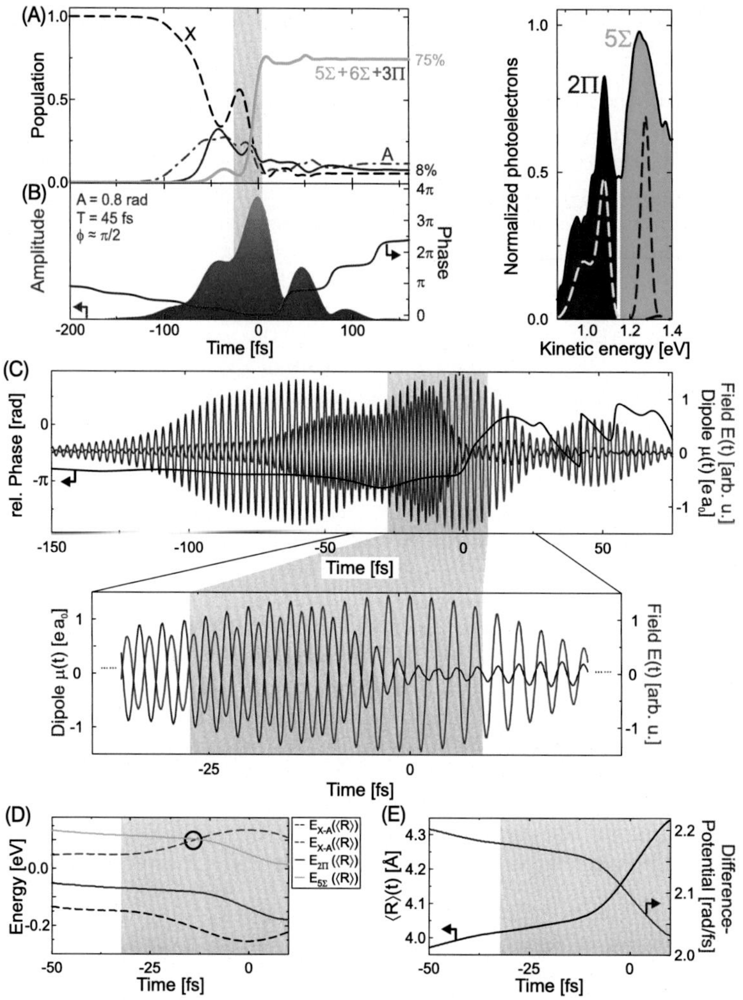

FIG. 3.7 Molecular dynamics simulations for the population of the upper target channel. The relevant time window for the population transfer is highlighted by a grey background. (A) Electronic population dynamics for the excitation with the shaped control pulse. The light green curve shows the sum over all state populations in the upper target channel. (B) The electric field of the control pulse is decomposed into its amplitude (red) and the time-dependent optical phase (blue). (C) Oscillations of the electric field (red) and the induced dipole moment (blue), as well as the relative phase of the oscillations (black). (D) Energy of the LIPs in the X–A-subsystem (black and red), the $5^1\Sigma_g^+$-state (green) and the $2^1\Pi_g$-state (blue) at $\langle R\rangle(t)$, each shifted by one photon energy. (E) Temporal evolution of the expectation value of the internuclear distance and the corresponding value of the difference potential between the $X^1\Sigma_g^+$- and the $A^1\Sigma_g^+$-state. In addition, a measured photoelectron spectrum containing the signals from the target channels, is shown in the upper right frame (cf. Fig. 3.5).

some population escapes to the high-lying target states. Around the same time, the dipole oscillation undergoes a phase jump of π due to the completion of the first Rabi half-cycle. This switches the charge oscillation in-phase with the field oscillation leading to the selective population of the lower LIP and suppressing further excitation of the upper target channel. After some complicated intermediate dynamics, dipole and field shift again in-phase around $t = 50$ fs. In this time window, the lower LIP reaches exact resonance with the lower target channel (black circle in Fig. 3.8D) which initiates the final population transfer to the low-lying target

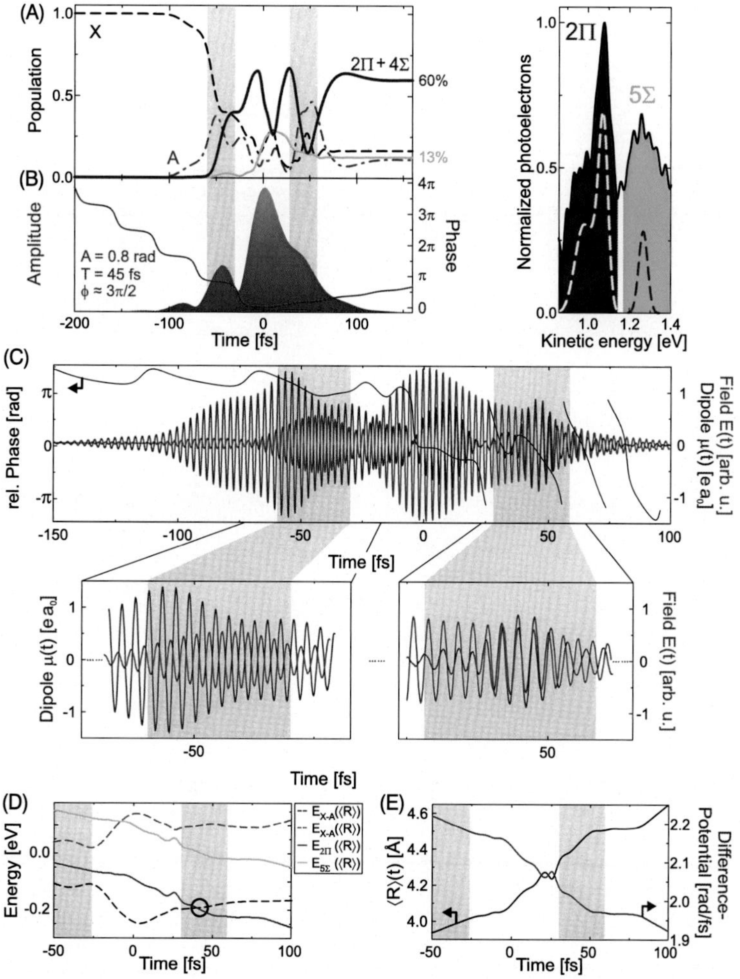

FIG. 3.8 Molecular dynamics simulations for the population of the lower target channel. The interesting time windows for the population dynamics are highlighted by a grey background. (A) Electronic population dynamics for the excitation with the shaped control pulse. The blue curve shows the sum over all state populations in the lower target channel. (B) The electric field of the control pulse is decomposed into its amplitude (red) and the time-dependent optical phase (blue). (C) Oscillations of the electric field (red) and the induced dipole moment (blue), as well as the relative phase of the oscillations (black). (D) Energy of the LIPs in the X–A-subsystem (black and red), the $5^1\Sigma_g^+$-state (green) and the $2^1\Pi_g$-state (blue) at $\langle R\rangle(t)$, each shifted by one photon energy. (E) Temporal evolution of the expectation value of the internuclear distance and the corresponding value of the difference potential of the $X^1\Sigma_g^+$- and the $A^1\Sigma_g^+$-state. In addition, a measured photoelectron spectrum containing the signals from the target channels, is shown in the upper right frame (cf. Fig. 3.5).

states. During this interaction, the internuclear distance increases by 18%. The Bohr frequency is reduced by 190 meV, which corresponds to a red-shift of the X–A-resonance to about 975 nm.

The complex non-perturbative population dynamics is the basis to break the initial anti-phase relation predetermined by the detuning. This is another clear indication for the strong-field character of the control scenario. To verify this finding, we performed the same simulation for a reduced intensity of $0.08 \cdot I_0$. The result is shown in the left column of Fig. 3.9. In this case, no Rabi oscillations are observed and field and dipole maintain their initial anti-phase relation throughout the entire interaction.

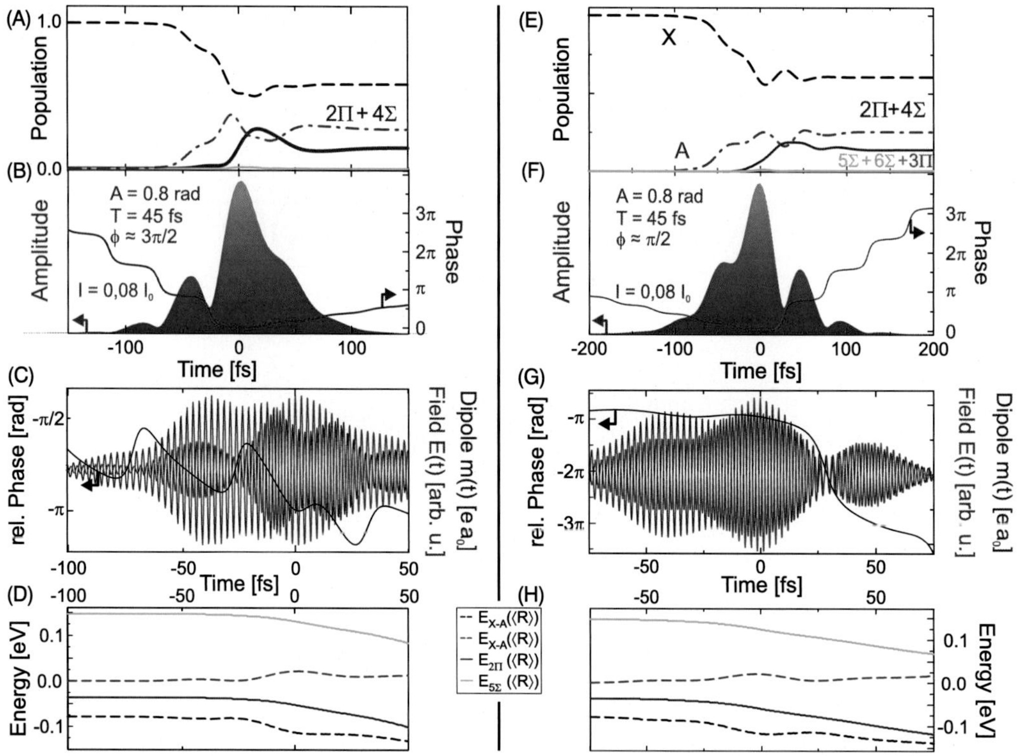

FIG. 3.9 (Left) Molecular dynamics simulations for the tailored laser pulse from Fig. 3.8B at decreased intensity. Under these excitation conditions, the induced dipole moment cannot be phase-shifted to oscillate in-phase with the driving laser field (see (C)). The relative phase changes between $-\pi/2$ and $-\pi$ and during the population transfer around $t = 0$ fs field and dipole oscillate in anti-phase. The upper target channel is energetically not accessible due to the low Stark splitting (see (D)) and only a small part of the population is transferred to the lower target channel (see (A)). (Right) Simulation results for the tailored laser pulse from Fig. 3.7B at decreased intensity. The relative phase of π between the electric field and the induced dipole moment is maintained during the major part of the interaction (see (G)). However, hardly any population is transferred to the upper target channel (see (E)) due to the insufficient Stark splitting (see (H)). Therefore, again only the lower target channel is significantly populated.

Global contrast minimum. The global minimum of the contrast landscape in Fig. 3.5 is located at large sine frequencies $T \geq 100$ fs. In this region, the subpulses of the pulse train are well-separated in time. According to the simulation results shown in Fig. 3.10, the dipole oscillation induced by the individual subpulses decays rapidly in between successive pulses. Due to the large subpulse separation, the vibrational wave packet $\psi_A(R,t)$ in the AΣ-state has sufficient time to propagate out of the Franck–Condon-window and, eventually, loses overlap with the more confined ground state wave packet $\psi_X(R,t)$. Because the amplitude of the dipole oscillation vanishes before the next subpulse arrives, it cannot be utilized to achieve selectivity among the target channels. Instead, each subpulse excites the molecule individually leading to a successive population and depopulation of the lower target channel by the weak subpulses, which are unable to address the upper target channel. Only the most intense main pulse is capable of opening up and transferring population to the upper target state. This holds true only for the ideal excitation conditions considered here. For lower intensities and molecular misalignment, the upper target channel is generally not accessible at all. This rationalizes the strongly negative photoelectron contrast observed in the experiment for large T. It should be noted, however, that despite vanishing dipole amplitude, the electronic phase coherence of the X–A-subsystem is maintained. Calculations performed by the group of Prof. R. de Vivie-Riedle suggest that at even larger time delays matching the recurrence

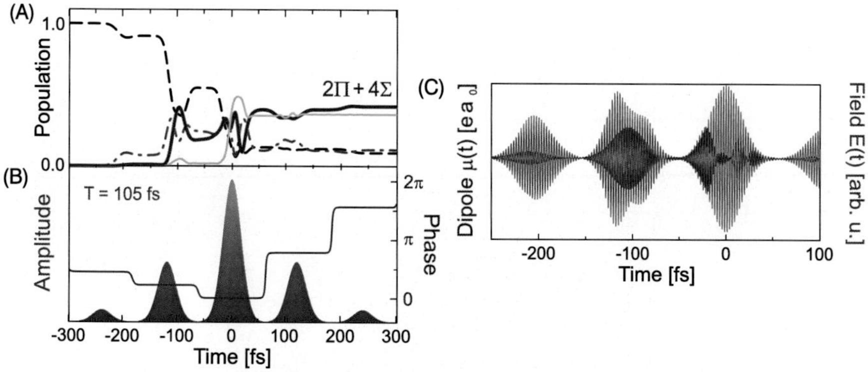

FIG. 3.10 Molecular dynamics simulations for a larger sine frequency of $T = 105$ fs. The shaped laser field is shown in (B). Due to the large temporal separation of the subpulses, the dipole moment induced by each subpulse decays before it can be utilized by the next pulse. As a result, each subpulse acts individually leading to a successive population of the lower target channel.

period of the $\psi_A(R,t)$ wave packet, the controllability of the population transfer is recovered (von den Hoff et al., 2011). At this time delay of about 550 fs, the overlap of the nuclear wave packets is restored and the dipole oscillation builds up again. In fact, the dipole amplitude may even exceed the initial amplitude due to the compression of the wave packet at the inner turning point. Seeing this effect in experiment would certainly be very interesting.

3.2.4.3 Target State Wave Packet Dynamics

In general, photo-excitation of different electronic target channels in molecules entails different nuclear dynamics leading to, e.g., vibration, fragmentation or isomerization of the molecule. Fig. 3.11 illustrates the nuclear wave packet dynamics in the target states of the potassium molecule, calculated for the excitation with the shaped IR pulses from Fig. 3.7B (selective population of the upper target channel; left column) and Fig. 3.8B (selective population of the lower target channel; right column). Already in this simple molecular model system, we observe distinctly different nuclear motions in both scenarios. Besides different vibrational periods, the wave packets cover a different range of internuclear distances during their propagation. In addition, the photoelectron energy distributions from the lower and the upper target channel, created by the same probe pulse, exhibit very different shapes.

Measured and calculated photoelectron spectra are displayed in the top row of Fig. 3.11. To rationalize the signal shape, the difference potentials between the ionic ground state $X\Sigma(K_2^+)$ and the corresponding target states are shown in addition. In the case of the upper target state 5Σ, shown in the left column, the difference potential is flat in the relevant R-window determined by the turning points of the $\psi_{5\Sigma}(R,t)$ wave packet oscillation. Therefore, photoelectrons released at different oscillation stages, associated with wave packet localization at different internuclear distances $\langle R \rangle$, are mapped onto the same kinetic excess energy in the ionization continuum. The resulting photoelectron signal is well-localized in the energy spectrum with a width essentially determined by the bandwidth of the probe pulse. In the case of the lower target state 2Π, shown in the right column, the difference potential decreases linearly in the relevant R-window. Here, photoelectrons created at different times are released with different kinetic energies. In particular, the two turning points of the $\psi_{2\Pi}(R,t)$ wave packet oscillation are mapped into different regions of the energy spectrum. In analogy to a classical pendulum, the probability density $|\psi_{2\Pi}(R,t)|^2$ – and hence the ionization probability – peaks at both turning points, leading to the observed double-hump structure of the corresponding photoelectron signal. The asymmetry of the signal, in favor of the high-energy hump, is explained by the more pronounced maximum of the density at the inner turning point as compared to the outer turning point. This disentanglement of different oscillation stages in the energy spectrum enables the time-resolved observation of the wave packet dynamics using a short, i.e. compressed probe pulse (Frohnmeyer et al., 1999).

3.2.5 Summary and Conclusion

In this section we presented a 2D strong-field approach being sensitive to selective population of dressed states (SPODS) in molecules. This is an extension of the well-known spin-locking pulse sequences in NMR to the op-

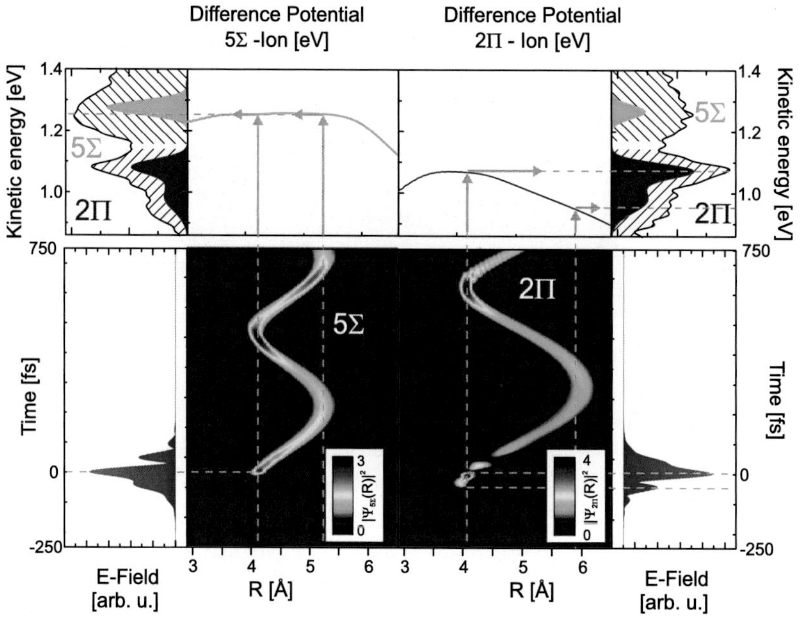

FIG. 3.11 Calculated wave packet dynamics in the $5^1\Sigma_g^+$- (left) and the $2^1\Pi_g$-state (right) for the excitation with the control pulses shown in Figs. 3.7B and 3.8B, respectively. The upper panel depicts the difference potentials between the ionic ground state and the neutral target state as well as the measured and simulated photoelectron signals. From Braun et al. (2014).

tical regime. Whereas spin-locking is realized in a two-level scheme, SPODS is realized in molecules taking the more complex electronic structure and the nuclear dynamics into account.

Inducing and controlling coherences in two-level schemes is important to increase sensitivity in spectroscopy where different pulse sequences developed in NMR are prominent examples to that end. Also different quantum technologies are currently emerging ranging from quantum computing to quantum sensing. In many cases they are based on different physical realizations of two-level schemes and the control or sensing of induced coherences. Considering the richer electronic structure of molecules in addition to the nuclear degrees of freedom, additional possibilities arise. One perspective is to induce and control coherences in molecular systems in order to steer photochemical processes selectively and efficiently. This is demonstrated here. By tailoring the optical phase of an intense IR laser pulse to the phase of the laser-induced charge oscillation and using the field amplitude to manipulate the potential energy landscape via resonant AC Stark splittings or non-resonant AC Stark shifts, the coupled electron-nuclear dynamics can be steered efficiently into preselected bound electronic target channels. Due to the non-perturbative character of the SPODS scenario, even target channels inaccessible in the weak-field regime can be addressed. In this spatiotemporal picture of the laser-induced dipole oscillation, the laser field initially induces an electronic coherence and, subsequently, aligns itself in-phase or in anti-phase to the corresponding charge oscillation, in order to minimize or maximize the interaction energy. In atomic systems, the eigenfrequency of the induced dipole, i.e. the Bohr frequency of the electronic transition, is constant. In molecular systems, however, the dipole is subject to additional phase dynamics induced by the motion of the nuclei which alters the electronic resonance. In addition, the amplitude of the dipole oscillation depends on the overlap of the nuclear wave packets in the electronic states.

A convenient way to implement the optical characteristics of such laser fields is sinusoidal spectral phase modulation in Fourier-transform-based pulse shaping approaches. In the temporal domain this modulation results in a pulse sequence with adjustable temporal separation, relative pulse ratio and phase. The generated pulse shapes range from clearly separated pulse trains down to complex-shaped laser fields when the individual subpulses start to merge.

In the presented experiment on the potassium dimer, we mapped 2D parameter spaces of the sine phase mask in terms of pulse separation and relative phase of the subpulses after fixing the pulse ratio. The final populations of the bound electronic target states were interrogated by a time-delayed VIS pulse employing two-color energy-resolved photoelectron spectroscopy. The experiments were supplemented by molecular dynamics simulations taking into account experimental conditions such as focal intensity and molecular orientation averaging. The good agreement between experimental and numerical results provides access to different aspects of the underlying neutral dynamics. The simulations verify that the attained control is based on non-perturbative laser–molecule interactions. Strong-field effects – such as AC-Stark splittings of several 100 meV in the resonantly driven X–A-subsystem, to shift the upper target states into resonance, or efficient multi-state Rabi cycling, to align dipole and field in-phase – are crucial to achieve selectivity among the target channels.

Transferred to even larger molecular systems (Wollenhaupt and Baumert, 2011), this 2D approach has potential applications from ultrafast control of valence bond chemistry including chemical sensing to quantum information processing (Vivie-Riedle and Troppmann, 2007).

3.3 ADIABATIC CONTROL SCENARIOS IN MOLECULES

Strong-field laser-matter-interactions are, in general, characterized by efficient population transfer between the involved quantum states. This is in strong contrast to quantum control scenarios in the weak-field regime, where by definition the population in the initial state remains constant. Adiabatic control mechanisms that allow for efficient population transfer (up to 100%) are of particular interest. They are largely insensitive to variations of the experimental parameters such as intensity or laser detuning. Famous examples for these kinds of robust excitation are the rapid adiabatic passage (RAP) or the stimulated Raman adiabatic passage (STIRAP) (Shore, 2008; Vitanov et al., 2001), see also Chapter 1 of this book (Marquardt and Quack, 2020). Especially for molecules an additional aspect is important: the modification of electronic potentials via the AC Stark-effect during the interaction with intense laser pulses. The nuclear motion can be strongly influenced by light-induced potentials. Frohnmeyer et al. (1999) and Sussman et al. (2006) show examples where the course of a reaction was controlled by modifying the potential surface after an initial excitation. The population of vibrational states can in addition be changed by the non-resonant interaction with excited vibrational wave packets (Goto et al., 2011). This non-resonant Stark control takes place on the time scale of the intensity envelope of the ultrashort laser pulse (Sussman, 2011).

In the previous Section 3.2, we discussed how a resonant femtosecond laser pulse can be utilized to control the induced coupled electron-nuclear dynamics and steer a molecule into a predefined neutral target state. The specifically tailored electrical fields are optimally adapted to the induced dynamics and energetic conditions in the molecule, allowing for efficient and selective population transfer. The reproduction of the experimental results by molecular dynamics simulations indicates that the ensemble average over the different excitation conditions has a huge influence onto the efficiency of the excitation and extent of the control attainable (Braun et al., 2014). Two different aspects are of relevance (see also Section 3.2.3.2): The spatial intensity distribution in the laser focus leads to different excitation intensities for the molecules distributed in the excitation area. Moreover, the intramolecular transitions in each molecule will be driven by different efficiencies depending on the orientation of the molecule with respect to the laser polarization. Both mechanisms have a significant impact on strong-field control scenarios which are sensitive to the laser intensity (Rabi-oscillations, photon locking). The additional degree of freedom of the nuclear motion in molecules complicates such excitation processes dramatically and has been barely investigated theoretically or experimentally.

The focus of this section is the development of adiabatic control scenarios for molecules. We use the results from Schneider et al. (2011) and Schneider (2014) as a basis. Starting from this previous work, we employ specifically phase-shaped femtosecond laser pulses to excite the molecules. So far, the experimental studies have been performed on large dye molecules in solution (porphyrazines and laser dyes, see Section 3.3.3). For a more transparent analysis of the light induced dynamics we now employ the potassium dimer (K_2) in the gas phase as a molecular model system.

3.3.1 Chirped Airy Pulses

The spectral phase of the femtosecond laser pulses used in the experiments on adiabatic molecular excitation is modulated by the third-order polynomial

$$\varphi_{\mathrm{mod}}(\omega) = \phi_2 \cdot (\omega - \omega_0)^2 + \phi_3 \cdot (\omega - \omega_0)^3. \quad (3.7)$$

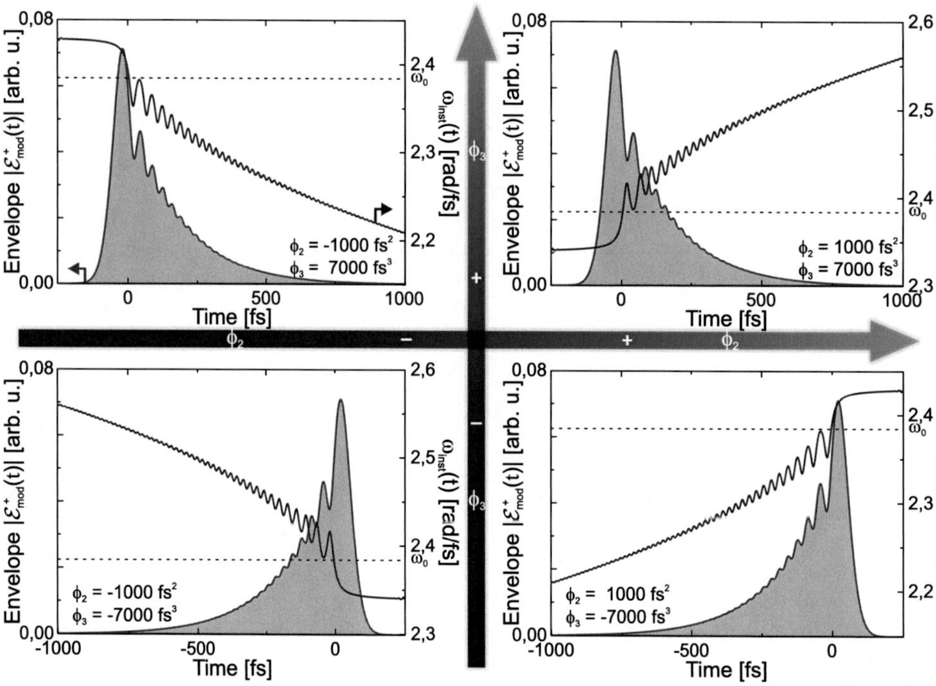

FIG. 3.12 Examples of chirped Airy pulses generated by combined second- and third-order spectral phase modulation. The sign of ϕ_3 determines the temporal asymmetry of the pulses. For $\phi_3 < 0$ (lower row), a pulse with a slowly increasing envelope $\mathcal{E}_{mod}(t)$ is generated which rapidly decays after the main pulse around $t = 0$. For $\phi_3 > 0$ (upper row), this behavior is mirrored: The resulting pulse starts with a steeply rising main pulse and slowly decays subsequently. The chirp parameter ϕ_2 determines the time evolution of the instantaneous frequency $\omega_{inst}(t)$ of the field. Negative ϕ_2 (left column) results in a decreasing instantaneous frequency. For $\phi_2 > 0$ (right column), the instantaneous frequency increases. The frequency's continuous oscillations are relics of discrete π-phase jumps obtained for unchirped Airy pulses.

Quadratic phase modulation, as described by the first term, yields linearly chirped laser pulses, which serve as prototypes for shaped laser fields with continuously varying instantaneous frequency. The cubic term in Eq. (3.7) yields shaped pulses with a temporal envelope described by an Airy function. Therefore, these pulses serve as prototypes for temporally asymmetric pulse shapes. A detailed analytical description of the temporal pulse shape resulting from combined second- and third-order spectral phase modulation is given in Schneider et al. (2011). Briefly, chirped Airy pulses exhibit a temporally asymmetric intensity profile with one slowly varying edge (onset or decay), as required for adiabatic excitation, and one rapidly varying edge capable of inducing impulsive excitations. The GDD smears out the typical subpulse-structure and also blurs the π-phase jumps from pure TOD modulation, thus smoothing the pulse envelope and providing a continuously varying instantaneous frequency, both of which are beneficial for adiabatic control scenarios. Examples of chirped Airy pulses for different combinations of ϕ_2 and ϕ_3 are illustrated in Fig. 3.12.

3.3.2 Adiabatic Control Scenarios

Methods for the population inversion of atomic systems are known and well-established. However, transfer of those schemes to molecules is demanding due to the vibrational degree of freedom. Consequently, theoretical approaches are sparse and experimental approaches almost non-existent. Experiments in the group of Prof. C. Shank (Cerullo et al., 1996) using chirped laser pulses motivated theoretical investigations on the applicability of this technique to an isotropically distributed ensemble of molecules in a solvent (Cao et al., 1998; Vala and Kosloff, 2001). These previous approaches for optimizing the population transfer in molecules mostly rely on the use of positively chirped pulses (positive GDD). In the limit of very long pulses (picoseconds) the population transfer can be attributed to adiabatic effects, for shorter pulses intrapulse–pump–dump wave packet dynamics are responsible (Cao et al., 2000). In the latter case, the instantaneous frequency of the laser pulse can follow the molecular resonance that will change during the interaction due to the nuclear wave packet dynamics. Thus, the population in the excited state can be dumped back to the ground state (down chirp). For a positive chirp, the excited state will be populated at the beginning of the pulse. However, during the following interaction the instantaneous frequency of the pulse and the resonance frequency of the molecule will evolve in opposite directions, thereby hindering the population return to the ground state. The population will remain in the excited state. A large spectral width is regarded as a vital prerequisite for an efficient population transfer, as the temporally varying instantaneous frequency may cover the whole bandwidth of the molecular excitation. In Zai et al. (2016) the authors report on the theoretical optimization of STIRAP-sequences for efficient population transfer in a molecule. For exact two-photon resonance, a complete population transfer is observed in the simulations.

3.3.3 Interaction of Chirped Airy Pulses With Porphyrazine Molecules

In Schneider et al. (2011) and Schneider (2014), experimental studies on adiabatic control of molecular dynamics were performed on sensitizer molecules using chirped Airy pulses. In the strong-field experiments, the porphyrazine Tetrakis(6,7,10,11-tetrakis(3,3,3-tritolylpropin-1-yl)dibenzo[f,h]chinoxalino)porphyrazinatozink(II) was used in its unprotonated (Sens3) and its protonated (Sens3H) form. After an intersystem crossing from the singlet to the triplet system, the molecules can be excited from the T_1- to the T_n-state by one 800 nm photon from a femtosecond laser source. For Sens3H the absorption band is detuned by ca. 20 nm in comparison to Sens3 and has a larger overlap with the employed laser spectrum. In both molecules the triplet–triplet transition was driven by chirped Airy pulses. The population in the T_n-state is detected by fluorescence. A simple model of the laser dye, consisting of two displaced harmonic potentials for the triplet states, was used to interpret the measured data. With the help of quantum mechanical wave packet simulations based on this simplified model the data could be reproduced qualitatively. The fluorescence landscapes measured as a function of the parameters GDD and TOD shown in Fig. 3.13 are taken from Schneider (2014). They compare the measured fluorescence from the excited triplet state T_n and the population transfer calculated using the harmonic model system. The data is presented for Sens3 and Sen3H and for two different pulse energies. We could identify different control scenarios some of which can be explained by an atom-like behavior of the molecules. The final populations are robust with regard to variations of the GDD- and TOD-parameter as well as the intensity. Fig. 3.14 presents an overview of these scenarios (from Schneider, 2014): (A) and (B) show the effect of a coupling of wave packet dynamics in the involved electronic states. During the propagation of the wave packets in the two potentials, the difference potential changes and can cross (in (A)) or evade (in (B)) the time-dependent laser frequency. In the first case, the shaped laser pulse adapts to the intramolecular dynamics such that an efficient transfer of electronic population to the excited state takes place. For case B, the excited state is populated only transiently and at the end of the interaction a wave packet is created in the ground state. The effect of wave packet coupling and the resulting joint dynamics in different potentials was termed JoMo – Joint Motion. If a vibrational eigenstate is excited in the molecule, the resonance frequency stays constant and may be crossed by the varying laser frequency (see Fig. 3.14 C). This scenario is reminiscent of the atomic RAP and results in an efficient transfer of population to the excited state. The measured parameter variations in Schneider (2014) verify that the effect is robust, i.e. small variations of the pulse parameters (energy, detuning with regard to the molecular resonance) barely influence the outcome of the interaction and the general form of the fluorescence landscape is maintained.

In addition, the interaction of chirped Airy pulses with the laser dyes IR125 and HITCI was investigated. These systems also show adiabatic behavior when excited by the shaped pulses. For example, plateau-type regions in the $\{\phi_2, \phi_3\}$ parameter space were identified, where population is transferred robustly to the excited state due to the chirped frequency of the shaped IR laser pulse. In different regions, the induced dynamics result in a coupling of wave packets and a coherent population return to the ground state. At the end of the interaction, a wave packet has been created in the ground state. Here the electronic states were modelled by a total of three harmonic potentials. The S_0 ground state is displaced with respect to the excited S_1 state along the reaction coordinate. The S_2-state, that is resonantly coupled to the S_1-state, has the same equilibrium coordinate as S_1. During the interaction, population from the S_2-state is transferred to a loss channel after 300 ps and is added to the population of S_1 at the end of the interaction. By this means, the internal conversion of higher-lying

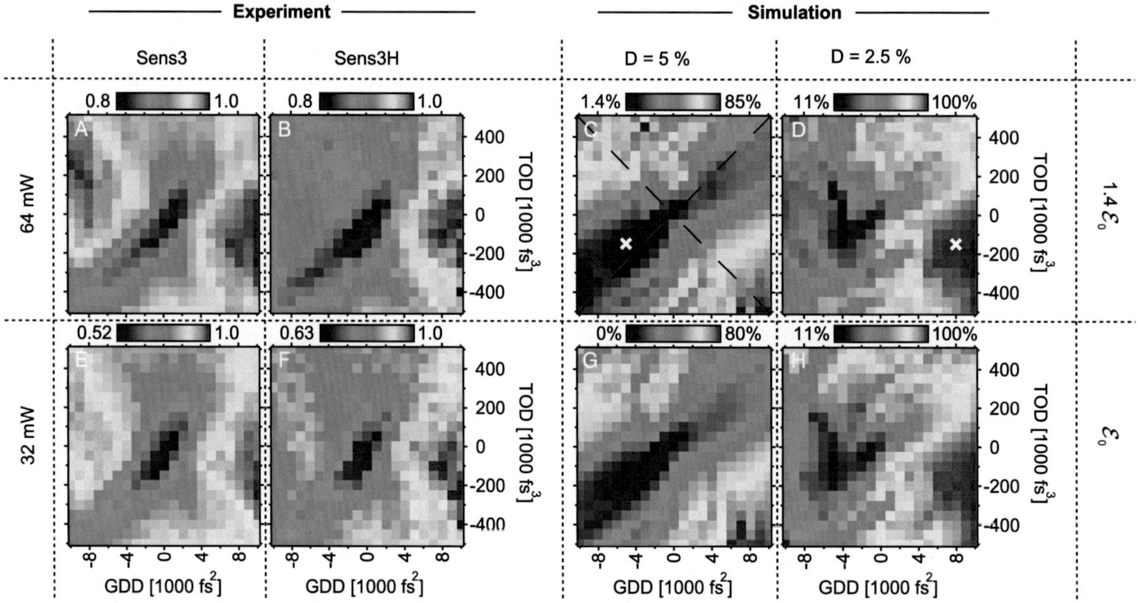

FIG. 3.13 (Left) Measured 2D fluorescence landscapes as a function of the phase modulation parameters ϕ_2 (GDD) and ϕ_3 (TOD) for the molecules Sens3 and Sens3H and two different pulse energies. (Right) Simulated 2D landscapes for the population transfer to the excited state in a two-state harmonic oscillator model. The simulations were performed for two different detunings of the laser spectrum with respect to the molecular resonance (left and right column) and two different pulse energies (upper and lower row). Data from Schneider et al. (2011).

excited states into the S_1-state can be modelled. This approach allows for a more realistic description of the $S_1 \rightarrow S_0$ fluorescence signal.

Modelling complex molecules like porphyrazines or laser dyes by only two or three harmonic potentials represents a substantial simplification, especially in the presence of a solvent environment. Although such simple models are suitable to identify basic control scenarios, more advanced simulations would be required to gain a detailed understanding of the laser-induced intramolecular dynamics. Alternatively, the control scheme based on chirped Airy pulses can be investigated on a less complex molecular prototype. The potassium dimer is well-suited to this end, as it can be accurately described in simulations and is readily available for experiments (von den Hoff et al., 2011; Petersen and Mitric, 2012; Wollenhaupt and Baumert, 2006).

The potassium molecule was already introduced in Section 3.2.1. Similar to the model for the porphyrazine molecules, K_2 has a resonant transition from the ground to the first excited state at around 800 nm. The $A\Sigma$-state couples strongly to the higher lying electronic states, such that efficient population transfer into further bound target states by a near-infrared femtosecond laser pulse is possible. In addition to selective population of the first excited state and population return to the ground state, an efficient population of higher-lying target states, e.g. 2Π and 5Σ, is possible. In the experiment, the populations in the electronic states will be mapped into the energy-resolved photoelectron spectrum by one-photon ionization of the molecules using a time-delayed VIS probe pulse.

3.3.4 Interaction of Chirped Airy Pulses With Potassium Molecules

3.3.4.1 Evaluation of Spectra

Fig. 3.15 shows the excitation schemes for potassium atoms and molecules interacting with a near-IR femtosecond laser pulse of $\lambda_0 = 790$ nm and a VIS probe pulse of $\lambda_{probe} = 570$ nm. The calculated signals from several excited states of the two species are also indicated in the photoelectron spectrum. Apparently, a multitude of signals is created by different combinations of photons from the control and the probe pulse due to frequency mixing and resonances. Of particular interest are the molecular two-color signals arising from excitation by the shaped IR pulse and time-delayed ion-

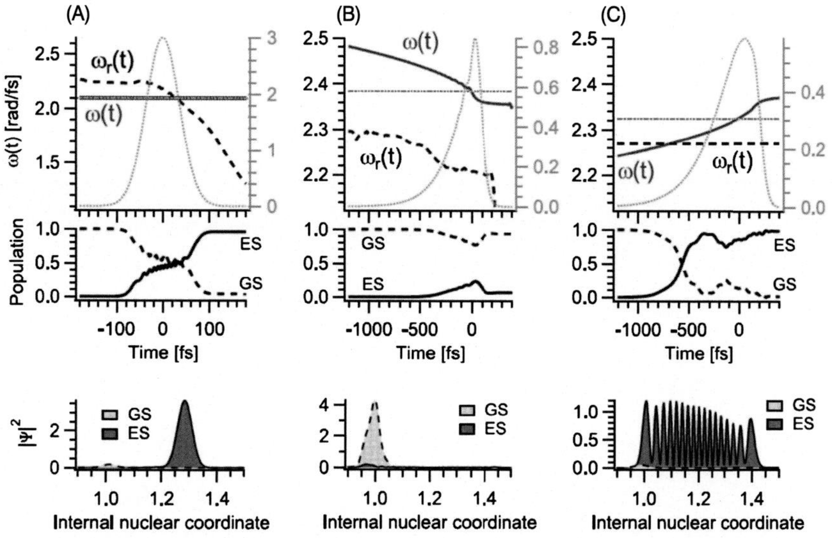

96% population transfer	6% population transfer	99% population transfer
Population transfer by crossing of ω(t) and ω_r(t) (JOMO-dynamics)	Population return due to non-crossing between ω(t) and ω_r(t) (JOMO-dynamics)	Population transfer due to crossing of ω(t) (chirp-dynamics) and ω_r(t)
JOMO allows for a meaningfull definition of ω_r(t) and the interpretation of the final populations in an atom-like picture.		ω_r(t) well-defined due to narrowband interaction (Franck-Condon window)

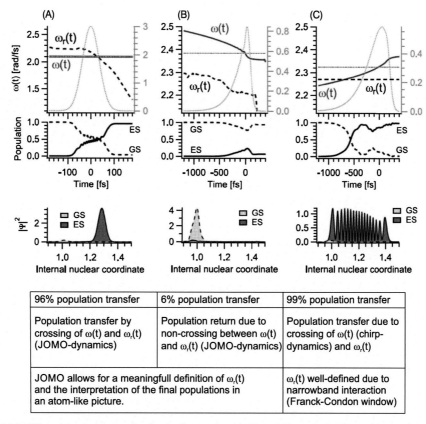

FIG. 3.14 Different control scenarios for the interaction of chirped Airy pulses with porphyrazine molecules, modelled by two harmonic potentials. The upper frames in each column shows the envelope of the laser pulse (blue) together with the instantaneous frequency $\omega(t)$ of the pulse (red) and the molecular resonance $\omega_\Gamma(t)$ (black) as a function of time. The population of the ground state and the excited state are plotted in the middle frames. The wave packets created in the ground and the excited state by the end of the interaction are depicted in the lower frames. The text table discusses (from top to bottom): The population transfer into the excited state. A short explanation of the dynamics on the basis of $\omega(t)$ and $\omega_\Gamma(t)$ (FC, Franck–Condon). Adapted from Schneider (2014).

ization by the VIS probe pulse. Fig. 3.16 shows simulated photoelectron spectra for atomic and molecular signals, as well as their combination. Photoelectrons from ionization of potassium atoms and molecules created by the control pulse, the probe pulse and frequency mixing of both pulses were calculated separately and subsequently superimposed incoherently. Their relative contributions do not necessarily reflect the experimental conditions, but serves to visualize the expected signal overlaps. The photoelectron signal from the $X\Sigma$ ground state of the potassium dimer, centered around 0.3 eV, is created by simultaneous absorption of two photons from the probe pulse. The two-color signal from the $A\Sigma$-state around 1.65 eV is created via two-photon ionization by the probe pulse after resonant excitation by the control pulse. Around 0.45 eV, an additional signal from state AΣ emerges due to the absorption of two photons from the control pulse. However, this signal overlaps with the one-photon ionization of the 2Π- and the 5Σ-state by the control pulse (not shown). The signals from the 2Π and the 5Σ-state, created by the probe pulse after excitation of potassium dimers by the control pulse, are located between 0.8 eV and 1.4 eV. Atomic single- and two-color photoelectron signals generated by the probe pulse are depicted in grey (4s ground state around 0.05 eV) and yellow (4p excited state around 1.6 eV), respectively. Direct 1+2 resonance-enhanced multiphoton ionization by the control pulse gives rise to the Autler–Townes-doublet centered around 0.4 eV (violet).

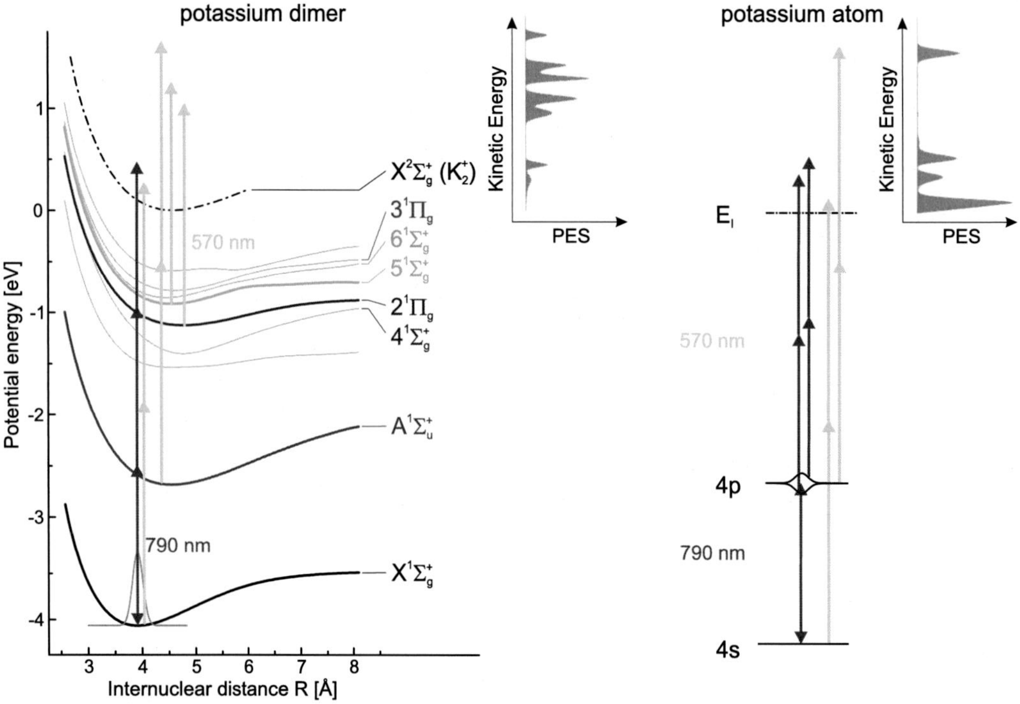

FIG. 3.15 Excitation scheme for potassium dimers (left) and atoms (right) for the interaction with an intense near-IR control pulse (790 nm) and a VIS probe pulse (570 nm). The top insets indicate the energy-resolved photoelectron signals from the various atomic and molecular states expected in the two-color experiment.

To disentangle atomic from molecular signals, which strongly overlap in the measured photoelectron spectra, we performed separate measurements on atoms only by interrupting the seed gas flow thus preventing the dimer formation. After determination of the atomic contributions, these signals were weighted and subtracted from the spectrum to retrieve the molecular signals from the XΣ- and the AΣ-state. The weighting factor, i.e. the relative contribution of atomic and molecular signals, was determined by the measurement of ion spectra using a Wiley–McLaren-type TOF mass spectrometer, as none of the atomic signals could be detected without molecular background when measuring with the molecular beam. For identical laser pulse parameters, the photoelectron signals are proportional to the particle density ϱ in the interaction area. By measuring the atomic signals in TOF mass spectra for the seeded and the unseeded beam, the relative abundance $\alpha = \varrho_{\text{seeded}}/\varrho_{\text{unseeded}}$ can be determined. The photoelectron spectra from the unseeded measurement is scaled by α and subtracted from the spectra of the seeded measurement. The resulting difference spectra only contain molecular photoelectron signals providing access to the final populations of the bound molecular states XΣ and AΣ. The signals from the high-lying states 2Π and 5Σ after excitation by the IR pump are detected background-free due to the choice of the VIS probe pulse wavelength (cf. Section 3.2.2).

The TOF mass spectra on the left side of Fig. 3.17 were obtained by ionization of the seeded (black) and the unseeded particle beam (green, to enhance the particle density, here potassium dispensers were used in addition) by bandwidth-limited, infrared femtosecond laser pulses. The spectrum of the seeded beam clearly shows signals of atomic potassium isotopes and potassium dimers. For the unseeded beam and additional use of the potassium dispensers at identical laser parameters, only atomic potassium ions are generated. The ratio of atomic signals in the two spectra is $\alpha = 3.8$. On the right sight of Fig. 3.17, the photoelectron spectra related to the respective TOF mass spectra are shown. They were generated by the interaction of the control and the probe pulse with the seeded (black) and the unseeded (red) particle beam. The purely atomic photoelectron spectrum (red) was already scaled by $\alpha = 3.8$. By subtracting the unseeded from the seeded photoelectron

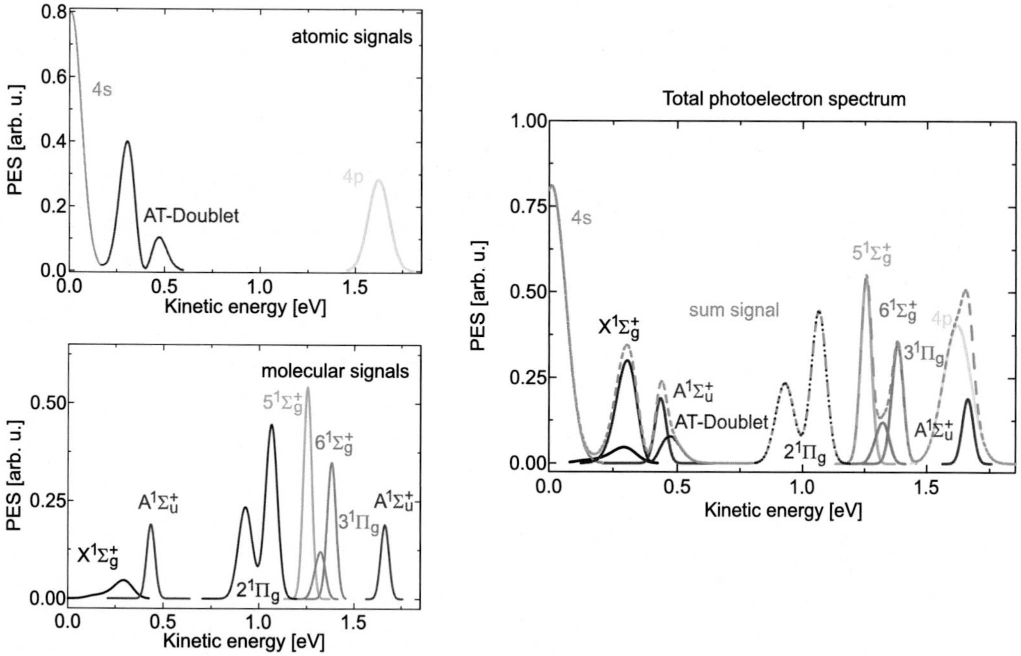

FIG. 3.16 (Left) Calculated photoelectron signals from excitation and ionization of potassium atoms (top) and dimers (bottom) with a 790 nm femtosecond laser pulse and a 570 nm probe pulse. Because each of the signals was calculated separately, their relative amplitudes are arbitrary and do not reflect a specific population distribution in the corresponding systems. (Right) Combination of the atomic and molecular photoelectron signals for assignment of contributions in the measured two-color spectra. Due to the arbitrary scaling of the signals, the simulated spectrum is not directly comparable to a measured spectrum. However, the general shape indicates a massive overlap of atomic and molecular signals. In particular, none of the atomic signals is detected without molecular background. Therefore, for disentangling atomic and molecular contributions the additional measurement of mass spectra is required.

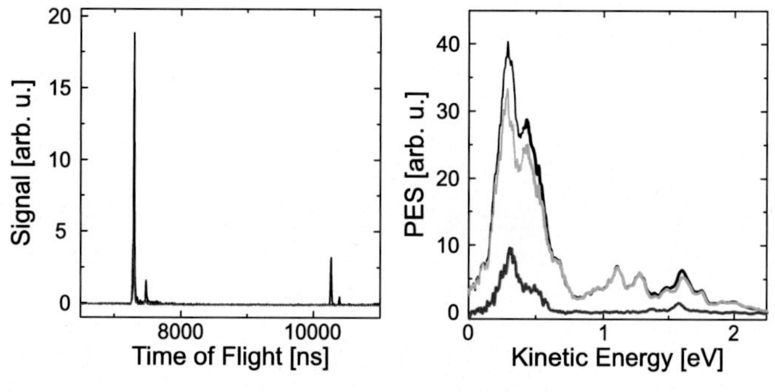

FIG. 3.17 (Left) TOF mass spectra that arise from the ionization of the seeded particle beam (black) and the unseeded particle beam with additional operation of the potassium dispenser source (red) by the bandwidth-limited IR pulse. While the spectrum of the seeded particle beam contains a clear signal of potassium dimers (centered at $\mathrm{TOF} = 10.250$ ns), no molecular signals are observed in the spectrum of the unseeded beam. This purely atomic spectrum displays ca. 3.8 times less ionic signal of potassium isotopes (centered around $\mathrm{TOF} = 7.250$ ns) than the spectrum of the seeded beam. (Right) Photoelectron spectra, generated by the interaction of the control and the probe pulse with the seeded (black) and the unseeded (red) particle beam. The depicted atomic photoelectron spectrum was already scaled by $\alpha = 3.8$. The difference of the two spectra yields the green spectrum containing only molecular signals.

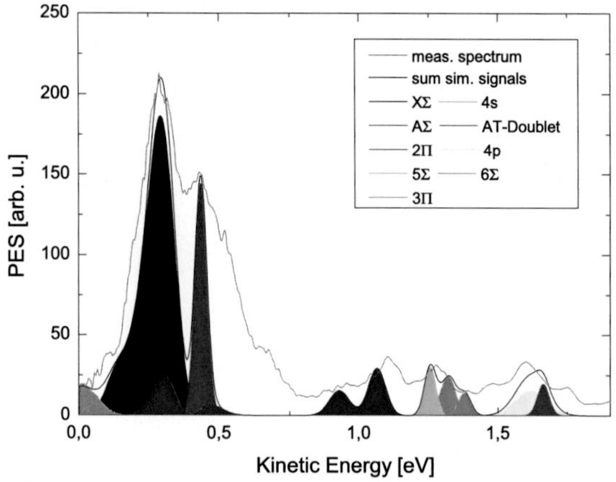

FIG. 3.18 Measured and simulated photoelectron spectra of potassium atoms and molecules. The simulated photoelectron signals of the atomic and molecular states (colored areas) were calculated individually and scaled arbitrarily. Comparison of the experimental result (magenta) and the calculated signals allows for an assignment the contributions in the measured spectrum.

spectrum, a spectrum only containing molecular photoelectron signals is created. A photoelectron spectrum with both atomic and molecular signals is shown in Fig. 3.18 together with simulated photoelectron signals, to identify the different contributions. For the simulated signals the contributions from the atomic and molecular states were scaled individually. The energetic position of all as well as the very specific form of some the signals enable an assignment of the electronic states. It is now guaranteed that the correct signals are integrated to obtain 2D control landscapes for specific states.

3.3.4.2 Experimental Results

The experiments on adiabatic population control in potassium dimers by chirped Airy pulses were performed using the experimental two-color setup described in Section 3.2.2. By systematic variation of the control parameters ϕ_2 and ϕ_3 and measurement of two-color photoelectron spectra, we mapped the control topology of the 2D parameter space. The two-color spectra were evaluated according to the procedure described in Section 3.3.4.1 to extract the signal yields from the molecular states AΣ, 2Π and 5Σ, indicative of the final population of the molecular target states. The signal yields are extracted by integration over the respective energy windows. Fig. 3.19 illustrates the experimental results in terms of 2D yield landscapes as a function of the parameters ϕ_2 and ϕ_3. Each 2D landscape was normalized individually to its maximum.

All three landscapes show a pronounced topological modulation, which is ascribed to different populations of the states after the interaction with the spectrally phase-modulated IR control pulses. The 5Σ-population is maximized by bandwidth-limited and moderately modulated pulses, as indicated by the distinct maximum located close to the landscape center ($\phi_2 = 0$, $\phi_3 = 0$). In fact, the maximum is slightly shifted to positive values of ϕ_3. This is not the case for the 2Π- and the AΣ-state whose maxima are situated clearly off-center. The AΣ-landscape exhibits a maximum at large positive values of ϕ_2 and large negative values of ϕ_3. In case of the 2Π-landscape, two maxima are observed at large negative ϕ_2, one located at large positive ϕ_3, the other one at large negative ϕ_3. Moreover both landscapes display valleys of the population transfer running along the positive diagonals of the respective landscapes

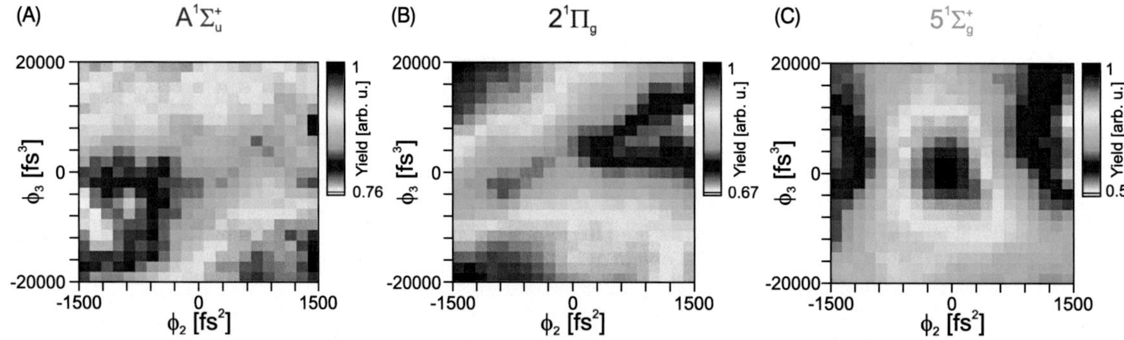

FIG. 3.19 2D landscapes of the photoelectron signal yields from the electronic states $A^1\Sigma_u^+$, $2^1\Pi_g$ and $5^1\Sigma_g^+$ as functions of the phase parameters ϕ_2 and ϕ_3. The yields were derived from the measured two-color spectra by integration of over the corresponding energy windows. Each landscaped is normalized to its maximum.

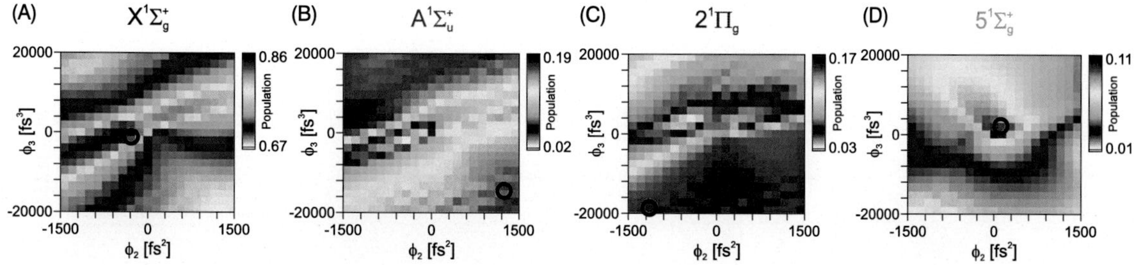

FIG. 3.20 Simulated final populations of the electronic states in the potassium dimer for the interaction with chirped Airy pulses as a function of the phase parameters ϕ_2 and ϕ_3. Focal intensity and orientation averaging was taken into account. The black circles highlight the parameter combinations given in Table 3.1.

(equal sign of ϕ_2 and ϕ_3). In the experiment no signals from the $X\Sigma$-state could be retrieved.

The measured landscapes are compared to calculated landscapes based on molecular dynamics simulations including both focal intensity and molecular orientation averaging (cf. Section 3.2.3.2). From the calculations, we derive the final population of the molecular electronic states at the end of the interaction with the control pulse in dependence of the internuclear distance R. In Fig. 3.20, the resulting 2D population landscapes are shown for the electronic states $A\Sigma$, 2Π and 5Σ evaluable in the experiment and, in addition, the $X\Sigma$-state. The simulated landscapes are in good agreement with the measured landscapes.

The diagonal valley structures observed experimentally in the final populations of the $A\Sigma$- and the 2Π-state, where only little population is transferred to those states, are reproduced in the corresponding simulated landscapes. Above and below the diagonal, a higher population transfer is observed, again similar to the experiment. In the simulations, the maximal population in the $A\Sigma$-state is found for $\phi_3 > 0$ and in the 2Π-state for $\phi_3 < 0$. However, in the measured data this behavior can only be seen as a tendency. The maximum of the 5Σ-landscape is less symmetrical than the experimental result, but also centered around the bandwidth-limited pulse, i.e. around $(\phi_2 = 0, \phi_3 = 0)$. In the simulation, efficient population transfer ($> 5\%$) is obtained for positive ϕ_3, whereas for negative ϕ_3 the population efficiency decreases much more rapidly with ϕ_3 than in the measured landscape. In the simulations, also the population dynamics of the $X\Sigma$-state is evaluated as a function of ϕ_2 and ϕ_3. The resulting landscape is shown in Fig. 3.9A. We find a pronounced maximum, corresponding to coherent population return to the ground state, stretching diagonally from negative (ϕ_2, ϕ_3) to positive (ϕ_2, ϕ_3). In the two other quadrants, the ground state is almost depleted and population is efficiently transferred to the excited electronic states.

While the experiment (cf. Fig. 3.19) was carried out using a fixed pulse energy, the simulations can be performed for several pulse energies. The resulting landscapes for increasing pulse energy, parametrized by the peak amplitude $\mathcal{E}_0$ of the unshaped input pulse, are shown in Fig. 3.21. Apparently, the topology of the landscapes and the qualitative position of maxima and minima remains constant. Only the absolute populations of the states change and for higher $\mathcal{E}_0$ a tendency for larger population transfer from the ground to the excited states is observed. Furthermore, the landscapes gradually develop plateaus where the final populations are almost unaffected by small variations of the phase parameters ϕ_2 and ϕ_3. The simulations thus suggest a certain robustness of the population transfer to the electronic states in the potassium molecule induced by chirped Airy pulses.

3.3.4.3 Wave Packet Dynamics Analysis

The simulated landscapes, calculated for increasing pulse energy, point to an overall robust behavior of the populations of the electronic states. For distinctive points in the landscapes we will investigate to what extent the simulated wave packet dynamics can be used to identify control mechanisms. We choose parameter combinations that lead to high population transfer to the respective electronic states in the simulations as well as in the experiment, see Table 3.1.

The discussion in Schneider (2014) reveals that in addition to the final population of the electronic state also an inspection of the molecular resonance in combination with the instantaneous frequency and the type of the resulting wave packet may be of interest. The simulations are performed at the highest pulse energy that is contained in the landscapes in Fig. 3.20. The orientation of the molecules is fixed at 45° with respect to

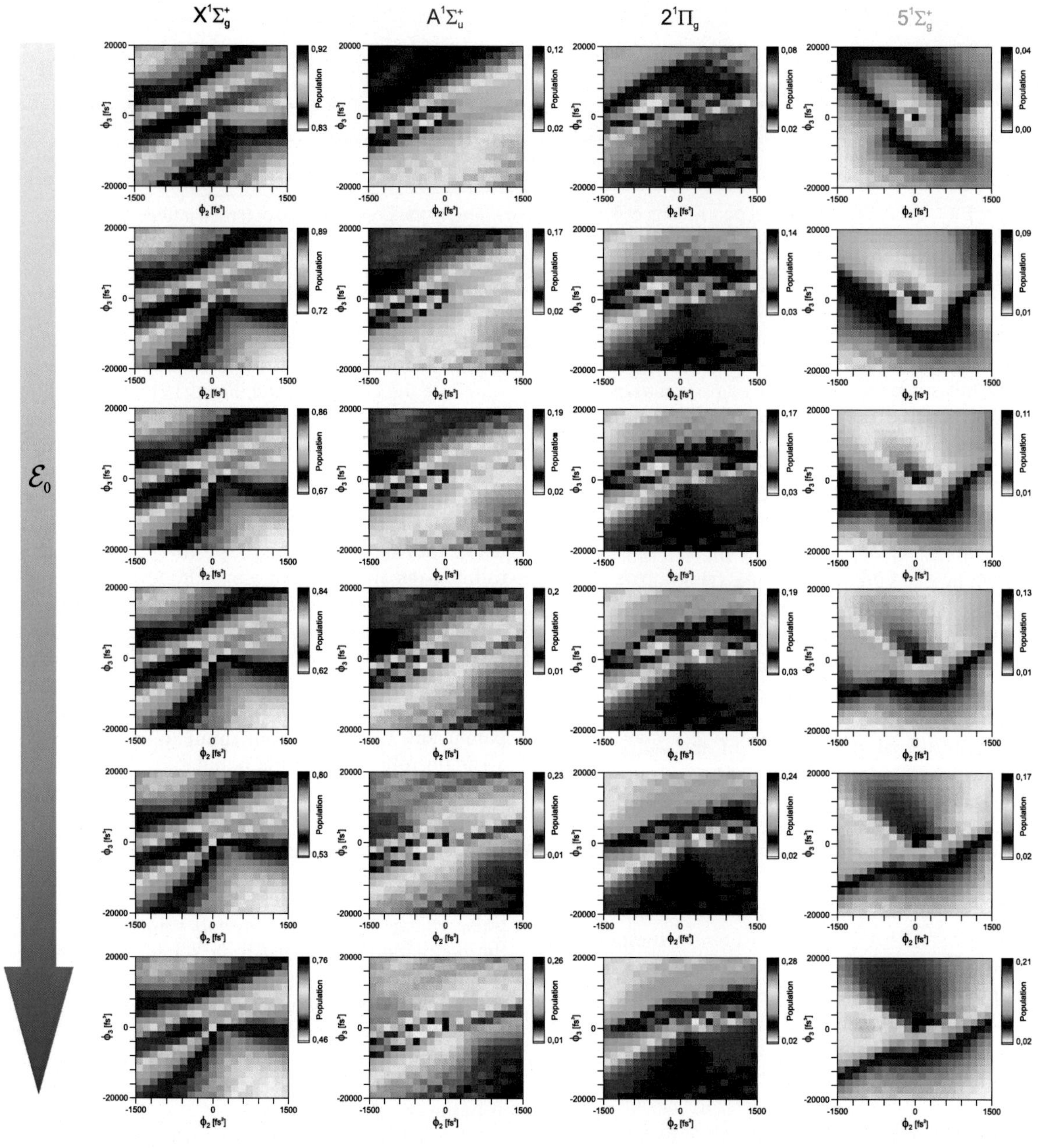

FIG. 3.21 Simulated ϕ_2–ϕ_3-landscapes of the final populations of the electronic states in the potassium dimer for increasing pulse energy (from top to bottom). Focal intensity and orientation averaging was taken into account. $\mathcal{E}_0$ marks the row already depicted in Fig. 3.20. The pulse energy was varied between $0.6 \cdot \mathcal{E}_0$ (top row) and $1.7 \cdot \mathcal{E}_0$ (bottom row). The landscapes display an overall robust behavior for excitation with increasing pulse energy. The general topology for each electronic state, as well as the qualitative position of minima and maxima in each landscape, remains constant.

TABLE 3.1
Parameter combinations (ϕ_2, ϕ_3) (middle and right column) for the selective excitation of different electronic target states (left column). The corresponding landscape points are marked by black circles in Fig. 3.20.

El. state	ϕ_2	ϕ_3
$X^1\Sigma_g^+$	$-200\ fs^2$	$-1000\ fs^3$
$A^1\Sigma_u^+$	$1250\ fs^2$	$-15.000\ fs^3$
$2^1\Pi_g$	$-1200\ fs^2$	$-20.000\ fs^3$
$5^1\Sigma_g^+$	$400\ fs^2$	$2000\ fs^3$

the polarization of the laser pulse, such that transitions of type Σ–Σ and Σ–Π are driven with the same field strength. Fig. 3.22 shows the results of the molecular dynamics simulations for the efficient population of the XΣ- and the 5Σ-state. We used the phase modulation parameters from Table 3.1. In comparison, the dynamics for the efficient population of the two different states show some similarities: The main population transfer takes place impulsively during the most intense part of the laser pulse. For the population of the XΣ-state, the ground state is initially nearly depopulated at $t = 0$ fs and the population transferred to the 5Σ-state. However, the population does not stay in that electronically excited state, but is transferred back to the ground state at the end of the main pulse.

Also the control pulse for the efficient population of the 5Σ-state initially depopulates the ground state. During the following part of the pulse the 5Σ-state is rapidly populated and even in the trailing edge of the pulse the population remains in this state.

Column B in Fig. 3.22 shows how the molecular resonance develops in the X–A-subsystem in analogy to Fig. 3.7E. The wave packets in the participating electronic states stay strongly localized during the whole interaction. Hence, the molecular resonance, evaluated at the expectation value of the internuclear distance $\langle R\rangle(t)$ in the X–A-subsystem is a meaningful quantity. In addition, the instantaneous frequency of the modulated laser pulses (blue) is depicted. In both cases of efficient population of either the XΣ- or the 5Σ-state the molecular resonance avoids the instantaneous frequency of

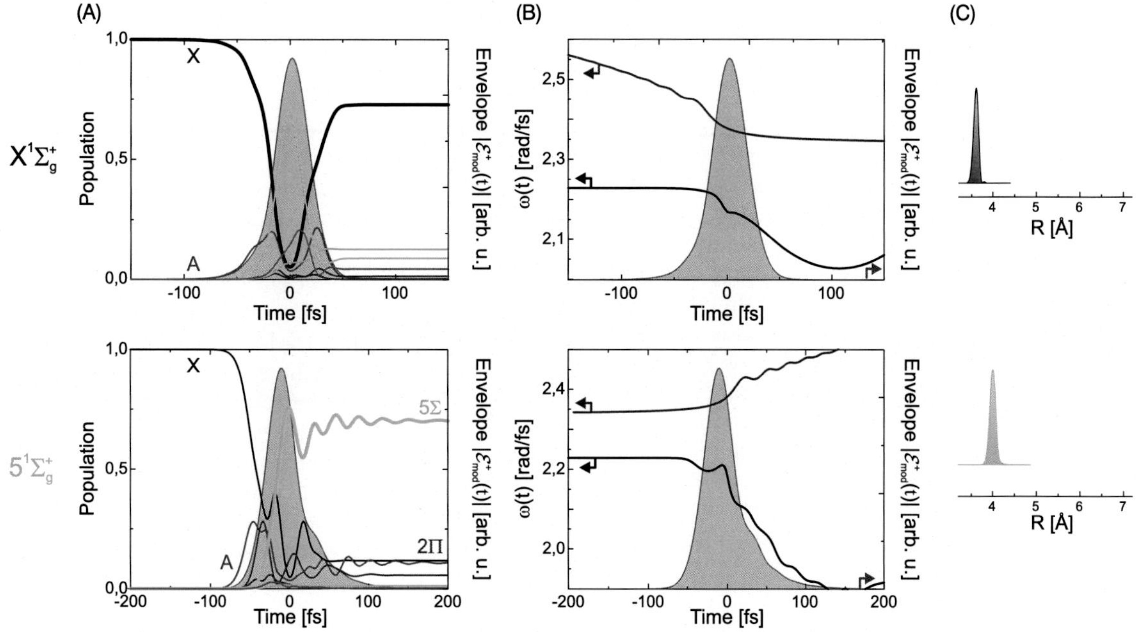

FIG. 3.22 Results for the molecular dynamics simulations for the efficient population of the $X^1\Sigma_g^+$- (top) and $5^1\Sigma_g^+$-state (bottom). (A) Simulated electronic population dynamics. (B) Time evolution of the instantaneous frequency of the laser pulse (blue) together with the molecular resonance of the X–A-subsystem (black). In addition, (A) and (B) display the envelope of the shaped laser pulse. (C) Nuclear wave packet in the respective electronic state by the end of the laser–molecule interaction.

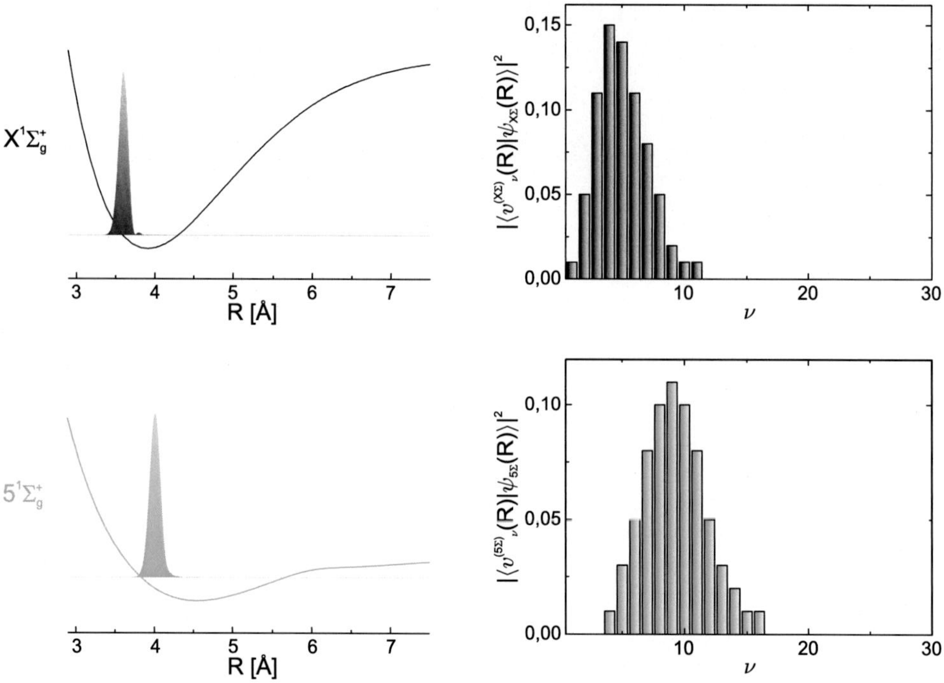

FIG. 3.23 (Left) Nuclear wave packets from Fig. 3.22C and corresponding electronic potentials. (Right) Underlying vibrational population distribution obtained by decomposition of the wave packet into vibrational eigenstates of the respective electronic state.

the laser pulse and no crossing of the two frequencies takes place. The joint movement of the wave packets in the XΣ- and the AΣ-state in the direction of larger internuclear distances is responsible for this behavior. For this movement the difference potential becomes smaller and the resonance frequency drops. Although these characteristics are similar in (B), the 5Σ-state is only transiently populated in the upper case and at the end of the interaction the main part of the population finds itself in the $X\Sigma$-state. In the lower case the population stays in the 5Σ-state after the excitation.

Column C displays the wave packet that is created in the respective electronic state. For both potentials we find a strongly localized wave packet. The populations $|v_\nu|^2 = |\langle\psi_m|v_\nu^{(m)}\rangle|^2$ of the vibrational eigenstates $|\, v_\nu^{(m)}\rangle$ of the electronic state $|\,\psi_m\rangle$ are presented in Fig. 3.23. After the interaction with the laser pulse and when neglecting decoherence and relaxation processes these $|v_\nu^{(m)}|^2$ stay constant. In both cases we obtain a broad distribution of populated vibrational eigenfunctions. (The wave function $\psi_m(R)$ is not normalized before the decomposition into vibrational eigenfunctions. Hence the populations of the vibrational eigenstates do not add up to unity.) While the $X\Sigma$-state is also populated before the interaction with the shaped laser pulse, after the interaction a nuclear wave packet is created in the ground state. In the 5Σ-state even higher vibrational eigenfunctions participate in the excited wave packet.

The results for the efficient population of the AΣ- and the 2Π-state (achieved with the phase modulation parameters ϕ_2 and ϕ_3 according to Table 3.1) are displayed in Fig. 3.24. Column A depicts the population dynamics, (B) the instantaneous frequency of the laser pulse and (C) the wave packet that was created in the electronic state at the end of the interaction. During the dynamics discussed for the efficient population of the XΣ- and the 5Σ-state the wave packets in the participating states stay localized. In contrast, the situation for the efficient excitation of the AΣ- and the 2Π-state is different. During the interaction with the chirped Airy pulses (shown in the background of Fig. 3.24A and B) only few vibrational eigenstates are excited in the $A\Sigma$- as well as in the 2Π-state. The excitation happens vertically from the XΣ ground state to the respective vibrational eigenstates of the AΣ state, which have spatial overlap with the wave packet in the ground state,

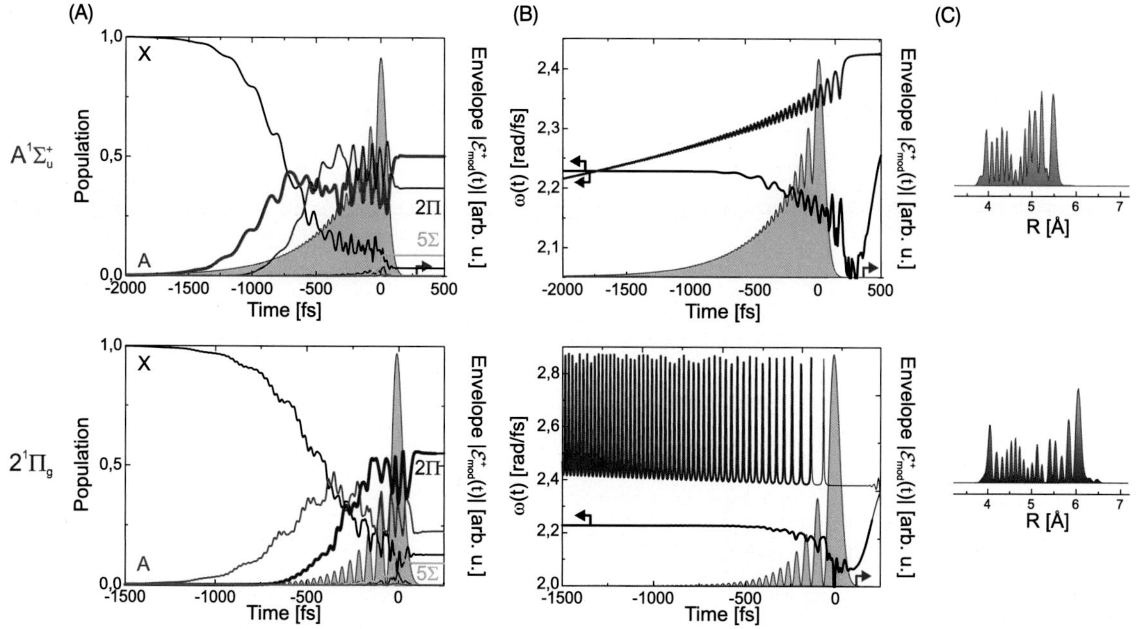

FIG. 3.24 Same as Fig. 3.22 for the efficient population of the $A^1\Sigma_g^+$- (top) and $2^1\Pi_g$-state (bottom).

according to the Franck–Condon-principle – see also Section 6.5.2 of this book, Vaníček and Begušić, 2020. No localized wave packet is created, but the participating vibrational eigenstates that are distributed over the whole potential overlap. Therefore, the resonance of the transition is given by the XΣ–AΣ difference potential at the expectation value $\langle R\rangle(t)$ in the XΣ-state that is also shown in (B). The wave packet in the ground state barely moves during the interaction with the depicted laser pulse, hence the resonance frequency stays constant during the major part of the interaction. Again, parallels in the induced dynamics can be found for the efficient population of the AΣ- and the 2Π-state. Both chirped Airy pulses, which transfer the population into the two electronic states, are strongly asymmetric and start with a slowly rising edge. The dynamics evoked by the pulse resemble each other especially at the beginning of the two pulses. Over the whole rising edge of the pulses population is initially transferred from the ground state into the AΣ state. Once ca. 25% of population is in this excited state, also the 2Π-state starts being slowly populated. At the end of the interaction with the specific laser pulses, wave packets have been created in both target states that are composed of few, higher lying vibrational eigenfunctions. The wave packets can be seen in Fig. 3.24C. Additionally, the populations of the vibrational eigenstates for these wave packets are depicted in Fig. 3.25. In both cases only few vibrational stats are populated. A major part of the population can be found in a single vibrational eigenstate.

The population dynamics that lead to the result for the AΣ state can be understood by considering the time evolution of the molecular resonance and the instantaneous frequency of the laser pulse in Fig. 3.24B. By changing the spectral phase not only the temporal amplitude but also the instantaneous frequency becomes time-dependent. Initially, this frequency is lower than the molecular resonance. At a later time, it crosses the molecular resonance and finally at the end of the pulse it is higher. Overall the course of the two frequencies is reminiscent of a RAP. In a RAP the electronic population is transferred to the excited state by virtue of a time-dependent frequency change, the slow change of the pulse envelope and the high intensity. In the potassium dimer the 2Π-state lies in one-photon resonance with the AΣ-state. After the latter has been populated efficiently ($>25\%$), the higher lying 2Π-state gets populated, too. During the following interaction Rabi oscillations take place between the two states, especially during the intense main pulse. A glance at Fig. 3.21 illustrates that the two states are populated or not populated together also for higher pulse energies. A clear selectivity is barely possible.

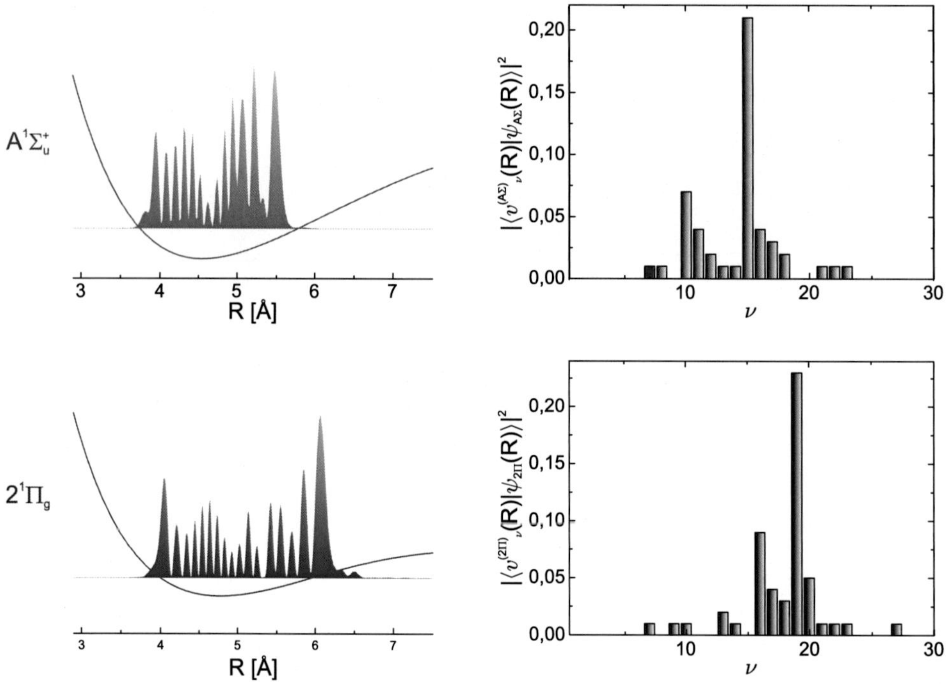

FIG. 3.25 Same as Fig. 3.23 for the nuclear wave packets from Fig. 3.24C.

The dynamics for the efficient population of the 2Π-state cannot be discussed similarly. The molecular resonance and the instantaneous frequency of the laser pulse do not cross for this case. The laser frequency stays higher than the molecular resonance during the entire interaction. Population from the ground state can therefore be excited into higher lying vibrational states of the AΣ-state. From these states a subsequent transfer to the 2Π-state takes place. Altogether the curves of the population dynamics are less smooth than in the case of efficient population of the $A\Sigma$-state and are modulated beyond the structure determined by the subpulses.

3.3.4.4 Discussion

With the help of the wave packet dynamics, discussed in the previous section, the population dynamics can be understood qualitatively. The control mechanisms are more complex than those used in the data interpretation in Schneider et al. (2011) and Schneider (2014). Similar to the scenarios that were presented in Section 3.3.3, the joint motion of wave packets can also be observed in the potassium dimer in the $X\Sigma$ ground state and the first excited state $A\Sigma$. The time evolution of the expectation value $\langle R \rangle$ towards larger internuclear distances leads to a decrease of the difference potentials of the two states and a time-dependent variation of the molecular resonance. According to Schneider et al. (2011) and Schneider (2014), a crossing of the molecular resonance and the instantaneous frequency under adiabatic conditions leads to population transfer to the excited state. A parallel progression of the two frequencies should result only in a transient population transfer and at the end of the interaction the molecule should be back in its ground state. The latter case is observed in the potassium dimer: by impulsive excitation a nuclear wave packet can be created in the ground state that comprises a broad distribution of vibrational eigenfunctions. However, for a different pulse shape a parallel development of the laser frequency and the molecular resonance results in efficient population transfer to the 5Σ-state.

In the previous experiments on porphyrazines the excitation of a single vibrational eigenstate was connected with a scenario similar to a RAP: due to the transfer of population from the ground state to one vibrational eigenstate of the AΣ-potential the molecular resonance stays constant during the interaction with the chirped Airy pulse. By virtue of the chirp the instantaneous frequency of the laser pulse can cross this resonance during the progression of the pulse. In combination with the slowly rising edge of the laser pulse these dynamics result in an efficient population transfer to the excited state. In the potassium molecule, this scenario is identified as well for the efficient population of the $A\Sigma$-state. The shaped laser pulse starts with a slowly rising edge and its time-dependent frequency crosses the constant molecular resonance. Approximately, only a single vibrational eigenstate is excited in the $A\Sigma$-state. Although the 2Π-state, which lies in one-photon resonance, leads to more complex population dynamics, the situation is still comparable to the one discussed in Schneider et al. (2011) and Schneider (2014). For the overall very similar evolution in the case of efficient population of the 2Π-state the progression of the relevant values cannot be interpreted in this form. The envelope of the chirped Airy pulse resembles the pulse that populates the $A\Sigma$-state efficiently, however, the instantaneous frequency does not cross the molecular resonance. Nevertheless, only few vibrational eigenstates are excited transiently in the $A\Sigma$- state and also finally in the 2Π-state.

The potassium dimer is less complex than the porphyrazines previously investigated and its theoretical description can be done more precisely. Still it is more complex due to the multitude of additional electronic states that are taken into account. These additional states do not only represent yet further targets for optimization of populations transfer, but do also influence the transient populations and the dynamics of the wave packets due to their coupling to the X–A-subsystem. It is not surprising that the control concepts, which were developed in a simplified model of two harmonic potentials, are not identical to those found in the potassium dimer. The control concepts emerge in the X–A-subsystem, in which the states can be efficiently populated by either coupling of wave packets or the excitation of one vibrational eigenstate. An efficient population of the higher-lying states 2Π and 5Σ is possible as well, but the underlying mechanisms are quite complex. Furthermore, the simulation of control landscapes at higher pulse energies indicates that the efficient population of the electronic states is robust with respect to fluctuations in the pulse energy and small variations of the phase modulation parameters ϕ_2 and ϕ_3. This hints towards adiabatic control of electronic population transfer in the molecule by chirped Airy pulses.

3.3.5 Conclusion and Outlook

In this section we presented a 2D strong-field approach being sensitive to adiabatic population transfer in molecules. This is an extension of the well-known RAP techniques, developed to invert two-level systems, to multilevel molecules including nuclear dynamics. There is high interest in the development of adiabatic control scenarios for molecules as efficient and robust population transfer to an excited state enhances the subsequent photochemistry. Selective and efficient state preparation may also help to simplify spectroscopy. Important natural pulse parameters to study RAP-type strong-field effects are frequency sweeps and slowly varying temporal intensity profiles. With the help of optical Fourier transform pulse shaping, these attributes are obtained by modulating the spectral phase with a polynomial of third order containing quadratic components. In this way temporally asymmetric chirped Airy pulses are generated that combine adiabatic with impulsive characteristics.

Following up on our experiments on strong-field excitation of dye molecules by chirped Airy pulses, in this section we investigated the same scenario on the simpler potassium dimer prototype to gain a transparent physical picture of the exerted control. Moreover, the higher-lying electronic states of the potassium dimer represent additional target states for optimization.

In the experiments, we employ two-color energy-resolved photoelectron spectroscopy and measure photoelectrons released from the excited molecular target states by a time-delayed VIS probe pulse. The two-color photoelectron spectra are recorded as a function of the control parameters ϕ_2 (second order polynomial) and ϕ_3 (third order polynomial). Comparison of the measured 2D landscapes to simulations indicates that chirped Airy pulses are able to robustly excite the molecules into predefined target states. The underlying molecular dynamics simulations allow for an investigation of the induced neutral dynamics for specifically tailored laser pulses that lead to efficient population transfer into different target states. The selected dynamics can be interpreted in connection with the evolution of the molecular resonance and the instantaneous frequency of the shaped laser field. For the subsystem consisting of the ground and the first excited state the creation of a ground state wave packet and the excitation of few vibrational eigenstates in the excited state is observed. These effects can be understood by an interplay between the pulse envelope (slowly rising or impulsive) and the crossing or non-crossing of the instantaneous laser frequency and the time-dependent molecular resonance due to the induced wave packet propagation. The higher lying electronic states are in one-photon resonance with the first excited state. Their efficient population and the respective population dynamics cannot be interpreted in direct analogy to the subsystem of ground and first excited state. For a deeper understanding of the underlying excitation dynamics, further experimental studies are required. Still, the numerical simulations for the interaction of the potassium dimer with chirped Airy pulses at different pulse energies indicate a robust behavior for all electronic states investigated.

Follow-up investigations of robust molecular excitation by applying this 2D approach to larger molecules in the gas phase as well as in the liquid phase in order to achieve selective adiabatic population transfer are promising for the reasons mentioned above. Note that for the liquid phase it was shown in simulations that the influence of electronic decoherence is softened when taking the nuclear dynamics into account (Schneider et al., 2011).

3.4 SUMMARY

To induce and control coherent electronic excitation is at the heart of many quantum technologies, where a vast number of different two-level systems are in use. Mainly the non-perturbative regime of light matter interaction is employed to that end. 2D spectroscopy to unveil molecular dynamics by separating contributions that would be overlapping in a one-dimensional measurement is often performed in the perturbative light–matter interaction regime when carried out in the optical domain.

In this contribution we demonstrated on the potassium dimer prototype – both experimentally and with the help of quantum dynamics simulation – an extension of impulsive spin-locking techniques and adiabatic RAP-type techniques to molecules. We make use of 2D parameterizations of the spectral phase, taking essential physical mechanisms into account in order to identify areas of electronic coherences and of efficient population transfer into preselected target states.

Sinusoidal spectral phase modulation results in a pulse sequence with adjustable temporal separation, relative pulse ratio and phase. The generated pulse shapes range from clearly separated pulse trains down to complex-shaped laser fields when the individual subpulses start to merge. By 2D scanning of the pulse separation and the phase after fixing the amplitude ratio, we were able to find areas where the induced electronic coherence can be brought in phase and out of phase with the driving electric field and, as a consequence, populate different final states in the molecule.

If the spectral phase is modulated by a polynomial of third order (containing quadratic components), temporally asymmetric chirped Airy pulses are generated that combine adiabatic with impulsive characteristics. By 2D scanning of the cubic and the quadratic part we identify regimes of robust and efficient population transfer.

We suggest that such physically motivated 2D parameterizations are a versatile tool to study and control the coupled electron-nuclear dynamics also on larger molecules in the gas phase as well as in solution on the femtosecond time scale with attosecond precision. These methods were originally developed in the area of coherent control of chemical reactions and are now ready for cross-fertilizing neighboring disciplines. For the induced electronic coherence part further applications are envisioned, for example, in the area of quantum computing using molecules, molecular quantum sensing or highly sensitive molecular identification schemes. With respect to the latter we currently work on enantiomeric excess determination combining this approach with circular dichroism techniques. For the efficient and robust population transfer in molecules a whole variety of photophysical processes could be optimized and by selective and efficient preparation of individual target states also molecular spectroscopy in large molecules or ensembles might be simplified.

REFERENCES

Autler, S.H., Townes, C.H., 1955. Stark effect in rapidly varying fields. Phys. Rev. 100 (2), 703–722.

Baumert, T., Engel, V., Meier, C., Gerber, G., 1992. High laser field effects in multiphoton ionization of sodium dimer. Experiment and quantum calculations. Chem. Phys. Lett. 200 (5), 488–494.

Bayer, T., Braun, H., Sarpe, C., Siemering, R., von den Hoff, P., Vivie-Riedle, R. de, Baumert, T., Wollenhaupt, M., 2013. Charge oscillation controlled molecular excitation. Phys. Rev. Lett. 110, 123003.

Bayer, T., Wollenhaupt, M., Braun, H., Baumert, T., 2014. Ultrafast and efficient control of coherent electron dynamics via SPODS. Adv. Chem. Phys., 1–50.

Bayer, T., Wollenhaupt, M., Braun, H., Baumert, T., 2016. Ultrafast and efficient control of coherent electron dynamics via SPODS. In: Brumer, P., Rice, S.A., Dinner, A.R. (Eds.), Advances in Chemical Physics, vol. 159. John Wiley & Sons, Inc, Hoboken, NJ, USA, pp. 235–282.

Braun, H., Bayer, T., Sarpe, C., Siemering, R., Vivie-Riedle, R. de, Baumert, T., Wollenhaupt, M., 2014. Coupled electron-nuclear wavepacket dynamics in potassium dimers. J. Phys. B, At. Mol. Opt. Phys. 47 (12), 124015.

Braun, H., von den Hoff, P., Bayer, T., Siemering, R., Vivie-Riedle, R. de, Wollenhaupt, M., Baumert, T., 2013. Efficient attosecond control of electron dynamics in molecules. EPJ Web Conf. 41, 2026.

Brixner, T., Gerber, G., 2001. Femtosecond polarization pulse shaping. Opt. Lett. 26 (8), 557–559.

Brixner, T., Krampert, G., Pfeifer, T., Selle, R., Gerber, G., Wollenhaupt, M., Graefe, O., Horn, C., Liese, D., Baumert, T.,

2004. Quantum control by ultrafast polarization shaping. Phys. Rev. Lett. 92 (20), 208301.

Cao, J., Bardeen, C.J., Wilson, K.R., 1998. Molecular pi pulses for total inversion of electronic state population. Phys. Rev. Lett. 80 (7), 1406–1409.

Cao, J.S., Bardeen, C.J., Wilson, K.R., 2000. Molecular pi pulses: population inversion with positively chirped short pulses. J. Chem. Phys. 113 (5), 1898–1909.

Cerullo, G., Bardeen, C.J., Wang, Q., Shank, C.V., 1996. High-power femtosecond chirped pulse excitation of molecules in solution. Chem. Phys. Lett. 262, 362–368.

Corkum, P.B., Krausz, F., 2007. Attosecond science. Nat. Phys. 3, 381–387.

von den Hoff, P., Kowalewski, M., Vivie-Riedle, R. de, 2011. Searching for pathways involving dressed states in optimal control theory. Faraday Discuss. Chem. Soc. 153, 159–171.

von den Hoff, P., Siemering, R., Kowalewski, M., Vivie-Riedle, R. de, 2012. Electron dynamics and its control in molecules: from diatomics to larger molecular systems. IEEE J. Sel. Top. Quantum Electron. 18 (1), 119–129.

Faraday Discussion vol. 153, 2011.

Frohnmeyer, T., Hofmann, M., Strehle, M., Baumert, T., 1999. Mapping molecular dynamics (Na2) in intense laser fields: another dimension to femtochemistry. Chem. Phys. Lett. 312 (5–6), 447–454.

Goto, H., Katsuki, H., Ibrahim, H., Chiba, H., Ohmori, K., 2011. Strong-laser-induced quantum interference. Nat. Phys. 7, 383–385.

Greenman, L., Koch, C.P., Whaley, K.B., 2015. Laser pulses for coherent XUV Raman excitation. Phys. Rev. A 92 (1), 13407.

Hartmann, S.R., Hahn, E.L., 1962. Nuclear double resonance in the rotating frame. Phys. Rev. 128 (2042), 2053.

Herek, J.L., Wohlleben, W., Cogdell, R., Zeidler, D., Motzkus, M., 2002. Quantum control of energy flow in light harvesting. Nature 417, 533–535.

Kling, M.F., Vrakking, M.J.J., 2008. Attosecond electron dynamics. Annu. Rev. Phys. Chem. 59, 463–499.

Köhler, J., Wollenhaupt, M., Bayer, T., Sarpe, C., Baumert, T., 2011. Zeptosecond precision pulse shaping. Opt. Express 19, 11638–11653.

Krausz, F., Ivanov, M., 2009. Attosecond physics. Rev. Mod. Phys. 81 (1), 163–234.

Marquardt, R., Quack, M., 2020. Foundations of time dependent quantum dynamics of molecules under isolation and in coherent electromagnetic fields. In: Marquardt, R., Quack, M. (Eds.), Molecular Spectroscopy and Quantum Dynamics. Elsevier, Amsterdam. Chapter 1 (this book).

Meier, C., Engel, V., 1994. Interference structure in the photoelectron spectra obtained from multiphoton ionization of Na_2 with a strong femtosecond laser pulse. Phys. Rev. Lett. 73 (24), 3207–3210.

Meshulach, D., Silberberg, Y., 1998. Coherent quantum control of two-photon transitions by a femtosecond laser pulse. Nature 396 (6708), 239–242.

Nuernberger, P., Ruetzel, S., Brixner, T., 2015. Multidimensional electronic spectroscopy of photochemical reactions. Angew. Chem., Int. Ed. 54 (39), 11368–11386.

Palacios, A., Bachau, H., Martín, F., 2007. Excitation and ionization of molecular hydrogen by ultrashort VUV laser pulses. Phys. Rev. A 75, 13408.

Petersen, J., Mitric, R., 2012. Electronic coherence within the semiclassical field-induced surface hopping method: strong field quantum control in K_2. Phys. Chem. Chem. Phys. 14 (23), 8299–8306.

Präkelt, A., Wollenhaupt, M., Assion, A., Horn, C., Sarpe-Tudoran, C., Winter, M., Baumert, T., 2003. Compact, robust and flexible setup for femtosecond pulse shaping. Rev. Sci. Instrum. 74 (11), 4950–4953.

Prokhorenko, V.I., Nagy, A.M., Waschuk, S.A., Brown, L.S., Birge, R.R., Miller, R.J.D., 2006. Coherent control of retinal isomerization in bacteriorhodopsin. Science 313, 1257–1261.

Remacle, F., Levine, R.D., 2006. An electronic time scale in chemistry. Proc. Natl. Acad. Sci. USA 103 (18), 6793–6798.

Roeding, S., Brixner, T., 2018. Coherent two-dimensional electronic mass spectrometry. Nat. Commun. 9 (1), 2519.

Ruge, M., Wilcken, R., Wollenhaupt, M., Horn, A., Baumert, T., 2013. Coherent control of colloidal semiconductor nanocrystals. J. Phys. Chem. C 117, 11780–11790.

Schneider, J., 2014. Kontrolle von Sensitizerdynamik mit dediziert geformten Femtosekunden-Laserpulsen. Dissertation, Kassel.

Schneider, J., Wollenhaupt, M., Winzenburg, A., Bayer, T., Köhler, J., Faust, R., Baumert, T., 2011. Efficient and robust strong-field control of population transfer in sensitizer dyes with designed femtosecond laser pulses. Phys. Chem. Chem. Phys. 13, 8733–8746.

Shapiro, M., Brumer, P., 2011. Quantum Control of Molecular Processes. Wiley-VCH, Berlin.

Shore, B.W., 2008. Coherent manipulations of atoms using laser light. Acta Phys. Slovaca 58 (3), 243–486.

Shore, B.W., 2011. Coherent Manipulation of Quantum Structures Using Laser Pulses. Cambridge University Press, Cambridge.

Siegmann, A.E., 1986. Lasers. University Science Books, Mill Valley.

Sinha, M.P., Caldwell, C.D., Zare, R.N., 1974. Alignment of molecules in gaseous transport: alkali dimers in supersonic nozzle beams. J. Chem. Phys. 61 (2), 491–503.

Sussman, B.J., 2011. Five ways to the nonresonant dynamic Stark effect. Am. J. Phys. 79 (5), 477–484.

Sussman, B.J., Townsend, D., Ivanov, M.Y., Stolow, A., 2006. Dynamic Stark control of photochemical processes. Science 314, 278–281.

Thallmair, S., Siemering, R., Kölle, P., Kling, M., Wollenhaupt, M., Baumert, T., Vivie-Riedle, R. de, 2014. The interplay of nuclear and electron wavepacket motion in the control of molecular processes: a theoretical perspective. In: Gatti, F. (Ed.), Molecular Quantum Dynamics. Springer, Heidelberg, New York, Dordrecht, London, pp. 213–248.

Vala, J., Kosloff, R., 2001. Coherent mechanism of robust population inversion. Opt. Express 8, 238–245.

Vaníček, J., Begušić, T., 2020. Ab initio semiclassical evaluation of vibrationally resolved electronic spectra with thawed gaussians. In: Marquardt, R., Quack, M. (Eds.), Molecular Spectroscopy and Quantum Dynamics. Elsevier, Amsterdam. Chapter 6 (this book).

Vitanov, N.V., Halfmann, T., Shore, B.W., Bergmann, K., 2001. Laser-induced population transfer by adiabatic passage techniques. Annu. Rev. Phys. Chem. 52, 763–809.

Vivie-Riedle, R. de, Troppmann, U., 2007. Femtosecond lasers for quantum information technology. Chem. Rev. 107, 5082–5100.

Weiner, A.M., 2000. Femtosecond pulse shaping using spatial light modulators. Rev. Sci. Instrum. 71 (5), 1929–1960.

Wollenhaupt, M., Assion, A., Baumert, T., 2012. In: Springer Handbook of Lasers and Optics. Springer.

Wollenhaupt, M., Assion, A., Bazhan, O., Horn, C., Liese, D., Sarpe-Tudoran, C., Winter, M., Baumert, T., 2003. Control of interferences in an Autler–Townes doublet: symmetry of control parameters. Phys. Rev. A 68, 15401.

Wollenhaupt, M., Baumert, T., 2006. Ultrafast strong field quantum control on K2 dimers. J. Photochem. Photobiol. A 180, 248–255.

Wollenhaupt, M., Baumert, T., 2011. Ultrafast laser control of electron dynamics in atoms, molecules and solids. Faraday Discuss. Chem. Soc. 153, 9–26.

Wollenhaupt, M., Krug, M., Köhler, J., Bayer, T., Sarpe-Tudoran, C., Baumert, T., 2009. Photoelectron angular distributions from strong-field coherent electronic excitation. Appl. Phys. B 95, 245–259.

Wollenhaupt, M., Liese, D., Präkelt, A., Sarpe-Tudoran, C., Baumert, T., 2006a. Quantum control by ultrafast dressed states tailoring. Chem. Phys. Lett. 419, 184–190.

Wollenhaupt, M., Präkelt, A., Sarpe-Tudoran, C., Liese, D., Baumert, T., 2005. Quantum control and quantum control landscapes using intense shaped femtosecond pulses. J. Mod. Opt. 52 (16), 2187–2195.

Wollenhaupt, M., Präkelt, A., Sarpe-Tudoran, C., Liese, D., Bayer, T., Baumert, T., 2006b. Femtosecond strong-field quantum control with sinusoidally phase-modulated pulses. Phys. Rev. A 73, 63409.

Zai, J.-B., Zhan, W.-S., Wang, S., Dang, H.-P., Han, X., 2016. Population transfer of a NaH molecule via stimulated Raman adiabatic passage. Laser Phys. 26 (9), 96002.

Zewail, A.H., 2000a. Femtochemistry: atomic-scale dynamics of the chemical bond. J. Phys. Chem. 104 (24), 5660–5694.

Zewail, A.H., 2000b. Femtochemistry: atomic-scale dynamics of the chemical bond using ultrafast lasers. Angew. Chem., Int. Ed. 39 (15), 2586–2631.

CHAPTER 4

Attosecond Molecular Dynamics and Spectroscopy

DENITSA BAYKUSHEVA*,† • HANS JAKOB WÖRNER*
*Laboratorium für Physikalische Chemie, ETH Zürich, Zürich, Switzerland
†Stanford PULSE Institute, SLAC National Accelerator Laboratory, Menlo Park, CA, United States

Abstract

This chapter presents an overview of the state of the art in attosecond time-resolved spectroscopy. The theoretical foundations of strong-field light–matter interaction and attosecond pulse generation are described. The enabling laser technologies are reviewed from chirped-pulse amplification and carrier-envelope-phase stabilization to the generation and characterization of attosecond pulses. The applications of attosecond pulses and pulse trains in electron- or ion-imaging experiments are presented, followed by attosecond electron spectroscopy in larger molecules. After this, high-harmonic spectroscopy and its applications to probing charge migration on attosecond time scales is reviewed. The rapidly evolving field of molecular photoionization delays is discussed. Finally, the applications of attosecond transient absorption to probing molecular dynamics are presented.

4.1 INTRODUCTION

Time-resolved spectroscopy, or the investigation of fundamental natural processes in real time, is a scientific field that lies at the intersection of physics, chemistry, and biology. Historically, the application of Schlieren photography by Alfred Toepler to study the instantaneous density profiles of fluids and propagating shockwaves by means of two subsequent light flashes (Krehl and Engemann, 1995) can be considered as the first time-resolved study of a microscopic process on a subsecond time-scale. These experiments laid out the methodology of the pump–probe technique, whereas further progress has remained coupled to the state of technological development setting the temporal limit to the cross-correlation of available pump/probe pulses. The nanosecond time scale was reached already in the 19th century, when Abraham and Lemoine (Abraham and Lemoine, 1899) demonstrated that the Kerr cell shutter can be operated with a time resolution of less than 10^{-8} s. In the first half of the 20th century, the development of the relaxation methods and the flash photolysis to study chemical reactions occurring on the micro- to nanosecond timescales allowed for the identification of transient reaction intermediates, a work whose importance was acknowledged by the Nobel prize in chemistry in 1967 awarded to M. Eigen, N. Porter and W. Norrish (Eigen, 1954; Norrish and Porter, 1949). Subsequently, the development of ultrafast laser technology, starting with the invention of the laser in 1960 and the subsequent introduction of pulsed-laser operation through Q-switching, modelocking, and frequency up-conversion techniques based on non-linear optical processes led to the improvement of the achievable time resolution by nearly six orders of magnitude. By the mid-1980s, pulses of few-femtosecond (fs) duration containing only few cycles of the fundamental carrier light wave became accessible using dye lasers (Fork et al., 1987). Stimulated by the work of A. Zewail (Nobel Prize in Chemistry 1999) (Zewail, 2000), these technologies have enabled the real-time tracking of detailed chemical reactions, including the monitoring of rotational and vibrational motion, dissociation processes occurring on bound and repulsive potential energy surfaces, and dynamics at conical intersections.

Since pulsed lasers operate mainly in the visible (VIS) and the (near-)infrared (NIR) part of the spectrum, the associated duration of a single optical cycle (e.g., 2.67 fs for 800-nm-carrier wave) imposes a lower limit to the pulse duration. Further extension of the achievable time resolution to the attosecond (10^{-18} s) frontier hence requires a different approach, namely the transfer of the ultrafast technology to shorter wavelengths via non-linear frequency up-conversion. Historically, this requirement entailed up-scaling of the energy of the femtosecond pulses by three orders of magnitude from the nJ- to the mJ-energy scale in order to reach the necessary laser field intensities. A multitude of techniques were developed toward meeting this goal, among those are the in-

Molecular Spectroscopy and Quantum Dynamics. https://doi.org/10.1016/B978-0-12-817234-6.00009-X

troduction of titanium–sapphire (Ti:Sa) as a broadband solid-state laser medium (Moulton, 1986), the Kerr-lens mode-locking (KLM) (Keller et al., 1991; Spence et al., 1991), the invention of semiconductor saturable absorbers (Sutter et al., 1999), and the chirped-pulse amplification (CPA) scheme (Maine et al., 1988; Strickland and Mourou, 1985). The latter development had far-reaching implications for fields not bearing direct relationship to fundamental research (i.e., industry/medicine) and was honored with the Nobel prize in physics in 2018. At the same time, the discovery of high-harmonic generation (HHG) (McPherson et al., 1987), to be discussed in detail in one of the subsequent sections, a highly non-linear non-perturbative process involving the up-conversion of the frequency of VIS or NIR laser pulses to a multiple of their frequency, provided a fully coherent, bright, broadband source of radiation in the vacuum- and extreme-ultraviolet (VUV-XUV) to soft-X-ray (SXR) ranges. Further developments, which will be treated in detail in Section 4.3, consolidated the role of HHG-based methods for the generation of attosecond pulse trains or even isolated attosecond pulses and the measurement and the control of the subcycle electric field evolution. These developments represent the tools necessary for the real-time observation and steering of electron dynamics on a sub-fs time scale.

In this chapter the technical advances and the theoretical developments to realize and rationalize attosecond-science experiments that reveal a new dynamical time scale (10^{-15}–10^{-18} s) are addressed, with a particular emphasis on molecular systems and the implications of attosecond processes for chemical dynamics. After a brief outline of the theoretical framework for treating non-perturbative phenomena in Section 4.2, we introduce the physical mechanisms underlying high-harmonic generation and attosecond technology. The relevant technological developments and experimental schemes are covered in Section 4.3. Throughout the remainder of the chapter, we report on selected applications in molecular attosecond physics, thereby addressing specific phenomena mediated by purely electronic dynamics: charge localization in a simple molecule (H_2), charge migration and delays in molecular photoionization. A particular focus will be placed on the description of the corresponding experimental methodology. An in-depth discussion of the state-of-the-art theoretical developments can be found in a recent review (Nisoli et al., 2017) as well as in a recent monograph (Vrakking and Lepine, 2019). Further reviews on attosecond science and technology can be found in (Kling and Vrakking, 2008; Krausz and Ivanov, 2009; Wörner and Corkum, 2011; Gallmann et al., 2012; Krausz, 2016; Kraus and Wörner,2018a, 2018b).

At this point, it is important to outline the new insights that attosecond metrology adds to the field of molecular physics. First and foremost, the majority of the experimental schemes enabling access to sub-fs dynamics rely on strong-field processes, i.e., highly non-linear light–matter interactions. The resulting novel phenomena pose a fundamental challenge to the well-established perturbative framework of quantum mechanics and make advanced theoretical methods indispensable for the rationalization of the experimental outcomes. Second, recalling that (within the simplest version of Bohr's model) the orbiting period of an electron in the hydrogen atom is about 152 as, attosecond techniques allow one to directly address the electronic degrees of freedom. One direction of research follows closely the extension of the operating principle of femtochemistry to the attosecond regime. Generally, femtochemistry studies nuclear rearrangements taking place on a timescale of tens to hundreds of femtoseconds (a vibration associated with the H–H bond has a fundamental period of 7.6 fs) and has further inspired the notion of coherent control, or the steering of reactions into predefined pathways. The attosecond analogue of this scheme (termed "attochemistry"; Lépine et al., 2014; Remacle and Levine, 2006) exploits the fact that (as far as the concept of an electronic state retains its validity) the potential energy surface (PES) dictating the nuclear motion is formed by the fast-moving electrons. Thus, perturbing the PES in a very specific manner can potentially give rise to a rearrangement of the nuclear framework, providing a steering mechanism solely based on electron dynamics. Third, attosecond technology allows one to address fundamental questions in chemical dynamics. One of them is the possibility to measure the finite duration of a photoionization event: Delays in photoemission ranging from several to several tens of attoseconds have successfully been resolved in atoms (Schultze et al., 2010; Klünder et al., 2011; Guénot et al., 2014; Sabbar et al., 2015; Heuser et al., 2016; Jordan et al., 2017), surfaces (Cavalieri et al., 2007; Neppl et al., 2012; Okell et al., 2015) and, more recently, in molecules (Huppert et al., 2016). Another exciting aspect concerns the role of many-body interactions, in particular the influence of electron correlation phenomena beyond the mean-field approximation. A particular example that has been widely explored theoretically and experimentally is the charge migration process (Cederbaum et al., 1986; Wörner et al., 2017; see also Section 4.5 as well as Gokhberg et al., 2020, Chapter 5 of this book), or the ultrafast electronic reorganization along the nuclear framework following a prompt ionization event. In general, the study of coherent electron dynamics and its implications for chemical reactivity have become possible only with the advancement of attosecond technology, specifically the generation of (isolated) pulses of attosecond duration. The significant bandwidths of the latter enable the coherent excitation of several electronic states; simultaneously, the short pulse durations ensure that the interaction time is confined within a temporal window below the typical lifetime of excited electronic states.

4.2 THEORETICAL DESCRIPTION OF STRONG-FIELD PHENOMENA

4.2.1 Overview of the Basic Terminology

The term "strong-field physics" is broadly used to classify the regime in which the treatment of laser–matter interactions in terms of perturbative non-linear optics is no longer applicable. Within the perturbative picture, the induced non-linear response is quantified in terms of a multipole expansion with respect to the driving field amplitude. For the case of transverse electromagnetic (TEM) fields, this treatment is applicable when the condition $U_\mathrm{p} < \omega_0$ holds, whereby ω_0 denotes the angular frequency of the oscillating field and U_p stands for the ponderomotive energy. The latter is defined as the cycle-averaged energy associated with the periodic motion of a free electron in a plane-wave field in the frame of reference in which it is minimal. In the commonly encountered case of a linearly polarized laser field with a peak electric-field amplitude F_0, U_p is given by

$$U_\mathrm{p} = \left(\frac{F_0}{2\omega_0}\right)^2. \tag{4.1}$$

The latter implies that the ponderomotive energy grows quadratically with increasing wavelength of the driving field. Here and in what follows, unless explicitly stated otherwise, we use atomic units for convenience. In the system of atomic units, the unit of mass is the electron mass, the unit of charge is the elementary charge, the unit of length is the bohr, symbol a_0 ($1\ a_0 \approx 53$ pm), and the unit of energy is the hartree, symbol E_h ($1\ E_\mathrm{h} \approx 4.610^{-18}$ J). When using atomic units, one usually reports relative quantities, i.e., a length r is given as r/a_0, an energy E is given as E/E_h (see also Cohen et al., 2008). The Keldysh parameter γ_K, which relates the electron's binding energy in its ground state, E_b, with the ponderomotive energy accumulated in the field via the expression

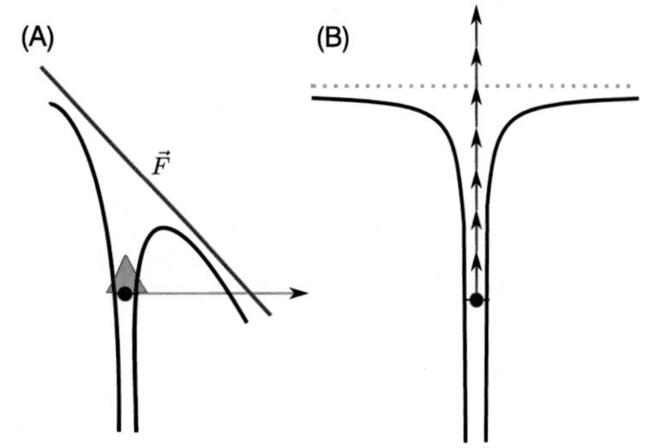

FIG. 4.1 Schematic illustration of the strong-field ionization process in the tunneling (A) and the multiphoton (B) regimes.

$$\gamma_\mathrm{K} = \sqrt{\frac{|E_\mathrm{b}|}{2U_\mathrm{p}}} = \frac{\omega_0}{F_0}\sqrt{2|E_\mathrm{b}|}, \tag{4.2}$$

is commonly employed in order to distinguish between perturbative and non-perturbative regimes in the context of strong-field ionization. As such, it has found a wide acceptance, although its universality has been subject to debate (Reiss, 2008a, 2008b). Strong-field phenomena occur in the regime of $\gamma_\mathrm{K} \ll 1$, also known as the "tunneling" or quasistatic regime and characterized by low laser frequencies (ω_0) and/or strong electric field amplitudes F_0. The opposite regime is known as the multiphoton ionization (MPI) limit. The reasoning behind this convention can be inferred from the schematic depiction of the two processes presented in Fig. 4.1. Tunneling ionization (TI) is typically modelled on the basis of a static Coulomb-like potential well modified by the strong, quasistatic field. A strong enough field can suppress/lower one side of the Coulomb potential, leading to the formation of a barrier of finite width, see Fig. 4.1A. A prerequisite for this situation to arise is that the time scale of the electric-field oscillation has to be slow enough compared to the time required for the electron to cross the tunneling barrier (in a naïve and incorrect classical picture) to permit adiabatic tunneling. Conversely, in weak and/or rapidly oscillating fields, the MPI picture (Fig. 4.1B) may be more appropriate. For the case of Ti:Sa wavelengths and the typical field strengths encountered under realistic conditions, $\gamma_\mathrm{K} \sim 1$, implying that most experiments reported to date take place in a regime that is best described as intermediate between the two limiting cases.

The nomenclature "Keldysh parameter" derives its origin from the 1965 publication of Keldysh (1965) on the topic of the quasistatic limit of strong-field ionization of a one-electron atom. This analysis was subsequently generalized to cover both tunnel and multiphoton limits in the works of Perelomov, Popov, and Terent'ev (also known as the PPT model Perelomov et al., 1966). Another popular approach for the quantitative treatment of the limiting case of tunnel ionization rates is the Ammosov, Delone, and Krainov model (Ammosov et al., 1986) (ADK), initially formulated for atoms and subsequently extended to molecules (Tong et al., 2002), or the work of Yudin and Ivanov (2001) addressing the aspect of non-adiabatic effects in few-cycle fields. Since the subject has been extensively covered in various reviews (Ivanov et al., 2005; Popruzhenko, 2014; Smirnova et al., 2007), we refrain from a more detailed mathematical description and only emphasize the following central result. With exponential accuracy, the total tunnel ionization rate is given by:

$$w_{\mathrm{TI}} \propto \exp\left(-\frac{2(2\,|\,E_{\mathrm{b}}\,|)^{3/2}}{3|E_0|}\right). \tag{4.3}$$

The exponential dependence of the TI rate on the strength of the applied field implied by the above equation has far-reaching consequences for the intensity requirements for technological applications of attosecond physics (cp. Section 4.3).

4.2.2 Electric-Dipole Approximation and Gauge Invariance

The theoretical descriptions of strong-field phenomena often employ the electric-dipole approximation (EDA, also simply referred to as "dipole approximation"). The latter amounts to neglecting the spatial variation of the propagating TEM field, which for a plane-wave treatment is governed by the phase factor

$$\omega_0 t - \vec{k}_{\mathrm{p}} \cdot \vec{r} \approx \omega_0 t, \tag{4.4}$$

where $\vec{k}_{\mathrm{p}}$ denotes the wave vector of the field. This assumption is justified in the case where the wavelength of the TEM field significantly exceeds the size of the system it interacts with. The direct consequence of Eq. (4.4) is the vanishing of the spatial dependence of the vector potential $\vec{A}(\vec{r},t) \approx \vec{A}(\vec{0},t) \equiv \vec{A}(t)$ associated with the TEM field as well as its magnetic-field component: $\vec{B}(\vec{r},t) = \vec{\nabla} \times \vec{A}(\vec{r},t) \approx 0$. These approximations are well-justified for most realistic experimental conditions, generally, the breakdown of the EDA is reached at sufficiently high intensities ($I \sim 10^{15}$ W/cm^2 for Ti:Sa wavelengths) or extremely long driving wavelengths (Ludwig et al., 2014).

At this point, we wish to address one specific aspect of the theoretical description of non-perturbative phenomena, namely the gauge invariance (Bandrauk et al., 2013; Reiss, 1980). Whereas the choice of the gauge is of no particular importance when describing the perturbative response, it can have far-reaching consequences for the theoretical or numerical results when dealing with strong-field phenomena. There are two commonly employed theoretical frameworks, referred to as length or velocity gauge, both of which represent special cases of the EM Coulomb gauge ($\vec{\nabla} \cdot \vec{A}(\vec{r},t) = 0$). Below we briefly discuss these two approaches in the context of the Schrödinger equation for a free particle in an electromagnetic field. The latter problem relates to the foundation of the strong-field approximation (Lewenstein et al., 1994) (SFA), which forms the starting point for the interpretation of a plethora of attosecond phenomena, including strong-field ionization, high-harmonic generation and above-threshold ionization. The SFA will be covered in detail for the case of HHG in Section 4.2.4.

In its essence, the velocity gauge is the Coulomb gauge from electromagnetism, whereby the EDA has been imposed: $\vec{A}(\vec{r},t) \to \vec{A}(t)$. The term "velocity" originates from the fact that the associated Hamiltonian contains the term $\vec{p} + \vec{A}(t)$, which is proportional to the kinematic momentum $\vec{k}(t)$ and thus to the velocity. Here and in what follows, the symbol $\vec{p}$ will be used to denote the canonical momentum, which is a conserved quantity. The corresponding Hamiltonian for a free electron in an EM field reads in the velocity gauge as

$$\mathrm{i}\partial_t|\Psi^{\mathrm{V}}\rangle = \frac{1}{2}\left[-\mathrm{i}\vec{\nabla} + \vec{A}(t)\right]^2|\Psi^{\mathrm{V}}\rangle. \tag{4.5}$$

The associated solution can be obtained in a straightforward manner by switching to the momentum-space representation

$$\begin{aligned}\Psi^{\mathrm{V}}(\vec{r},t) &\equiv \langle \vec{r}\,|\,\Psi^{\mathrm{V}}\rangle \\ &= \frac{1}{(2\pi)^{3/2}}\exp\left\{\mathrm{i}\vec{p}\cdot\vec{r} - \frac{\mathrm{i}}{2}\int_{-\infty}^{t}[\vec{p} + \vec{A}(t')]^2\mathrm{d}t'\right\}\end{aligned} \tag{4.6}$$

and is also known as the Volkov solution.

Within the length gauge, which is defined within the EDA, a transverse field is expressed in terms of the scalar potential as

$$\varphi(\vec{r},t) = -\vec{r} \cdot \vec{F}(t). \tag{4.7}$$

The above equation implies that the (transverse) plane-wave laser field is treated as a quasistatic, longitudinal

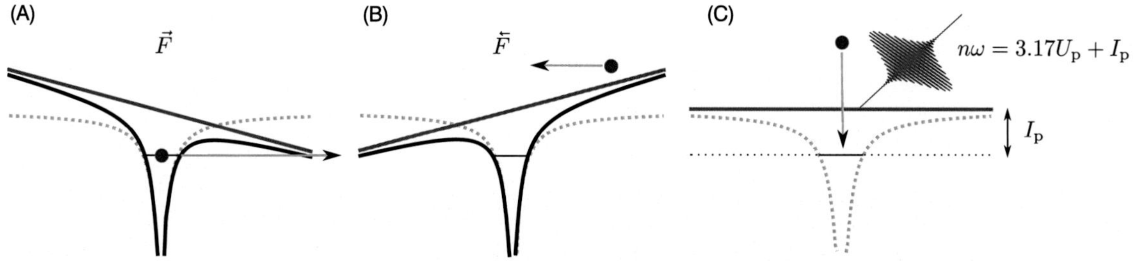

FIG. 4.2 Schematic illustration of each of the three processes constituting the three-step model: (A) ionization, (B) propagation, and (C) recombination. The approximate location of the cutoff is indicated in panel C.

field. The length gauge gives rise to many intuitive interpretations of strong-field phenomena. For instance, the fact that the laser field and the scalar potential reduce to additive scalars in a one-dimensional treatment gives rise to the tunneling picture illustrated in Fig. 4.1A. The free-electron Schrödinger equation in this gauge becomes

$$\mathrm{i}\partial_t|\psi^\mathrm{L}\rangle = \left[\frac{1}{2}(-\mathrm{i}\vec{\nabla})^2 - \vec{r}\cdot\vec{F}\right]|\psi^\mathrm{L}\rangle. \qquad (4.8)$$

Since there is only a scalar potential in the length gauge, one cannot establish a direct relationship between the above equation and the Volkov solutions defined by Eq. (4.6), which are defined for a transverse field. In order to circumvent this problem, one resorts to a gauge transformation of the velocity-gauge solution $\psi^\mathrm{V}(\vec{r},t)$. For this purpose, the velocity-gauge solution in Eq. (4.6) is multiplied by the transformation factor

$$\exp\left(\mathrm{i}\vec{r}\cdot\vec{A}(t)\right), \qquad (4.9)$$

which yields the length-gauge Volkov wavefunctions

$$\psi^\mathrm{L}(\vec{r},t) = \frac{1}{(2\pi)^{3/2}}\exp\left\{\mathrm{i}\left(\vec{p}+\vec{A}(t)\right)\cdot\vec{r} - \frac{\mathrm{i}}{2}\int_{-\infty}^{t}[\vec{p}+\vec{A}(t')]^2\mathrm{d}t'\right\}. \qquad (4.10)$$

4.2.3 The Three-Step Model of High-Harmonic Generation

Having defined the basic theoretical framework, we proceed with the description of the high-harmonic generation process, first in semiclassical terms (Section 4.2.3), and then within the quantum-mechanical framework of the strong-field approximation. The emphasis on HHG in this short overview is dictated by the central importance of this process for the development of attosecond science and technology, mainly through its application as a table-top source of bright, coherent radiation in the XUV and the soft-X-ray ranges, and in the generation of attosecond pulses. HHG is a parametric process which occurs in the course of the interaction of a strong ultrashort ($<$ ps) laser field (of frequency ω_0) with matter and leads to the emission of high-frequency radiation. The emitted intensity spectrum features three characteristic regions: (1) an abrupt decrease spanning 2–3 orders of magnitude for the first few harmonics, followed by (2) a "plateau" region, where the intensity of subsequent harmonic orders is nearly constant, and (3) a rapid falloff ("cut-off-region"). In centrosymmetric media, the harmonic comb consists of odd harmonic multiples ($n\omega_0, n \in \mathcal{N}_+$) of the driving field. Most of the underlying physics can be explained with the aid of a simple, non-perturbative approach introduced by P. Corkum, K. C. Kulander and K. Schafer in 1993 (Corkum, 1993; Kulander et al., 1993; Schafer et al., 1993), which became known as the "three-step model" (TSM).

The TSM model is summarized graphically in Fig. 4.2. First, the strong field, implicitly treated in the quasi-static limit, distorts ("bends") the Coulomb barrier, leading to the formation of a potential barrier of a finite width. The subsequent tunneling process can release an initially bound electron into a state belonging to the continuum. In the second step, the dynamics of the liberated electron is essentially governed by the laser field, while the Coulomb potential is assumed to represent a minor perturbation. The electron is accelerated and driven away from the core, after which the field switches its sign (for a linearly polarized field, this occurs after 1/4 of an optical cycle), it decelerates and is driven back to the origin. In the ensuing recombination, a pulse of high-energy radiation is emitted, which forms a part of the harmonic comb. The generation of a highly energetic photon is only one (and by far not the most probable one) of several strong-field processes that take place in parallel, among which laser-induced electron diffraction, rescattering, non-sequential double ionization, etc. Depending on the exact subcycle timing of the ionization event, some electrons are driven away by the field instead of being accelerated back.

Despite its simplicity, the intuitive picture implied by the TSM enables one to explain some characteristic experimental observations associated with HHG from a purely classical perspective. By treating the electron motion in the second step as a classical laser-driven acceleration of an initially stationary electron with zero velocity, the trajectories of the accelerated electrons can be obtained from integration of Newton's equation of motion by applying the constraint that the electron returns to its original position ($\vec{r}(t') = 0$) at a certain recombination time t. For a monochromatic field linearly polarized along the $\hat{x}$-direction, i.e., $\vec{F}(t) = F_0 \cos(\omega_0 t)\hat{x}$, the classical trajectory $\vec{r}(t) \equiv x(t)$ is given by:

$$\ddot{x}(t) = -F_0 \cos(\omega_0 t), \tag{4.11a}$$

$$x(t) = \frac{F_0}{\omega_0^2}\big(\cos(\omega_0 t) - \cos(\omega_0 t')\big) + \frac{F_0}{\omega_0}\sin(\omega_0 t')(t - t'). \tag{4.11b}$$

The above classical interpretation accounts for the observed position of the harmonic cutoff at $\approx 3.2U_\mathrm{p} + I_\mathrm{p}$. Further, it allows one to map the kinetic energy at the return time, $\frac{\dot{x}^2(t)}{2}$ (and thus the photon energy of the emitted radiation), to a given electron trajectory and thus to a given travel time of the electron in the continuum. For the classical case, restricted to recombination events within one cycle, this mapping is not unique, as there is a pair of trajectories that map to the same photon energy, as evident from Fig. 4.3. These two "branches" of solutions are associated with families of "long" and "short" trajectories, which refers to the length of their transit time in the continuum. The short trajectories originate from ionization after $t'/T_0 \sim 0.05$, whereas the long ones begin prior to this instant. The harmonic emission associated with these two trajectories is characterized by distinctly different properties. In particular, it implies that the emitted energy changes with time on a sub-cycle scale, giving rise to an intrinsic chirp of the high-harmonic emission, termed "atto-chirp". As evident from Fig. 4.3, the long and the short branches have opposite chirps (negative and positive, respectively). The two contributions can be differentiated in an experiment by exploiting macroscopic phase matching considerations owing to their different divergence properties. The latter enables a unique mapping of the harmonic energy to a given transit time of the electron in the continuum. As a result, ultrafast dynamics taking place between the two instants will be mapped onto the properties of the high-harmonic emission, i.e., intensity, polarization and ellipticity. This constitutes the self-probing aspect of high-harmonic spectroscopy (HHS). Moreover, recombining trajectories are characterized by durations of a fraction of a laser cycle, or several hundreds of attoseconds for a Ti:Sa-wavelength driver, thus endowing HHG with a subcycle temporal resolution in spite of the fact that the process is driven by multicycle (femtosecond) pulses.

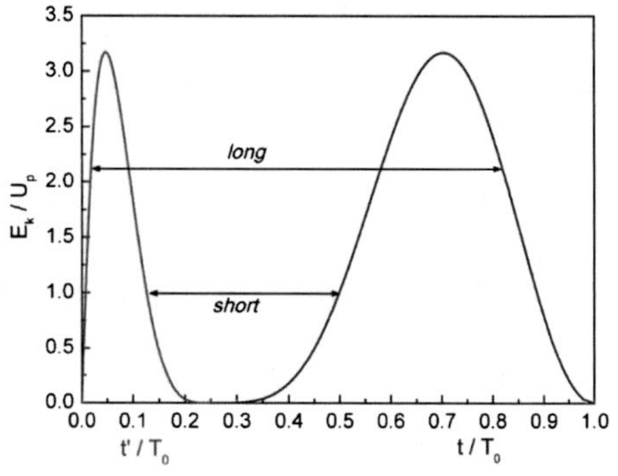

FIG. 4.3 Kinetic energy (in units of U_p) of the recolliding electron at the instant of recombination, as a function of the ionization time t' (red curve, in units of the laser period T_0) and the recombination time t (blue curve). The ionization/recombination times of the two trajectory branches are indicated with black arrows. Figure adapted from Nisoli et al. (2017).

4.2.4 High-Harmonic Generation Within the Strong-Field Approximation

Despite its remarkable success, the TSM does not allow for a more quantitative analysis of the HHG spectra in terms of its specific features, e.g., the intensity distribution of individual harmonics. We therefore conclude this section by outlining the quantum-mechanical treatment of HHG within the strong-field approximation. The starting point of the derivation of the HHG amplitude is the semiclassical length-gauge Schrödinger equation within the single-active electron approximation (SAE):

$$\mathrm{i}\partial_t |\Psi(t)\rangle = \hat{H}(t)|\Psi(t)\rangle = \left[\frac{1}{2}(-\mathrm{i}\vec{\nabla})^2 + \hat{V}(\vec{r}, t)\right]|\Psi(t)\rangle. \tag{4.12}$$

It will prove useful to separate the Hamiltonian into a field-free part $\hat{H}_0 = \frac{\vec{p}^2}{2} + U(\vec{r})$ and the perturbation induced by the time-dependent classical field $V_L(t) = -\vec{r}\cdot\vec{F}(t)$ as

$$\begin{aligned} i\partial_t|\Psi(t)\rangle &= \left[\frac{\vec{p}^2}{2} + U(\vec{r}) + V_L(t)\right]|\Psi(t)\rangle \\ &= \left[\hat{H}_0 + V_L(t)\right]|\Psi(t)\rangle, \end{aligned} \quad (4.13)$$

where $U(\vec{r})$ denotes the effective SAE potential. The time evolution of the eigenstates of the system in the presence of the field ($|\Psi(t)\rangle$) from t' to t can be expressed with the aid of the propagation operator $U(t,t')$,

$$|\Psi(t)\rangle = \hat{U}(t,t')|\Psi(t')\rangle. \quad (4.14)$$

Employing the Dyson series expansion, one arrives at the following formal expression for the propagator:

$$\begin{aligned} \hat{U}(t,t') =& \hat{U}_0(t,t') \\ &- i\int_{t'}^{t} \hat{U}(t,t'')V_L(t'')\hat{U}_0(t'',t')dt'', \end{aligned} \quad (4.15)$$

where $\hat{U}_0(t,t')$ denotes the propagator associated with the system in the absence of the external field. The formal solution for the perturbed wavefunction $|\Psi(t)\rangle$ then becomes

$$\begin{aligned} |\Psi(t)\rangle =& \hat{U}_0(t,t_0)|\Psi(t_0)\rangle \\ &- i\int_{t_0}^{t} \hat{U}(t,t')V_L(t')\hat{U}_0(t',t_0)|\Psi(t_0)\rangle dt', \end{aligned} \quad (4.16)$$

which is exact in terms of the single-active-electron treatment. The first assumption that constituted the SFA is to truncate the full set of bound-state wavefunctions to a single component only, i.e., $|\Psi(t_0)\rangle \equiv |\Psi_0(t_0)\rangle = e^{-iE_b t_0}|\Psi_0\rangle$. Next, the core ansatz that constitutes the essence of the SFA is the substitution of the full propagator $\hat{U}(t,t')$ in Eq. (4.16) by the propagator pertaining to a free particle in an EM field (cf. Eq. (4.6)), the so-called "Volkov propagator" $\hat{U}_V(t,t')$,

$$\hat{U}_V(t,t') = \int d^3\vec{p}\,|\Psi_V(\vec{p},t)\rangle\langle\Psi_V(\vec{p},t')|, \quad (4.17)$$

where $|\Psi_V(\vec{p},t)\rangle$ is the length-gauge Volkov function. The spatial representation of the latter can also be cast in the form

$$\begin{aligned} \Psi_V^{\vec{p}}(\vec{r},t) &= \langle\vec{r}\mid\Psi_V(\vec{p},t)\rangle = e^{i[\vec{p}+\vec{A}(t)]\cdot\vec{r} - i\vec{S}(t)} \\ &\equiv \langle\vec{r}\mid\vec{p}+\vec{A}(t)\rangle e^{-iS(t)} \end{aligned} \quad (4.18)$$

with the dependence on the canonical momentum denoted explicitly. Here, $S(t) = \frac{1}{2}\int^t[\vec{p}+\vec{A}(t')]^2dt'$ is the equivalent of the classical action. After inserting the Volkov resolvent into the expression for the time-dependent induced HHG dipole,

$$\vec{d}(t) = \langle\Psi(t)|\vec{r}|\Psi(t)\rangle, \quad (4.19)$$

and neglecting the continuum–continuum couplings, one arrives at

$$\begin{aligned} \vec{d}(t) \propto& -i\int_{-\infty}^{t} dt' \\ &\times \int d^3\vec{p}\, e^{-iS(\vec{p},t,t')}\langle\Psi_0|\vec{r}\mid\vec{p}+\vec{A}(t)\rangle \\ &\times \langle\vec{p}+\vec{A}(t')|\vec{r}\cdot\vec{F}(t')|\Psi_0\rangle, \end{aligned} \quad (4.20)$$

where the phase factor originating from the field-free propagation of the ground state has been incorporated into the semiclassical action (by means of the relation $I_p = -E_b$) as

$$S(\vec{p},t,t') = \frac{1}{2}\int_{t'}^{t}[\vec{p}+\vec{A}(t'')]^2dt'' + I_p(t-t'). \quad (4.21)$$

In the above, I_p stands for the ionization potential. According to Eq. (4.20), the dipole response can be reduced to an integral involving transition matrix elements between the ground state and the plane-wave continuum. Indeed, the terms in the bra-kets can be related to the dipole couplings associated with the ionization (resp., recombination) steps of the three-step model:

$$d_{ion}(\vec{p}+\vec{A}(t)) \equiv \langle\vec{p}+\vec{A}(t)|\vec{r}\cdot\vec{F}(t)|\Psi_0\rangle \quad (4.22a)$$

and

$$\vec{d}_{rec}(\vec{p}+\vec{A}(t)) \equiv \langle\Psi_0|\vec{r}\mid\vec{p}+\vec{A}(t)\rangle. \quad (4.22b)$$

Despite this significant reduction in complexity, Eq. (4.20) is still not (easily) amenable to a numerical treatment, due to the presence of the rapidly oscillating factor associated with the semiclassical action. As an alternative route, one can deform the integration path in the complex plane and utilize the steepest-descent method. This approach results in the saddle-point treatment of the SFA, whereby the integral in Eq. (4.20) reduces to a coherent summation over the contributions

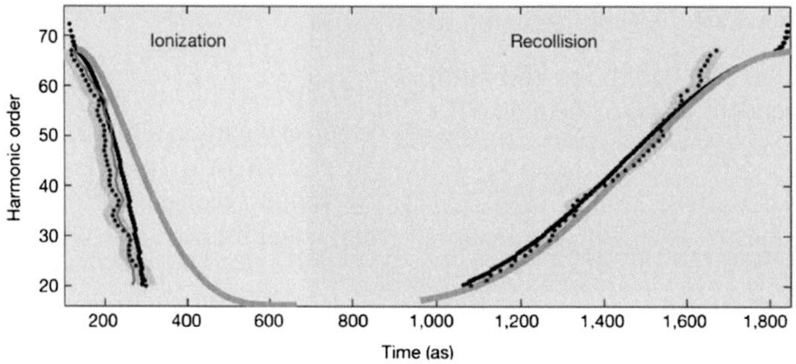

FIG. 4.4 Comparison of the experimentally extracted ionization and recombination times (red dots) with the predictions of the semiclassical model (grey) and the quantum-trajectory solutions (black). Figure adapted from Shafir et al. (2012) with permission.

of a finite set of electron trajectories defined in terms of the saddle points $(\overrightarrow{p_{st}}^{(j)}, t^{(j)}, t'^{(j)})$ of the semiclassical action (cp. Eq. (4.21)). In the language of Feynman's quantum-path formalism (Salières et al., 2001), these trajectories are often referred to as "quantum orbits". The HHG amplitude, expressed as the Fourier transform of the time-dependent dipole in Eq. (4.20), then reads:

$$\vec{\tilde{d}}(\omega) = \sum_j \left(\frac{2\pi}{t^{(j)} - t'^{(j)} + i\epsilon} \right)^{3/2} \times \left[\det\left(\frac{\partial^2 S}{\partial t'^{(j)2}} \right) \right]^{-1/2} \left[\det\left(\frac{\partial^2 S}{\partial t^{(j)2}} \right) \right]^{-1/2} \times \vec{d}_{rec}(\vec{p}_{st}^{(j)} + \vec{A}(t^{(j)})) \times \vec{d}_{ion}(\vec{p}_{st}^{(j)} + \vec{A}(t'^{(j)})) e^{-iS(\vec{p}_{st}^{(j)}, t^{(j)}, t'^{(j)}) + i\omega t^{(j)}}, \tag{4.23}$$

where the additional multiplicative factors result from the contributions of the Hessian determinant. With the above approximation one can easily recover the physical picture contained in the three-step model. It implies that the dipole emission can be decomposed in terms of the contributions from electrons emitted at the ionization instant t' and recombining with the parent system at time t, whereby the weight of each trajectory is proportional to the dipole-matrix element terms. One essential difference, however, is the complex character of the quantum-mechanical (QM) saddle points, which originate from the contribution of the tunneling process in the first step of the process. These quantum corrections have been confirmed and quantified experimentally using high-harmonic generation in a two-color field (Shafir et al., 2012) (see Fig. 4.4).

4.3 ATTOSECOND TECHNOLOGY

In this section we provide a concise overview of the recent experimental progress in attosecond source development. First, we review the principal methods for generating few-cycle laser pulses with high peak power and then proceed with the description of the various schemes for attosecond pulse generation. Afterwards, we present the operating principle of several dedicated techniques for attosecond spectroscopy experiments. The latter will be covered in a greater detail in the subsequent sections, where we focus on selected applications.

4.3.1 Chirped-Pulse Amplification

The invention of the chirped pulse amplification (CPA) technique in 1985 (Maine et al., 1988; Strickland and Mourou, 1985) has had a profound impact on the development of ultrafast laser physics. Within the CPA scheme, the low-energy sub-1ps pulses delivered by a mode-locked oscillator are first chirped and stretched in the time domain to a considerably longer duration (e.g., several hundreds of ps or even ns) by means of a dispersive element such as a grating pair or a long fiber. In the subsequent amplification stage, the energy after passing the gain medium is increased to a mJ- or even to a J-level. At the same time, the peak power is kept below the damage threshold of the optical components and the non-linear pulse distortion limit. Afterwards, a dispersive element (e.g., a second grating pair) is used to compensate the chirp and to recompress the pulse to a value close to the original pulse duration, thereby enabling peak powers reaching the TW-range and beyond. The CPA technique, in conjunction with the Ti:Sa technology, has undergone multiple improvements throughout the following decades. To date, peak powers up to the PW level are available at large-scale facilities, whereas table-top high-power systems delivering pulses with energies reaching 200 J and 10 fs in duration have been developed (see, e.g., Chu et al., 2015; Gan et al., 2017, or Danson et al., 2015 for recent reviews).

4.3.2 Carrier-Envelope Phase Stabilization

A further challenge associated with the realization of spectroscopic techniques relying on isolated attosecond pulses is the control over the carrier-envelope phase of the laser pulse. For a few-cycle pulse with an electric field amplitude profile $E_0(t)$, the instantaneous electric field assumes the form

$$E(t) = E_0(t)\cos(\omega_0 t + \phi). \tag{4.24}$$

Thereby, the phase offset ϕ is referred to as carrier-envelope phase (CEP) and is formally defined as $\phi = \Delta t\, \omega_0$, i.e., the offset between the maximum of the envelope and the position of the nearest field antinode. For pulse durations approaching three cycles or less, the slowly-varying-envelope approximation, commonly employed to simplify the description of ultrashort pulse propagation, breaks down as the CEP starts to influence the pulse propagation and interaction with other media. In a typical Ti:Sa oscillator or amplifier, the CEP strongly fluctuates from shot to shot: whereas the envelope travels with the group velocity through the active medium, the carrier wave is propagated at the phase velocity. The resulting velocity mismatch gives rise to a phase slippage $\Delta\phi$ between carrier and envelope between consecutive pulses separated by the cavity round-trip time. This circumstance is additionally aggravated by energy fluctuations (e.g., in the pump laser, or by Kerr non-linearities in the active material), which cause a variation of $\Delta\phi$ as a function of time. A variety of methods for (i) measuring and (ii) controlling the pulse-to-pulse CE phase-slip have been developed to address this problem. The earliest approaches for quantifying the CEP slippage occurring in an oscillator were based on interferometric cross-correlation between two consecutive pulses in a pulse train (Xu et al., 1996) but had only limited precision. The adaptation of radio-frequency heterodyne detection schemes (Telle et al., 1999) enabled significant progress in this respect. The most common implementations of the heterodyne scheme include the f–$2f$-interferometry, which is based on heterodyning the fundamental and the second harmonic, or difference-frequency generation (DFG, "0–f-interferometry") between different parts of the same frequency comb (see, e.g., the work by Fuji et al., 2005). Thereby, the frequency comb corresponds to the mode-locked pulse train. In case the oscillator output is not octave-spanning, external spectral broadening is necessary, which can be conveniently realized in a photonic crystal fiber medium (PCF) (Dudley et al., 2006). These methods are not directly applicable to amplified pulses, as the repetition rates of the latter lie in the kHz-range or below. In this case, CEP characterization is performed using spectral interferometry (Kakehata et al., 2001; Lepetit et al., 1995; Mehendale et al., 2000) by analyzing the spectra of two different harmonics that overlap temporally. A different method, termed "stereo-above-threshold-ionization" (stereo-ATI) was pioneered by G. Paulus and coworkers (Kreß et al., 2006; Paulus et al., 2001; Sayler et al., 2011).

The control of the CEP, on the other hand, is achieved using either *active* (i.e., based on electronic feedback loops) or *passive* (based on all-optical approaches such as DFG) methods. In its most widespread implementation, the active scheme consists of two dedicated loops, a "fast" and a "slow" one, designed to stabilize the CEP at the output of the oscillator and after the amplification medium. The "fast" loop is based on measuring and stabilizing the phase slippage $\Delta\phi$ using an electronic feedback on an acousto-optic modulator (AOM), thereby adjusting the oscillator pump power and selecting only pulses sharing the same CEP (Calegari et al., 2015). A different method utilizes an acousto-optic frequency-shifter (AOFS) for splitting the oscillator output by diffraction, whereby the zeroth order is used for measuring the CEP, whereas the first order (60–70%) is shifted to compensate the offset and subsequently used for amplification (Koke et al., 2010). The conventional implementation of the "slow loop" is based on the f–$2f$-characterization scheme. Thereby, a portion of the amplifier output is focused into a sapphire plate to generate an octave-spanning supercontinuum and subsequently frequency-doubled in a β-barium-borate (BBO) crystal. The extracted CEP variation $\Delta\phi$ is then used to control the AOM in the oscillator or compensated with the aid of wedges in front of the amplifier. Another scheme using acousto-optic programmable dispersive filter (AOPDF) has been described by Canova et al. (2009).

The passive scheme, introduced by Baltuška et al. (2002), is based on the all-optical process of DFG between pulses sharing the same CEP variation. Upon frequency-mixing in a non-linear crystal, the CEP of the generated wave cancels out (up to a constant phase factor). Most commonly, the passive scheme is exploited for CEP stabilization of the idler pulse produced in the process of optical parametric amplification (see Section 4.3.4).

4.3.3 Pulse Postcompression Techniques

On the basis of the technologies described in the preceding paragraphs, present-day commercial Ti:Sa-based systems routinely deliver CEP-stable pulses of 20 fs duration, with a central wavelength around 800 nm and at an energy in the mJ range. Generation of isolated pulses for HHG with high peak intensities requires further pulse compression to the few-cycle regime (i.e., <8 fs) in a postcompression stage. The latter is most commonly realized by means of spectral broadening via self-phase modulation (SPM) in a gas-filled hollow-core fiber (HCF), followed by recompression in a chirped-mirror assembly for ultrabroadband dispersion compensation. Owing to the long interaction lengths employed (~1 m) and the efficient coupling (achieved by using HCF inner diameters on the order of 200–300 μm), significant SPM levels can be maintained even at moderate gas pressures of 200–300 mbar. First demonstrated in 1996 (Nisoli et al., 1996), when it was used to compress 140 fs pulses of 0.66 mJ energy down to 10 fs (0.24 mJ), HCF-based pulse compression is now routinely employed to yield sub-5 fs pulse durations. Thereby, the optical cycle of the carrying wave (2.66 fs) represents a fundamental lower bound on the realizable pulse compression. In the initial version, the available energies were limited to the mJ-level. A further increase requires the implementation of a differential pumping scheme and a pressure gradient (Suda et al., 2005) to mitigate the detrimental effects of ionization. Other schemes rely on the usage of circularly-polarized drivers to suppress the onset of ionization. Using a combination of a pressure gradient and chirped pulses, the generation of 5 fs pulses of 5 mJ energy was reported in 2010 (Bohman et al., 2010). High-energy pulses of up to 13.7 mJ and with a pulse duration of 11.4 fs were obtained using an HCF filled with helium at a low gas pressure (Dutin et al., 2010). Spectral broadening in an HCF has also been extensively used for post-compressing pulses with a central energy red-shifted compared to the typical Ti:Sa output, such as the ones generated using the OPA scheme. In the spectral range around 1.8 μm and above, one can conveniently exploit the negative dispersion properties of bulk materials (e.g., fused silica) in order to shorten the pulse duration down to 2–3 cycles (Schmidt et al., 2011, 2010), thus circumventing the necessity for employing chirped mirrors.

Another related postcompression technique relies on filamentation (Hauri et al., 2004) by loose focusing of an intense laser pulse in a gas-filled cell, whereby argon is conventionally used as the generation medium. As a consequence of the dynamical equilibrium between self-focusing (due to the Kerr effect) and defocusing (due to ionization and plasma generation), the generated filament (typically 100 μm in diameter) can propagate over distances up to several tens of centimeters, thereby undergoing self-compression.

A conceptually different technique for generating optical pulses (i.e., covering the VIS region) with sub-fs duration was presented by Wirth et al. (2011). In this scheme, termed "light-wave synthesizer", a broad supercontinuum (260–1100 nm) generated in a hollow-core-fiber is first decomposed into three spectral regions (UV, 350–500 nm; VIS, 500–700 nm; and NIR, 700–1100 nm) with the aid of dichroic beam-splitters. After compressing each spectral region with the aid of dedicated chirped mirrors, the individual components are recombined. By using active stabilization in combination with dispersive elements controlling the CEP in each arm, the electric field of the light wave can be modulated on a subcycle time scale. In a later version of the synthesizer (Hassan et al., 2016), a fourth arm covering the deep UV-region (DUV, 260–350 nm) was added, which allowed for the synthesis of optical attosecond pulses with a duration of 415 as, as confirmed by the streaking measurements (see Section 4.3.6.1).

4.3.4 Attosecond Sources in the Mid-Infrared

Ti:Sa-based femtosecond sources are limited around 800 nm. Under typical experimental conditions, HHG driven by pulses in this wavelength range (NIR) yields access to photon energies of only up to 100 eV, thus precluding the investigation of electron dynamics of core-excited electrons. The demand for attosecond pulses with higher photon energies is mainly dictated by the high biological relevance of the spectral region between 280 and 543 eV, where the absorption edges of the main constituents of organic matter, carbon (K-edge, 284 eV), oxygen (K-edge, 543 eV), and nitrogen (K-edge, 397 eV) are located, whereas water (i.e., the cell environment) is transparent. On a different note, photon energies in the keV range are also required for the investigation of ultrafast processes in solid-state systems with X-ray absorption spectroscopy employing elements such as Ti, Cu, Al, Si (Seres et al., 2006).

One way to up-scale the photon energy involves increasing the laser intensity, thereby reaching the conditions of strong ionization. Using helium gas as a medium, soft-X-ray harmonics of energies up to 460 eV were reported in 1997 (Chang et al., 1997). By exploiting phase-matching techniques, bright high-harmonic generation reaching the keV range has been demonstrated (Popmintchev et al., 2009). Nevertheless, the photon flux achievable via these methods still does not meet the level required for practical applications, as a consequence of the phase mismatch caused by the generated electron plasma.

The quadratic dependence of the harmonic cutoff on the driving field wavelength (cp. ponderomotive term in Eq. (4.1)) provides an alternative route towards increasing the photon energy of the generated harmonics. This fact has therefore spawned the development of mid-IR-sources (MIR) for high-harmonic generation (Heyl et al., 2016; Ishii et al., 2014; Wolter et al., 2015). Conventionally, the optical parametric amplification (OPA), covered in detail by Cerullo and De Silvestri (2003), a wavelength down-conversion method in which a seed pulse is amplified by an ultrafast pump in a non-linear crystal, provides a straightforward route towards ultrashort pulses in the 1.1–3.0 μm range. Relying on the OPA technology, extension of high-harmonic generation to the water window was first demonstrated by Takahashi et al. (2008), whereby photon energies of up to 300 eV from Ne and 450 eV from He were generated using a 1.5 μm driver. However, the limited pulse energies (up to mJ), resp. peak powers (up to few GW), as well as the considerable output pulse durations (50 fs) pose fundamental limitations for the power-scalability of the OPA method.

Some of the inherent limitations of the OPA process can be overcome in a dual-chirp OPA scheme (DC-OPA), whereby properly optimized, chirped seed and pump pulses are utilized. First theoretically proposed in 2011 (Zhang et al., 2011), this scheme has been implemented to generate TW, sub-two-cycle IR pulses in the mid-IR region (Fu et al., 2015). Further, the generation of multi-TW, fs pulses with an energy exceeding 100 mJ was reported (Fu et al., 2018). In this aspect, optical parametric chirped-pulse amplification (OPCPA), first proposed in 1992 (Dubietis et al., 1992), a technology which unifies the OPA and the CPA concepts, has been increasingly gaining in popularity. OPCPA utilizes the main characteristics of the CPA scheme, i.e., stretching in the temporal domain, amplification, and recompression, however, the amplification stage is based on an optical parametric process involving energy transfer from a pump to a seed beam, instead of stimulated emission (as in Ti:Sa amplifiers). This method possesses an octave-spanning amplification bandwidth and is capable of generating sub-10-fs spectrum in the VIS range. Further, the detrimental influence of thermal distortions is considerably mitigated as the instantaneous nature of the OPA process does not require storage of the energy in the amplification medium. Utilizing this scheme, multiple groups have reported the generation of few-cycle pulses at varying central wavelengths: at 0.88 μm (Rothhardt et al., 2012), 2.1 μm (Deng et al., 2012), 3.2 μm (Chalus et al., 2010), or 3.9 μm (Andriukaitis et al., 2011). A comprehensive review of the state-of-the-art performance level of OPCPA systems including few-cycle table-top sources has been given by Vaupel et al. (2013). The OPCPA architecture itself has been reviewed in several papers (Butkus et al., 2004; Dubietis et al., 2006; Witte and Eikema, 2012). Another recent review on the topic of IR sources was given by Ciriolo et al. (2017). We conclude the discussion of MIR sources with the Fourier-domain optical parametric amplification (FOPA) method, a novel pulse amplification scheme that possesses the potential for overcoming the fundamental limitations of traditional methods (such as the gain bandwidth, or material damage threshold). Within FOPA, the dispersed spectral components of a transform-limited pulse are amplified independently of each other utilizing different non-linear crystals in the Fourier plane. In this manner, simultaneous up-scaling of energy and amplified bandwidth is achieved. The experimental realization of the FOPA concept utilizes a symmetric $4f$-scheme and has been first demonstrated via the generation of CEP-stable, 1.43 mJ pulses at 1.8 μm (Schmidt et al., 2014). A later work reported on the generation of two-cycle ($\sim$11 fs), 1.8 μm pulses with peak powers reaching 2.5 TW (Gruson et al., 2017).

4.3.5 Generation of Isolated Attosecond Pulses

As evident from the theoretical discussion in the preceding section, high-harmonic generation driven by sufficiently intense multicycle pulses involves the emission of attosecond pulse trains (APTs) consisting of XUV light bursts separated by half a period of the driving field. The first experimental observation of attosecond pulse trains was reported by Paul et al. (2001), whereas experiments by Mairesse et al. (2003) revealed the chirped nature of the high-harmonic emission, in agreement with the semi-classical prediction of the three-step model. The generation of isolated attosecond pulses (IAPs) thus requires confining the harmonic generation process to a single emission event within the driving pulse. This possibility was first theoretically investigated by Farkas and Tóth (1992), who considered the generation of pulses with durations of 30–70 as based on HHG. To date, the most popular methods for generating isolated attosecond pulses can be broadly grouped into *amplitude gating* and *temporal gating* techniques. Both gating schemes necessitate an electric field form which is reproducible from shot to shot, making CEP stabilization a prerequisite. The amplitude-gating approach resides on the spectral filtering of the high-harmonic emission: as the cut-off portion of the harmonic spectrum is generated in the most-intense half-cycle of the driving field, spectral isolation of this region corresponds to an attosecond pulse in the temporal domain. In order to ensure that the high-energy portion of the XUV spectrum is generated only within one half cycle of the driving pulse, strict requirements on the driver pulse duration (<5-fs) and intensity apply. IAPs with durations as short as 80 as have been produced using the amplitude-gating method (Goulielmakis et al., 2008).

The temporal-gating method, on the other hand, requires appropriate tailoring of the properties of the driving electric field in time (e.g., polarization) in order to confine the HHG event within a single half-cycle. This scheme is much more versatile, and here we limit the discussion to its most popular variants: polarization gating (PG), double-optical gating (DOG), and ionization gating (IG). The polarization-gating method originally proposed by Corkum et al. (1994) exploits the recollision nature of the HHG event (on a microscopic scale) to confine the emission to a single cycle by manipulating the polarization of the driving pulse. HHG driven by a monochromatic pulse exhibits a pronounced ellipticity dependence: the efficiency of the process drops by a factor of two for ellipticities above 10%. By temporally modulating the polarization of the driving field from circular to linear and then to circular again, one can thus restrict the XUV emission to the central part of the pulse, where the field is linearly polarized. An all-optical scheme based on the usage of two birefringent elements (a multiple-order quarter wave-plate (QWP) followed by a second zero-order QWP) was proposed by Tcherbakoff et al. (2003) and experimentally realized in 2006 with the generation of IAPs of durations reaching 130 as in the 25–50 eV spectral range (Sansone et al., 2006; Sola et al., 2006). The duration of two consecutive XUV bursts imposes an upper boundary for the duration of the linearly polarized segment of the pulse, consequently, driving pulses with durations of sub-2–3 optical periods are required (or ~7 fs for 800 nm driving fields). This requirement can be relaxed by breaking the symmetry of the driving field, e.g., by adding a second color with the appropriate pulse energy and phase delay. This is the principle behind the DOG scheme, first demonstrated by Chang (2007). An additional advantage of the DOG technique is the superior generation efficiency related to the lower depletion of the generation medium at the leading pulse edge, as well as the possibility to employ higher field intensities. Generation of XUV-supercontinuum with an FWHM of 200 eV using the DOG scheme in combination with a sub-8-fs 800 nm driver was reported by Mashiko et al. in 2009. Other variants of the polarization gating techniques, designed to mitigate the effects of ionization depletion, include generalized DOG (GDOG), whereby the leading and the trailing edges of the pulse have a lower ellipticity (0.5 instead of 1.0) compared to the DOG scheme. This technique allowed the generation of IAPs with durations of nearly 150 as by using 28 fs drivers at 800 nm (Feng et al., 2009). Interferometric polarization gating is another, more sophisticated approach based on the same principle, whereby up to 4 electric field components with properly optimized ellipticities, relative energies and phases are combined. In this manner, XUV-supercontinua supporting 260-as-pulses were generated using driving pulses longer than 50 fs. The high relevance of this method is the ability to employ a loose-focusing geometry, which translates into higher XUV flux (e.g., 20 nJ in the 20–70 eV spectral range Skantzakis et al., 2009; Tzallas et al., 2007).

As an alternative to the temporal gate provided by polarization gating, the ionization gate relies on the complete ionization depletion during the leading edge of the pulse, which suppresses HHG emission throughout the rest of the pulse due to phase mismatch. Typically, the temporal gate realized in this manner is not narrow enough to restrict the emission to a single-cycle event, therefore, additional bandpass-filtering is required (Bouhal et al., 1998; Jullien et al., 2008; Pfeifer et al., 2007). Alternatively, one can spatially filter the XUV beam in order to tailor its temporal characteristics (Ferrari et al., 2010). PASSAGE (Timmers et al., 2016), or polarization-assisted amplitude gating, is a technique which allows the generation of IAPs with a central frequency tunable in the range from 50 to 130 eV, and combines amplitude and polarization gates, presenting the additional advantage of relaxed requirements with regard to the pulse duration of the driver.

In the light of the discussion presented in Section 4.3.4, recent efforts center on transferring the technology described in the preceding for the case of Ti:Sa pulses to MIR drivers, thus targeting the generation of IAPs in the water window spectral range. IAPs with central energies near the carbon K-edge (248 eV), a pulse duration of 400 as and a spectral bandwidth supporting 30 as pulses were recently reported, whereby a spatiotemporal gating scheme based on wave-front rotation was employed (Silva et al., 2015). The current world record for the shortest pulse (43 as) was generated using a 1.8 μm driver in a xenon target (Gaumnitz et al., 2017). This result superseded the previously achieved pulse durations of 53 as for IAPs in the water-window range (Li et al., 2017).

As a concluding remark, we briefly address the recent development of novel schemes for generating IAPs that do not rely on gating. The lighthouse method first proposed in 2012 (Vincenti and Quéré, 2012) and experimentally demonstrated shortly thereafter (Kim et al., 2013; Wheeler et al., 2012) employs a driving field with a rotating wave front in order to generate high harmonics with a non-negligible pulse-front tilt. As each component of the attosecond pulse train is emitted in a different instantaneous direction, a far-field spatial separation of the individual attosecond pulses composing the train is achieved. The non-collinear optical gating (NOG) scheme, also called angular streaking, introduced in (Louisy et al., 2015), is based on a similar concept. Thereby, an intensity grating is created in the focus by superimposing two pulses at a small angle. As a result, the attosecond bursts are emitted along the bisector angle of the driving fields, and the emission direction corresponds to the orientation of the driving wavefront. By adjusting the delay between the two drivers, an effective wavefront rotation on a subcycle time scale can be achieved, which enables the spatial separation of the emitted XUV bursts in the far-field.

4.3.6 Attosecond Spectroscopic Techniques

The final paragraphs of this section are devoted to the presentation of several of the most commonly employed experimental configurations for conducting attosecond time-resolved pump–probe experiments. Currently, the low conversion efficiencies of the HHG process (lying in the 10^{-9}–10^{-5} range) preclude the application of high-harmonic sources to direct XUV-pump XUV-probe experiments. Therefore, the established approach for studying events occurring on the subcycle scale involves the combination of an XUV pulse with an ultrashort optical pulse. The most common realizations of this principle are the RABBITT (reconstruction of attosecond beating by interference of two-photon transitions) and the attosecond streaking schemes, which are reviewed in the following. In both cases, the XUV pump serves to initiate the electron dynamics, which are subsequently probed by the infrared pulse as a function of the mutual delay between both pulses.

4.3.6.1 Reconstruction of Attosecond Beating by Interference of Two-Photon Transitions

The RABBIT technique utilizes an APT in combination with a less intense and longer, multicycle NIR field, and was initially developed and utilized to characterize the duration of the XUV bursts emitted in the pulse train (Paul et al., 2001). As discussed in Section 4.2, the frequency-domain manifestation of an APT is given by a harmonic comb consisting of a finite number of peaks separated by twice the frequency of the driving field. In a RABBITT measurement, the XUV photons are detected indirectly through photoemission from a target using the photoelectric effect. Adding an NIR perturbing field (typically derived from the same field used for generating the APT) leads to the emergence of additional maxima in the photoelectron spectrum, shifted by the NIR photon energy with respect to the main harmonic peaks. These peaks, referred to as sidebands, result from the emission or absorption of an NIR photon under the influence of the dressing field in addition to the ionizing XUV photon. The non-distinguishability of the two pathways contributing to a given sideband will result in a modulation of the sideband intensity at a frequency given by twice the frequency of the NIR field. Denoting the (two-photon) amplitudes of the absorption, resp. the emission, pathway contributing to a given sideband by M_a, resp. M_e, the proability for the emission of a photoelectron at an energy $E_k = \omega_{XUV} + \omega_{IR}$ can be expressed mathematically as a sum of the squared amplitudes of the individual pathways:

$$P_k \approx |M_a + M_e|^2 = |M_a|^2 + |M_e|^2 + 2\mathcal{R}e\{M_a^* M_e\}. \tag{4.25}$$

The total probability thus depends on the relative phase ϕ between the two contributing quantum paths through the last term in Eq. (4.25). The latter can be controlled by changing the relative delay τ between the XUV and the IR, i.e., $\phi \equiv \omega_0\tau$. The dependence of the amplitudes of the individual pathways on the dressing field is given by $M_a \propto |E_{IR}|e^{i\omega_0\tau+\phi_a}$, resp. $M_e \propto |E_{IR}|e^{-i\omega_0\tau+\phi_e}$, implying that the last term in Eq. (4.25) will vary as $\propto \cos(2\omega_0\tau)$, i.e., at twice the frequency of the NIR pulse. If one considers the two steps in the NIR-assisted XUV photoionization as decoupled from each other, the relative delay can be further decomposed into its individual contributions as follows:

$$\tau = \tau_{GD} + \tau_{XUV} + \tau_{cc}. \tag{4.26}$$

The first term, τ_{GD}, corresponds to the relative group delay of the XUV relative to the NIR probe. The second and the third terms originate from the intrinsic phases of the individual matrix elements characterizing each pathway. Equation (4.26) illustrates the utility of the RABBIT technique in characterization measurements. In case the intrinsic phase difference can be calculated from *ab-initio* methods, which holds true for simple atomic systems, the relative phases between consecutive bursts in an APT can be inferred from a RABITT spectrogram, thus providing a direct characterization of the HHG spectrum. Inversely, in case the group delay can be measured by other means or eliminated by adopting a proper reference scheme, the sideband oscillation phases provide a direct access to the phases of the two-photon matrix elements. The latter implementation of the RABBIT scheme is most commonly utilized to infer photoionization delays associated with the absorption of a XUV photon.

4.3.6.2 Attosecond Streaking

The attosecond streak camera is closely related to the RABBITT scheme, with several major differences. First, the photoionization event is initiated with an isolated attosecond pulse instead of a pulse train, which imposes severe requirements on the NIR field in terms of pulse duration and CEP stability. Further, the dressing (or streaking) field, typically derived from the one used for the IAP generation in order to ensure optimal stabilization, is orders of magnitudes stronger than the typical field strengths employed in RABBITT. Consequently, the interaction of the system with the NIR is no longer perturbative as multiple photons of the IR are absorbed. For this reason, one more conveniently resorts to a semiclassical treatment of the dynamics of the liberated electron in the NIR field, which results in the following expression for the electron velocity at time instant t:

$$\begin{aligned} v(t) &= -\int_t E(t')\mathrm{d}t' + \left[v_0 + A(t_i)\right] \\ &= -A(t) + \left[v_0 + A(t_i)\right]. \end{aligned} \tag{4.27}$$

In the above, t_i denotes the instant of ionization, v_0 is the initial velocity, and $E(t)$, resp. $A(t)$, denote the electric field, resp. the associated vector potential. The first term of the right-most side represents the electron's quiver energy which goes to zero when the pulse is over. The motivation behind the term "streak camera" used for this technique can be inferred from the fact that the above equation represents a mapping of a given instant of ionization, t_i, or "birth in the continuum", onto a velocity displacement $v(t)$ that is a function of the vector potential of the NIR laser field (Kling and Vrakking, 2008). When recorded over a large set of delays between the IAP and the NIR, one recovers a cross-correlation between the EWP and the NIR field. The dependence on the electric field (through the vector potential $A(t)$) rather than the intensity yields sub-fs temporal resolution, despite the few-fs duration of the streaking field. The ability to reproduce the sub-cycle oscillations of the electric field amplitude renders the streaking technique the most accurate method for characterizing attosecond pulses. Initially introduced as a characterization method (Hentschel et al., 2001; Itatani et al., 2002), it was later implemented to infer the underlying attosecond dynamics of the ejected electron. Since an absolute reference for the exact emission time is lacking, one typically records streaking traces originating from electrons emitted from multiple (two or more) electronic shells of the system in a single measurement. The latter is possible given the large bandwidths of the employed APTs. Subsequently one analyzes the small phase shifts between the individual streaking traces, which are interpreted as temporal delays between the photoionization events associated with the underlying electronic shells. Although the spectral overlap resulting from the large IAP bandwidths has prevented widespread application of the streak camera for the photoionization dynamics in molecules, we include this brief description in the current review as this technique has been instrumental in the time-resolving of fundamental events in atomic systems (Drescher et al., 2002; Ossiander et al., 2018, 2017; Schultze et al., 2010; Uiberacker et al., 2007).

4.3.6.3 Photoelectron and Photoion Spectroscopy

Prior to concluding the current section, we address another unique aspect of attosecond spectroscopy: besides the unprecedented temporal resolution, the utilization of large-bandwidth pulses in the XUV/SXR ranges offers the possibility for spatial imaging with ångstrom resolution. The photon energies associated with the attosecond pulses derived from HHG exceed by far the ionization energies of valence electrons, leading to the emission of photoelectrons with very short de-Broglie wavelengths. Recording the angular distribution of the ejected electrons thus gives access to scattering dynamics in the molecular frame (for a recent perspective on the topic, see Vrakking, 2014). In general, in experimental configurations based on the detection of charged particles, i.e., photoelectron or photoion spectroscopy, the dynamics of the investigated process are encoded in the kinetic energy distributions of the detected species. The most commonly employed detection schemes are based on time-of-flight (TOF) electron or ion-mass spectrometry. The usage of magnetic-bottle spectrometers allows for superior detection efficiencies. Velocity-map imaging (VMI) techniques in combination with Abel inversion algorithms are employed in order to detect and retrieve the photoelectron angular distribution in the laboratory frame. In order to achieve a complete reconstruction of the scattering event in the

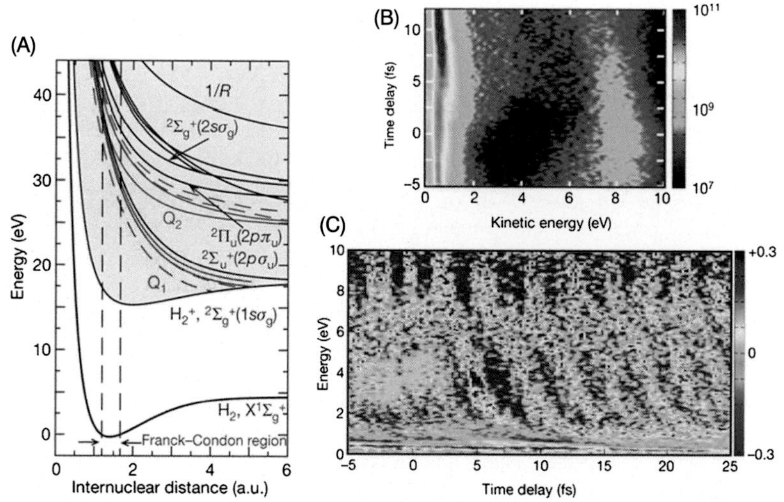

FIG. 4.5 (A) Potential energy curves of the states participating in the dissociative ionization of neutral hydrogen (bottom curve). Photoexcitation results in excitation of the Q_1 (red) and Q_2 (blue) doubly excited states and ionization to the $1s\sigma_g$ and $2p\sigma_u$ states. R denotes the internuclear distance, reported in atomic units (bohr). (B) Experimental kinetic energy distributions of the D^+-fragments as a function of the delay between the XUV and the NIR pulses. The color scale represents the fragment yield (arb. u.). (C) Experimentally extracted asymmetry parameter associated with the production of D^+-fragments in the dissociative ionization of D_2 as a function of the kinetic energy and the temporal delay between the two pulses. The asymmetry parameter exhibits pronounced oscillations as a function of time that vary strongly with kinetic energy. Figure adapted from Sansone et al. (2010) with permission.

molecular frame, a simultaneous detection of the emitted photoelectrons and photoions in coincidence is required. In this respect, the Cold Target Recoil Ion Momentum Spectroscopy (COLTRIMS) technique provides a way towards kinematically complete scattering experiments (Dörner et al., 2000; Ullrich et al., 2003, 1997).

4.4 ATTOSECOND ELECTRON/ION IMAGING SPECTROSCOPY

Given the high photon energy associated with typical attosecond pulses, the interaction of the latter with molecules results in ionization, leaving behind charged particles in potentially highly excited states, which can undergo further relaxation pathways specific to the target. The associated dynamics particularly lends itself to investigation using the techniques of attosecond electron/ion spectroscopy introduced in Section 4.3.

The examples covered in this section are seminal in attosecond spectroscopy as they constitute the first extension of attosecond pump–probe techniques to molecular systems, thereby addressing the fundamental role of vibronic interactions on the sub-fs localization of the electronic charge distributions in small molecules. The majority of the studies address the paradigmatic case of molecular hydrogen (H_2) or deuterium (D_2).

In 2010, Sansone et al. (2010) used the combination of isolated attosecond pulses and waveform-controlled, few-cycle intense NIR fields to track the charge localization dynamics in H_2 and D_2 following dissociative ionization (DI). Intense-field DI results in the production of H^+/D^+ fragments, and their velocity and angular distributions are detected in the experiment. Pronounced angular asymmetries in the fragment momentum distributions induced by the presence of the NIR field encode the spatial localization of the charge distribution and its dynamics.

In the experiment, ionization is induced using intense isolated attosecond EUV pulses with a bandwidth extending from 20 to 40 eV. Given the large spectral width of the IAP, the kinetic energy distribution (KED) spectra contain contributions from several ionization channels. The potential energy surfaces of the two lowest electronic states of the cation ($^2\Sigma_g^+(1s\sigma_g)$ and $^2\Sigma_u^+(2p\sigma_u)$) that are of relevance for this study are reproduced in Fig. 4.5A, together with the autoionizing (AI) states (Q_1 and Q_2) located in this energy region.

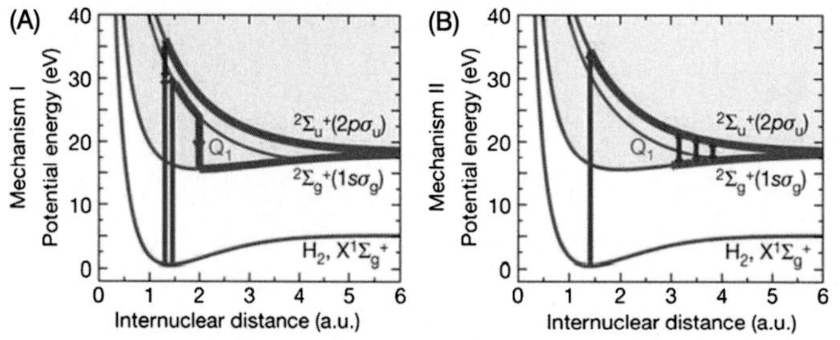

FIG. 4.6 Mechanisms underlying the observed asymmetry in XUV-NIR dissociative ionization. (A) Asymmetry caused by the interference of a wave packet initially launched in the $2p\sigma_u$ state by direct XUV ionization (blue arrows) and by rapid ionization of the $Q_1{}^1\Sigma_u^+$ doubly excited states by the IR pulse (red arrows) with a wavepacket in the $1s\sigma_g$-state resulting from autoionization of the $Q_1{}^1\Sigma_u^+$ states. Purple lines signify intrinsic molecular dynamics. (B) Asymmetry caused by the interference of a wave packet prepared directly in the $2p\sigma_u$ state by direct XUV ionization with a wave packet in the $1s\sigma_g$-state resulting from the stimulated emission during dissociation. Figure adapted from Sansone et al. (2010) with permission.

Direct ionization is the dominant process at photon energies up to 25 eV and leads to the population of the $^2\Sigma_g^+(1s\sigma_g)$-state of the ion, releasing a small fraction of low-KE fragments (see spectrum in Fig. 4.5B). Excitation to the doubly-excited $Q_1\,{}^1\Sigma_u^+$-states embedded in the continuum prevails at photon energies in the range between 25 and 36 eV, whereby the symmetry of the transition dipole moment (parallel) preferentially selects molecules aligned parallel to the laser polarization axis. These transitions are characterized by large oscillator strengths; subsequent AI leads to the population of the $^2\Sigma_g^+(1s\sigma_g)$ state and releases ionic fragments at characteristic kinetic energies between 2 and 7 eV (Fig. 4.5B). However, due to the large cone opening angle of the spectrometer employed in the experiment, the measured spectra above 31 eV contain contributions from the perpendicular transition, which involves the $Q_2{}^1\Pi_u$-doubly-excited states. The ionization channel resulting in the population of the $^2\Sigma_u^+(2p\sigma_u)$ state of the ion opens at 30 eV. At these photon energies, the ionization process can thus leave the ion in either the $^2\Sigma_g^+(1s\sigma_g)$- or the $^2\Sigma_u^+(2p\sigma_u)$-state, leading to different fragment energies.

A time-delayed, intense 6-fs NIR pulse subsequently redistributes the charge within the molecular ion, leaving its imprint in the form of an asymmetric angular distribution of the ejected fragments with respect to the laser polarization axis. This asymmetry is quantified via the laboratory-frame asymmetry parameter

$$A(E_k,\tau) = \frac{N_\uparrow(E_k,\tau) - N_\downarrow(E_k,\tau)}{N_\uparrow(E_k,\tau) + N_\downarrow(E_k,\tau)}, \qquad (4.28)$$

where $N_\uparrow(E_k,\tau)$ and $N_\downarrow(E_k,\tau)$ denote the fragments emitted in opposite directions (within 45°) relative to the polarization axis. This parameter bears a direct relationship to the localization of the charge after the ionization event. As evident from Fig. 4.5C, the asymmetry parameter displays distinct oscillations as a function of the delay between the XUV and the NIR: a shift between the two pulses by half the period of the IR results in a reversal of the charge localization site.

In absence of the IR, the electronic states $1s\sigma_g$ and $2p\sigma_u$ that can be accessed via a single-photon absorption of a XUV photon possess opposite, but well-defined parities, which results in symmetric distribution of electrons emitted "up" ($N_\uparrow(E_k,\tau)$) or "down" ($N_\downarrow(E_k,\tau)$). A coherent superposition of the $1s\sigma_g$ and $2p\sigma_u$ states is therefore a prerequisite for the emergence of charge localization. The dipole selection rules governing the single-photon excitation determine the angular momentum symmetry ($l_{g,u}$) of the outgoing photoelectron (*gerade* (l_g) in the case of $2p\sigma_u$ and *ungerade* (l_u) for $1s\sigma_g$). The two photoelectrons are thus distinguishable, erasing the coherence between the states. This situation changes under the influence of the NIR pulse, which can coherently couple the two states and prepare a coherent superposition. Two coupling mechanisms, illustrated schematically in Fig. 4.6, A and B, that can lead to this scenario are invoked. The first one involves an interference between the autoionization into the $1s\sigma_g$ state proceeding through the $Q_1\,{}^1\Sigma_u^+$ doubly-excited states that is accompanied by the emission of a photoelectron of p-symmetry, and the direct ionization of the WP launched on the $2p\sigma_u$-potential energy surface leading to the emission of an s-electron. In this scenario, the role of the NIR absorption consists in changing the angular momentum character of the outgoing electron, thereby erasing the link between the parity of

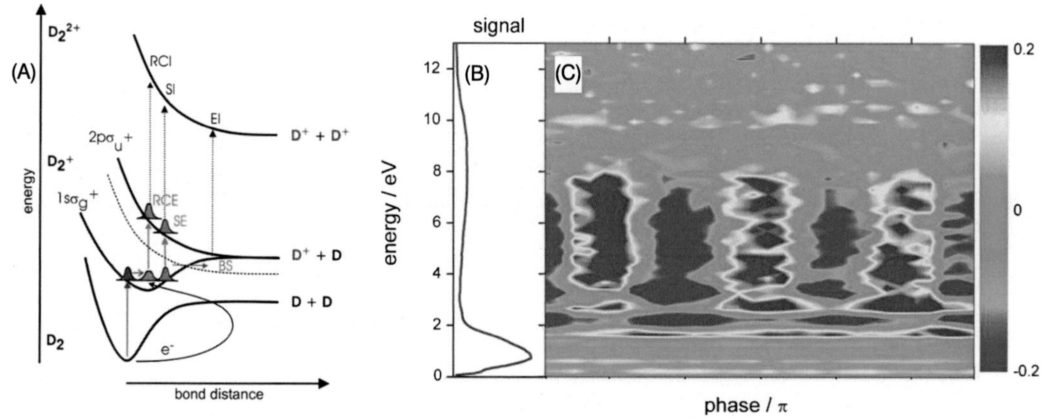

FIG. 4.7 Possible pathways for the production of D^+-fragments from D_2 either by photodissociation of the molecular ion (through bond softening (BS), sequential excitation (SE) or recollision excitation (RCE) or by Coulomb explosion (through RCI, SI, EI). (B) Kinetic energy distribution of the D^+-fragments originating from the dissociation of D_2 induced by a 5-fs, 10^{14} W/cm^2 laser pulses without CEP stabilization. (C) Asymmetry parameter as a function of the D^+ kinetic energy and the CEP phase offset. Figure adapted from Kling et al. (2006) with permission.

the cationic wavefunction and the angular momentum character of the photoelectron. Interaction with the NIR thus leaves the molecular ion and the electron in an entangled state. In the second mechanism, the NIR acts upon the molecular ion in the course of its dissociation on the $2p\sigma_u$-potential, thereby coupling it to the $1s\sigma_g$-surface, forming a coherent superposition between the two states. The asymmetric dissociation is then determined by the relative phase of the two molecular wave functions of opposite parity. This mechanism is particularly efficient at large internuclear distances as the energy separation between the relevant curves approaches one NIR photon.

Prior to this study, electron dynamics evolving on the attosecond timescale have been observed by Kling et al. (2006) in the dissociative ionization of D_2 using intense, CEP-controlled NIR (800 nm) few-cycle pulses of duration 5 fs and intensities up to 10^{14} W/cm^2. As in the preceding study, electron localization is inferred by measuring the D^+-fragment kinetic energy and angular distributions in a VMI apparatus, whereby in the current case the asymmetry of the fragments is monitored as a function of the CEP of the pulse. At the field intensities employed in the experiment, ionization of D_2 leaves the molecular ion in the $1s\sigma_g$-state, whereas the dynamics of the outgoing photoelectron is (almost completely) governed by the oscillatory laser field. Depending on the instant of the ionization within the NIR laser cycle, the electron can revisit the core in a manner reminiscent of the three-step-model for HHG introduced in Section 4.2.3. One of the possible outcomes of the recollision event is the excitation of the molecular ion to the higher-lying $2p\sigma_u$-state (cp. Fig. 4.7A). Due to the repulsive nature of this state, this scenario leads to the ejection of high-kinetic-energy D^+-fragments (up to 12 eV), which are observed as a broad peak (8–10 eV) in the kinetic energy release (KER, cp. Fig. 4.7B). The recollision nature of the mechanism leading to the production of these fragments is further confirmed by conducting the experiment using a circularly polarized (CPL) driver, in which case the suppressed recombination probability leads to the disappearance of the features in the energy range above 8 eV. The central outcome of the experiment is the observation of pronounced (a modulation depth of ≈50%, see Fig. 4.7C) periodic oscillations of the asymmetry parameter associated with these recollision fragments as a function of the CEP of the ionizing pulse. In a similar fashion to the case study presented above, the mechanistic origin of the observed modulation resides in the formation of a coherent wave packet with no well-defined parity, leading to an asymmetry of the fragment ejection that depends on the exact phase of the electric field. In this case, the D_2^+-molecular ion is formed in its ground state $1s\sigma_g$ around the maximum of the electric field. After an excursion time of up to 1.7 fs, the electron recollides with the core, promoting the ion to the strongly repulsive $2p\sigma_u$- state, which promptly dissociates, resulting in D/D^+ fragments with kinetic energies of up to 10 eV. During the dissociation event, the NIR field can induce a partial population transfer back to the ground state, thereby forming a dissociative wavepacket with a large excess kinetic energy. The coherent superposition of these two pathways underlies the observed time-dependent localization of the electron density within the molecule that can be controlled by switching the polarity of the field on a subcycle scale by varying the CEP.

In the experiments discussed in the preceding, the observed molecular attosecond dynamics are associated with the relaxation pathways of the ionized particles generated by the action of the attosecond pulse. A similar experimental principle has been transferred by Neidel et al. (2013) to a series of small-to-medium sized molecules in order to realize a "molecular attosecond Stark spectroscopy". The experimental approach is based on using a two-color pump–probe sequence, whereby a 1 kHz, moderately intense (10^{13} W/cm^2) 800-nm-NIR field of 30 fs duration serves as the pump, whereas the probe pulses are provided by a XUV attosecond pulse train derived from harmonic generation in argon. Both pulses are focused with a toroidal mirror in the active region of a VMI apparatus, in a molecular beam of N_2, CO_2, or C_2H_4. The collected TOF spectra of the molecular ion yields reveal pronounced periodic modulations as a function of the pump–probe delay. These oscillations are interpreted in terms of a time-dependent dipole moment induced during the interaction of the molecules with the NIR field in the pump step, which gives rise to a time-dependent modification of the electronic density and thus to a modulation of the XUV photoionization yield of the molecule. The amplitude of the XUV yield oscillations was found to correlate with the magnitude of the component of the polarizability tensor along the internuclear axis α_z.

The possibility of attosecond control in the molecular photoionization of H_2 using an APT in the place of an isolated pulse has been further explored in a work by Kelkensberg et al. (2011). Additional experiments analogous to those reported in the first part of this section have been conducted in other systems such as O_2 (Siu et al., 2011).

4.5 ATTOSECOND ELECTRON SPECTROSCOPY IN BIORELEVANT MOLECULES

The current section is dedicated to the investigation of charge transfer (CT) and related phenomena using attosecond spectroscopy. In general, CT describes the spatial redistribution of electronic charge in a molecular system, e.g., the backbone of an extended molecular chain, or between two or more molecules, and is of fundamental importance for photoinduced reactions in biochemistry and biophysics (for a recent review on the topic, see Wörner et al., 2017 and references therein). The CT process is predominantly mediated and influenced by molecular nuclear dynamics. Charge migration (CM), on the other hand, is used to denote the charge transfer driven specifically by electronic coherence and/or electron correlation. As the primary mechanism involves the electronic degrees of freedom, CM takes place much faster, typically on a subfemtosecond timescale, although the influence of the nuclear motion comes into play provided that the observation time scale is sufficiently long. Whereas CT phenomena have been studied extensively in the past employing various techniques at different temporal resolutions, CM shifted to focus only recently (Calegari et al., 2016), after progress in attosecond technology enabled the generation of attosecond pulses in the EUV regime. Prior to that, CM has been thoroughly investigated from a theoretical perspective, with a particular emphasis on the role of electronic correlations (Breidbach and Cederbaum, 2005, 2003; Hennig et al., 2005; Remacle and Levine, 2006; Gokhberg et al., 2020, Chapter 5 of this book). These efforts were motivated by the experimental work of Weinkauf et al. on the fragmentation of small peptides based on natural amino acids (Weinkauf et al., 1995). In these experiments, the initial hole was localized on a given specific site of the molecule, however, analysis of the produced fragments indicated that the fragmentation has occurred at a remote site. The interpretation suggested by the authors involved charge migration of the hole between the two sites of the molecule due to a fast CT mechanism (Weinkauf et al., 1996). However, the poor temporal resolution of the experiments prevented the estimation of the time scale of the process.

Here, the recent implementation of an XUV-pump-NIR probe experiment realized by Calegari et al. (2014) to investigate ultrafast dynamics in the small amino acid phenylalanine will be discussed. The topic of CM will be revisited again in Section 4.6.3, where the observation of attosecond charge migration using high-harmonic spectroscopy will be reviewed (Kraus et al., 2015a).

The experimental approach utilizes an XUV pump-NIR probe scheme, whereby a single isolated, sub-300 as XUV pulse with a bandwidth spanning from 15 to 35 eV is used to trigger ultrafast dynamics via ionization, and the ensuing dynamics is interrogated by a waveform-controlled NIR pulse centered at 1.77 eV with a duration of $\approx$4 fs. The two pulses are focused into a clean plume of isolated phenylalanine

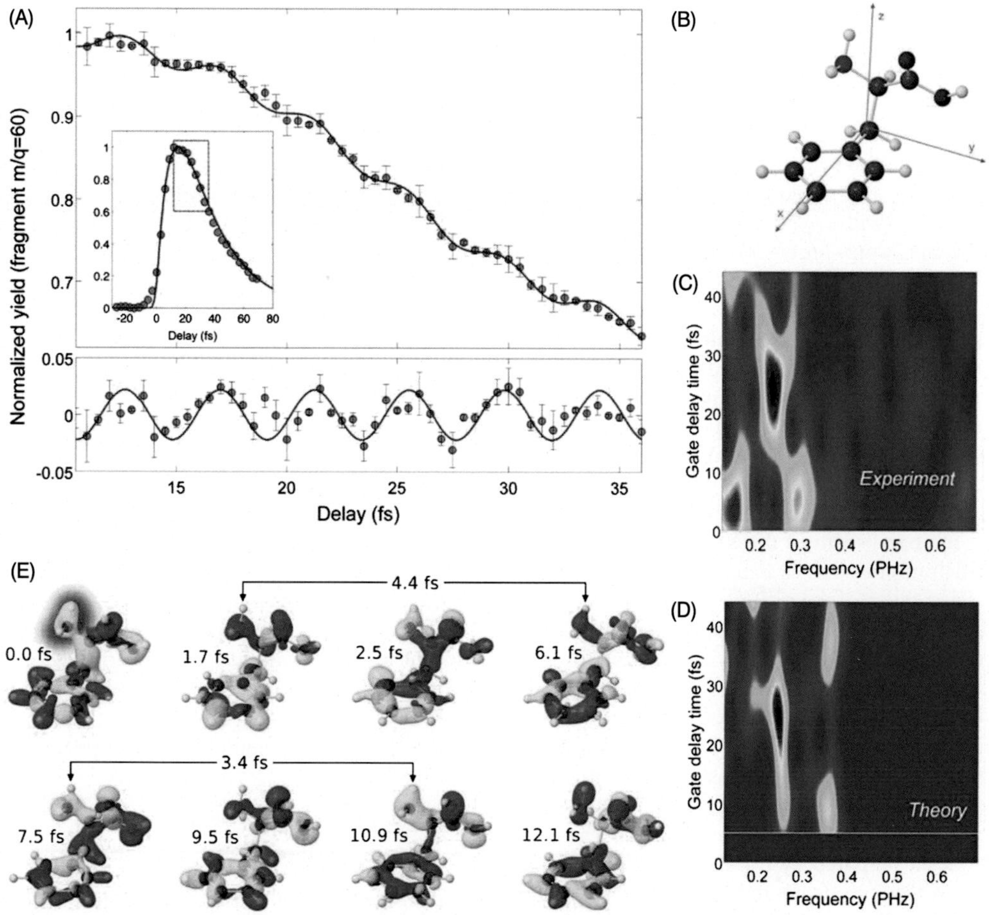

FIG. 4.8 (A) Yield of the doubly-charged immonium ion as a function of the pump–probe delay, measured with a temporal resolution of 3 fs (inset of the top panel) and 0.5 fs (top panel). The bottom panel displays the difference between the experimental data and the exponential fitting curve shown in the inset of the top panel. The error bars indicate the standard error obtained extracted from four measurements. (B) Molecular structure of the most abundant conformer of phenylalanine, obtained from DFT calculations with a B3LYP functional. (C) Sliding-Fourier-transform spectrograms of the data presented in the bottom panel of (A). (D) Same as (C), but extracted from theoretical calculations. (E) Relative variation of the hole density with respect to its time-averaged value as a function of time. The cutoffs of the isosurfaces are set at $+10^{-4}$ (arb.u, yellow) and -10^{-4} (arb.u, purple). The time axis is referenced to the end of the XUV pulse. Figure adapted from Calegari et al. (2014) with permission.

gas-phase molecules (Fig. 4.8B), generated in-situ via laser-induced evaporation from a thin metal foil. Due to the elevated temperatures of 430 K required for this process, the experiment samples over six equilibrium conformers of phenylalanine. The molecular fragments produced by the probe pulse are collected and analyzed in a linear TOF spectrometer. The analysis focusses on the doubly-charged immonium ion produced via secondary electron ejection by the NIR pulse, whose yield is monitored as a function of the pump–probe delay, with varying step sizes (0.3–3 fs).

The observed dynamics are summarized in Fig. 4.8A. The long-time evolution, monitored over a temporal window of 100 fs, displays a steep rise time of 10 ± 2 fs, followed by an exponential decay (with a characteristic constant of 25 ± 2 fs). The results recorded with a step size of 0.3 fs, centering on the first 25 fs, are displayed in more detail in the inset resp. the bottom part of panel A, and reveal an oscillation with a beating period of ≈4.3 fs. More insight can be obtained by performing a sliding-window Fourier-transform (FT) analysis, resulting in the time-frequency map reproduced in Fig. 4.8C. Two initially present components at 0.14 and 0.3 PHz dominate the response over the first 15 fs. Afterwards, a strong and broad component at 0.24 PHz forms and decays within 35 fs, with a spectral width increasing as the pump–probe delay progresses. The observed fast motion is not compatible with the timescale of typical nuclear dynamics, which, although it cannot be excluded to influence the dynamics at longer delays, evolves on a timescale of 9 fs for the fastest X–H stretching, and takes even longer for the skeleton vibrations. The authors therefore attribute the periodic beatings to CM dynamics mediated by electronic dynamics. A further indication for the participation of electron dynamics is the decrease in the yield of the immonium dication when the bandwidth of the XUV pulse is reduced to 3 eV, which implies that CM entails the participation of higher-lying states of the cation.

Extensive theoretical calculations were conducted to further corroborate the electron-dynamics hypothesis. The electron wave packet, modelled as a coherent superposition of many 1-hole ($1h$) cationic states calculated using time-dependent density-functional theory, was propagated for a delay of 500 as after excitation by the XUV pulse using a density-matrix formalism. The NIR probe was not included in the analysis. The initial wave packet prepared by the large-bandwidth IAP contains contributions from multiple open channels (35 per conformer), and the ionization amplitudes were determined by static-exchange and first-order density functional theories. After averaging over all spatial orientations, the hole density, defined as the difference of the electronic densities of the neutral and the cation was extracted and analyzed in the frequency domain. The temporal modulations of the hole density are found to maximize around the amine group, even though the initially created hole distribution is highly delocalized, preventing an intuitive description in terms of a simple CM from one molecular site to another (cp. Fig. 4.8E). The resulting frequency spectra, while depending on the choice of the conformer, all contain the three dominant peaks between 0.15 and 0.4 PHz, and correctly reproduce the appearance of the dominant peak at 0.25 PHz and the major features of its temporal evolution (Fig. 4.8D). These results agree with the experimentally observed beating dynamics in the immonium fragment yield. However, the agreement is only qualitative as the relative intensities and time dependencies of the individual frequency components deviate strongly. The remaining incongruence was attributed to the neglect of the influence of the NIR pulse and the freezing of the nuclear degrees of freedom (Lünnemann et al., 2008), which erases the influence of nuclear motion on longer timescales.

In fact, it remains unclear how the modulation of the electron–hole density at the amino group relates to the experimentally observed yield of the immonium ion. The latter is formed by breaking the C–COOH bond of phenylalanine, rather than the C–NH_2 bond that is more likely to be weakened by an increased electron–hole density at the amino group. An alternative theoretical analysis would consist in analyzing the bond order of the C–COOH group as a function of time in a more directed attempt to explain the experimental results. Therefore, additional theoretical work is desirable to relate the experimental and theoretical results.

Following this study, related experimental schemes based on the use of ionizing attosecond XUV pulses in combination with an optical probe were employed to elucidate the response of DNA building blocks to highly-energetic radiation. Examples include the study of ultrafast dynamics in thymidine and thymine (Månsson et al., 2017), attosecond pump–probe experiments in tryptophan (Lara-Astiaso et al., 2018), and the ultrafast hydrogen migration in glycine (Castrovilli et al., 2018).

4.6 HIGH-HARMONIC SPECTROSCOPY

The principle behind high-harmonic spectroscopy (HHS) and its self-probing character has been introduced in Section 4.2.3. The current section illustrates the capability of HHS to resolve, albeit indirectly, electron and nuclear dynamics with sub-fs resolution. We will cover the following aspects: the resolution of structural rearrangements occurring on a sub-fs-timescale, the laser-induced modification of electronic structure on the generic examples of two small polar polyatomic molecules, and, finally, the reconstruction and laser control of attosecond charge migration.

4.6.1 Observation of Sub-Fs Nuclear Dynamics Using High-Harmonic Spectroscopy

The physical mechanism underlying the sensitivity of HHG to nuclear dynamics can be best illustrated by considering the following expression for the HHG intensity ($I_{\text{HHG}}(N\omega)$), first proposed and validated by Lein (2005) for the case of the H_2 molecule:

$$I_{\text{HHG}}(N\omega) \propto |\, c(\tau(N\omega)\,|^2 \, I^{\text{FN}}_{\text{HHG}}(N\omega), \tag{4.29}$$

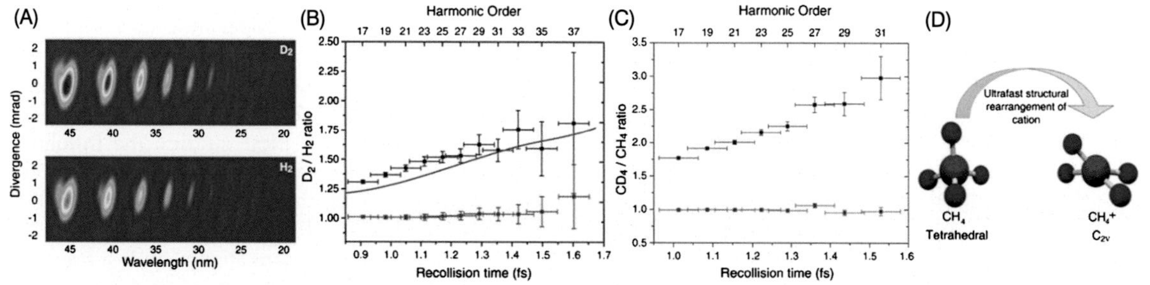

FIG. 4.9 (A) Raw CCD images of the HHG spectra recorded in D_2 (top) and H_2 (bottom), revealing that at all harmonic orders observed, harmonic emission in H_2 is weaker. (B) Ratio of harmonic peak intensities for D_2 and H_2 (black). The red lines represent the outcome of a control experiment in H_2. The blue line corresponds to the ratio obtained from theoretical calculations. (C) Ratio of harmonic signals in CD_4 and CH_4 (black), as well as the control ratio of two harmonic spectra from CD_4 (red). (D) Structures of CH_4 and CH_4^+ at equilibrium. Upon removal of an electron, a rapid structural rearrangement takes place. Figure adapted from Baker et al. (2006) with permission.

where $I_{\mathrm{HHG}}^{\mathrm{FN}}(N\omega)$ denotes the HHG intensity pertaining to the molecule with "frozen" nuclei, and $c(\tau(N\omega)$ is the nuclear autocorrelation function, i.e., the overlap of the nuclear parts $\chi(R,t)$ of the total molecular wavefunction in the ground state and at the instant of recombination, after a continuum excursion of duration τ, $c(\tau) = \int \chi(R,0)\chi(R,\tau)\mathrm{d}R$. The unique association of a given electron travel time with the Nth component of the harmonic comb, denoted explicitly through the argument of τ in Eq. (4.29), implies that the nuclear dynamics can potentially be inferred from a spectrum recorded in a single laser shot.

In 2006, Baker et al. (2006), used laser pulses centered at 775 nm to generate high-harmonic radiation in a pulsed gas jet of H_2 or D_2 molecules. Given the fact that the rotational period of the hydrogen molecule is $\approx$274 fs (and twice that value for D_2), the pulses (initially of 30 fs duration) were compressed down to 8 fs in a hollow-core fiber followed by a chirped-mirror assembly in order to preclude the onset of rotational motion from influencing the HHG process. Placing the focus 9 mm before the gas jet isolated the contribution of the short trajectories, which were subsequently detected in a grazing-incidence, angularly-resolving flat-field spectrometer. The experimentally determined ratio of the HHG responses of the two isomers as a function of the transit time in the continuum is reported in Fig. 4.9B. According to the time-to-frequency mapping discussed in Section 4.3, the experimental time resolution is set by the difference of the recollision times associated with two successive harmonic orders and amounts to 100 as at the wavelength employed ($\approx$800 nm). The HHG efficiency in D_2 exceeds the one in H_2 (cp. also Fig. 4.9A) over the entire energy range covered in the experiment, and the $r(D_2{:}H_2)$ ratio exhibits a monotonic increase at longer transit times. This result has been interpreted in the light of the faster nuclear motion in the lighter isomer. The latter interpretation has been further validated with the aid of SFA calculations which incorporate the autocorrelation functions derived from accurate potential energy curves for the two species and additionally include the two-center-interference effect (Lein et al., 2002a, 2002b; Vozzi et al., 2006, 2005), known to affect the HHG emission from the fraction of molecules aligned parallel to the driving field. The good quantitative agreement with the experimental results serves as a confirmation of the validity of the time-to-frequency mapping and has been used as a starting point for reconstructing the temporal variation of the mean internuclear distance on a subcycle time scale based on the measured data. This approach has been subsequently transferred to the more complex case of the methane molecule (CH_4 vs. CD_4). The HHG intensity ratio $r(CD_4{:}CH_4)$ depicted in Fig. 4.9C shows an even more pronounced isotope effect than the H_2/D_2 pair despite the presence of heavier nuclei. The faster nuclear dynamics is a direct consequence of the nonadiabatic vibronic couplings (the Jahn–Teller effect) in the methane cation, which are absent in the electronically simple case of H_2. Whereas the equilibrium geometry of methane is characterized by a tetrahedral geometry, the cation is unstable with respect to a Jahn–Teller distortion and adopts a C_{2v} structure in equilibrium (Frey and Davidson, 1988; Knight et al., 1984; Vager et al., 1986), with significantly smaller bond angles ($<60°$ vs. $109.4°$, cp. Fig. 4.9D). As the equilibrium structures of the ground state and the ion are highly disparate, this rearrangement has been suggested to take place on an ultrafast time scale. This result, as well as subsequent experiments on the H_2O/D_2O isotopomer pair (Farrell et al., 2011), established that the speed of the nuclear dynamics probed in an HHS experiment is mainly determined by the difference in the potential energy surfaces of the neutral and the ground state. The implications for the Jahn–Teller dynamics induced by nonadiabatic couplings (see Chapter 1 of this book, Marquardt and Quack, 2020) in the methane cation have been investigated theoretically in further detail (Mondal and Varandas, 2015, 2014; Patchkovskii, 2009), and, more recently, by Patchkovskii and Schuurman (2017). The frequency of the employed driving field sets an upper limit for the temporal window that can be covered in this type of experiment, which amounts to ≈1.6 fs for an 800-nm-field. Longer driving wavelengths are thus a prerequisite for observing nuclear dynamics of heavier nuclei. In later work, experiments on the H_2/D_2-system have been extended to 1.3 μm (Mizutani et al., 2011), with an emphasis on the role of the two-center interference. Other works have exploited HHS-based methods to gain access to the phase of the nuclear wave packet (Haessler et al., 2009a; Kanai et al., 2008).

A further specificity of the HHG-based approach for studying nuclear dynamics is that strong-field ionization, the initiating step of the HHG process, prepares a nuclear wave packet, whose dynamics differ from the ones prepared by single-photon ionization (Kjeldsen and Madsen, 2005; Urbain et al., 2004). A theoretical description based on the Franck–Condon factors and tunneling rates was presented and validated against experimental results on the umbrella mode in ammonia in Kraus and Wörner (2013). These results point out the capability of HHS to probe PES areas far from equilibrium, which are hardly accessible by other methods, thereby interrogating the dynamics with high temporal resolution.

4.6.2 Observation of Laser-Induced Modification of the Electronic Structure

The next two selected case studies showcase the sensitivity of HHG to both static electronic structure and electron dynamics and their modification in the presence of a strong electric field. The sensitivity to electronic structure is mediated by the laser-driven recombination step, which gives access to the bound-continuum matrix elements probed via photoionization and photoelectron spectroscopies, but with the additional benefit of the subcycle time resolution. HHS has been extensively used to characterize the field-free electronic structure, notable examples include Cooper minima (Wörner et al., 2009; Bertrand et al., 2012; Higuet et al., 2011; Wong et al., 2013), giant resonances (Shiner et al., 2011), shape resonances (Kraus et al., 2014; Ren et al., 2013). However, in all of these studies, the influence of the laser field present in the probe step has tacitly been omitted from consideration. This assumption has been questioned in the work of Kraus et al. (2015b), where harmonic emission from aligned and/or oriented polar molecules (CH_3F and CH_3Br) was shown to encode the modifications of the electronic structure induced by the strong laser electric field. The experimental approach combined high-harmonic generation driven by moderately intense fields (≈10^{14} W/cm^2) and techniques for field-free molecular alignment/orientation based on non-resonant impulsive Raman scattering (see Stapelfeldt and Seideman, 2003 and references therein). A moderately strong, temporally stretched fs laser pulse is used for inducing alignment, whereas orientation is achieved by means of a phase-controlled superposition of 800 and 400 nm pulses (two-color field, cp. De et al., 2009; Kraus et al., 2012; Frumker et al., 2012; Kraus et al., 2014). Imposing orientational order breaks the inversion symmetry of the sample, leading to the emission of even harmonics in addition to the odd ones. The choice of CH_3F and CH_3Br as generic examples is motivated by their large permanent dipole moments (1.85 and 1.81 D, respectively) which ensure a substantial degree of coupling with the laser field. The ensembles are probed with an HHG-driving pulse around the first alignment orientational revival time, under otherwise field-free conditions, whereby fine control over the rotational distribution is achieved by varying the pump–probe delay.

Quantification of the field-induced electronic-structure modifications is performed with respect to a carefully selected set of robust experimental observables. One of these is the ratio of the emission from aligned vs. isotropically distributed molecules. The photon-energy dependence of this quantity for the two investigated species has been determined at different driving wavelengths (800 nm and 1275, resp. 1330, nm) and intensities (0.8–1.2×10^{14} W/cm^2) and is displayed in Fig. 4.10 for CH_3F (panel A), resp. CH_3Br (panel B). The aligned-to-isotropic ratios exhibit a reversal between 16 and 18 eV for CH_3F, whereas in the case of methyl bromide the crossing is located between 18 and 21 eV. The insensitivity of the position of the crossing to variations of the laser parameters is a signature of the static electronic structure of the molecules and does not result from an interference effect due to participation of multiple electronic states in the emis-

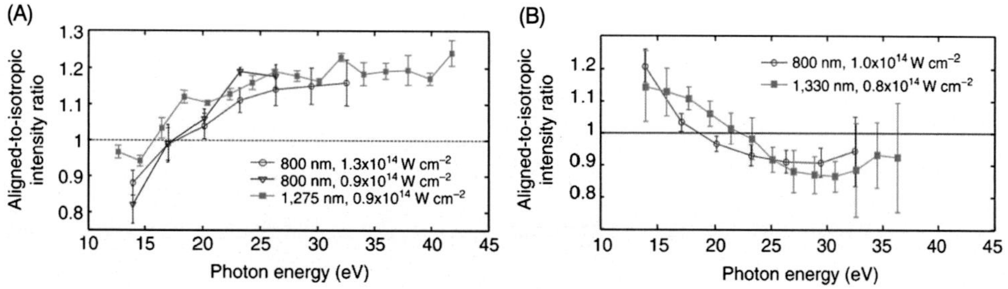

FIG. 4.10 (A) Measured intensity ratio of emission from aligned vs. isotropic molecules at different wavelengths and intensities for CH_3F. (B) Same as (A) for CH_3Br. Figure adapted from Kraus et al. (2015b) with permission.

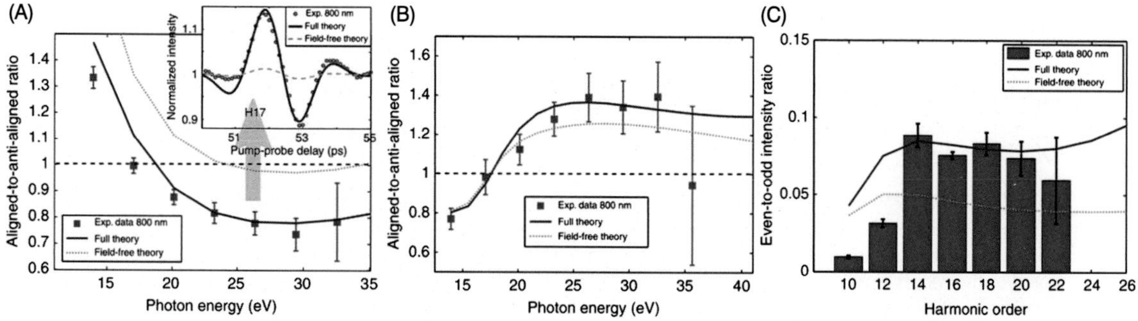

FIG. 4.11 Comparison of the experimental high-harmonic intensity ratios with theoretical results. (A) Ratios of aligned vs antialigned CH_3Br molecules compared with the predictions of the full theory and the field-free model (see text). The inset shows the comparison between the experimentally observed alignment pattern with the theoretical predictions. (B) Same as (A) for CH_3F. (C) Comparison of the even-to-odd ratios for CH_3F with the predictions of the two models. Figure adapted from Kraus et al. (2015b) with permission.

sion process. An additional observable for CH_3F is the even-to-odd ratio, i.e., the ratio of the intensities of the emitted even harmonics relative to the averaged intensities of two adjacent odd ones, determined at the pump–probe delay optimizing the degree of orientation. This observable, plotted in Fig. 4.11C, exhibits a steep increase until it reaches a maximum at H14, followed by a smooth decline. The theoretical interpretation of these observations required improving the theoretical description of HHG by including the influence of the electric field on each of the steps of the HHG process in a consistent manner. In the case of polar molecules, the treatment of the SFI rates, which is difficult by itself due to the exponential sensitivity to the asymptotic tail of the molecular potential, is additionally complicated due to the presence of a large, angle-dependent Stark-shift caused by the permanent dipoles of the neutral and the cation. This challenge is addressed by employing the recently developed weak-field asymptotic theory, consistently extended to the case of polar molecules (Tolstikhin et al., 2011). In the propagation step, an additional, orientation-dependent phase shift originating from the Stark effect (Dimitrovski et al., 2010; Etches and Madsen, 2010) was introduced, whereas the calculation of the matrix elements governing the recombination step was performed using molecular orbitals that take the distortion due to the static electric field into account (Śpiewanowski et al., 2013). As evident from the different curves shown in Figs. 4.11A–B, quantitative agreement with the experimental results is achieved only after all field-induced effects are accounted for, whereas the field-free model is qualitatively incorrect and even makes erroneous predictions in the case of CH_3Br. This pronounced difference in the level of agreement is interpreted as an evidence for the field-induced modifications of the electronic structure, which take place on the subcycle time scale of the HHG process.

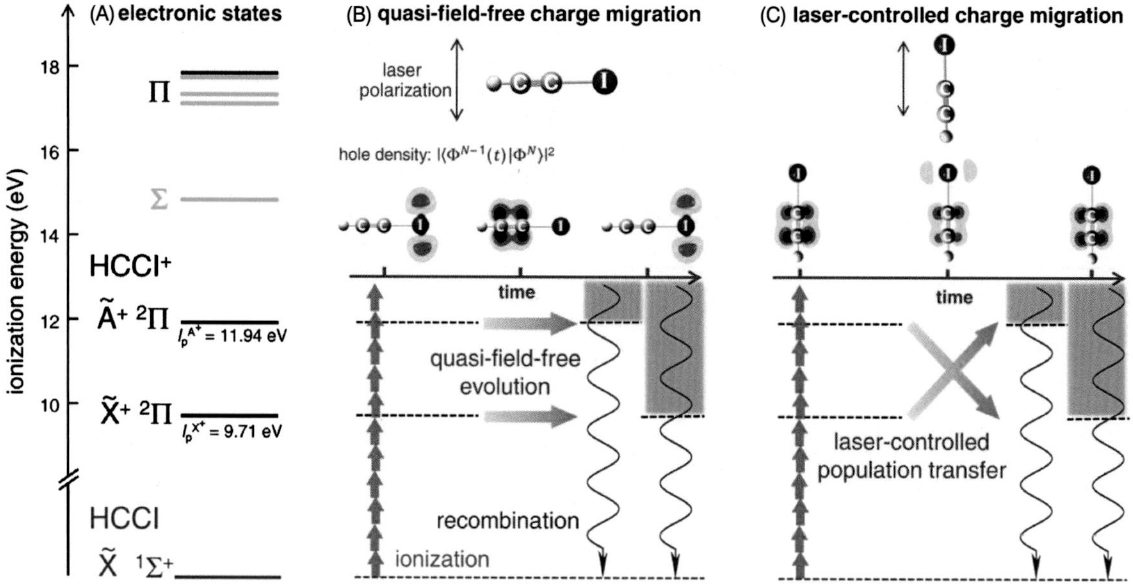

FIG. 4.12 (A) Energy level diagram of the relevant electronic states of HCCl. (B) Schematic illustration of the quasi field-free charge migration and the laser-controlled charge migration. Strong field ionization by the NIR field (red arrows) prepares the electron hole. Its evolution is encoded in the high-harmonic emission (violet) at the recombination instant. For perpendicularly aligned molecules (B), the hole populations are quasi time-independent. For molecules aligned parallelly to the field, the laser field drives a strong population transfer between the $\tilde{X}^+$ and the $\tilde{A}^+$ states. Figure adapted from Kraus et al. (2015a) with permission.

4.6.3 Measurement and Laser Control of Charge Migration in Ionized Iodoacetylene

In the studies reviewed so far, HHS has been used to address the static electronic structure of the system under study. The capability of HHS to probe electron dynamics occurring on a subcycle time scale has been exploited in the interpretation of intensity minima in CO_2 (Rupenyan et al., 2013; Smirnova et al., 2009a; Smirnova et al., 2009b), or the tomographic imaging of orbital wave functions in N_2^+ (Haessler et al., 2010). This section illustrates this feature on the example of the reconstruction of the attosecond charge migration (CM) dynamics (Kraus et al., 2015a) in a spatially oriented polar molecule with a temporal resolution of 100 as from the detailed analysis of the harmonic emission at 800 and 1300 nm from impulsively oriented iodoacetylene (HCCI) molecules. The iodoacetylene molecule represents an ideal system for studying CM (cp. energy diagram in Fig. 4.12A). Its energy structure is characterized by two closely-lying cationic states ($\tilde{X}^{+2}\Pi$ and $\tilde{A}^{+2}\Pi$), separated by 2.2 eV, which can be simultaneously populated by strong-field ionization at the employed intensities, while the population of higher-lying excited cationic states plays a negligible role. The two states are coupled via a strong electric transition dipole moment (≈3.5 D), which is oriented parallel to the internuclear axis in the molecular frame, thus resulting in a strong alignment dependence of the coupling to the external field (Fig. 4.12, B and C). All these characteristics make HCCI a very attractive system for studying field-free as well as field—steered charge migration.

In a manner similar to the experiment presented in the preceding paragraph, a two-color pump pulse is used to induce spatial control (alignment or orientation), and the ensemble is subsequently probed at delays around the first revival time either with an 800 nm or with a 1300 nm pulse. The durations of the electron trajectories at these two driving wavelengths define two observation windows spanning 0.9–1.5 fs, resp. 1.3–2.2 fs.

Contrary to the situation encountered in Section 4.6.2, the ratios of the emissions from molecules oriented perpendicular vs. parallel and the spectral intensity ratio of even and odd harmonics for the two employed probe wavelengths as a function of photon

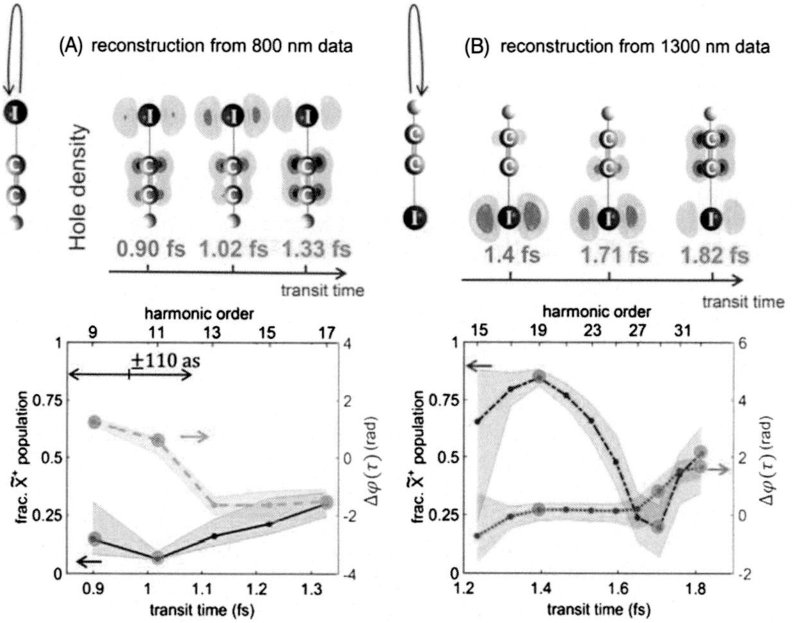

FIG. 4.13 (A) Electron–hole densities for HCCl molecules aligned along the laser polarization axis, reconstructed from high-harmonic emission driven by an 800 nm field. (B) Same as (A) for a 1300 nm driver. The lower panels display the fractional populations of the $\tilde{X}^+$-state (left vertical axis) as well as the relative phase between the $\tilde{X}^+$ and the $\tilde{A}^+$ states (right vertical axis). Figure adapted from Wörner et al. (2017) with permission.

energy are sensitive to the wavelength of the driver: for instance, the position of the minimum of the ratio shifts from 23.2 eV at 800 nm to 35.3 eV at 1300 nm, hinting at subcycle dynamics taking place between ionization and recombination. An additional observable that was detected in the experiment is the phase of the harmonic radiation as a function of the alignment angle using two-source interferometry between the harmonics emitted from aligned and spatially isotropic molecules (Rupenyan et al., 2012). The variation of the phase as a function of the alignment angle depends not only on the photon energy, but also on the driving wavelength, further consistent with laser-induced dynamics.

The complete reconstruction of the quantum-mechanical evolution of the coherent superposition of states prepared by SFI until the recombination instant requires knowledge of the relative populations of the two states and their relative phases, as a function of time. These unknowns were retrieved from the expериential observables (parallel-to-perpendicular resp. even-to-od intensity ratios and the measured phase differences) using a Levenberg–Marquardt inversion procedure. The underlying theoretical model incorporates the angular dependence of the SFI step via the WFAT theory, accurate quantum-mechanical scattering states, the effect of the nuclear motion, and the rotational anisotropy. The total emitted dipole is given by a coherent superposition of the emissions from the contributing electronic states at each energy $N\omega$, mapped onto time using the saddle-point trajectories.

The reconstruction for perpendicularly aligned molecules is represented in terms of the hole densities in Fig. 4.12B. The hole created in the SFI step is initially localized on the iodine atom, then delocalizes over the internuclear axis and localizes at the acetylene end after 930 as. These dynamics correspond to field-free charge migration as the orientation of the transition dipole between the two cationic states $\tilde{X}^+$ and $\tilde{A}^+$ precludes coupling to the field in this orientation.

The reconstruction for parallelly aligned molecules displayed in Fig. 4.13, on the other hand, shows much richer dynamics. In this case, the electron can tunnel via two possible sites, either via the hydrogen or via the iodine atom, whereby the dynamics is qualitatively similar. Here we focus on the CM using an 800 nm driver, depicted in Fig. 4.13A. For tunneling via the I-atom, which is the dominant case, the hole is first created on the acetylene fragment, and the relative phase difference between the $\tilde{X}^+$ and the $\tilde{A}^+$ states is $\approx \pi$. The reconstructed fractional populations indicate that the $\tilde{A}^+$ state is strongly depopulated at early transit times, reaching a minimum shortly after $\approx$1 fs and subsequently increasing again. This depopulation is accompanied by a jump in the relative phase. The dynamics in the case of tunneling via the H-atom, on the other hand, reveal a strong depopulation at early delays, and an earlier onset of repopulation. The observed rapid variation of the population of the $\tilde{X}^+$-state is a clear signature of laser-driven charge migration. The results at 1300 nm, shown only for the case of tunneling via the H-atom, reveal a gradual increase of the $\tilde{X}^+$-population until 1.53 fs, followed by a rapid decrease and minimum at 1.73 fs and subsequent rise. These results suggest that both the wavelength and the tunneling site offer a possibility to exert control over the CM process.

This study provided evidence that HHS can be used to conduct a spatiotemporal reconstruction of CM triggered by SFI, with a temporal resolution of 100 as. In addition, it highlighted the possibility for laser control through the orientational dependence and the driving wavelength.

4.7 ATTOSECOND TIME DELAYS IN MOLECULAR PHOTOIONIZATION

The recent progress in attosecond chronoscopy, in particular the development of interferometric XUV + IR two-color techniques such as the attosecond streak camera (cp. Section 4.3.6.2) or RABBIT (cp. Section 4.3.6.1), has enabled access to the photoionization dynamics in atomic and molecular gases, as well as solids (cp. (Pazourek et al., 2015) and (Gallmann et al., 2017) for recent reviews). Probing these dynamics in the time-domain provides complementary information to the observables accessible via frequency-domain (i.e., time-integrated) measurements (e.g., partial cross sections or phase shifts) and can thus assist the derivation of a complete quantum-mechanical description of the photoionization process (Jordan et al., 2017). Whereas atomic and solid-phase systems have been a subject to investigation using both streaking and RABBIT techniques (Cavalieri et al., 2007; Gaumnitz et al., 2017; Jain et al., 2018; Jordan et al., 2017; Klünder et al., 2011; Schultze et al., 2010), the relatively small amount of experimental data on valence-shell photoionization dynamics in molecules reported to date has been limited to the latter technique. This is due to the presence of multiple close-lying electronic states in molecular cations that fall within the bandwidth of a typical IAP, consequently, application of streaking to molecules is limited to well separated electronic states, such as core-ionized states (Biswas et al., 2020). Therefore, in this section we restrict our discussion to recent experimental studies employing APTs phase-locked to multicycle IR pulses, where the time-domain observables can be retrieved from the spectral phase information with sufficient energy resolution. The selected examples cover aspects such as the influence of autoionizing Rydberg states (Haessler et al., 2009b) on the phase of side-band oscillations, the manifestation of shape resonances (Huppert et al., 2016) in the measured attosecond photoionization delays, as well as the intricate dependence of the latter on the details of the molecular structure and the anisotropy of the molecular potential (Vos et al., 2018). The final case study addresses the subtle attosecond delays between electrons emitted in the forward vs. backward directions in the two enantiomers of a chiral molecule (Beaulieu et al., 2017). Although this latter study did not make use of the RABBIT technique, we include it here as it illustrates how self-referenced photoelectron interferometry can be used to access dynamics on attosecond time scales without the necessity of employing attosecond pulses/pulse trains, while noting that the interpretation of such experiments is more complicated than in the RABBIT framework.

4.7.1 Phase-Resolved Near-Threshold Photoionization of Molecular Nitrogen

The experimental characterization of the phase of the two-color two-photon near-threshold photoionization of molecular N_2 reported in the study of Haessler et al. (2009b) represents an important preparatory step towards the measurements of photoionization delays in molecules. Notably, this study, as well as the subsequently published theoretical analysis (Caillat et al., 2011), were carried out before the relation between the observables of RABBIT and photoionization delays was established (Klünder et al., 2011). Instead, Haessler et al. explored the manifestations of autoionizing Rydberg states on the phase of side-band oscillations observed in RABBIT measurements. Whereas the experimental publication only reported the measured phases as a function of the side-band order, the theoretical analysis (Caillat et al., 2011) additionally discussed the relation of these phases with the "formation time" of the side-band states, obtained as the derivative of the phases with respect to the photon energy of the dressing (and driving) fields. We note that these formation times are however distinct from the photoionization delays that are discussed in Sections 4.7.2 and 4.7.3. The variation of the two-color two-photon phases as a function of the photon energy in the neighborhood of a resonance has

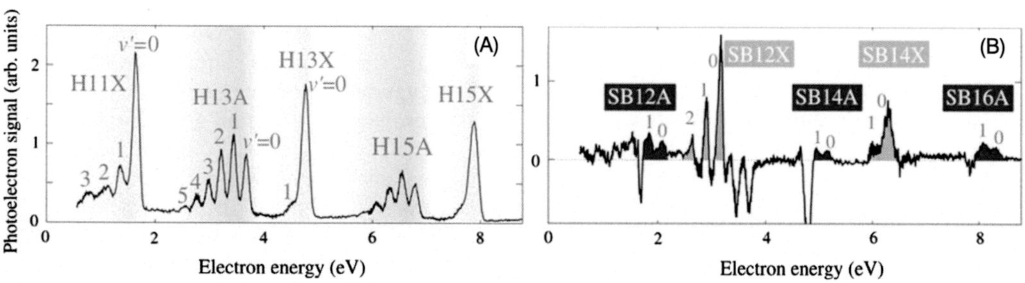

FIG. 4.14 (A) Photoionization spectrum of N_2 obtained with the high-harmonic (XUV) comb alone. (B) Difference between XUV-only spectra and spectra obtained in the presence of the perturbing NIR field. Figure adapted from Haessler et al. (2009b) with permission.

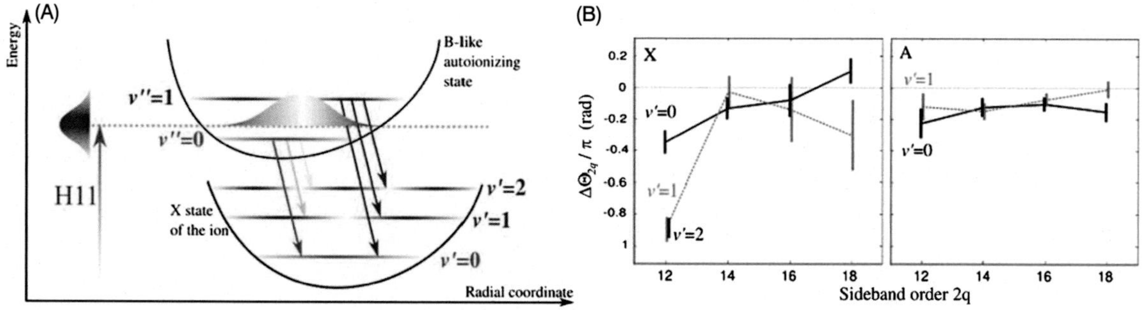

FIG. 4.15 (A) Schematic representation of the photoionization process taking place via vibrational levels of the autoionizing state. (B) Experimental molecular phase difference as a function of the sideband order for the X and A ionization channels. Figure adapted from Haessler et al. (2009b) with permission.

also been experimentally observed in helium (Swoboda et al., 2010) prior to the publication of (Klünder et al., 2011). The experiment of Haessler et al. employed an APT derived from a 20 Hz, 50 mJ system delivering 50 fs pulses which are subsequently spatially separated into a pump (HHG, outer part) and a probe (the dressing IR, inner part) arms. The XUV is generated by focusing the outer annular part of the beam in a pulsed argon gas jet, and, after filtering out the residual IR part, is recombined with the probe beam with the aid of a grazing-incidence Au-coated toroidal and focused in an effusive stream of N_2 molecules. The relative XUV–IR delay τ is adjusted with the aid of a piezoelectric transducer, and the generated photoelectrons are collected with a magnetic-bottle time-of flight (TOF) assembly. The accent of the study lies on the spectroscopically rich region just above the ionization potential of N_2, and, in particular, the "complex" resonance at around 17.12 eV, which corresponds to an autoionizing state belonging to the Rydberg series converging to the $B^2\Sigma_u^+$ state of the ion (Dehmer et al., 1984). The latter couples most efficiently to the continuum associated with the $X^2\Sigma_u^+$ ground states of N_2^+. As the fundamental IR (ω) frequency is centered at 1.565 eV, the resonance falls within the bandwidth of the 11th harmonic (17.21 eV, cp. also Fig. 4.15A).

The XUV photoelectron spectrum in the absence of the dressing IR field is shown in Fig. 4.14A. The spectral bandwidth of the attosecond pulse train results in photoemission from two close-lying photoionization channels associated with the $X^2\Sigma_g^+$ (15.58 eV) and the $A^2\Pi_u$ (16.69 eV) cationic states, which appear as two distinct photoelectron bands with a superimposed vibrational structure. The difference in the shapes of the PE bands reflects the different equilibrium distances in the two states: 1.12 Å vs 1.17 Å. In order to accentuate the differences in the details of the spectra induced by the perturbing IR and mitigate the spectral overlap of the two channels, the two-color spectrum presented in Fig. 4.14B has been background corrected by subtracting the XUV-only response. The absorption, resp. emission, of IR photons creates photoelectrons with energies corresponding to the even harmonics $2q$ of the fundamental frequency ω ("sidebands", label SB), which appear as positive contributions in contrast to the negative areas reflecting the loss of population at the positions of the main peaks. As there are two interfering pathways contributing to each sideband, the intensities of these peaks evolve as a function of the delay between the two colors:

$$SB_{2q} \propto A_{2q} \cos(2\omega\tau - \Delta\phi_{\mathrm{mol}} - \Delta\phi_{2q}). \quad (4.30)$$

In the experiment, the total phase of each sideband is extracted by means of an FFT analysis of the oscillatory signal. In order to isolate the non-trivial, molecule-specific phases denoted by $\Delta\phi_{\mathrm{mol}}$, the contribution of the phase of the XUV field ($\Delta\phi_{2q}$) to the total phase has been determined by a separate measurement in Ar and subsequently subtracted. This procedure invokes the implicit assumption that the $\Delta\phi_{\mathrm{at}}$ can be reliably calculated from theory (Mairesse et al., 2003), so that the harmonic phases can be directly read off from the phase of the SB oscillations. This analysis is performed for each sideband and each ionization channel and the extracted phases for each vibrational quantum number are plotted as a function of energy (sideband order) in Fig. 4.15B. Whereas most of the measured values for both channels are contained within $\pm 0.2\ \pi$ rad, the phase difference for SB12 in the X-state shows a noticeable exception to this trend, assuming a value of $-0.35\ \pi$ rad for the $v' = 0$ state and reaching up to $-0.9\ \pi$ rad for the $v' = 1, 2$ states. This behavior has been interpreted as a manifestation of the resonance in the vicinity of H11 and can be linked to the phase jump in the complex two-photon photoionization matrix element for the pathways involving absorption of H11. In addition, a remarkable dependence of the magnitude of $\Delta\phi_{\mathrm{mol}}(12)$ on the vibrational level is observed. To aid the rationalization of this behavior, a one-dimensional model is developed. The crucial insight is that the effect of the resonance on the molecular phase is most pronounced when the harmonic energy is situated below the resonance energy. With this knowledge, and given the fact that the H11 is positioned energetically between the $v'' = 0$ and $v'' = 1$ levels of the autoionizing B-like state, this behavior can be explained in terms of the Franck–Condon overlap with the vibrational levels of the X-cationic state. The latter show that whereas population of the $v' = 0$ state of the ion can efficiently take place from both the $v'' = 0$ (below resonance) and $v'' = 1$ (above resonance) states of the B-like state, the population of the higher-lying $v' = 1, 2$ states occurs mainly from the $v'' = 1$ state (above resonance). Thus, the phase shift $\Delta\phi_{\mathrm{mol}}(12)$ for the $v' = 0$-sideband contains contributions from both direct ($v'' = 0$) as well as resonant ($v'' = 1$) pathways, which diminishes the magnitude of the total phase shift. In contrast, the higher-lying vibrational states are almost entirely dominated by the resonant contribution and thus exhibit a strong phase shift ($\approx \pi$). The lack of sudden variations of the phases associated with the A-channel is attributed to the decreased efficiency of the coupling to the autoionizing state. This work demonstrates the sensitivity of the observables accessible via RABBIT measurements to the location of the one-photon ionization threshold, to the vibrational structure of the PE spectrum, and to the presence of autoionizing resonances. Several of these aspects will be further addressed and explored in more details in the case studies presented in the remaining part of this section.

4.7.2 Attosecond Photoionization Delays in the Nitrous Oxide and Water Molecules

The first measurements of time delays in molecular photoionization were reported by Huppert et al. (2016). While demonstrating the applicability of the RABBIT technique to molecules, the work by Haessler et al. did not report time delays for at least two reasons. First, the relation between the observables of RABBIT and time delays in photoionization (Klünder et al., 2011) had not yet been established when the study was published. Second, Haessler et al. investigated the effect of a resonance whose width is much smaller than the energetic separation of two neighboring harmonics. Although such a resonance manifests itself as an additional phase shift in the two-color two-photon matrix elements, the phase shift does not smoothly translate into a time delay because the phase varies too rapidly with energy to validate the finite-difference approximation (Dahlström et al., 2012; Klünder et al., 2011) that is used to relate phase shifts to time delays. The study by Haessler et al. simultaneously revealed some of the potential impediments originating from the molecular electronic structure, like the spectral congestion related to the participation of several ionization channels in the valence-shell photoemission process, which is inconvenient given the bandwidth of the employed APTs, or the spectral overlap of the photoelectron spectra associated with different harmonic orders. These limitations have been resolved in the work of Huppert et al. (2016) by spectrally filtering the attosecond pulse train, thereby removing the problematic spectral overlap that previously limited the reliability of molecular attosecond interferometry. The experiment is performed in an actively-stabilized attosecond beamline (Huppert et al., 2015) and a magnetic TOF spectrometer. High-harmonic generation takes place in an argon-filled gas cell (10 mbar)

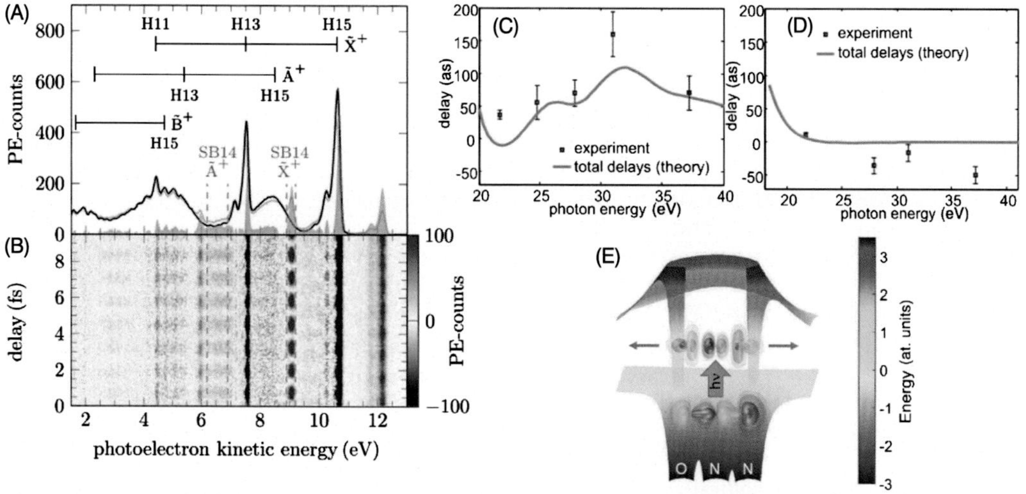

FIG. 4.16 (A) Photoelectron spectrum of N_2O generated by an APT transmitted through an Sn spectral filter without (black) and in the presence (orange) of the perturbing IR field. The red and the blue spectra correspond to difference spectra obtained by subtracting the XUV-only from the XUV + IR spectra and vice versa, respectively. (B) Difference spectrum as a function of the XUV–IR delay and the photoelectron energy. (C) Experimental and theoretical photoionization delays between photoelectrons leaving N_2O^+ in its X^+ or A^+ states. (D) Same as (C), but for H_2O. (E) Illustration of the shape resonance of sigma symmetry in the photon energy range of 25–30 eV and associated with the A^+ state of N_2O^+. The lower surface corresponds to the calculated molecular potential comprising electrostatic and exchange interactions. The upper surface corresponds to the total potential obtained after adding the molecular and the centrifugal potentials for partial wave quantum number $l = 5$. The color-coded isosurfaces indicate the wave functions of the bound orbital and the shape-resonant state. Figure adapted from Huppert et al. (2015) with permission.

driven by laser pulses centered at 800 nm (1.5 mJ, 30 fs). The separation and subsequent recombination of the XUV and the dressing IR pulse is performed with the aid of two perforated off-axis parabolic mirrors. The refocusing of the XUV is done with the aid of a toroidal mirror, and the time delay is adjusted by changing the length of the IR beam path. The complications posed by the spectral overlap are mitigated by introducing several significant experimental advances, most notably via the introduction of thin metal filters (Sn, Ti, Cr) for spectral isolation of harmonics, the single-shot acquisition setup and the insertion of a chopper wheel in the IR beam path which allows for recording high-fidelity (XUV-only-XUV+IR) difference spectrograms. Fig. 4.16A shows a typical photoionization spectrum of N_2O generated with an Sn-filtered APT.

The study investigates the relative photoionization delays between the two outermost valence shells of two triatomic molecules, nitrous oxide (N_2O, linear) and (gaseous) water (H_2O, bent). The selection of these two species allows one to survey the effect of a shape resonance on the molecular photoionization delays in a comparative manner. The spectrum of N_2O (cp. Fig. 4.16A) is dominated by photoelectrons corresponding to the first two (bound) cationic states $\tilde{X}^+$ and $\tilde{A}^+$ with ionization energies of 12.89 and 16.38 eV, respectively. The difference spectrogram as a function of the two-color delay is shown in Fig. 4.16B, whereby the oscillating positive contributions (red) correspond to the SBs of order 12 and 14 for the $\tilde{X}^+$ state and 14 for the $\tilde{A}^+$ state. Delays of up to 35 ± 6 as between photoionization events originating from the two valence orbitals ($\Delta\tau = \tau(\tilde{A}^+) - \tau(\tilde{X}^+)$) can be extracted after a Fourier analysis of the oscillation of SB14. Varying the metal filter allows one to access different energy regions; and the resulting delays as a function of the sideband order are summarized in Fig. 4.16C. Analogous measurements have been performed in H_2O, whereby the PE spectrum contains contributions from the $\tilde{X}^+$ and the $\tilde{A}^+$ states located at 12.62 and 14.74 eV, respectively.

A brief inspection of the energy variation of the relative delays between the $\tilde{X}^+$ and the $\tilde{A}^+$ states in the two species (cp. Fig. 4.16, C and D) reveals a pronounced enhancement of the magnitude of the attosecond delays in N_2O with respect to the ones determined for H_2O, with a maximum value of 160 ± 34 as around 31 eV.

Overall, the energy variation of $\Delta\tau$ in H_2O is relatively structureless, decaying monotonically as a function of the SB order.

In order to rationalize these results, the authors resort to a theoretical treatment that takes as its starting point the photoionization matrix elements derived from quantum scattering calculations (Gianturco et al., 1994; Natalense and Lucchese, 1999). The details of the theoretical work were published separately (Baykusheva and Wörner, 2017). Similar to the atomic case, the molecular photoionization delay $\Delta\tau$ can be decomposed into two contributions, a molecule-specific part (τ_W) and a continuum–continuum (τ_{cc}, measurement-induced) contribution. Whereas the continuum–continuum ("cc") part is mainly sensitive to the Coulomb part of the asymptotic potential and can thus be approximated by the analytical solution for a hydrogen-like potential, τ_W is strictly system-specific and encodes the signatures of the molecular potential. The subscript "W" stands for "Wigner" time delay, following the established nomenclature (Pazourek et al., 2015) for the system-specific part of the photoionization delays. It does not reduce, however, to the delay associated with the photoionization following the absorption of a single XUV photon. Due to the loss of spherical symmetry in molecular systems, the photoionization matrix elements contain contributions from multiple partial waves of varying angular momenta, which are subsequently coupled by the IR field in the dressing step in a non-trivial manner. The calculated results exhibit an overall good qualitative agreement with the experimental data (cp. Fig. 4.16, C and D), correctly predicting the local maximum in N_2O around 31 eV and the smooth variation in the H_2O-case. Moreover, the analysis in terms of the quantum-mechanical photoionization matrix elements allows the correlation of the enhancement of the photoionization delay for N_2O with the presence of several shape resonances in the investigated energy range. There are two shape resonances (one of σ- and one of π-symmetry) embedded in the energy interval between 21.7 and 37.2 eV that previous theoretical studies have linked to continua belonging to the $\tilde{A}^+$ state (Braunstein and McKoy, 1987; Rathbone et al., 2005). Two further shape resonances (of σ- and π-symmetry) have been predicted for the $\tilde{X}^+$ continuum (Huppert et al., 2016). Generally, a shape resonance is a single-electron phenomenon that occurs when the combined molecular and centrifugal potential experienced by the electron features a finite barrier, through which the ionizing electron can escape by tunneling, yielding a local enhancement in the photoionization cross section (cp. Fig. 4.16E). In the context of photoionization delays, the SR associated with the $\tilde{A}^+$ state leads to enhancement of the delays around 30 eV. This experiment demonstrated that attosecond interferometry, by providing access to the relative phase differences between different ionization continua, can serve as a probe of the time-domain manifestation of shape resonances.

4.7.3 Stereo-Wigner Time Delays in Molecular Photoionization of Carbon Monoxide

The photoionization matrix elements, whose phase behavior is intrinsically linked to the attosecond time delays probed in the conventional RABBIT experiments, exhibit a strong anisotropy both in the molecular frame as well as with respect to the mutual orientation of the molecule and the XUV polarization direction. The results presented so far, however, represent averages over all possible molecular orientations. Gaining access to the angular dependence of the delays would provide a direct measurement of the spatial localization of the escaping electron wavepacket (EWP) within the molecular potential. In a simple diatomic heteronuclear molecule like carbon monoxide (CO), one could then introduce a molecular photoionization "stereo-Wigner" time delay (Chacon et al., 2014), i.e., the relative time delay between the two possible escape sites:

$$\tau_{SW} = \tau_W(\text{C-side}) - \tau_W(\text{O-side}). \tag{4.31}$$

This possibility, theoretically proposed in Chacon et al. (2014) was experimentally realized only very recently by Vos et al. (2018) in a configuration combining the RABBIT excitation scheme with the COLTRIMS detection technique briefly described in Section 3.2.4. Photoionization dynamics is initiated by an APT (extending from harmonic 15 to 27 and centered at 34 eV) generated in an Ar-gas cell with a 30 fs pulse with a central wavelength of 776 nm. A portion of the generating beam is split prior to the HHG step and serves as a perturbation field, whereby the two-color delay is controlled by a piezo-controlled stage. The two pulses are collinearly focused in a supersonic jet expansion of CO molecules using a toroidal mirror, where dissociative photoionization takes place. The thereby generated electrons and fragment ions are extracted by a combination of (dc) electric and magnetic fields and detected in coincidence, whereby the acceptance angle of 4 π for both particles allows for a full 3D reconstruction of the fragment momentum distributions.

Out of the three ionization channels lying within the APT bandwidth (one direct channel, $CO + \omega \rightarrow$

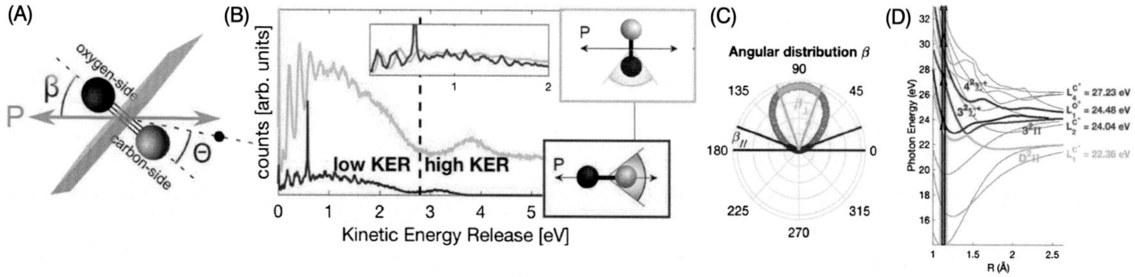

FIG. 4.17 (A) The laboratory frame is defined by the laser polarization axis P (blue arrow). The orientation of the molecular axis with respect to the field is denoted by the angle β, whereas θ_e corresponds to the emission direction of the electron (black dot) in the molecular frame. (B) Decomposition of the KER spectra in terms of contributions from $\beta_{\parallel}$ (blue) and $\beta_{\perp}$ (yellow), averaged over an emission cone with an opening angle of ±20° (cp. side panels). The inset indicates the vibrational structure observed in the low-KER (<2 eV) region. (C) Polar plot of the angular distribution of the recoil angle with respect to the field polarization axis beta. The selection areas corresponding to $\beta_{\perp}$ and $\beta_{\parallel}$ are indicated in orange and blue, respectively. (D) Potential energy curves of the most relevant electronic states of the CO^+ molecular ion as a function of the internuclear distance. Figure adapted from Vos et al. (2018) with permission.

$CO^+ + e^-$, and two dissociative channels leading to the production of C^+ or O^+ ions), only the channel associated with the production of C^+-fragments is studied in detail. Due to the substantial degree of spectral congestion (Fig. 4.17D) because of many close-lying states, a detection in state-resolved manner could not be implemented. Therefore, partial electronic state selection in the kinetic-energy release (KER) spectrum is imposed by discriminating between dissociation events where the recoil axis is oriented perpendicular ($\beta_{\perp}$) or parallel ($\beta_{\parallel}$) to the laser polarization (see Fig. 4.17). The angle β denotes the relative angle between the internuclear axis and the photoionizing laser field. KER spectra for the two orientations are presented in Fig. 4.17B, whereby the spectrum for perpendicularly oriented molecules is further subdivided into a high-KER region (>2.8 eV) and a low-KER region (<2.8) eV. With the aid of theoretical calculations of the nuclear dissociation dynamics based on non-adiabatic QM treatment derived from potential energy curves calculated with the MRCI/def2-TZVPP methods, the electronic states contributing to the $\beta_{\perp/\parallel}$-events could be determined. Photoelectron spectra of molecules oriented parallel to laser field are dominated by $^2\Sigma^+$-states ($3^2\Sigma^+$ vs. $4^2\Sigma^+$ at high/low energies, respectively). The spectra of the perpendicularly oriented molecules feature contributions from the $D^2\Pi$- and $3^2\Sigma^+$-states at low KER, and contributions from the $3^2\Pi$-state at high energies. As in the preceding examples, the SB intensity is monitored as a function of the two-color delay and is given by Eq. (4.30), whereby the oscillation phase can be written as $\Delta\phi_{\text{tot}} = \Delta\phi_{2q} + \Delta\phi_{\text{W}} + \Delta\phi_{\text{cc}}$.

In this case, however, one monitors the relative delay between electrons emitted from different locations of a single molecule. This implies that both the XUV-($\Delta\phi_{2q}$) and the measurement-induced ($\Delta\phi_{\text{cc}}$) contributions cancel out, yielding a quantity directly proportional to the stereo Wigner time delay (SWTD),

$$\tau_{\text{SW}} = \frac{\Delta\phi_{\text{W}}(C - side) - \Delta\phi_{\text{W}}(O - side)}{2\omega}. \quad (4.32)$$

The experimental results for $\beta_{\perp/\parallel}$, obtained after Fourier analysis of the sideband oscillations (averaged over an energy region of 1 eV centered at each SB), are displayed in Fig. 4.17. In the case of perpendicularly-oriented molecules, the SWTDs lie within ±35 as and are close to zero, resp. slightly positive, in the investigated energy interval. This behavior is expected from symmetry considerations and is further confirmed by the results of theoretical calculations based on two different models (see below). In the case of parallel-oriented CO molecules, the SWTDs exhibit an evolution from strongly negative (−165 as at 5.0 eV) to slightly positive (+30 at 14.4 eV) values. This behavior implies that the SWTDs changes sign as a function of energy, i.e., at low energies, the EWP escapes into the continuum via the C-site, whereas the opposite applies for higher kinetic energies. In order to track the origin of this trend, theoretical calculations based on the following two independent approaches have been considered: (i) the TD-RIS method described in detail in (Spanner and Patchkovskii, 2013, 2009), which accounts for the XUV-initiated dynamics in a fully quantum-mechanical

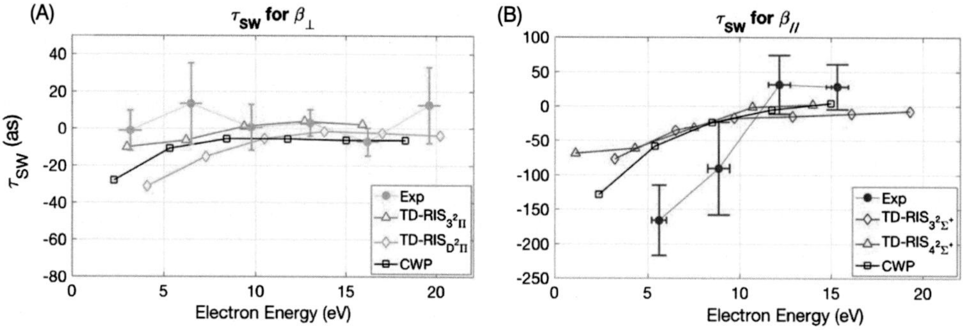

FIG. 4.18 (A) SWTDs for molecules oriented perpendicularly with respect to the laser field, integrated over the entire KER region. The experimental results, shown in orange, suggest that the SWTDs amount to zero within the error bars. This trend is reproduced in the theoretical delays calculated with different models (cp. legend). (B) Same as (A), but for the SWTDs of molecules aligned parallelly to the field. Figure adapted from Vos et al. (2018) with permission.

manner, but does not explicitly treat the IR-induced dynamics; and (ii) the classical Wigner propagation (CWP) method (Vos et al., 2018) based on propagating the Wigner function of the photoelectron in a classical manner, with the IR-field included. Both calculations (cp. Fig. 4.18) capture qualitatively the experimentally observed trend for $\beta_{\perp}$ (panel A) and $\beta_{\parallel}$ (panel B), thus revealing the sensitivity of the measured delay to the molecular orientation. There are two potential contributions that could give rise to the observed asymmetry in the SWTDs: either an asymmetry in the initial localization of the EWP, or an asymmetry in the molecular potential experienced by the receding EWP. The contribution of the molecular potential can be ruled out after analysis of the dipole moments of the ionic states associated with each channel, thus identifying the role of the initial localization, i.e., the mean position of the electron along the C–O bond) of the ionization event as the determining factor for the observed SWTD. This last conclusion is further supported by analyzing the Wigner function of the dipole matrix element of the Dyson orbitals for each of the contributing states. This quantity represents a coordinate and momentum representation of the EWP at the instant of the photoionization event, and its anisotropy directly reflects the asymmetry of the observed SWTD.

4.7.4 Phase-Resolved Two-Color Multiphoton Ionization of Chiral Molecules

A complementary approach to studying photoionization dynamics in the time domain has been introduced in (Zipp et al., 2014). Instead of employing single-photon ionization by XUV pulses as in the RABBIT technique, this measurement scheme relies on multiphoton ionization by UV pulses centered at 400 nm. Similar to the RABBIT technique, an IR dressing field centered at 800 nm is used to create photoelectron sidebands that can be accessed through two interfering pathways. In both cases, temporal information on the sub-femtosecond time scale is extracted by analyzing the phase of the side-band-intensity oscillations as a function of the delay between the ionizing and dressing pulses. This measurement scheme has been applied to rare gas atoms in Zipp et al. (2014) and Gong et al. (2017). A combination of a strong UV (400 nm) femtosecond pulse and an 800 nm perturbing field was used in Zipp et al. (2014) to extract the intrinsic and measurement-induced phase delays in the ATI of argon atoms in the region 2–15 eV, revealing a strong dependence of the observed phase shifts on the UV field intensity and thus on the ponderomotive potential. In Gong et al. (2017), on the other hand, a combination of orthogonally polarized 400 nm + 800 nm fields of equal intensities was used to study the photoemission dynamics of the Freeman resonance via the field-dressed 5p or 4f Rydberg states of Ar. While seemingly similar to the RABBIT technique, it is important to emphasize that the multiphoton-ionization scheme does not have any known straightforward relationship to the photoionization dynamics, in pronounced contrast with the RABBIT technique (Dahlström et al., 2012; Klünder et al., 2011). Experimentally, this fact becomes apparent through the strong dependence of the measured delays on the intensity of the ionizing laser pulse, which was reported in Zipp et al. (2014) and is absent in the RABBIT scheme. Theoretically, the origin of the time delays measured by the multiphoton technique

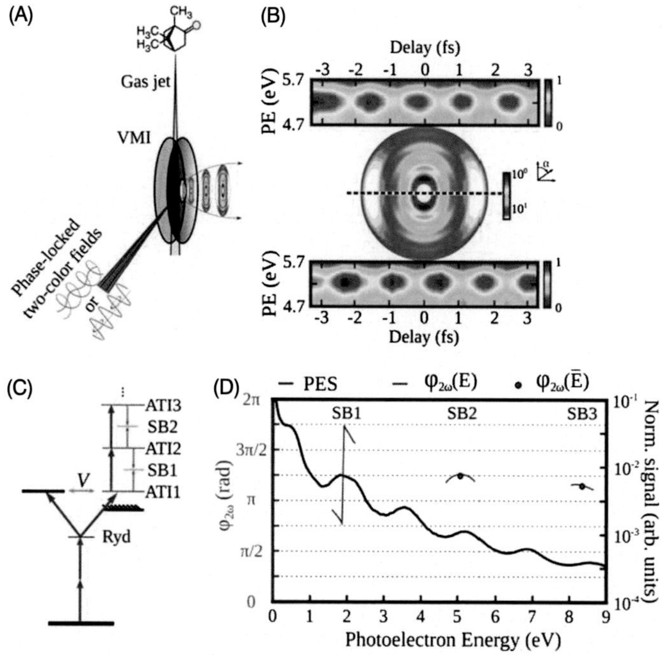

FIG. 4.19 (A) Schematic drawing of the experimental setup, comprising two phase-locked femtosecond laser pulses with either linear or circular polarization which are focused into a jet of enantiopure camphor molecules in the interaction zone of a VMI spectrometer. (B) Typical results for the photoelectron angular distribution as well as the oscillations of the second sideband as a function of the two-color delay. The phase shift (π) between the signals in the upper resp. lower half of the distribution reflect the up-down asymmetry of the ionizing field. (C) Schematic illustration of the physical mechanism underlying photoelectron interferometry. The UV field (blue arrows) induces multiphoton ionization, leading to ATI, whereby the first ionization transition is located close to an autoionizing resonance. The IR field (red arrows) subsequently induces additional transitions, leading to the creation of sidebands between the main ATI peaks. (D) Angle-integrated photoelectron spectrum comprising the main ATI peaks as well as the sidebands of orders 1–3. The oscillation phases (2ω) of each band, integrated over the emission angles, are shown as a function of energy (blue lines) as well as spectrally averaged (red dots, only for the non-resonant SBs 2 and 3). Figure adapted from Beaulieu et al. (2017) with permission.

has been traced to retrapped resonant ionization (Song et al., 2018) and different numerical values of the time delays compared to those measured by the RABBIT technique have been reported (see Fig. S7 therein).

The multiphoton measurement scheme has been applied to the photoionization of chiral molecules by Beaulieu et al. (2017). Specifically, they addressed the topic of the differential response between the two enantiomers of a given molecule when subject to multiphoton ionization. Two-color multiphoton ionization enables access to certain time-domain aspects of the chiral response, which is given in this case by photoemission into the forward or the backward directions (with respect to the polarization of the ionization field).

The detection scheme of Beaulieu et al. is based on the photoelectron circular dichroism (PECD) effect (Powis, 2008; Ritchie, 1975), which is defined as the asymmetry in the photoelectron angular distribution (typically recorded in a VMI apparatus, cp. Section 3.2.4) with respect to the light propagation direction and takes place when a chiral species is subject to photoionization in a circularly polarized laser field (see illustration in Fig. 4.19A).

The interferometric technique employed in this study, relies on the technique introduced in (Zipp et al., 2014). In the experiment, a UV laser field centered at 400 nm is used to induce ATI in camphor, a bicyclic ketone with an ionization potential of 8.76 eV. The electronic structure of camphor ($C_{10}H_{16}O$) is characterized by Rydberg states around 6.2 eV, and the ionization process takes place via a 2+n−REMPI pathway, whereby n denotes the order of the ATI band. Superimposing the ionizing UV radiation with a weak IR component at 800 nm leads to the appearance of sidebands between the ATI comb. The two-color field is asymmetric with respect to the PE emission axis, consequently, the electrons ejected in the forward or backward directions are modulated with opposite phases (cp. Fig. 4.19B). This implies that the phases evaluated for the upper and the lower halves of each recorded PAD image are shifted by π prior to comparison. The angularly integrated phases of the three sidebands resolved in the experiment are displayed as a function of the photoelectron kinetic energy in Fig. 4.19D. Whereas the second and the third sidebands (SB2 and SB3) exhibit a smooth phase variation across the ATI bandwidth, the first SB (SB1) is characterized by an abrupt discontinuity ($\approx \pi$ in magnitude), which is interpreted as a signature of the autoionizing resonance located at 1.9 eV. Studying the phases of the individual sidebands thus gives access to the chiral photoionization dynamics in two different regimes (resonant vs. non-resonant). In the non-resonant case, the phases reported are averaged over the bandwidth of the corresponding SB, whereas this treatment is not applicable for the first SB. The experimental results were analyzed in a framework analogous to Eq. (4.30), although the latter is not applicable to the ATI measurement scheme (see, e.g., Song et al., 2018):

$$\tau_{\mathrm{tot}} = \tau_{\mathrm{UV}} + \tau_{\mathrm{W}} + \tau_{\mathrm{cc}}. \tag{4.33}$$

Here τ_{UV} corresponds to the instant of the ionization event triggered off by the UV pulse, τ_{W} is the delay resulting from the influence of the molecular potential, and τ_{cc} is the additional delay induced by the absorption of the weak IR perturbing field, and mainly sensitive to the asymptotic tail of the molecular potential (see below). In order to eliminate the measurement-induced contribution encoded in τ_{UV}, one can define and extract a quantity corresponding to the relative photoemission delay $\Delta\tau^{\mathrm{f/b}}$ between electrons emitted in forward (f) or backward (b) directions, $\Delta\tau^{\mathrm{f/b}} = \Delta\tau^{\mathrm{f}} - \Delta\tau^{\mathrm{b}}$. Disentangling the remaining two contributions associated with $\Delta\tau_{\mathrm{W}}^{\mathrm{f/b}}$, resp. $\Delta\tau_{\mathrm{cc}}^{\mathrm{f/b}}$, is performed by measuring $\Delta\tau^{f/b}$ using two different combinations of linear (LP) vs. circularly-polarized (CP) light in the pump (UV), resp. probe (IR), steps. Choosing the polarization of the UV field to be circular while setting the IR to linear essentially restricts the chiral discrimination to the ionization step, implying that $\Delta\tau^{\mathrm{f/b}} = \Delta\tau_{\mathrm{W}}^{\mathrm{f/b}}$. In contrast, employing a linearly polarized UV pulse erases the asymmetry in the Wigner time delay and allows one to isolate the chiral contribution to the continuum–continuum delay, $\Delta\tau^{\mathrm{f/b}} = \Delta\tau_{\mathrm{cc}}^{\mathrm{f/b}}$. This treatment is based on the tacit assumption that the transitions induced by the UV and the IR fields are strictly independent, in which case the contributions to $\Delta\tau^{\mathrm{f/b}}$ can be treated in an additive manner as implied by Eq. (4.33). They are additionally based on the assumption of a specific time ordering of the interactions, i.e., UV interaction followed by IR interaction. Whereas this approximation is well justified in the RABBIT scheme (see Dahlström et al., 2012), it has no rigorous justification in the multiphoton scheme because of the similar frequencies of the employed laser pulses. Experimentally, $\Delta\tau^{\mathrm{f/b}}$ are measured for each of the two camphor enantiomers and the obtained values are subsequently averaged for the sake of improved accuracy. The values $\Delta\tau^{\mathrm{f/b}}$ switch sign on changing the enantiomer or, equivalently, the helicity of the CPL field, as expected from symmetry reasons.

In Section 4.7.2, it was argued that the continuum–continuum contribution is essentially insensitive to the molecular potential as it probes mainly the Coulombic tail of the asymptotic potential. This is also reflected in the current experimental results for angle-averaged $\Delta\tau_{\mathrm{cc}}^{\mathrm{f/b}}$ for SB2 and SB3 (Fig. 4.20A), which are equal to zero within the experimental accuracy of ±2 as. The chiral signatures can be isolated only after resolving the angular dependence of the photoionization process by integrating the PAD in slices of 10° centered around different ejection angles α, measured with respect to the polarization plane of the IR (cp. insets in Fig. 4.20, B and D). Even in this case, the chiral signal is weak, and maximal time delays are observed for SB2 for an ejection angle of 25° (5 ± 2 as). For SB3, even in the angular resolved case, no difference between forward and backward emission is detected. This is consistent with the notion that the IR-induced transitions take place away from the core region and $\Delta\tau_{\mathrm{cc}}^{\mathrm{f/b}}$ is thus less sensitive to the chiral features of the system. For the Wigner-like contribution to the f/b delay, $\Delta\tau_{\mathrm{W}}^{\mathrm{f/b}}$, (measured by breaking the f/b symmetry in the ionization step by employing a CPL UV pulse), the angle-averaged delays amount to ≈7 as in the case of SB2 (Fig. 4.20C), whereas the photoemission-angle-resolved delays reach up to 24 as for an ejection angle of 60–70° (Fig. 4.20D). Similar to the cc-contribution, the Wigner part $\Delta\tau_{\mathrm{W}}^{\mathrm{f/b}}$ for SB3 is not sensitive to the chiral character of the PI process. This behavior reflects the diminished sensitivity

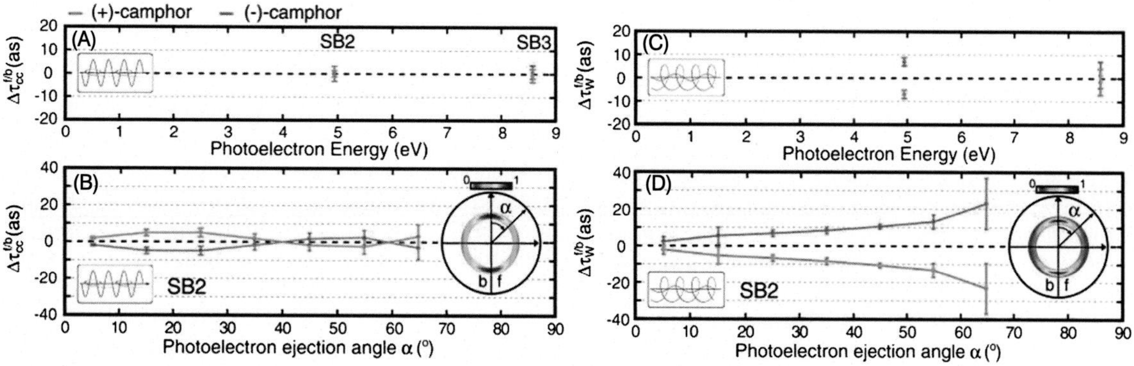

FIG. 4.20 (A) Forward/backward differential delays in non-resonant photoionization of camphor for the experimental scheme employing a linearly-polarized UV field and a left circularly-polarized IR field. (B) Angularly-resolved differential delays corresponding to the results presented in (A) for the second sideband. The inset shows the normalized PAD for the SB2 as well as the definition of the ejection angle alpha. (C)–(D) Same as for (A)–(B) for the experimental configuration employing a left-circularly polarized UV field and a linearly polarized IR. Figure adapted from Beaulieu et al. (2017) with permission.

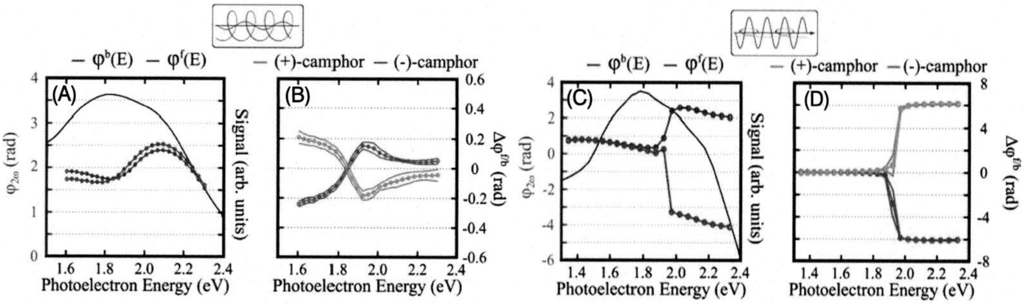

FIG. 4.21 (A) and (C): Spectral amplitudes (black lines) and spectral phases ($\varphi^f(E)$, blue; $\varphi^b(E)$, red) of the resonant sideband SB1 in $(1R)$–(+)-camphor. In (A), the UV is left-circularly polarized and the IR is linear, whereas in (C) the polarization of the fields is reversed. (B) and (D) Forward/backward spectral phase asymmetry between the two enantiomers of camphor, using a left circularly-polarized UV field and a linearly polarized IR in (B) and a linearly polarized UV/left circularly-polarized IR in (D). Figure adapted from Beaulieu et al. (2017) with permission.

of electrons with high-kinetic energy to the asymmetric character of the molecular potential.

Finally, we briefly discuss the resonant case, whereby the angle-integrated phase delays for each of the two polarization combinations are presented in Fig. 4.21 as a function of the photoelectron energy. The spectral phase exhibits a phase jump with a magnitude of $\approx$0.75 rad around 2.1 eV when a CPL UV is employed (panel A), with a slight difference between electrons emitted in the forward and the backward directions that is mirrored for the two enantiomers (panel B). Contrary to the non-resonant case, breaking the f/b symmetry with a CPL IR field leads to a very distinct signature in the spectral phase – a phase jump of nearly 0.9 π, in opposite direction for forward/backward electrons (Fig. 4.21, C and D). This result shows that in the vicinity of a resonance, the chiral character of the photoemission process can be probed with the aid of a weak IR-induced cc-transition.

This work, although restricted to the particular example of camphor, outlines a general approach for tracking the temporal evolution of chiral phenomena with attosecond temporal resolution. Additional theoretical work is required to establish the relation between the measured quantities and the concept of Wigner delays in the photoionization of chiral molecules.

4.8 ATTOSECOND TRANSIENT ABSORPTION SPECTROSCOPY

In this section, we review the current advances in attosecond transient absorption spectroscopy (ATAS), a technique which has now become well-established in the realm of femtosecond laser spectroscopy and has been only recently extended to the attosecond domain (for recent reviews, see Beck et al., 2015; Gallmann et al., 2013; Kraus et al., 2018; Kraus and Wörner, 2018a; Leone and Neumark, 2016; Ramasesha et al., 2016). In many respects, ATAS can be viewed as a complementary technique to the HHS method presented in Section 4.6 as well as the NIR+XUV methods covered in Sections 4.4–4.5. HHS possesses a temporal resolution of a fraction of the optical cycle, determined by the duration of an electron's continuum trajectory. However, due to the exponential dependence of the SFI rate on the ionization potential, it is mainly sensitive to valence-state dynamics, and, moreover, relies heavily on the availability of theoretical models for retrieving the electronic structure information from experimental observables. The two-color XUV-pump-NIR-probe techniques (XUV+NIR) presented in Section 4.4 represent a reliable and versatile approach for directly probing the attosecond dynamics, but also have inherent limitations. The majority of the experiments realized so far rely on the detection of charged particles, providing insight only to the dynamics in the ionized system, typically in a highly-excited state. Moreover, charged particle detection is accompanied by a multitude of drawbacks, such as the susceptibility to space-charge effects, stray magnetic and electric fields, limited detection speed and the presence of background signals due to the strong fields employed. ATAS, on the other hand, relies on photon detection, thus obviating the need for any additional electric or magnetic fields, and can benefit from high detection sensitivities due to the advances in charge-coupled device detector technology. In addition, high spectral resolution can be achieved, without compromising the bandwidth of the XUV/SXR pulses. The basic experimental scheme does not entail ionization, hence ATAS can serve as a sensitive probe for bound-continuum transitions and resonances.

4.8.1 Dynamics of Rydberg and Valence States in Molecular Nitrogen Probed by ATAS

The first example we will cover (Warrick et al., 2016) represents one of the few ATAS studies addressing sub-cycle electron dynamics in molecular systems (Cheng et al., 2014, 2013; Reduzzi et al., 2013). The work of Warrick et al. (2016) centers on molecular nitrogen (N_2) to highlight the possibility of investigating quantum-beat-dynamics between multiple bound and autoionizing electronic states of valence or Rydberg character, a task which has become accessible only with the advent of attosecond laser technology in the view of the wide energy separations between individual electronic states. Prior to this point, coherent wavepacket dynamics were investigated in the vibrational and the rotational domain using narrow-bandwidth femto- or picosecond sources. The electronic wavepacket dynamics in this simple molecule reveal many similarities to the extensively studied case of noble-gas atoms, but also contains novel features unique to diatomic molecules.

The experimental scheme employs 25 fs pulses derived from a 2 mJ, 1 kHz CEP-stabilized system which are subsequently spectrally broadened in an HCF filled with Ne-gas and compressed in a chirped-mirror assembly to a duration of 6 fs at 780 nm (0.8 mJ). The more intense part (70%) is reshaped in a double-optical-gating scheme (using two quartz plates and a BBO crystal) and used for attosecond pulse generation based on HHG in a static gas cell filled with Xe. An indium foil (200 nm) is used to block the residual IR. The XUV pump and the NIR probe (less-intense portion of the few-cycle pulse) are focused using a toroidal mirror, resp. spherical mirror, and recombined collinearly by means of an annular mirror into the interaction region consisting of a 1-mm-long gas cell filled with N_2. The temporal resolution of the piezoelectric stage used to delay the NIR is ~100 as. A second indium filter is employed to block the NIR after the target cell, and the transmitted XUV is dispersed and detected using a flat-field spectrometer equipped with an X-ray CCD camera. The experimental data is subject to a Fourier-filtering procedure and presented as absorbance. The static absorption lines in the energy region from 12.5 to 16.7 eV corresponding to the various transitions are indicated in panel B of Fig. 4.22, whereas the transient absorption spectrum covering pump–probe delays up to 350 fs is displayed in panel C.

The indium filter employed in the pump arm limits the excitation bandwidth of the XUV to 11–17 eV. The electronic structure of N_2 in this spectral region (cp. Fig. 4.22A) is characterized by the presence of two valence states ($b^1\Pi_u$ and $b'^1\Sigma_u^+$) as well as Rydberg series built on the ground state of the ion (*n*p series converging to $X^2\Sigma_g^+$ of N_2^+) and its first two excited states $A^2\Pi_u$ (*n*s and *n*d series) and $B^2\Sigma_u^+$ (3s series). As evident from Fig. 4.22C, the spectrum at negative delays is dominated by static absorption features associated with the dipole-allowed transitions to the vibrational levels of the valence $b^1\Pi_u$ and $b'^1\Sigma_u^+$ states (12.5–15 eV), whereby the

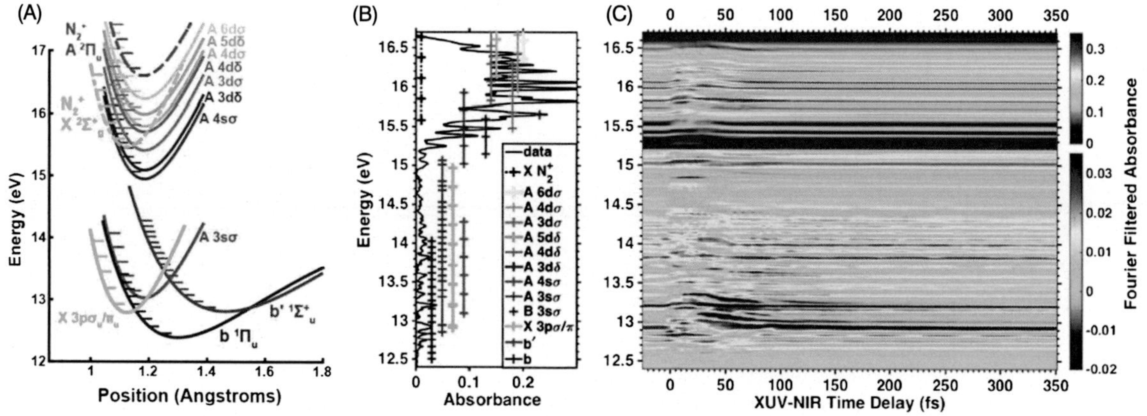

FIG. 4.22 (A) Potential energy curves of molecular nitrogen in the investigated energy region. (B) Energy levels relevant for the static absorption of nitrogen. The diagram lists the positions of the vibrational levels of the valence $b^1\Pi_u$ (blue) and $b'^1\Sigma_u^+$ (red) states and of the Rydberg series built on the $X^2\Sigma_g^+ N_2^+$ (green) and $A^2\Pi_u N_2^+$ cores. (C) Transient absorption spectrum of nitrogen recorded as a function of photon energy (12.5–16.7 eV) and time delay between XUV and IR fields, represented as a Fourier-filtered absorbance. The NIR pulse arrives after the XUV for positive time delays. Figure adapted from Warrick et al. (2016) with permission.

presence of the first Rydberg states as well as the onset of predissociation cause irregularities in the spectral features. The transitions to the Rydberg series converging to the ionic states start to play a role in the region above the first ionization limit to the $v = 0$-state of $X^2\Sigma_g^+ N_2^+$ located at 15.559 eV. At negative delays, the static features are not affected by the NIR due to its low intensity. In the temporal vicinity of the NIR/XUV-overlap, the features shift to higher energies, and in addition, become weaker or negative (i.e., emissive) at the central transition energy. The temporal duration of the shift affecting states between 15.3 and 16.7 eV is not uniform as a function of energy and shifts to longer values (from 20 to 80 fs) as the transition energy increases. The long-time behavior of some of the transition lines displays oscillatory character, whereby the beating period varies strongly with energy: 50 fs at 12.85 eV, 5–10 fs at 14.5 and 16.5 eV, and 1.3 fs at 13.8 eV. The analysis of these features employs a many-level non-perturbative model that treats each vibrational level or Rydberg level as a separate state. The absorption strength measured in a TAS experiment is directly proportional to the imaginary part of the time-dependent dipole moment in the frequency domain:

$$d_1(t) \propto e^{-\frac{t}{t_1}} \mu_{1g}^2 \Bigg[\left|A_1^1(t-\tau)\right| \sin(E_1 t + \varphi_{A_1^1}) + \sum_{n,n\neq 1} \frac{\mu_{ng}}{\mu_{1g}} \left|A_n^1(t-\tau)\right| \sin(E_1 t + \varphi_{A_n^1} + \Delta E_{n1}\tau) \Bigg]. \tag{4.34}$$

In the above, t_1 denotes the lifetime, μ_{ng} is the dipole coupling between state n and the ground state (g), τ is the NIR–XUV delay, E_n is the energy of the nth state, and $\Delta E_{n1} = E_n - E_1$ is the energy difference. The terms A_1^1 and A_n^1 capture the depletion of state 1 through ionization resp. population transfer to other states n. The physical model behind Eq. (4.34) is essentially based on the single-electron approximation discussed in Section 4.2. Its application to molecular systems can be justified only for the case of Rydberg states, which are essentially decoupled from the molecular core, and does not capture non-adiabatic effects due to coupling with the nuclear motion.

The second term in Eq. (4.34) encodes the population transfer coupling the initial state E_1 with another state of energy E_n. In the frequency domain, the sinusoidal dependence of this term on the time delay τ and the energy separation ΔE_{n1} between the two levels translates into a persistent oscillation of the absorption feature. In this framework, the sub-cycle quantum beats with a period of 1.3 fs (half the period of the NIR) can be assigned to interferences between two quantum paths transferring population between two levels spaced by twice the NIR photon energy. This population transfer pathway involves the absorption of two

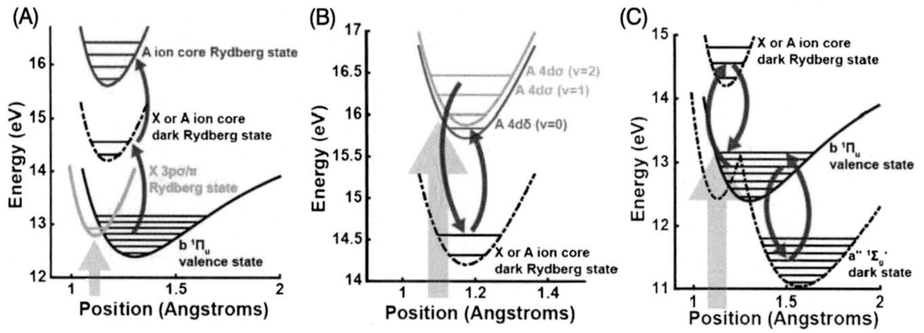

FIG. 4.23 (A) Scheme illustrating the origin of the subcycle oscillations caused by interferences between two levels spaced by double the NIR frequency. A dark state built on the $A^2\Pi_u N_2^+$ core is shown to illustrate the Franck–Condon overlap with the $b^1\Pi_u$ vibrational levels. (*B*) Scheme illustrating the origin of the slow oscillations in the $4d\delta v = 0$ state at 15.85 eV. The black curve shows a dark state built on the A-core. (C) Scheme illustrating the origin of the slow oscillations in the vibrational levels of the $b^1\Pi_u$ valence state. The black curves indicate two potential intermediate dark states: a Rydberg state built on the A core at 14 eV as well as the valence a″ dark state at 11 eV. In all panels, the population transfer is illustrated by the red arrows, whereas the Franck–Condon overlap with the ground state of nitrogen is indicated with the purple arrow. Figure adapted from Warrick et al. (2016) with permission.

NIR photons and involves the mediation of an XUV dipole-forbidden (dark) state. In molecular N_2, the potential candidates for mediating the two-photon coupling pathway are several dark Rydberg states built on the A and X ion-state cores (cp. scheme in Fig. 4.23A). The most dominant spectral features arising through this pathway are located around 15.8 eV and result from the population transfer to the $A4d\delta(v = 0)$-state and the $A3d\sigma(v = 2)$ state from the $3p\sigma/\pi$-Rydberg states (built on the $X^2\Sigma_g^+ N_2^+$ core at 12.95 eV). Additional examples are the coupling between the $A4d\delta(v = 2)$-state (16.4 eV) and the $A3s\sigma(v = 0)$ at 13.1 eV. Besides the population transfer pathways between Rydberg states of different ion cores and orbital symmetries, which have also been observed in atoms, the current experiment reveals the presence of interferences between states of Rydberg and valence character: the oscillations assigned to the $v = 2 - 4b^1\Pi_u$-state levels.

The few-fs (5–10 fs) oscillations, on the other hand, originate from the population transfer between two close-lying resonances via a dark state located in energy above or below the resonances via a "Lambda" or "V"-like coupling scheme. An essential requirement for the realization of this scheme, which has been also observed in atomic systems, is the presence of two NIR photons of differing energy, which is fulfilled given the substantial bandwidth of the few-cycle NIR (1.4–2.2 eV). In molecular nitrogen, the "V"-coupling scheme is realized by the A-Rydberg states above 15 eV that are coupled to the dark X or A ion-core Rydberg states around 14 eV (cp. Fig. 4.23B). One of the most prominent features in the current experiment are the beatings with frequencies of 11.5 and 7 fs, assigned to the $A4d\delta$ $(v = 0)$-state at 15.85 eV. These oscillations result from the population transfer with the $A4d\sigma$ $(v = 1)$ and $(v = 2)$ levels and represent the first observation of coherent beating between vibrational levels belonging to different electronic potential energy curves, a feature unique to molecular systems. The beatings between vibrational states belonging to the same electronic states give rise to the longer-period oscillations with a vibrational period of 51 fs, matching the vibrational period of the b-state. The vibrational beatings thus involve levels of the b-state separated by $\Delta v = \pm 1$ that are coupled by the NIR field (cp. scheme in Fig. 4.23C).

Finally, we briefly comment on the origin of the energy shifts of the absorption features observed in the vicinity of time zero. Following a theoretical analysis developed for atoms, the authors attribute the observed dynamics to the ac-Stark effect induced by the few-cycle NIR pulse. This effect is modeled by the introduction of the laser-imposed phase term $\varphi_{A_n^1}$ in Eq. (4.34). The NIR-mediated coupling between a resonant excited state and other dark states leads to an energy shift of the level that can be incorporated as a phase-shift of the time-dependent dipole moment. The increasing time duration regarding the shifts of the absorption features is explained by the increasing overlap of the energy levels in the high-density region of the A-Rydberg states as they approach the ionization potential.

In summary, the observed quantum beating dynamics were attributed to the energy repartitioning between electronic and nuclear degrees of freedom mediated by the perturbative NIR laser field. Whereas the results could to a major extent be interpreted using the theoretical framework built upon the single-electron response in atomic systems, several unique features arising from the nuclear degrees of freedom were identified.

4.8.2 Time-Resolved X-Ray Absorption Spectroscopy Using a Table-Top High-Harmonic Source

The rest of this section is dedicated to the recent progress in the development of table-top time-resolved X-ray absorption techniques (TR-XAS). Broadband soft-X-ray (SXR) continua with photon energies beyond the XUV range (10–124 eV) represent a powerful diagnostic tool due to their sensitivity to the local chemical structure (via the excitation from a specific core level of a given atom) including spin and oxidation state. Coupled with time-resolved NIR or optical ultrafast excitation schemes, this fingerprinting-capability has the potential to track the dynamics of oxidation and spin states with ultrafast temporal resolution and element specificity. Until recently, time-resolved studies employing XAS were limited to large-scale synchrotron facilities or free-electron lasers and were mostly confined to the study of sub-ps dynamics in the condensed phase. The advent of HHG-based light sources capable of producing sub-50 fs SXR pulses with full spatial and temporal coherence permitted the realization of time-resolved X-ray absorption measurements in a table-top manner.

We will illustrate this recent progress on the example of two pioneering studies (Pertot et al., 2017; Attar et al., 2017) reporting the realization of TR-XAS experiments with femtosecond temporal resolution employing SXR supercontinua extending beyond 160 eV and up to 350 eV, i.e., partially covering the chemically and biologically highly relevant energy range of 282–533 eV where water becomes transparent ("water window") and where the K-edge of C is located (282 eV). Prior to these two studies, XAS experiments were limited to $\approx$100 eV, covering only the element-specific edges in the XUV. These experiments enabled the tracking of the dynamics following strong-field ionization in a variety of systems such as Xe atoms (Loh et al., 2007), Kr (Goulielmakis et al., 2010) and thin-film silicon (Schultze et al., 2014). Studies centering on molecular systems have so far addressed the SFI-induced vibrational wavepacket dynamics in Br_2 (Hosler and Leone, 2013), the SFI-induced dissociation of dibromomethane (Chatterley et al., 2016a) and ferrocene (Chatterley et al., 2016b), the evolution of transition state during the SFI-induced dissociation of CH_3I (Attar et al., 2015), and the ring-opening reaction of selenophene (Lackner et al., 2016) following ionization.

In the experiment of Pertot et al. (2017), a high-power laser source based on a cryogenically-cooled Ti:Sa system is employed to generate 40 fs pulses (800 nm) of exceptional power (reaching 18 mJ at 1 kHz). The major part of the energy is down-converted in an OPA to yield MIR pulses centered at 1800 nm with an energy of 2.5 mJ, which are subsequently used to drive HHG in Ne, thereby resulting in an SXR supercontinuum extending up to 350 eV. A portion of the NIR beam (800 nm) is used to strong-field-ionize tetrafluoromethane (CF_4) molecules. The CF_4^+ cation is unstable with respect to a Jahn–Teller distortion and has an energy minimum at a lower symmetry (C_{2v}) compared to the tetrahedral neutral species. This symmetry lowering lifts the degeneracy of the initially triply degenerate $5t_2$-orbital of the neutral (cp. calculated spectra in Fig. 4.24A). In addition, the cation is unstable and undergoes a dissociation reaction to yield the trigonal-planar (D_{3h}) CF_3^+ and F (Fig. 4.24B). The geometry changes accompanying the C–F bond breakage and the structural rearrangement from C_{2v} to the D_{3h}-structure in CF_3^+ are tracked at each NIR-pump–SXR-probe delay by monitoring the X-ray absorption lines at the C-edge. The static absorption spectrum of the tetrahedral CF_4 is characterized by two intense lines (cp. Fig. 4.24C) which can be assigned to transitions to the $5t_2$ and $6t_2$ orbitals. Symmetry arguments dictate that the transition of the C1s $\rightarrow$ t_2-type associated with the initially tetrahedral carbon atom will split into two C1a $\rightarrow$ a_2''-lines and one 1s $\rightarrow$ e'-line as the environment around C progressively changes to trigonal-planar. The experimentally recorded transient-absorption spectra shown in Fig. 4.24C reflect this situation (cp. labels in panel C). With progressing time, the spectrum splits into multiple bands, with one band shifting down to 288 eV (by $\approx$10 eV), two further bands shifting up/down by 1 eV, and a fourth line appearing at 302 eV as a shoulder structure in the absorption spectrum. Thus, the TAS spectrum reflects the changes in the geometry as the C–F bond breakage takes its course. Further, the intensity evolution of the $5e'$- and the $6e'$-transitions (cp. Fig. 4.24A) is interpreted as an evidence of the mixing of Rydberg and valence character of these orbitals in the context of the fluorine "cage effect" (Dehmer et al., 1979). The $5e'$ orbital, initially a valence-type, well-localized orbital in the tetrahedral neutral species, develops a partial Rydberg character, whereas the $6e'$ orbital (of Rydberg character, localized outside of the region defined by the four F atoms in CF_4) develops a partial valence character. The difference

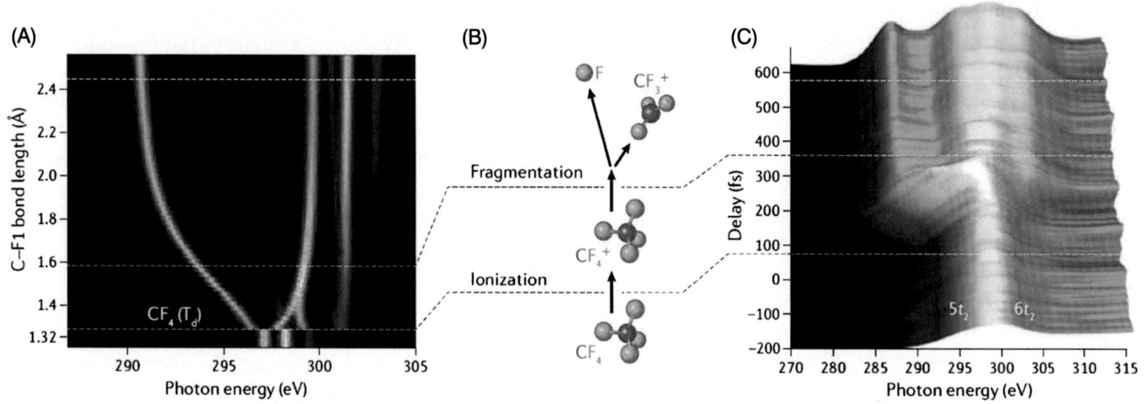

FIG. 4.24 (A) Calculated X-ray absorption spectra as a function of photon energy and C–F bond distance for the light-induced dissociation $CF_4{}^+ \rightarrow CF_3{}^+ + F$. (B) Schematic illustration of the light-induced dissociation of CF_3^+ after strong-field ionization of CF_4. (C) Experimental transient absorption spectrum at the carbon K-edge as a function of photon energy and NIR–XUV time delay. Figure adapted from Kraus et al. (2018) with permission.

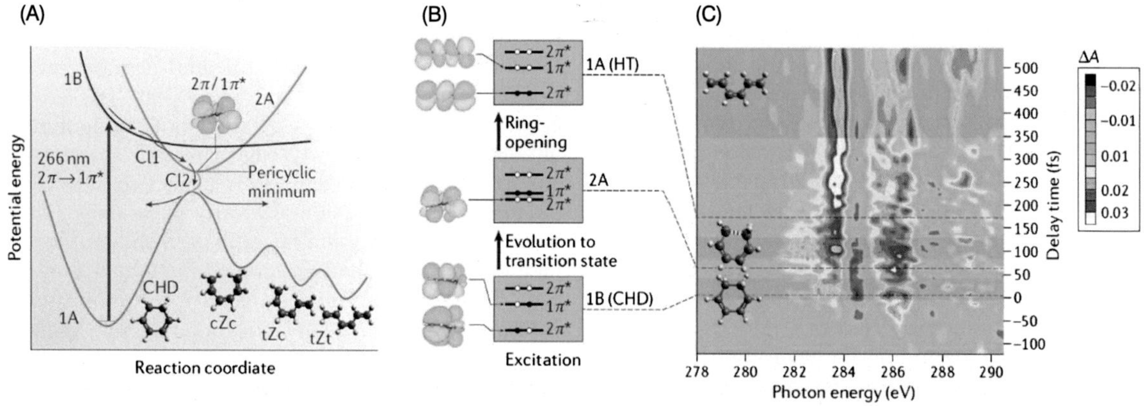

FIG. 4.25 (A) Schematic view of the potential energy surfaces relevant for the electrocyclic ring-opening of cyclohexadiene, which leads to the formation of three stereoisomers of 1,3,5-hexatriene (HT): s- cis,Z,s- cis (cZc); s- trans,Z,s- cis (tZc), and s- trans,Z,s- trans (tZt). (B) Orbital diagram and electronic configurations of the relevant transition states. (C) Experimental transient absorption spectrum at the carbon K-edge as a function of photon energy and NIR–XUV time delay. The black arrows highlight the connection of the transient intensity modulations in the absorption spectrum with the populations of various transition states (panel B) during the reaction. Figure adapted from Kraus et al. (2018) with permission.

in the overlap with the highly localized core orbital accounts for the evolution of the intensity distribution as the reaction progresses. This study first demonstrated the feasibility of table-top-based TR-XAS experiments and highlighted the potential of this technique in elucidating the dynamics of chemical reactions.

In a subsequent study, Attar et al. (2017) used a 266 nm, 100 fs pulse to induce the isomerization reaction of ring-closed cyclohexadiene (CHD) to the open-chain 1,3,5-hexatriene (HT). Mechanistically, the UV photoexcitation promotes the system to the 1B excited state via a strong symmetry-allowed ($\pi \rightarrow \pi^*$) transition. The ring-opening reaction proceeds near the conical intersection (CI) of the 1B state with the dark 2A excited state and reaches a "pericyclic minimum" in the transition state of 2A symmetry (cp. scheme in Fig. 4.25A). The latter is characterized by an electron configuration with a doubly-occupied antibonding π^*-orbital (see orbital scheme in Fig. 4.25B). After reaching a second conical intersection between the excited 2A and the ground 1A states, the wavepacket can bifurcate either to the open-chain hexatriene or back towards the cyclic isomer. The reaction progress has been tracked by monitoring the time-resolved XAS-signal near the C K-edge. With the aid of advanced TD-DFT calculations, the reaction constants pertaining to the major steps in the mechanistic pathway were determined: the intermediate transition-state minimum is reached within 60 ± 20 fs, whereas the decay to the open photoproduct occurs in further 110 ± 60 fs.

Although the highlighted examples involving molecular systems are still limited to the femtosecond domain, extension to the attosecond range is in near sight given the recent progress in the generation and characterization of IAP-based SXR continua in the water window (Cousin et al., 2017; Silva et al., 2015).

REFERENCES

Abraham, H., Lemoine, J., 1899. Disparition instantanée du phénomène de Kerr. C. R. Math. Acad. Sci. Paris 129, 206–208.

Ammosov, M.V., Delone, N.B., Krainov, V.P., 1986. Tunnel ionization of complex atoms and of atomic ions in an alternating electromagnetic field. Sov. Phys. JETP 64, 1191–1194.

Andriukaitis, G., Balčiūnas, T., Ališauskas, S., Pugžlys, A., Baltuška, A., Popmintchev, T., Chen, M.-C., Murnane, M.M., Kapteyn, H.C., 2011. 90 GW peak power few-cycle mid-infrared pulses from an optical parametric amplifier. Opt. Lett. 36, 2755–2757.

Attar, A.R., Bhattacherjee, A., Leone, S.R., 2015. Direct observation of the transition-state region in the photodissociation of CH_3I by femtosecond extreme ultraviolet transient absorption spectroscopy. Chem. Phys. Lett. 6, 5072–5077.

Attar, A.R., Bhattacherjee, A., Pemmaraju, C.D., Schnorr, K., Closser, K.D., Prendergast, D., Leone, S.R., 2017. Femtosecond X-ray spectroscopy of an electrocyclic ring-opening reaction. Science 356, 54–59.

Baker, S., Robinson, J.S., Lein, M., Chirila, C.C., Bandulet, H.C., Comtois, D., Villeneuve, D., Kieffer, J.-C., Tisch, J.W.G., Marangos, J.P., 2006. Probing proton dynamics in molecules on an attosecond timescale. Science 312, 424–427.

Baltuška, A., Fuji, T., Kobayashi, T., 2002. Controlling the Carrier-Envelope phase of ultrashort light pulses with optical parametric amplifiers. Phys. Rev. Lett. 88, 133901.

Bandrauk, A.D., Fillion-Gourdeau, F., Lorin, E., 2013. Atoms and molecules in intense laser fields: gauge invariance of theory and models. J. Phys. B At. Mol. Opt. Phys. 46.

Baykusheva, D., Wörner, H.J., 2017. Theory of attosecond delays in molecular photoionization. J. Chem. Phys. 146, 124306.

Beaulieu, S., Comby, A., Clergerie, A., Caillat, J., Descamps, D., Dudovich, N., Fabre, B., Géneaux, R., Légaré, F., Petit, S., Pons, B., Porat, G., Ruchon, T., Taïeb, R., Blanchet, V., Mairesse, Y., 2017. Attosecond-resolved photoionization of chiral molecules. Science 358, 1288–1294.

Beck, A.R., Neumark, D.M., Leone, S.R., 2015. Probing ultrafast dynamics with attosecond transient absorption. Chem. Phys. Lett. 624, 119–130.

Bertrand, J.B., Wörner, H.J., Hockett, P., Villeneuve, D.M., Corkum, P.B., 2012. Revealing the Cooper minimum of N_2 by molecular frame high-harmonic spectroscopy. Phys. Rev. Lett. 109, 143001.

Biswas, S., Förg, B., Ortmann, L., Schötz, J., Schweinberger, W., Zimmermann, T., Pi, L., Baykusheva, D., Masood, H.A., Liontos, I., Kamal, A.M., Kling, N.G., Alharbi, A.F., Alharbi, M., Azzeer, A.M., Hartmann, G., Wörner, H.J., Landsman, A.S., Kling, M.F., 2020. Probing molecular environment through photoemission delays. Nat. Phys.. In press.

Bohman, S., Suda, A., Kanai, T., Yamaguchi, S., Midorikawa, K., 2010. Generation of 5.0 fs, 5.0 mJ pulses at 1 kHz using hollow-fiber pulse compression. Opt. Lett. 35, 1887–1889.

Bouhal, A., Salières, P., Breger, P., Agostini, P., Hamoniaux, G., Mysyrowicz, A., Antonetti, A., Constantinescu, R., Muller, H.G., 1998. Temporal dependence of high-order harmonics in the presence of strong ionization. Phys. Rev. A 58, 389–399.

Braunstein, M., McKoy, V., 1987. Shape resonances in the photoionization of N_2O. J. Chem. Phys. 87, 224–228.

Breidbach, J., Cederbaum, L.S., 2003. Migration of holes: formalism, mechanisms, and illustrative applications. J. Chem. Phys. 118, 3983–3996.

Breidbach, J., Cederbaum, L.S., 2005. Universal attosecond response to the removal of an electron. Phys. Rev. Lett. 94, 033901.

Butkus, R., Danielius, R., Dubietis, A., Piskarskas, A., Stabinis, A., 2004. Progress in chirped pulse optical parametric amplifiers. Appl. Phys. B 79, 693–700.

Caillat, J., Maquet, A., Haessler, S., Fabre, B., Ruchon, T., Salières, P., Mairesse, Y., Taïeb, R., 2011. Attosecond resolved electron release in two-color near-threshold photoionization of N_2. Phys. Rev. Lett. 106, 093002.

Calegari, F., Ayuso, D., Trabattoni, A., Belshaw, L., De Camillis, S., Anumula, S., Frassetto, F., Poletto, L., Palacios, A., Decleva, P., Greenwood, J.B., Martín, F., Nisoli, M., 2014. Ultrafast electron dynamics in phenylalanine initiated by attosecond pulses. Science 346, 336–339.

Calegari, F., Sansone, G., Nisoli, M., 2015. Optical Technologies for Extreme-Ultraviolet and Soft X-Ray Coherent Sources, 1st ed. Springer-Verlag Berlin Heidelberg.

Calegari, F., Trabattoni, A., Palacios, A., Ayuso, D., Castrovilli, M.C., Greenwood, J.B., Decleva, P., Martín, F., Nisoli, M., 2016. Charge migration induced by attosecond pulses in bio-relevant molecules. J. Phys. B At. Mol. Opt. Phys. 49, 142001.

Canova, L., Chen, X., Trisorio, A., Jullien, A., Assion, A., Tempea, G., Forget, N., Oksenhendler, T., Lopez-Martens, R., 2009. Carrier-envelope phase stabilization and control using a transmission grating compressor and an AOPDF. Opt. Lett. 34, 1333–1335.

Castrovilli, M.C., Trabattoni, A., Bolognesi, P., O'Keeffe, P., Avaldi, L., Nisoli, M., Calegari, F., Cireasa, R., 2018. Ultrafast hydrogen migration in photoionized glycine. Chem. Phys. Lett. 9, 6012–6016.

Cavalieri, A.L., Müller, N., Uphues, T., Yakovlev, V.S., Baltuska, A., Horvath, B., Schmidt, B., Blümel, L., Holzwarth, R., Hendel, S., Drescher, M., Kleineberg, U., Echenique, P.M., Kienberger, R., Krausz, F., Heinzmann, U., 2007. Attosecond spectroscopy in condensed matter. Nature 449, 1029–1032.

Cederbaum, L.S., Domcke, W., Schirmer, J., von Niessen, W., 1986. Correlation effects in the ionization of molecules: breakdown of the molecular orbital picture. Adv. Chem. Phys. LXV, 115.

Cerullo, G., De Silvestri, S., 2003. Ultrafast optical parametric amplifiers. Rev. Sci. Instrum. 74, 1–18.

Chacon, A., Lein, M., Ruiz, C., 2014. Asymmetry of Wigner's time delay in a small molecule. Phys. Rev. A 89, 053427.

Chalus, O., Thai, A., Bates, P.K., Biegert, J., 2010. Six-cycle mid-infrared source with 3.8 μJ at 100 kHz. Opt. Lett. 35, 3204–3206.

Chang, Z., 2007. Controlling attosecond pulse generation with a double optical gating. Phys. Rev. A 76, 51403.

Chang, Z., Rundquist, A., Wang, H., Murnane, M.M., Kapteyn, H.C., 1997. Generation of coherent soft X rays at 2.7 nm using high harmonics. Phys. Rev. Lett. 79, 2967–2970.

Chatterley, A.S., Lackner, F., Neumark, D.M., Leone, S.R., Gessner, O., 2016a. Tracking dissociation dynamics of strong-field ionized 1, 2-dibromoethane with femtosecond XUV transient absorption spectroscopy. Phys. Chem. Chem. Phys. 18, 14644–14653.

Chatterley, A.S., Lackner, F., Pemmaraju, C.D., Neumark, D.M., Leone, S.R., Gessner, O., 2016b. Dissociation dynamics and electronic structures of highly excited ferrocenium ions studied by femtosecond XUV absorption spectroscopy. J. Phys. Chem. A 120, 9509–9518.

Cheng, Y., Chini, M., Wang, X., Wu, Y., Chang, Z., 2013. Probing hydrogen and deuterium molecular dynamics using attosecond transient absorption. In: Frontiers in Optics 2013. Optical Society of America, p. LW2H.2.

Cheng, Y., Chini, M., Wang, X., Wu, Y., Chang, Z., 2014. Attosecond transient absorption in molecular hydrogen. In: CLEO: 2014. Optical Society of America, p. FM2B.3.

Chu, Y., Gan, Z., Liang, X., Yu, L., Lu, X., Wang, C., Wang, X., Xu, L., Lu, H., Yin, D., Leng, Y., Li, R., Xu, Z., 2015. High-energy large-aperture Ti: sapphire amplifier for 5 PW laser pulses. Opt. Lett. 40, 5011–5014.

Ciriolo, G.A., Negro, M., Devetta, M., Cinquanta, E., Faccialà, D., Pusala, A., De Silvestri, S., Stagira, S., Vozzi, C., 2017. Optical parametric amplification techniques for the generation of high-energy few-optical-cycles IR pulses for strong field applications. Appl. Sci. 7.

Cohen, E.R., Cvitaš, T., Frey, J.G., Holmström, B., Kuchitsu, K., Marquardt, R., Mills, I., Pavese, F., Quack, M., Stohner, J., Strauss, H.L., Takami, M., Thor, A.J., 2008. Quantities, Units and Symbols in Physical Chemistry, 3rd ed. IUPAC and RSC Publishing, Cambridge.

Corkum, P.B., 1993. Plasma perspective on strong field multiphoton ionization. Phys. Rev. Lett. 71, 1994–1997.

Corkum, P.B., Burnett, N.H., Ivanov, M.Y., 1994. Subfemtosecond pulses. Opt. Lett. 19, 1870–1872.

Cousin, S.L., Di Palo, N., Buades, B., Teichmann, S.M., Reduzzi, M., Devetta, M., Kheifets, A., Sansone, G., Biegert, J., 2017. Attosecond streaking in the water window: a new regime of attosecond pulse characterization. Phys. Rev. X 7, 041030.

Dahlström, J.M., L'Huillier, A., Maquet, A., 2012. Introduction to attosecond delays in photoionization. J. Phys. B At. Mol. Opt. Phys. 45, 183001.

Danson, C., Hillier, D., Hopps, N., Neely, D., 2015. Petawatt class lasers worldwide. High Power Laser Sci. Eng. 3, e3.

De, S., Znakovskaya, I., Ray, D., Anis, F., Johnson, N.G., Bocharova, I.A., Magrakvelidze, M., Esry, B.D., Cocke, C.L., Litvinyuk, I.V., Kling, M.F., 2009. Field-free orientation of CO molecules by femtosecond two-color laser fields. Phys. Rev. Lett. 103, 153002.

Dehmer, J.L., Dill, D., Wallace, S., 1979. Shape-resonance-enhanced nuclear-motion effects in molecular photoionization. Phys. Rev. Lett. 43, 1005–1008.

Dehmer, P.M., Miller, P.J., Chupka, W.A., 1984. Photoionization of N_2 $X^1\Sigma^+{}_g$, $v'' = 0$ and 1 near threshold. Preionization of the Worley–Jenkins Rydberg series. J. Chem. Phys. 80, 1030–1038.

Deng, Y., Schwarz, A., Fattahi, H., Ueffing, M., Gu, X., Ossiander, M., Metzger, T., Pervak, V., Ishizuki, H., Taira, T., Kobayashi, T., Marcus, G., Krausz, F., Kienberger, R., Karpowicz, N., 2012. Carrier-envelope-phase-stable, 1.2 mJ, 1.5 cycle laser pulses at 2.1 µm. Opt. Lett. 37, 4973–4975.

Dimitrovski, D., Martiny, C.P.J., Madsen, L.B., 2010. Strong-field ionization of polar molecules: stark-shift-corrected strong-field approximation. Phys. Rev. A 82, 053404.

Dörner, R., Mergel, V., Jagutzki, O., Spielberger, L., Ullrich, J., Moshammer, R., Schmidt-Böcking, H., 2000. Cold target recoil ion momentum spectroscopy: a 'momentum microscope' to view atomic collision dynamics. Phys. Rep. 330, 95–192.

Drescher, M., Hentschel, M., Kienberger, R., Uiberacker, M., Yakovlev, V., Scrinzi, A., Westerwalbesloh, T., 2002. Time-resolved atomic inner-shell spectroscopy. Nature 419, 803–807.

Dubietis, A., Butkus, R., Piskarskas, A.P., 2006. Trends in chirped pulse optical parametric amplification. IEEE J. Sel. Top. Quantum Electron. 12, 163–172.

Dubietis, A., Jonušauskas, G., Piskarskas, A., 1992. Powerful femtosecond pulse generation by chirped and stretched pulse parametric amplification in BBO crystal. Opt. Commun. 88, 437–440.

Dudley, J.M., Genty, G., Coen, S., 2006. Supercontinuum generation in photonic crystal fiber. Rev. Modern Phys. 78, 1135–1184.

Dutin, C.F., Dubrouil, A., Petit, S., Mével, E., Constant, E., Descamps, D., 2010. Post-compression of high-energy femtosecond pulses using gas ionization. Opt. Lett. 35, 253–255.

Eigen, M., 1954. Methods for investigation of ionic reactions in aqueous solutions with half-times as short as 10^{-9} sec. Application to neutralization and hydrolysis reactions. Discuss. Faraday Soc. 17, 194–205.

Etches, A., Madsen, L.B., 2010. Extending the strong-field approximation of high-order harmonic generation to polar molecules: gating mechanisms and extension of the harmonic cutoff. J. Phys. B At. Mol. Opt. Phys. 43, 155602.

Farkas, G., Tóth, C., 1992. Proposal for attosecond light pulse generation using laser induced multiple-harmonic conversion processes in rare gases. Phys. Lett. A 168, 447–450.

Farrell, J.P., Petretti, S., Förster, J., McFarland, B.K., Spector, L.S., Vanne, Y.V., Decleva, P., Bucksbaum, P.H., Saenz, A., Gühr, M., 2011. Strong field ionization to multiple electronic states in water. Phys. Rev. Lett. 107, 083001.

Feng, X., Gilbertson, S., Mashiko, H., Wang, H., Khan, S.D., Chini, M., Wu, Y., Zhao, K., Chang, Z., 2009. Generation of isolated attosecond pulses with 20 to 28 Femtosecond lasers. Phys. Rev. Lett. 103, 183901.

Ferrari, F., Calegari, F., Lucchini, M., Vozzi, C., Stagira, S., Sansone, G., Nisoli, M., 2010. High-energy isolated attosecond pulses generated by above-saturation few-cycle fields. Nat. Photonics 4, 875.

Fork, R.L., Cruz, C.H.B., Becker, P.C., Shank, C.V., 1987. Compression of optical pulses to six femtoseconds by using cubic phase compensation. Opt. Lett. 12, 483–485.

Frey, R.F., Davidson, E.R., 1988. Potential energy surfaces of CH_4^+. J. Chem. Phys. 88, 1775–1785.

Frumker, E., Kajumba, N., Bertrand, J.B., Wörner, H.J., Hebeisen, C.T., Hockett, P., Spanner, M., Patchkovskii, S., Paulus, G.G., Villeneuve, D.M., Naumov, A., Corkum, P.B., 2012. Probing polar molecules with high harmonic spectroscopy. Phys. Rev. Lett. 109, 233904.

Fu, Y., Midorikawa, K., Takahashi, E.J., 2018. Towards a petawatt-class few-cycle infrared laser system via dual-chirped optical parametric amplification. Sci. Rep. 8, 7692.

Fu, Y., Takahashi, E.J., Midorikawa, K., 2015. High-energy infrared femtosecond pulses generated by dual-chirped optical parametric amplification. Opt. Lett. 40, 5082–5085.

Fuji, T., Rauschenberger, J., Apolonski, A., Yakovlev, V.S., Tempea, G., Udem, T., Gohle, C., Hänsch, T.W., Lehnert, W., Scherer, M., Krausz, F., 2005. Monolithic carrier-envelope phase-stabilization scheme. Opt. Lett. 30, 332–334.

Gallmann, L., Cirelli, C., Keller, U., 2012. Attosecond science: recent highlights and future trends. Annu. Rev. Phys. Chem. 63, 447–469.

Gallmann, L., Herrmann, J., Locher, R., Sabbar, M., Ludwig, A., Lucchini, M., Keller, U., 2013. Resolving intra-atomic electron dynamics with attosecond transient absorption spectroscopy. Mol. Phys. 111, 2243–2250.

Gallmann, L., Jordan, I., Wörner, H.J., Castiglioni, L., Hengsberger, M., Osterwalder, J., Arrell, C.A., Chergui, M., Liberatore, E., Rothlisberger, U., Keller, U., 2017. Photoemission and photoionization time delays and rates. Struct. Dyn. 4, 061502.

Gan, Z., Yu, L., Li, S., Wang, C., Liang, X., Liu, Y., Li, W., Guo, Z., Fan, Z., Yuan, X., Xu, L., Liu, Z., Xu, Y., Lu, J., Lu, H., Yin, D., Leng, Y., Li, R., Xu, Z., 2017. 200 J high efficiency Ti: sapphire chirped pulse amplifier pumped by temporal dual-pulse. Opt. Express 25, 5169–5178.

Gaumnitz, T., Jain, A., Pertot, Y., Huppert, M., Jordan, I., Ardana-Lamas, F., Wörner, H.J., 2017. Streaking of 43-attosecond soft-X-ray pulses generated by a passively CEP-stable mid-infrared driver. Opt. Express 25, 27506–27518.

Gianturco, F.A., Lucchese, R.R., Sanna, N., 1994. Calculation of low-energy elastic cross sections for electron-CF_4 scattering. J. Chem. Phys. 100, 6464–6471.

Gokhberg, K., Kuleff, A., Cederbaum, L.S., 2020. Electronic decay cascades in chemical environment. In: Marquardt, R., Quack, M. (Eds.), Molecular Spectroscopy and Quantum Dynamics. Elsevier, Amsterdam. Chapter 5 (this book).

Gong, X., Lin, C., He, F., Song, Q., Lin, K., Ji, Q., Zhang, W., Ma, J., Lu, P., Liu, Y., Zeng, H., Yang, W., Wu, J., 2017. Energy-resolved ultrashort delays of photoelectron emission clocked by orthogonal two-color laser fields. Phys. Rev. Lett. 118, 143203.

Goulielmakis, E., Loh, Z.-H., Wirth, A., Santra, R., Rohringer, N., Yakovlev, V.S., Zherebtsov, S., Pfeifer, T., Azzeer, A.M., Kling, M.F., Leone, S.R., Krausz, F., 2010. Real-time observation of valence electron motion. Nature 466, 739–743.

Goulielmakis, E., Schultze, M., Hofstetter, M., Yakovlev, V.S., Gagnon, J., Uiberacker, M., Aquila, A.L., Gullikson, E.M., Attwood, D.T., Kienberger, R., Krausz, F., Kleineberg, U., 2008. Single-cycle nonlinear optics. Science 320, 1614–1617.

Gruson, V., Ernotte, G., Lassonde, P., Laramée, A., Bionta, M.R., Chaker, M., Di Mauro, L., Corkum, P.B., Ibrahim, H., Schmidt, B.E., Légaré, F., 2017. 2.5 TW, two-cycle IR laser pulses via frequency domain optical parametric amplification. Opt. Express 25, 27706–27714.

Guénot, D., Kroon, D., Balogh, E., Larsen, E.W., Kotur, M., Miranda, M., Fordell, T., Johnsson, P., Mauritsson, J., Gisselbrecht, M., Varjù, K., Arnold, C.L., Carette, T., Kheifets, A.S., Lindroth, E., L'Huillier, A., Dahlström, J.M., 2014. Measurements of relative photoemission time delays in noble gas atoms. J. Phys. B At. Mol. Opt. Phys. 47, 245602.

Haessler, S., Boutu, W., Stankiewicz, M., Frasinski, L.J., Weber, S., Caillat, J., Taïeb, R., Maquet, A., Breger, P., Monchicourt, P., Carré, B., Salières, P., 2009a. Attosecond chirp-encoded dynamics of light nuclei. J. Phys. B At. Mol. Opt. Phys. 42, 134002.

Haessler, S., Caillat, J., Boutu, W., Giovanetti-Teixeira, C., Ruchon, T., Auguste, T., Diveki, Z., Breger, P., Maquet, A., Carré, B., Taïeb, R., Salières, P., 2010. Attosecond imaging of molecular electronic wavepackets. Nat. Phys. 6, 200–206.

Haessler, S., Fabre, B., Higuet, J., Caillat, J., Ruchon, T., Breger, P., Carré, B., Constant, E., Maquet, A., Mével, E., Salières, P., Taïeb, R., Mairesse, Y., 2009b. Phase-resolved attosecond near-threshold photoionization of molecular nitrogen. Phys. Rev. A 80, 011404.

Hassan, M.T., Luu, T.T., Moulet, A., Raskazovskaya, O., Zhokhov, P., Garg, M., Karpowicz, N., Zheltikov, A.M., Pervak, V., Krausz, F., Goulielmakis, E., 2016. Optical attosecond pulses and tracking the nonlinear response of bound electrons. Nature 530, 66.

Hauri, C.P., Kornelis, W., Helbing, F.W., Heinrich, A., Couairon, A., Mysyrowicz, A., Biegert, J., Keller, U., 2004. Generation of intense, carrier-envelope phase-locked few-cycle laser pulses through filamentation. Appl. Phys. B 79, 673–677.

Hennig, H., Breidbach, J., Cederbaum, L.S., 2005. Charge transfer driven by electron correlation: a non-Dyson propagator approach. J. Chem. Phys. 122, 134104.

Hentschel, M., Kienberger, R., Spielmann, C., Reider, G.A., Milosevic, N., Brabec, T., Corkum, P., Heinzmann, U., Drescher, M., Krausz, F., 2001. Attosecond metrology. Nature 414, 509.

Heuser, S., Jiménez Galán, Á., Cirelli, C., Marante, C., Sabbar, M., Boge, R., Lucchini, M., Gallmann, L., Ivanov, I., Kheifets, A.S., Dahlström, J.M., Lindroth, E., Argenti, L., Martín, F., Keller, U., 2016. Angular dependence of photoemission time delay in helium. Phys. Rev. A 94, 63409.

Heyl, C.M., Arnold, C.L., Couairon, A., L'Huillier, A., 2016. Introduction to macroscopic power scaling principles for high-order harmonic generation. J. Phys. B At. Mol. Opt. Phys. 50, 13001.

Higuet, J., Ruf, H., Thiré, N., Cireasa, R., Constant, E., Cormier, E., Descamps, D., Mével, E., Petit, S., Pons, B., Mairesse, Y., Fabre, B., 2011. High-order harmonic spectroscopy of the Cooper minimum in argon: experimental and theoretical study. Phys. Rev. A 83, 053401.

Hosler, E.R., Leone, S.R., 2013. Characterization of vibrational wave packets by core-level high-harmonic transient absorption spectroscopy. Phys. Rev. A 88, 023420.

Huppert, M., Jordan, I., Baykusheva, D., von Conta, A., Wörner, H.J., 2016. Attosecond delays in molecular photoionization. Phys. Rev. Lett. 117, 093001.

Huppert, M., Jordan, I., Wörner, H.J., 2015. Attosecond beamline with actively stabilized and spatially separated beam paths. Rev. Sci. Instrum. 86, 123106.

Ishii, N., Kaneshima, K., Kitano, K., Kanai, T., Watanabe, S., Itatani, J., 2014. Carrier-envelope phase-dependent high harmonic generation in the water window using few-cycle infrared pulses. Nat. Commun. 5, 3331.

Itatani, J., Quéré, F., Yudin, G.L., Ivanov, M.Y., Krausz, F., Corkum, P.B., 2002. Attosecond streak camera. Phys. Rev. Lett. 88, 173903.

Ivanov, M.Y., Spanner, M., Smirnova, O., 2005. Anatomy of strong field ionization. J. Modern Opt. 52, 165–184.

Jain, A., Gaumnitz, T., Bray, A., Kheifets, A., Wörner, H.J., 2018. Photoionization delays in xenon using single-shot referencing in the collinear back-focusing geometry. Opt. Lett. 43, 4510–4513.

Jordan, I., Huppert, M., Pabst, S., Kheifets, A.S., Baykusheva, D., Wörner, H.J., 2017. Spin-orbit delays in photoemission. Phys. Rev. A 95, 13404.

Jullien, A., Pfeifer, T., Abel, M.J., Nagel, P.M., Bell, M.J., Neumark, D.M., Leone, S.R., 2008. Ionization phase-match gating for wavelength-tunable isolated attosecond pulse generation. Appl. Phys. B 93, 433.

Kakehata, M., Takada, H., Kobayashi, Y., Torizuka, K., Fujihira, Y., Homma, T., Takahashi, H., 2001. Single-shot measurement of carrier-envelope phase changes by spectral interferometry. Opt. Lett. 26, 1436–1438.

Kanai, T., Takahashi, E.J., Nabekawa, Y., Midorikawa, K., 2008. Observing the attosecond dynamics of nuclear wavepackets in molecules by using high harmonic generation in mixed gases. New J. Phys. 10, 025036.

Keldysh, L.V., 1965. Ionization in the field of a strong electromagnetic wave. J. Exp. Theor. Phys. 20, 1307.

Kelkensberg, F., Siu, W., Pérez-Torres, J.F., Morales, F., Gademann, G., Rouzée, A., Johnsson, P., Lucchini, M., Calegari, F., Sanz-Vicario, J.L., Martín, F., Vrakking, M.J.J., 2011. Attosecond control in photoionization of hydrogen molecules. Phys. Rev. Lett. 107, 043002.

Keller, U., 'tHooft, G.W., Knox, W.H., Cunningham, J.E., 1991. Femtosecond pulses from a continuously self-starting passively mode-locked Ti: sapphire laser. Opt. Lett. 16, 1022–1024.

Kim, K.T., Zhang, C., Ruchon, T., Hergott, J.-F., Auguste, T., Villeneuve, D.M., Corkum, P.B., Quéré, F., 2013. Photonic streaking of attosecond pulse trains. Nat. Photonics 7, 651.

Kjeldsen, T.K., Madsen, L.B., 2005. Vibrational excitation of diatomic molecular ions in strong field ionization of diatomic molecules. Phys. Rev. Lett. 95, 073004.

Kling, M.F., Verhoef, A.J., Khan, J.I., Schultze, M., Ni, Y., Uiberacker, M., Drescher, M., Krausz, F., Vrakking, M.J.J., Siedschlag, C., Verhoef, A.J., Khan, J.I., Schultze, M., Uphues, T., Ni, Y., Uiberacker, M., Drescher, M., Krausz, F., Vrakking, M.J.J., 2006. Control of electron localization in molecular dissociation. Science 312, 246–248.

Kling, M.F., Vrakking, M.J.J., 2008. Attosecond electron dynamics. Annu. Rev. Phys. Chem. 59, 463–492.

Klünder, K., Dahlström, J.M., Gisselbrecht, M., Fordell, T., Swoboda, M., Guénot, D., Johnsson, P., Caillat, J., Mauritsson, J., Maquet, A., Taïeb, R., L'Huillier, A., 2011. Probing single-photon ionization on the attosecond time scale. Phys. Rev. Lett. 106, 143002.

Knight, L.B., Steadman, J., Feller, D., Davidson, E.R., 1984. Experimental evidence for a C_{2v} ($^2B_1$) ground-state structure of the methane cation radical: ESR and ab initio CI investigations of methane cation radicals (CH_4^+ and $CD_2H_2^+$) in neon matrixes at 4 K. J. Am. Chem. Soc. 106, 3700–3701.

Koke, S., Grebing, C., Frei, H., Anderson, A., Assion, A., Steinmeyer, G., 2010. Direct frequency comb synthesis with arbitrary offset and shot-noise-limited phase noise. Nat. Photonics 4, 462.

Kraus, P.M., Baykusheva, D., Wörner, H.J., 2014. Two-pulse field-free orientation reveals anisotropy of molecular shape resonance. Phys. Rev. Lett. 113, 023001.

Kraus, P.M., Mignolet, B., Baykusheva, D., Rupenyan, A., Horný, L., Penka, E.F., Grassi, G., Tolstikhin, O.I., Schneider, J., Jensen, F., Madsen, L.B., Bandrauk, A.D., Remacle, F., Wörner, H.J., 2015a. Measurement and laser control of attosecond charge migration in ionized iodoacetylene. Science 350, 790.

Kraus, P.M., Rupenyan, A., Wörner, H.J., 2012. High-harmonic spectroscopy of oriented ocs molecules: emission of even and odd harmonics. Phys. Rev. Lett. 109, 233903.

Kraus, P.M., Tolstikhin, O.I., Baykusheva, D., Rupenyan, A., Schneider, J., Bisgaard, C.Z., Morishita, T., Jensen, F., Madsen, L.B., Wörner, H.J., 2015b. Observation of laser-induced electronic structure in oriented polyatomic molecules. Nat. Commun. 6, 7039.

Kraus, P.M., Wörner, H.J., 2013. Attosecond nuclear dynamics in the ammonia cation: relation between high-harmonic and photoelectron spectroscopies. Chem. Phys. Chem. 14, 1445–1450.

Kraus, P.M., Wörner, H.J., 2018a. Perspectives of attosecond spectroscopy for the understanding of fundamental electron correlations. Angew. Chem., Int. Ed. Engl. 57, 5228–5247.

Kraus, P.M., Wörner, H.J., 2018b. Perspektiven für das Verständnis fundamentaler Elektronenkorrelationen durch Attosekundenspektroskopie. Angew. Chem. 130, 5324–5344.

Kraus, P.M., Zürch, M., Cushing, S.K., Neumark, D.M., Leone, S.R., 2018. The ultrafast X-ray spectroscopic revolution in chemical dynamics. Nat. Rev. Chem. 2, 82–94.

Krausz, F., 2016. The birth of attosecond physics and its coming of age. Phys. Scr. 91, 063011.

Krausz, F., Ivanov, M., 2009. Attosecond physics. Rev. Modern Phys. 81, 163–234.

Krehl, P., Engemann, S., 1995. August Toepler – the first who visualized shock waves. Shock Waves 5, 1–18.

Kreß, M., Löffler, T., Thomson, M.D., Dörner, R., Gimpel, H., Zrost, K., Ergler, T., Moshammer, R., Morgner, U., Ullrich, J., Roskos, H.G., 2006. Determination of the carrier-envelope phase of few-cycle laser pulses with terahertz-emission spectroscopy. Nat. Phys. 2, 327.

Kulander, K.C., Schafer, K.J., Krause, J.L., 1993. Dynamics of short-pulse excitation, ionization and harmonic conversion - super-intense laser-atom physics. In: Piraux, B., L'Huillier, A., Rzążewski, K. (Eds.), Super-Intense Laser-Atom Physics. In: NATO ASI Series (Series B: Physics), vol. 316. Springer US, Boston, MA, pp. 95–110.

Lackner, F., Chatterley, A.S., Pemmaraju, C.D., Closser, K.D., Prendergast, D., Neumark, D.M., Leone, S.R., Gessner, O., 2016. Direct observation of ring-opening dynamics in strong-field ionized selenophene using femtosecond inner-shell absorption spectroscopy. J. Chem. Phys. 145, 234313.

Lara-Astiaso, M., Galli, M., Trabattoni, A., Palacios, A., Ayuso, D., Frassetto, F., Poletto, L., De Camillis, S., Greenwood, J., Decleva, P., Tavernelli, I., Calegari, F., Nisoli, M., Martín, F., 2018. Attosecond pump–probe spectroscopy of charge dynamics in tryptophan. Chem. Phys. Lett. 9, 4570–4577.

Lein, M., 2005. Attosecond probing of vibrational dynamics with high-harmonic generation. Phys. Rev. Lett. 94, 1–4.

Lein, M., Hay, N., Velotta, R., Marangos, J.P., Knight, P.L., 2002a. Role of the intramolecular phase in high-harmonic generation. Phys. Rev. Lett. 88, 183903.

Lein, M., Hay, N., Velotta, R., Marangos, J.P., Knight, P.L., 2002b. Interference effects in high-order harmonic generation with molecules. Phys. Rev. A 66, 23805.

Leone, S.R., Neumark, D.M., 2016. Attosecond science in atomic, molecular, and condensed matter physics. Faraday Discuss. 194, 15–39.

Lepetit, L., Chériaux, G., Joffre, M., 1995. Linear techniques of phase measurement by femtosecond spectral interferometry for applications in spectroscopy. J. Opt. Soc. Amer. B 12, 2467–2474.

Lépine, F., Ivanov, M.Y., Vrakking, M.J.J., 2014. Attosecond molecular dynamics: fact or fiction? Nat. Photonics 8, 195–204.

Lewenstein, M., Balcou, P., Ivanov, M.Y., L'Huillier, A., Corkum, P.B., 1994. Theory of high-harmonic generation by low-frequency laser fields. Phys. Rev. A 49, 2117–2132.

Li, J., Ren, X., Yin, Y., Zhao, K., Chew, A., Cheng, Y., Cunningham, E., Wang, Y., Hu, S., Wu, Y., Chini, M., Chang, Z., 2017. 53-attosecond X-ray pulses reach the carbon K-edge. Nat. Commun. 8, 186.

Loh, Z.H., Khalil, M., Correa, R.E., Santra, R., Buth, C., Leone, S.R., 2007. Quantum state-resolved probing of strong-field-ionized Xenon atoms using femtosecond high-order harmonic transient absorption spectroscopy. Phys. Rev. Lett. 98, 143601.

Louisy, M., Arnold, C.L., Miranda, M., Larsen, E.W., Bengtsson, S.N., Kroon, D., Kotur, M., Guénot, D., Rading, L., Rudawski, P., Brizuela, F., Campi, F., Kim, B., Jarnac, A., Houard, A., Mauritsson, J., Johnsson, P., L'Huillier, A., Heyl, C.M., 2015. Gating attosecond pulses in a noncollinear geometry. Optica 2, 563–566.

Ludwig, A., Maurer, J., Mayer, B.W., Phillips, C.R., Gallmann, L., Keller, U., 2014. Breakdown of the dipole approximation in strong-field ionization. Phys. Rev. Lett. 113, 243001.

Lünnemann, S., Kuleff, A.I., Cederbaum, L.S., 2008. Charge migration following ionization in systems with chromophore-donor and amine-acceptor sites. J. Chem. Phys. 129, 104305.

Maine, P., Strickland, D., Bado, P., Pessot, M., Mourou, G., 1988. Generation of ultrahigh peak power pulses by chirped pulse amplification. IEEE J. Quantum Electron. 24, 398–403.

Mairesse, Y., de Bohan, A., Frasinski, L.J., Merdji, H., Dinu, L.C., Monchicourt, P., Breger, P., Kovačev, M., Taïeb, R., Carré, B., Muller, H.G., Agostini, P., Salières, P., 2003. Attosecond synchronization of high-harmonic soft X-rays. Science 302, 1540–1543.

Månsson, E.P., De Camillis, S., Castrovilli, M.C., Galli, M., Nisoli, M., Calegari, F., Greenwood, J.B., 2017. Ultrafast dynamics in the DNA building blocks thymidine and thymine initiated by ionizing radiation. Phys. Chem. Chem. Phys. 19, 19815–19821.

Marquardt, R., Quack, M., 2020. Foundations of time dependent quantum dynamics of molecules under isolation and in coherent electromagnetic fields. In: Marquardt, R., Quack, M. (Eds.), Molecular Spectroscopy and Quantum Dynamics. Elsevier, Amsterdam. Chapter 1 (this book).

McPherson, A., Gibson, G., Jara, H., Johann, U., Luk, T.S., McIntyre, I.A., Boyer, K., Rhodes, C.K., 1987. Studies of multiphoton production of vacuum-ultraviolet radiation in the rare gases. J. Opt. Soc. Amer. B 4, 595–601.

Mehendale, M., Mitchell, S.A., Likforman, J.-P., Villeneuve, D.M., Corkum, P.B., 2000. Method for single-shot measurement of the carrier envelope phase of a few-cycle laser pulse. Opt. Lett. 25, 1672–1674.

Mizutani, H., Minemoto, S., Oguchi, Y., Sakai, H., 2011. Effect of nuclear motion observed in high-order harmonic generation from D_2/H_2 molecules with intense multi-cycle 1300 nm and 800 nm pulses. J. Phys. B At. Mol. Opt. Phys. 44, 081002.

Mondal, T., Varandas, A.J.C., 2014. On extracting subfemtosecond data from femtosecond quantum dynamics calculations: the methane cation. J. Chem. Theory Comput. 10, 3606–3616.

Mondal, T., Varandas, A.J.C., 2015. Structural evolution of the methane cation in subfemtosecond photodynamics. J. Chem. Phys. 143, 14304.

Moulton, P.F., 1986. Spectroscopic and laser characteristics of Ti:Al_2O_3. J. Opt. Soc. Amer. B 3, 125–133.

Natalense, A.P.P., Lucchese, R.R., 1999. Cross section and asymmetry parameter calculation for sulfur 1s photoionization of SF_6. J. Chem. Phys. 111, 5344–5348.

Neidel, C., Klei, J., Yang, C.H., Rouzée, A., Vrakking, M.J.J., Klünder, K., Miranda, M., Arnold, C.L., Fordell, T., L'Huillier, A., Gisselbrecht, M., Johnsson, P., Dinh, M.P., Suraud, E., Reinhard, P.G., Despré, V., Marques, M.A.L., Lépine, F., 2013. Probing time-dependent molecular dipoles on the attosecond time scale. Phys. Rev. Lett. 111, 033001.

Neppl, S., Ernstorfer, R., Bothschafter, E.M., Cavalieri, A.L., Menzel, D., Barth, J.V., Krausz, F., Kienberger, R., Feulner, P., 2012. Attosecond time-resolved photoemission from core and valence states of magnesium. Phys. Rev. Lett. 109, 87401.

Nisoli, M., De Silvestri, S., Svelto, O., 1996. Generation of high energy 10 fs pulses by a new pulse compression technique. Appl. Phys. Lett. 68, 2793–2795.

Nisoli, M., Decleva, P., Calegari, F., Palacios, A., Martín, F., 2017. Attosecond electron dynamics in molecules. Chem. Rev. 117, 10760–10825.

Norrish, R.G.W., Porter, G., 1949. Chemical reactions produced by very high light intensities. Nature 164, 658.

Okell, W.A., Witting, T., Fabris, D., Arrell, C.A., Hengster, J., Ibrahimkutty, S., Seiler, A., Barthelmess, M., Stankov, S., Lei, D.Y., Sonnefraud, Y., Rahmani, M., Uphues, T., Maier, S.A., Marangos, J.P., Tisch, J.W.G., 2015. Temporal broadening of attosecond photoelectron wavepackets from solid surfaces. Optica 2, 383–387.

Ossiander, M., Riemensberger, J., Neppl, S., Mittermair, M., Schäffer, M., Duensing, A., Wagner, M.S., Heider, R., Wurzer, M., Gerl, M., Schnitzenbaumer, M., Barth, J.V., Libisch, F., Lemell, C., Burgdörfer, J., Feulner, P., Kienberger, R., 2018. Absolute timing of the photoelectric effect. Nature 561, 374–377.

Ossiander, M., Siegrist, F., Shirvanyan, V., Pazourek, R., Sommer, A., Latka, T., Guggenmos, A., Nagele, S., Feist, J., Burgdörfer, J., Kienberger, R., Schultze, M., 2017. Attosecond correlation dynamics. Nat. Phys. 13, 280.

Patchkovskii, S., 2009. Nuclear dynamics in polyatomic molecules and high-order harmonic generation. Phys. Rev. Lett. 102, 253602.

Patchkovskii, S., Schuurman, M.S., 2017. Full-dimensional treatment of short-time vibronic dynamics in a molecular high-order-harmonic-generation process in methane. Phys. Rev. A 96, 053405.

Paul, P.M., Toma, E.S., Breger, P., Mullot, G., Augé, F., Balcou, P., Muller, H.G., Agostini, P., 2001. Observation of a train of attosecond pulses from high harmonic generation. Science 292, 1689–1692.

Paulus, G.G., Grasbon, F., Walther, H., Villoresi, P., Nisoli, M., Stagira, S., Priori, E., De Silvestri, S., 2001. Absolute-phase phenomena in photoionization with few-cycle laser pulses. Nature 414, 182–184.

Pazourek, R., Nagele, S., Burgdörfer, J., 2015. Attosecond chronoscopy of photoemission. Rev. Modern Phys. 87.

Perelomov, A.M., Popov, V.S., Terent'ev, M.V., 1966. Ionization of atoms in an alternating electrical field. Sov. Phys. JETP 23, 924.

Pertot, Y., Schmidt, C., Matthews, M., Chauvet, A., Huppert, M., Svoboda, V., von Conta, A., Tehlar, A., Baykusheva, D., Wolf, J.-P., Wörner, H.J., 2017. Time-resolved X-ray absorption spectroscopy with a water window high-harmonic source. Science 355, 264–267.

Pfeifer, T., Jullien, A., Abel, M.J., Nagel, P.M., Gallmann, L., Neumark, D.M., Leone, S.R., 2007. Generating coherent broadband continuum soft-x-ray radiation by attosecond ionization gating. Opt. Express 15, 17120–17128.

Popmintchev, T., Chen, M.-C., Bahabad, A., Gerrity, M., Sidorenko, P., Cohen, O., Christov, I.P., Murnane, M.M., Kapteyn, H.C., 2009. Phase matching of high harmonic generation in the soft and hard X-ray regions of the spectrum. Proc. Natl. Acad. Sci. USA 106, 10516–10521.

Popruzhenko, S.V., 2014. Keldysh theory of strong field ionization: history, applications, difficulties and perspectives. J. Phys. B At. Mol. Opt. Phys. 47, 204001.

Powis, I., 2008. Photoelectron circular dichroism in gas phase chiral molecules. Adv. Chem. Phys. 138, 267–329.

Ramasesha, K., Leone, S.R., Neumark, D.M., 2016. Real-time probing of electron dynamics using attosecond time-resolved spectroscopy. Annu. Rev. Phys. Chem. 67, 41–63.

Rathbone, G.J., Poliakoff, E.D., Bozek, J.D., Toffoli, D., Lucchese, R.R., 2005. Photoelectron trapping in N_2O $7\sigma \rightarrow k\sigma$ resonant ionization. J. Chem. Phys. 123, 14307.

Reduzzi, M., Feng, C., Chu, W.-C., Dubrouil, A., Calegari, F., Nisoli, M., Frassetto, F., Poletto, L., Lin, C.-D., Sansone, G., 2013. Attosecond absorption spectroscopy in molecules. In: CLEO: 2013. Optical Society of America, p. QF2C.1.

Reiss, H.R., 1980. Gauges for intense-field electrodynamics. Phys. Rev. A 22, 770–772.

Reiss, H.R., 2008a. Limits on tunneling theories of strong-field ionization. Phys. Rev. Lett. 101, 043002.

Reiss, H.R., 2008b. Erratum: limits on tunneling theories of strong-field ionization [Phys. Rev. Lett. 101, 043002 (2008)]. Phys. Rev. Lett. 101, 159901.

Remacle, F., Levine, R.D., 2006. An electronic time scale in chemistry. Proc. Natl. Acad. Sci. 103, 6793–6798.

Ren, X., Makhija, V., Le, A.-T., Troß, J., Mondal, S., Jin, C., Kumarappan, V., Trallero-Herrero, C., 2013. Measuring the angle-dependent photoionization cross section of nitrogen using high-harmonic generation. Phys. Rev. A 88, 43421.

Ritchie, B., 1975. Theory of the angular distribution of photoelectrons ejected from optically active molecules and molecular negative ions. Phys. Rev. A 13, 1411–1415.

Rothhardt, J., Demmler, S., Hädrich, S., Limpert, J., Tünnermann, A., 2012. Octave-spanning OPCPA system delivering CEP-stable few-cycle pulses and 22 W of average power at 1 MHz repetition rate. Opt. Express 20, 10870–10878.

Rupenyan, A., Bertrand, J.B., Villeneuve, D.M., Wörner, H.J., 2012. All-optical measurement of high-harmonic amplitudes and phases in aligned molecules. Phys. Rev. Lett. 108, 33903.

Rupenyan, A., Kraus, P.M., Schneider, J., Wörner, H.J., 2013. High-harmonic spectroscopy of isoelectronic molecules: wavelength scaling of electronic-structure and multielectron effects. Phys. Rev. A 87, 033409.

Sabbar, M., Heuser, S., Boge, R., Lucchini, M., Carette, T., Lindroth, E., Gallmann, L., Cirelli, C., Keller, U., 2015. Resonance effects in photoemission time delays. Phys. Rev. Lett. 115, 133001.

Salières, P., Carré, B., Le Déroff, L., Grasbon, F., Paulus, G.G., Walther, H., Kopold, R., Becker, W., Milošević, D.B., Sanpera, A., Lewenstein, M., 2001. Feynman's path-integral approach for intense-laser-atom interactions. Science 292, 902–905.

Sansone, G., Benedetti, E., Calegari, F., Vozzi, C., Avaldi, L., Flammini, R., Poletto, L., Villoresi, P., Altucci, C., Velotta, R., Stagira, S., De Silvestri, S., Nisoli, M., 2006. Isolated single-cycle attosecond pulses. Science 314, 443–446.

Sansone, G., Kelkensberg, F., Pérez-Torres, J.F., Morales, F., Kling, M.F., Siu, W., Ghafur, O., Johnsson, P., Swoboda, M., Benedetti, E., Ferrari, F., Lépine, F., Sanz-Vicario, J.L., Zherebtsov, S., Znakovskaya, I., L'Huillier, A., Ivanov, M.Y., Nisoli, M., Martín, F., Vrakking, M.J.J., 2010. Electron localization following attosecond molecular photoionization. Nature 465, 763–766.

Sayler, A.M., Rathje, T., Müller, W., Rühle, K., Kienberger, R., Paulus, G.G., 2011. Precise, real-time, every-single-shot, carrier-envelope phase measurement of ultrashort laser pulses. Opt. Lett. 36, 1–3.

Schafer, K.J., Yang, B., DiMauro, L.F., Kulander, K.C., 1993. Above threshold ionization beyond the high harmonic cutoff. Phys. Rev. Lett. 70, 1599–1602.

Schmidt, B.E., Béjot, P., Giguère, M., Shiner, A.D., Trallero-Herrero, C., Bisson, É., Kasparian, J., Wolf, J.-P., Villeneuve, D.M., Kieffer, J.-C., Corkum, P.B., Légaré, F., 2010. Compression of 1.8 μm laser pulses to sub two optical cycles with bulk material. Appl. Phys. Lett. 96, 121109.

Schmidt, B.E., Shiner, A.D., Lassonde, P., Kieffer, J.-C., Corkum, P.B., Villeneuve, D.M., Légaré, F., 2011. CEP stable 1.6 cycle laser pulses at 1.8 μm. Opt. Express 19, 6858–6864.

Schmidt, B.E., Thiré, N., Boivin, M., Laramée, A., Poitras, F., Lebrun, G., Ozaki, T., Ibrahim, H., Légaré, F., 2014. Frequency domain optical parametric amplification. Nat. Commun. 5, 3643.

Schultze, M., Fieß, M., Karpowicz, N., Gagnon, J., Korbman, M., Hofstetter, M., Neppl, S., Cavalieri, A.L., Komninos, Y., Mercouris, T., Nicolaides, C.A., Pazourek, R., Nagele, S., Feist, J., Burgdörfer, J., Azzeer, A.M., Ernstorfer, R., Kienberger, R., Kleineberg, U., Goulielmakis, E., Krausz, F., Yakovlev, V.S., 2010. Delay in photoemission. Science 328, 1658–1662.

Schultze, M., Ramasesha, K., Pemmaraju, C.D., Sato, S.A., Whitmore, D., Gandman, A., Prell, J.S., Borja, L.J., Prendergast, D., Yabana, K., Neumark, D.M., Leone, S.R., 2014. Attosecond band-gap dynamics in silicon. Science 346, 1348–1353.

Seres, E., Seres, J., Spielmann, C., 2006. X-ray absorption spectroscopy in the keV range with laser generated high harmonic radiation. Appl. Phys. Lett. 89, 181919.

Shafir, D., Soifer, H., Bruner, B.D., Dagan, M., Mairesse, Y., Patchkovskii, S., Ivanov, M.Y., Smirnova, O., Dudovich, N., 2012. Resolving the time when an electron exits a tunnelling barrier. Nature 485, 343–346.

Shiner, A.D., Schmidt, B.E., Trallero-Herrero, C., Wörner, H.J., Patchkovskii, S., Corkum, P.B., Kieffer, J.-C., Légaré, F., Villeneuve, D.M., 2011. Probing collective multi-electron dynamics in xenon with high-harmonic spectroscopy. Nat. Phys. 7, 464–467.

Silva, F., Teichmann, S.M., Cousin, S.L., Hemmer, M., Biegert, J., 2015. Spatiotemporal isolation of attosecond soft X-ray pulses in the water window. Nat. Commun. 6, 6611.

Siu, W., Kelkensberg, F., Gademann, G., Rouzée, A., Johnsson, P., Dowek, D., Lucchini, M., Calegari, F., De Giovannini, U., Rubio, A., Lucchese, R.R., Kono, H., Lépine, F., Vrakking, M.J.J., 2011. Attosecond control of dissociative ionization of O_2 molecules. Phys. Rev. A 84, 063412.

Skantzakis, E., Tzallas, P., Kruse, J., Kalpouzos, C., Charalambidis, D., 2009. Coherent continuum extreme ultraviolet radiation in the sub-100-nJ range generated by a high-power many-cycle laser field. Opt. Lett. 34, 1732–1734.

Smirnova, O., Mairesse, Y., Patchkovskii, S., Dudovich, N., Villeneuve, D., Corkum, P., Ivanov, M.Y., 2009a. High harmonic interferometry of multi-electron dynamics in molecules. Nature 460, 972–977.

Smirnova, O., Patchkovskii, S., Mairesse, Y., Dudovich, N., Ivanov, M.Y., 2009b. Strong-field control and spectroscopy of attosecond electron-hole dynamics in molecules. Proc. Natl. Acad. Sci. 106, 16556–16561.

Smirnova, O., Spanner, M., Ivanov, M., 2007. Anatomy of strong field ionization II: to dress or not to dress? J. Modern Opt. 54, 1019–1038.

Sola, I.J., Mével, E., Elouga, L., Constant, E., Strelkov, V., Poletto, L., Villoresi, P., Benedetti, E., Caumes, J.-P., Stagira, S., Vozzi, C., Sansone, G., Nisoli, M., 2006. Controlling attosecond electron dynamics by phase-stabilized polarization gating. Nat. Phys. 2, 319.

Song, X., Shi, G., Zhang, G., Xu, J., Lin, C., Chen, J., Yang, W., 2018. Attosecond time delay of retrapped resonant ionization. Phys. Rev. Lett. 121, 103201.

Spanner, M., Patchkovskii, S., 2009. One-electron ionization of multielectron systems in strong nonresonant laser fields. Phys. Rev. A 80, 63411.

Spanner, M., Patchkovskii, S., 2013. Molecular strong field ionization and high harmonic generation: a selection of computational illustrations. Chem. Phys. 414, 10–19.

Spence, D.E., Kean, P.N., Sibbett, W., 1991. 60-fsec pulse generation from a self-mode-locked Ti: sapphire laser. Opt. Lett. 16, 42–44.

Śpiewanowski, M.D., Etches, A., Madsen, L.B., 2013. High-order-harmonic generation from field-distorted orbitals. Phys. Rev. A 87, 043424.

Stapelfeldt, H., Seideman, T., 2003. Colloquium: aligning molecules with strong laser pulses. Rev. Modern Phys. 75, 543–557.

Strickland, D., Mourou, G., 1985. Compression of amplified chirped optical pulses. Opt. Commun. 56, 219–221.

Suda, A., Hatayama, M., Nagasaka, K., Midorikawa, K., 2005. Generation of sub-10-fs, 5-mJ-optical pulses using a hollow fiber with a pressure gradient. Appl. Phys. Lett. 86, 111116.

Sutter, D.H., Steinmeyer, G., Gallmann, L., Matuschek, N., Morier-Genoud, F., Keller, U., Scheuer, V., Angelow, G., Tschudi, T., 1999. Semiconductor saturable-absorber mirror–assisted Kerr-lens mode-locked Ti: sapphire laser producing pulses in the two-cycle regime. Opt. Lett. 24, 631–633.

Swoboda, M., Fordell, T., Klünder, K., Dahlström, J.M., Miranda, M., Buth, C., Schafer, K.J., Mauritsson, J., L'Huillier, A., Gisselbrecht, M., 2010. Phase measurement of resonant two-photon ionization in Helium. Phys. Rev. Lett. 104, 103003.

Takahashi, E.J., Kanai, T., Ishikawa, K.L., Nabekawa, Y., Midorikawa, K., 2008. Coherent water window X ray by phase-matched high-order harmonic generation in neutral media. Phys. Rev. Lett. 101, 253901.

Tcherbakoff, O., Mével, E., Descamps, D., Plumridge, J., Constant, E., 2003. Time-gated high-order harmonic generation. Phys. Rev. A 68, 43804.

Telle, H.R., Steinmeyer, G., Dunlop, A.E., Stenger, J., Sutter, D.H., Keller, U., 1999. Carrier-envelope offset phase control: a novel concept for absolute optical frequency measurement and ultrashort pulse generation. Appl. Phys. B 69, 327–332.

Timmers, H., Sabbar, M., Hellwagner, J., Kobayashi, Y., Neumark, D.M., Leone, S.R., 2016. Polarization-assisted amplitude gating as a route to tunable, high-contrast attosecond pulses. Optica 3, 707–710.

Tolstikhin, O.I., Morishita, T., Madsen, L.B., 2011. Theory of tunneling ionization of molecules: weak-field asymptotics including dipole effects. Phys. Rev. A 84, 053423.

Tong, X.M., Zhao, Z.X., Lin, C.D., 2002. Theory of molecular tunneling ionization. Phys. Rev. A 66, 33402.

Tzallas, P., Skantzakis, E., Kalpouzos, C., Benis, E.P., Tsakiris, G.D., Charalambidis, D., 2007. Generation of intense continuum extreme-ultraviolet radiation by many-cycle laser fields. Nat. Phys. 3, 846.

Uiberacker, M., Uphues, T., Schultze, M., Verhoef, A.J., Yakovlev, V., Kling, M.F., Rauschenberger, J., Kabachnik, N.M., Schröder, H., Lezius, M., Kompa, K.L., Muller, H.-G., Vrakking, M.J.J., Hendel, S., Kleineberg, U., Heinzmann, U., Drescher, M., Krausz, F., 2007. Attosecond real-time observation of electron tunnelling in atoms. Nature 446, 627.

Ullrich, J., Moshammer, R., Dorn, A., Dörner, R., Schmidt, L.P.H., Schmidt-Böcking, H., 2003. Recoil-ion and electron momentum spectroscopy: reaction-microscopes. Rep. Progr. Phys. 66, 1463–1545.

Ullrich, J., Moshammer, R., Dörner, R., Jagutzki, O., Mergel, V., Schmidt-Böcking, H., Spielberger, L., 1997. Recoil-ion momentum spectroscopy. J. Phys. B At. Mol. Opt. Phys. 30, 2917–2974.

Urbain, X., Fabre, B., Staicu-Casagrande, E.M., De Ruette, N., Andrianarijaona, V.M., Jureta, J., Posthumus, J.H., Saenz, A., Baldit, E., Cornaggia, C., 2004. Intense-laser-field ionization of molecular hydrogen in the tunneling regime and its effect on the vibrational excitation of H_2^+. Phys. Rev. Lett. 92, 163004.

Vager, Z., Kanter, E.P., Both, G., Cooney, P.J., Faibis, A., Koenig, W., Zabransky, B.J., Zajfman, D., 1986. Direct determination of the stereochemical structure of CH_4^+. Phys. Rev. Lett. 57, 2793–2795.

Vaupel, A., Bodnar, N., Webb, B., Shah, L., Richardson, M.C., 2013. Concepts, performance review, and prospects of table-top, few-cycle optical parametric chirped-pulse amplification. Opt. Eng. 53, 051507.

Vincenti, H., Quéré, F., 2012. Attosecond lighthouses: how to use spatiotemporally coupled light fields to generate isolated attosecond pulses. Phys. Rev. Lett. 108, 113904.

Vos, J., Cattaneo, L., Patchkovskii, S., Zimmermann, T., Cirelli, C., Lucchini, M., Kheifets, A., Landsman, A.S., Keller, U., 2018. Orientation-dependent stereo Wigner time delay and electron localization in a small molecule. Science 360, 1326–1330.

Vozzi, C., Calegari, F., Benedetti, E., Berlasso, R., Sansone, G., Stagira, S., Nisoli, M., Altucci, C., Velotta, R., Torres, R., Heesel, E., Kajumba, N., Marangos, J.P., 2006. Probing two-centre interference in molecular high harmonic generation. J. Phys. B At. Mol. Opt. Phys. 39, S457–S466.

Vozzi, C., Calegari, F., Benedetti, E., Caumes, J.P., Sansone, G., Stagira, S., Nisoli, M., Torres, R., Heesel, E., Kajumba, N., Marangos, J.P., Altucci, C., Velotta, R., 2005. Controlling two-center interference in molecular high harmonic generation. Phys. Rev. Lett. 95, 153902.

Vrakking, M.J.J., 2014. Attosecond imaging. Phys. Chem. Chem. Phys. 16, 2775–2789.

Vrakking, M.J.J., Lepine, F., 2019. Attosecond Molecular Dynamics. Theoretical and Computational Chemistry Series. The Royal Society of Chemistry.

Warrick, E.R., Cao, W., Neumark, D.M., Leone, S.R., 2016. Probing the dynamics of Rydberg and valence states of molecular nitrogen with attosecond transient absorption spectroscopy. J. Phys. Chem. A 120, 3165–3174.

Weinkauf, R., Schanen, P., Metsala, A., Schlag, E.W., Bürgle, M., Kessler, H., 1996. Highly efficient charge transfer in peptide cations in the gas phase: threshold effects and mechanism. J. Phys. Chem. 100, 18567–18585.

Weinkauf, R., Schanen, P., Yang, D., Soukara, S., Schlag, E.W., 1995. Elementary processes in peptides: electron mobility and dissociation in peptide cations in the gas phase. J. Phys. Chem. 99, 11255–11265.

Wheeler, J.A., Borot, A., Monchocé, S., Vincenti, H., Ricci, A., Malvache, A., Lopez-Martens, R., Quéré, F., 2012. Attosecond lighthouses from plasma mirrors. Nat. Photonics 6, 829.

Wirth, A., Hassan, M.T., Grguraš, I., Gagnon, J., Moulet, A., Luu, T.T., Pabst, S., Santra, R., Alahmed, Z.A., Azzeer, A.M., Yakovlev, V.S., Pervak, V., Krausz, F., Goulielmakis, E., 2011. Synthesized light transients. Science 334, 195–200.

Witte, S., Eikema, K.S.E., 2012. Ultrafast optical parametric chirped-pulse amplification. IEEE J. Sel. Top. Quantum Electron. 18, 296–307.

Wolter, B., Pullen, M.G., Baudisch, M., Sclafani, M., Hemmer, M., Senftleben, A., Schröter, C.D., Ullrich, J., Moshammer, R., Biegert, J., 2015. Strong-field physics with mid-IR fields. Phys. Rev. X 5, 21034.

Wong, M.C.H., Le, A.-T., Alharbi, A.F., Boguslavskiy, A.E., Lucchese, R.R., Brichta, J.-P., Lin, C.D., Bhardwaj, V.R., 2013. High harmonic spectroscopy of the cooper minimum in molecules. Phys. Rev. Lett. 110, 33006.

Wörner, H.J., Arrell, C.A., Banerji, N., Cannizzo, A., Chergui, M., Das, A.K., Hamm, P., Keller, U., Kraus, P.M., Liberatore, E., Lopez-Tarifa, P., Lucchini, M., Meuwly, M., Milne, C., Moser, J.-E., Rothlisberger, U., Smolentsev, G., Teuscher, J., van Bokhoven, J.A., Wenger, O., 2017. Charge migration and charge transfer in molecular systems. Struct. Dyn. 4, 61508.

Wörner, H.J., Corkum, P.B., 2011. Attosecond spectroscopy. In: Quack, M., Merkt, F. (Eds.), Handbook of High-Resolution Spectroscopy. John Wiley and Sons Ltd., Chichester, pp. 1781–1803. Chapter 85.

Wörner, H.J., Niikura, H., Bertrand, J.B., Corkum, P.B., Villeneuve, D.M., 2009. Observation of electronic structure minima in high-harmonic generation. Phys. Rev. Lett. 102, 103901.

Xu, L., Spielmann, C., Poppe, A., Brabec, T., Krausz, F., Hänsch, T.W., 1996. Route to phase control of ultrashort light pulses. Opt. Lett. 21, 2008–2010.

Yudin, G., Ivanov, M., 2001. Nonadiabatic tunnel ionization: looking inside a laser cycle. Phys. Rev. A 64, 013409.

Zewail, A.H., 2000. Femtochemistry: atomic-scale dynamics of the chemical bond. J. Phys. Chem. A 104, 5660–5694.

Zhang, Q., Takahashi, E.J., Mücke, O.D., Lu, P., Midorikawa, K., 2011. Dual-chirped optical parametric amplification for generating few hundred mJ infrared pulses. Opt. Express 19, 7190–7212.

Zipp, L.J., Natan, A., Bucksbaum, P.H., 2014. Probing electron delays in above-threshold ionization. Optica 1, 361.

CHAPTER 5

Electronic Decay Cascades in Chemical Environment

KIRILL GOKHBERG • ALEXANDER I. KULEFF • LORENZ S. CEDERBAUM
Theoretische Chemie, Physikalisch-Chemisches Institut, Universität Heidelberg, Heidelberg, Germany

Abstract

Core-excited and core-ionized states created by absorption of X-ray photons carry an enormous energy of hundreds to thousands electronvolts. In the gas phase this energy is dissipated primarily in the Auger decay process or Auger decay cascades, whereby one or several electrons are emitted into the continuum producing multiply charged ions. Two interatomic electronic decay processes, interatomic Coulombic decay (ICD) and electron-transfer mediated decay (ETMD), ensure the efficient dissipation of energy beyond the Auger decay should X-ray absorption take place in a chemical medium characterized by weak van der Waals or hydrogen bonds. Numerous experiments and theoretical results have demonstrated that the Auger–interatomic decay cascades represent a common de-excitation mechanism of core vacancies in medium, whose length and complexity increase with increasing photon energy. Such cascades offer a means for very fast dissipation of the energy which is deposited by the photon into the system. They are also responsible for massive radiation damage to the molecules around the photoabsorption site. Surprisingly, ICD cascades remain a potent mechanism of energy dissipation also when the energy is invested in a cluster by a laser not as a single high energy photon but as multiple low energy ones. The recent research of the electronic decay in chemical media presented in this chapter strives to clarify the physical and chemical consequences of weakly subjecting bound clusters to the highly energetic light provided by modern light sources.

5.1 INTRODUCTION

X-rays have a long history of being employed for investigating structural or electronic properties of atoms, molecules, or condensed matter (Shi, 2014; Stöhr, 1992). Their ability to probe matter relies on their being either absorbed or scattered by the chemical medium. Absorption of X-ray photons by atoms and molecules mostly perturbs tightly bound core electrons. As a result, these electrons can be excited either to the valence orbitals or into the continuum, whereby highly energetic core-excited or core-ionized electronic states are produced. Such states are unstable and decay within a few femtoseconds emitting electrons, photons, and positive ions. The decay of these core-excited or core-ionized states of atoms and molecules may proceed by a variety of pathways. Once known, the decay mechanisms help explain the spectra of emitted particles, which in turn provide the insight into the electronic structure of irradiated species, and into the electron and nuclear dynamics which follow photoabsorption. The absorption of X-rays, even by extended chemical medium, is highly localized, and the energies deposited are tremendous compared to energies common in chemistry. Thus, recent research in its investigation of different decay pathways addresses a more general question. Namely, how the energy deposited locally by X-rays is dissipated throughout the system and what are the possible physico-chemical effects of this energy flow (Nagaya et al., 2016b; Erk et al., 2014, 2013; Motomura et al., 2015; Ouchi et al., 2017).

X-ray scattering forms the basis of the X-ray diffraction technique, the primary tool of structural biology. However, during the irradiation of chemical compounds by X-rays with energies below 10 keV, the energy range at which the majority of current experiments are done, photoabsorption dominates. Therefore, decay processes accompany the accumulation of diffraction images damaging the sample and distorting its molecular structure (Taberman, 2018). The same decay mechanisms are responsible for the radiation damage of living cells, where they constitute the first link in a long chain of physical, chemical, and biological processes which define the cell response to radiation (Goodhead, 1994). Thorough knowledge of the decay processes initiated by X-ray absorption, their characteristic lifetimes and the decay products is the first step in a detailed understanding of radiation damage of complex chemical and biological systems. This knowledge is particularly important in light of the coming of age of new powerful X-ray sources such as the X-ray free electron lasers (XFELs) (Pellegrini et al., 2016). Their immense light intensi-

Molecular Spectroscopy and Quantum Dynamics. https://doi.org/10.1016/B978-0-12-817234-6.00010-6

ties and short pulses promise to revolutionize structural biology by enabling one to obtain the images of native structures of biomolecules (Neutze et al., 2000; Chapman et al., 2011). This is possible if the diffraction image can be accumulated before the sample is damaged, which in turn requires prior knowledge of the electronic decay pathways and their timescales.

For atoms and molecules in the gas phase, central among these electronic relaxation mechanisms of core-ionized states is the Auger decay process (Åberg and Howat, 1982). Here an electron from a higher energy shell fills the core vacancy, while the excess energy is transferred to another electron which is emitted into the continuum. The coupling between the two electrons is furnished by electron correlation. This coupling in atoms and molecules is usually very strong which explains very short Auger decay times which lie between 10 and 0.1 fs (Krause and Oliver, 1979). Auger decay of the 1s vacancy in Ne may serve as a typical example. Removing a 1s electron requires photon energies above 870 eV, and the lifetime of the core vacancy is 2.3 fs (Avaldi et al., 1995). The Auger electrons are emitted at several well defined energies lying in the range between 747 and 804 eV. These energies correspond to the final states of Ne which are predominantly Ne^{2+} ions with two holes residing in the 2s and 2p electronic shells (Körber and Mehlhorn, 1964; Kelly, 1975).

The energy invested in the Ne atom by the X-rays in the 1s ionization event is sufficient to produce ions with charges up to seven. However, in the majority of the decay events almost all of it is carried away by a single Auger electron with Ne^{2+} formed as the result. The explanation is that the probability amplitudes corresponding to the emission of two, three, or more electrons appear in the second, third, etc., orders of perturbation theory (Amusia et al., 1992), and are much smaller than the normal Auger amplitude. Thus in Ne the branching ratio of the double Auger decay is 4.8% and of the triple Auger decay is only 0.3% (Morgan et al., 1997). These probabilities have been shown to increase when one goes to heavier atoms where electron correlation is stronger. In Ar the double Auger branching ratio is 13% (Viefhaus et al., 2004) and becomes even more prominent in the Kr and Xe atoms. Thus for 3d vacancies in Kr it is 25% to 30%, and for 4d vacancies in Xe this ratio is 20% (see Kolorenč et al., 2016).

Two mechanisms were put forward as being responsible for the emission of multiple Auger electrons: the direct mechanism, where the electrons are emitted simultaneously, and the cascade mechanism where electrons are emitted consecutively, when the system decays through a sequence of metastable states. The characteristic signatures of these mechanisms for the double Auger decay can be resolved in two-dimensional Auger spectra where the energies of the two Auger electrons serve as the axes. In the direct Auger decay the emitted electrons share the total available energy in a continuous fashion, and the electron spectra appears, therefore, as a continuous diagonal line along which the total energy of the two electrons is conserved (Viefhaus et al., 2004). Decay cascades, on the contrary, appear as a set of bright spots along this diagonal, each spot corresponding to electrons emitted with well defined energies in a specific two-step transition through some intermediate state (Viefhaus et al., 2005). Experimental evidence suggests that no cascades are involved in the double Auger decay of $Ne^{+}(1s^{-1})$, while they contribute 25% to the double Auger decay of $Ar^{+}(2p^{-1})$, and completely dominate it in $Kr^{+}(3d^{-1})$ (Kolorenč et al., 2016).

Auger decay cascades become even more prominent if deep lying shells of heavier atoms are ionized. Such initial states carry immense energy and result in multistep decay cascades which emit several Auger electrons with varying energies and leave behind multiply charged ions. For example, removing a 1s electron of Ar, whose binding energy is 3206 eV, results in an Auger cascade during which up to 6 Auger electrons are emitted, and the Ar^{q+} ions are produced with q ranging between 1 and 7. The most probable ionic charge is 4 and it is reached in a three step cascade (Guillemin et al., 2011, 2018). The length of a cascade, the number of emitted electrons and the ionic charge in the final state all grow with the atomic number and the energy of the core vacancy. Thus, the most probable ion in Auger decay cascades of the $Kr^{+}(2p_{3/2}^{-1})$ and $Xe^{+}(1s^{-1})$ core-ionized states are Kr^{5+} (Morishita et al., 2006b) and Xe^{8+} (Pomplun et al., 1987), respectively.

Although the Auger decay process and Auger decay cascades are the major mechanisms by which core-ionized or core-excited states in atoms and molecules dissipate their energy, the complete de-excitation pathway comprises also photon emission and, in the case of molecules, dissociation or isomerization. Photon emission is prominent in the initial relaxation step of the 1s core vacancies, and in heavier elements it even quenches the Auger decay step (Hubbell et al., 1994). Moreover, the final state of the electronic decay often consists of electronically excited ions which can only relax further by emitting a photon. For example, all Ne^{2+} ions produced in the Auger decay of the 1s vacancy in the Ne atom are in electronically excited states, and continue decaying radiatively. Auger decay or Auger decay cascades which take place in molecules produce multiply charged molecular ions, where the positive charge redistributes throughout the molecule causing it to undergo the Coulomb explosion (Erk et al., 2013, 2014; Motomura et al., 2015; Nagaya et al., 2016b; Rudenko et al., 2017). Summing up, we see that deposition of energy by X-rays in an *isolated* atom or molecule leads to complicated relaxation pathways, whereby energy is dissipated by producing energetic electrons, photons, and positive ions, and in the case of molecules often results in their disintegration.

The picture presented above is derived from the gas phase experiments and theoretical consideration of isolated atoms and molecules. The question, however, remains by how much it will be modified if we assume that the irradiated atom or molecule is not isolated but is instead a part of a weakly bound medium: embedded in a van der Waals cluster or in a solvent such as water. The comparatively large distances (2–4 Å) and weak interactions between the constituents of such a medium suggest that its effect should amount to a small perturbation of the original cascade. However, such a medium is also the source of relatively weakly bound electrons, and this leads to the appearance of surprisingly potent interatomic decay processes. These are interatomic Coulombic decay (ICD) (Cederbaum et al., 1997) and electron-transfer mediated decay (ETMD) (Zobeley et al., 2001). These processes complement and sometimes modify the Auger cascades in such a way that the knowledge of a relaxation pathway in the gas phase is of limited utility for understanding the relaxation of the same species in a chemical medium. In this chapter we introduce in detail the interatomic decay processes and discuss their role in a number of relaxation cascades observed experimentally and studied theoretically in atomic and molecular clusters. As will be seen in the following, these processes are responsible for a fast redistribution of energy and charge in weakly bound clusters and solutions. Within a few tens of femtoseconds they produce multiple highly reactive chemical species and set the stage for fast nuclear dynamics; to a large degree they determine the earliest response of the medium around the photoabsorption site to the impact of an X-ray photon.

5.2 INTERATOMIC DECAY PROCESSES

5.2.1 Interatomic Coulombic Decay (ICD)

Interatomic Coulombic decay or ICD is a process, whereby an electronic excitation localized on an atom or a molecule relaxes by transferring the excess energy to a weakly bound neighbor and ionizing it (see Fig. 5.1). Originally, the process was predicted theoretically to occur in the hydrogen bonded HF dimers and trimers following the creation of a 2s vacancy in the fluorine atom (Cederbaum et al., 1997). The Auger decay is energetically forbidden for the inner-valence ionized HF due to the high energy cost of accommodating two holes on the same monomer. However, the ICD threshold lies lower, since in the ICD final state the positive charge is delocalized over two HF molecules. Subsequent theoretical investigations showed that the ICD lifetime in the HF dimer is extremely short and lies in the range of 10 to 14 fs (Santra et al., 1999), i.e., not far from the 3.3 fs Auger lifetime of the 1s vacancy in the fluorine. This surprisingly high efficiency of ICD is due to

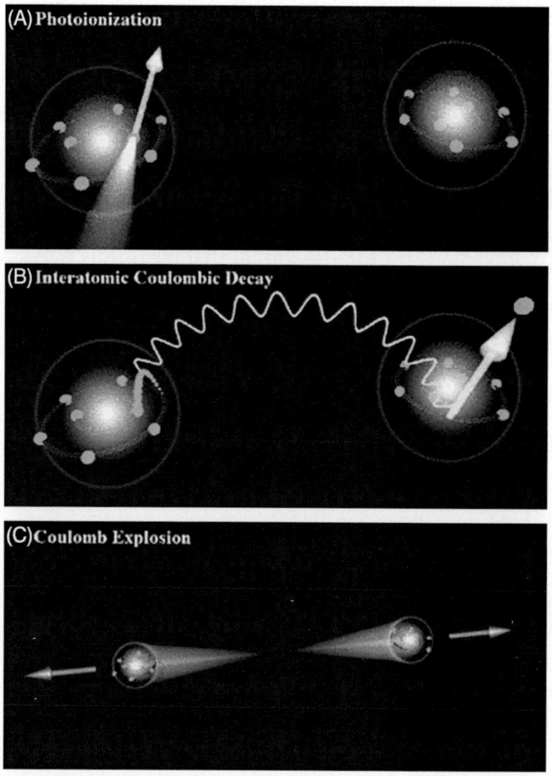

FIG. 5.1 Schematic description of ICD following the 2s ionization of a Ne atom in the Ne dimer. Absorption of a VUV photon leads to the emission of a 2s electron from the corresponding Ne atom and to the creation of the excited $Ne^+(2s^{-1})$ ion. This ion relaxes by transferring the excess energy to the Ne neighbor and ionizing it. After ICD is complete the doubly ionized $(Ne^+)_2$ dimer undergoes Coulomb explosion. From Jahnke et al. (2004); with permission.

several factors. First, the decay is driven by the correlation between electrons localized on different molecules. The Coulomb potential responsible for the coupling is long range and remains effective even though the distance between the neighboring F atoms for the equilibrium geometry of the dimer is about 2.7 Å. Second, the initially excited state is embedded in the electronic continuum, and therefore, it is always in resonance with the final states so that no vibrational motion is necessary to facilitate energy transfer. Third, in larger clusters the initial excitation is coupled to several neighbors increasing the number of ICD channels. Indeed, it was shown by Zobeley et al. (1998) that the ICD lifetime in the HF trimer was 4.5–6.5 fs, and it became even shorter in the tetramer.

Van der Waals clusters represent a different class of weakly bound systems where ICD can be observed. Typical representatives are the rare gas clusters which offer a flexible and tractable model of condensed matter. Their average size can be continuously varied in an experiment from two to many thousand atoms. They might contain a single atomic species (pure clusters) or two different rare gas atoms (mixed clusters). Dopants, such as metal atoms, molecules or small metal clusters, can be adsorbed on their surfaces or absorbed into the bulk (Rosso et al., 2007; Toennies and Vilesov, 2004). Importantly, weak interactions in such clusters facilitate both the interpretation of experimental results and their theoretical modeling which motivated the intensive investigation of these clusters by the electron spectroscopy methods (Rühl, 2003).

It was a theoretical demonstration that ICD in the rare gas van der Waals clusters occurs with high efficiency which paved the way for its experimental investigation. First *ab initio* computations for such clusters showed that the 2s vacancy of Ne in small Ne clusters decays by ICD (Santra et al., 2000). The ICD lifetime in the dimer was found to be about 100 fs, and it became considerably shorter as the number of Ne neighbors around the 2s-ionized Ne increased (Santra et al., 2001). These results allowed to estimate ICD lifetimes for a bulk and surface Ne atoms in a larger cluster as 3 and 10 fs, respectively. Moreover, the theory predicted that ICD would be manifested in the dimer by the appearance of electrons with kinetic energy below 2 eV and by a pair of Ne^+ ions with the total kinetic energy of about 4.5 eV (Scheit et al., 2004).

Experimental confirmation of ICD was first given by the publication of the electron spectra of large Ne clusters (Marburger et al., 2003; Öhrwall et al., 2004). They unequivocally showed the presence of a peak at about 1.2 eV which was absent in uncondensed Ne atoms and appeared in coincidence with 2s photoelectrons (Marburger et al., 2003). Moreover, the experimentalists extracted ICD lifetimes from the photoelectron spectra obtaining the values of 6 fs for the bulk atoms and 30 fs for the surface atoms in good agreement with the theory (Öhrwall et al., 2004). However, more direct experimental demonstration of ICD was achieved in Ne dimers using the COLTRIMS (cold target recoil ion momentum spectroscopy) technique (Ullrich et al., 2003) which allows coincident observation of all charged particles produced in the decay and the determination of their momenta and energies. In the experiment by Jahnke et al. (2004) ICD was observed as a slow (< 2 eV) electron and two Ne^+ ions detected in coincidence and shown to share energy following 2s ionization of Ne (see Fig. 5.2).

The coincidence experiments have remained the most powerful approach for the experimental study of the interatomic decay processes. Two observables are particularly important for unraveling the details of ICD: electron energy spectra, and sum kinetic energy or kinetic energy release of the nuclei (KER). In dimers slow electrons and two positive ions produced in the interatomic decay share energy (see Fig. 5.3). Therefore, two dimensional maps of the electron kinetic energy vs. KER allow to identify the ICD events as diagonal lines along which the total energy of all charged particles is conserved (see Fig. 5.2). If the repulsive potential energy curve in the final state is known (and it often has a simple a/R form), the KER spectrum, obtained by integrating over electron energies, provides information of the relative probability of decaying at a certain interatomic distance R. From the electron spectrum one learns which electronic transitions are involved in the decay. The sum of the electron and KER spectra, which is approximately equal to the energy difference between the decaying and the final state at infinite R, helps to identify the ICD states if the final states are known. The use of coincidence techniques in rare-gas dimers allows to investigate in great detail not only ICD but also a whole range of other interatomic electronic processes, such as electron-transfer mediated decay (ETMD) considered in the next section, and different types of charge transfer mechanisms. The picture provided by such techniques becomes less clear in larger clusters where inelastic scattering of slow electrons off atoms or molecules and energy transfer to unobservable neutral fragments make the experimental spectra more difficult to interpret. Similarly, a more tractable picture of the electronic structure and dynamics is obtained in *ab initio* calculations in dimers than in larger clusters. These considerations make the rare gas dimers benchmark systems in the study of the interatomic decay processes.

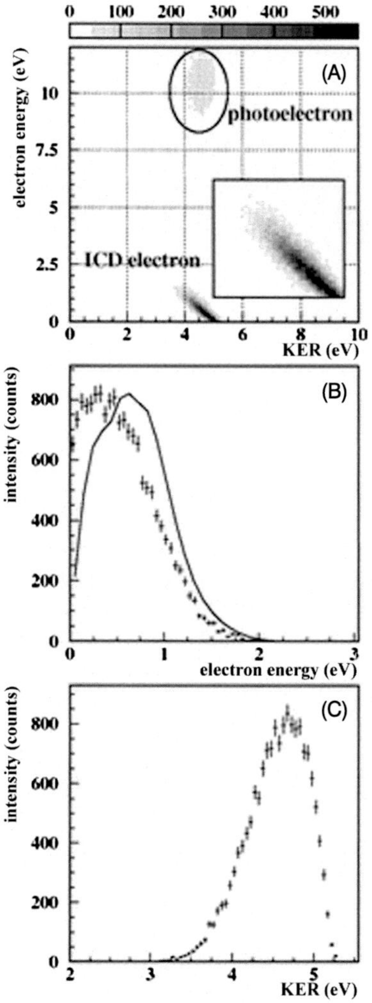

FIG. 5.2 (A) Map of the events where an electron is coincidentally observed with two Ne^+ ions following ionization of Ne_2 plotted as the function of the KER of the ions and the electron energy. ICD signature is visible as the diagonal streak in the lower half of the plot. The sum energy along the streak is approximately constant and equal to 5.1 eV – the energy of the system in the ICD state. (B)–(C) ICD electron and KER spectra obtained by integrating the map (A) along the complementary coordinate. Solid curve in (B) shows a theoretical spectrum. From Jahnke et al. (2004); with permission.

Already in the original publication by Cederbaum et al. (1997) it was asserted that ICD is a general phenomenon not limited to the decay of inner-valence ionized states. Indeed, any localized excited electronic state whose excess energy is larger than the ionization potential of the surrounding chemical medium will undergo

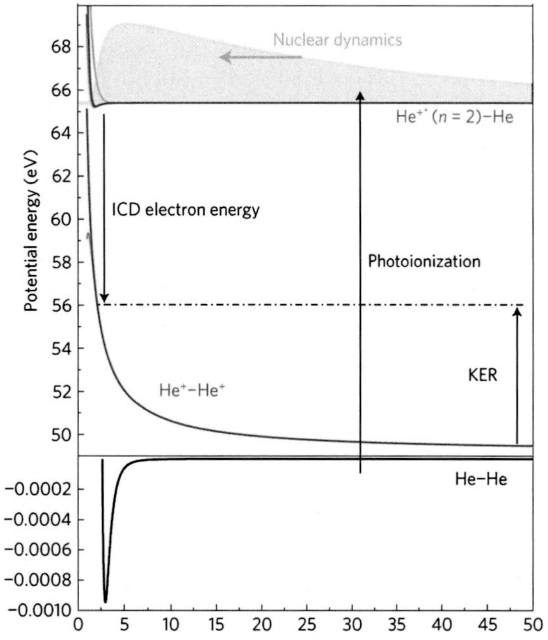

FIG. 5.3 Schematic representation of the nuclear dynamics during ICD as illustrated by ICD in the He dimer. The initial vibrational wavepacket is vertically transferred into the decaying He^{+*}–He state in a photoionization event. The wavepacket propagates in the ICD state while decaying into the repulsive He^+–He^+ final state and emitting an ICD electron of specific energy. The positive ions in the final state accelerate gaining kinetic energy which is measured in the experiment together with the energy of the ICD electron. The sum of the kinetic energy release of the nuclei (KER) and the ICD electron energy is constant and in the rare gas dimers approximately equal to the energy difference between the decaying and final state at the infinite interatomic separation. From Sisourat et al. (2010); with permission.

ICD. A striking demonstration of this fact, and also of the high potency of ICD to transfer energy over large distances, was achieved in the He dimer (Havermeier et al., 2010; Sisourat et al., 2010) (see Fig. 5.3). This system is extremely weakly bound with average interatomic distance being 52 Å. Ionization of He with photons, whose energy is larger than 66 eV, leads to the appearance of the ionized excited He^{+*} species, which then decay via ICD with the neighboring He atom.

Resonant excitation of neutral clusters can also lead to their fast de-excitation via ICD provided that it is energetically allowed. ICD was demonstrated both for the case of the inner-valence (Barth et al., 2005; Aoto et al., 2006; Gokhberg et al., 2006; Knie et al., 2014) and outer-valence (Gokhberg et al., 2005; Najjari et al., 2010; Trinter et al., 2013b) excited states. An interesting situation arises in clusters irradiated by intense laser light whose frequency lies close to the resonance frequency of the constituent atoms. Absorption of many photons leads to multiple excitations within the cluster, and ICD occurs between two excited atoms (Kuleff et al., 2010; Demekhin et al., 2013; Takanashi et al., 2017). The energy released in the de-excitation of one atom, while insufficient to ionize a ground state neighbor, is usually more than enough to remove a weakly bound Rydberg electron from an electronically excited atom. This mechanism, which we discuss in detail later in the chapter, is very important for cluster ionization in moderately strong laser fields.

A particularly interesting aspect of studying an interatomic decay process is the determination of its rate. As we noted above, in the case of ICD it was found already at the very beginning that it was an ultrafast process whose lifetime lies in the femtosecond range. However, accurate determination of ICD rates for specific systems is a difficult task both theoretically and experimentally. Difficulties in theory stem from the fact that the ICD state is a metastable resonance state which lies among the continuum of the final states. The latter cannot be directly handled by the efficient electronic structure methods, whose numerical implementation relies on the square-integrable Gaussian-type basis functions. The experimental approaches devised to determine the ICD rate do so by either observing the broadening of spectral lines due to the finite lifetime of the ICD state (Öhrwall et al., 2004, 2010; Trinter et al., 2013b) or by clocking the decay in time domain (Schnorr et al., 2013; Trinter et al., 2013a). Experimentally, accurate determination of the ICD rates from the widths of the peaks in the electron spectra is hampered by the limited energy resolution of the electron detectors. The difficulties in synchronizing the pump and the probe pulses impair the accuracy when it is the temporal evolution of an ultrafast process which is being observed.

The computational difficulties can be overcome in two ways. One is to use analytic continuation of the electronic Hamiltonian into the complex plane (Reinhardt, 1982; Moiseyev, 1998). The approach which is particularly suitable for use with the *ab initio* quantum chemistry methods consists in adding a suitable local complex potential to the electronic Hamiltonian, $\hat{H}_{\mathrm{el}}$, (Riss and Meyer, 1993, 1995; Santra et al., 1999; Santra and Cederbaum, 2002; Thomas and Robin, 2001; Gokhberg et al., 2005; Vaval and Cederbaum, 2007; Kunitsa et al., 2017; Jagau et al., 2014). The decaying states are then identified as certain solutions of the resulting complex symmetric eigenvalue problem. They have complex energy eigenvalues of the form $E = E_0 - \mathrm{i}\Gamma/2$, where Γ is the decay width, which is proportional to the decay rate and inversely proportional to the lifetime of the decaying state $\tau = \hbar/\Gamma$. Since the addition of complex potential discretizes the continuum, the use of L^2 basis sets becomes admissible (Riss and Meyer, 1993).

Alternatively, one may use the approach due to U. Fano (1961) and H. Feshbach (1958, 1962) to construct the decaying and the final states using the real symmetric electronic Hamiltonian, $\hat{H}_{\mathrm{el}}$. The width is then given as

$$\Gamma = 2\pi \sum_{\beta} \left| \langle \Phi | \hat{H}_{\mathrm{el}} | \chi_{\beta,\epsilon} \rangle \right|^2 \tag{5.1}$$

where $|\Phi\rangle$ is the decaying state, $|\chi_{\beta,\epsilon}\rangle$ is a continuum state of the free electron of energy ϵ in the decay channel β. Since the continuum states are constructed using standard *ab initio* methods which employ Gaussian basis sets, they have wrong normalization. Their renormalization to energy, and the following computation of the decay width can be done by Stieltjes imaging technique (Reinhardt, 1979; Langhoff, 1973; Hazi, 1978; Averbukh and Cederbaum, 2005; Kolorenč et al., 2008; Kopelke et al., 2011). At large interatomic distances the expression in Eq. (5.1) can be simplified and, provided the transition on the excited species is dipole allowed, the ICD width acquires a form characteristic for the dipole–dipole energy transfer (Averbukh et al., 2004)

$$\Gamma = \frac{3\hbar}{4\pi} \left(\frac{c}{\omega}\right)^4 \frac{\tau_A^{-1} \sigma_B}{R_{AB}^6} \tag{5.2}$$

where τ_A^{-1} is the radiative decay rate of the excited species, and σ_B the photoionization cross section of the neighbor evaluated at the frequency of the transferred "virtual" photon, ω. This asymptotic formula is usually accurate at large interatomic distances where orbital overlap between the excited species and its neighbors becomes negligible (Averbukh et al., 2004). Experience shows that it represents a *lower bound* to the ICD width at the distances at which the decay actually occurs.

For completeness we mention here that the ICD lifetimes can be directly deduced from the time-evolution of the electron density, which is traced using the multielectron wavepacket propagation method (Kuleff et al., 2005; see also Marquardt and Quack, 2020, Chapter 1 of this book). This technique was successfully used for studying the ICD process in NeAr initiated by the 2s ionization of Ne (Kuleff and Cederbaum, 2007).

If more than one neighbor is present, in the first approximation the total decay width can be found by summing up the widths of the individual excited species–neighbor pairs. In this case the total width $\Gamma \propto N$, where N is the number of equivalent neighbors. This pairwise approximation works well in van der Waals clusters where the neighbors interact weakly with each other (Kryzhevoi et al., 2007; Vaval and Cederbaum, 2007). However, in microsolvated clusters with their shorter distances and stronger interactions between the neighbors it can be off by a few tens of percentage points (Stumpf et al., 2016a).

The *ab initio* calculations carried out to date show that the ICD lifetimes in rare gas dimers are tens to hundreds of femtoseconds. Formula in Eq. (5.2) indicates that faster decay can be achieved by decreasing the interatomic distance or by adding more neighbors. The latter effect is demonstrated by ICD lifetimes in larger Ne clusters which are an order of magnitude shorter than in the Ne dimer (Santra et al., 2001; Öhrwall et al., 2004), and in $Ne@C_{60}$ where the ICD lifetime of the $Ne^{+}(2s^{-1})$ state was found numerically to be about 2 fs (Averbukh and Cederbaum, 2006). The effect of shortening interatomic distances and of polarizing the neighbor is particularly pronounced in microsolvated metal cations, such as Na^{+} and Mg^{2+}, where the ICD lifetimes were found to be a few femtoseconds only (Öhrwall et al., 2010; Stumpf et al., 2016a).

Despite the instrumental difficulties, ICD lifetimes were measured in a number of systems. Björneholm et al. extracted ICD widths from the photoelectron spectral lines which correspond to the ICD states either by fitting the Voigt profiles to them (Öhrwall et al., 2004), or by comparing their widths with the spectral linewidths which correspond to electronically stable states under the assumption of constant homogeneous broadening in the electron spectrum (Öhrwall et al., 2010). The results in large Ne clusters (Öhrwall et al., 2004) and microsolvated clusters of metal cations (Öhrwall et al., 2010) are in good agreement with the theory (Santra et al., 2001; Stumpf et al., 2016a). The ICD lifetime of the 2s vacancy in the Ne dimer was measured by Schnorr et al. in a pioneering pump–probe experiment using XUV FEL light (Schnorr et al., 2013). The global experimental lifetime of 150 ± 50 fs which they obtained matched well the theoretical value of about 100 fs at the equilibrium geometry of Ne_2. An ingenious approach by Trinter et al., called post collision interaction (PCI) streaking (Trinter et al., 2013a), allowed to observe ICD in real time in the He dimer. The technique relies on the fact that the observed kinetic energy loss by the slow photoelectron depends on the time at which the faster ICD electron is emitted. A good agreement was found between the experimental and theoretical survival probabilities as a function of time, and both show that the decay is strongly non-exponential.

The origin of this non-exponential behavior is due to two factors. First, many vibrational states in the decaying electronic state are populated in a sudden, or broadband, excitation from the initial state of the system. Second, strong variation of the electronic ICD width with the interatomic coordinate leads to large differences in the decay widths of the individual vibrational states in the ICD state. The joint experimental and theoretical investigation of ICD in photoexcited HeNe dimer showed that the ICD width may vary among the vibronic states by an order of magnitude (5 to 0.6 meV) (Trinter et al., 2013b; Jabbari et al., 2014).

The results of both theory and experiment show that ICD is a general de-excitation mechanism of localized excited electronic states in weakly bound chemical medium. Whenever energetically allowed, it usually occurs on the femtosecond timescale and is more efficient than competing relaxation processes: photon emission (Förstel et al., 2013), or nuclear rearrangements such as dissociation (Trinter et al., 2014) or proton transfer (Jahnke et al., 2010). One can expect ICD to play an important role whenever energy, sufficient to excite electrons, is absorbed by a chemical medium. While most of the discussion ahead will focus on the excitations produced as the result of photoabsorption, ICD has been also demonstrated following the impact by charged particles such as electrons (Yan et al., 2014; Ren et al., 2016, 2017; Yan et al., 2018; Ren et al., 2018) and ions (Kim et al., 2011; Titze et al., 2011).

5.2.2 Electron-Transfer Mediated Decay

In ICD the excess electronic energy is transferred between an excited atom or molecule and its weakly bound environment. However, it was shown that there exists another interatomic decay process in which electronic relaxation of an excited cation is mediated by electron transfer to the cation from its neighbor (Zobeley et al., 2001). In this electron-transfer mediated decay (ETMD) the electron is transferred from the environment to the embedded cation, while the energy released in the transfer is transferred back to and used to ionize the environment (see Fig. 5.4). One distinguishes between the ETMD(2) and ETMD(3) processes. In the former the electron donor is doubly ionized after the decay, while in the latter two neighbors are singly ionized. Originally, ETMD was investigated for inner-valence ionized states in heteroatomic rare

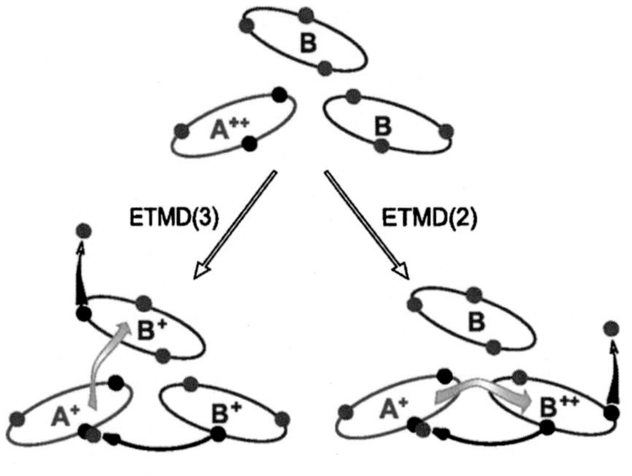

FIG. 5.4 Schematic representation of the ETMD(2) and ETMD(3) processes. A singly or multiply charged ion is created in a cluster in a prior physical process (Auger decay, photoionization, electron impact ionization, etc.). An electron is transferred from a neighbor to the ion, while the energy released in the transfer is used to ionize the donor or a second neighbor. As the result the charge on the initial ion is reduced by one, while the environment is doubly ionized. From Stumpf et al. (2013); with permission.

gas and rare gas–metal dimers (Zobeley et al., 2001; Averbukh and Cederbaum, 2005; Scheit et al., 2006). Since such initial states can also decay by ICD, and the partial decay width of ETMD was found to be two orders of magnitude smaller than that of ICD, ETMD was dismissed as a minor decay channel. However, it was noted that should there be an initial state where ICD is forbidden on energetic grounds while ETMD is allowed, the latter might become the major relaxation mechanism (Averbukh and Cederbaum, 2005). The inner-valence 3s ionized state of Ar in Ar–Kr mixed clusters (Pernpointner et al., 2008; Förstel et al., 2011) and the $Ar^{3+}(3p^{-3}\,^2D)$–Ar triply ionized state of the Ar dimer (Sakai et al., 2011) were identified as such states and were used for the first experimental observation of ETMD.

The most general class of systems where no ICD is possible, while ETMD channel remains open are atomic ions with one or several vacancies in the outer-valence shell embedded in a chemical medium. Such ions are either in their electronic ground state or in the lowest excited states whose excess energy is insufficient to ionize the environment. However, they possess large electron attachment energy so that electron transfer from a neighbor followed by the ionization of the same or a different neighbor is energetically allowed. Such ions can be efficiently produced by outer-valence ionization or in the Auger decay process. We will discuss the Auger–ETMD cascade in greater detail below. However, to illustrate the ETMD process we briefly consider two systems where it proceeds directly following the ionization of a single electron.

An interesting example is provided by He–Mg cluster (Stumpf et al., 2014). Ionization of the He atom by a photon with energy just above its ionization potential of 24.6 eV produces He^+ in the ground electronic state. While ICD is energetically forbidden, since the ion carries no excess energy, ETMD(2) with Mg is allowed due to high electron attachment energy of He^+ (24.6 eV), which is larger than the double ionization potential of Mg (22.7 eV). As the result of ETMD, He reverts to its neutral ground state, while Mg becomes doubly ionized. It was argued that the ETMD pathway should be responsible for the dramatic enhancement of the single-photon double ionization of atoms, clusters, and molecules embedded in He nanodroplets, where the photons would predominantly ionize the He atoms which in turn would undergo ETMD with the dopant. Indeed, the efficient double ionization of Mg clusters embedded in He nanodroplets was demonstrated experimentally by LaForge et al. who showed that it takes place mostly via the ETMD pathway (LaForge et al., 2016). Similar situation arises in water solutions of LiCl where ETMD occurs following photoionization of the Li^+ ion (Müller and Cederbaum, 2005). Experimental measurements show the presence of ETMD when the respective salt solutions are irradiated by X-rays with energy above the ionization potential of Li^+, and that the structure of the Li ion's solvation shell could be reconstructed from the ETMD electron spectra (Unger et al., 2017).

ETMD has been investigated to the much lesser extent than ICD. However, the following general picture emerges from the *ab initio* calculations. ETMD is a general interatomic decay process for ions embedded in a chemical environment. It is energetically allowed whenever the electron attachment energy of the ion is larger than the double ionization potential of the environment. As the result of ETMD, the charge on the initial ion is reduced by one, while the environment is doubly ionized and slow electrons are emitted into the continuum. The ETMD width also strongly depends on the interatomic distance. However, since it is an electron transfer process, whose rate is proportional to the degree of orbital overlap between the neighbors, the width falls off exponentially with the distance between them. The ETMD lifetimes for the equilibrium geometries of rare gas and rare gas–metal clusters range from several to several hundred picoseconds. Importantly, they become much shorter due to nuclear dynamics, and are generally shorter than the characteristic lifetimes of the alternative electron transfer processes. Indeed, as the following discussion will show, ETMD is an efficient mechanism responsible for the prompt redistribution of a high positive charge localized on an atom throughout a larger chemical system.

5.2.3 Radiative Charge Transfer and Charge Transfer Through Curve Crossing

To round off the discussion of the electronic relaxation and charge redistribution mechanisms present in chemical media we briefly introduce two additional electron-transfer processes. These processes are the radiative charge transfer and the charge transfer through curve crossing. Although slower than the electronic decay discussed above, these electron-transfer processes may appear in the relaxation cascades which follow the X-ray absorption whenever electronic decay becomes energetically forbidden (Ouchi et al., 2017).

In a diatomic AB the latter process occurs in the gas phase whenever there is a suitable crossing between the potential energy curves (PECs) corresponding to the states of the $A^{q+}B$ and $A^{(q-p)+}B^{p+}$ characters (Janev and Winter, 1985). In solutions at normal temperature such charge transfer is described by the Markus–Hush model, where the Gibbs (free) energy and generalized electron transfer coordinates take the place of PECs and interatomic distances (Marcus, 1956, 1965; Hush, 1961). In both cases nuclear motion is necessary to bring the initial and final electronic states into a resonance and ensure energy conservation in the electron-transfer step. This explains the slower rate of electron transfer compared to electronic decay.

In the radiative charge transfer (RCT) the excess energy which is released in electron transfer is carried away by a photon. The process was originally observed between a doubly charged rare gas ions and neutral rare gas atoms in drift-tube experiments (Johnsen and Biondi, 1978). The first observation in clusters was achieved in the Ar dimer by Saito et al. (2007) who measured the KER of two Ar^+ ions which originated from the decay of the $Ar^{2+}Ar$ Auger final states. These states are electronically stable and their potential energy curves do not cross the Ar^+–Ar^+ curves, which leaves RCT as the only allowed relaxation mechanism. Later experiments showed that RCT dominates the relaxation of the $Ne^{2+}Ne$ and $Ar^{2+}Ar$ states produced in the Auger decay (Kreidi et al., 2008b), direct double photoionization (Higuchi et al., 2010), or electron impact ionization (Yan et al., 2018; Ren et al., 2016, 2017, 2014).

In all these experiments RCT was observed indirectly via coincidence measurements of slow electrons (e.g., photoelectrons) and nuclei. Recently, a direct observation of RCT photons was achieved in Ne clusters following the excitation of 1s electrons of Ne and the successive Auger decay which produces Ne^{2+} ions (Hans et al., 2018). RCT took place between Ne^{2+} and neutral Ne, and the corresponding photons were collected. Unlike emission from excited atomic ions which appears as a collection of sharp atomic lines, RCT spectra appear as a band which bears the imprint of the cluster's vibrational motion. Comparison of the measured spectrum and the shape of the emission lines with the theoretical calculations led to the conclusion that RCT is a slow process. It is slower than intracluster vibrational relaxation (Haberland, 1985) or the charge transfer via curve crossing, and its theoretical lifetimes range from 1 ns to 1 µs. Indeed, the photon emission spectra indicated that vibrationally relaxed $Ne^{2+}Ne$ dimers were first formed in the Ne cluster and only then they underwent RCT (Hans et al., 2018).

5.3 DECAY CASCADES IN WEAKLY BOUND ATOMIC AND MOLECULAR SYSTEMS

5.3.1 Auger–ICD Cascades

The above discussion of the ICD and ETMD processes shows that they are triggered whenever a localized electronic excitation or multiple positive charge is created in a weakly bound system. Both types of localized electronic states are produced in the Auger decay process. Indeed, the initial core-hole state is strongly localized on the atom which absorbs the X-ray photon. The following decay involves predominantly the electrons from the higher shells whose wavefunctions have a non-negligible overlap with the core hole (Hergenhahn et al., 2003). For a weakly bound cluster it means that the decay occurs, and the resulting final Auger states are localized predominantly on the atom or molecule which carries the initial vacancy. These states may further undergo ICD or ETMD with a neighbor in the next decay step.

The existence of an electronic decay cascade consisting of an Auger step and an ICD step was put forward by Santra and Cederbaum (2003). They considered possible electronic decay processes in the Ne dimer triggered by the creation of a 1s vacancy in one of the neon atoms. The vacancy decays mostly in a normal Auger process which populates dicationic states of the $Ne^{2+}Ne$ character where both holes reside in the valence electronic shell. Of these states about 23% correspond to the excited electronic states $Ne^{2+}(2s^{-1}2p^{-1}\,{}^1P)Ne$, and $Ne^{2+}(2s^{-2}\,{}^1S)Ne$ which carry enough excess energy to decay by ICD with Ne (Kelly, 1975). Therefore, 23% of the initial 1s vacancies will relax in the cascade which leads to the triple ionization of the dimer as exemplified by the following scheme

$$\begin{aligned} Ne^{+}(1s^{-1})Ne &\rightarrow Ne^{2+}(2s^{-1}2p^{-1}\,{}^{1}P)Ne + e_{Auger} \\ &\rightarrow Ne^{2+}(2p^{-2})Ne^{+}(2p^{-1}) + \\ &+ e_{Auger} + e_{ICD}. \end{aligned} \tag{5.3}$$

One can see from this scheme that the energy invested in the dimer by a photon is carried away by the Auger and ICD electrons, as well as by the Ne^{+} ions. The computed ICD lifetime for the $Ne^{2+}(2s^{-1}2p^{-1}\,{}^{1}P)Ne$ state was found to be 80 fs; the value comparable with that of the inner-valence ionized Ne dimer. Since both the Auger decay step and the Coulomb explosion are much faster than ICD, 80 fs define a timescale within which the dimer becomes triply ionized and disintegrates.

Precisely such an Auger–ICD cascade was first demonstrated experimentally by Morishita et al. (2006a) for a 2p ionized Ar atom in Ar dimers. In a coincident measurement of a slow electron together with the Ar^{2+} and Ar^{+} ions they identified ICD events in which these three particles share energy. From the electron kinetic energy–KER map they were able to identify the ICD state as the ionization satellite $Ar^{2+}(3s^{-1}3p^{-1}\,{}^{1}P + 3p^{-3}3d\,{}^{1}P)Ar$. Moreover, the KER spectrum showed that the decay occurred mostly at the equilibrium geometry. This result indicates that little nuclear dynamics take place during the decay, and, consequently, that the ICD lifetime is less than the characteristic vibrational period in the decaying state, or a few hundred femtoseconds.

Following these two publications demonstrating its existence, the Auger–ICD cascade was investigated in a number of rare gas dimers both theoretically and experimentally. The results of coincidence experiments in the mixture of Ar_2, ArKr, and Kr_2 dimers obtained in the coexpansion of the Ar and Kr gases were reported in (Morishita et al., 2008; Ueda et al., 2008). The photon energy was set at 263 eV so that it lied both above the argon's 2p and krypton's 3d ionization thresholds. Following the irradiation of the dimer AB the $A^{+}B^{+}$ and $A^{2+}B^{+}$ ionic pairs were observed in coincidence with slow electrons. The $A^{2+}B^{+}$ pairs originate from the Auger decay of the initial core vacancy followed by ICD. The KER spectra corresponding to the $Ar^{2+}Ar^{+}$, $Ar^{2+}Kr^{+}$, $Kr^{2+}Ar^{+}$, and $Kr^{2+}Kr^{+}$ pairs appear as nearly symmetric Gaussian peaks. This indicates electronic decay which is much faster than the vibrational motion so that the initial distribution of interatomic distances given by the vibrational wavefunction in the neutral ground state remains unchanged throughout the decay. Another observation was that the electron spectrum coincident with the $Ar^{2+}Kr^{+}$ pair was shifted by 2 eV towards higher energies relative to the spectrum coincident with the $Ar^{2+}Ar^{+}$ pair. This shift towards higher energies is due to the fact that the excited electronic state on Ar which relaxes via ICD is the same in both dimers, while the ionization potential of Kr is about 2 eV lower than the ionization potential of Ar. One can see that the ICD electron spectrum obtained in the decay of a specifically prepared electronic state reflects the nature of the nearest neighbors and can be used for structural analysis of clusters.

A detailed investigation of the relaxation pathways of core-hole states observed experimentally and analyzed theoretically in the Ne_2 dimer allowed to ascertain the role of the Auger–ICD cascade among a host of possible relaxation cascades. The decay of the 1s vacancy in the Ne dimer was shown to produce Ne^{+}/Ne^{+} and Ne^{2+}/Ne^{+} pairs at the ratio 3 to 1 (Kreidi et al., 2008b; Stoychev et al., 2008). The decay into the "symmetric" $Ne^{+}Ne^{+}$ channel occurs mostly via the Auger decay followed by an electron transfer process. The PECs of the lowest electronic states populated in the Auger decay $Ne^{2+}(2p^{-2}\,{}^{1}D, {}^{1}S)Ne$ lie isolated in the spectrum, and in the absence of suitable crossings decay via RCT. The state $Ne^{2+}(2s^{-1}2p^{-1}\,{}^{3}P)Ne$ decays by transferring charge to the neighbor via the crossing with $Ne^{+}(2p^{-2}3d\,{}^{3}P)Ne^{+}$ states. In the "asymmetric" $Ne^{2+}Ne^{+}$ channel electron kinetic energy vs. KER spectra show strong lines which can be explained by the interatomic decay of energetic states populated in the Auger decay step: $Ne^{2+}(2s^{-1}2p^{-1}\,{}^{1}P)Ne$, and $Ne^{2+}(2s^{-2}\,{}^{1}S)Ne$. The former decays either by ICD to the $Ne^{2+}(2p^{-2}\,{}^{1}D)Ne^{+}$ or by the slower exchange ICD process[1] to the $Ne^{2+}(2p^{-2}\,{}^{3}P)Ne^{+}$ state. The latter decays by exchange ICD to $Ne^{2+}(2p^{-2}\,{}^{3}P)Ne^{+}$ or by ETMD in $Ne^{2+}(2p^{-2})Ne^{+}(2s^{-1}\,{}^{1}S)$ state. The expected share of ICD-active states produced in the decay of the 1s vacancy is 23% and matches well the experimental ratio of the decay into "symmetric" and "asymmetric" channels.

Thus, the results in the Ne dimer show that the Auger step might be followed by an ICD, ETMD or electron transfer step; the cascade culminates in Coulomb explosion of the dimer and the energy is shared between the electrons, nuclei and/or photons. Moreover, they establish an order of efficiency among different processes

[1] In exchange ICD (Jahnke et al., 2007) an electron is transferred to the excited system from a neighbor and an electron is emitted into continuum from the excited system. Exchange ICD appears when the radiative transition on the excited species becomes forbidden, for example, due to the spin selection rules, which bars energy transfer. Since electron transfer is involved in exchange ICD the latter is usually slower than its direct counterpart.

which follow the Auger step: ICD, ETMD, electron transfer through curve crossing, RCT. So that a particular interatomic process takes place only if the faster processes are energetically forbidden. The picture remains essentially the same for the mixed NeAr clusters (Demekhin et al., 2009; Ouchi et al., 2011, 2017); the difference being that lowering of the ionization potential of the neighbor allows ICD for some Auger final states, e.g., $Ne^{2+}(2s^{-1}2p^{-1}\,{}^3P)$, for which it is forbidden in Ne_2. Also the measurement of the photoelectron angular distributions in the "symmetric" and "asymmetric" decay channels of Ne_2 in (Yamazaki et al., 2008) allowed to answer the question whether the initial 1s core hole is localized on one Ne atom or delocalized over both of them due to the inversion symmetry. Indeed, in the $Ne^{2+}Ne^{+}$ channel which appears due to the Auger–ICD cascade such angular distribution relative to the dimers axis lacks the inversion symmetry, which indicates a degree of the core-hole localization.

One could glean more detailed information on the Auger–ICD cascade in the NeAr cluster by comparing experimental and theoretical KER spectra of the $Ne^{2+}(2s^{-1}2p^{-1}\,{}^1P)$Ar state (Demekhin et al., 2009; Ouchi et al., 2011). Their appearance as we noted above depends on whether nuclear dynamics take place during the decay and is sensitive to the ratio of the electronic decay lifetime to the vibrational period. The former rate was found to be 30 fs at the dimer's equilibrium distance, while the latter was found to be about 250 fs. The ratio of the rates suggests that practically no nuclear dynamics occur in the decaying state, and the spectrum should have the nearly Gaussian shape of the initial vibrational wavepacket, since no dynamics occur at the Auger step as well. The experimental peak, however, showed a marked tail towards higher energies which indicated the contraction of the interatomic distance during the decay (see Fig. 5.5). To obtain good correspondence between the experimental and theoretical spectra the theoretical ICD rate should be uniformly decreased by 30%. The tendency of the *ab initio* ICD rates computed by the Fano–Stieltjes method to overestimate the experimental ones by a few tens of percentage points seems to be general whenever lower level *ab initio* methods are used to obtain electronic structure (Stumpf et al., 2016a; Trinter et al., 2013a).

Increasing the cluster size or creating deeper lying core vacancies complicates the structure of the de-excitation cascades. Thus, experimental results show that removing an Ar 2p electron in Ar_3 leads not only to the production of Ar^{+}/Ar^{+} and Ar^{2+}/Ar^{+} pairs but also of three Ar^{+} ions (Liu et al., 2007). This becomes possible due to the coupling of the Auger final state

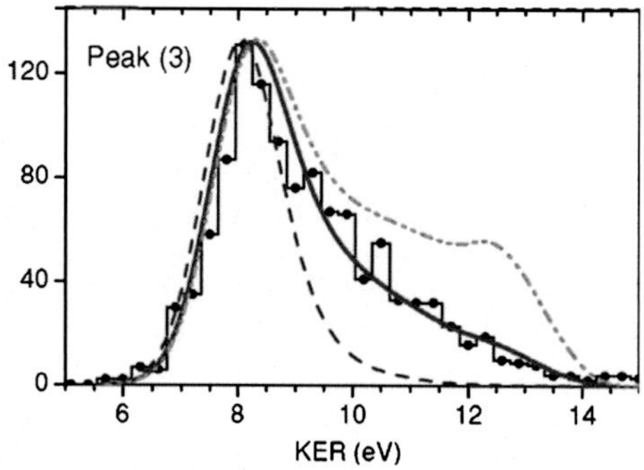

FIG. 5.5 KER peak corresponding to the decay of the $Ne^{2+}(2s^{-1}2p^{-1}\,{}^1P)$Ar state into the $Ne^{2+}(2p^{-2}\,{}^1D)Ar^{+}(3p^{-1})$ final state. (Black dotted line) experimental curve; (blue broken line) theoretical KER peak obtained using the *ab initio* ICD width $\Gamma(R)$; (red solid line) theoretical KER peak obtained with $\Gamma(R)$ scaled by factor 0.4; (green double dotted line) theoretical KER peak obtained with $\Gamma(R)$ scaled by factor 0.25. Note the appearance of the tail and the shoulder at larger energies as the ICD width is decreased, and the nuclear wavepacket has sufficient time to propagate towards shorter interatomic distances during the decay. The plot also illustrates the way to determine the error of the theoretical computations of a decay width without having to measure it directly. From Ouchi et al. (2011); with permission.

$Ar^{2+}(3s^{-1}3p^{-1})$–Ar_2 to the dicationic two-site satellite states $Ar^{+}(3p^{-2}nl)$–Ar^{+}–Ar, which in their turn decay by ICD. This result indicates that in larger clusters the Auger–ICD cascade may involve two or more neighbors in addition to the situation discussed above where only one neighbor was ionized. The study of photofragmentation of Ar dimers which follows the removal of the 1s electron demonstrates that the de-excitation cascades become more complicated when the photon energy is increased (Keshavarz et al., 2014). In the dimer, where the excited atom has a single neighbor, the cascades starting with two or more Auger steps produce multiply charged Ar^{q+} $(q = 1 - 7)$ ions and culminate with a single interatomic step. We will see later that in larger clusters the accumulation of high positive charge on an atom as the result of an Auger cascade triggers a sequence of interatomic decay steps.

5.3.2 Resonant Auger–ICD Cascades

Core ionization followed by an Auger–ICD cascade has a property which can be important for controlling radiation damage or for spectroscopic determination of composition and structure of weakly bound systems. Core orbitals do not participate directly in chemical bonding and retain their atomic character even in covalently bound systems. Therefore, core-ionized states are localized on the atom which absorbed the X-ray photon. Moreover, in weakly bound systems the positive charge produced in the Auger decay remains localized on the same species (Kreidi et al., 2008a). When ICD follows, it occurs mostly with the nearest neighbor and results in its ionization and the emission of a slow electron whose energy depends on the initial excitation and the nature of the neighbor. In radiotherapy this site-selectivity of X-ray absorption is utilized for delivering the radiation dose at the location where the damage to cancer cells is maximized, e.g., to a heavy atom marker attached to their DNA (Howell, 2008). The photoabsorption by the marker results first in an Auger cascade and the emission of Auger electrons which cause molecular damage throughout the system. Importantly, the ICD steps which follow the Auger decay damage the environment close to the photoabsorption site first by directly ionizing it, and additionally by producing slow reactive electrons which were demonstrated to be highly efficient in causing bond breaks (Martin et al., 2004). Also, the emission of an ICD electron with characteristic energy from a neighbor provides information as to the chemical composition and structure of the weakly bound system (Lundwall et al., 2007) close to the selected absorption site.

A major difficulty in utilizing this site-selectivity of the Auger–ICD cascade lies in the relatively low number of ICD-active electronic states populated in the Auger decay. We have seen that in the case of the Ne dimer only about 25% of the final Auger states may undergo ICD; the rest of the states decay either via ETMD (see next section) or in slow charge transfer processes (Kreidi et al., 2008b). This situation can be reversed if in the X-ray absorption a core electron is not removed from the system, but rather is excited into an unoccupied orbital. Such core-excited states undergo resonant Auger (RA) decay (Armen et al., 2000; Ueda, 2006), which proceeds along two pathways: the dominant spectator decay, or the minor participator decay. In the spectator decay the Auger process involves only the electrons in the occupied orbitals, so that one electron fills the initial core vacancy and another is emitted into the continuum, while the initially excited electron remains in the space of the unoccupied orbitals. In the participator decay the initially excited electron takes part in the process. It either recombines with the core hole, while another electron is emitted into the continuum, or leaves the system following a valence-to-core transition. In the neutral atoms or molecules the spectator Auger decay populates mostly ionization satellite states, i.e., ionic states of two-hole one-particle character; the participator decay results in the ionic main states, i.e., one-hole states. Ionization satellites usually have enough excess energy to ionize any neutral neighbor, therefore, one may expect a large number of ICD events to follow the resonant Auger step in a chemical medium.

In addition to populating a significant proportion of ICD-active states, core excitation and resonant Auger decay have other important differences from core ionization and normal Auger decay. First, resonant core excitations have much larger dipole transition probabilities than core ionizations. Therefore, the energy is delivered to a specific location in a cluster or in a solution more efficiently. Second, by selecting specific resonant frequencies one may distinguish not only between chemically different atoms, but also between identical atoms which occupy non-equivalent positions in a molecule or a cluster (Bolognesi et al., 2010). Thus, energy may be deposited with a higher selectivity than in the case of core ionization. Third, populating different core-excited states in the excitation step leads to the population of different satellite states following the resonant Auger step. Their interatomic decay, in turn, produces ICD electrons of different energies. Therefore, by selecting different core-excited states at the X-ray absorption step one may control the energies of the emitted ICD electrons. The combination of these three features suggests that the resonant Auger–ICD cascade provides the means to exercise some control over radiation damage in an extended chemical system through selecting where X-ray absorption takes place and controlling the energy of the damaging slow ICD electrons by exciting different core-excited states (see Fig. 5.6).

The cascade and its control possibilities can be illustrated by the example of a mixed rare gas dimer ArKr (Gokhberg et al., 2014; Miteva et al., 2014b). The lowest core excitations of Ar are the $2p_{3/2}4s$, $2p_{1/2}4s$, and $2p_{3/2}3d$ which lie at 244.39, 246.51, and 246.93 eV. The decay of the first two core excitations predominantly populates low-lying satellite states of $Ar^{+*}(3p^{-2}4s)$ and $Ar^{+*}(3p^{-2}3d)$ character. The resonant Auger decay of the last excitation is accompanied by a pronounced shake-up process which populates higher lying $Ar^{+*}(3p^{-2}4d)$ states (de Gouw et al., 1995). A large proportion of the states populated in the resonant Auger decay undergo ICD in the ArKr dimer, whereby the excited Ar^{+} ions de-excite and Kr is ionized. Coulomb explosion follows which allows recording KER spectra in coincidence with the ICD electron spectra.

The decay proceeds to the final state of $Ar^{+}(3p^{-1}\,{}^{2}P)Kr^{+}(4p^{-1}\,{}^{2}P)$ character whose PECs behave very nearly as $1/R$ and form a tight band at in-

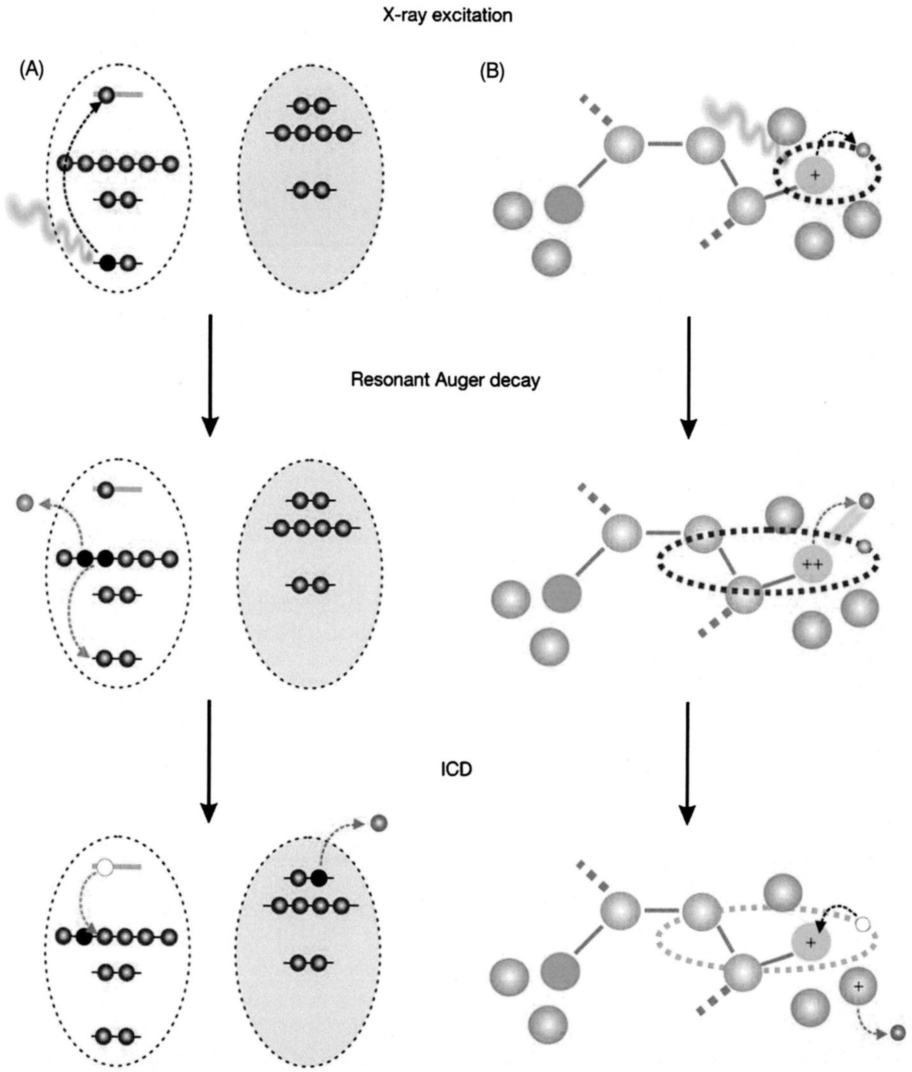

FIG. 5.6 Schematic representation of the resonant Auger–ICD cascade illustrating its site selectivity and electron energy control capabilities. Resonant X-ray absorption takes place on a specific atom populating a specific core-excited state. This state undergoes spectator resonant Auger decay emitting a fast Auger electron and populating a number of ionization satellites localized in the vicinity of the original excitation. The latter decay via ICD with the weakly bound neighbors such as solvent molecules emitting slow ICD electrons. By varying the energy of the X-ray photons one can select both the location of energy absorption and control the energy of the satellite states and, consequently, the kinetic energy of the ICD electrons. From Gokhberg et al. (2014); with permission.

teratomic distances of interest. Therefore, the electron spectra reflect different populations of the satellite states which decay by ICD. If one assumes that ICD is much faster than the vibrational motion, nuclear dynamics in the ICD states can be neglected and the interatomic distance can be fixed at the equilibrium geometry of the dimer ($R = 3.9$ Å). Then the ICD electron emitted in the RA–ICD cascade following $2p_{1/2}4s$ excitation has a kinetic energy in the 0 to 4 eV range and the spectrum consists of two peaks: a stronger at 0.7 eV and a weaker at 3 eV (see Fig. 5.7). The two peaks reflect the decay from the two groups of the low-lying satel-

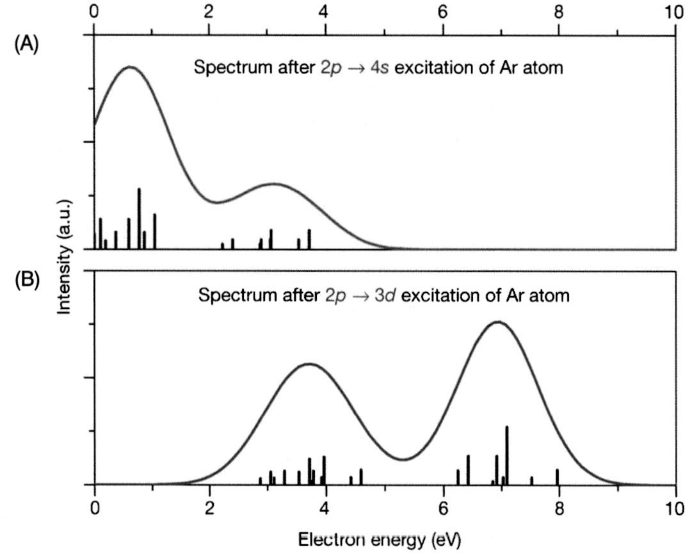

FIG. 5.7 Theoretical ICD electron spectra obtained for the RA–ICD cascade in ArKr initiated by the resonant excitation of either the $2p_{1/2}4s$ or $2p_{3/2}3d$ states of Ar. The spectra were obtained by assuming the electronic decay to be much faster than the vibrational motion and, therefore, takes place at the equilibrium geometry of the dimer. Control of ICD electron energies is possible in this system since the resonant Auger decay of the two core-excited states populates different groups of ICD states. The latter have markedly different energies, and their subsequent decay by ICD produces electrons of differing energies. From Gokhberg et al. (2014); with permission.

lites: one comprising the $Ar^{+*}(3p^{-2}[^3P,^1D]4s)Ar$ and $Ar^{+*}(3p^{-2}[^3P]3d)Ar$ states, while another includes the $Ar^{+*}(3p^{-2}[^1S]4s)Ar$ and $Ar^{+*}(3p^{-2}[^1D]3d)Ar$ states which lie at higher energy. Exciting Ar into the $2p_{3/2}3d$ state which lies only 0.4 eV above the $2p_{1/2}4s$ state populates higher lying satellite states of Ar which decay into the same manifold of final states. This results in faster ICD electrons and the ICD electron spectrum which consists of two peaks which lie at 3.5 and 7 eV (see Fig. 5.7), and can be attributed to the respective ICD of the $Ar^{+*}(3p^{-2}3d)Ar$ and $Ar^{+*}(3p^{-2}4d)Ar$ states. Thus, the energy of the emitted electron can be controlled by tuning the energy of the X-ray photon. Since the KER spectra reflect the distances at which the decay takes place, assuming the absence of nuclear dynamics it consists of a single peak at 3.7 eV.

The effects of nuclear dynamics and of the neighbor's IP on the ICD spectra were studied by Miteva et al., Miteva et al. (2014a, 2014b) in Ar_2 and ArKr using *ab initio* methods. The calculations showed that the low-lying $Ar^{+*}(3p^{-2}4s)Ar$ and $Ar^{+*}(3p^{-2}3d)Ar$ satellites have ICD lifetimes at the equilibrium geometry of Ar_2 between 28 fs and 1.35 ps, while the ICD lifetimes of the higher lying $Ar^{+*}(3p^{-2}5s)Ar$ and $Ar^{+*}(3p^{-2}4d)Ar$ states lie between 0.6 and 6.6 ps. Similar calculations in ArKr produced the ICD lifetimes for the low lying satellites between 15 and 670 fs, and for the higher lying states between 0.4 and 3.3 ps. The increase in the ICD lifetime with the principal quantum number of the excited electron is analogous to the increase of atomic autoionization lifetimes (Codling et al., 1967; Fano and Cooper, 1968), while the difference between the lifetimes of different multiplets within each group of ICD states is correlated with the overlap between the hole density and the density of the excited electron.

Characteristic vibrational period for the lower ICD states is between 0.6 and 1.3 ps in Ar_2 (290 fs and 2.2 ps in ArKr), while for the higher ICD states it is between 180 and 250 fs ($\approx$ 300 fs). This shortening of the vibrational period is due to the increased depth of the PEC in the higher excited states, which in turn can be explained by the diminished repulsion between the Rydberg electron and the neighboring neutral atom (Miteva et al., 2014c). Thus, we see that ICD of the lower energy satellites occurs almost instantaneously with the nuclei barely having time to move. In the case of the higher energy satellites ICD is slower than the nuclear dynamics. Thus, the initial nearly Gaussian vibrational

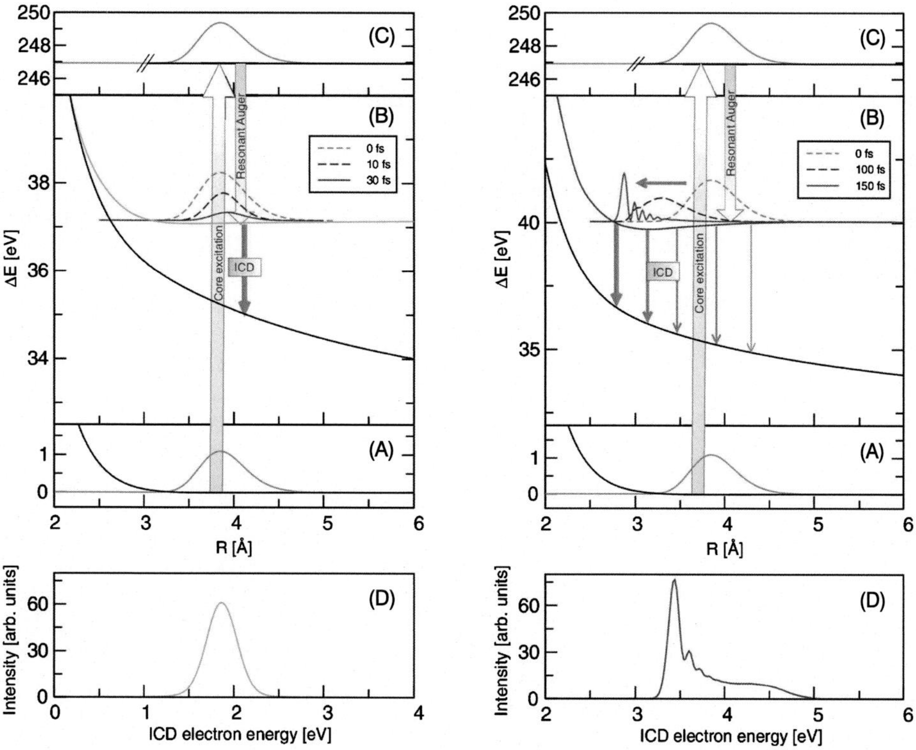

FIG. 5.8 Illustration of the impact the nuclear dynamics have on the resonant Auger–ICD cascade and on the ICD spectra in Ar_2. Due to their very short duration both the core excitation and the resonant Auger decay process can be considered to occur instantaneously. Therefore, the initial vibrational wave packet is transferred vertically from the neutral initial state to the ICD state. If the ICD rate is much larger than the vibrational frequency the wavepacket does not have time to move on the PEC of the ICD state and is vertically transferred to the dissociative PEC of the final state (left panel). The resulting electron spectrum appears as a nearly Gaussian peak (corresponding KER spectra are the mirror image of the electron ones Chiang et al., 2011). If the ICD rate is smaller than or comparable to the vibrational frequency, nuclear dynamics accompanies the decay which takes place mostly at the inner turning point where there is a considerable probability of finding the system and the ICD rate is the largest (right panel). The electron spectrum acquires a strongly asymmetric shape with a maximum corresponding to the decay at the inner turning point. From Miteva et al. (2014a); with permission.

wavepacket moves on the decaying PEC and becomes highly asymmetric with a pronounced maximum at the classical inner turning point (see Fig. 5.8). As the result, ICD mostly takes place at the inner turning point of the corresponding potential energy curve, where, in addition to the large probability of finding the system, the decay rate is the largest. This leads to the shift of the corresponding KER and ICD electron peaks towards higher and, respectively, lower energies compared to the case of the nuclei frozen at the ground state equilibrium distance (see Fig. 5.8).

In addition, comparison of the ICD thresholds in Ar_2 and ArKr shows that choosing a chemically "softer" neighbor with lower ionization potential leads to two effects. First, the decay of the ICD states generates faster electrons. Second, more ICD channels become open. Thus, the appearance of the electron spectra close to 0 eV (or at the ICD threshold) changes little, while the spectrum as the whole extends towards higher energies and its integrated intensity increases. It also means that in the presence of "softer" neighbors the share of the RA–ICD cascade in the de-excitation of the initial core excitation increases.

The RA–ICD cascade was confirmed experimentally in the $(N_2)_2$ and $(CO)_2$ molecular dimers (Trinter et al., 2014). The cascade was triggered by exciting 1s electrons of N and C, respectively, and slow electrons were observed in coincidence with pairs of the ground state molecular cations N_2^+ and CO^+. ICD following the resonant Auger decay was found to be the only mechanism responsible for the emission of the slow electrons. The alternative mechanism via the knock-out of electrons form the neighbor by the fast Auger electron could be ruled out due to the observed isotropic distribution of slow electrons and not the one aligned along the intermolecular axis. An interesting observation was made that the ICD states are highly excited dissociative electronic states of molecular ions. Therefore, ICD competes with molecular dissociation, which can be as short time as 20 fs in the case of CO^{+*}, after the resonant Auger step. The fact that a clear ICD signature is observed indicates that it proceeds at the comparable rate.

The coincidence experiments in the rare gas dimers again provided the clearest picture of the RA–ICD cascade. It was investigated in Ar_2, ArKr, and ArXe dimers following $2p_{3/2}4s$ and $2p_{3/2}3d$ excitations of Ar (Kimura et al., 2013b, 2013a). The dependence of the ICD electron spectra on the X-ray photon energy was clearly observed which confirmed the viability of the control mechanism. The effect of nuclear dynamics was also clearly visible in the ICD spectra, with the peaks corresponding to higher excited satellites, $Ar^{+*}(3p^{-2}5s)$ and $Ar^{+*}(3p^{-2}4d)$, shifted to smaller energies in the electron spectra and to larger energies in the KER spectra relative to their expected positions in the frozen-nuclei approximation. Similar conclusions were reached in (O'Keeffe et al., 2013) who investigated the RA–ICD cascade in the Ar_2 and ArNe clusters for a number of core-excited states of Ar which converge to the $Ar^+(2p_{3/2}^{-1})$ threshold. They showed that the ICD electron spectrum shifts to higher energies when a core electron is excited to the 3d and 4d orbitals. Moreover, they demonstrated that should the excited center have two different neighbors one could make ICD to occur either with only one or with both of them by selecting a suitable resonant core excitation. This finding offers another possibility of controlling the flow of electronic energy after the resonant Auger step in mixed clusters.

5.3.3 Auger–ETMD Cascade

As we noted above, normal Auger decay or Auger decay cascades in atoms and molecules result to a large degree in doubly or multiply charged positive ions in the ground state electronic configuration. In a weakly bound cluster or in solution the same ions will be produced surrounded by neutral neighbors due to the local nature of the Auger decay process. The pent-up Coulomb energy of these ions is released differently in atoms and molecules. While smaller molecules release some of this energy by dissociating (Nagaya et al., 2016b; Pedersen et al., 2013; Tavernelli et al., 2008), the atomic cations redistribute the localized charge via electron-transfer processes. Often they have electron attachment energies which are larger than the double ionization threshold of the surrounding chemical medium. This makes ETMD energetically allowed and leads to the fast charge redistribution and partial neutralization of the original cation.

For example, in the Ne atom the Auger decay of the 1s vacancy mostly populates $Ne^{2+}(2p^{-2}\,{}^1D, {}^1S)$ (66%) states. Attaching an electron to them, such that $Ne^+(2p^{-1})$ is formed, releases between 44 and 48 eV of energy. This energy is enough to doubly ionize a Xe atom which makes the NeXe cluster one of the few diatomic, weakly bound clusters where ETMD takes place. Taking a leaf out of the ICD book, where detailed understanding of the process had been obtained from the study of rare gas dimers, ETMD(2) after Auger decay was theoretically investigated in NeXe (Stumpf et al., 2013).

The expected process where ETMD(2) is followed by the Coulomb explosion is summarized in the following scheme

$$\begin{aligned} Ne^{2+}(2p^{-2})Xe &\rightarrow Ne^{+}(2p^{-1})Xe^{2+}(2p^{-2}) + e_{ETMD} \\ &\rightarrow Ne^{+}(2p^{-1}) + Xe^{2+}(2p^{-2}) + e_{ETMD} \end{aligned} \tag{5.4}$$

There are four decaying states (two Σ-terms, one Π-term, and one Δ-term) which originate from the $Ne^{2+}(2p^{-2}\,{}^1D, {}^1S)$ multiplets. The *ab initio* PECs of the decaying and final states of ETMD (see Fig. 5.9) show that this decay is energetically allowed at the equilibrium geometry of the dimer ($R_0 = 3.9$ Å). The corresponding ETMD lifetimes at R_0 of the decaying states in the dimer lie between 2 ps and 15.5 ps. The variation is explained by the varying degree of the overlap between the hole density on Ne^{2+} and valence electron density on Xe. Thus the Δ term, which originates from the $Ne^{2+}(2p^{-2}\,{}^1D)Xe$ multiplet, has hole density located off the molecular axis and exhibits a considerably smaller ETMD width than the Σ and Π terms where part of the hole density is located on the axis (see Fig. 5.10). The magnitude of the lifetimes indicates that ETMD will be accompanied by nuclear motion in the strongly attractive $Ne^{2+}Xe$ potential which will lead to the contraction of the interatomic distance. This and the fact that the ETMD rate decreases exponentially with the increasing distance (see Fig. 5.10) ensures that the decay will mostly take place at interatomic distances shorter than R_0.

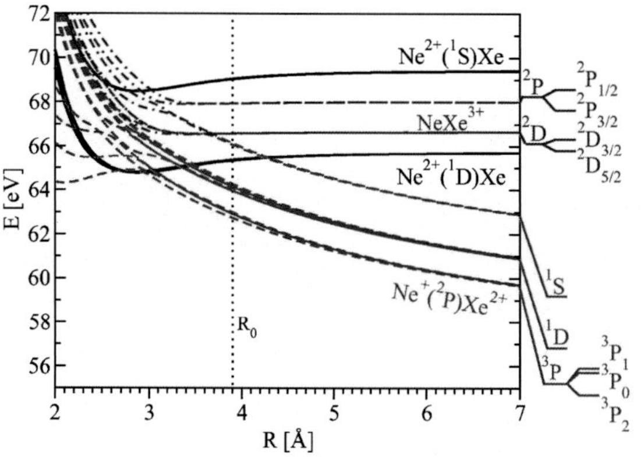

FIG. 5.9 Non-relativistic *ab initio* diabatic PECs of the decaying $Ne^{2+}(2p^{-2}\,{}^1D)Xe$ and $Ne^{2+}(2p^{-2}\,{}^1S)Xe$ states (solid lines) and adiabatic PECs of the final Ne^+Xe^{2+} states (dashed lines) of ETMD. The crossings between these two types of states indicate the positions of the ETMD thresholds. One can see that ETMD from the $Ne^{2+}(2p^{-2}\,{}^1D)Xe$ state becomes energetically forbidden at R shorter than 3.0 Å, while the decay from the $Ne^{2+}(2p^{-2}\,{}^1S)Xe$ state is allowed at all interatomic distances accessible by the dimer. The $NeXe^{3+}$ states (dashed–dotted lines), while not accessible through the ETMD process, can be populated through the avoided crossings with the ETMD final Ne^+Xe^{2+} states as the result of nuclear dynamics in the final states; R_0 marks the equilibrium distance of the NeXe dimer. From Stumpf et al. (2017); with permission.

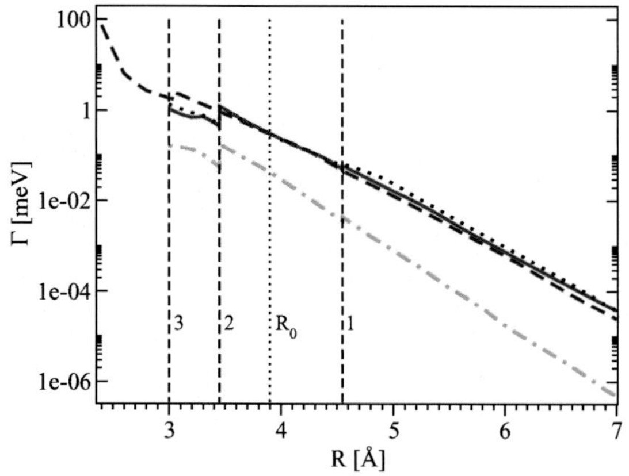

FIG. 5.10 Total ETMD widths of the terms deriving from the $Ne^{2+}(2p^{-2}\,{}^1D)Xe$ and $Ne^{2+}(2p^{-2}\,{}^1S)Xe$ decaying states: (dashed line) ${}^1\Sigma$ term derived from the $Ne^{2+}(2p^{-2}\,{}^1S)Xe$ state, (dotted line) ${}^1\Sigma$ term, (solid line) ${}^1\Pi$ term, (dashed–dotted line) ${}^1\Delta$ term derived from the $Ne^{2+}(2p^{-2}\,{}^1D)Xe$ state. The numbered dashed vertical lines enumerate different channel thresholds in the decay of the $Ne^{2+}(2p^{-2}\,{}^1D)Xe$ multiplet. The respective widths decrease at the threshold so that they remain on average constant between the equilibrium distance R_0 of the NeXe ground state (dotted vertical line) and $R = 3$ Å. Note the expected exponential behavior of the widths at $R > 5$ Å. From Stumpf et al. (2017); with permission.

The electronic energy diagram in Fig. 5.9 makes it clear that the decay will be markedly different for the $Ne^{2+}(2p^{-2}\,{}^1D)Xe$ and $Ne^{2+}(2p^{-2}\,{}^1S)Xe$ states. In the latter state ETMD is allowed at all interatomic distances at which nuclear dynamics take place during the decay. Therefore, we expect it to occur mostly at the inner turning point of the potential where the ETMD lifetime becomes as short as 10 fs! In the former state the ETMD channels become closed one after another as R decreases, so that no decay takes place between 3 Å and the inner turning point at 2.5 Å. The closing of the ETMD channels is also evident in the behavior of the corresponding ETMD widths which exhibit sudden jumps at the thresholds. As a result, the ETMD widths of each of the three multiplets which stem from the $Ne^{2+}(2p^{-2}\,{}^1D)Xe$ state remain approximately constant for the interatomic distances between R_0 and 3 Å below which ETMD becomes forbidden. As we will shortly see, the different behavior of the widths will result in the completely different appearance of the corresponding ETMD spectra.

Another short observation can be made from the electronic energy diagram in Fig. 5.9. The triply ionized $NeXe^{3+}$ states lie energetically low and interact with the ETMD Ne^+Xe^{2+} final states. These states are inaccessible in the ETMD process, since two electrons ought to be simultaneously transferred from Xe to Ne^{2+}. However, ETMD can be followed by charge transfer during the Coulomb explosion in the final Ne^+Xe^{2+} state resulting in a slow electron and a Xe^{3+} ion of 0.2 to 0.4 eV kinetic energy which can be observed in coincidence.

The difference in the behavior of the ETMD widths for the decaying states has a strong effect on the respective KER and electron spectra (see Fig. 5.11). Since the KER spectra reflect the probability to decay at a particular interatomic distance, the KER spectrum due to the ETMD of the $Ne^{2+}(2p^{-2}\,{}^1S)Xe$ state consists of a single peak at 12 eV which corresponds to the decay at the inner turning point of 2.45 Å. The KER spectrum of the $Ne^{2+}(2p^{-2}\,{}^1D)Xe$ shows that ETMD occurs with approximately equal probability anywhere between 3.8 and 3 Å, with the peaks at 8.5 and 10 eV marking the geometries where two ETMD channels become closed at 3.5 and 3 Å. This difference in the spectra is the direct consequence of the difference between the ETMD widths: one which increases dramatically towards the inner turning point in the $Ne^{2+}(2p^{-2}\,{}^1S)Xe$

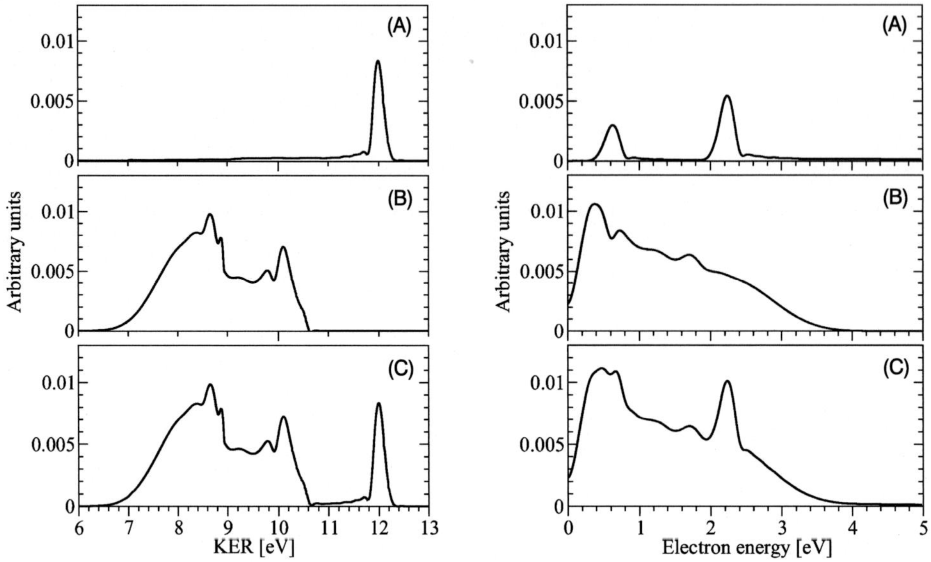

FIG. 5.11 KER (left panel) and electron kinetic energy (right panel) ETMD spectra of $Ne^{2+}Xe$ states. (A) Spectrum corresponding to the decay of the $Ne^{2+}(2p^{-2}\,{}^1S)Xe$ state. (B) Spectrum corresponding to the decay of the $Ne^{2+}(2p^{-2}\,{}^1D)Xe$ state. (C) Total spectrum obtained by summing the spectra in (A) and (B). From Stumpf et al. (2017); with permission.

state, and the others which remain nearly constant with an occasional drop at a channel threshold in the $Ne^{2+}(2p^{-2}\,{}^1D)Xe$ states. The electron spectra of the $Ne^{2+}(2p^{-2}\,{}^1S)Xe$ state show two sharp peaks at 0.6 eV and 2.2 eV which correspond to the decay into the two ETMD channels open at the inner turning point. Again, since the decay is almost equiprobable at all interatomic distances where the nuclear wavepacket does not vanish, the spectrum of the $Ne^{2+}(2p^{-2}\,{}^1D)Xe$ state consists of a broad peak between 0 and 4 eV with the ETMD thresholds visible at 0.4 and 1.9 eV.

The following conclusions can be drawn from the results presented above. Whenever allowed and provided that no competing ICD channel is open, ETMD will promptly follow an Auger step partially neutralizing a multiply charged ion. Nuclear dynamics in the decaying state will accelerate ETMD, dynamics in the final state might cause additional charge transfer between the ion and the environment. In rare gas clusters one expects the decay lifetimes to be few hundred femtoseconds to several picoseconds which makes ETMD the dominant electron transfer process.

Similar conclusions were drawn from the investigation of the ETMD(3) process in the $NeKr_2$ cluster following Auger decay of a 1s vacancy in Ne (Stumpf et al., 2013). The same final Auger states, $Ne^{2+}(2p^{-2}\,{}^1D,{}^1S)$, decay via ETMD(3) according to the following scheme

$$\begin{aligned} Ne^{2+}(2p^{-2})Kr_2 &\rightarrow Ne^{+}(2p^{-1})Kr^{+}(4p^{-1})Kr^{+}(4p^{-1}) \\ &\quad + e_{ETMD} \\ &\rightarrow Ne^{+} + Kr^{+} + Kr^{+} + e_{ETMD}. \end{aligned} \tag{5.5}$$

The ETMD(3) lifetimes at the equilibrium geometry of the cluster lie in the 7.5 to 32 ps range and are longer than the corresponding ETMD(2) lifetimes in NeXe, since the electron transfer from one neighbor is accompanied by energy transfer to another one. The decay is again accompanied by nuclear dynamics. It was shown that the shortening of the Ne–Kr distances and the following sharp increase in the decay rate leads to the substantial acceleration of ETMD. Thus, $Ne^{2+}(2p_x^{-1}2p_z^{-1})Kr_2$ state, whose lifetime at the equilibrium geometry is 7.5 ps, has an effective lifetime of 1 ps once nuclear dynamics are taken into account.

The experimental verification of the Auger–ETMD cascade was carried out in large mixed NeKr clusters (You et al., 2017). The mixed rare gas clusters usually have a shell structure with the heavier element forming the core and the lighter one the outer layers (Fasshauer et al., 2014). Therefore, the ETMD process in question can occur only in the interface region where Ne and Kr atoms are close. The clusters were irradiated by X-rays with energy just above the 1s ionization threshold of Ne. In the following decay the coincident observation of the Ne^{+} and two Kr^{+} ions together with a slow electron in the 0 to 5 eV range indicated the occurrence of the ETMD(3) of the $Ne^{2+}(2p^{-2}\,{}^{1}D, {}^{1}S)$ ions with Kr. The full electron spectrum stretching from 0 to 20 eV and multiple ion coincidences other than Ne^{+} and two Kr^{+} which were also observed suggest more complicated decay than the one in Eq. (5.5). For example, about 23% of the Auger final states are the excited ions $Ne^{2+}(2s^{-1}2p^{-1}\,{}^{3}P, {}^{1}P)$ which on the interface may undergo ICD with Kr. Since ICD is an energy transfer process the total charge of the Ne^{2+} cation will remain the same. Contrary to the situation in dimers or small clusters, the neighbors in a large cluster will hinder the Coulomb explosion. As a result $Ne^{2+}(2p^{-2})$ may remain at the interface long enough to undergo an additional ETMD step with Kr neighbors. Thus, in larger clusters interatomic ICD–ETMD cascades become a possibility after Auger decay. Some of these cascades, which follow the Auger decay step and explain observed ion coincidences and electron energies, are shown in Fig. 5.12. One can see that all of them consist of one or two ICD steps and culminate with the ETMD(3) step. This particular order of the interatomic decay processes reflects their relative efficiency, and suggests a scheme, according to which localized electronic energy is dissipated. Local processes precede the non-local ones, and energy transfer processes precede charge transfer ones.

In the particular case of the NeKr clusters the occurrence of ICD steps prior to ETMD explains the appearance of electrons faster than 7 eV in coincidence with the Ne^{+} and two Kr^{+} ions. As follows from the integration of the observed electron spectrum, a single ETMD step given by Eq. (5.5) remains the most probable process to follow the Auger decay of Ne 1s vacancy at the Ne–Kr interface. However, as we will see in the next section, creating multiply charged atomic ions in the Auger decay in larger clusters makes the chain of interatomic decay steps the rule and not an exception.

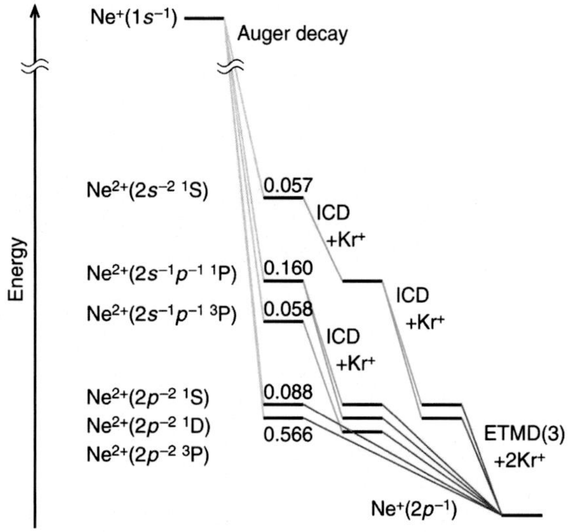

FIG. 5.12 Schematic representation of some decay cascades initiated in large NeKr clusters at the Ne–Kr interface as uncovered by coincident electron–ion spectroscopy following 1s ionization of Ne. The majority of Ne^{2+} ions are produced in the ground state electronic configuration and undergo ETMD with neighboring Kr atoms. Electronically excited Ne^{2+} ions promptly relax in one or two ICD steps into the $Ne^{2+}(2p^{-2})$ state, after which they also undergo ETMD. This is the first experimental demonstration that cascades of interatomic decay steps are possible in clusters after the Auger decay of a localized core-hole state. From You et al. (2017); with permission.

5.3.4 Electronic Decay Cascades in Microsolvated Clusters

The discussion in the last section showed that multiply charged atoms surrounded by neutral neighbors may undergo decay cascades consisting of several interatomic decay steps. An efficient way to create a multiply charged ion embedded in a cluster is to initiate a multistep Auger decay cascade in a neutral atom. Such cascades are already possible for third period atoms as demonstrated by the example of the 1s ionized Ar atom which we considered before. However, theoretical investigation of such cascades and, in particular, determining the electronic structure and populations of their final states quickly becomes a very difficult task with increasing number of decay steps due to the explosive growth in the number of intermediate states and the possible transitions between them (Son and Santra, 2012). To circumvent this problem one might start with a positive ion which will become multiply charged already after a single Auger step. Cations of alkali or alkaline earth metals such as (Na^{+}, Mg^{2+}, K^{+}, Ca^{2+}, etc.) appear as natural candidates due to the following reasons. First, they are isoelectronic with the well studied rare gas atoms (Ne, Ar, etc.) and are, therefore, amenable in the equal degree to *ab initio* calculations. Second, there is a well-studied class of clusters, called microsolvated clusters, which consist of such ions surrounded by solvent molecules (Barran et al., 2000; Rodriguez-Cruz et al., 1998, 1999), and they can be used to study experimentally the decay cascades which are triggered by core ionization of the metal ion. Third, these ions are naturally abundant. They are important for proper functioning of living cells, with Mg^{2+} being, for example, the most abundant metal found in enzymes (Maguire and Cowan, 2002). Therefore, the results obtained in a model system should reveal the kind of physico-chemical processes one would expect in a realistic setting, e.g., in an enzyme irradiated by X-rays during a crystallographic measurement.

Such cascades were studied in microsolvated Mg^{2+} clusters (Stumpf et al., 2016b). Mg^{2+} can coordinate up to six H_2O molecules in the first solvation shell forming an octahedron with the equilibrium Mg^{2+}–O distance in the fully coordinated cluster being 2.078 Å. This distance is much shorter than the typical distances in the rare gas clusters (3.1 to 4 Å) which has an enormous effect on the rates of the interatomic decay. Also, although Mg^{2+} is isoelectronic with Ne, the larger atomic charge of Mg compared to Ne implies larger 1s ionization energy (1317 eV vs. 870 eV) and faster Auger decay (1.9 fs vs. 2.3 fs) in the former. However, the computed branching ratios of the Mg^{4+} final states of the normal Auger process in Mg^{2+} were found to be equal to the ratios of their isoelectronic counterparts in Ne. The final states in question are $Mg^{4+}(2p^{-2}\,{}^1D,{}^1S)$ (63%), $Mg^{4+}(2s^{-1}2p^{-1}\,{}^1P)$ (21%), $Mg^{4+}(2s^{-1}2p^{-1}\,{}^3P)$ (9%), $Mg^{4+}(2s^{-2}\,{}^1S)$ (7%). As we will presently see, their de-excitation and neutralization sustain a complicated decay cascade of interatomic processes.

We will first look into the course of the electronic decay by assuming the cluster nuclear configuration to be frozen at the equilibrium geometry. The initial 1s vacancy relaxes first in a local Auger decay step populating mostly the $Mg^{4+}(2p^{-2}\,{}^1D,{}^1S)$ states. Their excess energy relative to the $Mg^{4+}(2p^{-2}\,{}^3P)$ ground state is 4 to 9 eV, and that is not sufficient to ionize a water ligand which rules out ICD as the next step. However, these states can efficiently decay by both ETMD(2) and ETMD(3), partially neutralizing the Mg^{4+} cation to $Mg^{3+}(2p^{-1}\,{}^2P)$ and producing H_2O^{2+} or two H_2O^{+} cations (see Fig. 5.13). The ETMD lifetime is only 16 fs, much shorter than the few picoseconds lifetimes in the rare gas clusters considered previously. This is due to shorter Mg–O distances and a larger number of nearest neighbors. In a trimer ETMD(3) is usually less efficient than ETMD(2), since the transferred and emitted electrons originate on different neighbors and interact weaker than when they come from the same neighbor. However, ETMD(3) scales quadratically with the number of neighbors, and the ETMD(3)/ETMD(2) branching ratio in the fully coordinated $Mg^{2+}(H_2O)_6$ was found to be 1.6/1. Electrons emitted in the decay have energies between 11 and 26 eV for ETMD(2) and 23 and

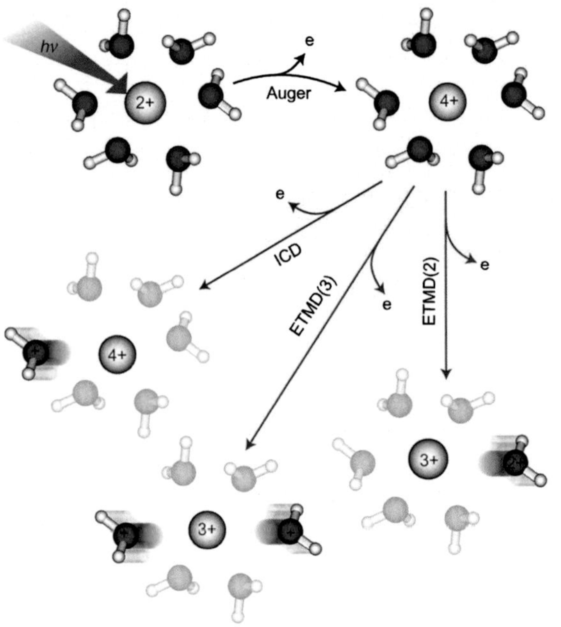

FIG. 5.13 Schematic representation of electronic decay steps realized in a microsolvated Mg^{2+} cluster following 1s ionization of the metal ion. They include the ICD process, whereby the Mg ion de-excites without altering its charge, while the solvation shell is singly ionized. They also include the ETMD process, whereby the Mg ion is partially neutralized, while the solvation layer is doubly ionized. From Stumpf et al. (2016b); with permission.

40 eV for ETMD(3). The larger energies of the ETMD(3) electrons are due to the larger charge delocalization in the ETMD(3) final states which lowers their energies relative to the ETMD(2) ones.

The $Mg^{4+}(2s^{-1}2p^{-1}\,{}^1P)$ states have enough excess energy to decay by ICD. The ICD lifetime found *ab initio* was as short as 0.7 fs at the equilibrium geometry. Surprisingly, it is shorter than the Auger lifetime of the 1s hole in Mg^{2+}. Similar ultrashort ICD lifetimes were obtained experimentally for $Mg^{3+}(2s^{-1})$ ions in aqueous solutions (Öhrwall et al., 2010). *Ab initio* calculations in microsolvated clusters confirmed these values and showed that the large number of neighbors, relative proximity of the excited moiety to the ionizable neighbors, and polarization of the neighbors by the multiply charged Mg ion are the reasons responsible for the extraordinarily short ICD lifetimes (Stumpf et al., 2016a). As a result of the ICD step, one water molecule will be singly ionized and an electron with energy between 0 and 7 eV will be emitted (see Fig. 5.13).

Such high ICD rates mean that ICD will completely quench ETMD which is also energetically allowed for this state. Since ICD is an energy transfer process, the charge of Mg in the final state remains unaltered and one obtains the already familiar $Mg^{4+}(2p^{-2}\,{}^1D, {}^1S)$ ions. Computations for a cluster in which the H_2O^+ cation produced in the ICD step is replaced by a positive point charge show that ETMD with the remaining neutral ligands remains energetically allowed. The effect of the positive charge is to close some of the ETMD channels and to increase the lifetime of the decaying state from 16 to 22 fs. Thus, electronically excited states of multiply charged ions in microsolvated clusters will first shed their excess energy via ICD and then will be partially neutralized in the following ETMD step. The same pattern of very fast energy transfer followed by slower electron transfer is repeated in the decay of the $Mg^{4+}(2s^{-1}2p^{-1}\,{}^3P)$ and $Mg^{4+}(2s^{-2}\,{}^1S)$ states, so that all Auger final states decay interatomically to produce Mg^{3+} ions surrounded by a few intact and a few ionized water molecules. The metal ions will be mostly in the $Mg^{3+}(2p^{-1}\,{}^2P)$ state, although a small percentage will be in the excited $Mg^{3+}(2s^{-1}\,{}^2S)$ state. The latter de-excites in the fast ICD step to $Mg^{3+}(2p^{-1}\,{}^2P)$ which undergoes the last ETMD step upon which the metal reverts to its initial Mg^{2+} state and the electronic decay stops.

The complete de-excitation cascade, schematically shown in Fig. 5.14, comprises branches consisting of interatomic decay cascades that are up to four steps long. These steps are responsible for the quick dissipation of up to 170 eV of energy retained by the metal atom after the Auger decay and emission of a fast Auger electron ($\sim$ 1150 eV). This excess energy is used for ionizing the molecules coordinated around the metal and for initiating nuclear dynamics such as Coulomb explosion or proton transfer. Calculations show that all electronic decay steps shown in Fig. 5.14 are allowed even if the water cations accumulated in the previous decay steps are retained close to the metal ion. The energy transfer to and the electron transfer from the surrounding solvation shell leads to its multiple ionization (4.3 times per 1s vacancy) and to the emission of 2.4 slow ICD and ETMD electrons. Moreover, the ultrashort lifetimes of the interatomic decay processes ensure that the decay of 90% of the 1s vacancy is completed within 220 fs.

These times are bound to become even shorter if one would take into account the nuclear motion. The original hexa-coordinated cluster is stable with respect to electron transfer from water and dissociation (Carl and Armentrout, 2013). Following ICD or ETMD, the nearest water neighbors are ionized which initiates nuclear dynamics. In a cluster with only the first solvation shell any ionization of the water molecules will lead to the Coulomb explosion and the detachment of the H_2O^+ or H_2O^{2+} cations with energies of up to 40 eV. Once the second solvation shell is formed, the proton transfer between the first and farther shells becomes dominant (Barran et al., 2000). Both the Coulomb explosion and proton transfer occur on the timescale of the electronic decay, and their effect is to remove positive charge away from the metal cation, which opens additional ICD and ETMD channels and accelerates the de-excitation cascade.

In our classification of the decay cascades we make a distinction between the ones initiated by core ionization followed by the Auger decay and the ones initiated by core excitation followed by the resonant Auger decay. The distinction washes out in larger clusters. Thus higher Rydberg excitations of bulk atoms form a conduction band, so that an excited electron delocalizes throughout the cluster (Kivimäki et al., 2005). The rate of this delocalization measured in Ar multilayers was found to be comparable to the Auger rate of a few fs (Wurth et al., 1993). Such "internal ionization" of core excited atoms and molecules is further accelerated in hydrogen bonded systems where it might become shorter than a femtosecond (Nordlund et al., 2007). The excited electrons in the core-excited states of ions also tend to delocalize even though they occupy a valence orbital (Miteva et al., 2016). Therefore, the core excited states in larger clusters will decay by both resonant Auger–ICD and Auger–ICD type cascades whose respective contribution and the final product mix will depend on the delocalization rate.

5.3.5 Interatomic Coulombic Decay Cascades in Multiply Excited Clusters

The interatomic decay cascades discussed in the previous sections were driven by the large amount of energy deposited locally in a weakly bound cluster by a single X-ray photon. Such energy can also be absorbed nonlocally in the form of many lower energy photons, and its dissipation will again lead to the multiple ionization of the cluster. The role which ICD plays in the multiple ionization of clusters in intense laser fields was first investigated theoretically by Kuleff et al. (2010). They considered a Ne cluster irradiated by photons whose energy lied below the ionization potential of the cluster and was equal to the energy of the $2p \to 3s$ transition in the Ne atom. In strong fields some atoms in the cluster will be ionized by absorbing two photons in the process known as multiphoton ionization (MPI). However, a large fraction of the Ne atoms will be resonantly ex-

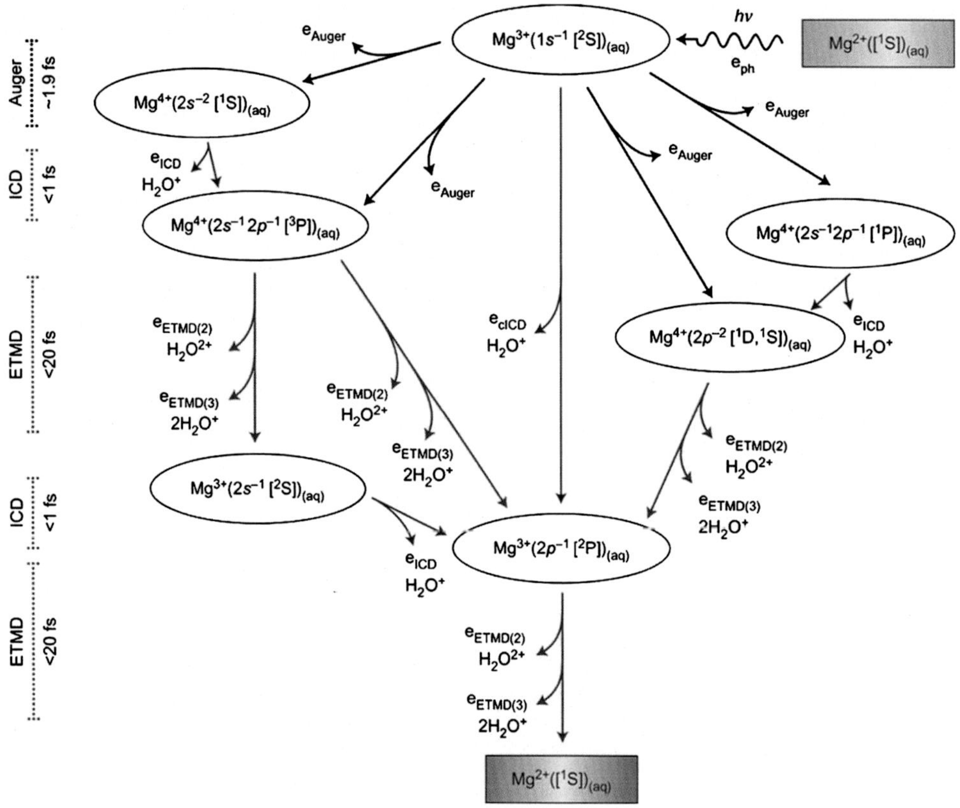

FIG. 5.14 Schematic representation of the electronic decay cascade in the microsolvated Mg^{2+} cluster. Black arrows denote the Auger steps, blue arrows denote the ICD and red arrows the ETMD steps. The timescale on the left shows the lifetimes of different processes. Removal of the 1s electron of Mg^{2+} leads to the Auger decay and population of several Mg^{4+} states. Electronically excited states de-excite via ICD within one femtosecond ionizing water molecules in the solvation shell; at the same time the charge of the metal cation is retained. The Mg^{4+} which have the ground state electronic configuration partially neutralize via ETMD with the solvating molecules doubly ionizing the solvation shell. This ICD–ETMD, de-excitation–neutralization, sequence of steps continues till the metal ion reverts to its original charge state. Multiple ionization of the solvating layer takes place in this interatomic decay cascade, followed by the release of reactive radicals in subsequent dissociation of water cations and slow ICD and ETMD electrons. From Stumpf et al. (2016b); with permission.

cited. ICD is energetically allowed between two excited Ne atoms (see Fig. 5.15) and is very efficient whenever they are within 1 nm distance from one another. Therefore, in the model considered by Kuleff et al. (2010) the cluster could be ionized via two distinct pathways: by a direct multiphoton ionization of a constituent atom, or through ICD in a pair of two resonantly excited Ne atoms.

The important question is what are the conditions under which one or the other pathway dominates. On the one hand, *ab initio* calculations in the Ne dimer show that the ICD lifetime at the equilibrium distance is about 120 fs, and it equals the rate of competing photon emission only at the interatomic separation ~ 1 nm. More importantly, the ICD rate grows quadratically with the number of excited Ne atoms, and, therefore, is expected to be considerable in a multiply excited larger Ne cluster. Thus, should photoabsorption lead to the multiple excitation of the cluster, ICD would be the dominant ionization mechanism. On the other hand, stronger fields favor MPI of individual Ne atoms over multiple excitations in the cluster. The probability of two-photon ionization scales quadratically with the field strength, while the probability of electronic excitation scales only

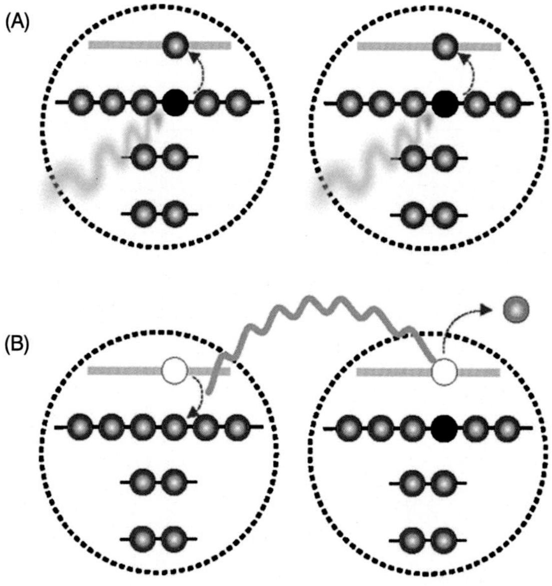

FIG. 5.15 Schematic representation of the ICD process between two resonantly excited atoms. (A) Two different atoms are excited by absorbing two photons. (B) One atom de-excites, transfers the excess energy to the neighbor and ionizes it. From Kuleff et al. (2010); with permission.

linearly. Solving a system of rate equations for varying field strength showed that multiple excitation of the cluster followed by ICD will be the prevalent ionization pathway for moderate field strengths of about 10^{12} W/cm^2, while the MPI's contribution dominates as the fields become stronger.

The characteristic dependence of these ionization mechanisms on the field intensity suggests a way of distinguishing them experimentally. It has been observed that clusters which are multiply ionized by XUV radiation disintegrate, and singly charged atomic ions are the most abundant product for moderate cluster sizes (Bostedt et al., 2008). Therefore, measuring their yield as a function of the field strength should indicate by what mechanism the ionization has taken place. The yield of Ne^+ ions produced by irradiating Ne_N clusters ($\langle N \rangle = 1000$) with XUV-FEL light was measured as the function of the field strength by Nagaya et al. (2013). The photon energy was set to 20 eV which matches the 2p → 3d transition energy in the Ne atom. The ion yield from the cluster clearly showed linear dependence on the field strength, while the similar yield from free Ne atoms showed the quadratic dependence characteristic of MPI. These results rule out MPI of Ne atoms in the cluster as the leading ionization mechanism, and suggest that ionization proceeds through multiply excited states of the cluster.

Absorption of multiple photons by a cluster in an intense laser field leads eventually to the formation of nanoplasma (Fennel et al., 2010). The underlying mechanism differs for the intense IR and intense XUV radiation. In the latter case, whenever single photon ionization is frustrated by rising charge of multiply ionized cluster, resonant excitation of many atoms follows, excited electrons delocalize throughout the cluster and move collectively in the field of positive ions. In the former case, strong fields cause tunneling ionization of constituent atoms and the appearance of quasifree electrons. Electrons evaporate from the heated plasma, and their energy shows typical exponential distribution with the maximum at zero energy. Ionization of clusters due to plasma heating cannot be distinguished from the one by ICD when the ion yield alone is measured. It depends in both cases on the number of excited atoms and, hence, linearly on the field strength. However, the different decay mechanisms can be recognized in photoelectron spectra, since the ICD electrons should appear as well defined peaks which are superimposed on the exponential background of the evaporated electrons.

Precisely such electron spectra were obtained following the irradiation of He clusters by XUV light at different intensities in the experiments reported in (Ovcharenko et al., 2014). They demonstrated the transition between the ICD and plasma evaporation regimes in cluster ionization as the filed intensity grew. The experiments were conducted in He droplets. The photon energy was chosen to coincide with the absorption maximum corresponding to the production of the 1s2p excited state of a He droplet at 21.4 eV. A clear peak, observed at 18 eV in smaller ($\langle N \rangle = 1000$) droplets at lower field intensities of 8.7×10^{10} W/cm^2, was assigned to ICD between two excited He atoms. In larger droplets ($\langle N \rangle = 50000$) and in stronger fields ($\gtrsim 10^{12}$ W/cm^2) this peak disappears. This behavior of the ICD peak was explained along the following lines. At lower intensities the few excited atoms formed in the cluster decay pairwise by ICD. In addition, the observation of ICD electrons is enhanced in smaller clusters, since they originate mostly at the cluster's surface and do not undergo inelastic collisions with other excited atoms. In larger clusters and at stronger fields the number of excited atoms grows and many electron processes, such as double-ICD,

$$\begin{aligned} &He^*(1s2p)He^*(1s2p)He^*(1s2p) \\ &\quad \rightarrow HeHe^+He^+ + 2e_{ICD}, \end{aligned} \tag{5.6}$$

as well as ICD followed by ionization of another He* via the ICD electron impact, become more probable than a simple ICD step. Since two electrons which are emitted in these processes share the excess energy in a continuous manner, the ICD peak disappears. At even higher intensities when the nanoplasma is formed, such collective autoionization processes become dominant, and the resulting electron energy spectrum exhibits the exponential behavior. Multiphoton ionization mechanism was found to be insignificant, since its signature was absent from the electron spectra at higher field intensities. Independent experiments in He droplets ($\langle N \rangle = 50000$) by LaForge et al. (2014) at the same excitation frequencies and comparable intensities showed that He^+ yield grew with the intensity slower than linearly. This result confirmed that cluster ionization proceeds via complicated collective autoionization pathways.

Clear demonstration of the ICD signal was achieved in Iablonskyi et al. (2016) in multiply excited Ne clusters. Since the 2p → 3s excitation energy of the surface and bulk atoms differ by some 0.5 eV they could be selectively excited by FEL light. The resulting photoelectron spectra are shown in Fig. 5.16. The peaks at 11.8 eV plotted for different laser intensities correspond to ICD in pairs of excited Ne atoms. These peaks are pronounced in the case of surface excitations, and are superimposed on the background due to collective autoionization which is much stronger for the case of the bulk excited states. The electron spectra in Fig. 5.16 make it clear that exciting surface atoms leads to higher ratio of ICD to collective autoionization events than in the case of the bulk, similar to the case of He clusters discussed above. Moreover, the ICD peaks are shifted to lower energies relative to their expected positions in the cluster such that they correspond to ICD between two isolated excited Ne atoms. This shift was explained as being due to the cluster dynamics prior to the decay. Indeed, the excitation energies of neutral rare gas atoms in clusters are shifted towards higher energies due to the repulsive interaction between the excited electron and surrounding ground state atoms. In the case of excited Ne atoms this interaction pushes the neighboring atoms away so that the excited Ne atom resides in a bubble (Leung et al., 1983). The bubble creation leads to the lowering of the excitation energy to the values which correspond to the isolated atom, and the known relaxation lifetimes in solid Ne are about 1 ps. Since nearest Ne neighbors cannot be simultaneously excited due to the Coulomb blockade, ICD takes place between excited non-nearest neighbors, and its lifetime was estimated to be also about 1 ps or longer. Therefore, one might expect that nuclear dynamics in the multiply excited Ne cluster and the formation of bubbles around excited Ne atoms precede or at least occur simultaneously with ICD which explains the position of the ICD electron peak.

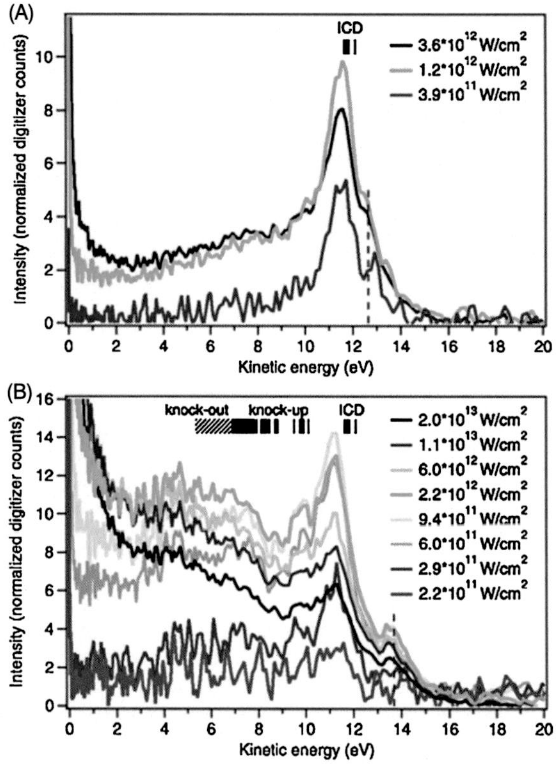

FIG. 5.16 Electron emission spectra of FEL irradiated Ne clusters. (A) 2p → 3s excitations of surface atoms at 17.12 eV. (B) 2p → 3s excitations of bulk atoms at 17.65 eV. The peaks at about 11.8 eV correspond to the ICD between two excited Ne atoms, while the plateaus at lower energies are due to the collective autoionization process. The dashed lines denote the kinetic energy of the electrons emitted due to ionization of uncondensed atoms by the respective second FEL harmonic. From Iablonskyi et al. (2016); with permission.

The discussion above shows that ICD is an important mechanism for cluster ionization by intense XUV light whose absorption leads to multiply excited clusters. Surprisingly, it also plays a notable role in the cluster ionization in strong IR fields where single photon resonant excitation of atoms is impossible. It was demonstrated by Schütte et al. (2015) that pronounced peaks superimposed on the exponential background appear in the electron spectrum of pure Ar and mixed ArKr clusters (see Fig. 5.17). The peaks close to the ionization

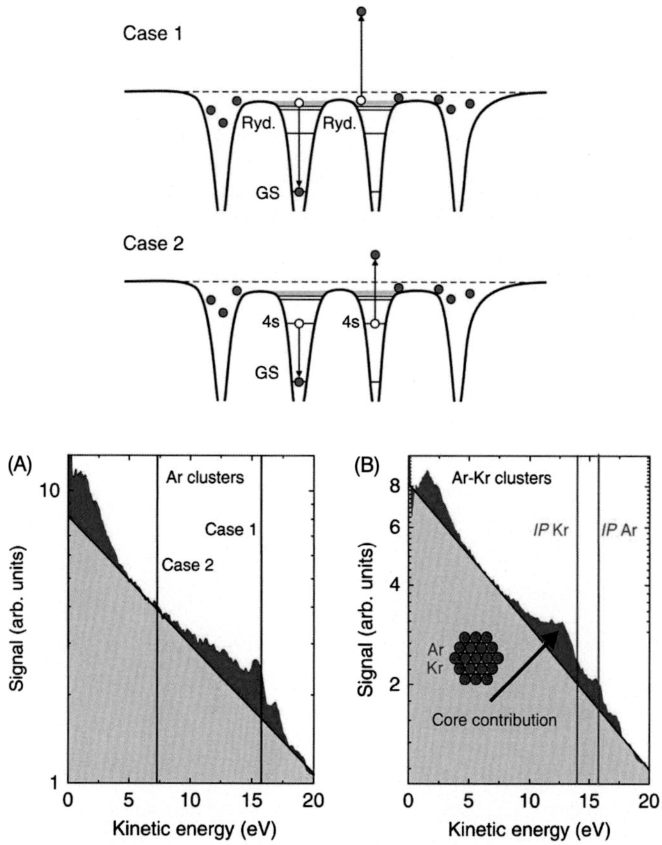

FIG. 5.17 Electron kinetic energy spectra of (A) Ar and (B) ArKr clusters following near IR ionization. The peak in (A) close to the Ar ionization threshold at 15.76 eV corresponds to the ICD in a pair of highly excited Rydberg Ar atoms (case 1). Its onset at about 7.5 eV (case 2) corresponds to the ICD in a pair of two Ar($3p^{-1}4s$) atoms. The signal in excess of the exponential background at energies below 5 eV is assigned to the ICD whereby a Rydberg atom de-excites into a lower Rydberg state and not in the atomic ground state. In (B) two peaks are visible converging to the respective Kr and Ar ionization thresholds. This indicates that ICD might take place between excited Ar atoms, Kr atoms, as well as excited Ar and Kr atoms. From Schütte et al. (2015); with permission.

potential of Ar in the Ar cluster and to the respective ionization potentials of Ar and Kr in the mixed ArKr cluster were interpreted as due to ICD between Rydberg atoms, such as $Ar^*(3p^{-1}nl)$–$Ar^*(3p^{-1}ml)$, where *nl* and *ml* stand for Rydberg orbitals from 4s and higher. The Rydberg atoms are produced following the creation of the nanoplasma when Coulomb repulsion and thermal pressure of the excited electrons cause the cluster to expand (Fennel et al., 2010). The quasifree plasma electrons are captured by atomic ions forming excited atoms. This expansion also strongly reduces the positive cluster potential, which would otherwise shift the ICD electron peaks from their observed position towards lower energies. So once again nuclear dynamics in the multiply excited cluster leads to the formation of atomic excited states which undergo ICD with one another. The ICD lifetime in Ar clusters was found to be 87 ps in a pump–probe experiment. This lifetime agreed within the order of magnitude with the theoretical estimates in the Ar dimer at the equilibrium geometry. Importantly, the modeling of the cluster dynamics showed that its expansion timescale is 10 ps, so that the Rydberg atoms can be formed and the cluster potential removed before ICD takes place.

The emitted electrons with energies just below the respective Ar and Kr ionization potentials originate in

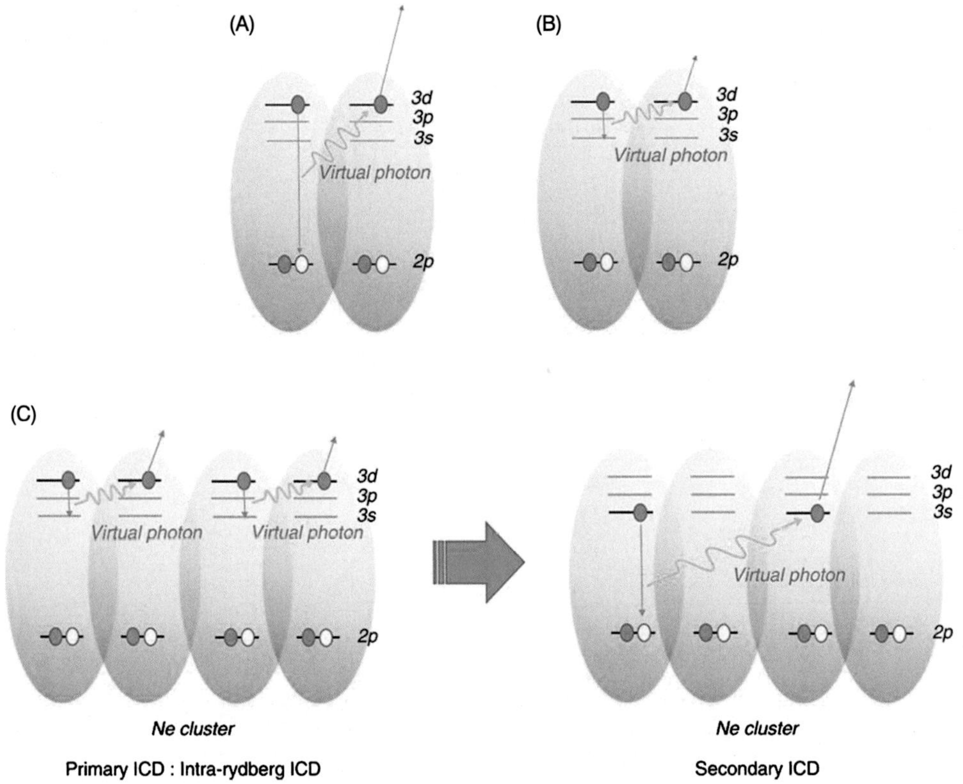

FIG. 5.18 Schematic diagram of ICD pathways in multiply 2p → 3d excited Ne clusters. (A) Direct ICD process; (B) intra-Rydberg ICD process; (C) two-step ICD cascade. From Nagaya et al. (2016a); with permission.

the ICD events where one Rydberg atom relaxes to its ground state while the other is ionized. Particularly interesting is the appearance in the electron spectra of the peaks in the range between 0 eV and 5 eV. They were assigned to the ICD events where the transitions on one excited atom took place between Rydberg states, so that the transferred energy and, consequently, the energy of the ICD electron were small. If the density of the excited atoms is high enough such transitions may constitute a de-excitation cascade, whereby a Rydberg atom loses its energy in a succession of several transitions between Rydberg states transferring energy and ionizing a different neighbor at each step.

The existence and the efficiency of such cascades were investigated in a joint experimental and theoretical study reported in Nagaya et al. (2016a). Ne clusters were irradiated by FEL light with photon energy of 20.3 eV, which corresponds to the 2p → 3d transition in the Ne atom. In a multiply excited cluster the $Ne^*(2p^{-1}3d)$ atom may de-excite either into the Ne ground state or into the $Ne^*(2p^{-1}4s)$ state and ionize an excited Ne neighbor in direct or intra-Rydberg ICD processes, respectively (see Fig. 5.18). The $Ne^*(2p^{-1}4s)$ atom may further de-excite to the Ne ground state by ionizing another excited Ne neighbor and forming, therefore, a two-step ICD cascade. Theoretical estimates of the ICD lifetimes for the direct and intra-Rydberg ICD of $Ne^*(2p^{-1}3d)$ produced 88 ns and 9.2 fs, respectively. Although the estimates were obtained using asymptotic formulas of the type given in Eq. (5.2), the expected errors of this approximation are much smaller than the seven orders of magnitude difference between the lifetimes. Therefore, one may conclude that the direct ICD process will be quenched by the ICD cascade. This quenching can be seen in the ICD electron spectra in Fig. 5.19. The cascade produces electrons with kinetic energy of about 2 and 11.6 eV, while the electrons due to direct ICD will appear at 18.6 eV. Indeed, the experimental spectrum shows that the electrons at the energies corresponding to the cascade predominate. The theoret-

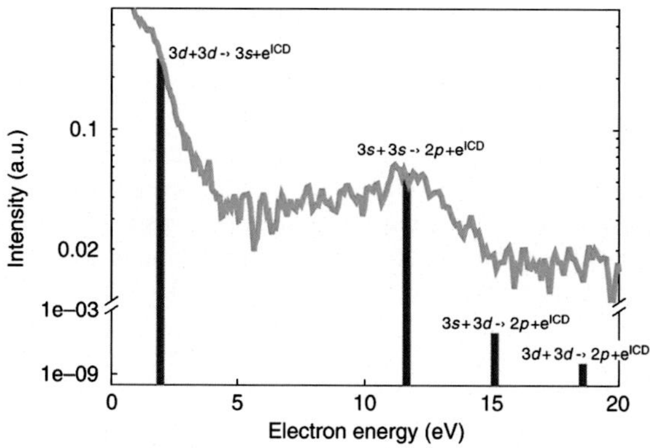

FIG. 5.19 Experimental electron kinetic energy spectrum obtained by irradiating Ne cluster by FEL light with photon energy of 20.3 eV. Vertical lines show the theoretical electron yields due to different ICD steps. The theoretical data is normalized to the experimental intensity at 12 eV. The spectrum clearly shows the predominance of the cascades of ICD processes in cluster's relaxation. From Nagaya et al. (2016a); with permission.

ical branching ratios between different decay pathways obtained by solving rate equations confirm the experimental results. Thus, the multiply highly excited Ne clusters de-excite in an ICD cascade which involves ultrafast intra-Rydberg ICD steps. The ICD cascade can be nearly two times more efficient in ionizing a cluster than the direct ICD process if high Rydberg states are produced, since the relaxation of a single atom leads to ionization of multiple excited neighbors. Moreover, since the energy is transferred in smaller packets in the ICD cascade among Rydberg atoms, slow free electrons will be predominantly produced.

5.4 CONCLUDING REMARKS

In their interaction with chemical media X-ray photons are strongly coupled to localized core electrons of atoms and molecules. Therefore, once the photon is absorbed, a core electron is excited, and the energy of hundreds or even thousands of electronvolts is deposited locally and is afterwards dissipated throughout the system. Since interelectron interaction is the strongest force in chemical systems, the initial stages of the decay are dominated by electronic decay. At first, this decay takes the form of Auger decay or Auger decay cascades, whereby electrons which occupy higher shells in the core excited atom or molecule fill the core vacancies, while other electrons from these shells are emitted in the continuum. The Coulomb interaction between electrons localized on an atom or a molecule which carries the core vacancy is very strong. This explains the ultra-short Auger lifetimes and the prevalence of the local Auger decay process even in van der Waals or hydrogen bonded clusters or liquid solutions. Whenever the Auger decay takes place in such systems, multiply charged positive ions in the ground or excited electronic states are produced among the neutral atomic or molecular neighbors. Due to the long-range nature of the Coulomb potential these ions de-excite and are neutralized in further electronic decay steps, although this time these steps are of interatomic character.

Two processes are important in this respect: interatomic Coulombic decay (ICD) and electron-transfer mediated decay (ETMD). In the former process, a localized electronic excited state embedded in a weakly bound chemical medium de-excites by ionizing a neighbor. In the second process, a positively charged ion is neutralized via electron transfer from the environment, while the released energy is transferred back to the environment and used to ionize it. Numerous theoretical and experimental studies provide one with a clear picture of these two processes. ICD is a completely general process and occurs whenever there is a localized electronic excitation in a weakly bound medium such that its energy lies above the ionization potential of the environment. The ICD lifetimes are about 100 fs in rare gas dimers; they become shorter with decreasing interatomic distances and increasing number of neighbors. As a result, they reach values of a few fs or even less than a femtosecond in microsolvated clusters, where short distances predominate, or in endohedral fullerenes, which provide a large number of nearest neighbors. ETMD is particularly important for neutralization of multiply charged ions embedded in clusters. Its lifetime in rare gas clusters is in the picosecond range but becomes one order of magnitude shorter once nuclear dynamics during the decay is accounted for. In microsolvated clusters of metal cations, where the distance between the ligands and the metal center is shorter than between two rare gas atoms ETMD lifetime becomes tens of femtoseconds.

Both processes were shown to follow Auger decay in rare gas clusters, demonstrating that the Auger decay common in the gas phase is replaced in a chemical medium by Auger–ICD or Auger–ETMD cascades. Excited electronic states produced in the Auger decay would de-excite via ICD whenever it is energetically allowed. Multiply charged ions in the ground state configurations are neutralized via ETMD. Thus local concentrations of energy and positive charge become efficiently distributed throughout the cluster. An interesting detail of such cascades is the possibility of controlling the location of ICD and the energy of emitted ICD electrons in the resonant Auger–ICD cascade which relies on the properties of the resonant X-ray absorption and resonant Auger decay.

Such cascades become complicated and involve multiple interatomic steps when the energy of the X-ray photon increases and highly charged ions appear in the medium. Investigation of the relaxation of microsolvated clusters of Mg^{2+} following X-ray photoabsorption by the metal showed that decay cascades of up to four steps appear following the Auger decay. As a result, several slow electrons are emitted from the surrounding hydration layer, which becomes massively ionized. The ionization of ligands triggers molecular dynamics leading to Coulomb explosion, dissociation, and proton transfer processes.

This particular cascade illustrates the important role of the interatomic decay processes in shaping the physico-chemical events which accompany X-ray absorption by chemical systems. It has been observed that an X-ray diffraction image deteriorates with the exposure length due to the damage incurred in the sample by the energetic photons (Taberman, 2018). The damage is the result of multiple ionization of the molecule either by the X-ray photon or by secondary electrons. It changes electron density, creates reactive radicals and slow electrons, and initiates nuclear dynamics, such that the diffraction picture does not correspond to the native structure of the molecule. Should a biomolecule have a metal center it is damaged at significantly lower doses than the rest of the molecule (Carugo and Carugo, 2005). This sensitivity can be explained by the cascades of Auger and interatomic electronic decay processes which are initiated by the photon absorption through the metal. Indeed, absorbing a singe photon by microsolvated Mg^{2+} leads to the ionization of more than four nearest neighbors and completely demolishes the first coordination shell of the metal. The promise of modern XFEL techniques is to be able to record molecular structure before the effects of the radiation damage appear (Chapman et al., 2014). Indeed, a highly resolved image of the photosystem II has been obtained by irradiating it with XFEL pulses of 10 fs duration (Suga et al., 2015). It was claimed that the irradiation time was too short for the damage to propagate, therefore the image was essentially of an undamaged molecule. However, we have seen that interatomic processes occur already within a few fs following the Auger step damaging the ligands and they impose tighter temporal bounds on the XFEL crystallography.

Interatomic cascades appear not only when a single high energy photon is absorbed but also when multiple photons with energies below the ionization threshold of constituent atoms are absorbed by clusters. Should a sufficient number of Rydberg atoms be produced in the cluster, some of them de-excite, whereby electronic transitions take place between their Rydberg states, while the excess energy is transferred to other Rydberg atoms ionizing them. This ICD cascade is responsible for the efficient ionization of clusters in moderately intense XUV light or in an expanding nanoplasma. Thus, irrespective of the way in which a large quantity of energy is deposited in a chemical medium by photons, the following energy redistribution will be shaped to a large degree by the cascades of the interatomic decay processes.

ACKNOWLEDGMENTS

The authors gratefully acknowledge financial support by the European Research Council (ERC) (Advanced Investigator Grant No. 692657) and by the Deutsche Forschungsgemeinschaft (DFG-FOR 1789).

REFERENCES

Åberg, T., Howat, G., 1982. The theory of the Auger effect. In: Mehlhorn, W. (Ed.), Encyclopedia of Physics, vol. 31. Springer, Berlin, pp. 469–630.

Amusia, M.Y., Lee, I.S., Kilin, V.A., 1992. Double Auger decay in atoms: probability and angular distribution. Phys. Rev. A 45, 4576–4587.

Aoto, T., Ito, K., Hikosaka, Y., Shigemasa, E., Penent, F., Lablanquie, P., 2006. Properties of resonant interatomic Coulombic decay in Ne dimers. Phys. Rev. Lett. 97, 243401.

Armen, G.B., Aksela, H., Åberg, T., Aksela, S., 2000. The resonant Auger effect. J. Phys. B 33 (2), R49.

Avaldi, L., Dawber, G., Camilloni, R., King, G.C., Roper, M., Siggel, M.R.F., Stefani, G., Zitnik, M., Lisini, A., Decleva, P., 1995. Measurement of Ne 1s and 1snl satellite photoelectron spectra near threshold. Phys. Rev. A 51, 5025–5028.

Averbukh, V., Cederbaum, L.S., 2005. Ab initio calculation of interatomic decay rates by a combination of the Fano ansatz, Green's-function methods, and the Stieltjes imaging technique. J. Chem. Phys. 123, 204107.

Averbukh, V., Cederbaum, L.S., 2006. Interatomic electronic decay in endohedral fullerenes. Phys. Rev. Lett. 96, 053401.

Averbukh, V., Müller, I.B., Cederbaum, L.S., 2004. Mechanism of interatomic Coulombic decay in clusters. Phys. Rev. Lett. 93, 263002.

Barran, P.E., Walker, N.R., Stace, A.J., 2000. Competitive charge transfer reactions in small $[Mg(H_2O)_n]^{2+}$ clusters. J. Chem. Phys. 112 (14), 6173–6177.

Barth, S., Joshi, S., Marburger, S., Ulrich, V., Lindblad, A., Öhrwall, G., Björneholm, O., Hergenhahn, U., 2005. Observation of resonant interatomic Coulombic decay in Ne clusters. J. Chem. Phys. 122, 241102.

Bolognesi, P., O'Keeffe, P., Ovcharenko, Y., Coreno, M., Avaldi, L., Feyer, V., Plekan, O., Prince, K.C., Zhang, W., Carravetta, V., 2010. Pyrimidine and halogenated pyrimidines near edge X-ray absorption fine structure spectra at C and N K-edges: experiment and theory. J. Chem. Phys. 133 (3), 034302.

Bostedt, C., Thomas, H., Hoener, M., Eremina, E., Fennel, T., Meiwes-Broer, K.-H., Wabnitz, H., Kuhlmann, M., Plönjes, E., Tiedtke, K., Treusch, R., Feldhaus, J., de Castro, A.R.B., Möller, T., 2008. Multistep ionization of argon clusters in intense femtosecond extreme ultraviolet pulses. Phys. Rev. Lett. 100, 133401.

Carl, D.R., Armentrout, P.B., 2013. Threshold collision-induced dissociation of hydrated magnesium: experimental and theoretical investigation of the binding energies for $Mg^{2+}(H_2O)_x$ complexes ($x = 2$–10). ChemPhysChem 14 (4), 681–697.

Carugo, O., Carugo, K.D., 2005. When X-rays modify the protein structure: radiation damage at work. Trends Biochem. Sci. 30 (4), 213–219.

Cederbaum, L.S., Zobeley, J., Tarantelli, F., 1997. Giant intermolecular decay and fragmentation of clusters. Phys. Rev. Lett. 79, 4778.

Chapman, H.N., Caleman, C., Timneanu, N., 2014. Diffraction before destruction. Philos. Trans. R. Soc. Lond. B 369 (1647).

Chapman, H.N., Fromme, P., Barty, A., White, T.A., Kirian, R.A., Aquila, A., Hunter, M.S., Schulz, J., DePonte, D.P., Weierstall, U., Doak, R.B., Maia, F.R.N.C., Martin, A.V., Schlichting, I., Lomb, L., Coppola, N., Shoeman, R.L., Epp, S.W., Hartmann, R., Rolles, D., Rudenko, A., Foucar, L., Kimmel, N., Weidenspointner, G., Holl, P., Liang, M., Barthelmess, M., Caleman, C., Boutet, S., Bogan, M.J., Krzywinski, J., Bostedt, C., Bajt, S., Gumprecht, L., Rudek, B., Erk, B., Schmidt, C., Hömke, A., Reich, C., Pietschner, D., Strüder, L., Hauser, G., Gorke, H., Ullrich, J., Herrmann, S., Schaller, G., Schopper, F., Soltau, H., Kühnel, K.-U., Messerschmidt, M., Bozek, J.D., Hau-Riege, S.P., Frank, M., Hampton, C.Y., Sierra, R.G., Starodub, D., Williams, G.J., Hajdu, J., Timneanu, N., Seibert, M.M., Andreasson, J., Rocker, A., Jönsson, O., Svenda, M., Stern, S., Nass, K., Andritschke, R., Schröter, C.-D., Krasniqi, F., Bott, M., Schmidt, K.E., Wang, X., Grotjohann, I., Holton, J.M., Barends, T.R.M., Neutze, R., Marchesini, S., Fromme, R., Schorb, S., Rupp, D., Adolph, M., Gorkhover, T., Andersson, I., Hirsemann, H., Potdevin, G., Graafsma, H., Nilsson, B., Spence, J.C.H., 2011. Femtosecond X-ray protein nanocrystallography. Nature 470, 73–77.

Chiang, Y.-C., Otto, F., Meyer, H.-D., Cederbaum, L.S., 2011. Interrelation between the distributions of kinetic energy release and emitted electron energy following the decay of electronic states. Phys. Rev. Lett. 107, 173001.

Codling, K., Madden, R.P., Ederer, D.L., 1967. Resonances in the photo-ionization continuum of Ne I (20–150 eV). Phys. Rev. 155, 26–37.

de Gouw, J.A., van Eck, J., Peters, A.C., van der Weg, J., Heideman, H.G.M., 1995. Resonant Auger spectra of the $2p^{-1}nl$ states of argon. J. Phys. B 28 (11), 2127.

Demekhin, P.V., Chiang, Y.-C., Stoychev, S.D., Kolorenč, P., Scheit, S., Kuleff, A.I., Tarantelli, F., Cederbaum, L.S., 2009. Interatomic Coulombic decay and its dynamics in NeAr following K–LL Auger transition in the Ne atom. J. Chem. Phys. 131, 104303.

Demekhin, P.V., Gokhberg, K., Jabbari, G., Kopelke, S., Kuleff, A.I., Cederbaum, L.S., 2013. Overcoming blockade in producing doubly excited dimers by a single intense pulse and their decay. J. Phys. B 46, 021001.

Erk, B., Boll, R., Trippel, S., Anielski, D., Foucar, L., Rudek, B., Epp, S.W., Coffee, R., Carron, S., Schorb, S., Ferguson, K.R., Swiggers, M., Bozek, J.D., Simon, M., Marchenko, T., Küpper, J., Schlichting, I., Ullrich, J., Bostedt, C., Rolles, D., Rudenko, A., 2014. Imaging charge transfer in iodomethane upon X-ray photoabsorption. Science 345 (6194), 288–291.

Erk, B., Rolles, D., Foucar, L., Rudek, B., Epp, S.W., Cryle, M., Bostedt, C., Schorb, S., Bozek, J., Rouzée, A., Hundertmark, A., Marchenko, T., Simon, M., Filsinger, F., Christensen, L., De, S., Trippel, S., Küpper, J., Stapelfeldt, H., Wada, S., Ueda, K., Swiggers, M., Messerschmidt, M., Schröter, C.D., Moshammer, R., Schlichting, I., Ullrich, J., Rudenko, A., 2013. Ultrafast charge rearrangement and nuclear dynamics upon inner-shell multiple ionization of small polyatomic molecules. Phys. Rev. Lett. 110, 053003.

Fano, U., 1961. Effects of configuration interaction on intensities and phase shifts. Phys. Rev. 124 (6), 1866–1878.

Fano, U., Cooper, J.W., 1968. Spectral distribution of atomic oscillator strengths. Rev. Mod. Phys. 40, 441–507.

Fasshauer, E., Förstel, M., Pallmann, S., Pernpointner, M., Hergenhahn, U., 2014. Using ICD for structural analysis of clusters: a case study on NeAr clusters. New J. Phys. 16, 103026.

Fennel, T., Meiwes-Broer, K.-H., Tiggesbäumker, J., Reinhard, P.-G., Dinh, P.M., Suraud, E., 2010. Laser-driven nonlinear cluster dynamics. Rev. Mod. Phys. 82, 1793–1842.

Feshbach, H., 1958. Unified theory of nuclear reactions. Ann. Phys. 5 (4), 357–390.

Feshbach, H., 1962. A unified theory of nuclear reactions. II. Ann. Phys. 19 (2), 287–313.

Förstel, M., Arion, T., Hergenhahn, U., 2013. Measuring the efficiency of interatomic Coulombic decay in Ne clusters. J. Electron Spectrosc. Relat. Phenom. 191, 16.

Förstel, M., Mucke, M., Arion, T., Bradshaw, A.M., Hergenhahn, U., 2011. Autoionization mediated by electron transfer. Phys. Rev. Lett. 106, 033402.

Gokhberg, K., Averbukh, V., Cederbaum, L.S., 2006. Interatomic decay of inner-valence-excited states in clusters. J. Chem. Phys. 124, 144315.

Gokhberg, K., Kolorenč, P., Kuleff, A.I., Cederbaum, L.S., 2014. Site- and energy-selective slow-electron production through intermolecular Coulombic decay. Nature 505, 661.

Gokhberg, K., Trofimov, A.B., Sommerfeld, T., Cederbaum, L.S., 2005. Ionization of metal atoms following valence-excitation of neighbouring molecules. Europhys. Lett. 72, 228.

Goodhead, D., 1994. Initial events in the cellular effects of ionizing radiations: clustered damage in dna. Int. J. Radiat. Biol. 65 (1), 7–17.

Guillemin, R., Bomme, C., Marin, T., Journel, L., Marchenko, T., Kushawaha, R.K., Trcera, N., Piancastelli, M.N., Simon, M., 2011. Complex decay patterns in atomic core photoionization disentangled by ion-recoil measurements. Phys. Rev. A 84, 063425.

Guillemin, R., Jänkälä, K., de Miranda, B.C., Marin, T., Journel, L., Marchenko, T., Travnikova, O., Goldsztejn, G., Ismail, I., Püttner, R., Céolin, D., Lassalle-Kaiser, B., Piancastelli, M.N., Simon, M., 2018. Interplay of complex decay processes after argon 1s ionization. Phys. Rev. A 97, 013418.

Haberland, H., 1985. A model for the processes happening in a rare-gas cluster after ionization. Surf. Sci. 156, 305–312.

Hans, A., Stumpf, V., Holzapfel, X., Wiegandt, F., Schmidt, P., Ozga, C., Reiß, P., Ltaief, L.B., Küstner-Wetekam, C., Jahnke, T., Ehresmann, A., Demekhin, P.V., Gokhberg, K., Knie, A., 2018. Direct evidence for radiative charge transfer after inner-shell excitation and ionization of large clusters. New J. Phys. 20 (1), 012001.

Havermeier, T., Jahnke, T., Kreidi, K., Wallauer, R., Voss, S., Schöffler, M., Schössler, S., Foucar, L., Neumann, N., Titze, J., Sann, H., Kühnel, M., Voigtsberger, J., Morilla, J.H., Schöllkopf, W., Schmidt-Böcking, H., Grisenti, R.E., Dörner, R., 2010. Interatomic Coulombic decay following photoionization of the helium dimer: observation of vibrational structure. Phys. Rev. Lett. 104, 133401.

Hazi, A.U., 1978. A purely L^2 method for calculating resonance widths. J. Phys. B 11 (8), L259–L264.

Hergenhahn, U., Rüdel, A., Maier, K., Bradshaw, A., Fink, R., Wen, A., 2003. The resonant Auger spectra of formic acid, acetaldehyde, acetic acid and methyl formate. Chem. Phys. 289, 57–67.

Higuchi, I., Ouchi, T., Sakai, K., Fukuzawa, H., Liu, X.-J., Ueda, K., Iwayama, H., Nagaya, K., Yao, M., Ding, D., Zhang, D., Tamenori, Y., Saito, N., 2010. Radiative charge transfer and interatomic Coulombic decay following direct double photoionization of neon dimers. J. Phys. Conf. Ser. 235, 012015.

Howell, R.W., 2008. Auger processes in the 21st century. Int. J. Radiat. Biol. 84 (12), 959–975.

Hubbell, J.H., Trehan, P.N., Singh, N., Chand, B., Mehta, D., Garg, M.L., Garg, R.R., Singh, S., Puri, S., 1994. A review, bibliography, and tabulation of K, L, and higher atomic shell X-ray fluorescence yields. J. Phys. Chem. Ref. Data 23 (2), 339–364.

Hush, N.S., 1961. Adiabatic theory of outer sphere electron-transfer reactions in solution. Trans. Faraday Soc. 57, 557–580.

Iablonskyi, D., Nagaya, K., Fukuzawa, F., Motomura, K., Kumagai, Y., Mondal, S., Tachibana, T., Takanashi, T., Nishiyama, T., Matsunami, K., Johnsson, P., Piseri, P., Sansone, G., Dubrouil, A., Reduzzi, M., Carpeggiani, P., Vozzi, C., Devetta, M., Negro, M., Calegari, F., Trabattoni, A., Castrovilli, M.C., Faccialá, D., Ovcharenko, Y., Möller, T., Mudrich, M., Stienkemeier, F., Coreno, M., Alagia, M., Schütte, B., Berrah, N., Kuleff, A.I., Jabbari, G., Callegari, C., Plekan, O., Finetti, P., Spezzani, C., Ferrari, E., Allaria, E., Penco, G., Serpico, C., De Ninno, G., Nikolov, I., Diviacco, B., Di Mitri, S., Giannessi, L., Prince, K.C., Ueda, K., 2016. Slow interatomic Coulombic decay of multiply excited neon clusters. Phys. Rev. Lett. 117, 276806.

Jabbari, G., Klaiman, S., Chiang, Y.-C., Trinter, F., Jahnke, T., Gokhberg, K., 2014. Ab initio calculation of ICD widths in photoexcited HeNe. J. Chem. Phys. 140, 224305.

Jagau, T.-C., Zuev, D., Bravaya, K.B., Epifanovsky, E., Krylov, A.I., 2014. A fresh look at resonances and complex absorbing potentials: density matrix-based approach. J. Phys. Chem. Lett. 5 (2), 310–315.

Jahnke, T., Czasch, A., Schöffler, M., Schössler, S., Käsz, M., Titze, J., Kreidi, K., Grisenti, R.E., Staudte, A., Jagutzki, O., Schmidt, L.P.H., Weber, T., Schmidt-Böcking, H., Ueda, K., Dörner, R., 2007. Experimental separation of virtual photon exchange and electron transfer in interatomic Coulombic decay of neon dimers. Phys. Rev. Lett. 99, 153401.

Jahnke, T., Czasch, A., Schöffler, M.S., Schössler, S., Knapp, A., Käsz, M., Titze, J., Wimmer, C., Kreidi, K., Grisenti, R.E., Staudte, A., Jagutzki, O., Hergenhahn, U., Schmidt-Böcking, H., Dörner, R., 2004. Experimental observation of interatomic Coulombic decay in neon dimers. Phys. Rev. Lett. 93, 163401.

Jahnke, T., Sann, H., Havermeier, T., Kreidi, K., Stuck, C., Meckel, M., Schöffler, M., Neumann, N., Wallauer, R., Voss, S., Czasch, A., Jagutzki, O., Malakzadeh, A., Afaneh, F., Weber, T., Schmidt-Böcking, H., Dörner, R., 2010. Ultrafast energy transfer between water molecules. Nat. Phys. 6, 139.

Janev, R., Winter, H., 1985. State-selective electron capture in atom-highly charged ion collisions. Phys. Rep. 117 (5), 265–387.

Johnsen, R., Biondi, M.A., 1978. Measurements of radiative charge-transfer reactions of doubly and singly charged rare-gas ions with rare-gas atoms at thermal energies. Phys. Rev. A 18, 996–1003.

Kelly, H.P., 1975. K Auger rates calculated for Ne^+. Phys. Rev. A 11, 556–565.

Keshavarz, E., Farrokhpour, H., Sabzyan, H., Noorisafa, Z., Kivimäki, A., Richter, R., 2014. Core photoionization of the argon dimer in the photon-energy range of 255–340 eV studied by a photoelectron–photoion–photoion coincidence technique. Phys. Rev. A 89, 053409.

Kim, H.-K., Titze, J., Schöffler, M., Trinter, F., Waitz, M., Voigtsberger, J., Sann, H., Meckel, M., Stuck, C., Lenz, U., Odenweller, M., Neumann, N., Schössler, S., Ullmann-Pfleger, K., Ulrich, B., Fraga, R.C., Petridis, N., Metz, D., Jung, A., Grisenti, R., Czasch, A., Jagutzki, O., Schmidt, L., Jahnke, T., Schmidt-Böcking, H., Dörner, R., 2011. Enhanced production of low energy electrons by alpha particle impact. Proc. Natl. Acad. Sci. 108 (29), 11821–11824.

Kimura, K., Fukuzawa, H., Sakai, K., Mondal, S., Kukk, E., Kono, Y., Nagaoka, S., Tamenori, Y., Saito, N., Ueda, K., 2013a. Efficient site-specific low-energy electron production via interatomic Coulombic decay following resonant Auger decay in argon dimers. Phys. Rev. A 87, 043414.

Kimura, K., Fukuzawa, H., Tachibana, T., Ito, Y., Mondal, S., Okunishi, M., Schöffler, M., Williams, J., Jiang, Y., Tamenori, Y., Saito, N., Ueda, K., 2013b. Controlling low-energy electron emission via resonant-Auger-induced interatomic Coulombic decay. J. Phys. Chem. Lett. 4, 1838.

Kivimäki, A., Sorensen, S.L., Tchaplyguine, M., Gisselbrecht, M., Marinho, R.R.T., Feifel, R., Öhrwall, G., Svensson, S., Björneholm, O., 2005. Resonant Auger spectroscopy of argon clusters at the 2p threshold. Phys. Rev. A 71, 033204.

Knie, A., Hans, A., Förstel, M., Hergenhahn, U., Schmidt, P., Reiss, P., Ozga, C., Kambs, B., Trinter, F., Voigtsberger, J., Metz, D., Jahnke, T., Dörner, R., Kuleff, A.I., Cederbaum, L.S., Demekhin, P.V., Ehresmann, A., 2014. Detecting ultrafast interatomic electronic processes in media by fluorescence. New J. Phys. 16, 102002.

Kolorenč, P., Averbukh, V., Feifel, R., Eland, J., 2016. Collective relaxation processes in atoms, molecules and clusters. J. Phys. B 49 (8), 082001.

Kolorenč, P., Averbukh, V., Gokhberg, K., Cederbaum, L.S., 2008. Ab initio calculation of interatomic decay rates of excited doubly ionized states in clusters. J. Chem. Phys. 129, 244102.

Kopelke, S., Gokhberg, K., Averbukh, V., Tarantelli, F., Cederbaum, L.S., 2011. Ab initio interatomic decay widths of excited states by applying Stieltjes imaging to Lanczos pseudospectra. J. Chem. Phys. 134, 094107.

Körber, H., Mehlhorn, W., 1964. The KLL Auger spectrum of neon. Phys. Lett. 13 (2), 129–130.

Krause, M.O., Oliver, J.H., 1979. Natural widths of atomic K and L levels, $K\alpha$ X-ray lines and several KLL Auger lines. J. Phys. Chem. Ref. Data 8 (2), 329–338.

Kreidi, K., Jahnke, T., Weber, T., Havermeier, T., Grisenti, R.E., Liu, X.-J., Morisita, Y., Schössler, S., Schmidt, L., Schöffler, M., Odenweller, M., Neumann, N., Foucar, L., Titze, J., Ulrich, B., Sturm, F., Stuck, C., Wallauer, R., Voss, S., Lauter, I., Kim, H.K., Rudloff, M., Fukuzawa, H., Prümper, G., Saito, N., Ueda, K., Czasch, A., Jagutzki, O., Schmidt-Böcking, H., Semenov, S.K., Cherepkov, N.A., Dörner, R., 2008a. Localization of inner-shell photoelectron emission and interatomic Coulombic decay in Ne_2. J. Phys. B 41, 101002.

Kreidi, K., Jahnke, T., Weber, T., Havermeier, T., Liu, X., Morisita, Y., Schössler, S., Schmidt, L., Schöffler, M., Odenweller, M., Neumann, N., Foucar, L., Titze, J., Ulrich, B., Sturm, F., Stuck, C., Wallauer, R., Voss, S., Lauter, I., Kim, H.K., Rudloff, M., Fukuzawa, H., Prümper, G., Saito, N., Ueda, K., Czasch, A., Jagutzki, O., Schmidt-Böcking, H., Stoychev, S., Demekhin, P.V., Dörner, R., 2008b. Relaxation processes following 1s photoionization and Auger decay in Ne_2. Phys. Rev. A 78, 043422.

Kryzhevoi, N.V., Averbukh, V., Cederbaum, L.S., 2007. High activity of helium droplets following ionization of systems inside those droplets. Phys. Rev. B 76, 094513.

Kuleff, A.I., Breidbach, J., Cederbaum, L.S., 2005. Multielectron wave-packet propagation: general theory and application. J. Chem. Phys. 123 (4), 044111.

Kuleff, A.I., Cederbaum, L.S., 2007. Tracing ultrafast interatomic electronic decay processes in real time and space. Phys. Rev. Lett. 98, 083201.

Kuleff, A.I., Gokhberg, K., Kopelke, S., Cederbaum, L.S., 2010. Ultrafast interatomic electronic decay in multiply excited clusters. Phys. Rev. Lett. 105, 043004.

Kunitsa, A.A., Granovsky, A.A., Bravaya, K.B., 2017. CAP-XMCQDPT2 method for molecular electronic resonances. J. Chem. Phys. 146 (18), 184107.

LaForge, A.C., Drabbels, M., Brauer, N.B., Coreno, M., Devetta, M., Di Fraia, M., Finetti, P., Grazioli, C., Katzy, R., Lyamayev, V., Mazza, T., Mudrich, M., O'Keeffe, P., Ovcharenko, Y., Piseri, P., Plekan, O., Prince, K.C., Richter, R., Stranges, S., Callegari, C., Möller, T., Stienkemeier, F., 2014. Collective autoionization in multiply-excited systems: a novel ionization process observed in helium nanodroplets. Sci. Rep. 4, 3621.

LaForge, A.C., Stumpf, V., Gokhberg, K., von Vangerow, J., Stienkemeier, F., Kryzhevoi, N.V., O'Keeffe, P., Ciavardini, A., Krishnan, S.R., Coreno, M., Prince, K.C., Richter, R., Moshammer, R., Pfeifer, T., Cederbaum, L.S., Mudrich, M., 2016. Enhanced ionization of embedded clusters by electron-transfer-mediated decay in helium nanodroplets. Phys. Rev. Lett. 116, 203001.

Langhoff, P., 1973. Stieltjes imaging of atomic and molecular photoabsorption profiles. Chem. Phys. Lett. 22 (1), 60–64.

Leung, C.H., Emery, L., Song, K.S., 1983. Self-trapped exciton and rare-gas impurity centers in solid Ne. Phys. Rev. B 28, 3474–3481.

Liu, X.-J., Saito, N., Fukuzawa, H., Morishita, Y., Stoychev, S., Kuleff, A., Suzuki, I.H., Tamenori, Y., Richter, R., Prümper, G., Ueda, K., 2007. Evidence of sequential interatomic decay in argon trimers obtained by electron–triple-ion coincidence spectroscopy. J. Phys. B 40, F1.

Lundwall, M., Pokapanich, W., Bergersen, H., Lindblad, A., Rander, T., Öhrwall, G., Tchaplyguine, M., Barth, S., Hergenhahn, U., Svensson, S., Björneholm, O., 2007. Self-assembled heterogeneous argon/neon core-shell clusters studied by photoelectron spectroscopy. J. Chem. Phys. 126, 214706.

Maguire, M.E., Cowan, J.A., 2002. Magnesium chemistry and biochemistry. Biometals 15 (3), 203–210.

Marburger, S., Kugeler, O., Hergenhahn, U., Möller, T., 2003. Experimental evidence for interatomic Coulombic decay in Ne clusters. Phys. Rev. Lett. 90, 203401.

Marcus, R.A., 1956. Electrostatic free energy and other properties of states having nonequilibrium polarization. I. J. Chem. Phys. 24 (5), 979–989.

Marcus, R.A., 1965. On the theory of electron-transfer reactions. VI. Unified treatment for homogeneous and electrode reactions. J. Chem. Phys. 43 (2), 679–701.

Marquardt, R., Quack, M., 2020. Foundations of time dependent quantum dynamics of molecules under isolation and in coherent electromagnetic fields. In: Marquardt, R., Quack, M. (Eds.), Molecular Spectroscopy and Quantum Dynamics. Elsevier, Amsterdam. Chapter 1 (this book).

Martin, F., Burrow, P.D., Cai, Z., Cloutier, P., Hunting, D., Sanche, L., 2004. DNA strand breaks induced by 0–4 eV electrons: the role of shape resonances. Phys. Rev. Lett. 93, 068101.

Miteva, T., Chiang, Y.-C., Kolorenč, P., Kuleff, A.I., Gokhberg, K., Cederbaum, L.S., 2014a. Interatomic Coulombic decay following resonant core excitation of Ar in argon dimer. J. Chem. Phys. 141, 064307.

Miteva, T., Chiang, Y.-C., Kolorenč, P., Kuleff, A.I., Gokhberg, K., Cederbaum, L.S., 2014b. The effect of the partner atom on the spectra of interatomic Coulombic decay triggered by resonant Auger processes. J. Chem. Phys. 141, 164303.

Miteva, T., Klaiman, S., Gromov, E.V., Gokhberg, K., 2014c. Ionization satellites of the ArHe dimer. J. Chem. Phys. 140 (20), 204320.

Miteva, T., Wenzel, J., Klaiman, S., Dreuw, A., Gokhberg, K., 2016. X-ray absorption spectra of microsolvated metal cations. Phys. Chem. Chem. Phys. 18, 16671–16681.

Moiseyev, N., 1998. Quantum theory of resonances: calculating energies, widths and cross-sections by complex scaling. Phys. Rep. 302 (5), 212–293.

Morgan, D.V., Sagurton, M., Bartlett, R.J., 1997. Single-photon multiple ionization of neon in the K-edge region. Phys. Rev. A 55, 1113–1118.

Morishita, Y., Liu, X.-J., Saito, N., Lischke, T., Kato, M., Prümper, G., Oura, M., Yamaoka, H., Tamenori, Y., Suzuki, I.H., Ueda, K., 2006a. Experimental evidence of interatomic Coulombic decay from the Auger final states in argon dimers. Phys. Rev. Lett. 96, 243402.

Morishita, Y., Saito, N., Suzuki, I.H., Fukuzawa, H., Liu, X.-J., Sakai, K., Prümper, G., Ueda, K., Iwayama, H., Nagaya, K., Yao, M., Kreidi, K., Schöffler, M., Jahnke, T., Schössler, S., Dörner, R., Weber, T., Harries, J., Tamenori, Y., 2008. Evidence of interatomic Coulombic decay in ArKr after Ar 2p Auger decay. J. Phys. B 41, 025101.

Morishita, Y., Tamenori, Y., Okada, K., Oyama, T., Yamamoto, K., Tabayashi, K., Ibuki, T., Moribayashi, K., Suzuki, I.H., 2006b. Formation mechanisms of multi-charged Kr ions through 2p shell photoionization using a coincidence technique. J. Phys. B 39 (6), 1323.

Motomura, K., Kukk, E., Fukuzawa, H., Wada, S.-I., Nagaya, K., Ohmura, S., Mondal, S., Tachibana, T., Ito, Y., Koga, R., Sakai, T., Matsunami, K., Rudenko, A., Nicolas, C., Liu, X.-J., Miron, C., Zhang, Y., Jiang, Y., Chen, J., Anand, M., Kim, D.-E., Tono, K., Yabashi, M., Yao, M., Ueda, K., 2015. Charge and nuclear dynamics induced by deep inner-shell multiphoton ionization of CH_3I molecules by intense X-ray free-electron laser pulses. J. Phys. Chem. Lett. 6 (15), 2944–2949.

Müller, I.B., Cederbaum, L.S., 2005. Electronic decay following ionization of aqueous Li^+ microsolvation clusters. J. Chem. Phys. 122, 094305.

Nagaya, K., Iablonskyi, D., Golubev, N.V., Matsunami, K., Fukuzawa, H., Motomura, K., Nishiyama, T., Sakai, T., Tachibana, T., Mondal, S., Wada, S., Prince, K.C., Callegari, C., Miron, C., Saito, N., Yabashi, M., Demekhin, P.V., Cederbaum, L.S., Kuleff, A.I., Yao, M., Ueda, K., 2016a. Interatomic Coulombic decay cascades in multiply excited neon clusters. Nat. Commun. 7, 13477.

Nagaya, K., Motomura, K., Kukk, E., Fukuzawa, H., Wada, S., Tachibana, T., Ito, Y., Mondal, S., Sakai, T., Matsunami, K., Koga, R., Ohmura, S., Takahashi, Y., Kanno, M., Rudenko, A., Nicolas, C., Liu, X.-J., Zhang, Y., Chen, J., Anand, M., Jiang, Y.H., Kim, D.-E., Tono, K., Yabashi, M., Kono, H., Miron, C., Yao, M., Ueda, K., 2016b. Ultrafast dynamics of a nucleobase analogue illuminated by a short intense X-ray free electron laser pulse. Phys. Rev. X 6, 021035.

Nagaya, K., Sugishima, A., Iwayama, H., Murakami, H., Yao, M., Fukuzawa, H., Liu, X.-J., Motomura, K., Ueda, K., Saito, N., Foucar, L., Rudenko, A., Kurka, M., Kühne, K.-U., Ullrich, J., Czasch, A., Dörner, R., Feifel, R., Nagasono, M., Higashiya, A., Yabashi, M., Ishikawa, T., Togashi, T., Kimura, H., Ohashi, H., 2013. Unusual under-threshold ionization of neon clusters studied by ion spectroscopy. J. Phys. B 46, 164023.

Najjari, B., Voitkiv, A.B., Müller, C., 2010. Two-center resonant photoionization. Phys. Rev. Lett. 105, 153002.

Neutze, R., Wouts, R., van der Spoel, D., Weckert, E., Hajdu, J., 2000. Potential for biomolecular imaging with femtosecond X-ray pulses. Nature 406, 752–757.

Nordlund, D., Ogasawara, H., Bluhm, H., Takahashi, O., Odelius, M., Nagasono, M., Pettersson, L.G.M., Nilsson, A., 2007. Probing the electron delocalization in liquid water and ice at attosecond time scales. Phys. Rev. Lett. 99, 217406.

Öhrwall, G., Ottosson, N., Pokapanich, W., Legendre, S., Svensson, S., Björneholm, O., 2010. Charge dependence of solvent-mediated intermolecular Coster–Kronig decay dynamics of aqueous ions. J. Phys. Chem. B 114, 17057.

Öhrwall, G., Tchaplyguine, M., Lundwall, M., Feifel, R., Bergersen, H., Rander, T., Lindblad, A., Schulz, J., Peredkov, S., Barth, S., Marburger, S., Hergenhahn, U., Svensson, S., Björneholm, O., 2004. Femtosecond interatomic Coulombic decay in free neon clusters: large lifetime differences between surface and bulk. Phys. Rev. Lett. 93, 173401.

O'Keeffe, P., Ripani, E., Bolognesi, P., Coreno, M., Devetta, M., Callegari, C., Di Fraia, M., Prince, K.C., Richter, R., Alagia, M., Kivimäki, A., Avaldi, L., 2013. The role of the partner atom and resonant excitation energy in interatomic Coulombic decay in rare gas dimers. J. Phys. Chem. Lett. 4, 1797.

Ouchi, T., Sakai, K., Fukuzawa, H., Higuchi, I., Demekhin, P.V., Chiang, Y.-C., Stoychev, S.D., Kuleff, A.I., Mazza, T., Schöffler, M., Nagaya, K., Yao, M., Tamenori, Y., Saito, N., Ueda, K., 2011. Interatomic Coulombic decay following Ne 1s Auger decay in NeAr. Phys. Rev. A 83, 053415.

Ouchi, T., Stumpf, V., Miteva, T., Fukuzawa, H., Sakai, K., Liu, X.-J., Mazza, T., Schöffler, M., Iwayama, H., Nagaya, K., Tamenori, Y., Saito, N., Kuleff, A.I., Gokhberg, K., Ueda, K., 2017. Ion pair formation in the NeAr dimer irradiated by monochromatic soft X-rays. Chem. Phys. 482, 178–184.

Ovcharenko, Y., Lyamayev, V., Katzy, R., Devetta, M., LaForge, A., O'Keeffe, P., Plekan, O., Finetti, P., Di Fraia, M., Mudrich, M., Krikunova, M., Piseri, P., Coreno, M., Brauer, N., Mazza, T., Stranges, S., Grazioli, C., Richter, R., Prince, K.C., Drabbels, M., Callegari, C., Stienkemeier, F., Möller, T., 2014. Novel collective autoionization process observed in electron spectra of He clusters. Phys. Rev. Lett. 112, 073401.

Pedersen, H.B., Domesle, C., Lammich, L., Dziarzhytski, S., Guerassimova, N., Treusch, R., Harbo, L.S., Heber, O., Jordon-Thaden, B., Arion, T., Förstel, M., Stier, M., Hergenhahn, U., Wolf, A., 2013. Photolysis of water-radical ions H_2O^+ in the XUV: fragmentation through dicationic states. Phys. Rev. A 87, 013402.

Pellegrini, C., Marinelli, A., Reiche, S., 2016. The physics of X-ray free-electron lasers. Rev. Mod. Phys. 88, 015006.

Pernpointner, M., Kryzhevoi, N.V., Urbaczek, S., 2008. Possible electronic decay channels in the ionization spectra of small clusters composed of Ar and Kr: a four-component relativistic treatment. J. Chem. Phys. 129, 024304.

Pomplun, E., Booz, J., Charlton, D.E., 1987. A Monte Carlo simulation of Auger cascades. Radiat. Res. 111 (3), 533–552.

Reinhardt, W.P., 1979. L^2 discretization of atomic and molecular electronic continua: moment, quadrature and J-matrix techniques. Comput. Phys. Commun. 17 (1), 1–21.

Reinhardt, W.P., 1982. Complex coordinates in the theory of atomic and molecular structure and dynamics. Annu. Rev. Phys. Chem. 33 (1), 223–255.

Ren, X., Al Maalouf, E.J., Dorn, A., Denifl, S., 2016. Direct evidence of two interatomic relaxation mechanisms in argon dimers ionized by electron impact. Nat. Commun. 7, 11093.

Ren, X., Miteva, T., Kolorenč, P., Gokhberg, K., Kuleff, A.I., Cederbaum, L.S., Dorn, A., 2017. Observation of fast and slow interatomic Coulombic decay in argon dimers induced by electron-impact ionization. Phys. Rev. A 96, 032715.

Ren, X., Wang, E., Skitnevskaya, A.D., Trofimov, A.B., Gokhberg, K., Dorn, A., 2018. Experimental evidence for ultrafast intermolecular relaxation processes in hydrated biomolecules. Nat. Phys. 14 (10), 1062–1066.

Riss, U., Meyer, H.-D., 1993. Calculation of resonance energies and widths using the complex absorbing potential method. J. Phys. B 26, 4503.

Riss, U., Meyer, H.-D., 1995. Reflection free complex absorbing potentials. J. Phys. B 28, 1475.

Rodriguez-Cruz, S.E., Jockusch, R.A., Williams, E.R., 1998. Hydration energies of divalent metal ions, $Ca^{2+}(H_2O)_n$ ($n =$ 5–7) and $Ni^{2+}(H_2O)_n$ ($n = 6$–8), obtained by blackbody infrared radiative dissociation. J. Am. Chem. Soc. 120 (23), 5842–5843.

Rodriguez-Cruz, S.E., Jockusch, R.A., Williams, E.R., 1999. Binding energies of hexahydrated alkaline earth metal ions, $M^{2+}(H_2O)_6$, M = Mg, Ca, Sr, Ba: evidence of isomeric structures for magnesium. J. Am. Chem. Soc. 121 (9), 1986–1987.

Rosso, A., Pokapanich, W., Öhrwall, G., Svensson, S., Björneholm, O., Tchaplyguine, M., 2007. Adsorption of polar molecules on krypton clusters. J. Chem. Phys. 127 (8), 084313.

Rudenko, A., Inhester, L., Hanasaki, K., Li, X., Robatjazi, S.J., Erk, B., Boll, R., Toyota, K., Hao, Y., Vendrell, O., Bomme, C., Savelyev, E., Rudek, B., Foucar, L., Southworth, S.H., Lehmann, C.S., Kraessig, B., Marchenko, T., Simon, M., Ueda, K., Ferguson, K.R., Bucher, M., Gorkhover, T., Carron, S., Alonso-Mori, R., Koglin, J.E., Correa, J., Williams, G.J., Boutet, S., Young, L., Bostedt, C., Son, S.-K., Santra, R., Rolles, D., 2017. Femtosecond response of polyatomic molecules to ultra-intense hard X-rays. Nature 546, 129–132.

Rühl, E., 2003. Core level excitation, ionization, relaxation, and fragmentation of free clusters. Int. J. Mass Spectrom. 229 (3), 117–142.

Saito, N., Morishita, Y., Suzuki, I., Stoychev, S., Kuleff, A., Cederbaum, L., Liu, X.-J., Fukuzawa, H., Prümper, G., Ueda, K., 2007. Evidence of radiative charge transfer in argon dimers. Chem. Phys. Lett. 441 (1), 16–19.

Sakai, K., Stoychev, S., Ouchi, T., Higuchi, I., Schöffler, M., Mazza, T., Fukuzawa, H., Nagaya, K., Yao, M., Tamenori, Y., Kuleff, A.I., Saito, N., Ueda, K., 2011. Electron-transfer-mediated decay and interatomic Coulombic decay from the triply ionized states in argon dimers. Phys. Rev. Lett. 106, 033401.

Santra, R., Cederbaum, L.S., 2002. Complex absorbing potentials in the framework of electron propagator theory. I. General formalism. J. Chem. Phys. 117 (12), 5511–5521.

Santra, R., Cederbaum, L.S., 2003. Coulombic energy transfer and triple ionization in clusters. Phys. Rev. Lett. 90, 153401.

Santra, R., Cederbaum, L.S., Meyer, H.-D., 1999. Electronic decay of molecular clusters: non-stationary states computed by standard quantum chemistry methods. Chem. Phys. Lett. 303, 413.

Santra, R., Zobeley, J., Cederbaum, L.S., 2001. Electronic decay of valence holes in clusters and condensed matter. Phys. Rev. B 64, 245104.

Santra, R., Zobeley, J., Cederbaum, L.S., Moiseyev, N., 2000. Interatomic Coulombic decay in van der Waals clusters and impact of nuclear motion. Phys. Rev. Lett. 85, 4490.

Scheit, S., Averbukh, V., Meyer, H.-D., Moiseyev, N., Santra, R., Sommerfeld, T., Zobeley, J., Cederbaum, L.S., 2004. On the interatomic Coulombic decay in the Ne dimer. J. Chem. Phys. 121, 8393.

Scheit, S., Averbukh, V., Meyer, H.-D., Zobeley, J., Cederbaum, L.S., 2006. Interatomic Coulombic decay in a heteroatomic rare gas cluster. J. Chem. Phys. 124, 154305.

Schnorr, K., Senftleben, A., Kurka, M., Rudenko, A., Foucar, L., Schmid, G., Broska, A., Pfeifer, T., Meyer, K., Anielski, D., Boll, R., Rolles, D., Kübel, M., Kling, M.F., Jiang, Y.H., Mondal, S., Tachibana, T., Ueda, K., Marchenko, T., Simon, M., Brenner, G., Treusch, R., Scheit, S., Averbukh, V., Ullrich, J., Schröter, C.D., Moshammer, R., 2013. Time-resolved measurement of interatomic Coulombic decay in Ne_2. Phys. Rev. Lett. 111, 093402.

Schütte, B., Arbeiter, M., Fennel, T., Jabbari, G., Kuleff, A.I., Vrakking, M., Rouzée, A., 2015. Observation of correlated electronic decay in expanding clusters triggered by near-infrared fields. Nat. Commun. 6, 8596.

Shi, Y., 2014. A glimpse of structural biology through X-ray crystallography. Cell 159, 995.

Sisourat, N., Kryzhevoi, N.V., Kolorenč, P., Scheit, S., Jahnke, T., Cederbaum, L.S., 2010. Ultralong-range energy transfer by interatomic Coulombic decay in an extreme quantum system. Nat. Phys. 6, 508.

Son, S.-K., Santra, R., 2012. Monte Carlo calculation of ion, electron, and photon spectra of xenon atoms in X-ray free-electron laser pulses. Phys. Rev. A 85, 063415.

Stöhr, J., 1992. NEXAFS Spectroscopy. Springer, Berlin and Heidelberg.

Stoychev, S.D., Kuleff, A.I., Tarantelli, F., Cederbaum, L.S., 2008. On the interatomic electronic processes following Auger decay in neon dimer. J. Chem. Phys. 129, 074307.

Stumpf, V., Brunken, C., Gokhberg, K., 2016a. Impact of metal ion's charge on the interatomic Coulombic decay widths in microsolvated clusters. J. Chem. Phys. 145, 104306.

Stumpf, V., Gokhberg, K., Cederbaum, L.S., 2016b. The role of metal ions in X-ray-induced photochemistry. Nat. Chem. 8, 237.

Stumpf, V., Kolorenč, P., Gokhberg, K., Cederbaum, L.S., 2013. Efficient pathway to neutralization of multiply charged ions produced in Auger processes. Phys. Rev. Lett. 110, 258302.

Stumpf, V., Kryzhevoi, N.V., Gokhberg, K., Cederbaum, L.S., 2014. Enhanced one-photon double ionization of atoms and molecules in an environment of different species. Phys. Rev. Lett. 112, 193001.

Stumpf, V., Scheit, S., Kolorenč, P., Gokhberg, K., 2017. Electron-transfer mediated decay in NeXe triggered by K-LL Auger decay of Ne. Chem. Phys. 482, 192.

Suga, M., Akita, F., Hirata, K., Ueno, G., Murakami, H., Nakajima, Y., Shimizu, T., Yamashita, K., Yamamoto, M., Ago, H., Shen, J.-R., 2015. Native structure of photosystem II at 1.95 Å resolution viewed by femtosecond X-ray pulses. Nature 517, 99–103.

Taberman, H., 2018. Radiation damage in macromolecular crystallography—an experimentalist's view. Crystals 8 (4), 157.

Takanashi, T., Golubev, N.V., Callegari, C., Fukuzawa, H., Motomura, K., Iablonskyi, D., Kumagai, Y., Mondal, S., Tachibana, T., Nagaya, K., Nishiyama, T., Matsunami, K., Johnsson, P., Piseri, P., Sansone, G., Dubrouil, A., Reduzzi, M., Carpeggiani, P., Vozzi, C., Devetta, M., Negro, M., Faccialà, D., Calegari, F., Trabattoni, A., Castrovilli, M.C., Ovcharenko, Y., Mudrich, M., Stienkemeier, F., Coreno, M., Alagia, M., Schütte, B., Berrah, N., Plekan, O., Finetti, P., Spezzani, C., Ferrari, E., Allaria, E., Penco, G., Serpico, C., De Ninno, G., Diviacco, B., Di Mitri, S., Giannessi, L., Jabbari, G., Prince, K.C., Cederbaum, L.S., Demekhin, P.V., Kuleff, A.I., Ueda, K., 2017. Time-resolved measurement of interatomic Coulombic decay induced by two-photon double excitation of Ne_2. Phys. Rev. Lett. 118, 033202.

Tavernelli, I., Gaigeot, M., Vuilleumier, R., Stia, C., Hervé du Penhoat, M., Politis, M., 2008. Time-dependent density functional theory molecular dynamics simulations of liquid water radiolysis. ChemPhysChem 9 (14), 2099–2103.

Thomas, S., Robin, S., 2001. Efficient method to perform CAP/CI calculations for temporary anions. Int. J. Quant. Chem. 82 (5), 218–226.

Titze, J., Schöffler, M.S., Kim, H.-K., Trinter, F., Waitz, M., Voigtsberger, J., Neumann, N., Ulrich, B., Kreidi, K., Wallauer, R., Odenweller, M., Havermeier, T., Schössler, S., Meckel, M., Foucar, L., Jahnke, T., Czasch, A., Schmidt, L.P.H., Jagutzki, O., Grisenti, R.E., Schmidt-Böcking, H., Lüdde, H.J., Dörner, R., 2011. Ionization dynamics of helium dimers in fast collisions with He^{++}. Phys. Rev. Lett. 106, 033201.

Toennies, J.P., Vilesov, A.F., 2004. Superfluid helium droplets: a uniquely cold nanomatrix for molecules and molecular complexes. Angew. Chem., Int. Ed. 43 (20), 2622–2648.

Trinter, F., Schöffler, M.S., Kim, H.-K., Sturm, F.P., Cole, K., Neumann, N., Vredenborg, A., Williams, J., Bocharova, I., Guillemin, R., Simon, M., Belkacem, A., Landers, A.L., Weber, T., Schmidt-Böcking, H., Dörner, R., Jahnke, T., 2014. Resonant Auger decay driving intermolecular Coulombic decay in molecular dimers. Nature 505, 664.

Trinter, F., Williams, J.B., Weller, M., Waitz, M., Pitzer, M., Voigtsberger, J., Schober, C., Kastirke, G., Müller, C., Goihl, C., Burzynski, P., Wiegandt, F., Bauer, T., Wallauer, R., Sann, H., Kalinin, A., Schmidt, L., Schöffler, M., Sisourat, N., Jahnke, T., 2013a. Evolution of interatomic Coulombic decay in the time domain. Phys. Rev. Lett. 111, 093401.

Trinter, F., Williams, J.B., Weller, M., Waitz, M., Pitzer, M., Voigtsberger, J., Schober, C., Kastirke, G., Müller, C., Goihl, C., Burzynski, P., Wiegandt, F., Wallauer, R., Kalinin, A., Schmidt, L., Schöffler, M., Chiang, Y.-C., Gokhberg, K., Jahnke, T., Dörner, R., 2013b. Vibrationally resolved decay width of interatomic Coulombic decay in HeNe. Phys. Rev. Lett. 111, 233004.

Ueda, K., 2006. Core excitation and de-excitation spectroscopies of free atoms and molecules. J. Phys. Soc. Jpn. 75 (3), 032001.

Ueda, K., Fukuzawa, H., Liu, X.-J., Sakai, K., Prümper, G., Morishita, Y., Saito, N., Suzuki, I.H., Nagaya, K., Iwayama, H., Yao, M., Kreidi, K., Schöffler, M., Jahnke, T., Schössler, S., Dörner, R., Weber, T., Harries, J., Tamenori, Y., 2008. Interatomic Coulombic decay following the Auger decay: experimental evidence in rare-gas dimers. J. Electron Spectrosc. Relat. Phenom. 166–167, 3.

Ullrich, J., Moshammer, R., Dorn, A., Dörner, R., Schmidt, L.P.H., Schmidt-Böcking, H., 2003. Recoil-ion and electron momentum spectroscopy: reaction-microscopes. Rep. Prog. Phys. 66 (9), 1463.

Unger, I., Seidel, R., Thürmer, S., Pohl, M.N., Aziz, E.F., Cederbaum, L.S., Muchová, E., Slavíček, P., Winter, B., Kryzhevoi., N.V., 2017. Observation of electron-transfer-mediated decay in aqueous solution. Nat. Chem. 9, 708.

Vaval, N., Cederbaum, L.S., 2007. Ab initio lifetimes in the interatomic Coulombic decay of neon clusters computed with propagators. J. Chem. Phys. 126, 164110.

Viefhaus, J., Braune, M., Korica, S., Reinköster, A., Rolles, D., Becker, U., 2005. Auger cascades versus direct double Auger: relaxation processes following photoionization of the Kr 3d and Xe 4d, 3d inner shells. J. Phys. B 38 (21), 3885.

Viefhaus, J., Cvejanović, S., Langer, B., Lischke, T., Prümper, G., Rolles, D., Golovin, A.V., Grum-Grzhimailo, A.N., Kabachnik, N.M., Becker, U., 2004. Energy and angular distributions of electrons emitted by direct double Auger decay. Phys. Rev. Lett. 92, 083001.

Wurth, W., Rocker, G., Feulner, P., Scheuerer, R., Zhu, L., Menzel, D., 1993. Core excitation and deexcitation in argon multilayers: surface- and bulk-specific transitions and autoionization versus Auger decay. Phys. Rev. B 47, 6697–6704.

Yamazaki, M., Adachi, J.-I., Kimura, Y., Yagishita, A., Stener, M., Decleva, P., Kosugi, N., Iwayama, H., Nagaya, K., Yao, M., 2008. Decay channel dependence of the photoelectron angular distributions in core-level ionization of ne dimers. Phys. Rev. Lett. 101, 043004.

Yan, S., Zhang, P., Ma, X., Xu, S., Tian, S.X., Li, B., Zhu, X.L., Feng, W.T., Zhao, D.M., 2014. Dissociation mechanisms of the Ar trimer induced by a third atom in high-energy electron-impact ionization. Phys. Rev. A 89, 062707.

Yan, S., Zhang, P., Stumpf, V., Gokhberg, K., Zhang, X.C., Xu, S., Li, B., Shen, L.L., Zhu, X.L., Feng, W.T., Zhang, S.F., Zhao, D.M., Ma, X., 2018. Interatomic relaxation processes induced in neon dimers by electron-impact ionization. Phys. Rev. A 97, 010701(R).

You, D., Fukuzawa, H., Sakakibara, Y., Takanashi, T., Ito, Y., Maliyar, G.G., Motomura, K., Nagaya, K., Nishiyama, T., Asa, K., Sato, Y., Saito, N., Oura, M., Schöffler, M., Kastirke, G., Hergenhahn, U., Stumpf, V., Gokhberg, K., Kuleff, A.I., Cederbaum., L.S., Uleda, K., 2017. Charge transfer to ground-state ions produces free electrons. Nat. Commun. 8, 14277.

Zobeley, J., Cederbaum, L.S., Tarantelli, F., 1998. Highly excited electronic states of molecular clusters and their decay. J. Chem. Phys. 108, 9737.

Zobeley, J., Santra, R., Cederbaum, L.S., 2001. Electronic decay in weakly bound heteroclusters: energy transfer versus electron transfer. J. Chem. Phys. 115, 5076.

CHAPTER 6

Ab Initio Semiclassical Evaluation of Vibrationally Resolved Electronic Spectra With Thawed Gaussians

JIŘÍ VANÍČEK • TOMISLAV BEGUŠIĆ
Laboratory of Theoretical Physical Chemistry, Institut des Sciences et Ingénierie Chimiques, Ecole Polytechnique Fédérale de Lausanne (EPFL), Lausanne, Switzerland

Abstract

Vibrationally resolved electronic spectra of polyatomic molecules provide valuable information about the quantum properties of both electrons and nuclei. This chapter reviews the recent progress in *ab initio* semiclassical calculations of such spectra, based on the thawed Gaussian approximation and its extensions. After reviewing molecular quantum dynamics induced by the interaction with electromagnetic field and the most common semiclassical approximations to quantum dynamics, we explain details of the thawed Gaussian approximation and its variants. Next, we discuss the time-dependent approach to steady-state and time-resolved electronic spectroscopy, and review several standard models that facilitate interpreting vibrationally resolved electronic spectra. Finally, we present the on-the-fly *ab initio* implementation of the thawed Gaussian approximation and provide several examples of both linear and pump–probe spectra computed with this methodology, which, at a low additional cost and without sacrificing the ease of interpretation, outperforms the standard global harmonic approaches.

6.1 INTRODUCTION

Vibrationally resolved electronic spectroscopy provides a powerful tool for studying both electronic and nuclear motions in molecules: The measured spectrum contains precious information about the shapes of molecular potential energy surfaces, such as displacements between their minima, Duschinsky rotation between the normal modes of the ground-state and excited-state surfaces, couplings among vibrational modes, and non-adiabatic couplings between electronic states. The steady-state version of molecular electronic spectroscopy, used for many years, has been usually interpreted through the time-independent Franck–Condon picture, but an alternative, time-dependent approach (Heller, 1981b; Mukamel, 1999; Tannor, 2007; Heller, 2018), which examines the evolution of the nuclear wavepacket before evaluating the spectra, is more fundamental. The time-dependent approach is also much more natural for interpreting time-resolved spectra, and will be used exclusively in this chapter.

Accurate calculations of vibrationally resolved electronic spectra are difficult because they require both accurate potential energy surfaces, which necessitate expensive electronic structure calculations, and accurate nuclear quantum dynamics, impossible in high dimensions. As a result, such calculations are typically limited to accurate nuclear quantum dynamics calculations on potential energy surfaces which are either full-dimensional but approximate, such as those based on global harmonic models (Baiardi et al., 2013), or which take into account only a few degrees of freedom. To include all degrees of freedom as well as their anharmonicity without constructing global potential energy surfaces, various authors developed a number of trajectory-based on-the-fly *ab initio* (also called "direct dynamics" or "first-principles dynamics") quantum and semiclassical methods (Curchod and Martínez, 2018; Shalashilin and Child, 2004; Richings et al., 2015; Saita and Shalashilin, 2012; Šulc et al., 2013; Tatchen and Pollak, 2009; Ianconescu et al., 2013; Wong et al., 2011; Ceotto et al., 2009b; Gabas et al., 2017). Despite their success, which has, in fact, motivated the writing of this chapter, these methods are rather expensive when it comes to computing the vibrational structure of electronic spectra, since their convergence requires a large number of classical trajectories. In this chapter, we will, therefore, devote most attention to the on-the-fly *ab initio* version (Wehrle et al., 2014, 2015) of the thawed Gaussian approximation (TGA) (Heller, 1975), which provides a computationally efficient compromise between accurate but expensive multiple-trajectory semiclassical methods and cheap but restrictive and often

Molecular Spectroscopy and Quantum Dynamics. https://doi.org/10.1016/B978-0-12-817234-6.00011-8

flawed global harmonic methods. The TGA can be easily run in full dimensionality, and even its single-trajectory version captures some anharmonicity of the potential. Obviously, it cannot describe high resolution spectra, tunneling, wavepacket splitting, and very anharmonic dynamics in floppy molecules, yet, we will show several examples of surprisingly accurate molecular absorption, emission, and photoelectron spectra computed with the TGA.

Below, we define the notation and acronyms used in this chapter, the remainder of which is organized as follows: Section 6.2, devoted to the quantum molecular dynamics induced by the interaction with electromagnetic field, introduces the most common approximations for the molecule–field interaction. Section 6.3 reviews briefly the most popular semiclassical approaches for solving the time-dependent Schrödinger equation. One of these, the thawed Gaussian approximation, is discussed in more detail, together with its extensions in Section 6.4. In Section 6.5, the time-dependent approach to both steady-state and time-resolved spectra calculations is presented in the framework of approximations explained in Section 6.2. We also discuss rotational averaging, important in describing electronic spectra beyond the Condon approximation but often ignored in theoretical calculations. In Section 6.6, several "standard" models useful for interpreting molecular electronic spectra are defined and accompanied with numerical examples demonstrating both their advantages and shortcomings. Intricacies of the on-the-fly *ab initio* implementation of the thawed Gaussian approximation are discussed in Section 6.7 and several examples of absorption, emission, photoelectron, and pump–probe spectra computed with this approximation are presented in Section 6.8. Finally, a conclusion and outlook are given in Section 6.9.

6.1.1 Notation

Throughout this chapter, we consider a molecular system with S relevant electronic states and D vibrational degrees of freedom. Electronic states are indexed by a Greek letter, such as $\alpha = 1, \ldots, S$, while vibrational modes are indexed by a Latin letter, such as $j = 1, \ldots, D$. Three different vector spaces play an important role in such systems: the ambient 3-dimensional space $\mathbb{R}^3$, the nuclear D-dimensional real coordinate space $\mathbb{R}^D$, and the electronic S-dimensional complex Hilbert space $\mathbb{C}^S$. To distinguish vectors and operators on these spaces, 3-dimensional vectors are denoted with an arrow (e.g., $\vec{\epsilon}$), nuclear D-dimensional vectors or $D \times D$ matrices use no special notation (e.g., generalized nuclear coordinates q), and the electronic S-dimensional vectors or $S \times S$ matrices (such as $\boldsymbol{\mu}$ or $\boldsymbol{A}$) representing electronic states or operators are denoted with the **bold** font. Scalar and matrix products in both the 3-dimensional and nuclear D-dimensional spaces are denoted with a dot (as in $\vec{\mu}_{21} \cdot \vec{\epsilon}$ or $p^T \cdot m^{-1} \cdot p$), whereas the matrix product in the electronic Hilbert space use no special notation; it is expressed by a juxtaposition of matrices (as in $\boldsymbol{AB}$). Finally, the nuclear operators are denoted with a hat ^ (as in $\hat{p}$). The ground vibrational state g of the ground electronic state 1 is denoted by $|1, g\rangle$. Angular frequency of light ω is used in all expressions, while the numerical results are displayed in terms of the wavenumber $\tilde{\nu} = \omega/(2\pi c)$.

6.1.2 List of Acronyms

3TGA	three thawed Gaussians approximation
CASPT2	complete active space second-order perturbation theory
CCSD	coupled cluster singles and doubles
ETGA	extended thawed Gaussian approximation
MP2	Møller–Plesset second-order perturbation theory
n-TGA	n thawed Gaussians approximation
TGA	thawed Gaussian approximation

6.2 MOLECULAR QUANTUM DYNAMICS INDUCED BY THE INTERACTION WITH ELECTROMAGNETIC FIELD

Vibrationally resolved electronic spectra reflect the molecular motion following electronic excitations induced by the interaction with visible or ultraviolet electromagnetic field. Such an electronic excitation can induce non-adiabatic dynamics between different electronic states, but we will assume the validity of the Born–Oppenheimer approximation (see also Marquardt and Quack, 2020, Chapter 1 of this book), which will allow us to focus on adiabatic dynamics, i.e., dynamics on a single Born–Oppenheimer potential energy surface because many interesting phenomena in continuous-wave electronic spectroscopy depend only on dynamics on a single surface. Likewise, in time-resolved spectroscopy with well-separated ultrashort pulses, one can get rather far by considering only a sequence of such elementary steps, each of which takes place on a single surface. Therefore, we will not cover non-adiabatic dynamics here and instead refer the interested reader to various books and reviews of this vast subject (Domcke and Stock, 1997; Mukamel, 2000; Baer, 2006; Levine and Martínez, 2007; Domcke and Yarkony, 2012; Nakamura, 2012; Ryabinkin et al., 2017; Bircher et al., 2017).

6.2.1 Exact Dynamics, Electric Dipole Approximation, and Quasiresonant Condition

To justify our focus on electronically adiabatic dynamics we start the discussion with the full molecular wavefunction that involves both electronic and nuclear degrees of freedom. This will be useful because all our applications come from electronic spectroscopy, where the electromagnetic field induces the transition of the molecule to a different electronic state; this transition is followed by nuclear adiabatic dynamics on the corresponding, new potential energy surface.

The simplest electronic transition involves only two electronic states and, in this two-dimensional electronic basis, the time-dependent molecular wavepacket can be written as

$$\boldsymbol{\psi}(t) = \begin{pmatrix} |\psi_1(t)\rangle \\ |\psi_2(t)\rangle \end{pmatrix}, \tag{6.1}$$

where $|\psi_\alpha(t)\rangle$ is a time-dependent nuclear wavepacket moving on the αth potential energy surface. Evolution of $\boldsymbol{\psi}(t)$ is determined by the time-dependent molecular Schrödinger equation

$$i\hbar \frac{d}{dt}\boldsymbol{\psi}(t) = \hat{\boldsymbol{H}}_{\text{full}}(t)\boldsymbol{\psi}(t), \tag{6.2}$$

driven by the "full" time-dependent Hamiltonian

$$\hat{\boldsymbol{H}}_{\text{full}}(t) = \hat{\boldsymbol{H}} + \hat{\boldsymbol{V}}_{\text{int}}(t), \tag{6.3}$$

where $\hat{\boldsymbol{H}}$ is the time-independent molecular Hamiltonian and $\hat{\boldsymbol{V}}_{\text{int}}(t)$ the interaction potential of the molecule with the electromagnetic field. The exact solution of Eq. (6.2), which can be formally written as

$$\boldsymbol{\psi}(t) = \mathcal{T}\exp\left[-\frac{i}{\hbar}\int_{t_0}^{t} \hat{\boldsymbol{H}}_{\text{full}}(t')\,dt'\right]\boldsymbol{\psi}(t_0), \tag{6.4}$$

where $\mathcal{T}$ is the time-ordering operator, is greatly simplified by invoking the following approximations.

Within the *electric dipole approximation* (see, e.g., Schatz and Ratner, 2002; Marquardt and Quack, 2020, Chapter 1 of this book), the interaction potential is given by

$$\hat{\boldsymbol{V}}_{\text{int}}(t) = -\hat{\vec{\boldsymbol{\mu}}} \cdot \vec{E}(t), \tag{6.5}$$

where $\hat{\vec{\boldsymbol{\mu}}}$ is the molecular electric transition dipole moment and $\vec{E}(t)$ the electric field at the location of the molecule. This approximation requires the wavelength of the electromagnetic field to be longer than the size of the molecule, which is quite well justified for medium size molecules and visible light.

We further assume that there are no non-adiabatic or spin–orbit couplings between the two electronic states and, as a consequence, that the only electronic transitions are induced by the interaction with the electromagnetic field. This so-called *Born–Oppenheimer approximation* (see Chapters 1, Marquardt and Quack, 2020; and 2, Császár et al., 2020) implies that the molecular Hamiltonian $\hat{\boldsymbol{H}}$ is a diagonal S-dimensional matrix $\text{diag}(\hat{H}_1, \ldots, \hat{H}_S)$. To rigorously justify neglecting non-adiabatic or spin–orbit couplings, several criteria (Bircher et al., 2017; Baer, 2006; Domcke and Yarkony, 2012; Nakamura, 2012) can be used, starting from static criteria such as the size of the energy gap between electronic states or the strength of non-adiabatic couplings, to more dynamical criteria such as the population dynamics. Among the most rigorous dynamical criteria, "adiabaticity" (Zimmermann and Vaníček, 2010; MacKenzie et al., 2012; Zimmermann and Vaníček, 2012a, 2012b) is defined as the overlap of the adiabatically and non-adiabatically evolved molecular wavefunctions: If adiabaticity is close to 1, the non-adiabatic effects can be safely neglected, whereas if adiabaticity is much smaller than 1, they must be included in the simulation. Although rigorous, this criterion by itself would not be very practical; fortunately, there exist efficient approximate semiclassical methods to estimate the adiabaticity without solving the full Schrödinger equation (6.2) exactly (Zimmermann and Vaníček, 2010, 2012a, 2012b; Prlj et al., 2020).

A visible or ultraviolet electromagnetic field will excite electronic transitions, and if it is approximately in resonance with the transition from state 1 to state 2, we are allowed to retain only the off-diagonal elements of the transition dipole moment:

$$\hat{\vec{\boldsymbol{\mu}}} \approx \begin{pmatrix} 0 & \hat{\vec{\mu}}_{12} \\ \hat{\vec{\mu}}_{21} & 0 \end{pmatrix}. \tag{6.6}$$

This is a special case of the *quasiresonant condition* [see Chapter 1 and Quack (1978, 1979), Quack and Sutcliffe (1985), Marquardt and Quack (1989)].

6.2.2 Perturbation Theory, Zero-Temperature and Condon Approximations

If the electromagnetic field is strong, one must treat it explicitly and worry about the coupled dynamics on the two surfaces at least during the excitation process; in other words, one must evolve the two-component state $\boldsymbol{\psi}(t)$. For sufficiently weak fields or for short interaction times, one may employ the *time-dependent perturbation theory*. Whereas the first-order perturbation theory is often sufficient for linear spectroscopy, the second order is required, e.g., for resonance Raman spectroscopy and the third or higher order for more sophisticated non-linear and time-resolved spectroscopic techniques (Mukamel, 1999).

Within the first-order perturbation theory, the molecular state evolves as

$$\boldsymbol{\psi}(t)=\hat{\boldsymbol{U}}(t)\boldsymbol{\psi}(0) - \frac{\mathrm{i}}{\hbar}\int_{-\infty}^{t} \mathrm{d}t'\hat{\boldsymbol{U}}(t-t')\hat{\boldsymbol{V}}_{\mathrm{int}}(t')\hat{\boldsymbol{U}}(t')\boldsymbol{\psi}(0), \quad (6.7)$$

where $\hat{\boldsymbol{U}}(t)=\exp(-\mathrm{i}\hat{\boldsymbol{H}}t/\hbar)$ denotes the molecular evolution operator in the absence of the electromagnetic field. For electronic transitions, expression (6.7) simplifies; the only interesting part is the first-order term describing the wavepacket generated by the field on the second potential energy surface:

$$|\psi_2(t)\rangle = \frac{\mathrm{i}}{\hbar}\int_{-\infty}^{t} \mathrm{d}t'\hat{U}_2(t-t')\hat{\vec{\mu}}_{21}\cdot\vec{E}(t')\hat{U}_1(t')|\psi_1(0)\rangle. \quad (6.8)$$

This equation implies that the initial state first evolves freely on the first surface, then, at time t', interacts with the field, which induces instantaneously an electronic transition to the second electronic state, and, finally, evolves for the remaining time on the second surface. The total wavepacket generated in the excited state is obtained by integrating this elementary process over all possible interaction times t'. In practical terms, since $\hat{U}_\alpha(t)=\exp(-\mathrm{i}\hat{H}_\alpha t/\hbar)$ is the nuclear evolution operator on surface α, one only has to solve the nuclear Schrödinger equation

$$\mathrm{i}\hbar\frac{\mathrm{d}}{\mathrm{d}t}|\phi(t)\rangle = \hat{H}_\alpha|\phi(t)\rangle \quad (6.9)$$

with a time-independent Hamiltonian $\hat{H}_\alpha$ and general initial state $|\phi(0)\rangle$ instead of the more complicated molecular equation (6.2) with a time-dependent Hamiltonian. The solution of Eq. (6.9) can be written formally as

$$|\phi(t)\rangle = \exp\left(-\mathrm{i}\hat{H}_\alpha t/\hbar\right)|\phi(0)\rangle. \quad (6.10)$$

At room temperature, most of the molecules are typically in the vibrational ground state $|1,g\rangle$ of the ground electronic state, which is, in particular, an eigenstate of $\hat{H}_1$, hence, the first evolution operator $\hat{U}_1(t')$ yields only a phase factor $\exp(-\mathrm{i}E_{1,g}t/\hbar)$, the only effect of which is an overall shift of the electronic spectrum by the zero-point vibrational energy $E_{1,g}$ of the initial potential energy surface (we will show this explicitly below). As a result, in the case of electronic transitions, the only interesting dynamics occurs after time t', in the second electronic state, and hence, as promised, the problem reduces to adiabatic dynamics on the second surface. The assumption that the initial state is a vibrational ground state of $\hat{H}_1$ is usually referred to as the *zero-temperature approximation* and avoids the necessity of Boltzmann averaging over different initial states. It is a good approximation for vibrationally resolved electronic spectroscopy. [Note that we anticipated the low-temperature approximation already in Eq. (6.1) by focusing our attention to pure states of the molecule. This is justified as long as we are not interested in rotationally resolved spectra, which would, at room temperature, necessitate a density operator treatment.]

Finally, one also frequently makes the *Condon approximation* (Franck and Dymond, 1926; Condon, 1927, 1928), which amounts to ignoring the dependence of the transition dipole on nuclear coordinates: $\vec{\mu}_{12}(q)\approx \mathrm{const}=\vec{\mu}_{12}$. Note that removing the coordinate dependence from $\vec{\mu}_{12}(q)$ permits taking the product $\hat{\vec{\mu}}_{21}\cdot\vec{E}(t')$ outside of the integral over $\mathrm{d}t'$ in Eq. (6.8).

6.3 SEMICLASSICAL APPROXIMATION TO QUANTUM DYNAMICS

Approximations for treating the molecule-field interaction, which were discussed in the previous section, simplify the solution of the time-dependent Schrödinger equation (6.2). Unfortunately, even after making these approximations, the quantum propagation of a wavefunction of a large molecule is impossible due to the exponential scaling of the computational cost with dimensionality. In addition, quantum dynamics requires construction of global potential energy surfaces, which is another task scaling exponentially with dimensions.

Semiclassical trajectory-based methods circumvent these two challenges and thus provide an interesting alternative for molecular dynamics simulations. On the one hand, the propagation of classical trajectories requires only local knowledge of the potential energy surfaces, allowing on-the-fly evaluation of necessary *ab initio* data. On the other hand, and in contrast to even simpler classical molecular dynamics, semiclassical trajectories carry along phases, and as a result can approximately describe nuclear quantum effects, such as the zero-point energy and quantum coherence. Many semiclassical methods have been proposed in the literature; here we can mention only a few, and therefore we focus on those that have been combined with *ab initio* evaluation of the electronic structure.

Probably the simplest semiclassical method is the single-trajectory *thawed Gaussian approximation* (Heller, 1975, 2018), in which the quantum wavepacket is approximated by a single Gaussian whose center moves along a classical trajectory. Because it has a finite nonzero width and because it is allowed to rotate and stretch in phase space, the thawed Gaussian captures some quantum effects. The TGA, however, obviously cannot describe wavepacket splitting, and therefore is expected to be valid only for rather short times and describe only low or medium resolution spectra. Below, we will discuss in detail various extensions and present several on-the-fly *ab initio* applications of this method, and hopefully surprise the reader by the usefulness of the TGA in electronic spectroscopy.

Wavepacket splitting and full anharmonicity of the potential can be captured by multi-trajectory methods, such as the *initial value representation* (Miller, 1970, 2001) or the *frozen Gaussian approximation* (Heller, 1981a), in which the propagated wavepacket is represented by an ensemble of rigid Gaussians. The two concepts are merged in one of the most accurate semiclassical approximations, the *Herman–Kluk* initial value representation (Herman and Kluk, 1984; Miller, 2001; Kay, 2005; Lasser and Sattlegger, 2017), which can be derived from the stationary-phase approximation to the Feynman path integral propagator. In the Herman–Kluk initial value representation, the nuclear quantum evolution operator, $\hat{U}(t) = \exp(-\mathrm{i}\hat{H}t/\hbar)$, needed in evaluating Eq. (6.8), is approximated as

$$e^{-\mathrm{i}\hat{H}t/\hbar} \approx h^{-D} \int \mathrm{d}q_0 \mathrm{d}p_0 \, R_t(q_0, p_0) \times e^{\mathrm{i}S_t(q_0,p_0)/\hbar} |q_t p_t\rangle\langle q_0 p_0|, \tag{6.11}$$

where (q_t, p_t) denote the phase-space coordinates at time t of a point along the classical trajectory and $S_t(q_0, p_0)$ is the corresponding classical action. In the position representation, Glauber's canonical coherent states $|q_t p_t\rangle$ (Glauber, 1963) from Eq. (6.11) are Gaussians

$$\langle q|q_t p_t\rangle = \det\left(\frac{g}{\pi\hbar}\right)^{1/4} \exp\left\{\frac{1}{\hbar}\left[-\frac{1}{2}(q-q_t)^T \cdot g \cdot (q-q_t) + \mathrm{i}p_t^T \cdot (q-q_t)\right]\right\} \tag{6.12}$$

with a time-independent, real symmetric $D \times D$ width matrix g,

$$R_t(q_0, p_0) = \sqrt{\det\left[\frac{1}{2}\left(M_{t,qq}+g^{-1}\cdot M_{t,pp}\cdot g - \mathrm{i}M_{t,qp}\cdot g + \mathrm{i}g^{-1}\cdot M_{t,pq}\right)\right]} \tag{6.13}$$

is the Herman–Kluk prefactor, and symbols $M_{t,xy} = \partial x_t/\partial y_0$ represent the four $D \times D$ subblocks of the $2D \times 2D$ stability matrix

$$M_t = \begin{pmatrix} M_{t,qq} & M_{t,qp} \\ M_{t,pq} & M_{t,pp} \end{pmatrix} := \begin{pmatrix} \frac{\partial q_t}{\partial q_0} & \frac{\partial q_t}{\partial p_0} \\ \frac{\partial p_t}{\partial q_0} & \frac{\partial p_t}{\partial p_0} \end{pmatrix}, \tag{6.14}$$

where $:=$ denotes a definition of a new quantity. Note that the form of the coherent state wavefunctions in Eq. (6.12) differs from that of Klauder (Klauder and Skagerstam, 1985; Child and Shalashilin, 2003) by a factor $\exp(\mathrm{i}p_t q_t/2\hbar)$, which modifies the action S_t but leaves the classical equations of motion for q_t and p_t unchanged. The phase-space integral in Eq. (6.11) is usually evaluated by sampling the initial conditions of classical trajectories using Monte Carlo techniques; the subsequent propagation requires computing the potential energy V to evolve the action S, the force (i.e., $-\operatorname{grad} V$) to evolve positions and momenta, and the Hessian of V to evolve the stability matrix M_t.

Despite some progress, the application of the Herman–Kluk initial value representation to large systems is difficult. First, the oscillatory nature of the integrand in Eq. (6.11) implies that a very large number of trajectories is required to converge the results. Second, calculating the Hessian is much more expensive than evaluating the force needed in classical *ab initio* molecular dynamics. The computational cost can be reduced by invoking various additional approximations, such as the prefactor-free propagator (Zhang and Pollak, 2003), time averaging (Kaledin and Miller, 2003), and Filinov filtering (cellularization) (Makri and Miller, 1988; Heller, 1991; Walton and Manolopoulos, 1996), which has been used to improve Monte Carlo statistics (Walton and Manolopoulos, 1996; Wang et al., 2001; Church et al., 2017) and to derive new approximate semiclassical methods (Thoss et al., 2001; Šulc and Vaníček, 2012; Zambrano et al., 2013; Church et al., 2017; Antipov et al., 2015). Thanks to these acceleration techniques, the Herman–Kluk propagator has been successfully merged with on-the-fly dynamics and used to calculate vibrationally resolved spectra (Tatchen and Pollak, 2009; Ceotto et al., 2009a, 2009b; Wong et al., 2011) and internal conversion rates (Ianconescu et al., 2013). Time averaging has proved useful in on-the-fly simulations as a central ingredient of the multiple-coherent-states time-averaged semiclassical initial value representation (Ceotto et al., 2009a, 2009b, 2011). This method is especially well suited for the determination of vibrational frequencies and prediction of vibrational spectra (Ceotto et al., 2011, 2013; Buchholz et al., 2016; Gabas et al., 2017). A clever choice of the initial coherent states allows a drastic reduction of the number of trajectories required for convergence of desired spectral regions.

Both the thawed Gaussian approximation and various forms of the initial value representation are general methods for approximating quantum dynamics and are exact in arbitrary harmonic systems, but most require the knowledge of the Hessian of the potential energy along the trajectory. In the case of electronic spectroscopy, which is the main subject of this chapter, there is a class of methods that only require the force, and not the Hessian of the potential, but at a cost of being exact only for displaced harmonic systems. These closely related methods are known as the phase averaging (Mukamel, 1982, 1999), Wigner averaged classical limit (Egorov et al., 1998, 1999), linearized semiclassical initial value representation (Shi and Geva, 2005), or dephasing representation (Vaníček, 2004, 2006; Zambrano et al., 2013).

6.4 THAWED GAUSSIAN APPROXIMATION

The main goal of this chapter is to demonstrate the power of on-the-fly *ab initio* semiclassical dynamics in electronic spectroscopy on the example of the thawed Gaussian approximation. Let us, therefore, discuss this method and its extensions in more detail.

6.4.1 Thawed Gaussian Approximation

The TGA (Heller, 1975) relies on the fact that a Gaussian wavepacket propagated exactly quantum-mechanically in an arbitrary time-dependent harmonic potential remains Gaussian. The TGA improves on the frequently used global harmonic models (discussed below, in Section 6.6.1) by approximating the potential energy surface to the second order only locally, in what is known as the *local harmonic approximation*. Although approximating the initial state in electronic spectroscopy with a Gaussian is only reasonable within the Condon approximation, let us first discuss this simplest case because it will serve as a starting point for extensions to more general forms of the initial wavepacket needed to describe Herzberg–Teller spectra.

A general D-dimensional Gaussian wavepacket at time t can be written in the position representation as

$$\psi(q,t) = \exp\left\{\frac{i}{\hbar}\left[\frac{1}{2}(q-q_t)^T \cdot A_t \cdot (q-q_t) + p_t^T \cdot (q-q_t) + \gamma_t\right]\right\}, \tag{6.15}$$

where (q_t, p_t) are the phase-space coordinates of the center of the wavepacket, A_t is a symmetric complex width matrix, and γ_t a complex number whose real part is a dynamical phase and imaginary part guarantees the normalization at all times. The wavepacket is propagated with an effective Hamiltonian

$$\hat{H}_{\text{eff}}(t) \equiv H_{\text{eff}}(\hat{q},\hat{p},t) = \frac{1}{2}\hat{p}^T \cdot m^{-1} \cdot \hat{p} + V_{\text{eff}}(\hat{q},t), \tag{6.16}$$

where $m = \text{diag}(m_1, \ldots, m_D)$ is the diagonal mass matrix and V_{eff} an effective time-dependent potential given by the local harmonic approximation of the true potential V at the center of the wavepacket:

$$V_{\text{eff}}(q,t) = V|_{q_t} + (\text{grad}_q V|_{q_t})^T \cdot (q-q_t) + \frac{1}{2}(q-q_t)^T \cdot \text{Hess}_q V|_{q_t} \cdot (q-q_t). \tag{6.17}$$

In this equation, $(\text{grad}_q V)_i := \partial_{q_i} V$ is the gradient of V and $(\text{Hess}_q V)_{ij} := \partial_{q_i}\partial_{q_j} V$ is the symmetric Hessian matrix of V. Insertion of the wavepacket ansatz (6.15) and the effective potential (6.17) into the nuclear time-dependent Schrödinger equation (6.9) yields a system of ordinary differential equations for the time-dependent parameters of the Gaussian (Heller, 1975):

$$\dot{q}_t = m^{-1} \cdot p_t, \tag{6.18}$$

$$\dot{p}_t = -\text{grad}_q V|_{q_t}, \tag{6.19}$$

$$\dot{A}_t = -A_t \cdot m^{-1} \cdot A_t - \text{Hess}_q V|_{q_t}, \tag{6.20}$$

$$\dot{\gamma}_t = L_t + \frac{i\hbar}{2}\text{Tr}\left(m^{-1} \cdot A_t\right). \tag{6.21}$$

In the last equation, L_t denotes the Lagrangian

$$L_t = \frac{1}{2}\dot{q}_t^T \cdot m \cdot \dot{q}_t - V(q_t) = \frac{1}{2}p_t^T \cdot m^{-1} \cdot p_t - V(q_t). \tag{6.22}$$

The system of differential equations (6.18)–(6.21), which is within the local harmonic approximation equivalent to the solution of the Schrödinger equation

(6.9), implies that phase-space coordinates q_t and p_t follow classical Hamilton's equations of motion, while the propagation of the width A_t and phase γ_t involves propagating the $2D \times 2D$ stability matrix (6.14), and therefore requires evaluating the Hessians of the potential energy surface V.

6.4.2 Parameter Propagation of the Thawed Gaussian Wavepacket

Eqs. (6.18)–(6.19) for the position and momentum are classical Hamilton's equations of motion for Hamiltonian $H(q, p)$, so q_t and p_t can be readily propagated with the Verlet algorithm or other symplectic integrators. Solving Eqs. (6.20)–(6.21) for the width matrix A_t and generalized phase γ_t is more complicated. The approach by Heller (1976) and Hagedorn (1980) [see also Lee and Heller (1982), Faou et al. (2009), and Lasser and Lubich (2020)] suggests splitting the complex symmetric width matrix A_t into a product of two complex $D \times D$ matrices P_t and Q_t:

$$A_t = P_t \cdot Q_t^{-1}, \tag{6.23}$$

$$P_t = m \cdot \dot{Q}_t. \tag{6.24}$$

Differential equations for Q_t and P_t, obtained from Eqs. (6.20), (6.23), and (6.24),

$$\dot{Q}_t = m^{-1} \cdot P_t, \tag{6.25}$$

$$\dot{P}_t = -\mathrm{Hess}_q V|_{q_t} \cdot Q_t, \tag{6.26}$$

can be interpreted as Hamilton's equations of motion of the Hamiltonian

$$H_{\mathrm{sc}}(Q, P; q_t) = \frac{1}{2}\mathrm{Tr}\Big[P^\dagger \cdot m^{-1} \cdot P + Q^\dagger \cdot \mathrm{Hess}_q V|_{q_t} \cdot Q\Big], \tag{6.27}$$

which depends on time through the parameter q_t. The solution can then be written explicitly in terms of the stability matrix M_t:

$$\begin{pmatrix} Q_t \\ P_t \end{pmatrix} = M_t \begin{pmatrix} Q_0 \\ P_0 \end{pmatrix}; \tag{6.28}$$

In other words, Q_t and P_t are solutions of a linear Hamiltonian system, obtained by linearizing the original system about the center of the thawed Gaussian. Choosing the initial values as $Q_0 = (\mathrm{Im}\, A_0)^{-1/2}$ and $P_0 = A_0 \cdot Q_0$ guarantees, in addition, that the sum $H(q_t, p_t) + (\hbar/2)H_{\mathrm{sc}}(Q_t, P_t)$ equals the total energy of the thawed Gaussian wavepacket (Begušić et al., 2019).

As for γ_t, which is a generalization of classical action and represents both the dynamical phase and normalization, it is evaluated as

$$\gamma_t = \gamma_0 + \int_0^t L_\tau \mathrm{d}\tau + \frac{\mathrm{i}\hbar}{2}\int_0^t \mathrm{Tr}\left(m^{-1} \cdot A_\tau\right)\mathrm{d}\tau \tag{6.29}$$

$$= \gamma_0 + \int_0^t L_\tau \mathrm{d}\tau + \frac{\mathrm{i}\hbar}{2}\int_0^t \mathrm{Tr}\left(\dot{Q}_\tau \cdot Q_\tau^{-1}\right)\mathrm{d}\tau \tag{6.30}$$

$$= \gamma_0 + \int_0^t L_\tau \mathrm{d}\tau + \frac{\mathrm{i}\hbar}{2}\ln\left[\frac{\det Q_t}{\det Q_0}\right], \tag{6.31}$$

where the conditions imposed on Q and P are used in order to obtain the final expression. Because the determinant in Eq. (6.31) is complex, a proper branch of the logarithm must be taken in order to make γ_t continuous in time. If continuity were not imposed on γ_t, the wavepacket would show sudden jumps by $\pm\pi$ in the overall phase. Phase continuity is also important in the evaluation of the correlation function (to be discussed in Section 6.5), which requires taking a square root of a complex determinant $\det(A_t - A_0^*)$. The continuity of the correlation function is enforced by taking the appropriate branch of the square root.

6.4.3 Extended Thawed Gaussian Approximation (ETGA)

To tackle the propagation of more general initial states, Lee and Heller (1982) proposed an extension of the TGA that assumes the initial state to be a Gaussian multiplied by a polynomial (Lee and Heller, 1982; Patoz et al., 2018):

$$\phi(q, 0) = P\,(q - q_0)\,\psi(q, 0). \tag{6.32}$$

The basic idea of the *extended TGA* (ETGA) is the same as of the original TGA; it uses the local harmonic approximation (6.17) for the potential along the trajectory q_t, but makes no other approximation.

Because the only dependence of $\psi(q, 0)$ on p_0 comes from the exponent $p_0^T \cdot (q - q_0)$ [see Eq. (6.15)], the polynomial prefactor in Eq. (6.32) can be replaced by the *same* polynomial in the derivatives with respect to p_0:

$$\phi(q, 0) = P\left(\frac{\hbar}{\mathrm{i}}\frac{\partial}{\partial p_0}\right)\psi(q, 0). \tag{6.33}$$

Appendix D of Begušić et al. (2018a) provides a detailed proof that the local harmonic approximation implies that the ETGA wavepacket at time t has the simple form (Lee and Heller, 1982)

$$\phi(q, t) = P\left(\frac{\hbar}{\mathrm{i}}\frac{\partial}{\partial p_0}\right)\psi(q, t), \tag{6.34}$$

where the dependence of $\psi(q,t)$ on initial conditions q_0 and p_0 is taken into account. In particular, equations of motion (6.18)–(6.21) for q_t, p_t, A_t, and γ_t remain unchanged.

As for the parameters of the polynomial, the simplest possibility beyond the original TGA is to consider only the constant and linear terms, which will be the only ones required in the Herzberg–Teller approximation (6.66) for the transition dipole moment, used in Section 6.5.4 for calculations of electronic spectra beyond the Condon approximation. Appendix D of Begušić et al. (2018a) provides a detailed demonstration that in this case (Lee and Heller, 1982)

$$\phi(q,t) = \left[a_0 + b_t^T \cdot (q - q_t)\right] \psi(q,t), \tag{6.35}$$

where the linear parameter of the polynomial at time t is

$$b_t = \left(-A_t \cdot M_{t,qp} + M_{t,pp}\right) \cdot b_0 \tag{6.36}$$

$$= (Q_0 \cdot Q_t^{-1})^T \cdot b_0. \tag{6.37}$$

Because all ingredients needed in Eq. (6.36) are already evaluated for the propagation of the parameters of the Gaussian, the evaluation of b_t comes at almost no additional cost.

6.4.4 Multiple Thawed Gaussians (n-TGA)

Another natural way to generalize the TGA is by considering superpositions

$$\psi(q,t) = \sum_{i=1}^{n} c_i \psi_i(q,t), \tag{6.38}$$

of thawed Gaussian wavepackets $\psi_i(q,t)$, where each component $\psi_i(q,t)$ is normalized at time 0 and propagated with its own effective time-dependent Hamiltonian $\hat{H}_{\text{eff},i}(t)$ according to the original TGA prescription. Let us call this method *multiple thawed Gaussians* or *n-TGA*. The expansion (6.38) is useful only if the initial state $\psi(q,0)$ is accurately represented by a small number of Gaussians; below, we shall see an example with $n = 3$. If n is very large, it is more convenient to use the frozen Gaussians (Heller, 1981a), which are easier to propagate.

6.4.5 (Non)Conservation of Norm, Inner Product, and Energy

One of the basic properties of quantum evolution is the conservation of probability, which is reflected in the conservation of the norm of the wavefunction:

$$\|\psi_t\| = \|\hat{U}(t)\psi_0\|, \tag{6.39}$$

where $\|\psi\|^2 = \langle\psi|\psi\rangle$ and $\hat{U}(t) = \exp(-i\hat{H}t/\hbar)$ is a time-evolution operator for a time-independent Hermitian Hamiltonian $\hat{H}$. [In Eq. (6.39), we have and, in the rest of this short section, we shall use a shorthand notation ψ_t for $\psi(t)$.] It is easy to show that time-independent Hamiltonians conserve energy, defined as the expectation value

$$E_t := \langle\psi_t|\hat{H}|\psi_t\rangle \tag{6.40}$$

of the Hamiltonian, because

$$\begin{aligned} E_t &:= \langle\psi_0|\hat{U}(t)^\dagger \hat{H} \hat{U}(t)|\psi_0\rangle \\ &= \langle\psi_0|\hat{U}(t)^\dagger \hat{U}(t) \hat{H}|\psi_0\rangle = \langle\psi_0|\hat{H}|\psi_0\rangle = E_0, \end{aligned} \tag{6.41}$$

where we used the facts that $\hat{U}(t)$ commutes with $\hat{H}$ and that $\hat{U}$ is unitary, i.e.,

$$\hat{U}^\dagger \hat{U} = [\exp(-i\hat{H}t/\hbar)]^\dagger \exp(-i\hat{H}t/\hbar) = \hat{U}^{-1}\hat{U} = 1 \tag{6.42}$$

because $\hat{H}$ is Hermitian. The conservation of norm remains valid even for time-dependent Hermitian Hamiltonians $\hat{H}(t)$, for which the time-evolution operator is a function of both the initial and final times:

$$\hat{U}(t,t_0) = \mathcal{T} \exp\left(-i \int_{t_0}^{t} \hat{H}(\tau)d\tau/\hbar\right). \tag{6.43}$$

In contrast, the energy is not conserved because $[\hat{H}(t), \hat{U}(t,t_0)] \neq 0$ since even

$$[\hat{H}(t_1), \hat{H}(t_2)] \neq 0 \tag{6.44}$$

in general for two different times t_1 and t_2.

It is easy to show that if the evolution operator $\hat{U}(t,t_0)$ conserves not only the norm, but also the inner product $\langle\psi_t|\phi_t\rangle$, then $\hat{U}(t,t_0)$ must be linear. Conservation of the inner product implies, of course, the conservation of the norm, since $\|\psi\| = \langle\psi|\psi\rangle^{1/2}$. However, there are operators $\hat{U}(t,t_0)$ that conserve the norm but not the inner product; by the previous argument such operators must be non-linear.

The time-evolution operator of the TGA is a perfect example of such a "pathological" operator because it corresponds to a time-dependent effective Hamiltonian $\hat{H}_{\text{eff}}(t)$ that depends on the initial state ψ_0. Therefore

$$\hat{U}_{\text{TGA}} \equiv \hat{U}_{\text{TGA}}(t,t_0,\psi_0), \tag{6.45}$$

which implies that it is a non-linear operator. It comes as no surprise that $\hat{U}_{\text{TGA}}$ does not preserve the inner product between two different states:

$$\langle \psi_t | \phi_t \rangle = \langle \psi_0 | \hat{U}_{\text{TGA}}(t, t_0, \psi_0)^\dagger \hat{U}_{\text{TGA}}(t, t_0, \phi_0) | \phi_0 \rangle \neq \langle \psi_0 | \phi_0 \rangle . \tag{6.46}$$

In contrast, $\hat{U}_{\text{TGA}}$ still conserves the norm of any initial state:

$$\begin{aligned} \|\psi_t\|^2 &= \langle \psi_t | \psi_t \rangle \\ &= \langle \psi_0 | \hat{U}_{\text{TGA}}(t, t_0, \psi_0)^\dagger \hat{U}_{\text{TGA}}(t, t_0, \psi_0) | \psi_0 \rangle \\ &= \langle \psi_0 | \psi_0 \rangle = \|\psi_0\|^2 . \end{aligned}$$

Finally, the non-conservation of energy by the TGA follows from the time dependence of $\hat{H}_{\text{eff}}(t)$. In fact, one can go further and partition the energy of the thawed Gaussian wavepacket as $E_{\text{TGA}} = E_{\text{cl}} + E_{\text{sc}}$ into a "classical" energy of the guiding trajectory,

$$E_{\text{cl}} = H(q_t, p_t) = \frac{1}{2} p_t^T \cdot m^{-1} \cdot p_t + V(q_t), \tag{6.47}$$

and the "semiclassical" energy associated with the finite width of the wavepacket:

$$E_{\text{sc}} = \frac{1}{2} \operatorname{Tr}\left(m^{-1} \cdot \Pi_t^2\right) + \frac{1}{2} \operatorname{Tr}\left(\operatorname{Hess}_q V|_{q_t} \cdot \Sigma_t^2\right) \tag{6.48}$$

$$= \frac{\hbar}{2} H_{\text{sc}}(Q_t, P_t; q_t), \tag{6.49}$$

where

$$\begin{aligned} \Sigma_t^2 &:= \langle \psi_t | (\hat{q} - q_t) \otimes (\hat{q} - q_t)^T | \psi_t \rangle \\ &= \frac{\hbar}{2} (\operatorname{Im} A_t)^{-1} = \frac{\hbar}{2} Q_t \cdot Q_t^\dagger , \end{aligned} \tag{6.50}$$

$$\begin{aligned} \Pi_t^2 &:= \langle \psi_t | (\hat{p} - p_t) \otimes (\hat{p} - p_t)^T | \psi_t \rangle \\ &= \frac{\hbar}{2} A_t \cdot (\operatorname{Im} A_t)^{-1} \cdot A_t^\dagger = \frac{\hbar}{2} P_t \cdot P_t^\dagger \end{aligned} \tag{6.51}$$

are position and momentum width (covariance) matrices. Since the guiding trajectory (q_t, p_t) is propagated with the exact, time-independent Hamiltonian $H(q, p)$, the classical energy E_{cl} is conserved exactly. In contrast, the semiclassical contribution $E_{\text{sc}} = (\hbar/2) H_{\text{sc}}(Q_t, P_t; q_t)$ can be interpreted as a scaled classical energy of the trajectory (Q_t, P_t) guided by a Hamiltonian $H_{\text{sc}}(Q, P; q_t)$ that depends on time via q_t. As a result, E_{sc} depends on time and is responsible for the non-conservation of the total energy E_{TGA}. Interestingly, if the dependence of H_{sc} on q_t vanishes, which happens in a global harmonic potential or if one approximates $\operatorname{Hess}_q V|_{q_t}$ at all times by a constant, reference Hessian, the energy E_{TGA} is conserved exactly (Begušić et al., 2019).

The above considerations generalize easily to the ETGA. The corresponding evolution operator $\hat{U}_{\text{ETGA}}$ again depends on both the initial and final time as well as on the initial state ϕ_0. Expression (6.34) for $\phi(q, t)$ implies that $\hat{U}_{\text{ETGA}}$ is fully determined by these three parameters, so

$$\hat{U}_{\text{ETGA}} \equiv \hat{U}_{\text{ETGA}}(t, t_0, \phi_0). \tag{6.52}$$

As a consequence, the ETGA conserves the norm, but both the inner product and energy generally depend on time.

By now it should be obvious that multiple thawed Gaussians do not even preserve the norm of the initial state. It is a simple consequence of the non-conservation of the inner product by the TGA:

$$\begin{aligned} \|\psi_t\|^2 &= \sum_{i,j=1}^{n} c_i^* c_j \langle \psi_{i,t} | \psi_{j,t} \rangle \neq \sum_{i,j=1}^{n} c_i^* c_j \langle \psi_{i,0} | \psi_{j,0} \rangle \\ &= \|\psi_0\|^2 . \end{aligned} \tag{6.53}$$

Because they do not conserve the norm, the multiple thawed Gaussians do not conserve the inner product either. That the energy generally depends on time is clear.

6.5 TIME-DEPENDENT APPROACH TO ELECTRONIC SPECTROSCOPY

Quantum and semiclassical dynamics methods, discussed in Sections 6.3 and 6.4, yield the quantum or semiclassical molecular wavepacket at time t. In order to compare with experimental electronic spectra, it is necessary to translate the knowledge of the time-dependent molecular state to a cross-section that can be measured experimentally. To be specific, let us inspect in detail the linear electronic absorption spectrum and, at the end, mention briefly the main differences that show up in non-linear spectroscopy on the example of time-resolved stimulated emission spectrum.

6.5.1 Linear Absorption Spectra

Within the *electric-dipole approximation* and first-order *time-dependent perturbation theory*, the absorption cross section for a linearly polarized light of frequency ω can be expressed as the Fourier transform

$$\sigma(\vec{\epsilon}, \omega) \approx \frac{\omega}{2\hbar c \varepsilon_0} \int_{-\infty}^{\infty} C_{\mu\mu}(\vec{\epsilon}, t) e^{i\omega t} \mathrm{d}t \tag{6.54}$$

of the dipole time autocorrelation function

$$C_{\mu\mu}(\vec{\epsilon}, t) = \mathrm{Tr}\left[\hat{\rho}\hat{\mu}(t)\hat{\mu}\right], \tag{6.55}$$

where $\vec{\epsilon}$ is the unit polarization vector of the electric field, $\hat{\rho}$ is the density operator of the initial state, $\hat{\mu} := \vec{\epsilon} \cdot \hat{\vec{\mu}}$ is the projection of the molecular electric dipole operator on the direction $\vec{\epsilon}$ of the field, and

$$\hat{\mu}(t) = e^{\mathrm{i}\hat{H}t/\hbar}\hat{\mu}e^{-\mathrm{i}\hat{H}t/\hbar} \tag{6.56}$$

is this projected molecular dipole moment operator at time t (in the Heisenberg picture). Assuming the *zero temperature approximation*, the initial state is $\hat{\rho} = |1, g\rangle\langle 1, g|$, i.e., the ground vibrational state g of the ground electronic state 1; in particular, it is a pure state. Assuming, furthermore, that the incident radiation is in resonance only with a single pair of electronic states 1 and 2 (*quasiresonance condition*; see also Marquardt and Quack, 2020, Chapter 1 of this book), that those states are not vibronically coupled (*Born–Oppenheimer approximation*), the dipole time autocorrelation function reduces (Heller, 1981b) to

$$C_{\mu\mu}(\vec{\epsilon}, t) \approx \langle 1, g|e^{\mathrm{i}\hat{H}_1 t/\hbar}\hat{\mu}_{12}e^{-\mathrm{i}\hat{H}_2 t/\hbar}\hat{\mu}_{21}|1, g\rangle \tag{6.57}$$

where $\hat{H}_1$ and $\hat{H}_2$ are the nuclear Hamiltonian operators in the ground and excited electronic states, and $\hat{\mu}_{21}$ is the matrix element of the projected molecular transition dipole moment matrix $\hat{\mu}$.

To evaluate the autocorrelation function (6.57), let us rewrite it, without any further approximation, as

$$C_{\mu\mu}(\vec{\epsilon}, t) \approx C_{\phi\phi}(t)e^{\mathrm{i}E_{1,g}t/\hbar} \tag{6.58}$$

in terms of the vibrational zero-point energy $E_{1,g}$ of the ground electronic state and wavepacket autocorrelation function

$$C_{\phi\phi}(t) = \langle\phi(0)|\phi(t)\rangle \tag{6.59}$$

of the (unnormalized) initial wavepacket $|\phi(0)\rangle = \hat{\mu}_{21}|1, g\rangle$ propagated on the excited-state surface with the Hamiltonian $\hat{H}_2$:

$$|\phi(t)\rangle = e^{-\mathrm{i}\hat{H}_2 t/\hbar}|\phi(0)\rangle. \tag{6.60}$$

6.5.2 Condon Approximation

The electric transition dipole moment is, in general, a function of nuclear coordinates, yet, within the *Condon approximation* (Condon, 1926, 1927; see also Braun et al., 2020, Chapter 3 of this book), this moment is assumed to be independent of the molecular geometry and is commonly approximated with its value at the equilibrium geometry q_{eq} of the initial state:

$$\mu_{12}(q) \approx \mathrm{const} = \mu_{12}(q_{\mathrm{eq}}). \tag{6.61}$$

Within the Condon approximation, the spectrum can be written as the Fourier transform

$$\sigma(\vec{\epsilon}, \omega) \approx \frac{\omega}{2\hbar c\varepsilon_0}\mu_{12}^2 \int_{-\infty}^{\infty} C_{\psi\psi}(t)e^{\mathrm{i}(\omega + E_{1,g}/\hbar)t}\,\mathrm{d}t \tag{6.62}$$

of the wavepacket autocorrelation function

$$C_{\psi\psi}(t) = \langle\psi|\psi(t)\rangle = \langle\psi|e^{-\mathrm{i}\hat{H}_2 t/\hbar}|\psi\rangle \tag{6.63}$$

of a normalized initial state $|\psi\rangle = |1, g\rangle$.

The beauty of Eqs. (6.62) and (6.63) lies in their simple interpretation (see also Fig. 6.1, left): the absorption of a photon of frequency ω lifts the stationary vibrational ground state $|\psi\rangle = |1, g\rangle$ of the ground-state surface instantaneously to the excited-state surface, where this, now non-stationary state starts moving under the influence of the excited-state Hamiltonian *alone*. In particular, the explicit form of the electromagnetic field does not play any role and the linear absorption spectrum is determined solely by the field-free dynamics of the wavepacket $\psi(t)$ on the excited-state surface. Indeed, this fact is the essence of *linear response theory*, which is here equivalent to the first order time-dependent perturbation theory.

6.5.3 Connection to Fidelity Amplitude

Note that since the initial state $|\psi\rangle \equiv |1, g\rangle$ is an eigenstate of $\hat{H}_1$, the spectrum can be also written as

$$\sigma(\vec{\epsilon}, \omega) = \frac{\omega}{\hbar c\varepsilon_0}\mu_{12}^2 \,\mathrm{Re}\int_0^{\infty} f(t)e^{\mathrm{i}\omega t}\,\mathrm{d}t, \tag{6.64}$$

where

$$f(t) = \langle\psi_1(t)|\psi_2(t)\rangle = \langle\psi|e^{\mathrm{i}\hat{H}_1 t/\hbar}e^{-\mathrm{i}\hat{H}_2 t/\hbar}|\psi\rangle \tag{6.65}$$

is a wavepacket cross-correlation function, also known as the *fidelity amplitude* (Gorin et al., 2006), between two states $\psi_1(t)$ and $\psi_2(t)$, both starting from the same initial state ψ, but one evolved with $\hat{H}_1$ and the other with $\hat{H}_2$. As the name suggests, the fidelity amplitude measures the similarity between the quantum evolutions on the ground and excited surfaces. Not only is expression (6.64) "nicer" than Eq. (6.62) because it does not contain the zero-point energy in the exponent, it is also much more general because it remains valid for

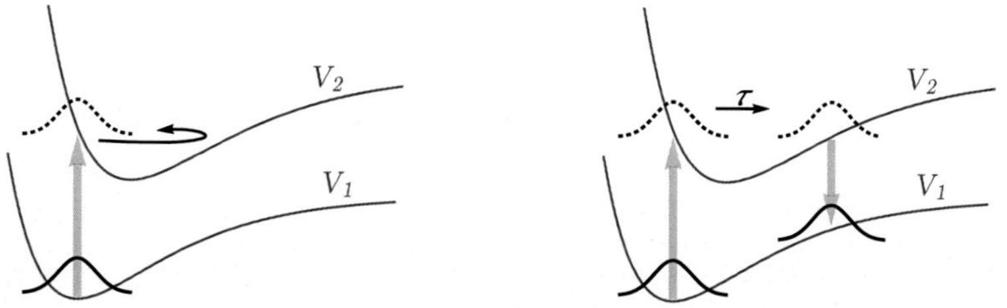

FIG. 6.1 Schematic representation of physical processes underlying two types of vibrationally resolved electronic spectra. (Left) Linear absorption. (Right) Time-resolved stimulated emission.

non-stationary initial states, such as arbitrary superpositions of vibrational eigenstates or states prepared by the pump pulse in time-resolved spectroscopy. Moreover, this alternative, although less known expression for an electronic spectrum is not just a mathematical curiosity; indeed, it is the direct outcome of the derivation of the spectrum using the first-order time-dependent perturbation theory, and it is only due to the *additional* assumption that ψ is a vibrational ground state (or another eigenstate) of $\hat{H}_1$ that one obtains the much better known expression (6.62) for the spectrum in terms of the wavepacket autocorrelation function (6.63).

The more general correlation function (6.65) has many important applications: Outside of electronic spectroscopy (Pollard et al., 1990b; Shemetulskis and Loring, 1992; Rost, 1995; Li et al., 1996; Egorov et al., 1998, 1999; Shi and Geva, 2005), it has proved useful, e.g., in NMR spin echo experiments (Pastawski et al., 2000) and theories of quantum computation (Gorin et al., 2006), decoherence (Gorin et al., 2006), and inelastic neutron scattering (Petitjean et al., 2007). In chemical physics, the fidelity amplitude was also used as a measure of the dynamical importance of diabatic (Zimmermann and Vaníček, 2010), non-adiabatic (Zimmermann and Vaníček, 2012b), or spin–orbit couplings (Zimmermann and Vaníček, 2012a), and of the accuracy of molecular quantum dynamics on an approximate potential energy surface (Li et al., 2009; Zimmermann et al., 2010).

6.5.4 Herzberg–Teller Approximation

The widespread use of the Condon approximation (6.61) is justified by its validity in many systems; it can describe most of the strongly symmetry-allowed transitions both qualitatively and quantitatively. However, a number of molecules exhibit "electronically forbidden" transitions, i.e., transitions $\alpha \leftarrow \beta$ with $\vec{\mu}_{\alpha\beta}(q_{\text{eq}}) = 0$, which cannot be described within the Condon approximation. Such systems, as well as systems in which the Condon term is small but not exactly zero, can be treated with the *Herzberg–Teller approximation* (Herzberg and Teller, 1933) that takes into account at least the gradient of the transition dipole moment with respect to nuclear degrees of freedom:

$$\vec{\boldsymbol{\mu}}(q) \approx \vec{\boldsymbol{\mu}}(q_{\text{eq}}) + \partial_q \vec{\boldsymbol{\mu}}|_{q_{\text{eq}}}^T \cdot (q - q_{\text{eq}}). \tag{6.66}$$

Although the Herzberg–Teller approximation is often discussed in terms of vibronic couplings between different electronic states (Seidner et al., 1992), this relation is not obvious from Eq. (6.66). It becomes, however, clear from a remarkable equality (Patoz et al., 2018; Begušić et al., 2018a)

$$\partial_{q_j} \vec{\boldsymbol{\mu}}(q) = \left[\vec{\boldsymbol{\mu}}(q), \boldsymbol{F}_j(q)\right] + \partial_{q_j} \vec{\mu}_{\text{nu}}(q)\mathbf{1} \tag{6.67}$$

satisfied by the gradient of the matrix representation $\vec{\boldsymbol{\mu}}(q)$ of the molecular dipole operator $\hat{\vec{\mu}}_{\text{mol}}$ at the nuclear configuration q. In this equation, matrix elements of $\vec{\boldsymbol{\mu}}(q)$ are defined as partial, electronic expectation values

$$\vec{\mu}_{\alpha\beta}(q) := \langle \alpha(q) | \hat{\vec{\mu}}_{\text{mol}} | \beta(q) \rangle, \tag{6.68}$$

elements of the matrix $\boldsymbol{F}_j(q)$ of non-adiabatic vector couplings are obtained as

$$F_{\alpha\beta,j} := \langle \alpha(q) | \partial_{q_j} \beta(q) \rangle, \tag{6.69}$$

and $\hat{\vec{\mu}}_{\text{nu}}$ is the nuclear component of $\hat{\vec{\mu}}_{\text{mol}}$.

Direct interpretation of relation (6.67), which is proven in Appendix B of Begušić et al. (2018a), explains the concepts of *vibronic transitions* and *intensity borrowing*. Namely, the gradient of the transition dipole moment between states α and β can be nonzero only if there exists an intermediate state γ that is non-adiabatically (i.e., *vibronically*) coupled to one of the states and electric-dipole coupled to the other state. Typically, the non-adiabatic couplings with the ground state ($\beta = 1$) can be neglected at the ground-state optimized geometry, around which the transition dipole moment is expanded. This leads to an expression (Li et al., 2010)

$$\partial_{q_j} \vec{\mu}_{\alpha 1} \approx -\sum_{\gamma} F_{\alpha\gamma, j} \vec{\mu}_{\gamma 1} \tag{6.70}$$

that explains the meaning of *intensity borrowing*, in which the symmetry forbidden transition to the formally dark state α occurs due to borrowing of the transition intensity from the neighboring bright electronic states γ that are vibronically coupled to the state α. Note that, despite introducing non-adiabatic couplings between excited electronic states, the original Born–Oppenheimer picture may be valid—the vibronic couplings that induce the transition do not necessarily influence the nuclear wavepacket dynamics. Although these non-adiabatic couplings are essential for describing the existence of symmetry-forbidden spectra, their contribution to the field-free Hamiltonian of the system may be negligible. The rather high resolution of the absorption spectrum of benzene (discussed in Section 6.8.2) supports these considerations—otherwise, significant population transfer would lead to the shortening of the excited-state lifetime and, consequently, to significant broadening of the spectral lines.

6.5.5 Rotational Averaging of the Spectrum

To compare with an experiment in gas phase or another isotropic medium, one has to average the vibronic spectrum (6.54) over all orientations of the molecule with respect to the polarization $\vec{\epsilon}$ of the electric field. This procedure assumes the rotations to be classical and, therefore, cannot be used to compute rotationally resolved spectra.

Within the Condon approximation, the transition dipole moment is independent of coordinates, and this averaging is trivial, namely

$$\overline{C_{\mu\mu}(\vec{\epsilon}, t)} = \frac{1}{3} C_{|\vec{\mu}||\vec{\mu}|}(t), \tag{6.71}$$

where $|\vec{\mu}|$ is the magnitude of the transition dipole moment, so only a single calculation is required—for the transition dipole moment aligned with the field.

For a general dipole moment (Andrews and Thirunamachandran, 1977; Gelin et al., 2017), it is useful to define the spectrum *tensor* $\overleftrightarrow{\sigma}(\omega)$, from which the spectrum (6.54) for a specific polarization $\vec{\epsilon}$, is obtained by "evaluation":

$$\sigma(\vec{\epsilon}, \omega) = \vec{\epsilon}^{\,T} \cdot \overleftrightarrow{\sigma}(\omega) \cdot \vec{\epsilon}. \tag{6.72}$$

The *rotational averaging* of the spectrum corresponds to the averaging of $\sigma(\vec{\epsilon}, \omega)$ over all unit vectors $\vec{\epsilon}$. Due to the isotropy of the 3-dimensional Euclidean space, the average over all orientations need not be performed numerically, and is, instead, reduced to an arithmetic average over only three arbitrary orthogonal orientations of the molecule with respect to the field:

$$\begin{aligned}\overline{\sigma(\vec{\epsilon}, \omega)} &= \frac{1}{3} \operatorname{Tr} \overleftrightarrow{\sigma}(\omega) \\ &= \frac{1}{3}\left[\sigma_{xx}(\omega) + \sigma_{yy}(\omega) + \sigma_{zz}(\omega)\right] \\ &= \frac{1}{3}\left[\sigma(\vec{e}_x, \omega) + \sigma(\vec{e}_y, \omega) + \sigma(\vec{e}_z, \omega)\right].\end{aligned} \tag{6.73}$$

Appendix A of Begušić et al. (2018a) contains an explicit proof of these equalities. Within the Condon approximation, in which $\vec{\mu}$ is coordinate-independent, the orientational average (6.73) simplifies further into the standard textbook recipe (6.71).

6.5.6 Time-Resolved Electronic Spectra

In the case of non-linear spectra, the autocorrelation picture is no longer valid. In contrast, the more general picture using fidelity amplitude (6.64) still applies. A wide variety of non-linear time-resolved spectra belong to the pump-probe scheme, in which an ultrashort pump pulse prepares a non-stationary nuclear wavepacket in an excited electronic state, and an ultrashort probe pulse measures the dynamics of this wavepacket after a certain time delay τ. There are many possible experimental setups depending on the polarization and mutual orientation of the pump and probe laser beams and on the direction in which the signal is detected (Mukamel, 1999), but, to be specific, we will only consider time-resolved stimulated emission here (see Fig. 6.1, right).

Besides the assumptions used for linear spectra, a simplified picture of time-resolved stimulated emission assumes that the pump and probe pulses can be treated independently (*non-overlapping pulses approximation*) and are short compared with the nuclear vibrational period and, at the same time, long compared with the time scale of the relevant electronic transition (*ultrashort pulse approximation*).

Assuming the validity of the zero-temperature, electric dipole, rotating-wave, and Condon approximations, and combining them with the third-order time-dependent perturbation theory, one obtains the differential time-resolved stimulated emission spectrum at frequency ω and time delay τ between the pump and probe pulses as the Fourier transform

$$\sigma(\omega, \tau) \propto \operatorname{Re} \int_0^{\infty} dt\, f(t, \tau)\, e^{i\omega t} \tag{6.74}$$

of the wavepacket correlation function (Pollard et al., 1990b; Wehrle et al., 2011; Šulc and Vaníček, 2012)

$$f(t,\tau)=\langle\psi_1(t,\tau)|\psi_2(t,\tau)\rangle, \tag{6.75}$$

where τ is the time delay between the pump and probe pulses, t denotes the time elapsed after the probe pulse, and

$$|\psi_\alpha(t,\tau)\rangle := e^{-i\hat{H}_\alpha t/\hbar}e^{-i\hat{H}_2\tau/\hbar}|\psi\rangle \tag{6.76}$$

stands for the initial state evolved for the delay time τ with the excited state Hamiltonian and subsequently for time t with either the ground or excited state Hamiltonian ($\alpha = 1, 2$).

As written, the correlation function $f(t,\tau)$ from Eq. (6.75) has an immediate interpretation as the quantum fidelity amplitude between states $\psi_1(t,\tau)$ and $\psi_2(t,\tau)$. This fidelity amplitude now corresponds to evolutions for time $t+\tau$ of the same initial state ψ with two Hamiltonians, a time-independent Hamiltonian equal to $\hat{H}_2$ and a time-dependent Hamiltonian equal to $\hat{H}_2$ until time τ, and equal to $\hat{H}_1$ at later times. Note that the correlation function $f(t,\tau)$ can be also interpreted as a correlation function (6.65) from linear spectroscopy, but applied to a non-stationary initial state $\exp(-i\tau\hat{H}_2/\hbar)|\psi\rangle$ prepared by the pump pulse (Pollard et al., 1990b; Shemetulskis and Loring, 1992).

6.6 "STANDARD MODELS" OF ELECTRONIC SPECTROSCOPY

Having reviewed the time-dependent approach to spectroscopy, we are ready to discuss several models commonly used for approximating molecular potential energy surfaces in calculations of vibrationally resolved electronic spectra. Due to their simplicity, these models facilitate approximate evaluations of spectra even of molecules for which obtaining more accurate surfaces would be impossible. Moreover, it is easier to analyze these models than the exact potential surfaces.

From now on, we assume that translations have been removed and that rovibrational coupling can be neglected within the accuracy provided by the methods used for dynamics. Therefore, we may assume that the nuclear Hamiltonian for the αth electronic state of a non-linear polyatomic molecule has been reduced from $3N$ to $D=3N-6$ dimensions, and is expressed in normal mode coordinates as

$$H_\alpha(q,p)=\frac{1}{2}p^T\cdot m^{-1}\cdot p+V_\alpha(q), \tag{6.77}$$

where q and p are D-dimensional vectors of position and momentum, and m is a D-dimensional diagonal mass matrix.

In *vibrational spectroscopy*, only one electronic state is involved, and therefore only one potential energy surface is needed. The simplest model is the harmonic potential,

$$V_1(q)=\frac{1}{2}(q-q_1)^T\cdot k_1\cdot(q-q_1), \tag{6.78}$$

where k_1 is a D-dimensional diagonal force constant matrix and where we assume that the minimum occurs at $q=q_1$ with energy $V_1(q_1)=0$. The harmonic model often provides a good approximation close to the minimum of a potential energy surface, but breaks down, of course, in anharmonic and floppy systems.

In *electronic spectroscopy*, the situation is much richer because energy surfaces describing different electronic states can have different minima, different force constants, and even different normal modes. To be specific, let us consider only two electronic states. The simplest model for the excited state surface is the *displaced harmonic potential*

$$\begin{aligned}V_2(q) &:= \Delta E + V_1(q-\Delta q)\\ &= \Delta E+\frac{1}{2}(q-q_2)^T\cdot k_1\cdot(q-q_2),\end{aligned} \tag{6.79}$$

where ΔE is the adiabatic excitation energy and $\Delta q = q_2 - q_1$ is the displacement from the minimum of the harmonic potential V_1 given by Eq. (6.78). Modes j with non-zero Δq_j are the modes *excited* by the electronic transition.

There are two natural ways to construct displaced harmonic models as approximations to an anharmonic system: the adiabatic shift and vertical gradient approximations. Both methods use Eqs. (6.78) and (6.79) with k_1 given by the Hessian of the ground surface at the equilibrium geometry, and thus avoid computing excited-state Hessians. In the *adiabatic shift approximation*, q_2 is given by the optimized excited-state geometry, whereas in the *vertical gradient approximation*, this optimization is avoided and the q_2 parameter in the model (6.79) is obtained by evaluating the gradient of the *ab initio* excited-state potential energy at the ground-state equilibrium geometry (Avila Ferrer and Santoro, 2012; Egidi et al., 2014; Fortino et al., 2019):

$$q_2 := q_1 - k_1^{-1}\cdot \mathrm{grad}_q V_{2,\mathrm{ab\ initio}}|_{q_1}. \tag{6.80}$$

An improvement that accounts for changes in the force constants in the excited state is provided by the

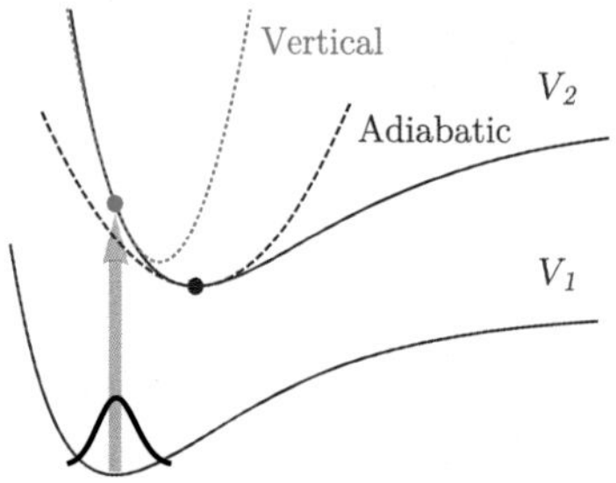

FIG. 6.2 In the vertical harmonic and adiabatic harmonic models, the excited-state potential energy is expanded to the second order about the minimum of the ground-state and excited-state potential energy surfaces, respectively.

displaced and distorted harmonic potential

$$V_2(q) := \Delta E + \frac{1}{2}(q - q_2)^T \cdot k_2 \cdot (q - q_2) \qquad (6.81)$$

with a diagonal matrix $k_2 \neq k_1$. Finally, one can also include the *Duschinsky rotation* between the ground-state and excited-state normal modes, to obtain the most general *harmonic model*, described still by Eq. (6.81), but allowing a general $D \times D$ symmetric force constant matrix k_2.

A harmonic model, given by Eqs. (6.78) and (6.81), which has been constructed as an approximation to an anharmonic system, is said to be a global *harmonic approximation*. Among all possible global harmonic approximations, two special ones stand out. In the *vertical harmonic approximation*, both V_1 and V_2 are expanded to the second order about the minimum of the ground state potential V_1 (see Fig. 6.2). This approach provides a good approximation in the Franck–Condon region, and is therefore expected to be accurate for short propagation times, which determine the initial decay of the autocorrelation function $C(t)$ and hence the overall envelope of the electronic spectrum. In the *adiabatic harmonic approximation*, in contrast, each V_α is expanded about its own minimum (see Fig. 6.2); in particular, k_2 is given by the Hessian of V_2 at the minimum of V_2. Often, but not always, this second approach provides a better approximation for the period of oscillations and hence for the vibrational peak spacing in the spectra.

Further improvements require including anharmonicity. The one-dimensional *Morse potential*

$$V_\alpha(q) = V_{\alpha,0} + d_\alpha(1 - e^{-a_\alpha(q-q_\alpha)})^2 \qquad (6.82)$$

is a good starting point for including anharmonicity of the degrees of freedom corresponding to bond stretches. In Eq. (6.82), q_α is again the location of the minimum $V_{\alpha,0}$ of the potential, while parameters a_α and d_α are both related to the curvature and anharmonicity of the potential. The adiabatic excitation energy is given by the difference $\Delta E := V_{2,0} - V_{1,0}$.

Anharmonicity in the symmetric torsional or umbrella motions can be represented by a *quartic* perturbation to the quadratic potential:

$$V_\alpha(q) = V_{\alpha,0} + \frac{1}{2}k_\alpha(q - q_\alpha)^2 + \frac{1}{24}\delta_\alpha(q - q_\alpha)^4. \quad (6.83)$$

Here, δ_α controls the magnitude of the perturbation, and we have, for simplicity, written out V_α explicitly only for one degree of freedom.

One can, of course, generate more accurate anharmonic potential energy surfaces by fitting more sophisticated analytical functions to *ab initio* energies evaluated at an appropriately chosen set of molecular configurations. In contrast, in the *on-the-fly approach*, the potential required for the dynamics is evaluated only along the trajectory; the fitting and construction of global surfaces are avoided. The on-the-fly approach has the advantage that the true molecular potential does not have to be approximated—instead, one can employ the best available electronic structure to obtain the energy at the point of interest. The drawback is, of course, that the on-the-fly approach works only in connection with semiclassical, mixed quantum-classical, or trajectory-based quantum methods, but not with methods that rely on global or semiglobal surfaces.

Finally, in electronic spectroscopy, one can choose to approximate one surface by a global harmonic model, while treating the other with the on-the-fly approach. This *combined global harmonic/on-the-fly approximation* is popular in linear spectroscopy, where the initial wavepacket is commonly the ground vibrational state of the initial electronic state. Because the harmonic approximation is valid near the minimum of the surface, it produces an accurate initial wavepacket. For wavepackets generated this way, the on-the-fly and combined methods are equivalent in the linear spectra calculations. In contrast, in the simulations of time-resolved spectra, the nuclear dynamics is performed in both electronic states; as a result, the global harmonic, combined, and on-the-fly approaches all give different results (Begušić et al., 2018b).

6.6.1 Several Few-Dimensional Examples

Let us demonstrate the basic features and shortcomings of various models by computing linear absorption spectra of several few-dimensional examples. The first system is a pair of two-dimensional harmonic potentials

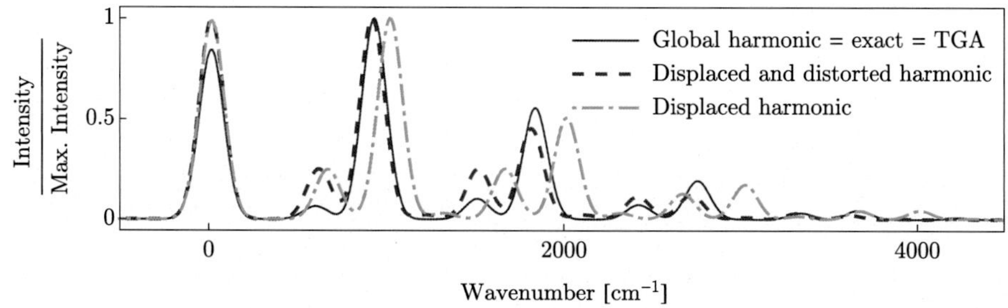

FIG. 6.3 Absorption spectrum of a pair of two-dimensional harmonic potentials. The excited-state surface is displaced, distorted, and rotated with respect to the ground-state surface. The relative dimensionless displacements $\Delta_i = \sqrt{m_i \omega_{1,i}/2\hbar}(q_{2,i} - q_{1,i})$ between the minima of the two potentials are $\Delta_a = 1$ and $\Delta_b = 0.5$, the wavenumbers of the two modes are $\tilde{\nu}_{1,a} = 1000$ cm^{-1} and $\tilde{\nu}_{1,b} = 650$ cm^{-1} in the ground state, and $\tilde{\nu}_{2,a} = 800$ cm^{-1} and $\tilde{\nu}_{2,b} = 600$ cm^{-1} in the excited state; the modes are coupled via the off-diagonal elements $k_{2,ab} = k_{2,ba} = k_{1,aa}/10$ of the symmetric force constant matrix of the excited state, where $k_{1,aa}$ is the element of the ground-state force constant corresponding to the mode a.

with a non-diagonal excited-state force constant, and thus exhibiting the Duschinsky effect. Fig. 6.3 shows the spectra computed using either the global harmonic model, which is exact here, or one of two approximate models. The simpler, displaced harmonic oscillator model does not account for the change in the force constant and, therefore, results in incorrect peak positions and intensities. Including the distortion of the surface in the displaced and distorted harmonic model fixes the peak positions; however, the intensities are still rather poor because this model neglects the mode-mixing present in the system.

Due to their simplicity, intuitive interpretation, and acceptable accuracy in many molecular systems, global harmonic models have served as the methods of choice for computing vibronic spectra. However, they tend to fail in floppy molecules and the thawed Gaussian approximation aims to correct this by including the anharmonicity of the potential at least partially. Let us see how this works in practice on two examples of one-dimensional quartic potentials (6.83) with different degrees of anharmonicity.

In the less anharmonic case (Fig. 6.4A), the vertical harmonic model yields an incorrect peak spacing due to incorrect description of the classical motion of the wavepacket on the excited-state surface. The peak spacing, directly related to the period between the recurrences of the autocorrelation function, is captured correctly with the adiabatic harmonic model, which approximates the classical dynamics much better. However, the envelope of the spectrum is determined not by the position but by the shape of recurrences. The shape, in turn, depends on the width and phase of the wavepacket in the Franck–Condon region. Because the adiabatic harmonic model is not a good representation of the potential in this region, some of the peak intensities are incorrect. In contrast, the vertical model, constructed from the Franck–Condon data, recovers the envelope of the progression almost perfectly. Nevertheless, in this weakly anharmonic system, the adiabatic harmonic model works, overall, better than the vertical model.

In the more anharmonic case (Fig. 6.4B), both global harmonic models fail to reproduce the exact spectrum. The adiabatic model no longer recovers the peak spacing and, therefore, describes well only the first few peaks of the progression. The vertical harmonic model again yields incorrect peak positions; however, it performs somewhat better than the adiabatic approach because it reproduces at least the envelope of the spectrum. A similar effect has been observed (Wehrle et al., 2015) in the absorption spectrum of ammonia (see Section 6.8 below).

Finally, the thawed Gaussian approximation reproduces the exact spectrum better than both global harmonic models because it makes only a local harmonic approximation, which describes the true potential more accurately at all times. In general, modes of different degrees of anharmonicity are excited when a molecule interacts with an external electromagnetic field; however, one cannot know in advance their relative contributions to the spectrum. Therefore, it is hard to decide, *a priori*, whether to use the vertical or adiabatic harmonic model. In contrast, the thawed Gaussian approximation is expected to be at least as accurate as the better of the two global harmonic approaches, and, in addition, requires less human input.

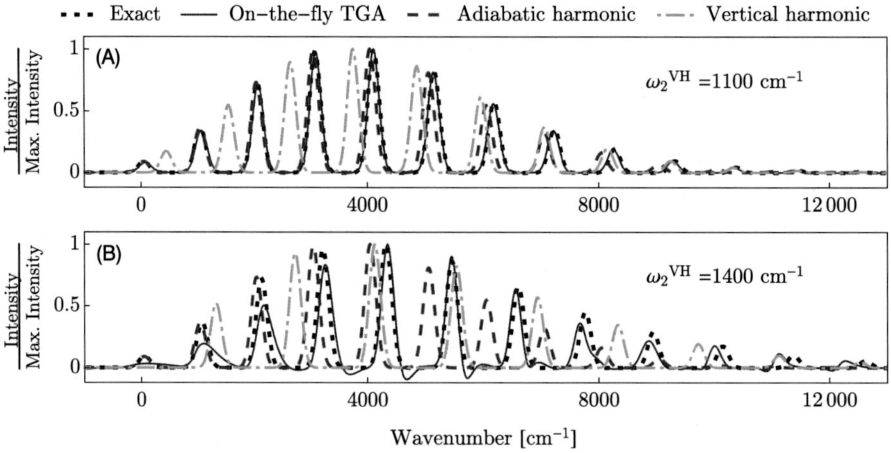

FIG. 6.4 Absorption spectra of two quartic potentials with different degrees of anharmonicity. In both cases, the ground state is a global harmonic potential of wavenumber $\tilde{\nu}_1 = 1000\ \mathrm{cm}^{-1}$ and the minimum of the excited-state quartic potential is shifted by a relative dimensionless displacement $\Delta = 3$ (see caption of Fig. 6.3 for definition of Δ). Quartic parameter δ_2 was chosen according to the formula $\delta_2 = 2(k_2^{\mathrm{VH}} - k_2^{\mathrm{AH}})/(q_{2,i} - q_{1,i})^2$ so that the adiabatic wavenumber was $\tilde{\nu}_2^{\mathrm{AH}} = 800\ \mathrm{cm}^{-1}$ and vertical wavenumber was either (A) $\tilde{\nu}_2^{\mathrm{VH}} = 1100\ \mathrm{cm}^{-1}$ (less anharmonic case) or (B) $\tilde{\nu}_2^{\mathrm{VH}} = 1400\ \mathrm{cm}^{-1}$ (more anharmonic case).

6.7 ON-THE-FLY *Ab Initio* IMPLEMENTATION OF THE THAWED GAUSSIAN APPROXIMATION

Because the center of the thawed Gaussian wavepacket follows a classical trajectory, the thawed Gaussian approximation requires only local information about the potential energy surface, and is perfectly suited for on-the-fly evaluation of energies, forces, and Hessians using an *ab initio* electronic structure code. As a consequence, any *ab initio* dynamics code can be easily extended to include the thawed Gaussian approximation.

The computational cost of the Hessian evaluation is often significantly higher than the cost of the corresponding gradients. However, the Hessians of the potential are not needed for the classical propagation of the center of the wavepacket; therefore, Hessians at different points along the classical trajectory can be evaluated in parallel after the full classical trajectory is known, significantly accelerating the calculation. Because the Hessian of the potential energy changes more slowly than its gradient, the computational cost is further reduced by evaluating the Hessians only once in every several steps of the trajectory (typically, every few femtoseconds). The Hessians at intermediate steps can be obtained by interpolation. Alternative approaches exist, such as those based on Hessian updates (Ceotto et al., 2013; Zhuang et al., 2013) or Gaussian process regression (Alborzpour et al., 2016; Laude et al., 2018), in which the frequency of *ab initio* Hessian calculations can be further reduced by using the energies and gradients to improve the accuracy of the interpolation or extrapolation procedure. In summary, if parallelization and interpolation are employed, the Hessian evaluation ceases to be a bottleneck of the thawed Gaussian approximation as it can be often performed even faster than the propagation of a single classical trajectory.

Once all the *ab initio* data are collected, the molecular geometries, gradients, and Hessians of the potential are transformed from the $3N$ Cartesian coordinates to the internal $3N - 6$ dimensional coordinate space. The mass-scaled normal mode coordinates are well-suited for vibrational dynamics of molecules: they provide an intuitive interpretation of the dynamics and spectra in terms of the vibrational modes of one electronic state. Although some electronic structure codes accept normal-mode and even general internal coordinates as input, working with the Cartesian coordinates to communicate between the dynamics and electronic structure codes is much more robust.

The coupling between the internal (vibrational) and external (translational and rotational) degrees of freedom is reduced by translating and rotating the molecular configuration to the Eckart frame. Let ξ^{ref} be the $3N$-dimensional Cartesian coordinate vector of a reference configuration and ξ a corresponding vector of a general configuration that we wish to transform. Furthermore, let X_A denote the 3-dimensional vector representing the coordinates of the Ath atom and $X := (X_1, X_2, \dots, X_N)$ the $3 \times N$ matrix containing the coordinates of all atoms. Finally, let M_A be the mass of the Ath atom and M the $N \times N$ diagonal matrix with N atomic masses on the diagonal.

The translations are separated and the translational Eckart condition,

$$\sum_{A=1}^{N} M_A X_A^{\text{trans}} = 0, \tag{6.84}$$

is satisfied in the center-of-mass coordinate frame, $X_A^{\text{trans}} := X_A - X_{\text{CM}}$, where

$$X_{\text{CM}} := \sum_{A=1}^{N} M_A X_A / \sum_{A=1}^{N} M_A \tag{6.85}$$

is the center-of-mass coordinate vector. Satisfying the rotational Eckart condition,

$$\sum_{A=1}^{N} M_A X_A^{\text{ref}} \times X_A^{\text{rot}} = 0, \tag{6.86}$$

where $X_A^{\text{rot}} := R \cdot X_A^{\text{trans}}$ and R denotes a rotation matrix, is equivalent to finding a 3×3 special orthogonal matrix R that minimizes the square distance

$$\left\| X^{\text{ref}} - X^{\text{rot}} \right\|^2 := \sum_{A=1}^{N} M_A \left| X_A^{\text{ref}} - X_A^{\text{rot}} \right|^2 \tag{6.87}$$

in the mass-scaled Cartesian coordinates between the reference configuration X^{ref} and the final translated and rotated configuration X^{rot} (Kudin and Dymarsky, 2005); this problem can be solved, e.g., with the Kabsch (Kabsch, 1978) or quaternion (Kearsley, 1989; Coutsias et al., 2004) algorithm. The Kabsch algorithm consists in computing the 3×3 cross-covariance matrix

$$S = X^{\text{trans}} \cdot M \cdot (X^{\text{ref}})^T, \tag{6.88}$$

finding its singular value decomposition $S = U \Sigma V^T$ (where U and V are real orthogonal matrices), and finally recovering the rotation matrix through

$$R = V \begin{pmatrix} 1 & 0 & 0 \\ 0 & 1 & 0 \\ 0 & 0 & \det(VU^T) \end{pmatrix} U^T. \tag{6.89}$$

To summarize the procedure of separating translations and rotations in a single equation, let us revert to the $3N$-dimensional vector space notation and define a $3N \times 3N$ block-diagonal rotation matrix R_ξ, composed of N identical 3×3 rotation matrices R on the diagonal, and a $3N$-dimensional center-of-mass configuration ξ_{CM} containing the N three-dimensional vectors X_{CM}. The final rotated and translated configuration is given by

$$\xi^{\text{rot}} = R_\xi \cdot (\xi - \xi_{\text{CM}}). \tag{6.90}$$

Let O be the orthogonal matrix that diagonalizes the mass-scaled Cartesian Hessian matrix evaluated at ξ^{ref}, i.e.,

$$O^T \cdot m^{-1/2} \cdot \text{Hess}_\xi V|_\xi^{\text{ref}} \cdot m^{-1/2} \cdot O = \Omega^2, \tag{6.91}$$

where m is the $3N \times 3N$ diagonal mass matrix and Ω a $3N \times 3N$ diagonal matrix with normal mode frequencies on the diagonal. Then, the transformation from Cartesian to mass-scaled normal mode coordinates is given by

$$q = L^T \cdot m^{\frac{1}{2}} \cdot (\xi^{\text{rot}} - \xi^{\text{ref}}), \tag{6.92}$$

where L is a $3N \times (3N - 6)$ submatrix of O from which the six translational and rotational degrees of freedom are dropped. Similarly, the gradients and Hessians are transformed using

$$\text{grad}_q V = L^T \cdot m^{-\frac{1}{2}} \cdot R_\xi \cdot \text{grad}_\xi V, \tag{6.93}$$

$$\text{Hess}_q V = L^T \cdot m^{-\frac{1}{2}} \cdot R_\xi \cdot \text{Hess}_\xi V \cdot R_\xi^T \cdot m^{-\frac{1}{2}} \cdot L. \tag{6.94}$$

6.8 EXAMPLES OF ON-THE-FLY *Ab Initio* CALCULATIONS OF ELECTRONIC SPECTRA

Having presented, in Section 6.6.1, simple examples demonstrating merits and shortcomings of commonly used models for describing electronic spectra, let us look at realistic examples based on the on-the-fly *ab initio* calculations of molecules of increasing size. The examples use the thawed Gaussian approximation and its extensions (the extended TGA and the 3TGA), described in Section 6.4, together with the on-the-fly *ab initio* implementation described in Section 6.7.

6.8.1 Absorption and Photoelectron Spectra of Ammonia

Due to its floppy nature, ammonia provides a challenging test of approximations. Absorption and pho-

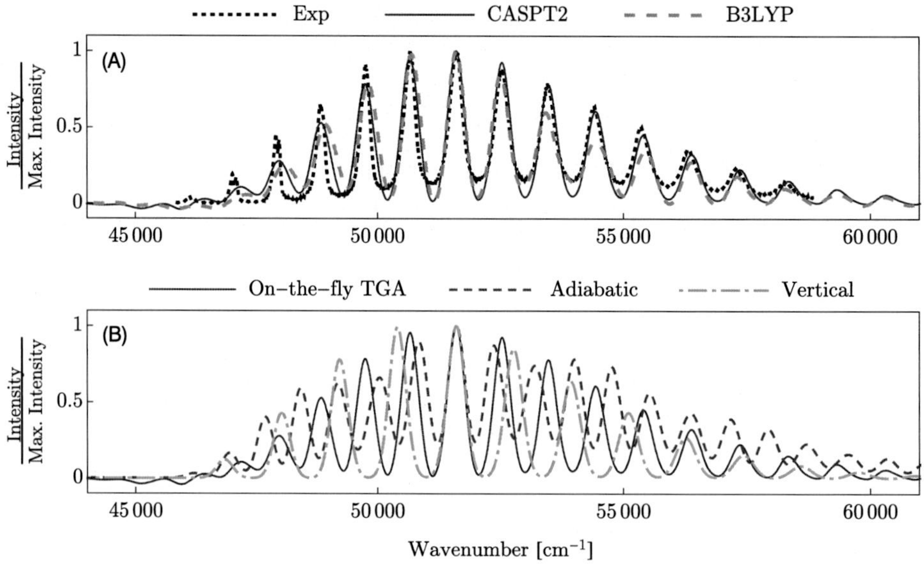

FIG. 6.5 Absorption spectrum of NH_3. (A) Comparison of the experimental spectrum recorded at the temperature of 175 K (Chen et al., 1999) with the spectra computed with the on-the-fly *ab initio* TGA using CASPT2 and B3LYP electronic structure methods. (B) Spectra evaluated using the on-the-fly, vertical harmonic, and adiabatic harmonic models (all using CASPT2). All spectra are rescaled and shifted so that the highest spectral peak in each spectrum is of unit intensity and positioned at the same wavenumber. Adapted with permission from Wehrle et al. (2015). Copyright 2015 American Chemical Society.

toelectron spectra of ammonia have, therefore, been used to validate new methods, such as the use of curvilinear internal coordinates (Capobianco et al., 2012) or the combination of the discrete variable representation for the large-amplitude motion and the global harmonic model for the modes orthogonal to it (Baiardi et al., 2017). The experimental $\tilde{A}^1A_2'' \leftarrow \tilde{X}^1A_1'$ ($S_1 \leftarrow S_0$) absorption spectrum of ammonia contains a single long progression due to the highly displaced umbrella mode. The nuclear configuration changes from non-planar (pyramidal) in the $\tilde{X}^1A_1'$ state to planar in the $\tilde{A}^1A_2''$ state, thus inducing a large-amplitude motion on the excited-state potential energy surface. Interestingly, the excitation of the totally symmetric N–H stretch mode, which also accompanies this electronic transition, is not observed as a separate progression (Tang et al., 1990)—this is an example of the missing-mode effect (Tutt et al., 1987).

The computation of the absorption spectrum

$$\sigma_{\text{abs}}(\omega) = \frac{\omega}{3\hbar c \varepsilon_0} |\vec{\mu}_{21}|^2 \times \text{Re} \int_0^\infty \langle \psi(0) | \psi(t) \rangle e^{\mathrm{i}(E_{1,g}/\hbar + \omega)t} \mathrm{d}t \quad (6.95)$$

involves the propagation of the initial vibrational wavepacket $|\psi(0)\rangle = |1, g\rangle$ on the excited-state potential energy surface $V_2(q)$. Eq. (6.95) assumes the Condon approximation and accounts for the orientational averaging of the spectrum, hence the factor of three in the denominator.

Fig. 6.5A compares the experimental spectrum with spectra calculated (Wehrle et al., 2015) with the on-the-fly *ab initio* TGA at two different levels of electronic structure theory: time-dependent density functional theory, with the B3LYP functional, and the CASPT2 method. Interestingly, the peak spacings, which are well described by the on-the-fly TGA, are not affected by the level of theory employed for the electronic structure. Unlike the on-the-fly approach, which partially includes the anharmonicity of the excited-state surface, the global harmonic methods fail to recover the correct shape of the spectrum (see Fig. 6.5B). The adiabatic harmonic model gives an inaccurate approximation of the excited-state surface, resulting in a double progression which is otherwise hidden in the experiment through the missing-mode effect. The vertical harmonic model captures at least the overall envelope of the spectrum but overestimates the frequency of the umbrella mode and, consequently, the spacings between the peaks.

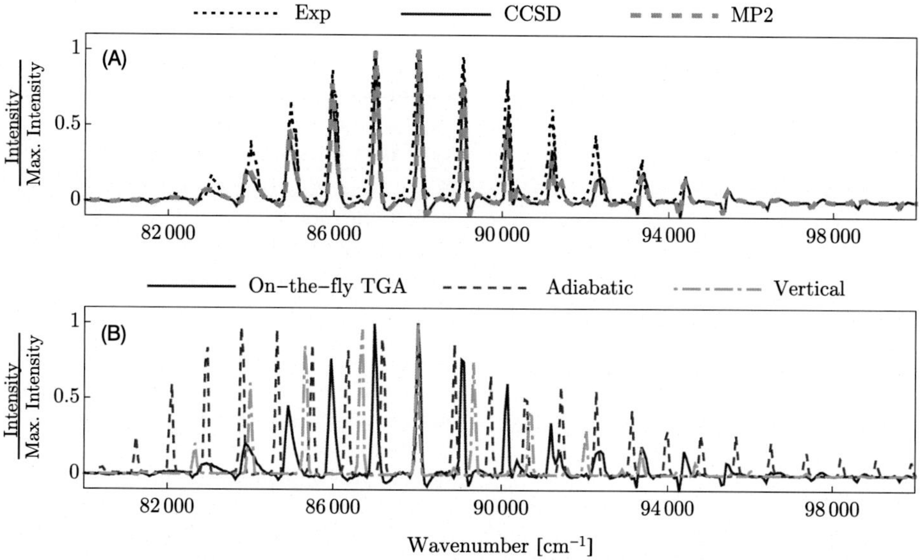

FIG. 6.6 Photoelectron spectrum of NH_3. (A) Comparison of the experimental spectrum recorded at the temperature of 77 K (Edvardsson et al., 1999) with the spectra computed with the on-the-fly *ab initio* thawed Gaussian approximation using either the CCSD or MP2 electronic structure method. (B) Comparison of the spectra evaluated with the on-the-fly, adiabatic harmonic, and vertical harmonic models (all using CCSD). All spectra are rescaled and shifted so that the highest spectral peak in each spectrum is of unit intensity and positioned at the same wavenumber. Adapted with permission from Wehrle et al. (2015). Copyright 2015 American Chemical Society.

Within the Condon approximation, a rotationally averaged photoelectron spectrum can be computed also with Eq. (6.95), but now the state 2 is the ground cationic state instead of an excited state of the neutral molecule (see, e.g., Lami et al., 2004). The photoelectron spectrum of ammonia (Fig. 6.6) is better resolved than the absorption spectrum and, therefore, provides a more stringent test of the on-the-fly *ab initio* TGA (Wehrle et al., 2015). Yet, the TGA performs rather well. As in the absorption spectrum, the choice of the electronic structure method (Fig. 6.6A) affects only slightly the on-the-fly result, whereas the adiabatic harmonic model contains spurious progressions and the vertical harmonic model completely misses the positions of the peaks (Fig. 6.6B).

6.8.2 Absorption Spectra Beyond Condon Approximation

In general, the transition dipole moment is a function of nuclear coordinates, and the absorption spectrum is calculated by propagating the modified initial wavepacket $|\phi(0)\rangle = \hat{\mu}_{21}|1, g\rangle$, which includes the transition dipole moment as a factor:

$$\sigma_{\text{abs}}(\vec{\epsilon}, \omega) = \frac{\omega}{\hbar c \varepsilon_0} \text{Re} \int_0^{\infty} \langle \phi(0)|\phi(t)\rangle e^{i(E_{1,g}/\hbar + \omega)t} dt. \tag{6.96}$$

As discussed in Section 6.5.4, within the Herzberg–Teller approximation, the transition dipole moment is a linear function (6.66) of nuclear coordinates; as a result, the initial nuclear wavefunction takes the form

$$\phi(q, 0) = \left[\mu_{21}(q_0) + \partial_q \mu_{21}|_{q_0}^{T} \cdot (q - q_0)\right] \psi(q, 0), \tag{6.97}$$

where $\psi(q, 0)$ is the wavefunction (6.15) of the initial Gaussian state $|\psi(0)\rangle = |1, g\rangle$. Because the projection of the transition dipole moment $\mu_{21}(q) = \vec{\mu}_{21}(q) \cdot \vec{\epsilon}$ changes depending on the orientation of the molecule with respect to the polarization $\vec{\epsilon}$ of the electric field, one has to carefully average the spectrum over all orientations, as discussed in Section 6.5.5.

Wavepacket $|\phi\rangle$ can be evolved with the extended thawed Gaussian approximation, described in Section 6.4.3 (Patoz et al., 2018). The ETGA wavepacket

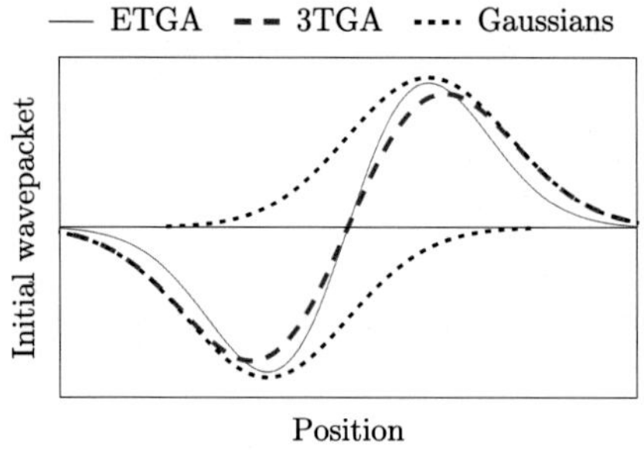

FIG. 6.7 Herzberg–Teller part of the wavepacket propagated with the ETGA is a Gaussian multiplied by a linear polynomial in nuclear coordinates. The 3TGA approximates the Herzberg–Teller part with two displaced Gaussians (black dotted lines). Reprinted from Begušić et al. (2018a).

(6.97) conserves its form and the corresponding parameters are propagated according to Eqs. (6.35) and (6.36). In contrast to the original TGA, the probability density at the center of the ETGA wavepacket is not maximum, but actually vanishes. Therefore, the local harmonic approximation about the wavepacket's center becomes questionable.

To address this issue, one can use another generalization of the TGA suitable for evaluating the Herzberg–Teller spectra, but which also accounts for the wavepacket splitting. Recall that in Section 6.4.4, we mentioned the possibility of using multiple thawed Gaussians to represent the initial wavepacket. Indeed, the Herzberg–Teller part of the initial wavepacket is well described by two displaced Gaussians of opposite sign (see Fig. 6.7) (Begušić et al., 2018a); the total initial wavepacket in the *three thawed Gaussians approximation* (3TGA), which also contains a Condon term, is

$$\phi^{3\mathrm{TGA}}(q,0) = \mu_{21}\, g_{q_0}(q) + f_d\left[g_{q_0+\Delta_d}(q) - g_{q_0-\Delta_d}(q)\right]. \quad (6.98)$$

In Eq. (6.98), $g_{q_c}(q)$ is a normalized Gaussian

$$g_{q_c}(q) := e^{\frac{i}{\hbar}\left[\frac{1}{2}(q-q_c)^T \cdot A_0 \cdot (q-q_c)+\gamma_0\right]} \quad (6.99)$$

centered at q_c, with zero initial momentum ($p_0 = 0$) and $\gamma_0 = -i\hbar \ln[\det(\mathrm{Im}A_0/\pi\hbar)]/4$. Δ_d is a displacement vector and f_d a scaling factor ensuring normalization. The two displaced Gaussians are centered at the extrema of the Herzberg–Teller part of the initial wavepacket; note that there are always exactly two local extrema, regardless of the number D of vibrational degrees of freedom (for a proof, see Appendix E of Begušić et al., 2018a).

An extreme case of Herzberg–Teller spectra occurs in molecules, in which the electronic transition is symmetry-forbidden, but vibronically allowed due to the non-zero elements of the gradient of the transition dipole moment (Herzberg, 1966). A prototypical example is the absorption spectrum of benzene $\tilde{A}^1B_{2u} \leftarrow \tilde{X}^1A_{1g}$ (Herzberg, 1966; Li et al., 2010), in which the non-zero elements, corresponding to the doubly degenerate e_{2g} vibrational modes, arise due to the non-adiabatic coupling between the first and higher excited electronic states (as discussed in Section 6.5.4 and in Li et al., 2010; Quack and Merkt, 2011; Begušić et al., 2018a). The absorption spectrum contains a single progression, attributed to the totally symmetric ring-breathing mode, and a number of weak hot bands. Because the ETGA and 3TGA, as described above, do not treat finite-temperature effects, we will only discuss the peaks of the main progression.

The spectra shown in Fig. 6.8A imply that in benzene the local harmonic approximation holds even for the Herzberg–Teller wavepacket because the 3TGA spectrum is only slightly red-shifted ($\approx 10\ \mathrm{cm}^{-1}$) with respect to the ETGA spectrum (Begušić et al., 2018a). The global harmonic models (Fig. 6.8B) are not as accurate as the on-the-fly approach (Patoz et al., 2018; Begušić et al., 2018a). Whereas the relative intensities of adiabatic harmonic peaks have errors of 20% to 50% and the vertical harmonic model fails completely, the relative intensities of the on-the-fly peaks lie within 5% of the experimental values. Finally, Fig. 6.8C confirms that including the Herzberg–Teller contribution in this system is absolutely essential because the purely electronic Condon spectrum is zero.

6.8.3 Emission Spectra of Large Systems: Quinquethiophene

Computational efficiency is, of course, one of the advantages of a single- or few-trajectory methods, permitting the treatment of larger systems, which are inaccessible to multi-trajectory semiclassical methods. Wehrle et al. (2014) applied the on-the-fly *ab initio* TGA to compute the emission spectra of oligothiophenes with up to five thiophene units, i.e., up to quinquethiophene, which, due to its size (105 vibrational degrees of freedom), poses a formidable challenge to both quantum dynamical and electronic structure methods.

An emission spectrum is obtained by taking the vibrational ground state, $|2, g\rangle$, of the excited-state surface as the initial wavepacket $|\psi(0)\rangle$ and by evolving $|\psi(t)\rangle$ on the ground-state surface. Within the Condon approximation and zero-temperature limit, the *spontaneous emission* (or *fluorescence*) spectrum, measured as

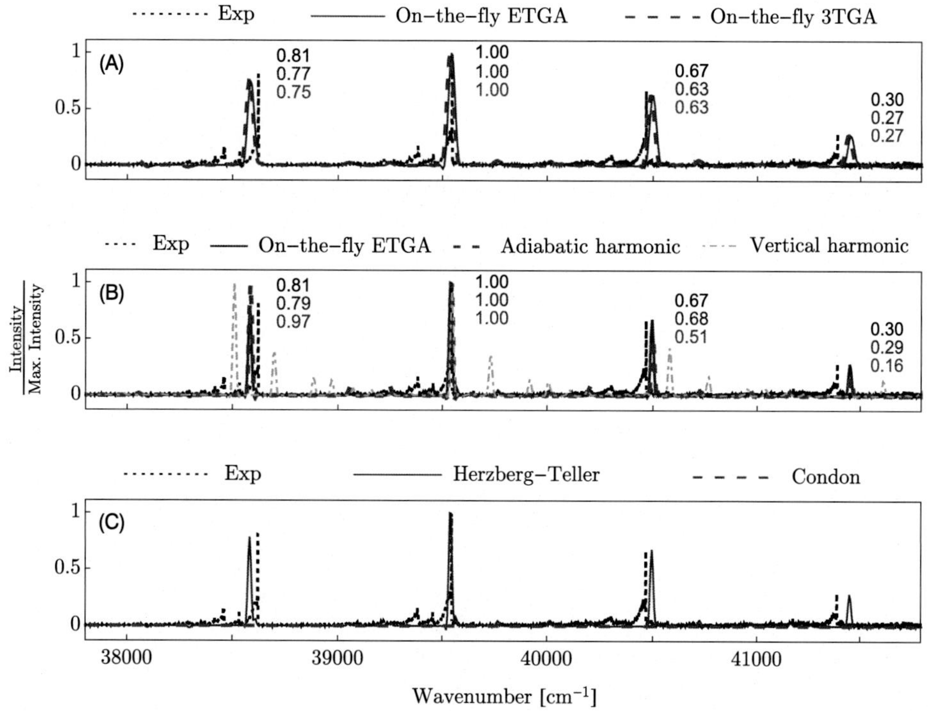

FIG. 6.8 Calculated absorption spectra of benzene $\tilde{A}^1B_{2u} \leftarrow \tilde{X}^1A_{1g}$ electronic transition compared to the experimental (Fally et al., 2009; Keller-Rudek et al., 2013) spectrum measured at 293 K. (A) Comparison of the on-the-fly *ab initio* ETGA and 3TGA spectra. (B) Comparison of the on-the-fly, adiabatic harmonic, and vertical harmonic spectra evaluated with ETGA, i.e., within the Herzberg–Teller approximation. (C) Comparison of the Condon and Herzberg–Teller approximations (both evaluated using the on-the-fly approach). All spectra are horizontally shifted and rescaled according to the highest peak. To clarify the difference between the adiabatic harmonic and on-the-fly spectra, we show the scaled intensities of the experimental, adiabatic harmonic, and on-the-fly peaks. The *ab initio* calculations are based on the B3LYP/6-31+G(d,p) electronic structure method. The ETGA requires only a single trajectory, which was propagated for longer times than the trajectories used in the 3TGA. Therefore, in (A), the broadening is determined by the length of the shorter 3TGA simulation, while in (B) and (C), where only the ETGA is considered, the spectra are broadened less. (A) Adapted from Begušić et al. (2018a). (B) and (C) Adapted with permission from Patoz et al. (2018). Copyright 2018 American Chemical Society.

the rate of emission per unit frequency between ω and $\omega + d\omega$, is a dimensionless quantity that can be computed as (see, e.g., Lami et al., 2004; Niu et al., 2010)

$$\sigma_{\mathrm{em}}(\omega) = \frac{\omega^3}{3\pi^2\hbar c^3 \varepsilon_0}|\vec{\mu}_{21}|^2 \times \mathrm{Re}\int_0^\infty \langle\psi(t)|\psi(0)\rangle e^{\mathrm{i}(\omega - E_{2,g}/\hbar)t}\,\mathrm{d}t, \quad (6.100)$$

where $E_{2,g}$ denotes the eigenenergy of the state $|2, g\rangle$ and $|\psi(t)\rangle = \exp(-\mathrm{i}\hat{H}_1 t/\hbar)|\psi(0)\rangle$.

Oligothiophenes, including quinquethiophene, undergo a change in symmetry upon the transition from the first excited to the ground electronic state: whereas the excited-state geometry is planar, the ground-state minimum is twisted along the inter-ring torsional degrees of freedom (see Fig. 6.9A). Therefore, the initial wavepacket, constructed at the minimum of the excited-state surface, starts its evolution at the top of a potential barrier of the ground-state surface. In this scenario, one could expect the exact wavepacket to split, and therefore the TGA to fail. However, due to the rather low magnitudes of the imaginary frequencies corresponding to the torsional modes, the simple spreading of the thawed Gaussian wavepacket (Fig. 6.9B) provides an adequate description of the dynamics along these modes, at least

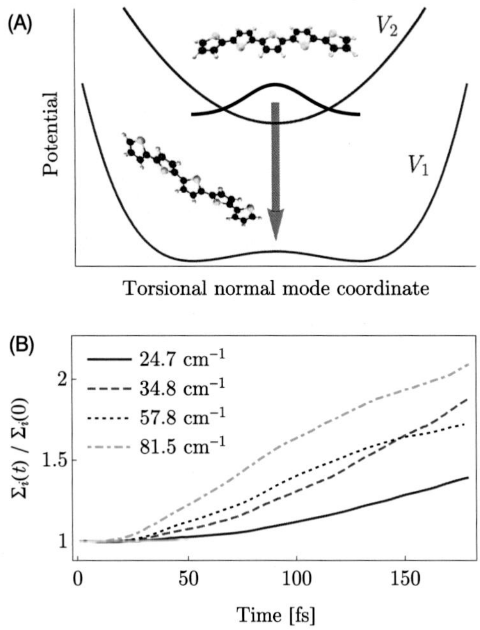

FIG. 6.9 Inter-ring torsional modes of quinquethiophene and the spreading of the wavepacket. (A) Schematic representation of a section of the potential energy surface along one of the four torsional normal mode coordinates; the excited-state minimum has a planar configuration, whereas the ground-state minimum is twisted. (B) Spreading of the wavepacket along the four inter-ring torsional degrees of freedom induced by the inverted potential. To distinguish the four modes, their excited-state wavenumbers are shown in the legend. The width of the wavepacket along the mode i is measured by the standard deviation of the probability density, $\Sigma_i(t) = (2\mathrm{Im}A_{ii,t}/\hbar)^{-1/2}$; the figure shows relative values with respect to the initial widths. (B) Adapted from Wehrle et al. (2014), with the permission of AIP Publishing.

during the short time required for simulating the low-resolution emission spectrum (Wehrle et al., 2014). This picture is confirmed in Fig. 6.10A, in which the experimental and calculated spectra are shown to agree both in the positions and intensities of the peaks. (Note that, in contrast to previous examples, here the computed spectra were not shifted; a small shift between the experimental and computed peaks is most likely due to the error of electronic structure method.)

The single TGA trajectory can be used not only to evaluate the spectrum, but also to understand the spectral features and to generate partially uncoupled models (Wehrle et al., 2014) that are accessible to more rigorous quantum approaches, such as the multi-layer multi-configurational time-dependent Hartree (ML-MCTDH) method (Wang and Thoss, 2003). The reason is that the stability matrix M_t, needed in TGA, provides information about the dynamical coupling within each pair of modes. The simplified system is obtained by neglecting the couplings below a chosen threshold (Wehrle et al., 2014); after the decoupling, the independent subsystems that contain no significantly excited modes are dropped from the analysis (Wehrle et al., 2014). Remarkably, the whole process of generating the partially uncoupled fewer-dimensional models is fully automated and thus completely avoids "chemical intuition." Comparison of the full- and reduced-dimensionality TGA calculations of quinquethiophene spectra in Fig. 6.10B proves that the main progression of the spectrum is generated by *only* four inter-ring-stretch and ring-squeeze modes. To account for the broadening of the peaks, it is necessary to include in the model only four additional vibrational modes, corresponding to the chain and C-H bond deformations. Overall, by including as few as eight vibrational modes, the computed spectrum reproduces very well the full 105-dimensional result. A related "divide-and-conquer" semiclassical approach, in which the decoupling is done differently, has been used by Ceotto and coworkers to evaluate vibrational spectra of fullerene and other large systems that would be otherwise inaccessible to semiclassical initial value representation (Ceotto et al., 2017).

6.8.4 Vibrationally Resolved Pump-Probe Spectra

In a pump–probe experiment, the difference between the pump-on and pump-off spectra is measured. Within the approximations made in Section 6.5.6, the differential absorption cross-section at a probe delay time τ (often simply called the "pump-probe spectrum") is evaluated as the half-Fourier transform (Pollard et al., 1990b; Mukamel, 1999; Domcke and Stock, 1997)

$$\sigma_{\mathrm{PP}}(\vec{\epsilon}^{\,\mathrm{pu}}, \vec{\epsilon}^{\,\mathrm{pr}}, \omega, \tau) = \frac{4\pi^2\omega|\tilde{E}^{\mathrm{pu}}(\omega_{21})|^2}{\hbar^3 c\varepsilon_0}\left|\mu_{21}^{\mathrm{pu}}\right|^2$$
$$\times \mathrm{Re}\int_0^{\infty}\Big[\left|\mu_{n2}^{\mathrm{pr}}\right|^2 C_{\mathrm{ESA}}^*(t,\tau) - \left|\mu_{21}^{\mathrm{pr}}\right|^2 C_{\mathrm{TRSE}}(t,\tau)$$
$$- \left|\mu_{21}^{\mathrm{pr}}\right|^2 C_{\mathrm{GSB}}(t,\tau)\Big]e^{i\omega t}\,dt, \tag{6.101}$$

where $\mu_{\alpha\beta}^{\mathrm{pu}}$ and $\mu_{\alpha\beta}^{\mathrm{pr}}$ are the transition dipole moments $\vec{\mu}_{\alpha\beta}$ [Eq. (6.68)] projected onto the polarization vectors $\vec{\epsilon}^{\,\mathrm{pu}}$ and $\vec{\epsilon}^{\,\mathrm{pr}}$ of the pump and probe electric fields, and $\tilde{E}^{\mathrm{pu}}(\omega_{21})$ is the Fourier transform of the pump electric field amplitude evaluated at the transition frequency ω_{21} between the ground and excited electronic states

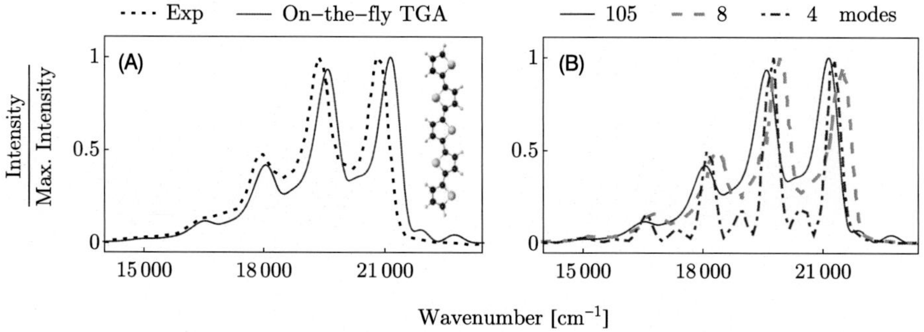

FIG. 6.10 Emission spectrum of quinquethiophene. (A) Comparison of the experimental spectrum and the spectrum computed with the full-dimensional on-the-fly *ab initio* TGA using all 105 normal modes. (B) Comparison of the spectra computed either with the full-dimensional on-the-fly *ab initio* TGA or with the TGA applied to the automatically generated reduced-dimensionality models. The spectra are labeled by the number of normal modes treated. Adapted from Wehrle et al. (2014), with the permission of AIP Publishing.

(Begušić et al., 2018b). The three terms in the integrand arise from three different physical processes: absorption to a higher-excited electronic state n ("excited-state absorption"),

$$C_{\mathrm{ESA}}(t,\tau) = \langle 1, g|\, e^{\mathrm{i}\hat{H}_2\tau/\hbar} e^{\mathrm{i}\hat{H}_n t/\hbar} e^{-\mathrm{i}\hat{H}_2(t+\tau)/\hbar}\, |1, g\rangle\,, \tag{6.102}$$

time-resolved stimulated emission,

$$C_{\mathrm{TRSE}}(t,\tau) = \langle 1, g|\, e^{\mathrm{i}\hat{H}_2\tau/\hbar} e^{\mathrm{i}\hat{H}_1 t/\hbar} e^{-\mathrm{i}\hat{H}_2(t+\tau)/\hbar}\, |1, g\rangle\,, \tag{6.103}$$

and ground-state bleach,

$$C_{\mathrm{GSB}}(t,\tau) = \langle 1, g|\, e^{\mathrm{i}\hat{H}_1 t/\hbar} e^{-\mathrm{i}\hat{H}_2 t/\hbar}\, |1, g\rangle\,. \tag{6.104}$$

In the following discussion, we concentrate on the time-resolved stimulated emission spectrum and only in the final result we add the ground-state bleach; the excited-state absorption signal can be computed in a similar way. Although the excited-state absorption is not negligible, it is commonly found at a different frequency than the other two contributions. Therefore, the excited state absorption can be analyzed separately, unlike the ground-state bleach and stimulated emission, which often overlap (Berera et al., 2009).

As shown in Section 6.5.6, to evaluate the time-resolved stimulated emission spectrum, one must first propagate the nuclear wavepacket for time τ in the excited state, and then propagate it for time t simultaneously in both the ground and excited states [Eqs. (6.75) and (6.76)]. Within the Condon and ultrashort pulse approximations, the ground-state bleach contribution to the spectrum is much simpler to evaluate because it is equal to a scaled linear absorption spectrum discussed in Section 6.5.1.

Fig. 6.11 shows the time-resolved stimulated emission spectrum of phenyl radical, computed with the on-the-fly *ab initio* TGA in Begušić et al. (2018b). Because the Herzberg–Teller contribution to the absorption spectrum of phenyl radical is negligible (Patoz et al., 2018; Begušić et al., 2018a), the Condon approximation was used for the time-resolved spectrum.

The spectrum exhibits oscillations over a broad range of wavenumbers, reflecting the dynamics of the wavepacket in the excited state: As the wavepacket leaves the Franck–Condon region, the energy gap decreases, shifting the spectrum towards lower wavenumbers [Eq. (6.113)]. Along with the changes in the position of the spectrum, the vibrational resolution also changes periodically as a function of the delay time (see Fig. 6.12). As the wavepacket moves away from the initial position, the spectra become broader and less resolved; the resolution of the spectrum is recovered when the wavepacket returns to the Franck–Condon region. The period of ≈ 36 fs observed in the time-resolved stimulated emission spectrum corresponds to the wavenumber of the most-displaced mode in the excited electronic state (924 cm^{-1}). The time-resolved stimulated emission component of the pump-probe spectrum in Fig. 6.11B changes significantly with the delay time τ, whereas the ground-state bleach is constant; a similar pattern has been observed in other systems, both experimentally (Fragnito et al., 1989; van der Veen et al., 2011) and in calculations based on harmonic models (Pollard et al., 1990b, 1990a; Pollard and Mathies, 1992).

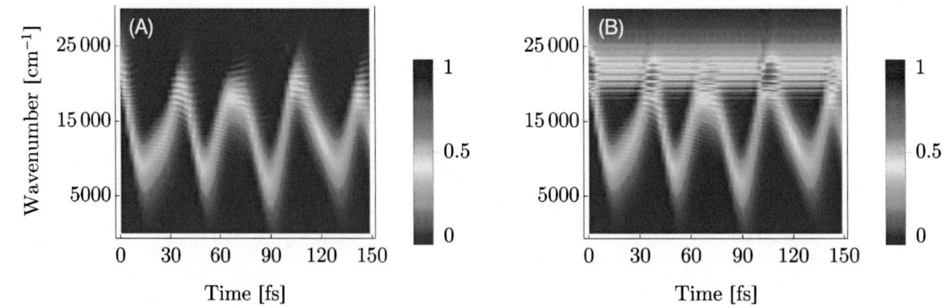

FIG. 6.11 Time-resolved stimulated emission spectrum (A) and the pump–probe spectrum (B) including both the stimulated emission and ground-state bleach of phenyl radical evaluated with the on-the-fly *ab initio* TGA. Both spectra were rescaled according to the maximum of the pump–probe spectrum in the right panel. *Ab initio* calculations are based on the B3LYP/SNSD (Baiardi et al., 2013) electronic structure method. Reprinted from Begušić et al. (2018b).

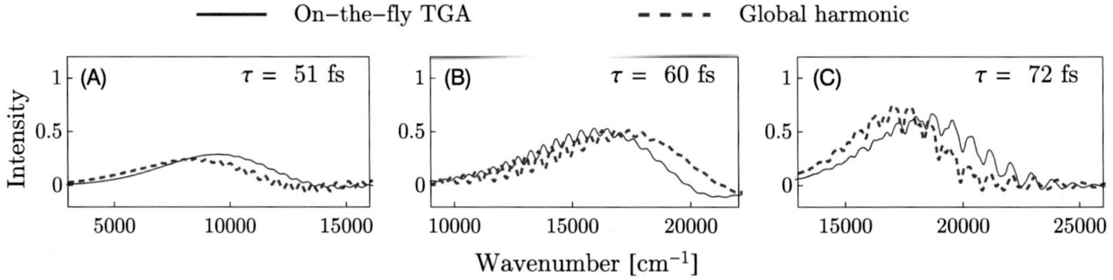

FIG. 6.12 Computed time-resolved stimulated emission spectra of the phenyl radical at several probe delay times τ. All spectra are scaled by the maximum of the on-the-fly spectrum at zero delay time. Adapted from Begušić et al. (2018b).

Useful measures for analyzing the pump–probe spectra as a function of the delay time τ are the mean $\langle\omega\rangle_\tau := \int \omega\sigma_0(\omega,\tau)d\omega$ and width (more precisely, standard deviation) $\Delta\omega_\tau := \sqrt{\langle\omega^2\rangle_\tau - \langle\omega\rangle_\tau^2}$ of the normalized spectral lineshapes (Pollard et al., 1990b; Ferrer et al., 2013)

$$\sigma_0(\omega,\tau) = \frac{\sigma_{\mathrm{PP}}(\omega,\tau)/\omega}{\int(\sigma_{\mathrm{PP}}(\omega,\tau)/\omega)d\omega}. \tag{6.105}$$

Both quantities can be expressed in terms of the energy gap operator $\Delta\hat{V} = \hat{V}_2 - \hat{V}_1$ and the excited-state wavepacket $|\psi(\tau)\rangle = |\psi_\alpha(0,\tau)\rangle$ [Eq. (6.76)] at time τ. The mean of the spectrum is determined by the expectation value of the energy gap (Pollard et al., 1990b),

$$\langle\omega\rangle_\tau = \frac{1}{\hbar}\langle\psi(\tau)|\Delta\hat{V}|\psi(\tau)\rangle, \tag{6.106}$$

while the width is given by the standard deviation of the energy gap (Pollard et al., 1990b):

$$\Delta\omega_\tau = \frac{1}{\hbar}\sqrt{\langle\psi(\tau)|\Delta\hat{V}^2|\psi(\tau)\rangle - \langle\psi(\tau)|\Delta\hat{V}|\psi(\tau)\rangle^2}. \tag{6.107}$$

For a thawed Gaussian wavepacket $\psi(\tau)$ and within the local harmonic approximation, the mean and width of the spectra can be evaluated analytically (Begušić et al., 2018b) as

$$\langle\omega\rangle_\tau^{\mathrm{LHA}} = \frac{1}{\hbar}[\Delta V_\tau + \frac{1}{2}\mathrm{Tr}(\Delta V_\tau'' \cdot \Sigma_\tau^2)], \tag{6.108}$$

$$\Delta\omega_\tau^{\mathrm{LHA}} = \frac{1}{\hbar}\sqrt{(\Delta V_\tau')^T \cdot \Sigma_\tau^2 \cdot \Delta V_\tau' + \frac{1}{2}\mathrm{Tr}(\Delta V_\tau'' \cdot \Sigma_\tau^2 \cdot \Delta V_\tau'' \cdot \Sigma_\tau^2)}, \tag{6.109}$$

where ΔV_τ, $\Delta V_\tau'$, $\Delta V_\tau''$ are, respectively, the energy gap, its gradient, and Hessian evaluated at the center q_τ of the wavepacket at time τ:

$$\Delta V_\tau = V_2|_{q_\tau} - V_1|_{q_\tau}, \tag{6.110}$$

$$\Delta V_\tau' = \mathrm{grad}_q\, V_2|_{q_\tau} - \mathrm{grad}_q\, V_1|_{q_\tau}, \tag{6.111}$$

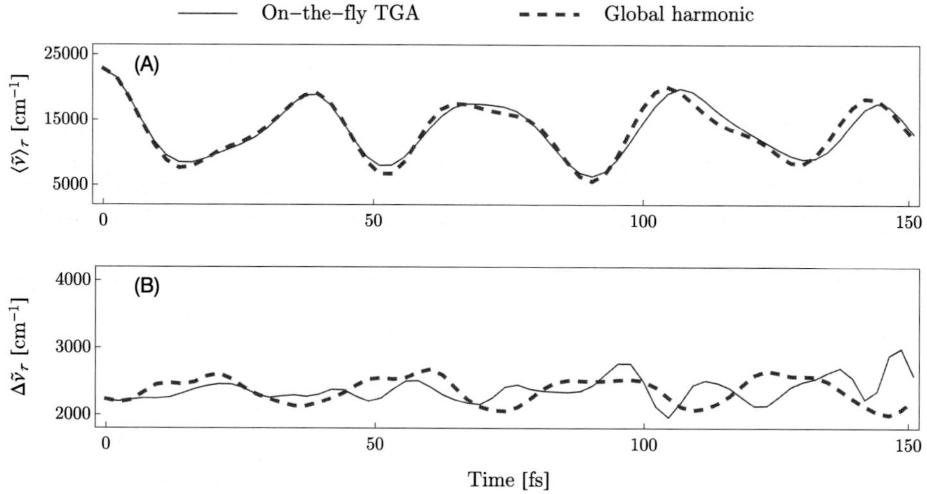

FIG. 6.13 Means (A) and widths (B) of the computed time-resolved stimulated emission spectra of the phenyl radical. Adapted from Begušić et al. (2018b).

$$\Delta V''_\tau = \mathrm{Hess}_q V_2|_{q_\tau} - \mathrm{Hess}_q V_1|_{q_\tau}. \tag{6.112}$$

Remarkably, evaluating the mean and width of the pump-probe spectrum from expressions (6.106)–(6.107) does not require running trajectories on the ground surface; this is only necessary if the full, frequency-resolved spectrum is wanted.

If the two potential surfaces have similar curvatures ($\Delta V''_\tau \approx 0$), then Eqs. (6.108) and (6.109) simplify further to:

$$\langle\omega\rangle_\tau^{\mathrm{appr}} = \Delta V_\tau/\hbar \tag{6.113}$$

$$\Delta\omega_\tau^{\mathrm{appr}} = \frac{1}{\hbar}\sqrt{(\Delta V'_\tau)^T \cdot \Sigma_\tau^2 \cdot \Delta V'_\tau}. \tag{6.114}$$

According to Eq. (6.113), the mean of the spectrum can be computed by evaluating the energy gap along a single excited-state classical trajectory. The widths of the spectra, however, require the full TGA propagation on the excited-state surface as well as the ground-state gradients along the excited-state trajectory.

Comparison between the global harmonic and the more accurate on-the-fly results (see Fig. 6.12) indicates the presence of anharmonicity effects. In spite of a very good description of the mean of the spectrum (see Fig. 6.13), the global harmonic approximation completely loses the accuracy of the frequency-resolved features. Moreover, the success of the global harmonic approximation in describing at least the mean of the spectra is due to a partial error cancellation arising from omitting anharmonicity in both potential energy surfaces.

6.9 CONCLUSION AND OUTLOOK

In this chapter, we have discussed one of many possible approaches to the calculation of vibrationally resolved electronic spectra of molecules: The on-the-fly *ab initio* semiclassical thawed Gaussian approximation avoids constructing a global potential energy surface by evaluating the potential energy, its gradient, and Hessian only where needed along a classical trajectory and, moreover, takes into account all degrees of freedom. At the other extreme are approaches using exact quantum propagation on reduced-dimensionality surfaces, globally harmonic surfaces, or other approximate surfaces such as those with the potential energy function expressed in the "sum-of-products" form.

The very simple TGA cannot compete with the accuracy of exact quantum methods if those are feasible in a given system. Yet, the TGA has two properties of central importance for molecular quantum dynamics and, in particular, for evaluating vibrationally resolved electronic spectra: First, the TGA is exact in a globally harmonic potential, which is often a decent starting point for approximating the exact potential energy surface of a rigid molecule. Moreover, the TGA performs well in nearly-harmonic potentials, outperforming the global harmonic approaches by including anharmonicity at least partially. Second, the TGA is exact in the short-time limit and, in practice, accurate at times before the wavepacket splitting takes place; this is often sufficient for describing low-resolution vibronic and ultrafast time-resolved spectra.

In Section 6.8, we have shown several examples of on-the-fly *ab initio* calculations of both steady-state and time-resolved molecular electronic spectra. Surprisingly, the Condon absorption and photoelectron spectra computed with the TGA are accurate even in the rather floppy molecule of ammonia. We have mentioned two extensions of the TGA for evaluating spectra beyond the Condon approximation. Both the extended TGA and the 3TGA allow the treatment of Herzberg–Teller spectra of fairly large molecules. In benzene and other molecules in which wavepacket splitting is negligible, the spectra are well described with the simpler, one-trajectory ETGA, at a cost of the original TGA. As shown on the emission spectrum of quinquethiophene, the on-the-fly *ab initio* TGA is feasible in rather large molecules. In this particular system, the accuracy of the spectrum is not corrupted by the double-well nature of the potential energy surface, which justifies using the Gaussian wavepacket ansatz in the simulation of low-resolution electronic spectra. On the quinquethiophene example, we have also shown that the single-trajectory TGA would be useful even if its accuracy were unsatisfactory; when this happens, the reduced-dimensionality models generated through the automatic decoupling procedure can be subsequently treated with more accurate semiclassical or exact quantum dynamical methods. A similar automated "divide and conquer" approach, based on somewhat different criteria for the decoupling, was used successfully by Ceotto and coworkers for computing vibrational spectra (Ceotto et al., 2017). The last of the examples, namely the pump–probe spectrum of the phenyl radical, proves that the on-the-fly *ab initio* calculations are feasible even when more trajectories are required and that the blind application of global harmonic models is dangerous: Due to cancellation of errors between the ground and excited states, the global harmonic models can yield accurate observables, such as means of spectra, even in anharmonic systems whose vibrationally resolved spectra are strongly affected by the anharmonicity.

Because the TGA and its extensions mentioned in this chapter require only a single or few classical trajectories, along with the corresponding Hessians, such methods allow an on-the-fly implementation in conjunction with rather high-level *ab initio* electronic structure calculations (Prlj et al., 2020). In contrast, many-trajectory approaches such as the variants of the Herman–Kluk initial value representation, on one hand result in more accurate nuclear dynamics, but, on the other hand, require many trajectories for convergence, which makes them practical only with less accurate electronic structure methods. Overall, it seems that until on-the-fly *ab initio* many-trajectory approaches become more efficient, the TGA will remain an interesting option for evaluating spectra of slightly anharmonic molecules. Therefore, it is also worthwhile to explore ways to eliminate the shortcomings of the TGA without increasing too much the computational cost.

The local harmonic approximation, employed in the TGA, holds as long as the wavepacket is compact because wavepacket spreading leads to incorrect description of the potential at its tails. Consequently, the single Gaussian wavepacket ansatz cannot describe wavepacket splitting or tunneling through barriers. Yet, when the initial wavepacket happens to land on the top of a potential barrier, which happens in many molecules, the splitting is inevitable. Typical examples are provided by the absorption spectrum of formaldehyde (Tatchen and Pollak, 2009; Bonfanti et al., 2018) and emission spectrum of ammonia (Tang et al., 1991). In ammonia, wavepacket splitting is captured at least qualitatively by representing the initial wavepacket as a sum of two Gaussians slightly displaced along the floppy mode. The resulting emission spectrum includes also the progression assigned to the low-frequency umbrella mode, whereas the single-trajectory TGA reproduces only the N–H stretch progression (Begušić et al., 2020).

Another possible target for improvement is a further reduction of the computational cost by exploring different approximations to the Hessians needed in the thawed Gaussian propagation, such as Hessian updating (Ceotto et al., 2013; Zhuang et al., 2013) or interpolation schemes. An extreme possibility is using only one Hessian for the whole propagation. Note that this is different from a global harmonic approximation because the single-Hessian approximation will affect only the wavepacket width and phase, while the center of the wavepacket will follow the exact, anharmonic on-the-fly classical trajectory, as in the original TGA. Such an approximation can be useful when dealing with very large systems, where the quadratic scaling of the Hessian calculations becomes the computational bottleneck. Remarkably, using only one Hessian for the thawed Gaussian propagation decouples the classical propagation of the center of the wavepacket [Eqs. (6.18) and (6.19)] from the semiclassical part [Eqs. (6.20) and (6.21)], resulting in rigorous conservation of the semiclassical energy E_{TGA} (Begušić et al., 2019), which is not true in the original TGA.

As for the time-resolved spectra calculations, the discussion in Sections 6.5.6 and 6.8.4 did not take into account finite pulse duration. If the pulses are much shorter than the vibrational period, this ultrashort pulse approximation is valid; however, when the time resolution is lower, the oscillations in the spectra corresponding to high-frequency vibrational modes disappear. In most experiments, the pulse duration is in the impulsive limit (i.e., satisfies the ultrashort pulse approximation) only for the low-frequency vibrational modes, while the dynamics in the faster modes is not observed. To account for pulse duration in the TGA and therefore permit a more direct comparison with experiment, all one has to do is extend the pump-probe methodology from Section 6.8.4 by evaluating the third-order response function (Begušić and Vaníček, 2020) and then convolving it with the pulse envelopes (Pollard and Mathies, 1992).

In conclusion, we believe that, despite its age and simplicity, it is too early to discard the TGA and that, thanks to their simplicity, on-the-fly *ab initio* calculations in general, and TGA in particular, can be useful not only for the interpretation of existing experimental data, but also for the design of new experiments. Finally, we would like to acknowledge the support from European Research Council (ERC) under the European Union's Horizon 2020 research and innovation programme (grant agreement No. 683069 – MOLEQULE).

REFERENCES

Alborzpour, J.P., Tew, D.P., Habershon, S., 2016. Efficient and accurate evaluation of potential energy matrix elements for quantum dynamics using Gaussian process regression. J. Chem. Phys. 145 (17), 174112.

Andrews, D.L., Thirunamachandran, T., 1977. On three-dimensional rotational averages. J. Chem. Phys. 67 (11), 5026–5033.

Antipov, S.V., Ye, Z., Ananth, N., 2015. Dynamically consistent method for mixed quantum-classical simulations: a semiclassical approach. J. Chem. Phys. 142, 184102.

Avila Ferrer, F.J., Santoro, F., 2012. Comparison of vertical and adiabatic harmonic approaches for the calculation of the vibrational structure of electronic spectra. Phys. Chem. Chem. Phys. 14 (39), 13549–13563.

Baer, M., 2006. Beyond Born-Oppenheimer: Electronic Nonadiabatic Coupling Terms and Conical Intersections, 1st ed. Wiley.

Baiardi, A., Bloino, J., Barone, V., 2013. General time dependent approach to vibronic spectroscopy including Franck–Condon, Herzberg–Teller, and Duschinsky effects. J. Chem. Theory Comput. 9 (9), 4097–4115.

Baiardi, A., Bloino, J., Barone, V., 2017. Simulation of vibronic spectra of flexible systems: hybrid DVR-harmonic approaches. J. Chem. Theory Comput. 13 (6), 2804–2822.

Begušić, T., Cordova, M., Vaníček, J., 2019. Single-Hessian thawed Gaussian approximation. J. Chem. Phys. 150 (15), 154117.

Begušić, T., Patoz, A., Šulc, M., Vaníček, J., 2018a. On-the-fly ab initio three thawed Gaussians approximation: a semiclassical approach to Herzberg–Teller spectra. Chem. Phys. 515, 152–163.

Begušić, T., Roulet, J., Vaníček, J., 2018b. On-the-fly ab initio semiclassical evaluation of time-resolved electronic spectra. J. Chem. Phys. 149, 244115.

Begušić, T., Vaníček, J. (2020). Unpublished.

Begušić, T., Zhang, Z.T., Choi, S., Vaníček, J. (2020). Unpublished.

Berera, R., van Grondelle, R., Kennis, J.T.M., 2009. Ultrafast transient absorption spectroscopy: principles and application to photosynthetic systems. Photosynth. Res. 101 (2–3), 105–118.

Bircher, M.P., Liberatore, E., Browning, N.J., Brickel, S., Hofmann, C., Patoz, A., Unke, O.T., Zimmermann, T., Chergui, M., Hamm, P., Keller, U., Meuwly, M., Wörner, H.J., Vaníček, J., Rothlisberger, U., 2017. Nonadiabatic effects in electronic and nuclear dynamics. Struct. Dyn. 4 (6), 061510.

Bonfanti, M., Petersen, J., Eisenbrandt, P., Burghardt, I., Pollak, E., 2018. Computation of the $S_1 \leftarrow S_0$ vibronic absorption spectrum of formaldehyde by variational Gaussian wavepacket and semiclassical IVR methods. J. Chem. Theory Comput. 14, 5310–5323.

Braun, H., Bayer, T., Wollenhaupt, M., Baumert, T., 2020. 2D strong-field spectroscopy to elucidate impulsive and adiabatic ultrafast electronic control schemes in molecules. In: Marquardt, R., Quack, M. (Eds.), Molecular Spectroscopy and Quantum Dynamics. Elsevier, Amsterdam. Chapter 3 (this book).

Buchholz, M., Grossmann, F., Ceotto, M., 2016. Mixed semiclassical initial value representation time-averaging propagator for spectroscopic calculations. J. Chem. Phys. 144 (9), 094102.

Capobianco, A., Borrelli, R., Noce, C., Peluso, A., 2012. Franck–Condon factors in curvilinear coordinates: the photoelectron spectrum of ammonia. Theor. Chem. Acc. 131 (3), 1–10.

Ceotto, M., Atahan, S., Shim, S., Tantardini, G.F., Aspuru-Guzik, A., 2009a. First-principles semiclassical initial value representation molecular dynamics. Phys. Chem. Chem. Phys. 11, 3861–3867.

Ceotto, M., Atahan, S., Tantardini, G.F., Aspuru-Guzik, A., 2009b. Multiple coherent states for first-principles semiclassical initial value representation molecular dynamics. J. Chem. Phys. 130 (23), 234113.

Ceotto, M., Di Liberto, G., Conte, R., 2017. Semiclassical "divide-and-conquer" method for spectroscopic calculations of high dimensional molecular systems. Phys. Rev. Lett. 119, 010401.

Ceotto, M., Tantardini, G.F., Aspuru-Guzik, A., 2011. Fighting the curse of dimensionality in first-principles semiclassical calculations: non-local reference states for large number of dimensions. J. Chem. Phys. 135 (21), 214108.

Ceotto, M., Zhuang, Y., Hase, W.L., 2013. Accelerated direct semiclassical molecular dynamics using a compact finite difference Hessian scheme. J. Chem. Phys. 138 (5), 054116.

Chen, F.Z., Judge, D.L., Wu, C.Y.R., Caldwell, J., 1999. Low and room temperature photoabsorption cross sections of NH3 in the UV region. Planet. Space Sci. 47 (1–2), 261–266.

Child, M.S., Shalashilin, D.V., 2003. Locally coupled coherent states and Herman–Kluk dynamics. J. Chem. Phys. 118 (5), 2061–2071.

Church, M.S., Antipov, S.V., Ananth, N., 2017. Validating and implementing modified Filinov phase filtration in semi-classical dynamics. J. Chem. Phys. 146, 234104.

Condon, E.U., 1926. A theory of intensity distribution in band systems. Phys. Rev. 28 (6), 1182–1201.

Condon, E.U., 1927. Coupling of electronic and nuclear motions in diatomic molecules. Proc. Natl. Acad. Sci. USA 13, 462–466.

Condon, E.U., 1928. Nuclear motions associated with electron transitions in diatomic molecules. Phys. Rev. 32, 858–872.

Coutsias, E.A., Seok, C., Dill, K.A., 2004. Using quaternions to calculate RMSD. J. Comput. Chem. 25 (15), 1849–1857.

Császár, A., Fábri, C., Szidarovszky, T., 2020. Stationary state quantum dynamics. In: Marquardt, R., Quack, M. (Eds.), Molecular Spectroscopy and Quantum Dynamics. Elsevier, Amsterdam. Chapter 2 (this book).

Curchod, B.F.E., Martínez, T.J., 2018. Ab initio nonadiabatic quantum molecular dynamics. Chem. Rev. 118 (7), 3305–3336.

Domcke, W., Stock, G., 1997. Theory of ultrafast nonadiabatic excited-state processes and their spectroscopic detection in real time. Adv. Chem. Phys. 100, 1–169.

Domcke, W., Yarkony, D.R., 2012. Role of conical intersections in molecular spectroscopy and photoinduced chemical dynamics. Annu. Rev. Phys. Chem. 63 (1), 325–352.

Edvardsson, D., Baltzer, P., Karlsson, L., Wannberg, B., Holland, D.M.P., Shaw, D.A., Rennie, E.E., 1999. A photoabsorption, photodissociation and photoelectron spectroscopy study of NH_3 and ND_3. J. Phys. B, At. Mol. Opt. Phys. 32 (11), 2583–2609.

Egidi, F., Bloino, J., Cappelli, C., Barone, V., 2014. A robust and effective time-independent route to the calculation of resonance Raman spectra of large molecules in condensed phases with the inclusion of Duschinsky, Herzberg–Teller, anharmonic, and environmental effects. J. Chem. Theory Comput. 10 (1), 346–363.

Egorov, S.A., Rabani, E., Berne, B.J., 1998. Vibronic spectra in condensed matter: a comparison of exact quantum mechanical and various semiclassical treatments for harmonic baths. J. Chem. Phys. 108 (4), 1407–1422.

Egorov, S.A., Rabani, E., Berne, B.J., 1999. Nonradiative relaxation processes in condensed phases: quantum versus classical baths. J. Chem. Phys. 110 (11), 5238–5248.

Fally, S., Carleer, M., Vandaele, A.C., 2009. UV Fourier transform absorption cross sections of benzene, toluene, meta-, ortho-, and para-xylene. J. Quant. Spectrosc. Radiat. Transf. 110 (9–10), 766–782.

Faou, E., Gradinaru, V., Lubich, C., 2009. Computing semiclassical quantum dynamics with Hagedorn wavepackets. SIAM J. Sci. Comput. 31 (4), 3027–3041.

Ferrer, F.J., Cerezo, J., Stendardo, E., Improta, R., Santoro, F., 2013. Insights for an accurate comparison of computational data to experimental absorption and emission spectra: beyond the vertical transition approximation. J. Chem. Theory Comput. 9 (4), 2072–2082.

Fortino, M., Bloino, J., Collini, E., Bolzonello, L., Trapani, M., Faglioni, F., Pedone, A., 2019. On the simulation of vibrationally resolved electronic spectra of medium-size molecules: the case of styryl substituted BODIPYs. Phys. Chem. Chem. Phys. 21 (7), 3512–3526.

Fragnito, H.L., Bigot, J.Y., Becker, P.C., Shank, C.V., 1989. Evolution of the vibronic absorption spectrum in a molecule following impulsive excitation with a 6 fs optical pulse. Chem. Phys. Lett. 160 (2), 101–104.

Franck, J., Dymond, E.G., 1926. Elementary processes of photochemical reactions. Trans. Faraday Soc. 21, 536–542.

Gabas, F., Conte, R., Ceotto, M., 2017. On-the-fly ab initio semiclassical calculation of glycine vibrational spectrum. J. Chem. Theory Comput. 13 (6), 2378.

Gelin, M.F., Borrelli, R., Domcke, W., 2017. Efficient orientational averaging of nonlinear optical signals in multichromophore systems. J. Chem. Phys. 147 (4), 044114.

Glauber, R.J., 1963. Theory of optical coherence. Phys. Rev. 130 (6), 2529–2539.

Gorin, T., Prosen, T., Seligman, T.H., Žnidarič, M., 2006. Dynamics of Loschmidt echoes and fidelity decay. Phys. Rep. 435 (2–5), 33–156.

Hagedorn, G.A., 1980. Semiclassical quantum mechanics. I. The $\hbar \to 0$ limit for coherent states. Commun. Math. Phys. 71 (1), 77–93.

Heller, E.J., 1975. Time-dependent approach to semiclassical dynamics. J. Chem. Phys. 62 (4), 1544–1555.

Heller, E.J., 1976. Classical s-matrix limit of wave packet dynamics. J. Chem. Phys. 65 (11), 4979–4989.

Heller, E.J., 1981a. Frozen Gaussians: a very simple semiclassical approximation. J. Chem. Phys. 75 (6), 2923–2931.

Heller, E.J., 1981b. The semiclassical way to molecular spectroscopy. Acc. Chem. Res. 14 (12), 368–375.

Heller, E.J., 1991. Cellular dynamics: a new semiclassical approach to time-dependent quantum mechanics. J. Chem. Phys. 94 (4), 2723–2729.

Heller, E.J., 2018. The Semiclassical Way to Dynamics and Spectroscopy. Princeton University Press, Princeton, NJ.

Herman, M.F., Kluk, E., 1984. A semiclassical justification for the use of non-spreading wavepackets in dynamics calculations. Chem. Phys. 91 (1), 27–34.

Herzberg, G., 1966. Molecular Spectra and Molecular Structure: III. Electronic Spectra of Polyatomic Molecules. D. Van Nostrand Company Inc..

Herzberg, G., Teller, E., 1933. Schwingungsstruktur der Elektronenübergänge bei mehratomigen Molekülen. Z. Phys. Chem. B 21 (5–6), 410.

Ianconescu, R., Tatchen, J., Pollak, E., 2013. On-the-fly semiclassical study of internal conversion rates of formaldehyde. J. Chem. Phys. 139 (15), 154311.

Kabsch, W., 1978. A discussion of the solution for the best rotation to relate two sets of vectors. Acta Crystallogr. A 34 (5), 827–828.

Kaledin, A.L., Miller, W.H., 2003. Time averaging the semiclassical initial value representation for the calculation of vibrational energy levels. J. Chem. Phys. 118 (16), 7174–7182.

Kay, K.G., 2005. Semiclassical initial value treatments of atoms and molecules. Annu. Rev. Phys. Chem. 56 (1), 255–280.

Kearsley, S.K., 1989. On the orthogonal transformation used for structural comparisons. Acta Crystallogr. A 45, 208–210.

Keller-Rudek, H., Moortgat, G.K., Sander, R., Sörensen, R., 2013. The MPI-Mainz UV/VIS spectral atlas of gaseous molecules of atmospheric interest. Earth Syst. Sci. Data 5 (2), 365–373.

Klauder, J.R., Skagerstam, B.-S., 1985. Coherent States. World Scientific Publishing, Singapore.

Kudin, K.N., Dymarsky, A.Y., 2005. Eckart axis conditions and the minimization of the root-mean-square deviation: two closely related problems. J. Chem. Phys. 122 (22), 224105.

Lami, A., Petrongolo, C., Santoro, F., 2004. Absorption, emission, and photoelectron continuous-wave spectra. In: Domcke, W., Yarkony, D.R., Köppel, H. (Eds.), Conical Intersections: Electronic Structure, Dynamics and Spectroscopy. World Scientific Publishing, Singapore, pp. 699–738. Chapter 16.

Lasser, C., Lubich, C., 2020. Computing quantum dynamics in the semiclassical regime. arXiv:2002.00624.

Lasser, C., Sattlegger, D., 2017. Discretising the Herman–Kluk propagator. Numer. Math. 137, 119–157.

Laude, G., Calderini, D., Tew, D.P., Richardson, J.O., 2018. Ab initio instanton rate theory made efficient using Gaussian process regression. Faraday Discuss. 212, 237–258.

Lee, S.-Y., Heller, E.J., 1982. Exact time-dependent wave packet propagation: application to the photodissociation of methyl iodide. J. Chem. Phys. 76 (6), 3035–3044.

Levine, B.G., Martínez, T.J., 2007. Isomerization through conical intersections. Annu. Rev. Phys. Chem. 58, 613–634.

Li, Z., Fang, J.-Y., Martens, C.C., 1996. Simulation of ultrafast dynamics and pump–probe spectroscopy using classical trajectories. J. Chem. Phys. 104 (18), 6919–6929.

Li, J., Lin, C.-K., Li, X.Y., Zhu, C.Y., Lin, S.H., 2010. Symmetry forbidden vibronic spectra and internal conversion in benzene. Phys. Chem. Chem. Phys. 12 (45), 14967–14976.

Li, B., Mollica, C., Vaníček, J., 2009. Efficient evaluation of accuracy of molecular quantum dynamics using dephasing representation. J. Chem. Phys. 131 (4), 041101.

MacKenzie, R., Pineault, M., Renaud-Desjardins, L., 2012. Optimizing adiabaticity in quantum mechanics. Can. J. Phys. 90 (2), 187.

Makri, N., Miller, W.H., 1988. Monte Carlo path integration for the real time propagator. J. Chem. Phys. 89, 2170–2177.

Marquardt, R., Quack, M., 1989. Infrared-multiphoton excitation and wave packet motion of the harmonic and anharmonic oscillators: exact solutions and quasiresonant approximation. J. Chem. Phys. 90 (11), 6320–6327.

Marquardt, R., Quack, M., 2020. Foundations of time dependent quantum dynamics of molecules under isolation and in coherent electromagnetic fields. In: Marquardt, R., Quack, M. (Eds.), Molecular Spectroscopy and Quantum Dynamics. Elsevier, Amsterdam. Chapter 1 (this book).

Miller, W.H., 1970. Classical S matrix: numerical application to inelastic collisions. J. Chem. Phys. 53 (9), 3578–3587.

Miller, W.H., 2001. The semiclassical initial value representation: a potentially practical way for adding quantum effects to classical molecular dynamics simulations. J. Phys. Chem. A 105 (13), 2942.

Mukamel, S., 1982. On the semiclassical calculation of molecular absorption and fluorescence spectra. J. Chem. Phys. 77 (1), 173–181.

Mukamel, S., 1999. Principles of Nonlinear Optical Spectroscopy, 1st ed. Oxford University Press, New York.

Mukamel, S., 2000. Multidimensional femtosecond correlation spectroscopies of electronic and vibrational excitations. Annu. Rev. Phys. Chem. 51, 691–729.

Nakamura, H., 2012. Nonadiabatic Transition: Concepts, Basic Theories and Applications, 2nd ed. World Scientific Publishing Company.

Niu, Y., Peng, Q., Deng, C., Gao, X., Shuai, Z., 2010. Theory of excited state decays and optical spectra: application to polyatomic molecules. J. Phys. Chem. A 114 (30), 7817–7831.

Pastawski, H.M., Levstein, P.R., Usaj, G., Raya, J., Hirschinger, J., 2000. A nuclear magnetic resonance answer to the Boltzmann–Loschmidt controversy? Physica A 283 (1–2), 166–170.

Patoz, A., Begušić, T., Vaníček, J., 2018. On-the-fly ab initio semiclassical evaluation of absorption spectra of polyatomic molecules beyond the Condon approximation. J. Phys. Chem. Lett. 9 (9), 2367–2372.

Petitjean, C., Bevilaqua, D.V., Heller, E.J., Jacquod, P., 2007. Displacement echoes: classical decay and quantum freeze. Phys. Rev. Lett. 98 (16), 164101.

Pollard, W.T., Fragnito, H.L., Bigot, J.Y., Shank, C.V., Mathies, R.A., 1990a. Quantum-mechanical theory for 6 fs dynamic absorption spectroscopy and its application to Nile blue. Chem. Phys. Lett. 168 (3–4), 239–245.

Pollard, W.T., Lee, S.-Y., Mathies, R.A., 1990b. Wave packet theory of dynamic absorption spectra in femtosecond pump–probe experiments. J. Chem. Phys. 92 (7), 4012–4029.

Pollard, W.T., Mathies, R.A., 1992. Analysis of femtosecond dynamic absorption spectra of nonstationary states. Annu. Rev. Phys. Chem. 43 (1), 497–523.

Prlj, A., Begušić, T., Zhang, Z.T., Fish, G.C., Wehrle, M., Zimmermann, T., Choi, S., Roulet, J., Moser, J.-E., Vaníček, J., 2020. Semiclassical approach to photophysics beyond Kasha's rule and vibronic spectroscopy beyond the Condon approximation. The case of azulene. J. Chem. Theory Comput. 16 (4), 2617–2626.

Quack, M., 1978. Theory of unimolecular reactions induced by monochromatic infrared radiation. J. Chem. Phys. 69, 1282–1307.

Quack, M., 1979. Nonlinear intensity dependence of the rate coefficient in unimolecular reactions induced monochromatic by infrared radiation. Chem. Phys. Lett. 65, 140–145.

Quack, M., Merkt, F., 2011. Handbook of High-Resolution Spectroscopy. John Wiley & Sons.

Quack, M., Sutcliffe, E., 1985. On the validity of the quasiresonant approximation for molecular infrared-multiphoton excitation. J. Chem. Phys. 83, 3805–3812.

Richings, G.W., Polyak, I., Spinlove, K.E., Worth, G.A., Burghardt, I., Lasorne, B., 2015. Quantum dynamics simulations using Gaussian wavepackets: the vMCG method. Int. Rev. Phys. Chem. 34 (2), 269–308.

Rost, J.M., 1995. Analytical total photo cross-section for atoms. J. Phys. B 28 (19), L601.

Ryabinkin, I.G., Joubert-Doriol, L., Izmailov, A.F., 2017. Geometric phase effects in nonadiabatic dynamics near conical intersections. Acc. Chem. Res. 50 (7), 1785–1793.

Saita, K., Shalashilin, D.V., 2012. On-the-fly ab initio molecular dynamics with multiconfigurational Ehrenfest method. J. Chem. Phys. 137 (22), 22A506.

Schatz, G.C., Ratner, M.A., 2002. Quantum Mechanics in Chemistry. Dover Publications.

Seidner, L., Stock, G., Sobolewski, A.L., Domcke, W., 1992. Ab initio characterization of the S1–S2 conical intersection in pyrazine and calculation of spectra. J. Chem. Phys. 96 (7), 5298–5309.

Shalashilin, D.V., Child, M.S., 2004. Real time quantum propagation on a Monte Carlo trajectory guided grids of coupled coherent states: 26D simulation of pyrazine absorption spectrum. J. Chem. Phys. 121 (8), 3563–3568.

Shemetulskis, N.E., Loring, R.F., 1992. Semiclassical theory of the photon echo: application to polar fluids. J. Chem. Phys. 97 (2), 1217–1226.

Shi, Q., Geva, E., 2005. A comparison between different semiclassical approximations for optical response functions in nonpolar liquid solutions. J. Chem. Phys. 122 (6), 064506.

Šulc, M., Hernández, H., Martínez, T.J., Vaníček, J., 2013. Relation of exact Gaussian basis methods to the dephasing representation: theory and application to time-resolved electronic spectra. J. Chem. Phys. 139 (3), 034112.

Tang, S.L., Abramson, E.H., Imre, D.G., 1991. Time-dependent study of the fluorescence spectrum of ammonia. J. Phys. Chem. 95 (13), 4969–4976.

Tang, S.L., Imre, D.G., Tannor, D., 1990. Ammonia: dynamical modeling of the absorption spectrum. J. Chem. Phys. 92 (10), 5919–5934.

Tannor, D.J., 2007. Introduction to Quantum Mechanics: A Time-Dependent Perspective. University Science Books, Sausalito.

Tatchen, J., Pollak, E., 2009. Semiclassical on-the-fly computation of the $S_0 \rightarrow S_1$ absorption spectrum of formaldehyde. J. Chem. Phys. 130 (4), 041103.

Thoss, M., Wang, H., Miller, W.H., 2001. Generalized forward-backward initial value representation for the calculation of correlation functions in complex systems. J. Chem. Phys. 114, 9220–9235.

Tutt, L.W., Zink, J.I., Heller, E.J., 1987. Simplifying the MIME: a formula relating normal mode distortions and frequencies to the MIME frequency. Inorg. Chem. 26 (5), 2158–2160.

van der Veen, R.M., Cannizzo, A., van Mourik, F., Vlček, A., Chergui, M., 2011. Vibrational relaxation and intersystem crossing of binuclear metal complexes in solution. J. Am. Chem. Soc. 133 (12), 305–315.

Vaníček, J., 2004. Dephasing representation: employing the shadowing theorem to calculate quantum correlation functions. Phys. Rev. E 70 (5), 055201.

Vaníček, J., 2006. Dephasing representation of quantum fidelity for general pure and mixed states. Phys. Rev. E 73 (4), 046204.

Šulc, M., Vaníček, J., 2012. Accelerating the calculation of time-resolved electronic spectra with the cellular dephasing representation. Mol. Phys. 110 (9–10), 945–955.

Walton, A.R., Manolopoulos, D.E., 1996. A new semiclassical initial value method for Franck–Condon spectra. Mol. Phys. 87 (4), 961–978.

Wang, H., Manolopoulos, D.E., Miller, W.H., 2001. Generalized Filinov transformation of the semiclassical initial value representation. J. Chem. Phys. 115 (14), 6317–6326.

Wang, H., Thoss, M., 2003. Multilayer formulation of the multiconfiguration time-dependent Hartree theory. J. Chem. Phys. 119 (3), 1289–1299.

Wehrle, M., Oberli, S., Vaníček, J., 2015. On-the-fly ab initio semiclassical dynamics of floppy molecules: absorption and photoelectron spectra of ammonia. J. Phys. Chem. A 119 (22), 5685.

Wehrle, M., Šulc, M., Vaníček, J., 2011. Time-resolved electronic spectra with efficient quantum dynamics methods. Chimia 65, 334–338.

Wehrle, M., Šulc, M., Vaníček, J., 2014. On-the-fly ab initio semiclassical dynamics: identifying degrees of freedom essential for emission spectra of oligothiophenes. J. Chem. Phys. 140 (24), 244114.

Wong, S.Y.Y., Benoit, D.M., Lewerenz, M., Brown, A., Roy, P.-N., 2011. Determination of molecular vibrational state energies using the ab initio semiclassical initial value representation: application to formaldehyde. J. Chem. Phys. 134 (9), 094110.

Zambrano, E., Šulc, M., Vaníček, J., 2013. Improving the accuracy and efficiency of time-resolved electronic spectra calculations: cellular dephasing representation with a prefactor. J. Chem. Phys. 139 (5), 054109.

Zhang, S., Pollak, E., 2003. Monte Carlo method for evaluating the quantum real time propagator. Phys. Rev. Lett. 91 (19), 190201.

Zhuang, Y., Siebert, M.R., Hase, W.L., Kay, K.G., Ceotto, M., 2013. Evaluating the accuracy of Hessian approximations for direct dynamics simulations. J. Chem. Theory Comput. 9 (1), 54–64.

Zimmermann, T., Ruppen, J., Li, B., Vaníček, J., 2010. Efficient evaluation of the accuracy of molecular quantum dynamics on an approximate analytical or interpolated ab initio potential energy surface. Int. J. Quant. Chem. 110 (13), 2426–2435.

Zimmermann, T., Vaníček, J., 2010. Communications: evaluation of the nonadiabaticity of quantum molecular dynamics with the dephasing representation of quantum fidelity. J. Chem. Phys. 132 (24), 241101.

Zimmermann, T., Vaníček, J., 2012a. Evaluation of the importance of spin-orbit couplings in the nonadiabatic quantum dynamics with quantum fidelity and with its efficient "on-the-fly" ab initio semiclassical approximation. J. Chem. Phys. 137 (22), 22A516.

Zimmermann, T., Vaníček, J., 2012b. Measuring nonadiabaticity of molecular quantum dynamics with quantum fidelity and with its efficient semiclassical approximation. J. Chem. Phys. 136 (9), 094106.

CHAPTER 7

Atomic and Molecular Tunneling Processes in Chemistry

MARTIN QUACK • GEORG SEYFANG
Laboratorium für Physikalische Chemie, ETH Zürich, CH-8093 Zürich, Switzerland, Martin@Quack.CH

Abstract

The article presents an overview of atomic and molecular tunneling processes, where tunneling of "heavy" particles (i.e., nuclei, not electrons) is important. After an introductory tour d'horizon including a brief history of the tunnel effect, we discuss some topics of current interest. These include tunneling and parity violation in chiral molecules, tunneling in weakly bound complexes, and tunneling processes in slightly asymmetric potentials with tunneling switching and the development of a molecular quantum switch. We then present a selection of recent results for isomerization reactions, bimolecular reactions, tunneling in ions, radicals and electronically excited states, and of the motions of molecules inside a cage. We conclude with some remarks on certain general problems related to tunneling.

7.1 INTRODUCTION

7.1.1 Aim and Overview of the Article

Tunneling is a fundamental quantum-mechanical phenomenon with a wide range of applications in chemistry and physics, on which numerous books have been written already. As far as atomic and molecular processes are concerned, one must distinguish phenomena where the "light" electrons are involved in tunneling (for instance, in redox reactions, tunneling ionization in strong fields, etc.) and processes, where the tunnel effect is important for the "heavy" particles, that is, the atomic nuclei, where we assign the attribute "heavy" also to "light nuclei" such as hydrogen and deuterium where tunneling processes are of particular relevance. The present review deals only with such processes involving "heavy-particle" tunneling, which is also often called "quantum atomic and molecular tunneling", which also forms the theme of a conference series (Eckert et al., 2015, 2017, 2019). The organization of the chapter is as follows: After an introductory "tour d'horizon" of the role of atomic and molecular tunneling in current chemical and molecular dynamics, we provide a brief sketch of the history of the tunnel effect, which is motivated by some interesting aspects of the early history, on the one hand, and by the frequent misrepresentations and misconceptions of that history in the literature, on the other hand. In the main part of the article we discuss in turn various topics of particular current interest, as already outlined in the abstract and table of contents, and we conclude with some remarks and an outlook. The present article is in part based on an earlier brief review (in German) (Seyfang and Quack, 2018).

7.1.2 The Quantum Mechanical Tunneling Process for "Heavy" Particles (Atoms and Molecules): a Tour d'Horizon

Tunneling processes with the motion of electrons are dominating many physical and chemical processes due to the small mass of the electron. The quantum-mechanical tunneling process was discovered, however, by Friedrich Hund (1927a; 1927c; 1927b) for the example of molecular transformations with the motion of "heavier" atomic particles, like the stereomutation of the enantiomers of chiral molecules or for the "textbook example" of the inversion motion of ammonia, NH_3. The early history is found in text- and handbooks of spectroscopy and kinetics (Herzberg, 1939, 1945; Bell, 1980), and historical articles (Quack, 2014), and also for recent developments, for example, in Quack and Merkt (2011). We shall give a brief summary below. Remarkable progress has been made recently by the development of new spectroscopic methods (Quack and Merkt, 2011), and especially by theoretical methods for the description of high-dimensional quantum mechanics of polyatomic molecules (Császár et al., 2012, 2020, Chapter 2, Carrington, 2011; Yamaguchi and Schaefer, 2011; Breidung and Thiel, 2011; Tennyson, 2011) in the framework of the equally high-dimensional underlying potential hypersurfaces (Marquardt and Quack, 2011).

Molecular Spectroscopy and Quantum Dynamics. https://doi.org/10.1016/B978-0-12-817234-6.00012-X

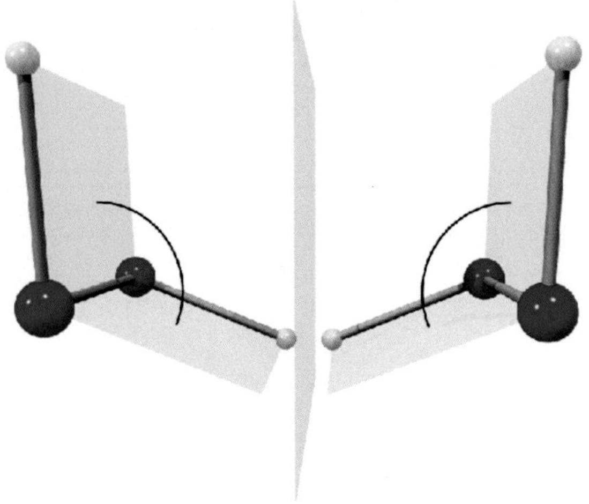

FIG. 7.1 H_2O_2 in its two enantiomeric equilibrium geometries: P-enantiomer on the left and M-enantiomer on the right connected by a fast tunneling reaction (after Quack, 1999, see also Fehrensen et al., 1999a,b, 2007).

The still today widespread classical molecular dynamics (MD) with empirical force fields (van Gunsteren et al., 2006; Riniker et al., 2012; Meier et al., 2013b; Karplus, 2014; Levitt, 2014; Warshel, 2014), as well as Car–Parrinello "ab initio" MD based on the potential hypersurfaces from density functional theory (Car and Parrinello, 1985) neglect the quantum mechanical tunnel effect which might be crucial for many processes with the motion of H-atoms and, as we know today, also for many processes with the motion of heavier atoms (like C, N, S, Cl, etc.), also in many biologically relevant reactions (Allemann and Scrutton, 2009). The development of exact procedures and "good" approximation procedures for the description of the tunneling process are therefore a field of research with vigorous development.

Thereby in the first place, Diffusion Quantum Monte Carlo (DQMC) methods have been applied for the full-dimensional description of tunneling processes in the simplest hydrogen-bond complexes like $(HF)_2$ (Quack and Suhm, 1991a, 1995; Klopper et al., 1998b), later on the more widely usable DVR-methods (discrete variable representation) (Meyer, 1970; Luckhaus and Quack, 1992; Bačić and Light, 1989; Light and Carrington, 2001; Carrington, 2011; Tennyson, 2011), which also allowed for a first full-dimensional time-independent and time-dependent description of stereomutation reaction between the two enantiomers of the hydrogen peroxide molecule HOOH (Fehrensen et al., 1999b, 2007) (see also Fig. 7.1). Today exact methods are implemented in software packages like GENIUSH (Fábri et al., 2011; Császár et al., 2012; Császár and Furtenbacher, 2016). This accurate approach has recently been combined with the accurate theoretical treatment of the interaction with coherent radiation fields (Quack, 1978) to provide a full-dimensional quantum picture of the control of the tunneling motion in ammonia isotopomers (including the chiral NHDT) (Fábri et al., 2019). Exact treatments can also be used for the validation of approximation methods like the quasiadiabatic channel RPH method (Fehrensen et al., 1999b, 2007, 1999a), (reaction path Hamiltonian, Miller et al., 1980), which also can be related to older methods such as the statistical adiabatic channel model (Quack and Troe, 1974, 1998; Merkel and Zülicke, 1987) or similar methods (Hofacker, 1963; Hougen et al., 1970), related to "adiabatic transition state theory". Approximation concepts related to semiclassical dynamics and transition state theory (Miller, 1975b) are also found in the instanton theory of the tunneling process (Benderski et al., 1994; Zimmermann and Vaníček, 2010; Richardson, 2016; Richardson et al., 2015; Kästner, 2014b,a), see also Chapter 9 (Cvitaš and Richardson, 2020). Additional newer developments are found in the RPD-method (ring polymer dynamics) (Richardson and Althorpe, 2009, 2011; Richardson, 2016). The development of the quantum mechanical path integral method for the description of tunneling processes has great potential for an extension of the program systems of the classical MD for the application to biomolecular tunneling processes (Ceriotti et al., 2011; Mátyus et al., 2016; Zimmermann and Vaníček, 2010; Buchowiecki and Vaníček, 2010; Marx, 2006; Ivanov et al., 2015, see also Marquardt and Quack, 2020, Chapter 1), where there are also a large number of experimental investigations (Roston and Kohen, 2013; Roston et al., 2014; Francis et al., 2016; Luk et al., 2013; Ruiz-Pernía et al., 2013; Chan-Huot et al., 2013; Shenderovich et al., 2015).

Another very versatile procedure for the description of the quantum dynamics of polyatomic molecules including tunneling is based on the MCTDH method (multiconfiguration time-dependent Hartee) (Meyer et al., 1990, 2009; Ansari and Meyer, 2016; Manthe, 2015). The "multimode" package has also been applied to tunneling processes, together with an RPH-like approximation (Carter et al., 2013; Wang and Bowman, 2013; Homayoon et al., 2014). An interesting theoretical development is also found in the investigation of the "electronic" and "nuclear flux" in the tunneling process (Grohmann et al., 2013; Liu et al., 2015; Bredtmann et al., 2015; Hermann et al., 2016; Bredtmann et al., 2016; Lu et al., 2016).

We have mentioned here a selection of recent theoretical methods for the description of the tunneling process which can now be applied to interesting physical-chemical processes. We shall turn to these after providing a short survey of the history of the discovery of the tunnel effect.

7.1.3 A Brief History of the Discovery of the Tunnel Effect and Further Developments

When one discusses the history of the tunnel effect, one must distinguish between the *experimental observation of phenomena which involve the tunnel effect and the theoretical understanding of the underlying effect based on quantum mechanics and wave mechanics*. Only the latter should be counted as part of the history of the discovery of the "tunnel effect" properly speaking. Otherwise one would have to tell the history of all kinds of wave phenomena, including even macroscopic surface wave phenomena (Stocker and Johnson, 1991) (which can be related to the tunnel effect) in the long and old history of science. Also, radioactive α-decay was discovered experimentally early on by Rutherford (1900), but without any understanding of the underlying quantum mechanical phenomena. Indeed, it is the discovery and theoretical understanding of this most striking phenomenon of transgressing a potential barrier without ever having sufficient energy to overcome that barrier – completely impossible, even unthinkable of in classical mechanics – which should be qualified as the discovery of the tunnel effect. This discovery was made by Friedrich Hund (1927a; 1927c; 1927b) in a theoretical investigation of certain spectroscopic and kinetic consequences of the quantum-mechanical treatment of the vibrational (vibrational-tunneling) dynamics of molecules, in particular chiral molecules and pyramidal molecules such as ammonia. In this early work Hund provided a careful explanation of the quantum mechanical phenomena and the basic equations for the relation of the "tunneling splitting" ΔE between energy eigenstates of different parity in such molecules and their time evolution, which we quote here as the tunneling period τ,

$$\tau = h/\Delta E, \tag{7.1}$$

and the transfer time for the motion of the molecular system from one well in a double minimum potential to the other well separated by a potential barrier much exceeding the total available energy of the molecule (see also Figs. 7.2 and 7.5 further below)

$$t_{\text{transfer}} = \tau/2 = h/(2\Delta E). \tag{7.2}$$

Hund then proceeds to discuss also the dependence of the tunneling time upon shape, with height and width, of the potential barrier. With simple quantum-mechanical estimates of the Born–Oppenheimer like potential barriers for substituted chiral methane derivatives, he calculated lifetimes longer than millions of years for such chiral molecules, therefore resolving what was later often called by others "Hund's paradox": The fact that for chiral molecules one does not observe the energy eigenstates of well defined parity (assuming parity conservation to be exact as was common then) but localized time-dependent states. The predicted long lifetimes would provide a resolution of the "paradox" (see also Janoschek, 1991). We shall see in Section 7.2 that the physics of the situation is changed today fundamentally. We might also note in passing that later (Pfeiffer, 1983) and Primas (1981) have discussed another paradox: Why can one synthesize and buy the chiral enantiomers but not the achiral superpositions which might be assumed to be more "natural" by being energy eigenstates? As pointed out in Quack (1989), this latter paradox has two solutions: first, by dominant parity violation making, in fact, the chiral states to be the energy eigenstates, and second, because the phase relations necessary for the superpositions or "parity isomers" are quickly destroyed by collisions.

Returning to the history of tunneling, the next effects to be treated along these lines were electron emission (Fowler and Nordheim, 1928) and α-decay (Gamov, 1928b,a; Gurney and Condon, 1928, 1929). Indeed, Gamov's work on α-decay is frequently referred to as the discovery of the tunnel effect in the literature, frequently also without any mention of Hund. A recent example among many is the article by Reiss (2014), although historical articles and books do mention the earlier work of Hund (Bell, 1980; Devault, 1984; Merzbacher, 2002). The work of Gamov and of Gurney and Condon was actually carried out independently and essentially simultaneously (the first paper by Gurney and Condon has a submission date 30 July 1928, published September 1928, and the paper by Gamov has a submission 2 August 1928, published shortly thereafter in 1928). Both papers date from about a year after the work of Hund, thus quite long afterwards. Indeed, Gurney and Condon (1928) make explicit reference to Hund's publication, noting also the similarly striking, very strong dependence of the tunneling rate upon barrier height and width. While this work carried out in another continent across the Atlantic thereby gives explicit reference and credit to the earlier work of Hund, the paper of Gamov submitted from Göttingen in the Zeitschrift für Physik, thus from the same place and in the same journal (also in German) just like Hund's (1927a; 1927c; 1927a) work makes no mention of Hund's publication. Gamov mentions earlier work by Oppenheimer (Oppenheimer, 1928) and Nordheim (Fowler and Nordheim, 1928) on electron emission. It seems unlikely that, while in Göttingen and being a reader of Zeitschrift für Physik, Gamov did not know about Hund's work. Thus the omission is striking. Similarly later Eckart (1930) and Bell (1933) make no mention of Hund's work in their treatment of the tunnel effect while Wigner (1932) does mention it in the very first instance as being important for chemical reaction. In these early papers the word "tunneling" is not used for the effect. Wigner (1932) calls it "non-mechanical transgression" (unmechanisches Überschreiten) and Born and Weisskopf (1931) in a treatment of adsorption use the words "breaking the energy barrier" ("Durchbrechung der Energieschwelle"). On the other hand, Cremer and Polanyi (1932) already use the expression "tunneling-theory" ("Tunneltheorie") and "tunnel-effect" ("Tunneleffekt") in their treatment of quantum effects in heterogeneous catalysis, see also Schottky (1931, writing about a "wave-mechanical tunnel effect" (einen wellenmechanischen Tunneleffekt)). Later much use is made of these expressions and also of the effect in chemical reactions, particularly in relation to non-classical "tunneling corrections" to transition state theory (Eyring, 1938a,b; Wigner, 1938; Eyring et al., 1944; Bell, 1980). We shall add just a sketch of further developments. The early work is usually restricted to one-dimensional treatments of tunneling both in terms of model problems for various processes (Heilbronner et al., 1956; Brickmann and Zimmermann, 1968, 1969; Löwdin, 1963; Zimmermann, 1964) and in applications to spectroscopic problems (Herzberg, 1939, 1945), where even the one-dimensional treatment was considered to be quite demanding numerically until about 1970, giving motivation to provide and use solutions in extended tables (see, e.g., Coon et al., 1966; Quack and Stockburger, 1972). Similarly in transition state theory, numerous one-dimensional tunneling corrections were discussed (see, e.g., the review by Truhlar et al., 1996) or mentioned in the state selective version of transition state theory in the framework of the statistical adiabatic channel model (SACM) (Quack and Troe, 1974, 1998). An important development towards multidimensional tunneling theory arose from Miller's semiclassical approach towards quantum dynamics (Miller, 1974, 1975a, 2014), which can be considered to be related to WKB theory (Wentzel, 1926; Kramers, 1926; Brillouin, 1926) which itself can be related to the "old" quantum theory of Bohr and Sommerfeld (see Merkt and Quack, 2011). Multidimensional tunneling is thus incorporated into transition state theory (Stratt et al., 1979; Hernandez and Miller, 1993; Clary, 2018; Shan et al., 2019) and in the so-called instanton theory of tunneling (Miller, 2014; Benderski et al., 1994; Goldanskii, 1976; Richardson, 2016, 2018; Richardson and Althorpe, 2011; Meisner and Kästner, 2018; Rommel et al., 2011). As mentioned, the approximate reaction-path Hamiltonian (RPH) is an alternative to multidimensional tunneling in transition state theory (Miller et al., 1980). It is essentially a version of adiabatic transition state theory, see also Hofacker (1963); Marcus (1964, 1965); Quack and Troe (1974, 1998). An extension is based on an idea of the statistical adiabatic channel model (SACM) to "diabatize" vibrationally adiabatic channels in such a way that certain channels may cross in order to allow them to maintain their physical nature, which leads to the quasiadiabatic channel RPH model (Fehrensen et al., 1999a,b, 2007; Prentner et al., 2015), which has been successfully applied to spectroscopic problems and tested against full-dimensional, "numerically exact" computations of stereomutation (Fehrensen et al., 2007; Quack et al., 2008). By now, exact quantum dynamics for tunneling has been carried out, for example, using diffusion quantum Monte Carlo (DQMC) techniques for the dimer $(HF)_2$, indeed, already some time ago (Quack and Suhm, 1991a) much extended later on, using other techniques (Vissers et al., 2003; Wu et al., 1995; Felker and Bačić, 2019). Full-dimensional vibration–rotation–tunneling, including time dependence under coherent laser excitation, was studied most recently for the prototype system of ammonia isotopomers (Fábri et al., 2019) which brings us to modern times to be discussed from here onwards.

7.2 TUNNELING AND PARITY VIOLATION IN CHIRAL MOLECULES

7.2.1 Exact and Approximate Studies of Tunneling in Prototypical Molecules: Hydrogen Peroxide and Ammonia Isotopomers

Chiral molecules have, as already mentioned, played a central role in the history of the discovery of the tunneling process (Hund, 1927a,c,b; Bell, 1980; Quack, 2014). Fig. 7.1 shows the H_2O_2 molecule with the two enantiomers in their equilibrium geometry. This was to our knowledge the first example where a reaction of stereomutation was described on a full-dimensional potential surface with all 6 internal degrees of freedom (Kuhn et al., 1998) and a quantum-mechanical treatment with exact DVR methods for the spectroscopic

TABLE 7.1
Tunneling splittings from the numerically exact solution ($\Delta\tilde{\nu}_i^{6D}$), from the quasiadiabatic channel RPH approximation ($\Delta\tilde{\nu}_i^{RPH}$) and from experiment ($\Delta\tilde{\nu}_i^{\exp}$). T^{6D} is the tunneling transfer time[a].

i	$\omega_i/\mathrm{cm}^{-1}$	$\tilde{\nu}_i^{\exp}/\mathrm{cm}^{-1}$	$\tilde{\nu}_i^{6D}/\mathrm{cm}^{-1}$	$\Delta\tilde{\nu}_i^{\exp}/\mathrm{cm}^{-1}$	$\Delta\tilde{\nu}_i^{6D}/\mathrm{cm}^{-1}$	$\Delta\tilde{\nu}_i^{RPH}/\mathrm{cm}^{-1}$	T^{6D}/ps
0	0.0	0.0	0.0	11.4	11.0	11.1	1.5
1	3778	3609.8	3617.7	8.2	7.6	8.4	2.2
2	1453	1395.9	1392.0	(2.4?)	6.1	5.0	2.7
3	889	865.9	850.5	12.0	11.1	10.8	1.5
4	392	254.6	259.3	116	118	120	0.14
5	3762	3610.7	3605.8	8.2	7.4	7.4	2.0
6	1297	1264.6	1236.5	20.5	20.8	21.8	0.8

[a] After Fehrensen et al., 1999b, 2007.

stationary states and the time-dependent wave packet dynamics of the tunneling process (Fehrensen et al., 1999b, 2007). Hydrogen peroxide is well suited as a prototype molecule for such investigations due to the large tunneling splitting of ca. 10 cm^{-1} in the ground state which leads to the result that effectively a quantum dynamics in the spirit of Hund is valid independent of the also existing, but for the dynamics negligible, parity violation (see below) (Bakasov et al., 1996, 1998; Berger and Quack, 2000a,b). In Table 7.1 the tunneling splittings from the numerically exact solution are compared to the approximate results from the quasiadiabatic channel RPH approach and to experimental results. The strongly mode selective tunneling times as a function of the excitation of the different degrees of freedom can be recognized. Thus by exciting various vibrational modes, say with a pulsed laser, one can control the stereomutation rates. For example, the excitation of the OH-stretching vibration results in a slowing down of the tunneling process even though the excitation energy is a multiple of the barrier height: it remains an effectively quasiadiabatic tunneling process with a slightly modified effective quasiadiabatic channel potential and moment of inertia ("quasi"-tunneling mass) slightly increased by the excitation of the OH-stretching vibration which explains qualitatively the slowing down of the process. Similar effects have been found for the inversion motion in the aniline isotopomers (with the chiral isotopomer C_6H_5NHD) (Hippler et al., 2011; Albert et al., 2016d), which shows such a mode selective "nonstatistical" tunneling process with a slowing down after NH-stretching excitation despite the very high density of states at high excitation.

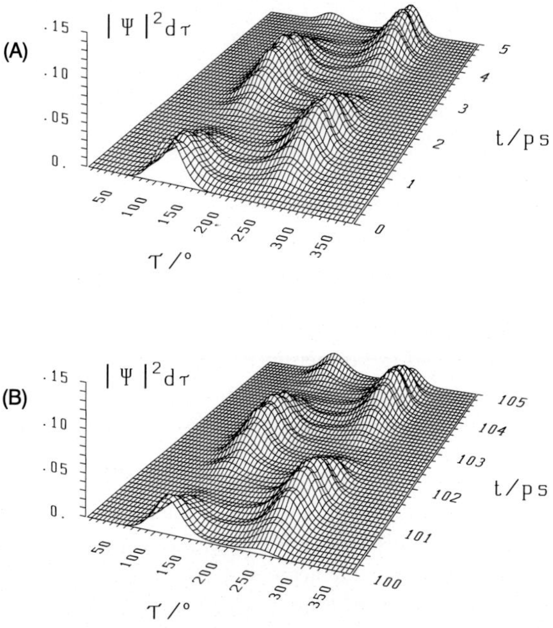

FIG. 7.2 Six-dimensional wave packet evolution for H_2O_2; $|\Psi|^2$ shows the time-dependent probability as a function of the torsional coordinate, where the probability density is integrated over all other coordinates: (A) shows the time interval 0–5 ps and (B) the time interval 100–105 ps with identical initial conditions at $t = 0$ as in (A) (Fehrensen et al., 1999b, 2007). The migration of the wave packet from the left to the right corresponds to a change from one enantiomer to the other in Fig. 7.1.

Fig. 7.2 shows as an example the "wave packet dynamics" in the ground state of the H_2O_2 molecule in terms of the 6-dimensional quantum-mechanical prob-

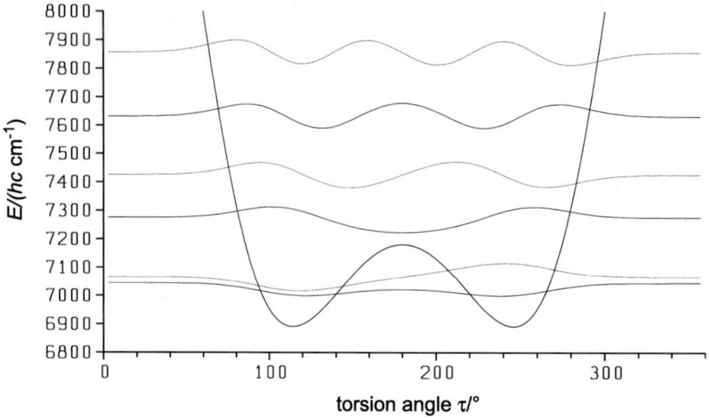

FIG. 7.3 The lowest torsional wavefunctions supported by the adiabatic channel corresponding to the antisymmetric bending fundamental ν_6 (calculated with the RPH model). (Full lines) Wavefunctions of positive parity (+); (Dotted lines) Wavefunctions of negative parity (-). The ordinate axis labels refer to the energies $E/(hc\ \text{cm}^{-1})$, the wavefunctions are shown without scale for illustration, centered at the positions of the corresponding energy levels (after Fehrensen et al., 2007).

ability density as a function of the torsional angle (integrated over the other 5 degrees of freedom). One can see the initially, at time $t = 0$, "left" localized density which shows an approximately Gaussian shape around the equilibrium geometry of the "left" enantiomer and which after 1.5 ps transforms to the "right" localized wavepacket close to the equilibrium geometry of the second ("right") enantiomer.

The detailed full-dimensional analysis of the wavepacket dynamics can be used to test the validity of the quasiadiabatic channel-RPH approximation, which turns out to be a remarkably good approximation for this example. Even at high energies one finds indeed quasiadiabatic tunneling above the barrier. The different tunneling velocities for various excitations of different channels in all possible degrees of freedom can be used for a "mode selective" tunneling control of the reaction velocity, for instance, with the help of laser excitation (Prentner et al., 2015; Fábri et al., 2019).

Fig. 7.3 shows a quasiadiabatic channel potential with one quantum excitation of the antisymmetric bending fundamental ν_6 in HOOH, together with the supporting torsional wavefunctions. As seen from Table 7.1, this excitation leads to an enhancement of the tunneling rate by a factor of 2, and the quasiadiabatic channel RPH result is within 5% of the exact result. The time-dependent wavepackets from the exact and approximate results are virtually indistinguishable by eye although the small numerical differences can, of course, be established easily (Fehrensen et al., 2007).

While rotation was treated approximately by Fehrensen et al. (1999b, 2007), in the recent work on tunneling in ammonia isotopomers all degrees of freedom (vibration and rotation) were treated numerically exactly in a 9-dimensional vibration–rotation–tunneling calculation (Fábri et al., 2019). In addition to tunneling wavepackets in the isolated ammonia isotopomers, the control of tunneling rates by well-designed electromagnetic (laser) radiation fields was studied in this work. Fig. 7.4 shows the *tunneling enhancement* achieved for the chiral isotopomer NHDT, where with a suitable laser excitation a transfer time of 40 ps can be obtained compared to a transfer time of 150 ps in the isolated NHDT molecule in its ground state.

7.2.2 Tunneling in Chiral Molecules Where Parity Violation Dominates Over Tunneling

At the time of Hund's discussion of chiral molecules (1927), parity conservation was assumed to be rigorous. The discovery of parity violation in particle physics (Lee and Yang, 1956; Wu et al., 1957; Garwin et al., 1957; Friedman and Telegdi, 1957; Schopper, 1957) leads to a fundamental change also in the tunneling dynamics of chiral molecules, as a consequence in molecular physics arising from high energy physics. Such tunneling processes and their theoretical analysis have gained recently a new fundamental interest in connection with the phenomenon of parity violation in chiral molecules (Quack, 2011b; Quack et al., 2008). Indeed, parity vi-

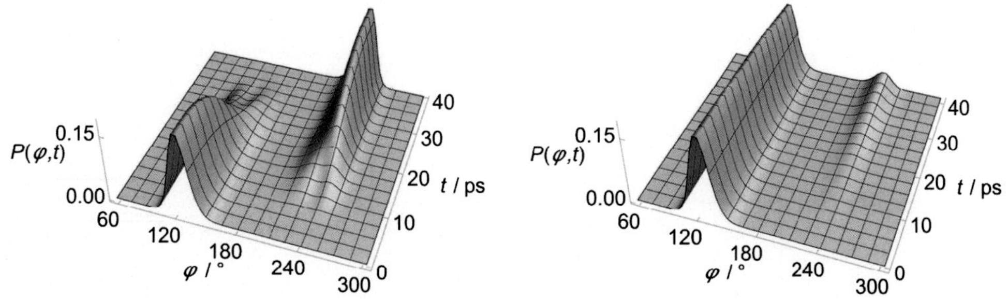

FIG. 7.4 Reduced probability density as a function of the inversion coordinate φ and time t for NHDT. The two enantiomeric structures correspond to $\varphi \simeq 120°$ and $\varphi \simeq 240°$, respectively. The left and right panels show reduced probability densities (probability densities integrated over all other coordinates) for the tunneling enhancement (left) and field-free dynamical schemes (right). The parameters of the laser pulse are $\tilde{\nu}_0 = 793.5.0$ cm^{-1}, $I_{\max} = 3.15$ GW cm^{-2}, and $t_p = 40$ ps (after Fábri et al., 2019).

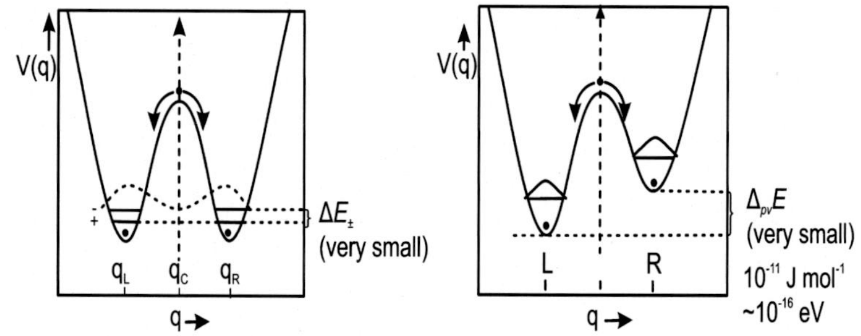

FIG. 7.5 Illustration of the symmetrical potential (left, as discussed by Hund 1927a; 1927c; 1927b) for chiral molecules and the asymmetry induced by the electroweak interaction in an originally symmetric double minimum potential (right, see Quack et al., 2008; Quack, 1986, 2015b).

olation creates an effectively asymmetric potential for tunneling as shown in Fig. 7.5, as calculated by "electroweak quantum chemistry" derived from the standard model of particle physics (SMPP) which allows one to obtain the very small extra potential showing the asymmetry (it should not be viewed as Born–Oppenheimer potential; Bakasov et al., 1996, 1998). While older theories of this effect were incorrect by one to two orders of magnitude, the new theoretical values (Bakasov et al., 1996, 1998; Berger and Quack, 2000a,b; Quack and Stohner, 2000) are now confirmed by several research groups (see reviews in Quack et al., 2008; Quack, 2011b, 2015b,a). The asymmetric potential distinguishing between the enantiomers is predicted to be in the sub-feV range (corresponding to about 100 pJ/mol); it dominates the dynamics in molecules with very small tunneling splitting (Prentner et al., 2015). Recent examples of this kind are the HOOH-analogue molecule ClOOCl (Prentner et al., 2015; Horný and Quack, 2015; Horný et al., 2016) and the 1,2-dithiine molecule (Albert et al., 2016a,b). A further example with a more complex tunneling dynamics is the HSSSH molecule, trisulfane (Fábri et al., 2015). In this molecule typical tunneling splittings $\Delta E_{\pm}$ on the order of less than $(\text{hc}) \cdot 10^{-20}$ cm^{-1} (corresponding to about 1 yeV, yoctoelectronvolt $= 10^{-24}$ eV) are calculated for the ground state, while the parity violating energy difference $\Delta_{\text{PV}}\text{E}$, corresponding to the effective asymmetry of the potential, is calculated in the range from $(\text{hc}) \cdot 10^{-11}$ to 10^{-12} cm^{-1} (corresponding to ca. 1 to 0.1 feV, see also Prentner et al., 2015; Horný and Quack, 2015; Horný et al., 2016; Albert et al., 2017, 2016a,b; Fábri et al., 2015). For these cases, as always when one has

$$\Delta_{\text{PV}} E \gg \Delta E_{\pm}, \tag{7.3}$$

parity violation is dominating the quantum dynamics, as demonstrated also quantitatively for ClSSCl (Berger et al., 2001). Indeed, in the work on ClSSCl a new extrapolation method was developed based on WKB theory and exact calculations in order to estimate tunneling

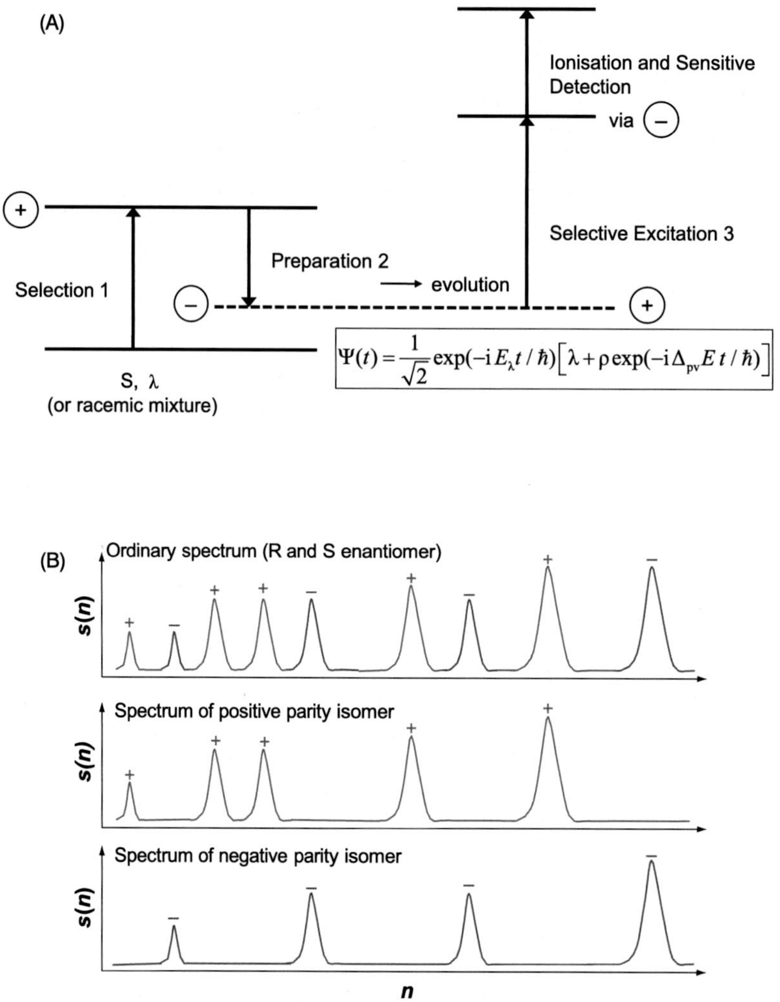

FIG. 7.6 (A) Sequence of the steps in the experiment to measure the parity violating energy difference $\Delta_{PV}E$ (Quack, 1986, 2015b). (B) Time evolution of the measured spectra under the influence of the parity violating energy difference $\Delta_{PV}E$ after preparation of a superposition state with initially negative parity (in the example, the reverse would be equivalent) (Quack et al., 2008; Quack, 2015b). ($n = (\nu - \nu_0)/\nu_0$ is a reduced frequency scale and $s(n)$ is a reduced absorbance signal.)

splittings $\Delta E_\pm \ll (\mathrm{hc})10^{-60}\ \mathrm{cm}^{-1}$ for the hypothetical symmetrical potential while $\Delta E_{\mathrm{PV}} \simeq (\mathrm{hc})10^{-12}\ \mathrm{cm}^{-1}$.

The extremely small values of $\Delta_{\mathrm{PV}}E$ can be measured according to a scheme shown in Fig. 7.6. After the selection of a parity state, a "parity isomer" of a chiral molecule is prepared which is at the same time an R and S (or P and M) enantiomer, therefore a superposition of both enantiomers, which shows a well-defined parity (a negative parity in our example of Fig. 7.6). Due to parity violation, this state develops into a state of opposite parity which shows a different spectrum (Fig. 7.6B), where the population p_+ (t) of the new state (initially forbidden) follows the equation

$$p_+(t) = \sin^2(\pi t \Delta_{\mathrm{PV}}E/h) = 1 - p_-(t), \tag{7.4}$$

and approximately at the beginning (for small values of t and the argument of the $\sin^2$ function)

$$p_+(t) \approx \pi^2 t^2 \Delta_{\mathrm{PV}}E^2/h^2. \tag{7.5}$$

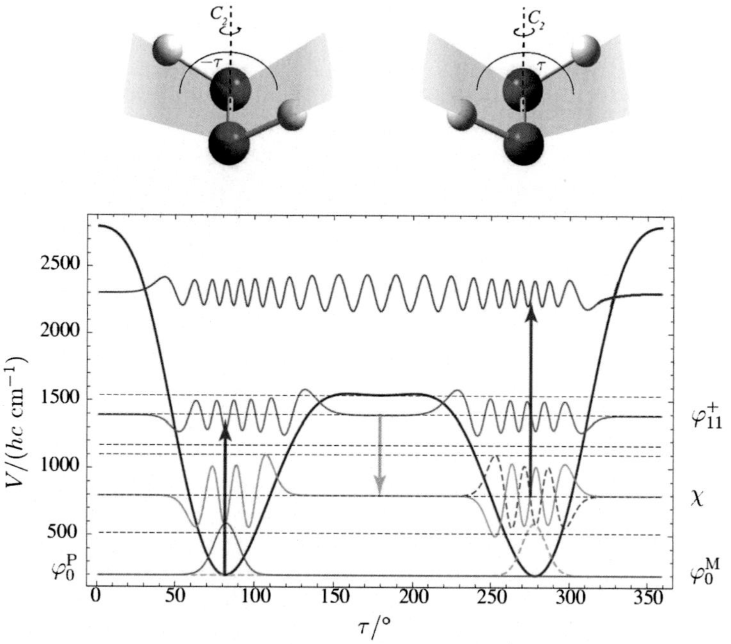

FIG. 7.7 Model calculations for the laser preparation (red) and of the laser detection (blue) for the measurement of the parity violating energy difference in an XYYX-molecule (Cl_2O_2). The wave functions of the levels populated by the laser excitation are shown (after Prentner et al., 2015).

This state of positive parity can then be detected very sensitively by multiphoton ionization as a function of time.

For the example of the achiral ammonia molecule NH_3, it could be shown recently in test experiments that a sensitivity can be reached which allows, in principle, for the measurement of values of $\Delta_{PV}E \simeq 100$ aeV or larger (Dietiker et al., 2015). A simulation for the realistic example of the chiral ClOOCl molecule shows that in principle for this molecule a measurement should be possible (Prentner et al., 2015) (see also Figs. 7.7 and 7.8). This simulation is a first example where tunneling processes including the weak parity violating interaction and coherent laser excitation could be addressed. In addition to the quasiadiabatic RPH approximation for the multidimensional tunneling process, also the rotational states have been considered explicitly with an approximate effective Hamilton operator. The time-dependent interaction with the coherent laser field has been calculated with and without the quasiresonant approximation (Quack, 1978; Quack and Sutcliffe, 1985, 1986; Marquardt and Quack, 1989, see also Chapter 1, Marquardt and Quack, 2020). However, there are no appropriate lasers available in the necessary frequency range for the ClOOCl molecule.

The laser sources for such a molecular beam experiment in the mid-infrared (Dietiker et al., 2015) are in principle available for molecules like 1,2-dithiine and trisulfane which should have appropriate molecular states above 2500 cm^{-1} for the parity selection. The relatively complex spectra of these molecules are currently subject of intense investigations (Albert et al., 2017, 2016b,a; Fábri et al., 2015). Fig. 7.9 shows the potential for the tunneling process and the parity violating potential in dithiine for which a very favorable, relatively high value for $\Delta_{PV}E$ is predicted (approximately 1 feV or $\Delta_{PV}E/hc \simeq 10^{-11}\ \mathrm{cm}^{-1}$). For the example of the trisulfane, due to the two hindered S–H-rotors, in principle, a two-dimensional tunneling problem has to be solved for which the 2-dimensional potential surface is shown in Fig. 7.10. However, trisulfane has the advantage that for such a 5-atomic molecule today, in principle an exact multidimensional calculation is feasible (9-dimensional) which will for a long time not be possible for larger molecules like 1,2-dithiine. These problems of the tunneling dynamics including parity violation are the subject of current studies at the limit of experimental and theoretical possibilities and of our knowledge of effects reaching from high energy physics to molecular quantum dynamics. Trisul-

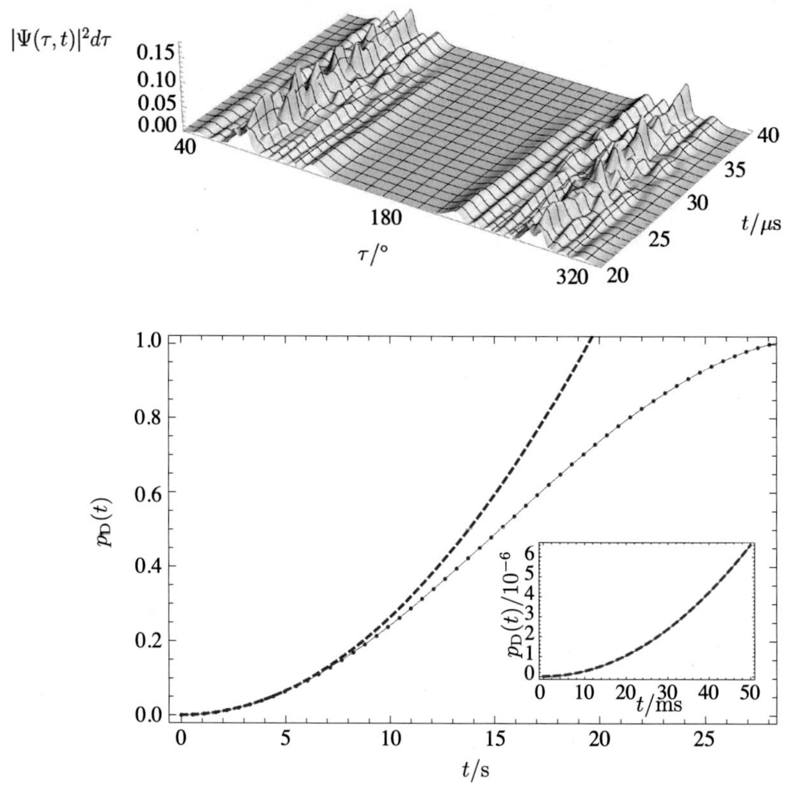

FIG. 7.8 (A) Time dependence of the wave packet for ClOOCl from Fig. 7.7 after the end of the preparation step. For $t = 40$ μs, a wave packet with well defined parity is obtained. (B) Time evolution of the level population $p_{+}(t)$ corresponding to $\Delta_{PV}E \gg \Delta E_{\pm}$ (blue) and $p_{+}(t) = \sin^2(\pi t \Delta_{PV}E/h) = 1 - p_{-}(t)$ or approximated by $p_{+}(t) = \pi^2 t^2 \Delta_{PV}E^2/h^2$ (red). The insert shows the evolution for the first 50 ms (after Prentner et al., 2015).

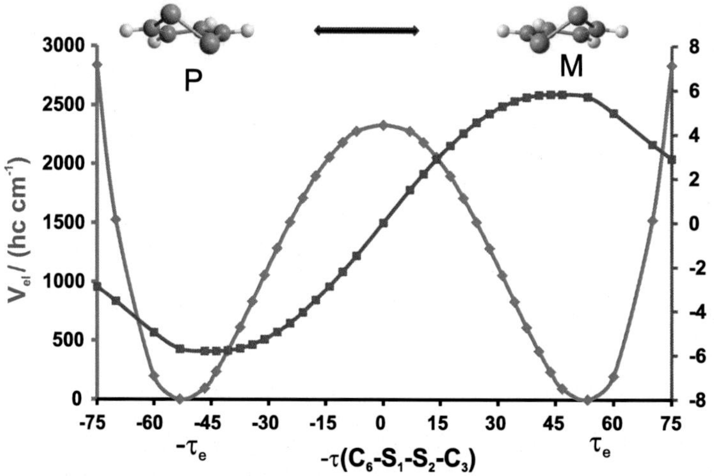

FIG. 7.9 Calculated symmetric Born–Oppenheimer potential (blue, scale to the left) and the parity violating antisymmetric potential (red, scale to the right) for 1,2-dithiine. The structure of the two enantiomers is displayed in the upper part of the figure (after Albert et al., 2016b).

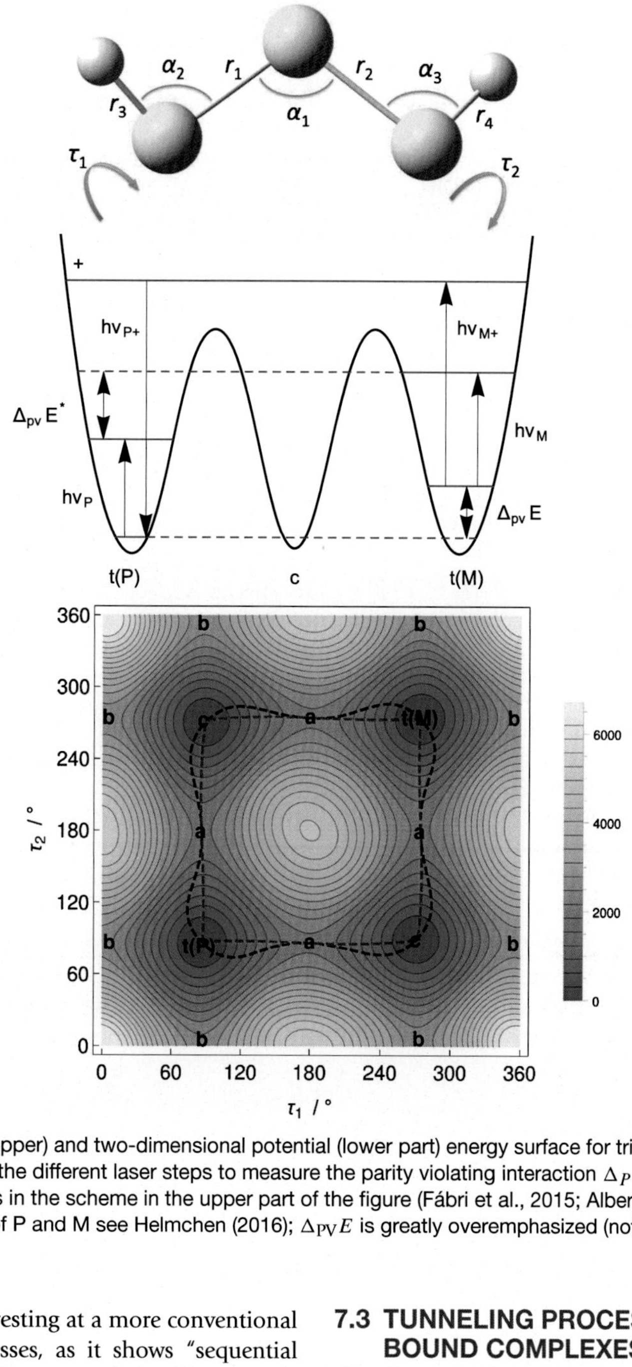

FIG. 7.10 One- (upper) and two-dimensional potential (lower part) energy surface for trisulfane, HSSSH. Also displayed are the different laser steps to measure the parity violating interaction Δ_{PV}E in trisulfane as given by the arrows in the scheme in the upper part of the figure (Fábri et al., 2015; Albert et al., 2017). For the nomenclature of P and M see Helmchen (2016); $\Delta_{PV}E$ is greatly overemphasized (not to scale) for this illustration.

fane, HSSSH, is also interesting at a more conventional level of tunneling processes, as it shows "sequential tunneling" trans(P) → cis → trans(M) with an intermediate cis structure of slightly higher energy than the two trans structures (P and M) for which the energy difference Δ_{PV}E is overemphasized (not to scale); see Fig. 7.10.

7.3 TUNNELING PROCESSES IN WEAKLY BOUND COMPLEXES

The dimer $(HF)_2$, Fig. 7.11, is prototypical for hydrogen bond complexes and is the first example where the tunneling process for the reorganization of the hydrogen bound has been investigated by high-resolution microwave (MW) spectroscopy (Dyke et al., 1972) and

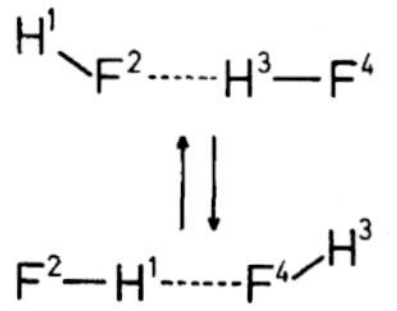

FIG. 7.11 Scheme of $(HF)_2$ conformations and inter conversions with definition of the numbering of the nuclei for the molecular symmetry group (after von Puttkamer and Quack, 1987).

later understood in detail by full-dimensional calculations (Quack and Suhm, 1991a; Klopper et al., 1998b). Here too, the mode selectivity showed pronounced non-classical and non-statistical behavior (Manca et al., 2008; Quack, 1981; von Puttkamer and Quack, 1989). Because $(HF)_2$ has been such an important prototype in our understanding of hydrogen bond tunneling dynamics, we add a little more detail here. This dimer is highly non-rigid with large amplitude motions, both by vibration and rotation, which greatly influence the tunneling rearrangement process shown in Fig. 7.11. Thus it is the permutation inversion group $S^*_{2,2} = S_2 \times S'_2 \times S^*$ and its subgroups such as M_{S4} which are relevant for analyzing the spectroscopy and quantum dynamics for the vibration–rotation–hydrogen-bond–tunneling motion. We use the notation of Quack (1977, 2011b); Merkt and Quack (2011) where S_2 is the symmetric group of the permutations of the two equivalent H atoms (or protons) 1 and 3, S'_2 corresponds to the F nuclei 2 and 4, and S^* is the inversion group. Table 7.2 provides the relevant character tables (from von Puttkamer and Quack, 1987). We note that the character table for $S^*_{2,2}$ is not given explicitly but is easily obtained, because of the direct product property $S^*_{2,2} = S_{2,2} \times S^*$, by taking the table for $S_{2,2}$ and adding E^* and all permutation inversions with a star such as $(13)^*$, etc. The symmetry species A_1, A_2, B_1, B_2 will have an exponent "+" for all characters of the starred operations being the same as the unstarred and will have an exponent "–" for all species where the characters of the starred operations have the opposite sign compared to the unstarred operation (thus + and – indicate positive and negative parity for the same species in the permutation group).

Because the barrier to exchanging protons attached to different F atoms is very high (Klopper et al., 1998a) the tunneling splitting corresponding to such exchanges is not observed and the subgroup of feasible permutation-inversion operations, according to Longuet-Higgins (1963), is the molecular symmetry group M_{S4} of order 4 given in Table 7.2 as well. This has four symmetry species A^+, A^-, B^+, B^- in the systematic notation of Quack (1977) (with an alternative notation being added in parenthesis), which assigns unique permutation species (symmetric A, antisymmetric B) and parity (positive + and negative –). This group describes the observable tunneling splittings for the rearrangement process of Fig. 7.11. It should be noted here that the concept of "feasible" symmetry operations of Longuet-Higgins (1963), which has occasionally been criticized, invokes the concept of "high barriers" in the Born–Oppenheimer approximation preventing "feasibility".

TABLE 7.2
Character tables for the symmetry groups of the dimer (HF-HF).

(a) Character table of the molecular symmetry group M_{S4} of $(HF)_2$ (after von Puttkamer and Quack, 1987)

Species	E	E^*	(ab) $(13)(24)$	$(ab)^*$ $(13)(24)^*$	$\Gamma(M_{S4}) \uparrow S^*_{2,2}$
$A^+(A_g)$	1	1	1	1	$A_1^+ + A_2^+$
$A^-(A_u)$	1	−1	1	−1	$A_1^- + A_2^-$
$B^+(B_u)$	1	1	−1	−1	$B_1^+ + B_2^+$
$B^-(B_g)$	1	−1	−1	1	$B_1^- + B_2^-$

(b) Character table for $S_{2,2}$

Species	E	(13)	(24)	(13)(34)	$(S_{2,2}) \downarrow M_{S2}$
$A \times A \equiv A_1$	1	1	1	1	A
$B \times B \equiv A_2$	1	−1	−1	1	A
$B \times A \equiv B_1$	1	−1	1	−1	B
$A \times B \equiv B_2$	1	1	−1	−1	B

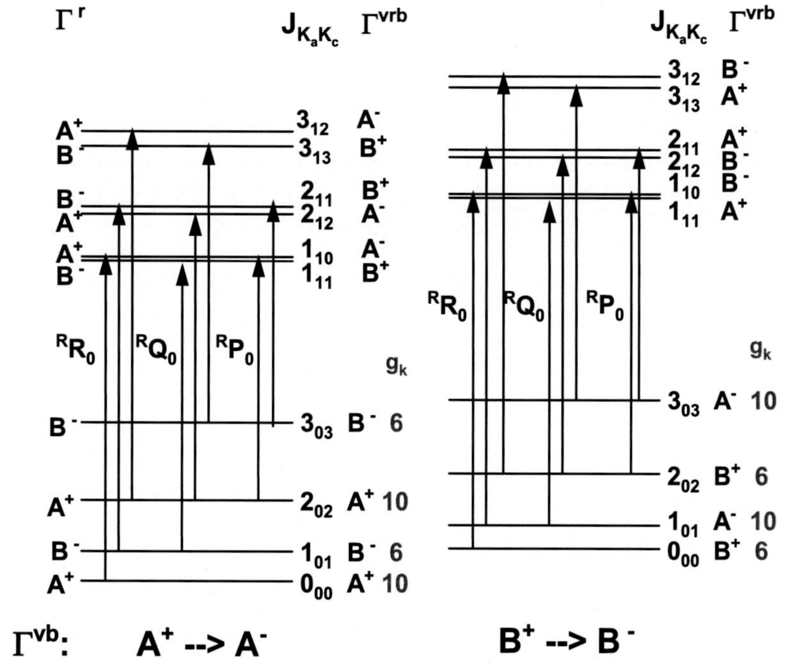

FIG. 7.12 Scheme for the observed rovibrational transitions assigned to the torsional (out of plane bending) fundamental of $(HF)_2$; $\Gamma^{vb}\Gamma^{r} = \Gamma^{vrb}$ is the vibrational, rotational symmetry species including the bending motion with the tunneling splitting; Γ^{r} is the rotational symmetry species (after von Puttkamer and Quack, 1987). The nuclear spin statistical weights g_k are given in red color and help the unique assignment of the high resolution spectra.

It can, however, be introduced much more rigorously relying only on exact theory or, in fact, directly on experiment, without invoking the Born–Oppenheimer or other related approximations (Quack, 1985, 2011b; Merkt and Quack, 2011; see also Chapters 1 and 2 and Section 7.9 of the present chapter). Fig. 7.12 shows in an exemplary way how these symmetry considerations leading to selection rules and nuclear spin statistics allow for an assignment of tunneling spectra.

In the investigations of $(HF)_2$ it turned out that tunneling depends in a highly mode-selective way on the excitation of various degrees of freedom. The "low frequency" motions such as strongly non-rigid K-rotation of the near symmetric rotor and the out of plane bending, "librational" or "torsional" mode promote tunneling appreciably (von Puttkamer and Quack, 1987; von Puttkamer et al., 1988). On the other hand, excitation of the high frequency (HF)-stretching vibration (Pine and Lafferty, 1983; Pine et al., 1984; Pine and Howard, 1986) as well as the even higher lying (HF)-stretching overtones (von Puttkamer and Quack, 1985, 1989; Hippler et al., 2007; Chuang et al., 1997; He et al., 2007; Yu et al., 2005; Manca et al., 2008) leads to a decrease of the tunneling splittings and a considerable lengthening of the tunneling times for the switching process in Fig. 7.11. This early work has revealed many interesting and fundamental facets of the hydrogen bond dynamics in such clusters, see the early review (Quack and Suhm, 1998). The possibilities for concerted proton exchange were studied systematically for the larger (cyclic) hydrogen fluoride clusters, $(HF)_n$ (with $n = 3$–6) (Klopper et al., 1998a). It was shown that pair potentials are quite insufficient to describe tunneling barriers, at least 3-body terms in the many body expansion must be included (see also Quack et al., 1993b, 2001; Quack and Suhm, 1997; Marquardt and Quack, 2011). The early results on simple prototype systems $(HF)_n$ can be seen as an introduction to much later work on more complex systems. $(H_2O)_n$ is an obvious candidate of interest, and we refer to the review by Cvitaš and Richardson in Chapter 9 (Cvitaš and Richardson, 2020). Indeed, many of the basic effects originally found in the quantum dynamics of hydrogen fluoride clusters were later also found in water clusters. These include the mode selective enhancement or inhibition of tunneling by rotational and vibrational excitation, or the importance of three-body terms in the development of the potential hypersurfaces to give just two important effects (Quack and Suhm, 1991a,b, 1995; Quack et al., 1997; Quack and Suhm, 1998; Quack et al., 1993b, 2001; Klopper et al., 1998a,b; Keutsch et al., 2001; Mukhopadhyay et al., 2018; Barclay et al., 2019; Luzar and Chandler, 1996; Luzar, 2000; Richard et al., 2014; Leforestier, 2012; Paesani, 2016; Cole et al., 2015, 2017, 2018), and for many body effects in water clusters see (Xantheas, 1996), for example. But also other and even more complex systems have found interest recently. Indeed, since then numerous additional hydrogen-bond complexes have been investigated. Tunneling processes are generally fast here, due to the low barrier for these processes. Many experimental studies have used microwave- or IR-spectroscopy (Grabow, 2011; Shipman and Pate, 2011; Caminati, 2011; Tanaka et al., 2011; Endo and Sumiyoshi, 2011; Albert et al., 2011; Snels et al., 2011; Havenith and Birer, 2011; Potapov and Asselin, 2014). Three relatively complex recent examples concern the structure and tunneling in methylsalicylate-monohydrate (Ghosh et al., 2015), benzoic-acid monhydrate (Schnitzler and Jäger, 2014), and the trimer of fluoroethanol (Thomas et al., 2015) and the H_2O_2 - formic acid complex (Li et al., 2018). Further complex tunneling systems have been investigated by low temperature UV–VIS and NMR spectroscopy with the aim to understand the influence of the polarity on the isomerization of carboxylate and similar phenomena (Koeppe et al., 2013; Pylaeva et al., 2015). Among the prototypical larger complexes recent studies of the trimer $(HF)_3$ could be mentioned (Asselin et al., 2014) where already earlier a full dimensional potential with a many body potential expansion has been formulated (Marquardt and Quack, 2011; Quack et al., 1993a,b, 1997, 2001). $(HF)_n$ clusters up to $n = 6$ have been studied by supersonic jet spectroscopy (Quack et al., 1993a; Luckhaus et al., 1995), see also Nesbitt (1994) and this species has also led to the first IR-spectroscopic observation of nanoclusters (Quack and Suhm, 1997; Quack et al., 1997; Maerker et al., 1997). Much effort has been devoted to understand the far infrared and microwave spectra of water clusters of different size theoretically (Richardson et al., 2011, 2013, 2016, see also Cvitaš and Richardson, 2020, Chapter 9 in the current volume). Quantum effects and tunneling processes in supercooled water have been investigated by neutron scattering and relaxation spectroscopy (Agapov et al., 2015). The GENUISH program was used to understand the spectra and tunneling processes in ionic complexes F^- (H_2O)) and F^- (D_2O)) (Sarka et al., 2016b) and for the important prototypical complex $(CH_4–H_2O)$ (Sarka et al., 2016a), the prototype of the methane-hydrates (see also the review by (Tanaka et al., 2011) for spectroscopic studies of complexes). Tunneling phenomena may even be of relevance in every day phenomena such as the slipperiness of ice. It was already established that this effect is related to the surface mobility of H_2O molecules by a rolling motion, so far studied with a classical molecular dynamics simulation in relation to experiments (Weber et al., 2018). However, since such a motion involves rearrangement of hydrogen bonding, tunneling may be important.

The double proton (or H-atom) transfer by tunneling in the dimer $(HCOOH)_2$ of formic acid has a long history as a simple straightforward example of this kind of processes (Ortlieb and Havenith, 2007; Havenith and Birer, 2011; Luckhaus, 2006; Zielke and Suhm, 2007; Xue and Suhm, 2009, see also Quack and Jans-Bürli, 1986; Blumberger et al., 2000). Recently an attempt was made to describe the process on a full dimensional potential hypersurface (Qu and Bowman, 2016). Also tunneling processes in the prototype system malonaldehyde still find much interest (Tautermann et al., 2002; Hargis et al., 2008; Lüttschwager et al., 2013; Vaillant et al., 2018, and references cited therein).

7.4 TUNNELING PROCESSES IN SLIGHTLY ASYMMETRIC POTENTIALS, TUNNELING SWITCHING, AND THE MOLECULAR QUANTUM SWITCH

The very weakly asymmetric potentials of chiral molecules under the influence of parity violation are an extreme case which is still today only proven by theory (Quack, 2011a,b). Somewhat larger asymmetries, but often still rather small ones, are found for isomerization reactions of isotopically substituted molecules. For such reactions the electronic Born–Oppenheimer potential for the tunneling process is symmetric, see also Chapters 1 and 2 (Marquardt and Quack, 2020; Császár et al., 2020). Through an appropriate substitution with isotopes the effective potential becomes asymmetric due to zero point energy effects. A straightforward example is the rearrangement of the hydrogen bonding in the dimer $(HF)_2$ (Dyke et al., 1972; Quack and Suhm, 1991a,b; Klopper et al., 1998a,b). As discussed in Section 7.3 this process exchanges the positions of the symmetrically equivalent (by the generalized Pauli principle identical) H-atoms in the dimer. The tunneling splitting $(\Delta E_T/hc)$ is approximately $0.7\ cm^{-1}$ in the ground

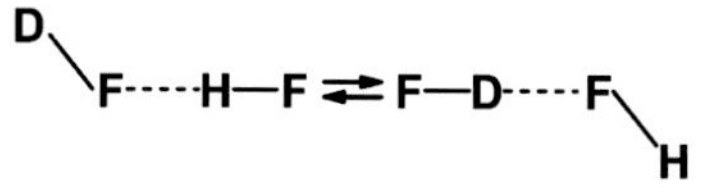

FIG. 7.13 If in the HF-Dimer one H is replaced by a D, then the structures are distinguishable and the zero point energy is different for the two structures.

state. If one H-atom is substituted by D the two structures become distinguishable (Fig. 7.13).

Even though the potential minima have the same energy within the Born–Oppenheimer approximation the zero point energy of the two structures is different which results in an energy difference of the ground states of $(\Delta E_{eff}/hc) \approx 70\ cm^{-1}$ (Klopper et al., 1998b). As there is

$$\Delta E_{eff} \gg \Delta E_T, \tag{7.6}$$

the energy levels of the corresponding isomers are localized in the ground state. For more highly excited levels one finds

$$\Delta E_{eff} \ll \Delta E_T \tag{7.7}$$

with delocalized eigenstates. This transition is called "tunneling switching" (Quack and Willeke, 2006; Albert et al., 2013). Another example is found in the mixed H/D isotopomers of the CH_4^+-cations (Signorell et al., 1999; Signorell and Merkt, 2000; Wörner et al., 2007; Wörner and Merkt, 2009). The high resolution analyses of tunneling processes of this kind in more complex and larger molecules are rare. It was predicted (Albert et al., 2013) that for phenol, marked with deuterium in the ortho or meta position, an energy difference on the order of $(\Delta E_{eff}/hc) \simeq 1\ cm^{-1}$ is to be expected, orders of magnitude larger than ΔE_T in symmetric phenol C_6H_5OH. This prediction and also the phenomena of tunneling switching after excitation could be confirmed recently in the high resolution analysis of the GHz- (mm wave) and THz (FTIR)-spectra with synchrotron radiation of meta-D-phenol (Albert et al., 2016c).

Fig. 7.14 shows a summary of the results for transitions between syn- and anti-isomers. For the ground state and the transition to the first excited state of the torsional vibration of the OH-group relative to the rigid aromatic ring structure, an effectively localized structure is found which could be proven by the different rotational constants of the two isomers in the rotational spectra and also of the corresponding rotational-vibrational spectra in comparison with the theory (Albert et al., 2016c). One finds separate transitions for both isomers, without transitions between the isomers. Above the torsional state excited with 2 quanta, one finds delocalized structures with correspondingly allowed transitions between the syn- and anti-structures, as is shown in the corresponding diagram of Fig. 7.14. The transitions considered are slightly weaker than the "main" transitions in the center of the spectrum, but the rotational fine structure could be precisely simulated and the corresponding spectroscopic parameters could be ascertained. A similar picture is obtained in ab initio predictions with the quasiadiabatic channel RPH model where the agreement with experiment (without any adjustment) is excellent within the uncertainty of the quantum chemical electronic structure methods for problems of this size. This first spectroscopic proof of such a tunneling switching is on the one hand of importance for the corresponding analysis of molecules like 1,2-diithine showing tunneling switching with parity violating potentials which then requires most highly resolved, currently available laser spectroscopy. On the other hand, tunneling switching in molecules like m-D-phenol can be used for quantum mechanical molecular switches and a possible quantum technology and quantum mechanical machine of this kind (Albert et al., 2016c; Fábri et al., 2018). Indeed, based on the accurate spectroscopic data and the analysis for m-D-phenol, a true molecular quantum switch could be recently demonstrated. Fig. 7.15 shows the wavepacket for the quantum switch transfer from a syn- to an anti-structure, which behaves quasiclassically. On the other hand, Fig. 7.16 shows the preparation of a "bistructural state" at the end of the laser pulse sequences, which is a superposition of syn- and anti-structures and shows time independent probability distributions, but time dependent spectra. This type of quantum switch may offer realistic opportunities for quantum machines and quantum computing in the future (Fábri et al., 2018).

Here we should point out also a fundamentally new isotope effect which is introduced by the parity violating interaction in "isotopically chiral" molecules (Quack, 1989). This effect has been calculated quantitatively for the first time (for $PF^{35}Cl^{37}Cl$) in Berger et al. (2005). It arises from the differences in the electroweak charges Q_W of the nuclei as available from the standard model of particle physics (SMPP) see Quack (2006) and Marquardt and Quack (2020), Chapter 1. A requirement for this effect and for stable isotopic chirality is a sufficiently small tunneling splitting (e.g., not fulfilled in the case of NHDT or hydrogen peroxide) for the stereomutation. The small effects of isotopic chirality have been discussed recently in connection with asymmet-

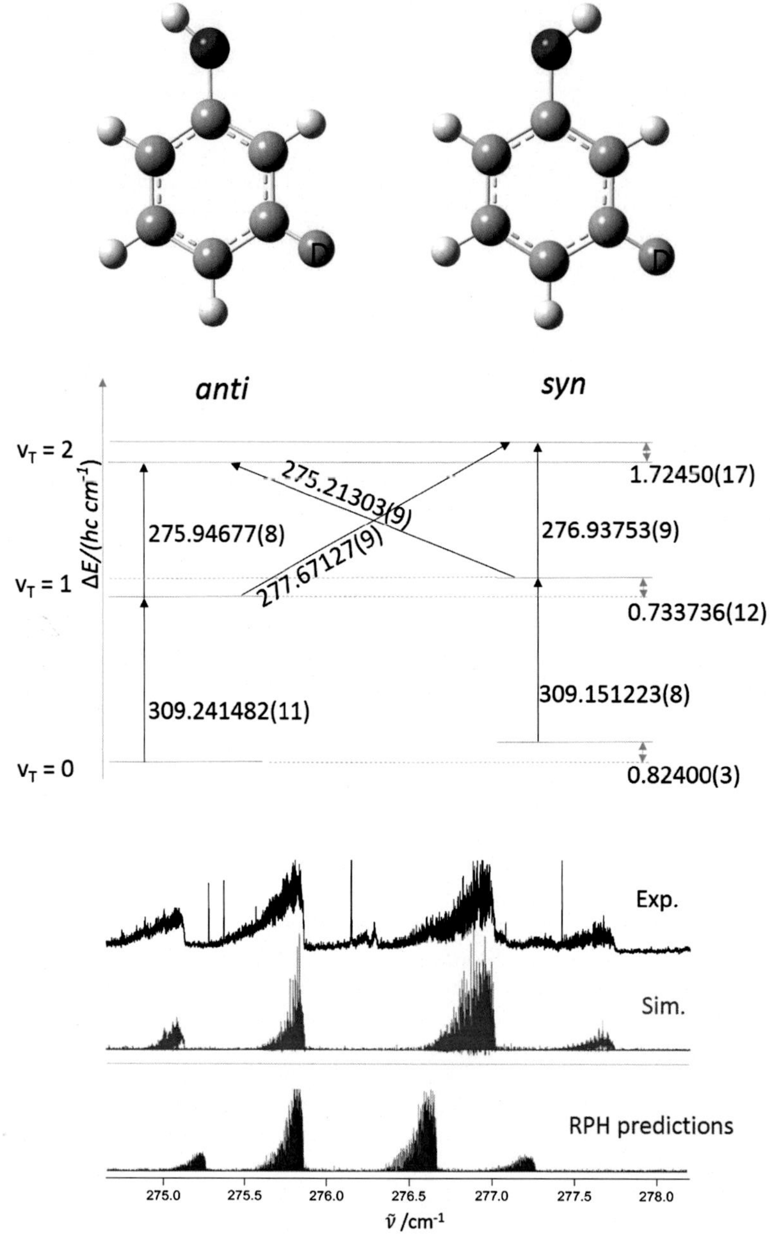

FIG. 7.14 (Top) Conformers of the meta-D-phenol. (Middle) Scheme of the energy levels with respect to the tunneling coordinate up to $v_T = 2$. (Bottom) Measured spectrum for the transition from $v_T = 1$ to $v_T = 2$, compared to a simulation with the fitted parameter of an effective Hamiltonian and with the parameters of the RPH-prediction. The transitions, which are crossing over between left and right in the level scheme, correspond to the weaker transitions becoming allowed by the tunneling switching (after Albert et al., 2016c).

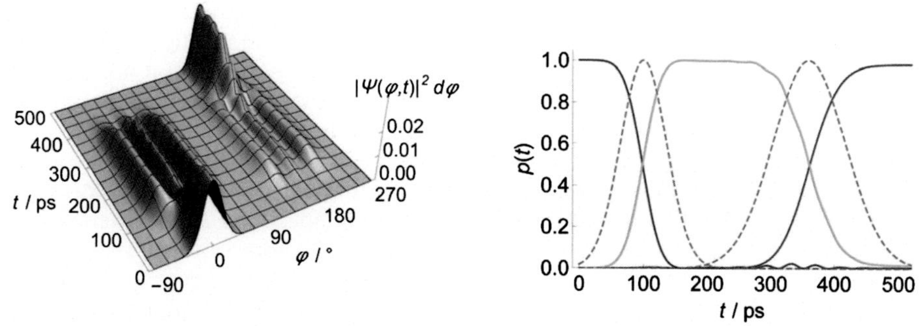

FIG. 7.15 Time-dependent wavepacket for meta-D-phenol (reduced probability density as a function of the torsional angle φ and time t) and populations ($|0_{\rm anti}\rangle$, blue; $|0_{\rm syn}\rangle$, red; $|2_{\rm lower}\rangle$, green) for the two-pulse excitation scheme $|0_{\rm anti}\rangle \rightarrow |2_{\rm lower}\rangle \rightarrow |0_{\rm syn}\rangle$. The parameters of the two laser pulses are $\tau_P = 50$ ps, $t_0 = 100$ ps, $I_{\max} = 150$ MW cm^{-2} and $\tilde{\nu}_0 = 582.82$ cm^{-1} (first pulse, resonant with the $|0_{\rm anti}\rangle \rightarrow |2_{\rm lower}\rangle$ transition), and $\tau_P = 80$ ps, $t_0 = 360$ ps, $I_{\max} = 150$ MW cm^{-2} and $\tilde{\nu}_0 = 582.00$ cm^{-1} (second pulse, resonant with the $|0_{\rm syn}\rangle \rightarrow |2_{\rm lower}\rangle$ transition). The Gaussian envelope functions (normalized to maximum intensity) of the two sequential laser pulses are shown by the dashed lines (after Fábri et al., 2018, where detailed definitions and conditions are given).

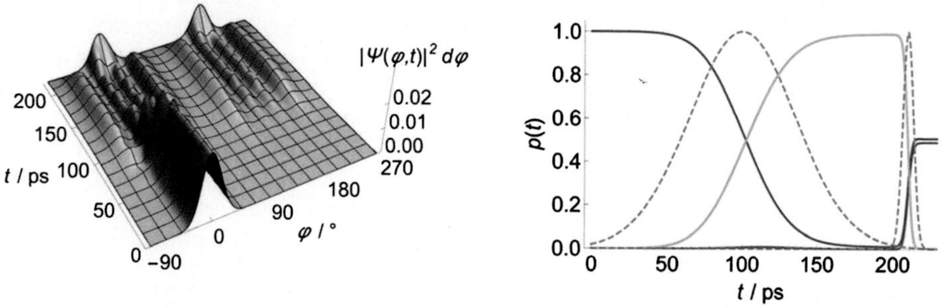

FIG. 7.16 Time-dependent wavepacket and populations for meta-D-phenol ($|0_{\rm anti}\rangle$, blue; $|0_{\rm syn}\rangle$, red; $|3_{\rm lower}\rangle$, green) for the two-pulse excitation scheme $|0_{\rm anti}\rangle \rightarrow |3_{\rm lower}\rangle \rightarrow |0_{\rm anti}\rangle + |0_{\rm syn}\rangle$. The parameters of the two laser pulses are $\tau_P = 50$ ps, $t_0 = 100$ ps, $I_{\max} = 1$ GW cm^{-2} and $\tilde{\nu}_0 = 813.72$ cm^{-1} (first pulse, resonant with the $|0_{\rm anti}\rangle \rightarrow |3_{\rm lower}\rangle$ transition), and $\tau_P = 5$ ps, $t_0 = 210$ ps, $I_{\max} = 60$ GW cm^{-2} and $\tilde{\nu}_0 = 813.10$ cm^{-1} (second pulse). The Gaussian envelope functions (normalized to maximum intensity) of the two sequential laser pulses are shown by the dashed lines (after Fábri et al., 2018, where detailed definitions and conditions are given).

ric induction by the Soai mechanism (Soai et al., 1995; Kawasaki et al., 2009; Sato et al., 2003; Matsumoto et al., 2016) which is of interest in the context of the evolution of biological homochirality (Quack, 2002, 2015b; Blackmond, 2004; Al-Shamery, 2011; Luisi, 2006; Lehn, 2002; Hawbaker and Blackmond, 2019). If asymmetric induction is possible by labeling with "heavy" atoms (like ^{14}N/^{15}N) this might be an indication for a possible asymmetric induction by the very small effects arising from the parity violating interaction (see also the recent report on the history of the nomenclature of chiral molecules Helmchen, 2016, and a more general overview in Jortner, 2006; Meierhenrich, 2008; Quack, 1989, 1999, 2015b,a).

7.5 ISOMERIZATION REACTIONS WHICH ARE SUBSTANTIALLY INFLUENCED BY TUNNELING

7.5.1 A Brief Overview Over Some Recent Studies

We provide here a brief review of examples for such reactions. Isomerization reactions with a reaction barrier between 30 and 60 kJmol^{-1} are ideally suited for investigations by infrared spectroscopy in a low temperature matrix (Schreiner et al., 2011; Schreiner, 2017). There the reaction can be initiated through the absorption of a photon in the infrared or visible spectral region. Examples for such isomerization reactions with significant contributions from tunneling are: 2,4-pentanedione

(Lozada-Garcia et al., 2012), tetrazol-5-yl-acetic acid (TAA) (Araujo-Andrade et al., 2014), glyoxalic acid (Gerbig and Schreiner, 2015), pyruvic acid (Reva et al., 2015), oxalic acid (Schreiner et al., 2015), carbonic acid (Wagner et al., 2016), trifluoromethyl-hydroxycarbene (Mardyukov et al., 2016). For the interpretation of the data often ab initio calculations are used in combination with the WKB (Wentzel–Kramers–Brillouin) theory of tunneling. The case of methyl-hydroxy-carbene is of interest as it shows two competing reactions: on the one hand, to vinyl alcohol through a [1,2] hydrogen shift along the C–C bond and, on the other hand, to acetaldehyde through a [1,2] hydrogen shift along the C–O bond. At low temperatures below 250 K, the reaction channel to the aldehyde is preferred with a higher but narrower barrier than for the competing channel to the alcohol which can be explained easily by theory (Kästner, 2013). An even more complex tunneling reaction is the proton transfer in triplet 2-formyl phenylazide which is produced by UV-photolysis. It reacts in the dark on a time scale of 7 hours to 6-imino-2,4-cyclohexadiene-1-ketene. The [1,4] H-atom migration from the nitrene to the imino ketene proceeds on the triplet surface before the molecule is transferred to the ground state (Nunes et al., 2016).

The tautomerism of porphycene has been studied by femtosecond pump–probe experiments in a molecular beam (Ciaćka et al., 2016). The rate of isomerization is an order of magnitude faster in the S_0-state compared to the S_1-state with a strong isotope effect upon N-deuteration. This study was extended to 19 differently substituted porphycenes (Ciaćka et al., 2015), as well as to experiments in He-droplets in a molecular beam (Mengesha et al., 2015) and on a metal surface Cu(111) (Böckmann et al., 2016) and to investigations of single molecules using an STM (Scanning Tunneling Microscope) (Ladenthin et al., 2016). Details on this interesting type of reactions have been summarized in a recent review article (Waluk, 2017).

The one-dimensional rotational motion of CH_2Cl about the C–Cl axis was investigated by infrared spectroscopy after photolysis of CH_2ClI in solid parahydrogen at 3.7 K. It was found that the nuclear spin conversion between the ortho and para nuclear spin species of the radical occurred on a time scale of a few hours (Miyamoto et al., 2013). The conformational composition and the change in conformational ratio induced by UV irradiation of β-alanine have been investigated in solid parahydrogen using FTIR spectroscopy. It could be shown that the conformational changes are significantly faster in p-H_2 in an Ar-matrix (Wong et al., 2015). Benzvalene, fulvene and Dewar benzene were obtained by exposing benzene to UV-radiation at 193 or 253.7 nm in solid parahydrogen. Neither tunneling reaction for any isomerization nor reaction with a hydrogen molecule from the matrix could be detected on the time scale of days (Toh et al., 2015).

Heavy atom tunneling is the dominant mechanism in ring-expansion and ring-opening reactions at cryogenic temperatures after activation by a visible or UV-photon. Photon activated 1H-bicyclo[3.1.0]-hexa-3,5-diene-2-one rearranges to 4-oxacyclo-hexa-2,5-dienylidene. The rate of rearrangement was found to be temperature independent for temperatures smaller than 20 K and was interpreted as a tunneling process (Ertelt et al., 2014). Through experiments in H_2, HD, and D_2 at cryogenic temperatures, is has been shown that tunneling governs the insertion of 1-azulenylcarbene into H_2 and D_2 (Henkel and Sander, 2015). Benzazirines were identified by IR-spectroscopy after photolysis of 4-methoxyphenyl azide and 4-methylthiophenyl azide. Despite a calculated barrier of 14 kJ mol^{-1}, the benzazirines rearranged to ketenimines at 10 K by a tunneling process (Inui et al., 2013). By a similar tunneling process the triplet 2-formyl phenylnitrene, obtained from the photolysis of 2-formyl phenylazide, is reacting to singlet 6-imino-2,4-cyclohexadiene-1-ketene (Nunes et al., 2016). The ring expansion of noradamantyl-methylcarbene could either proceed to 2-methyl-adamantene or to 3-vinylnoradamantane. It could be shown by DFT calculations that at 10 K the formation reaction of 3-vinylnoradamantane is by 8 orders of magnitude faster than the one to 2-methyladamantene (Kozuch et al., 2013). The SCI (small-curvature tunneling) approximation has been applied to estimate tunneling corrections to classically calculated rate constants for the ring expansions of noradamantyl-carbenes to form adamantenes. The calculation showed that below temperatures of 30–50 K the rearrangement is dominated by heavy atom tunneling (Kozuch, 2014b). Another system which can be treated as heavy atom tunneling is the isomerization of substituted pentalenes, heptalenes, and acepentalenes (Kozuch, 2014a). Particularly noteworthy are also the studies of Limbach and coworkers on proton tunneling in organic model systems (Limbach, 2007). Tunneling by the puckering motion in ring compounds has been studied extensively (Laane, 1999), and we refer here to the prototypical study of the cyclobutane as an example (Blake and Xantheas, 2006; Frey et al., 2011).

For the prototypical reaction of the degenerate isomerization of semibullvalene there is a prediction that at low temperatures a temperature independent rate constant of $2 \cdot 10^{-3}$ s^{-1} should be found (Zhang et al., 2010). This prediction has been tested by studies of the rate of equilibration after the synthesis of a mixture of 1,5-dimethyl-semibullvalene-2-D_1 and 4-D_1 (Ertelt et al., 2015). Another prototypical isomerization is the proton transfer in tropolone (Tanaka et al., 1999). This has been recently studied by making the effective potential for tunneling slightly asymmetric with various ^{13}C substitutions (Tanaka et al., 2019), similar to our study of phenols, discussed in Section 7.4 (Fábri et al., 2018; Albert et al., 2013).

Tunneling is not only relevant for conversion processes but may also influence the molecular structure, if the atomic wave function penetrates deeply into the classically forbidden region of the potential function. An extreme example is the very weakly bound He_2 molecule (Zeller et al., 2016). Here, it was shown in recent experiments that 80 % of the probability density in the ground state of the molecule is found within the classically forbidden region. This results in an extremely large "bond length", which, however, is not to be understood as a sharply defined quantity.

7.5.2 Ammonia as a Prototype for the Inversion at Nitrogen and Mode Selective Control of Tunneling Processes

Ammonia has for a long time been a fundamental molecule of physics and chemistry (Townes, 1964, 1965; Townes and Schawlow, 1975; Ertl, 2008) In addition to complex isomerization reactions in organic prototype molecules, the simplest tunneling processes in the ammonia molecule have further attracted interest with new results. For the ammonia molecule a number of full-dimensional potential energy surfaces were formulated in the last decade. A global potential hypersurface not only took into account the tunneling process of the inversion at low excitation energies, but also all possible dissociation reactions at higher energies (Marquardt et al., 2005; Marquardt and Quack, 2011; Marquardt et al., 2013). The inversion potential for NH_3 has been prototypical for this type of inversion at the nitrogen atom (Hund, 1927a,c,b; Fermi, 1932; Prelog and Wieland, 1944; Laane, 1999). Mode selective tunneling has been repeatedly discussed in the past for the ammonia molecule and its isotopomers, for example, in Herzberg (1945); Marquardt et al. (2003a,b, 2005); Snels et al. (2000, 2003, 2006b,a, 2011). Tunneling splittings have been discussed and analyzed in different recent publications (Marquardt et al., 2013, 2003a; Al Derzi et al., 2015; Fábri et al., 2014). The time dependent dynamics has also been discussed (Marquardt et al., 2003a; Sala et al., 2012) and was extended in most recent times to the complete series of isotopomers NH_3, NH_2D, NHD_2, ND_3, NHDT, NH_2Mu, ND_2Mu, and NHDMu, where the stereomutation of the chiral isotopomers is of special importance (Fábri et al., 2015; Fábri et al., 2019). Very highly resolved measurements of infrared spectra using a special OPO laser system resulted in hyperfine structure resolved spectra of the excited NH-stretching vibrations and have also been used as a test system for the preparation of an experiment to measure the parity violating energy difference in chiral molecules (Dietiker et al., 2015).

Because ammonia has been a prototypical system for tunneling by inversion, somewhat related to our discussion of $(HF)_2$ being prototypical for tunneling in hydrogen bonded systems, we shall discuss here also in some more detail the developments in our understanding of spectroscopy and tunneling dynamics in the NH_3 molecule. Indeed, after the early discussion of Hund (1927a,c,b) there was an analysis of the vibrational spectrum and potential by Fermi (1932) and at almost that same time as well a quantitative analysis of the tunneling dynamics in the framework of WKB theory (Dennison and Uhlenbeck, 1932). The early work typically dealt with the ground state tunneling – including notably the ammonia Maser (Gordon et al., 1955) – and used one-dimensional models for the tunneling process (Morse and Stückelberg, 1931; Manning, 1935; Wall and Glockler, 1937; Newton and Thomas, 1948; Costain and Sutherland, 1952; Swalen and Ibers, 1962). Much of the early history can be found in Herzberg (1939, 1945, 1966). Ammonia tunneling by inversion has been reviewed on the broader context of potential barriers for inversion at the nitrogen atom by Lehn (1970) and in relation to one dimensional potential functions for tunneling by Laane (1999). Strongly hindered inversion at nitrogen has been discussed in the context of molecular chirality (Prelog and Wieland, 1944; Felix and Eschenmoser, 1968), see also the review by Quack (1989). Ammonia has also been used as a testing ground for developing quantum chemical methods to compute barriers for isomerization (see Veillard et al., 1968; Császár et al., 1998; Klopper et al., 2001; Marquardt et al., 2005, 2013, and references therein). Finally, tunneling is also important in prototypical studies of the photoionization of ammonia isotopomers (Hollenstein et al., 2007; Merkt et al., 2011) and ammonia has been used as a test case for nuclear spin symmetry conservation (Wichmann et al., 2020). Here we shall discuss in some detail a recent extended investigation of spectroscopy and quantum dynamics of ammonia and its isotopomers in view of an understanding of its multidimensional tunneling and mode selective reaction

TABLE 7.3
Character table of the symmetry group S_3^* for NH_3 (ND_3) isotopomers (compared to D_{3h}). The first three columns provide species following three different notations: 1, [partition]^{parity}; 2, [S_3 species]^{parity}; and 3, D_{3h} point group species (see text).

		D_{3h} (class):	E	$2C_3$	$3C_2$	σ_h	$2S_3$	$3\sigma_v$		
		S_3^* (class):	E	2(123)	3(12)	E*	2(123)*	3(12)*		
species										
$\Gamma[S_3^*]$		$\Gamma[D_{3h}]$							$\Gamma(S_3^*)\downarrow S_2^*$	$\Gamma(S_3^*)\downarrow M_{S6}(C_{3v})$
$[3]^+$	A_1^+	A_1'	1	1	1	1	1	1	A^+	A_1
$[1^3]^+$	A_2^+	A_2'	1	1	−1	1	1	−1	B^+	A_2
$[2,1]^+$	E^+	E'	2	−1	0	2	−1	0	$A^+ + B^+$	E
$[3]^-$	A_1^-	A_1''	1	1	1	−1	−1	−1	A^-	A_2
$[1^3]^-$	A_2^-	A_2''	1	1	−1	−1	−1	1	B^-	A_1
$[2,1]^-$	E^-	E''	2	−1	0	−2	1	0	$A^- + B^-$	E

TABLE 7.4
Character table of the symmetry groups S_2^* for NH_2D and NHD_2 compared to C_{2v} (transition state structure group).

		C_{2v} (class):	E	C_2	σ_{yz}	σ_{xz}			
		S_2^* (class):	E	(12)	E*	(12)*	$\Gamma(S_2^*)\uparrow S_3^*$	$\Gamma(S_2^*)\downarrow M_{S2}(C_S)$	$\Gamma(S_2^*)\downarrow S^*$
species	$\Gamma(S_2^*)$	$\Gamma(C_{2v})$							
$[2]^+$	A^+	A_1	1	1	1	1	$A_1^+ + E^+$	A	A^+
$[2]^-$	A^-	A_2	1	1	−1	−1	$A_1^- + E^-$	B	A^-
$[1^2]^-$	B^-	B_1	1	−1	−1	1	$A_2^- + E^-$	A	A^-
$[1^2]^+$	B^+	B_2	1	−1	1	−1	$A_2^+ + E^+$	B	A^+

control of the corresponding isomerization by inversion at the nitrogen atom. As far as tunneling in vibrationally excited states is concerned, an extensive MARVEL analysis has reviewed the available data for the isotopomer $^{14}NH_3$ (Al Derzi et al., 2015).

Much less work is available on the deuterated isotopomers. However, in an extended spectroscopic project directed specifically also at the understanding of vibrationally mode-selective effects and control of tunneling rates the deuterated isotopomers ND_3, NHD_2, and NH_2D were investigated in some detail (Snels et al., 2000, 2003, 2006b,a) and combined with quantum-dynamical wavepacket calculations under coherent excitation including global potential and electric dipole hypersurfaces (Marquardt et al., 2003b,a; Fábri et al., 2019). Here we highlight some of the aspects with respect to mode selective tunneling dynamics as derived from the high resolution spectra. One important aspect concerns the symmetry of the rotation–vibration–tunneling levels, which can be understood in relation to the character tables (Tables 7.3, 7.4, 7.5, and 7.6) relevant for the different isotopomers. For ammonia and its isotopomers one has the special situation that one can use equivalently the point group of the planar transition structure (D_{3h} for NH_3, ND_3, C_{2v} for NH_2 D and NHD_2 and also C_s for NHDT, for which, however, no high resolution spectroscopic data seem to be available (Snels et al., 2006a; Quack, 2002)) or the corresponding isomorphous permutation inversion groups S_3^*, S_2^* and S^* with a one-to-one relation between group elements and irreducible representations (symmetry species). This procedure is not generally possible, for instance, for methane and other molecules this would not be possible (Quack, 2011b). However, for ammonia we can give the corresponding notations in Tables 7.3, 7.4, 7.5, and 7.6. The tables include also relevant induced and subduced

TABLE 7.5
Character table molecular symmetry group M_{S2} and point group C_s for NH_2D and NHD_2.

	C_S (class)		E	σ	
	M_{S2} (class)		E	$(12)^*$	
species	Γ (C_S)	Γ (M_{S2})			Γ (M_{S2})$\uparrow$ S_2^*
	A'	A	1	1	$A^+ + B^-$
	A"	B	1	−1	$B^+ + A^-$

TABLE 7.6
Character table of the subgroup M_{S6} of the molecular symmetry group M_{S12} (S_3^*) and induced representation Γ (M_{S6}) $\uparrow$ S_3^*; M_{S6} is isomorphous to C_{3v}[a].

C_{3v}	E	$2C_3$	$3\sigma_v$	
M_{S6}	E	2(123)	$3(12)^*$	$\Gamma(M_{S6}) \uparrow S_3^*$
A_1	1	1	1	$A_1^+ + A_2^-$
A_2	1	1	−1	$A_2^+ + A_1^-$
E	1	−1	0	$E^+ + E^-$

[a] After Dietiker et al., 2015.

representations. Table 7.6 gives also the character table for C_{3v}, isomorphous to the molecular symmetry group M_{S6}. The induced representation Γ (M_{S6}) $\uparrow$ S_3^* gives the symmetry structure of the tunneling sublevels in NH_3 and ND_3. We note furthermore that for NH_3 the Pauli principle excludes the A_1^+ and A_1^- levels in the corresponding tunneling doublets (thus hypothetical, but not real) whereas E^+ and E^- doublets exist and are observable by spectroscopy. On the other hand, all the rovibrational symmetry species $A_1^{\pm}$, $A_2^{\pm}$ and $E^{\pm}$ are Pauli allowed for ND_3 with appropriate nuclear spin functions and therefore all the tunneling doublets exist as observable levels. Fig. 7.17 summarizes one main result from the analysis of the tunneling spectra in ND_3 (Snels et al., 2000).

The ground state tunneling splitting in ND_3 is 0.053 cm^{-1}, as expected much smaller than in NH_3 (0.7 cm^{-1}). Thus the tunneling transfer time is rather long in the vibrational ground state (314 ps). However, exciting the symmetric ND_3 stretching fundamental at 2420.1 cm^{-1} one finds an increase of the tunneling splitting to 0.541 cm^{-1} corresponding to a "promotion" or enhancement of the tunneling rate by a factor of 10, with a shorter tunneling transfer time of 31 ps.

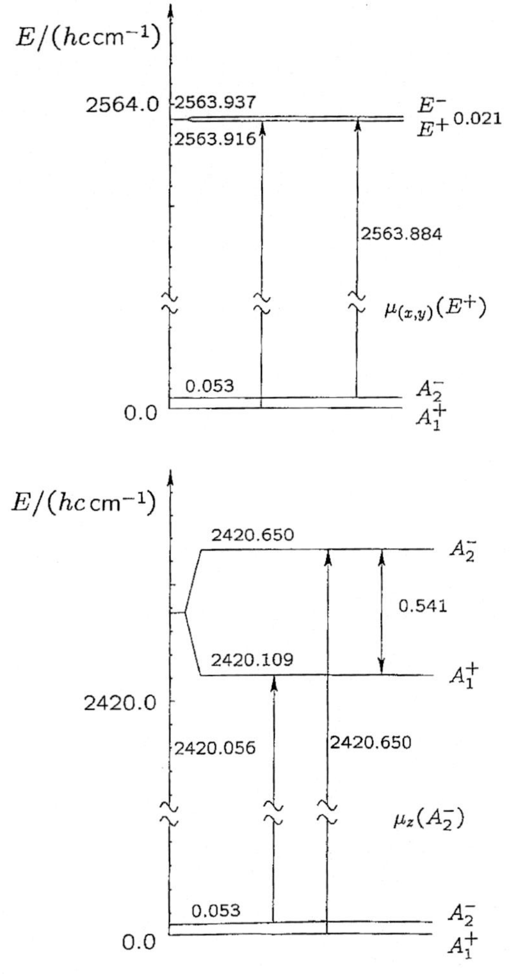

FIG. 7.17 Schematic drawing of observed transitions and vibrational tunneling levels in ND_3. The tunneling splittings are about to scale: the vibrational wavenumber scale is interrupted (after Snels et al., 2000). See also text.

While such an enhancement of tunneling with vibrational excitation might seem natural, one finds the reverse effect at even higher excitation with the degenerate ND_3 stretching fundamental centered at 2563.9 cm^{-1}, where the tunneling splitting is reduced to 0.021 cm^{-1}, by more than a factor of 2 compared to the ground state splitting and correspondingly an inhibition of the tunneling rate with an excitation well above the "tunneling barrier" and a very long tunneling transfer time of more than 600 ps. Of course, a very comparable situation has been found for H_2O_2 (see Table 7.1) and this is a further illustration of "quasiadiabatic above barrier tunneling". Indeed, a very rich variety of such en-

TABLE 7.7
Summary of tunneling times $\tau_{L \to R}$ (in ps) for NH_2D and NHD_2 ground states and fundamentals[a].

	NH_2D		NHD_2	
	$\tilde{\nu}$ (cm^{-1})	$\tau_{L \to R}$ (ps)	$\tilde{\nu}$ (cm^{-1})	$\tau_{L \to R}$ (ps)
ν_0	0	41	0	98
ν_1	2506	27	3404	214
ν_2	886	0.83	815	1.8
ν_{3a}	3366	7.1	2433	4.4
ν_{3b}	3439	99	2560	107
ν_{4a}	1598	1.1	1235	6.6
ν_{4b}	1390	28	1462	84

[a] After Snels et al., 2006a.

hancements and inhibitions of tunneling is also found in the extensive analysis on the NH_2D and NHD_2 isotopomers as summarized in Table 7.7. There the overall ranges of enhancements and inhibitions with various excitation of fundamental vibrations cover a factor of more than 100 with a shortest tunneling transfer time of about 1 ps and a longest tunneling transfer time of 214 ps. These experimental results can be compared to recent full-dimensional theoretical calculations (Fábri et al., 2015, 2019).

7.5.3 Methyl Group Internal Rotation as a Prototype for Tunneling

The methyl group ($-CH_3$) is a ubiquitous building block of organic molecules and biomolecular systems. Its torsional motion, a hindered internal rotation, shows a great variety of tunneling effects, which range from slightly anharmonic torsional vibration with only very small tunneling splittings to essentially free internal rotation, depending upon the molecular environment. The quantum dynamics of methyl group tunneling has been extensively studied in the solid, liquid and gas phases by NMR spectroscopy, neutron diffraction and optical spectroscopy, including high resolution studies from the microwave to the UV regions. It is a "classic" among the tunneling systems and has been extensively reviewed (Press, 1981; Heidemann et al., 1987; Prager and Heidemann, 1997; Horsewill, 1999; Bauder, 2011; Pratt, 2011). A great variety of systems has been studied in detail (Tan et al., 1989, 1991; de Haag et al., 1990; Barlow et al., 1992; Piecha-Bisiorek et al., 2014), including also $-CD_3$ group dynamics (Lalowicz et al., 1988; Börner et al., 1991). A particularly noteworthy aspect is the coupling of the tunneling motion with nuclear spin symmetry and the studies of the kinetics of the nuclear spin symmetry conversion (Haupt, 1972; Beckmann et al., 1977; Vandemaele et al., 1986; Buekenhoudt et al., 1990; Häusler, 1990; Würger, 1990; Hartmann et al., 1992; Plazanet et al., 2000). Due to the generalized Pauli principle (Quack, 2011b) the symmetries of the lowest tunneling levels for the hindered rotation are (in C_3) A combining with a total nuclear spin of the three protons $I_{3H} = 3/2$ in the ground state and E with $I_{3H} = 1/2$ in the first excited state, with only weak couplings between levels of different nuclear spin symmetry and correspondingly slow interconversion kinetics, depending also on the environment. The quantum dynamics of the methyl group tunneling in 4-methyl pyridine (γ-picoline, CH_3–C_5H_5N), for example, generates a tunneling splitting of about 130 GHz (4 cm^{-1}) resulting in long lived nuclear spin states (up to hours and more) and the so-called Haupt effect (Haupt, 1972). This effect has notably been used for a development of Haupt magnetic double resonance (Tomaselli et al., 2003, 2004) in view of sensitivity enhancements in NMR spectroscopy, see also Ludwig et al. (2010); Icker and Berger (2012); Ernst et al. (1987) and the theoretical treatment in Meier et al. (2013a); Dumez et al. (2015). We should mention here also the close relation of the symmetry effects in the long lived nuclear spin symmetry states in the methyl group tunneling with the effect of nuclear spin symmetry conservation observed by high resolution infrared spectroscopy of supersonic jet expansions of CH_3X molecules (Horká-Zelenková et al., 2019) and of NH_3 (Wichmann et al., 2020).

We shall discuss here as an example in some detail methanol (CH_3OH), which is among the simplest examples for the tunneling motion by hindered internal rotation of the methyl group, or strictly speaking the relative internal rotation of the methyl- and OH-fragments with respect to each other and with the corresponding reduced moment of inertia. The spectroscopy and quantum dynamics of methanol have been extensively studied and reviewed (Moruzzi et al., 2018). It shows many of the essential features of methyl group tunneling and also some interesting special effects. For this molecule a full-dimensional ab initio potential hypersurface has been developed by Qu and Bowman (2013), which includes the dissociation into $CH_3 + OH$ at high energies but also the low energy hindered internal rotation barriers. The mode selective multidimensional tunneling dynamics has been studied by Fehrensen et al. (2003) using the approximate quasiadiabatic channel reaction path Hamiltonian approach. Fig. 7.18 shows the molecular structure and coordinate definitions including the torsional angle τ defining in essence the tunneling reac-

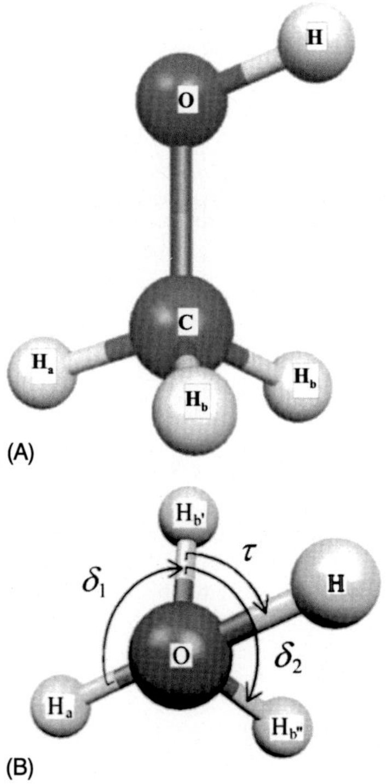

FIG. 7.18 (A) Equilibrium r_e structure of CH_3OH calculated with a basis of triple ζ quality on the MP2 level of theory (Fehrensen et al., 2003) The corresponding structure parameters are $r_{CO} = 142.3$ pm, $r_{OH} = 95.8$ pm, $r_{CHb} = 108.8$ pm, $r_{CHa} = 108.1$ pm, $\theta(\text{HOC}) = 107.9°$, $\theta(\text{OCH}_a) = 112.0°$, $\theta(\text{OCH}_b)) = 106.6°$, and $\delta(\text{HOCH}_b) = 61.5°$ (r = bond lengths, θ = bond angles, δ = dihedral angle). (B) Newman projection of methanol indicating the torsional angle τ, which is defined as the dihedral angle between the HOC plane and the $H_{b'}$CO plane; τ closely follows the reaction coordinate, but relaxation of the CH_3 group bond angles and lengths slightly breaks the C_3 symmetry along τ. A more accurate (but still approximate) representation of the reaction coordinate is the symmetry adapted combination $\tau + (\delta_1 - \delta_2)/3$; δ_1 is defined as dihedral angle between the planes spanned by $H_{b'}$, C, O and by C, O, H_a, δ_2 is the corresponding dihedral angle between the planes spanned by $H_{b'}$, C, O and by C, O, $H_{b''}$; τ, δ_1 and δ_2 have a clockwise orientation (after Fehrensen et al., 2003).

tion path coordinate, and Fig. 7.19 shows the effective tunneling potential with the lowest energy levels of A and E symmetry (Fehrensen et al., 2003).

The ground state splitting between the lowest levels is $\Delta\tilde{\nu}_{EA} = \tilde{\nu}_E - \tilde{\nu}_A = 9.12$ cm^{-1} from experiment, in good agreement with the ab initio predictions from the anharmonic quasiadiabatic channel RPH model (7.9 cm^{-1}). It is thus about twice as large as for the 4-methylpyiridine mentioned above. The first excited torsional state is located essentially at the top of the barrier with inverted order for symmetry reasons and has a $\Delta\tilde{\nu}_{EA}(1) = \tilde{\nu}_E(1) - \tilde{\nu}_A(1) = -85.5$ cm^{-1} from experiment (Moruzzi et al., 2018) and -81.1 cm^{-1} from the prediction, expressed here as a negative quantity in order to retain the definition of $\Delta\tilde{\nu}_{EA}$ and showing the inversion. Given the position at the barrier it is obviously a "borderline tunneling" problem at this level. At higher energies the levels approach the energy level patterns of free internal rotation in a well known manner, see, for example, the correlation scheme given in Bauder (2011). An interesting result concerns the interaction of the tunneling motion with the OH stretching vibration ν_1, which in fact acts as an inhibiting mode, slowing down the tunneling motion. For the first excited OH stretching state $v_1 = 1$ one finds $\Delta\tilde{\nu}_{EA}(v_1 = 1) = 6.3$ cm^{-1} (6.2 cm^{-1} from the prediction) and this trend continuous smoothly up to $\Delta\tilde{\nu}_{EA}(v_1 = 6) = 1.5$ cm^{-1} (1.5 cm^{-1} from the prediction). The agreement between the experimental result by Boyarkin et al. (1999) and the ab initio prediction from the quasiadiabatic channel RPH model is very good. Fig. 7.20 summarizes the trend for the tunneling splittings in conjunction with the trend for the quasiadiabatic channel potential barriers, which smoothly increase from less than 400 cm^{-1} at $v_1 = 0$ to more than 600 cm^{-1} at $v_1 = 6$. This increase of the effective barriers explains in part the decrease in tunneling splitting. An about equally important contribution to the decrease in tunneling splittings arises from the lengthening of the effective O–H bond length $\langle r_{OH} \rangle$ with excitation of the strongly anharmonic O–H stretching vibration. This lengthening of $\langle r_{OH} \rangle$ results in an increase of the effective moment of inertia which acts in a similar way as an effective "tunneling mass" (Fehrensen et al., 2003).

Another interesting effect was observed in experiments of the tunneling splittings of the asymmetric CH stretching fundamentals ν_2 and ν_9, where the tunneling doublets are inverted $\Delta\tilde{\nu}_{EA}(v_2 = 1) = -3.26$ cm^{-1} and $\Delta\tilde{\nu}_{EA}(v_9 = 1) = -5.48$ cm^{-1} (Xu et al., 1997; Wang and Perry, 1998). This effect is explained within the quasiadiabatic channel RPH model as arising from a Herzberg–Longuet-Higgins–Berry type phase from a conical intersection of the vibrationally adiabatic potential surface in the collinear C–OH structure where the modes ν_2 and ν_9 correlate with a degenerate mode of symmetry E in C_{3v}. The theoretical results give then correspondingly $\Delta\tilde{\nu}_{EA}(v_2 = 1) = -3.6$ cm^{-1} and $\Delta\tilde{\nu}_{EA}(v_9 = 1) = -3.2$ cm^{-1} in reasonable agreement with experiment (Fehrensen et al., 2003). This type of

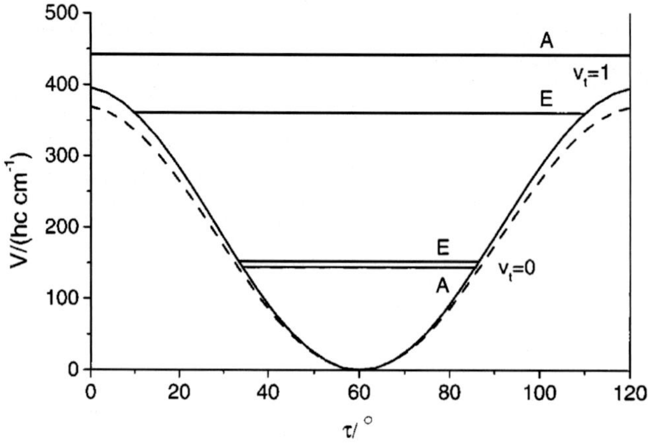

FIG. 7.19 Quasiadiabatic channel potential (full line) for the zero point level of $^{12}CH_3OH$ including the ground state doublet and the first excited torsional state doublet. For comparison the pure electronic potential $V_e(\tau)$ relative to its minimum energy is shown with a dashed line. The maximum of the adiabatic channel potential corresponds to 406.4 cm^{-1} and for the pure electronic potential V_e to 369.0 cm^{-1}, in each case relative to the corresponding minimum energy. The potential is periodic with period 120° (after Fehrensen et al., 2003).

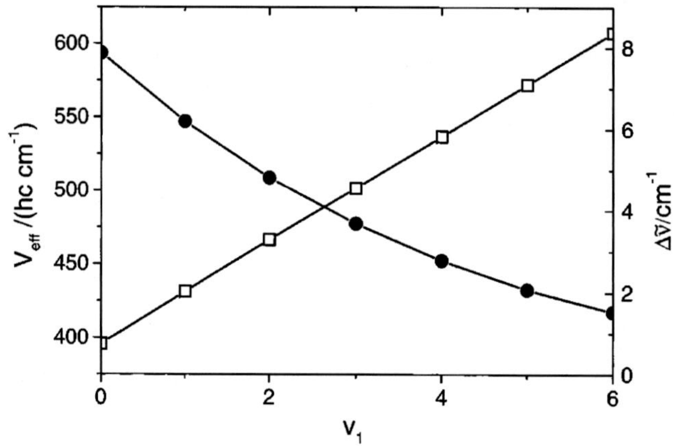

FIG. 7.20 Effective quasiadiabatic channel barrier height V_{eff} as a function of the OH stretching excitation v_1 (open squares) and the corresponding tunneling splittings $\Delta\tilde{\nu}_{EA} = \tilde{\nu}_E - \tilde{\nu}_A$ (filled circles, after Fehrensen et al., 2003).

symmetry effect arising from vibrational conical intersections has notably been further pursued by Perry and coworkers (Perry, 2008, 2009; Dawadi and Perry, 2014) and by Li Hong Xu and coworkers (Xu et al., 2010, 2011, 2014) using various models and perspectives and we shall not enter here the intricate discussions of the many detailed effects arising in the different model treatments (see also Fehrensen et al., 2003; Hänninen et al., 1999).

We should point out, however, a general aspect both of the "conical intersections" and the relative phases associated with them. Similar to the case of conical intersections of electronic potential energy hypersurfaces arising from the Born–Oppenheimer approximation (Yarkony, 1996; Domcke and Yarkony, 2012), these effects are always model dependent. If one omits the Born–Oppenheimer approximation and treats the electrons and nuclei "exactly" on a similar footing in a treatment, which has been called "pre-Born–Oppenheimer theory" (Simmen et al., 2013; see also Marquardt and Quack, 2020, Chapter 1, and Császár et al., 2020,

Chapter 2) there are no Born–Oppenheimer potential hypersurfaces, and thus no conical intersections and associated "Berry phases". This holds true in an analogous fashion for the vibrational-rotational problem in CH_3OH: If that is treated without making the vibrationally adiabatic approximations, say in a 12-dimensional (or 15-dimensional vibration-rotation) computation, no reference needs to be made to vibrational conical intersections and the associated phases. While such an "exact" treatment (similar to that quoted above for NH_3) would give numerically accurate results, there is still merit in the simpler models beyond just the enormous reduction in computational costs: The simple models such as the quasiadiabatic channel RPH model allow for an interpretation and deeper understanding of the plain numerical results (Quack, 2014). An example for such an understanding can be seen from Fig. 7.19 and Fig. 7.20 and, of course, there are many such cases.

We may mention here the discussion of the role of vibrational conical intersections in water dimers (Hamm and Stock, 2013) in relation to intramolecular vibrational redistribution (IVR) also in the chiral conformers for methanol isotopomers (Quack and Willeke, 1999) as intrinsically vibrationally non-adiabatic effects (Beil et al., 1997; Kushnarenko et al., 2018). In a more general context we may mention also the role of parity violation in the isotopomer CHD_2OH with transient chirality due to tunneling and CHDTOH with persistent chirality (Arigoni and Eliel, 1969; Lüthy et al., 1969; Berger et al., 2003). We conclude by mentioning also the methyl rotor tunneling studied also in the relatively complex molecules formic-acetic-acid anhydride (Bauder, 2013) and in para-halotoluenes (Shubert et al., 2013) showing most interesting dynamical effects. M. Schnell provided a detailed discussion of some complex cases with several methyl rotors (Schnell, 2011).

7.6 TUNNELING IN BIMOLECULAR REACTIONS

7.6.1 Direct Bimolecular Reactions

The prototype for tunneling in direct bimolecular reaction is certainly

$$H + H_2 \rightarrow H_2 + H \tag{7.8}$$

with its various isotopic variants. It has been in the focus of interest from the start of reaction dynamics, when it also was used for designing potential hypersurfaces for reaction dynamics and for experiments on the ortho–para conversion in H_2 (Eyring et al., 1944). Just as H_2, as a molecule, has been the prototype for fundamental tests of theory until today (Sprecher et al., 2013; Hölsch et al., 2019 and references therein), the $H + H_2$ reaction system continues to be of interest in the study of the tunnel effect in direct bimolecular reactions, as is also the H_3^+ system for complex forming reactions to be studied in the later Section 7.6.2.

A comprehensive review of the $H + H_2$ system would be far beyond our scope. We shall mention here a few highlights. Early studies started with classical trajectory calculations on ab initio potential hypersurfaces and simple tunneling corrections, if any; see Karplus (2014) for a review. The reaction may have been the first, where instanton theory was used to study tunneling in a bimolecular reaction (Chapman et al., 1975). It was the reaction where the so-called Marcus Coltrin path was discussed with "corner cutting" (Marcus and Coltrin, 1977). Among early experiments providing absolute cross-sections, we may mention the work of Wolfrum and coworkers (Buchenau et al., 1990; Wolfrum, 2001; Kneba and Wolfrum, 1980) and notably the recent work of Zare and coworkers on $H + D_2$ and $D + H_2$ and other isotopic variants (Jankunas et al., 2012, 2013a,b; Jambrina et al., 2016; Mukherjee et al., 2017; Aoiz and Zare, 2018; Althorpe et al., 2002; Aoiz et al., 2002; Ayers et al., 2003; Pomerantz et al., 2004; Ausfelder et al., 2004; Koszinowski et al., 2005 and references therein).

The reaction of muonium, the "lightest isotope of the H atom" (with a mass of 0.114 Da), with H_2 has recently received attention due to its large inverse kinetic isotope effect and the extremely large enhancement of the rate when H_2 is in its first excited vibrational state. The reaction rate constant for 300 K has been determined experimentally and by converged quantum dynamics calculations using a Born–Huang (BH) definition for the potential surface (Bakule et al., 2012). The importance of zero-point energy, tunneling and adiabaticity of the vibrational state for the understanding of this reaction has been summarized in a recent review article (Mielke et al., 2015). Different theoretical methods to consider the zero-point energy and tunneling effects within a classical or semiclassical framework were compared to accurate quantum calculations for the reaction D + HMu → DMu + H. It was found that RPMD (ring polymer molecular dynamics) calculations provide overall a better result than methods based on transition state theory (TST) like canonical variational theory (CVT) or semiclassical instanton theory (SCI) (de Tudela et al., 2013). In time independent quantum calculations and quasi classical trajectory calculations the rate constants for H-atom abstraction and Mu-transfer were calculated for the temperature range from 100 to 1000 K. The calculations showed that for the complete temperature range the rate constant for the Mu-transfer is found 1 to 2.5 orders of magnitude above the rate constant for H-atom abstraction (Aoiz et al., 2014).

The reaction of myonic helium with H_2 has been investigated experimentally and by accurate quantum dynamics calculations on a Born–Huang potential energy surface and compared to the reaction of myonium with H_2 in the temperature range from 300 to 500 K. For the higher temperatures the calculated rate constant agreed with experiment. However, for the lower temperatures the theoretical value was found to be substantially smaller than for the experiment (Fleming et al., 2011b,a).

The importance of tunneling in bimolecular reactions has been extensively studied in the past for the reaction $F + H_2$ and its deuterated modifications, where quantum effects are important already at room temperature (Lee, 1987). To model the temperature dependence of the rate constant for the temperature range from 10 to 350 K, a model function for the temperature dependence of the Arrhenius activation energy was derived by comparison with exact quantum calculations. It was found that the temperature dependence of the activation energy can be characterized by a "sub-Arrhenius" behavior (Aquilanti et al., 2012; Cavalli et al., 2014). Instanton and canonical variational transition state theories with optimized multidimensional tunneling were applied to calculate the reaction rates and kinetic isotope effects for the reaction $H_2 + OH \rightarrow H_2O + H$ and all its possible isotopic combinations down to temperatures of 50 K. It was found that atom tunneling is becoming important for temperatures below 250 K for proton transfer and below 200 K for deuterium transfer (Meisner and Kästner, 2016). In recent experiments He-droplets have been found to be an ideal environment for heavy atom tunneling. In a He-droplet the heliophilic Xe resides inside the droplet, while the heliophobic Rb stays outside at the surface of the droplet. The dissociation of this XeRb-complex could be described by a tunneling process (Poms et al., 2012).

Parahydrogen, forming quantum crystals at low temperatures, is a new matrix to study the physical and chemical processes of molecules at low temperatures. The large lattice constant and the large zero-point amplitude of the lattice vibrations in solid hydrogen provide more free space for the guest molecules than other matrices and the interactions with the surrounding host molecules become weak compared with those in noble gas matrices. An early experiment is the reaction of H_2 with differently deuterated methyl radicals at 5 K. For the deuterated modifications a rate constant around $10^{-6}\ s^{-1}$ was found which is increasing with the number of D atoms. For the reaction with CH_3 an upper limit for the rate constant of $8 \cdot 10^{-8}\ s^{-1}$ could be determined (Hoshina et al., 2004).

Bimolecular reactions on dust grains play an important role in interstellar chemistry (Bromley et al., 2014). An experimental investigation of O-atom diffusion and reactivity on water ice in the temperature range of 6–25 K showed that tunneling processes are important for O-atom diffusion at 6 K (Minissale et al., 2013; Congiu et al., 2014). The formation of O_2 and O_3 at low temperatures on an amorphous silicate surface is determined by the mobility and reactivity of the adsorbed O-atoms. The diffusion of O-atoms in the temperature range from 6.5 to 30 K could be explained by a Langmuir–Hinshelwood mechanism taking into account a non-negligible tunneling process (Minissale et al., 2014). A hybrid model of quantum and molecular mechanics has been combined with instanton theory to investigate the reaction H + HNCO on an amorphous solid water surface. Including contributions from tunneling the rate constant was calculated for temperatures down to 103 K and compared with results for the gas phase. The kinetic isotope effect (KIE) increased from 146 for the gas phase to 231 for the reaction on the cold surface (Song and Kästner, 2016). In a recent publication the reaction rate for the reaction $H + H_2O_2$ has been calculated for temperatures down to 50 K using the instanton theory based on electronic potentials from DFT calculations of different complexity. The calculations showed that at 114 K the reaction to $H_2O +$ OH is preferred by 2 orders of magnitude as compared to the reaction to $H_2 + HO_2$. Including the Eley–Rideal reaction mechanism it could be shown that the second reaction is at low temperature on surfaces of similar order of magnitude (Lamberts et al., 2016). The MCTDH (Multiconfiguration Time-Dependent Hartree) method was used to calculate the relaxation rates of the vibrational states for H-atoms adsorbed on a Pd(111) surface. The calculation showed that even at room temperature quantum effects cannot be neglected to describe the diffusion of H-atoms on a Pd(111) surface (Firmino et al., 2014). An interesting discussion of the importance of tunneling in interstellar chemistry and applications of "harmonic quantum transition state theory" as a reformulation of instanton theory can be found in (Bromley et al., 2014).

7.6.2 Bimolecular Reactions With Intermediate Complex Formation

Numerous bimolecular reactions do not occur by direct collision mechanisms but rather pass in a two step process from reactants to products via a more or less long lived and possibly strongly bound intermediate complex according to the general scheme

$$A + BC \rightleftharpoons ABC^* \rightleftharpoons AB + C. \qquad (7.9)$$

One thus has in a first step a (re)combination or capture and in the second step a (re)dissociation, which is a unimolecular reaction. Indeed, by microscopic reversibility all these processes can be understood as unimolecular reaction dynamics of the intermediate complex ABC^*. The reactions are frequently very fast, occurring without barriers (or very small ones), when they involve free radicals such as the reaction (Farrar and Lee, 1974), Fig. 7.21

$$F + C_6H_5Cl \rightarrow C_6H_5FCl^* \rightarrow C_6H_5F + Cl, \qquad (7.10)$$

or (Quack and Troe, 1974, 1975a,b, 1976, 1977a,b,c; Glänzer et al., 1976, 1977; Quack, 1979)

$$H + CH_3 \rightarrow CH_4^* \rightarrow CH_3 + H, \qquad (7.11)$$

or when they involve ions such as the prototypical ion molecule reaction (and its isotopic variants)

$$H^+ + H_2 \rightarrow H_3^+ \rightarrow H_2 + H^+, \qquad (7.12)$$

which has no barrier for the "capture" process forming the strongly bound molecular ion H_3^+ with a deep minimum in the electronic potential, quite in contrast to the direct reaction (7.8) involving neutral H_3 which has a pronounced barrier. When two ions of opposite charge are concerned one has charge recombination quite specifically. Numerous reactions of this general type have been studied, both for radical and ion molecule collisions and we provide just a small selection of some early references (Herschbach, 1973; Farrar and Lee, 1974; Schultz et al., 1972; Rice, 1975; Chesnavich and Bowers, 1979; Herschbach, 1987; Lee, 1987; Polanyi, 1987). Fig. 7.21 shows a schematic energy diagram for an example of such reactions (Quack, 1980).

For barrierless reactions one might assume that tunnel effects are unimportant, except for a small quantum effect with a threshold law that slightly reduces the cross-section compared to the classical reaction cross-section (Quack and Troe, 1974, 1975a,b, 1981, 1998). However, in such reactions there are important effects from angular momentum conservation (Zare, 1988; Quack and Troe, 1975a, 1998). Thus in collisions with capture one has to take into account effective potentials, which have "centrifugal barriers" due to extra rotational energy to be added to the purely attractive electronic

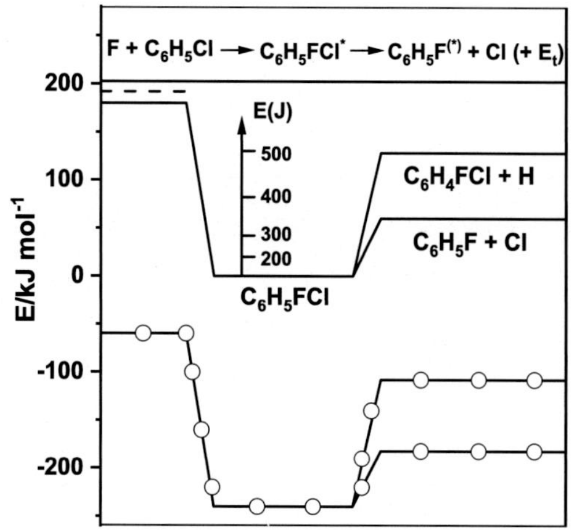

FIG. 7.21 Schematic energy diagram for a typical complex forming reaction. $E(J)$ gives the minimum rotational energy for a total angular momentum J of C_6H_5FCl as indicated by the numbers. The open circles represent the electronic energy without the zero point energy (after Quack, 1980).

potential. Tunneling through these effective potential barriers then becomes important at threshold as shown in Fig. 7.22.

These effective barriers for tunneling have been widely calculated using treatments where the relative rotation of the collision partners is considered to be free, with a quasidiatomic quasiclassical treatment for the centrifugal barriers, such as in phase space theory (Pechukas and Light, 1965; Nikitin, 1965), Gorin type models for radical recombinations (Gorin, 1938), or Langevin models for ion–molecule reactions (Gioumousis and Stevenson, 1958; Levine and Bernstein, 1989; Levine, 2005). We shall address here an important aspect for such effective tunneling potentials, which can be understood in the framework of the quasiadiabatic channel model (Quack and Troe, 1998). There the effective tunneling potential includes effects from the quantized energy of hindered relative rotation of the diatomic or polyatomic collision partners as illustrated with the potential labeled $V_{acm}(J, a)$ in Fig. 7.22. These zero point energy effects are quite important and have been calculated quantitatively in a numerically exact way by the quasiadiabatic channel Diffusion Quantum Monte Carlo (DQMC) method of Quack and Suhm (1991a) for the collisions

$$2HF \rightarrow (HF)_2 \qquad (7.13)$$

forming the hydrogen bonded complex $(HF)_2$. Fig. 7.23 shows such results for the lowest quasiadiabatic channel potentials with $J = 0$, where the effective barriers

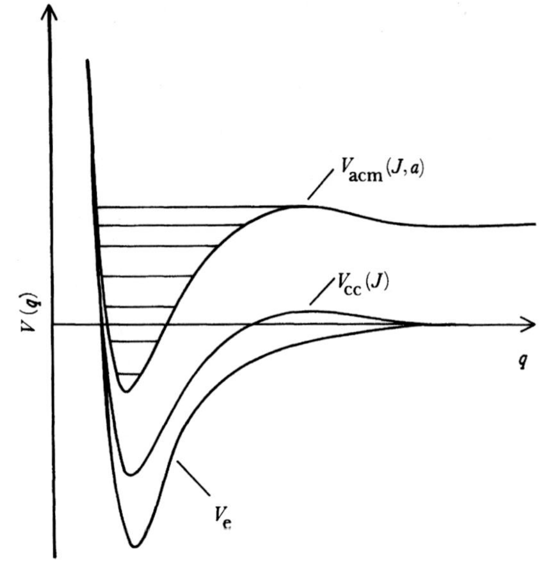

FIG. 7.22 Effect of zero point energy and angular momentum on polyatomic reaction dynamics as illustrated by the adiabatic channel model. V_e is the electronic (Born–Oppenheimer) potential in the reaction coordinate q, $V_{cc}(J)$ the effective, classical centrifugal potential (for a total angular momentum quantum number J, and $V_{acm}(J,a)$ the effective adiabatic channel potential including zero point energy for J and channel a (after Quack, 1990).

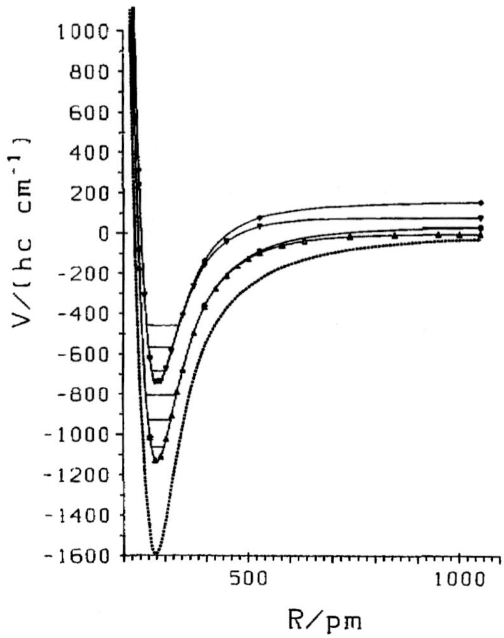

FIG. 7.23 Lowest quasiadiabatic DQMC dissociation channels of A^+ (△), A^- (▽), B^+ (○), B^- (◇) symmetry and 1D stretching levels for $(HF)_2$. Note that for the A^- and B^- channels, the symbol is about the size of the error bar, while it is twice the error bar for B^+ and 5–20 times the error bar for A^+, depending on $R_{ab'}$. The channel potentials V_a (R_{ab}) are referred to V_0 ($R_{ab} \to \infty$) $\equiv 0$. The electronic potential is shown as a dotted line, defining V_{el} ($R_{ab} \to \infty$) $\equiv 0$ in this case (after Quack and Suhm, 1991a).

are not visible, and Fig. 7.24 shows the lowest channels for higher total angular momenta (J = 10, 30, 60), where they are easily visible. These full-dimensional quasiadiabatic channel Quantum Monte Carlo calculations carried out on the full-dimensional SQSBDE potential hypersurface have demonstrated the great quantitative importance of the zero point energy effects. For instance, in the $J = 60$ channel the effective pure rotational energy barrier for capture occurs at $R'_{ab} = 15a_0$ with an energy $E'_B = (hc)33.6\ \text{cm}^{-1}$, whereas with inclusion of the correct quantum zero point energy by DQMC it occurs at $R'_{ab} = 11a_0$ with a much higher energy $E'_B = (hc)123\ \text{cm}^{-1}$. This demonstrates that clearly the full quantum dynamical treatment of the effective potentials must be carried out before doing calculations for capture in such systems with or without tunneling.

Notably, Troe and coworkers have applied the SACM (statistical adiabatic channel model), sometimes including further approximations to a variety of capture processes (Auzinsh et al., 2013; Troe, 1987, 1992, 1996; Nikitin and Troe, 2010a,b; Dashevskaya et al., 2003, 2010, 2011).

The fast reaction with intermediate complex formation have received further recent attention due to developments in studying reactions at very low temperatures (Smith, 2006; Hoshina et al., 2004; Ospelkaus et al., 2010; Perreault et al., 2017; Kilaj et al., 2018; Hall et al., 2013), see also Bergmann et al. (2019) for state selective preparation by STIRAP, and references therein. Besides the possible role of tunneling through small centrifugal barriers in such reactions, there are also interesting threshold effects due to symmetry, which we can illustrate again with the $(HF)_2$ system. Fig. 7.25 illustrates the symmetry correlations for the lowest quasiadiabatic channels (Quack and Suhm, 1991a). We can use again the group $S^*_{2,2}$ as discussed in Section 7.3 or the smaller group M_{S4} if the HF monomer units remain intact, because the barrier for exchanging protons between the different F-atom partners in the HF units is very high. One sees in Fig. 7.25 that the ground state tunneling doublet correlates with the ground state products A^+ and an excited vibrational state with B^+-symmetry and

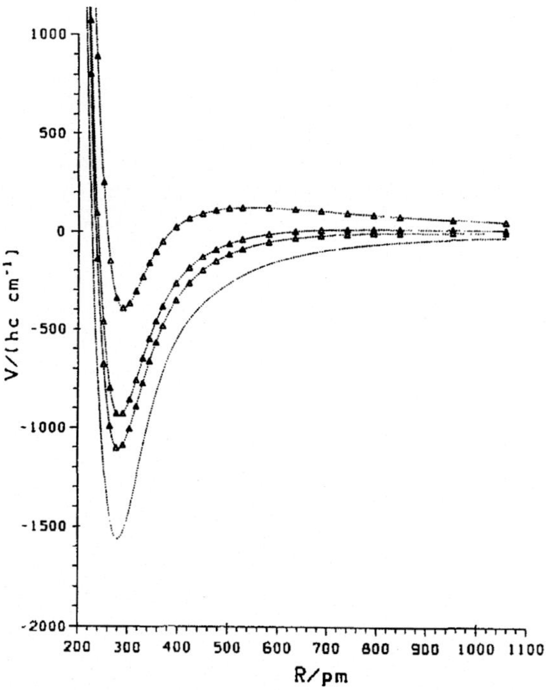

FIG. 7.24 Lowest quasiadiabatic DQMC dissociation channels for $J = 10, 30, 60$ together with the minimum energy path potential of dissociation (after Quack and Suhm, 1991a).

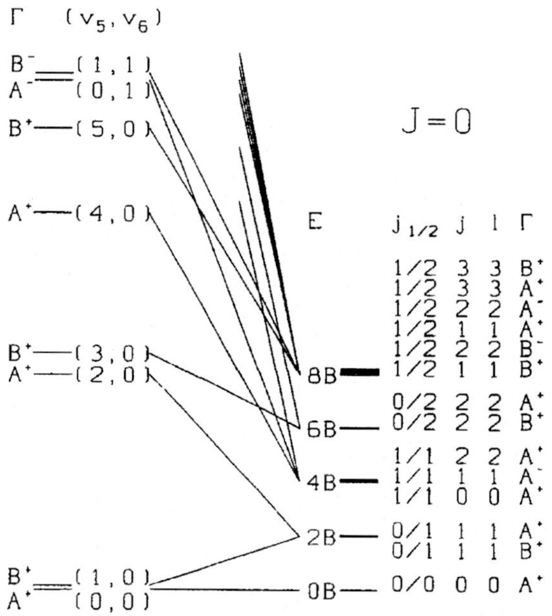

FIG. 7.25 Correlation diagram for the lowest rotationless states of $(HF)_2$ and the corresponding monomer fragments; $\upsilon_3 > \upsilon_6$ is assumed (after Quack and Suhm, 1991a).

an energy corresponding to 2B (with B being the rotational constant of HF). The lowest A^- channel involves excitation of the out of plane bending (vibrational, torsional) mode ν_6 neat 400 cm^{-1} in $(HF)_2$ and correlates at the product side with the lowest possible A^- state with energy corresponding to $4B$. This implies, for instance, that the excited levels of $(HF)_2$ with A^- symmetry and $J = 0$ cannot predissociate even if their energy is above the lowest threshold (energy = 0 in Fig. 7.25) but below $(hc*4B)$, unless one invokes parity violation (Quack, 2011b). These excited states would therefore be extremely long lived. The upper tunneling level of $\upsilon_6 = 1$ has B^- symmetry and correlates with a product B^- channel of energy $(hc*8B)$. Thus B^- excited levels (with $J = 0$) cannot predissociate above threshold unless they reach an energy $(hc * 8B)$ or more. In the energy range between $hc*4B$ to $hc*8B$ they could predissociate with violation of nuclear spin symmetry and below $hc * 4B$ again only with parity violation, thus being extremely long lived. Of course, these correlations also appear in the effective channel potentials in Fig. 7.24.

As a fully state selective version of transition state theory the statistical adiabatic channel model (SACM) allowed for proper explicit inclusion of all conserved quantum numbers, such as angular momentum, parity and nuclear spin symmetry, as also of other approximately conserved quantum numbers. At the same time it allowed for a consistent treatment of barrierless reactions by a kind of transition state approximation, a unique property at the time of its development, as the usual versions of transition state theory relied on a well defined transition structures at a saddle point (in the multidimensional hypersurface), which is absent in barrierless reactions. The latter feature was later included in the so-called "variational transition state theory" (Truhlar and Garrett, 1984; Truhlar et al., 1996), but which has been shown to be only an approximation to the full SACM (Quack and Troe, 1977c,a,b). Variational transition state theory generally gives somewhat too high values for the rate constant of barrierless reactions, although the results are much better than in the older versions of transition state theory or RRKM (Rice–Ramsperger–Kassel–Marcus) theory. As a early crucial experimental test we may mention the prediction of an increase of the barrierless methyl radical recombination rate by a factor of 20 by the then available RRKM calculations

$$CH_3 + CH_3 \rightarrow C_2H_6, \quad (7.14)$$

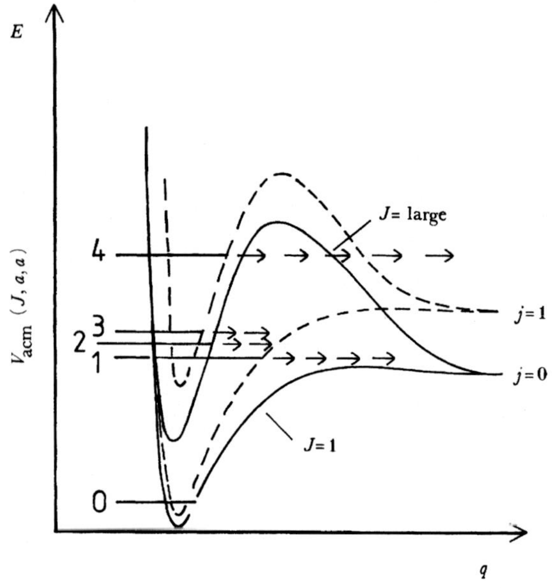

FIG. 7.26 Schematic drawing of adiabatic channel potentials for the predissociation of H_3^+ (see detailed discussion in the text, the electronic potential is omitted) (after Quack, 1990).

whereas the SACM predicts a constant or slightly decreasing rate coefficient over the range 300 to 1300 K. The SACM prediction (Quack and Troe, 1974) was confirmed by experiment (Glänzer et al., 1976, 1977).

7.7 TUNNELING IN IONS AND IN ELECTRONICALLY EXCITED STATES

We have already mentioned the prototypical ion H_3^+ in the context of barrierless recombination reactions. It can also be used as a prototype for studying tunneling and symmetry in unimolecular dissociation of ions. Indeed, its high resolution infrared predissociation spectrum has remained for decades without assignment and analysis (Carrington et al., 1982; Carrington and Kennedy, 1984). Symmetry and tunneling in this predissociation can be analyzed as illustrated in Fig. 7.26 (Quack, 1990).

For the symmetry analysis one can in practise use the group $S_3^* = S_3 \otimes S^*$ as the direct product of the symmetric group S_3 of the three protons and the inversion group S^*. We consider here the case of the totally symmetric electronic ground state. The $j = 1$ and $j = 0$ product rotational channels in

$$H_3^+ + h\nu \rightarrow (H_3^+)^* \rightarrow H_2(j) + H^+ \qquad (7.15)$$

are separated by about 120 cm^{-1} (1.4 kJ mol^{-1}) for a low total angular momentum channel with J = 1 with no or with nearly zero barrier, or a high angular momentum channel with large J and a large centrifugal barrier as indicated in Fig. 7.26.

Let us consider a predissociating state labeled (1) in Fig. 7.26. It may dissociate to give $j = 0$ product with an "orbital" angular momentum quantum number $l = 1$ for the relative motion of the fragments ($l = J = 1$ in this case with $j = 0$, i.e., para-H_2). This channel has negative parity and E^- symmetry with total nuclear spin I = 1/2, as A_1^- is Pauli forbidden. Thus if $|1\rangle$ is of A_2^+ symmetry, it can only predissociate by parity violation not by tunneling, thus it will be extremely long lived. The same conclusion is true for an E^+ level of H_3^+ at an energy of $|1\rangle$. On the other hand, an E^- state at that energy can predissociate by tunneling, the lifetime will then only depend on the tunneling probability and may be short enough to be found in predissociation. An A_2^- level at the energy of $|1\rangle$ could only predissociate by nuclear spin symmetry change, thus it would be long lived (but not as long lived as an A_2^+ level). More conclusions can be obtained in a similar way, see Quack (1990). One can reverse this reasoning: When detecting the product state of H_2 in reaction (7.15) (ortho or para) in conjunction with the predissociation spectral line, one can assign the symmetry of the predissociating state of H_3^+ observed through that line. We may note here that for the reaction

$$H_2^+ + H_2 \rightarrow H_3^+ + H, \qquad (7.16)$$

which proceeds without barrier through a very short lived H_4^+, the symmetry selection rules (Quack, 1977) were experimentally confirmed by Uy et al. (1997).

Tunneling processes are also of relevance in other positively charged ions where H_5^+ is an important example. The GENUISH program has been used to calculate the rotational-vibrational energy levels for H_5^+ and its deuterated modifications for a reduced and full dimensional model (Sarka and Császár, 2016). The "astructural" highly non-rigid ion CH_5^+ has found much interest as well (Fábri et al., 2017 and references therein). The calculations showed that energy separations of the rotational and vibrational levels are in each case on the same order of magnitude, reflecting the flatness of the potential hypersurface and therefore there is no good way to separate "rotational" from "vibrational" spectral structures, while, however, symmetries can be properly included following (Fábri et al., 2017).

Tunneling processes in the ground state of the ion had to be included to explain the ZEKE (zero-kinetic energy) spectra of propen-H_6 and -D_6 (Vasilatou et al.,

2010). A modified Hamiltonian for the Jahn–Teller effect has been derived to enable a joint treatment of the Jahn–Teller effect and the internal hindered rotation to explain the zero-kinetic energy photoelectron spectrum of C_2H_4 and C_2D_4 (Schulenburg and Merkt, 2010). Threshold photoelectron–photonion coincidence spectra on the two isotopic modifications of dimethyl ether, (CH_3OCH_3, DME-$H_6)_n$ and (CD_3OCD_3, DME-$D_6)_n$, $n = 1$ and 2 did not show any isotope effect. The assumption of a barrierless proton transfer across weak CH$\cdots$O hydrogen bonds was supported by DFT calculations (Yoder et al., 2015). The tunneling process of the pseudo rotation in the $1^2E'$ ground state of Na_3, a classical example for the Jahn–Teller effect, has been revisited in a recent publication (Hauser et al., 2015). The ground state of Na_3 is defined by the interplay of Jahn–Teller-, spin-orbit-, rovibrational and hyperfine interactions. In the investigation the rovibrational levels, theoretically obtained from an effective Hamiltonian have been compared to the results from highly resolved microwave spectra (Hauser et al., 2015).

Rotational and vibrational predissociation is a prominent tunneling process in the electronically excited states of small molecules and ions (Herzberg, 1966; Merkt and Quack, 2011; Merkt et al., 2011; Wörner and Merkt, 2011; Amano, 2011; Pratt, 2011; Schmitt and Meerts, 2011; Guennoun and Maier, 2011). In the vacuum ultraviolet spectrum of the $C_1 \leftarrow X_0^+$ and $D_0^+ \leftarrow X_0^+$ band system of ArXe a second state of $\Omega = 1$ symmetry was assumed to explain several tunneling/predissociation resonances found $\sim$200 cm^{-1} above the C_1-state (Pittico et al., 2012). For the benzoic acid dimer a tunneling time of 111 ps is found experimentally for the double proton exchange in the electronic ground state and of 666 ps in the electronically excited state. In the electronically exited state the predicted excitonic splitting of the S_1- and S_2-state is of the same order of magnitude. For the electronically excited states an exciton splitting between the S_1- and S_2-state is predicted theoretically. However, an exciton splitting of 0.94 cm^{-1} is found experimentally, significantly smaller than predicted (Ottinger and Leutwyler, 2012; Ottinger et al., 2012), see also the reviews of Willitsch (2011) and Müller-Dethlefs and Riese (2011) for further experimental studies of ions and predissociation by tunneling.

At this point we should also mention an alternative theoretical approach for treating proton dynamics in polyatomic ions, not together with the other nuclei on a Born–Oppenheimer potential hypersurface but rather together with the electrons as "light" particles. This interesting approach was named nuclear-electron orbital (NEO) approach by Sharon Hammes-Schiffer and developed for a number of applications which automatically should include "quantum tunneling" of the protons implicitly as also other quantum effects related to proton motion, see Yang et al. (2017); Brorsen et al. (2017).

7.8 TUNNELING OF MOLECULES INSIDE A CAGE

The quantum dynamics of molecules in cages have found much recent interest in the context of metal organic frameworks (MOFs) (Yaghi, 2016; Yaghi et al., 2019; Yaghi, 2019; Zhou et al., 2016; Schoedel et al., 2016; Rungtaweevoranit et al., 2016; Nguyen et al., 2014; Kundu et al., 2016) and Zeolites, for example, (Sauer, 2016; Tuma and Sauer, 2015; Piccini et al., 2018), which are important examples for heterogeneous catalysis often involving presumably tunneling reactions (Sauer and Freund, 2015).

The inside volumes of C_{60} and other fullerenes provide an ideal environment for the quantum dynamics of coupled translational and rotational motions of H_2 and similar molecules which are accessible to inelastic neutron scattering. Using a 5D translational-rotational Hamiltonian, the eigenstates could be obtained to enable the assignment of the INS (inelastic neutron scattering) spectrum for different temperatures. From the assigned INS spectra a special selection rule could be derived for INS spectra of the motion of H_2 in a near-spherical potential (Xu et al., 2013, 2019; Felker et al., 2019; Lauvergnat et al., 2019). The new selection rule is also found for HF in C_{60} (Xu et al., 2019). With the same theoretical approach the 6D translational-rotational (TR) eigenstates of para- and ortho-H_2O inside C_{60} could be calculated (Felker and Bačić, 2016). The nuclear spin symmetry conversion between the ortho and para species of the endohedral H_2O inside C_{60} was studied in the solid phase by low temperature NMR. The experimental data are consistent with a second-order kinetics, indicating a bimolecular spin conversion process. Numerical simulations showed a spin diffusion process allowing neighboring ortho and para molecules to exchange their angular momenta (Mamone et al., 2014). This should be compared to nuclear spin symmetry conservation in H_2O in supersonic jet expansions (Manca Tanner et al., 2013, 2018), where nuclear spin symmetry conversion may occur with cluster formation.

Hydrogen clathrate hydrates are inclusion compounds where hydrogen molecules are trapped inside closely packed polyhedral cavities within the ice-like lattice of hydrogen-bonded water molecules. The translational-rotational eigenstates and INS spectra were calculated by a quantum-mechanical method for H_2 inside spherical clathrate domains of gradually increasing radius and increasing number of water molecules (Powers et al., 2016). In an additional study the diffusion of H_2 and D_2 through clathrate hydrates was investigated. Path integral based molecular dynamics simulations were used to calculate the free-energy profiles for the H_2 and D_2 diffusion rates as a function of temperature. For temperatures above 25 K, zero point energies of vibrations perpendicular to the reaction path lead to a decrease of the quantum mechanical (calculated) rates compared to classically calculated rates. For low temperatures, tunneling becomes dominant (Cendagorta et al., 2016). Using neutron scattering and ab initio calculations a new "quantum tunneling state" of the water molecule confined in 500 pm wide channels in the mineral beryl has been discovered, which is characterized by extended proton and electron delocalization. The maxima in the neutron scattering data could be uniquely assigned to a new tunneling state of water (Kolesnikov et al., 2016).

Metal–organic frameworks (MOFs) represent a class of porous materials that have been considered to be promising for applications in H_2 storage. The inelastic scattering of neutrons from adsorbed H_2 is an ideal method for obtaining information on the type and nature of H_2 binding sites in porous materials. Energy levels and transition energies may be obtained from quantum dynamics calculations with an accuracy limited by the quality of the potential energy surface and can be directly compared to the experiment. In a recent review the results from some quantum and classical mechanical calculations have been summarized (Pham et al., 2016b). An example for such a porous material is $[Mg_3(O_2CH)_6]$, a metal–organic framework (MOF) that consists of a network of Mg_2^+-ions coordinated to formate ligands. In simulations and two-dimensional quantum calculations of the rotational levels it was found that different groups of maxima in the inelastic neutron scattering experiments correspond to particular adsorption sites in the material. In addition, 8 different rotational tunneling transitions could be identified experimentally and theoretically as well and their barrier height could be determined (Pham et al., 2016a).

7.9 CONCLUDING REMARKS ON TUNNELING

In many theories and mechanistic discussions in chemistry atomic and molecular tunneling processes are neglected even though chemical examples had been known from the beginning of the discovery of the tunnel effect in the early times of quantum mechanics (Hund, 1927a; Herzberg, 1939, 1945; Bell, 1980; Quack and Merkt, 2011). The present survey shows that just now in recent years a large number of phenomena in chemistry based on the tunnel effect with contributions from atoms or atomic groups in molecules and not only from electrons have been discovered and understood in detail. Here, for the future still important applications have to be expected, and we can highlight as just one example the possible use of tunneling in the design of molecular quantum switches and quantum machines (Fábri et al., 2018), discussed here in Sect 7.4. Numerous further examples of quantum atomic and molecular tunneling systems in molecular kinetics and spectroscopy have been illustrated in the present review. These provide a view of "molecules in motion" (Quack, 2001) from the true quantum perspective of the tunnel effect. By necessity a review of such a vast field cannot provide a comprehensive coverage of the literature, and we extend our apologies to those authors whose work may have been less than adequately covered in our selection of examples of recent interest and developments in the field.

To conclude we shall address a few conceptual points, which we have until now omitted but which are of some current relevance. The first concerns a remark on nomenclature for "tunneling sublevels". These are often labeled as 0^+, 0^-, 1^+, 1^-, v_n^+, v_n^-, where the number v_n give some vibrational (e.g., torsional, inversional, etc.) quantum number in the "high barrier" limit whereas the exponent "+" indicates the lower and "−" the upper tunneling sublevel. This widely used nomenclature is related to the intuitive notion that the lower sublevel is often symmetric and the upper sublevel antisymmetric with respect to some symmetry operation. This type of symmetry assignment is, however, not generally true and it obviously fails in slightly asymmetric potentials where at low energies the wavefunctions may be localized, whereas at higher energies one may have, indeed, delocalized "tunneling sublevels". A more general useful label would be simply "l" for the "lower" and "u" for the "upper" sublevel, which is a neutral nomenclature leaving the question of symmetry and localization and delocalization open and perhaps to be specified by some further nomenclature. Indeed, when the tunneling sublevels arise from a process related to the concept of "feasibility" according to Longuet-Higgins (1963), one can then use a symmetry label from the appropriate symmetry group. There has been a somewhat unfortunate multitude of conventions and nomenclatures in this field and we should strongly advocate the use of symmetry labels which give a unique symbol for symmetries in the restricted permutation group (A for symmetric, B for antisymmetric with indices A_i, B_i as may be needed, E for doubly degenerate, F for triply degenerate species, and then with increasing degeneracy systematically G, H, etc.) or else the partition in the full permutation group S_n as is conventional in mathematics. The parity of the species can be indicated by an exponent "+" for positive parity and "−" for negative parity. This notation was systematically introduced in Quack (1977) and has been recommended in Cohen et al. (2011) and Stohner and Quack (2011) for example. It differs sometimes from the various less systematic conventions used also in Bunker and Jensen (1998), which is among the frequently used sources for conventions, but there are also quite a few further ones (for instance, Oka, 2011, similar but not identical to our recommended one). A critical discussion and summary has been given in Quack (2011b), to which we refer for further details, and which also leads to the recommendation we give here (see also for some of the history and further references Mills and Quack, 2002). Obviously all this is a matter of convention and not fundamental. In principle any convention and nomenclature can be used, if it is well defined. Nevertheless, a good systematic nomenclature can be helpful in scientific work.

There is a second point of current interest, which, in fact, has led to some debate: The notion of the "tunneling time". This particular time is sometimes (not always) defined as the "time the particle spends in the classically forbidden region under the barrier during the tunneling process". We shall not enter here the details of the debate around the question of the "tunneling time", but we refer to some of the key references from which the reader can obtain a perspective on this debate (MacColl, 1932; Hauge and Stovneng, 1989; Landauer and Martin, 1994; Lépine et al., 2014; Dahlström et al., 2012; Telle et al., 1999; Gallmann and Keller, 2011; Gallmann et al., 2012; Landsman et al., 2014; Landsman and Keller, 2014, 2015; Cattaneo et al., 2016; Hofmann et al., 2019; Pollak, 2017; Petersen and Pollak, 2017, 2018; Wörner and Corkum, 2011). The discussions have sometimes led to considerations of fundamental physics extending into cosmology (Nimtz and Haibel, 2004; Hawking, 1975; Vilenkin, 1988; Low, 1998; Nimtz, 2011). We shall not dig deeply into these considerations here but conclude with the following remark relevant to "heavy particle" tunneling in molecules in the framework of potential barriers arising in the framework of Born–Oppenheimer or related potential hypersurfaces. In the limit where the non-relativistic dynamics as described by the time-dependent Schrödinger equation can be used ($i = \sqrt{-1}$),

$$i\frac{h}{2\pi}\frac{\partial\Psi(\vec{q},t)}{\partial t} = \hat{H}\,\Psi(\vec{q},t), \tag{7.17}$$

one can certainly state that experimentally observable quantities can be calculated for any tunneling process, and thus if one can define some kind of "tunneling time" which can be measured experimentally, then one can also compute it by solving the Schrödinger equation and thereby "understand" it, in that sense there is no real enigma. Similar statements hold true if we use the Heisenberg picture with the Heisenberg equations of motion or considering the time evolution of the density operator according to the Liouville–von Neumann equation (Merkt and Quack, 2011; Marquardt and Quack, 2020, Chapter 1).

Finally, we shall address a fundamental aspect related to the concept of the "potential function", which appears in the usual discussions of tunneling. Indeed, at first sight it may seem that the concept of "tunneling" necessarily involves the existence of such a potential function. A closer look shows, however, that things are not quite as simple. We shall start here by considering tunneling within the quasiadiabatic channel reaction path Hamiltonian and "quasiadiabatic above barrier tunneling". The lowest levels and wavefunctions shown in Fig. 7.3 show a tunneling sublevel structure and tunneling wavepacket dynamics as given by the effective one-dimensional quasiadiabatic channel potential, although their energy is high above the barrier in the multidimensional Born–Oppenheimer surface. One might then be tempted to say that it is not a "true" tunneling process, as it results from an approximate model. However, the "exact" 6-dimensional vibration-tunneling calculation on the complete hypersurface shows about the same level structure and wavepacket dynamics as the approximate calculations, and therefore one must conclude that the full quantum dynamics shows "tunneling behavior" even above the barrier. It is furthermore true that also the tunneling sublevel structure and dynamics at energies below the 6-dimensional saddle point result from an approximation. Without the Born–Oppenheimer approximation (see also Marquardt and Quack, 2020; Császár et al., 2020, Chapters 1 and 2) the "Born–Oppenheimer potential barriers and saddle points" disappear. If one treats hydrogenperoxide or other molecules with Born–Oppenheimer barriers "exactly" by means of quantum dynamics of a collection of the electrons and nuclei in the molecule, the "true" potential arises in essence from the Coulomb-interaction of these particles in a very high dimensional space. But even then without any Born–Oppenheimer barriers to tunnel through, the "exact" quantum dynamics will show very much the same sublevel structure and tunneling dynamics as it is described by the approximate theory and confirmed by experiment. This suggests a new definition of quantum tunneling dynamics that does not depend on the concept of potential barriers, as we have pointed out already in the context of the definition of the "molecular symmetry group" of non-rigid molecules. While Longuet-Higgins has, indeed, motivated this group by the existence of high Born–Oppenheimer potential barriers separating symmetrically equivalent isomers and thereby generating systematically degenerate level structures, we pointed out that in a rigorous discussion the symmetry groups should not be defined by an approximation but rather by the induced representation corresponding to the degenerate sublevels, which one can define from an exact theory without any approximation and also by experiment (Quack, 1977, 1985, 2011b). When the perfect degeneracy is lifted by tunneling leading to observable tunneling splittings, one can consider this effect then as the breaking of an approximate symmetry without having to refer to tunneling through potential barriers from the Born–Oppenheimer approximation. Thus the tunnel effect can be understood as a quantum dynamical phenomenon without making reference to approximate concepts such as Born–Oppenheimer potential hypersurfaces. One can say, however, that the Born–Oppenheimer approximation and also the quasiadiabatic channel reaction path Hamiltonian with their effective potentials provide us with simple models (Quack, 2014), which allow us in the first place to qualitatively understand the phenomena and in the second place also compute the phenomena approximately without too much effort (see also the general discussion by Roald Hoffmann (1998) on qualitative understanding versus computation).

The discussion can be continued similarly at an even deeper level: From the standard model of particle physics we understand even the Coulomb potential as not being a "fundamental preexisting" potential but rather as arising from photons as field particles mediating the electromagnetic interaction (Quack, 2006; Marquardt and Quack, 2020, Chapter 1). Similarly the parity violating interaction leading to the slight asymmetry effect in the effective potential as illustrated in Fig. 7.5 arises from the Z-bosons as field particles mediating the weak interaction between electrons and nuclei (or protons, neutrons, quarks, etc.). The small effective "parity violating potential" (really an extra effective potential hypersurface which is antisymmetric with respect to inversion and should not at all be interpreted as a "Born–Oppenheimer hypersurface") arises from an approximation in carrying out the computation in electroweak quantum chemistry (Bakasov et al., 1998; Berger and Quack, 2000a,b; Quack and Stohner, 2000; Quack, 2006, 2011b). While there would be no need to make the approximation, it nevertheless corresponds to a useful model which allows us to understand the small symmetry violation and to practically compute it with feasible effort. Conceptually, however, exact tunneling dynamics in this effective asymmetric potential should be understood on the basis of the quantum sublevel structure arising from the symmetry breaking removal of a degeneracy, which can be observed by experiment or could also be derived from an exact theory without making any reference to the parity violating potential hypersurface.

A brief note is also useful concerning the role of relativistic effects. As is well known, these are important for the dynamics of electron motion, whenever the heavier elements are involved in the molecules considered. These effects can be calculated by relativistic quantum chemistry (Reiher and Wolf, 2009) and can lead, indeed, to dramatic changes in the effective Born–Oppenheimer potential barriers for tunneling. Once these effects are included, the tunneling motion of atoms and molecules can be computed and understood in very much the same way using the Schrödinger equation as discussed for non-relativistic potentials. If the molecules move at relativistic speeds, one has to consider the changes in the definition of time which is then to be measured by an atomic clock moving at relativistic speed. Indeed, molecular tunneling systems such as ammonia can be and have been used as molecular clocks, and one has the well understood (and, in fact, experimentally observed) relativistic effects, such as an atomic and molecular "twin paradox" (Einstein, 1922). Further considerations arise when considering violations of time reversal symmetry and possibly a hypothetical violation of CPT symmetry and we refer to Quack (1999, 2003, 2004, 2006, 2011b,a) where one can also find a discussion of the "42 open problems", some of which are related to tunneling.

ACKNOWLEDGMENTS

We gratefully acknowledge support, help from and discussions with Ziqiŭ Chen, Csaba Fábri, Roberto Marquardt, Frédéric Merkt, Eduard Miloglyadov, Robert Prentner, Jürgen Stohner, Martin Willeke, and Gunther Wichmann as well as financial support from ETH

Zürich, the laboratory of Physical Chemistry and an Advanced Grant of the European Research Council ERC as well as the COST Project MOLIM. Many further coworkers have contributed to our work (see the lists in Quack, 2003; Ernst et al., 2013).

REFERENCES

Agapov, A.L., Kolesnikov, A.I., Novikov, V.N., Richert, R., Sokolov, A.P., 2015. Quantum effects in the dynamics of deeply supercooled water. Phys. Rev. E 91, 022312.

Al Derzi, A.R., Furtenbacher, T., Tennyson, J., Yurchenko, S.N., Császár, A.G., 2015. MARVEL analysis of the measured high-resolutions spectra of $^{14}NH_3$. J. Quant. Sepctrosc. Rad. Transf. 161, 117–130.

Al-Shamery, K., 2011. Moleküle aus dem All? Wiley-VCH, Weinheim.

Albert, S., Arn, F., Bolotova, I., Chen, Z., Fábri, C., Lerch, P., Quack, M., Seyfang, G., Wokaun, A., Zindel, D., 2016a. Synchrotron-based highest resolution terahertz spectroscopy of the ν_{24} band system of 1, 2-Dithiine ($C_4H_4S_2$): a candidate for measuring the parity violating energy difference between enantiomers of chiral molecules. J. Phys. Chem. Lett. 7, 3847–3853.

Albert, S., Bolotova, I., Chen, Z., Fábri, C., Horný, L., Quack, M., Seyfang, G., Zindel, D., 2016b. High resolution GHz and THz (FTIR) spectroscopy and theory of parity violation and tunneling for 1, 2-Dithiine ($C_4H_4S_2$) as a candidate for measuring the parity violating energy difference between enantiomers of chiral molecules. Phys. Chem. Chem. Phys. 18, 21976–21993.

Albert, S., Bolotova, I., Chen, Z., Fábri, C., Quack, M., Seyfang, G., Zindel, D., 2017. High-resolution FTIR spectroscopy of trisulfane HSSSH: a candidate for detecting parity violation in chiral molecules. Phys. Chem. Chem. Phys. 19 (19), 11738–11743.

Albert, S., Chen, Z., Fábri, C., Lerch, P., Prentner, R., Quack, M., 2016c. A combined Gigahertz and Terahertz (FTIR) spectroscopic investigation of meta-D-phenol: observation of tunnelling switching. Mol. Phys. 19, 2751–2768.

Albert, S., Keppler Albert, K., Quack, M., 2011. High resolution Fourier transform infrared spectroscopy. In: Quack, M., Merkt, F. (Eds.), Handbook of High Resolution Spectroscopy, Vol. 2. Wiley, Chichester, New York, pp. 965–1019 (Chapter 26).

Albert, S., Lerch, P., Prentner, R., Quack, M., 2013. Tunneling and tunneling switching dynamics in phenol and its isotopomers from high-resolution FTIR spectroscopy with synchrotron radiation. Angew. Chem., Int. Ed. 52 (1), 346–349.

Albert, S., Lerch, P., Quack, M., 2016d. Tunneling dynamics of aniline. In: Stohner, J., Yeretzian, C. (Eds.), Proceedings of the 19th Symposium on Atomic, Cluster and Surface Physics 2016 (SASP 2014). Davos, Switzerland. Innsbruck University Press (IUP), Innsbruck, pp. 169–177.

Allemann, R.K., Scrutton, N.S., 2009. Quantum Tunnelling in Enzyme-Catalysed Reactions. RSC Publishing, Cambridge.

Althorpe, S.C., Fernández-Alonso, F., Bean, B.D., Ayers, J.D., Pomerantz, A.E., Zare, R.N., Wrede, E., 2002. Observation and interpretation of a time-delayed mechanism in the hydrogen exchange reaction. Nature 410, 67–70.

Amano, T., 2011. High-resolution microwave and infrared spectroscopy of molecular cations. In: Quack, M., Merkt, F. (Eds.), Handbook of High-Resolution Spectroscopy, Vol. 2. Wiley, Chichester, New York, pp. 1267–1290 (Chapter 33).

Ansari, N., Meyer, H.-D., 2016. Isotope effects of ground and lowest lying vibrational states of $H_{3-x}D_xO_2^-$ complexes. J. Chem. Phys. 144, 054308.

Aoiz, F.J., Aldegunde, J., Herrero, V.J., Sáez-Rábanos, V., 2014. Comparative dynamics of the two channels of the reaction of D + MuH. Phys. Chem. Chem. Phys. 16, 9808–9818.

Aoiz, F.J., Banares, L., Castillo, J.F., Sokolovski, D., Fernández-Alonso, F., Bean, B.D., Ayers, J.D., Pomerantz, A.E., Zare, R.N., 2002. Observation of scattering resonances in the H + D_2 reaction: direct probe of the HD_2 transition-state geometry. In: Douhal, A., Santamaria, J. (Eds.), Femtochemistry and Femtobiology - Ultrafast Dynamics in Molecular Science. World Scientific, Singapore, pp. 61–72.

Aoiz, F.J., Zare, R.N., 2018. Quantum interference in chemical reactions. Phys. Today 71 (2), 70–71.

Aquilanti, V., Mundim, K.C., Cavalli, S., De Fazio, D., Aguilar, A., Lucas, J.M., 2012. Exact activation energies and phenomenological description of quantum tunneling for model potential energy surfaces. The F + H_2 reaction at low temperature. Chem. Phys. 398, 186–191.

Araujo-Andrade, C., Reva, I., Fausto, R., 2014. Tetrazole acetic acid: tautomers, conformers, and isomerization. J. Chem. Phys. 140, 064306.

Arigoni, D., Eliel, E.L., 1969. Chirality due to the presence of hydrogen isotopes at noncyclic positions. In: Allinger, N.L., Eliel, E.L. (Eds.), Topics in Stereochemistry, Vol. 4. John Wiley, Chichester, New York, pp. 127–243.

Asselin, P., Soulard, P., Madebène, B., Goubet, M., Huet, T.R., Georges, R., Pirali, O., Roye, P., 2014. The cyclic ground state structure of the HF trimer revealed by far infrared jet-cooled Fourier transform spectroscopy. Phys. Chem. Chem. Phys. 16, 4797–4806.

Ausfelder, F., Pomerantz, A.E., Zare, R.N., Althorpe, S.C., Aoiz, F.J., Banares, L., Castillo, J.F., 2004. Collision energy dependence of the HD(v = 2) product rotational distribution of the H + D_2 reaction in the range 1.30–1.89 eV. J. Chem. Phys. 120, 3255–3264.

Auzinsh, M., Dashevskaya, E.I., Nikitin, E.E., Troe, J., 2013. Quantum capture of charged particles by rapidly rotating symmetric top molecules with small dipole moments: analytical comparison of the flywheel and adiabatic channel limits. Mol. Phys. 111, 2003–2011.

Ayers, J.D., Pomerantz, A.E., Fernández-Alonso, F., Ausfelder, F., Bean, B.D., Zare, R.N., 2003. Measurement of the cross section for H + D_2 → HD(v′=3, J′=0) + D as a function of angle and energy. J. Chem. Phys. 119, 4662–4670.

Bačić, Z., Light, J.C., 1989. Theoretical methods for rovibrational states of floppy molecules. Annu. Rev. Phys. Chem. 40, 469–498.

Bakasov, A., Ha, T.K., Quack, M., 1996. Ab initio calculation of molecular energies including parity violating interactions. In: Chela-Flores, J., Raulin, F. (Eds.), Chemical Evolution, Physics of the Origin and Evolution of Life. Proc. of the 4th Trieste Conference (1995). Kluwer Academic Publishers, Dordrecht, pp. 287–296.

Bakasov, A., Ha, T.K., Quack, M., 1998. Ab initio calculation of molecular energies including parity violating interactions. J. Chem. Phys. 109 (17), 7263–7285.

Bakule, P., Fleming, D.G., Sukhorukov, O., Ishida, K., Pratt, F., Momose, T., Torikai, E., Mielke, S.L., Garrett, B.C., Peterson, K.A., Schatz, G.C., Truhlar, D.G., 2012. State-selected reaction of muonium with vibrationally excited H_2. J. Phys. Chem. Lett. 3, 2755–2760.

Barclay, A.J., McKellar, A.R.W., Moazzen-Ahmadi, N., 2019. Spectra of the D_2O dimer in the O-D fundamental stretch region: vibrational dependence of tunneling splittings and lifetimes. J. Chem. Phys. 150, 164307.

Barlow, M.J., Clough, S., Horsewill, A.J., Mohammed, M.A., 1992. Rotational frequencies of methyl group tunneling. Solid State Nucl. Magn. Reson. 1, 197–204.

Bauder, A., 2011. Fundamentals of rotational spectroscopy. In: Quack, M., Merkt, F. (Eds.), Handbook of High-Resolution Spectroscopy, Vol. 1. Wiley, Chichester, New York, pp. 57–116 (Chapter 2).

Bauder, A., 2013. Microwave spectrum of formic acetic anhydride. Mol. Phys. 111, 1999–2002.

Beckmann, P., Clough, S., Hennelf, J.W., Hill, J.R., 1977. The Haupt effect: coupled rotational and dipolar relaxation of methyl groups. J. Phys. C, Solid State Phys. 10, 729–742.

Beil, A., Luckhaus, D., Quack, M., Stohner, J., 1997. Intramolecular vibrational redistribution and unimolecular reaction: concepts and new results on the femtosecond dynamics and statistics in CHBrClF. Ber. Bunsenges. Phys. Chem. 101 (3), 311–328.

Bell, R.P., 1933. The application of quantum mechanics to chemical kinetics. Proc. Roy. Soc. A 139, 466–474.

Bell, R.P., 1980. The Tunnel Effect in Chemistry. Chapman and Hall, London.

Benderski, V.A., Makarov, D.E., Wight, C., 1994. Chemical dynamics at low temperatures. Adv. Chem. Phys. 88, 1–385.

Berger, R., Gottselig, M., Quack, M., Willeke, M., 2001. Parity violation dominates the dynamics of chirality in dichlorodisulfane. Angew. Chem., Int. Ed. 40 (22), 4195–4198; Angew. Chem. 113, 2001 4342–4345.

Berger, R., Laubender, G., Quack, M., Sieben, A., Stohner, J., Willeke, M., 2005. Isotopic chirality and molecular parity violation. Angew. Chem., Int. Ed. 44 (23), 3623–3626; Angew. Chem. 117, 2005 3689–3693.

Berger, R., Quack, M., 2000a. Multi-configuration linear response approach to the calculation of parity violating potentials in polyatomic molecules. J. Chem. Phys. 112, 3148–3158.

Berger, R., Quack, M., 2000b. Electroweak quantum chemistry of alanine: parity violation in gas and condensed phases. Chem. Phys. Chem. 1 (1), 57–60.

Berger, R., Quack, M., Sieben, A., Willeke, M., 2003. Parity-violating potentials for the torsional motion of methanol (CH_3OH) and its isotopomers CD_3OH, $^{13}CH_3OH$, CH_3OD, CH_3OT, CHD_2OH, and CHDTOH. Helv. Chim. Acta 86 (12), 4048–4060.

Bergmann, K., Nägerl, H.C., Panda, C., Gabrielse, G., Miloglyadov, E., Quack, M., Seyfang, G., Wichmann, G., Ospelkaus, S., Kuhn, A., Longhi, S., Szameit, A., Pirro, P., Hillebrands, B., Zhu, X.F., Zhu, J., Drewsen, M., Hensinger, W.K., Weidt, S., Halfmann, T., Wang, H.L., Paraoanu, G.S., Vitanov, N.V., Mompart, J., Busch, T., Barnum, T.J., Grimes, D.D., Field, R.W., Raizen, M.G., Narevicius, E., Auzinsh, M., Budker, D., Pálffy, A., Keitel, C.H., 2019. Roadmap on STIRAP applications. J. Phys. B: At. Mol. Phys. 52, 202001.

Blackmond, D.G., 2004. Asymmetric autocatalysis and its implications for the origin of homochirality. Proc. Nat. Acad. Sci. 101, 5732–5736.

Blake, T.A., Xantheas, S.S., 2006. Structure, vibrational spectrum, and ring puckering barrier of cyclobutane. J. Phys. Chem. A 110, 10487–10494.

Blumberger, J., Ha, T.K., Paff, J., Quack, M., Seyfang, G., 2000. In: Proceedings 12th SASP. Folgaria, Trento 2000, pp. PB-2, 1-4.

Böckmann, H., Liu, S., Mielke, J., Gawinkowski, S., Waluk, J., Grill, L., Wolf, M., Kumagai, T., 2016. Direct observation of photoinduced tautomerization in single molecules at a metal surface. Nano Lett. 16, 1034–1041.

Born, M., Weisskopf, V., 1931. Quantenmechanik der Adsorptionskatalyse. Z. Phys. Chem. B 12, 206–227.

Börner, K., Diezemann, G., Rössler, E., Vieth, H.M., 1991. Low-temperature methyl group dynamics of hexamethylbenzene in crystalline and glassy matrices as studied by ^2H NMR. Chem. Phys. Lett. 181, 563–568.

Boyarkin, O.V., Rizzo, T.R., Perry, D.S., 1999. Intramolecular energy transfer in highly vibrationally excited methanol. III. Rotational and torsional analysis. J. Chem. Phys. 110, 11359–11367.

Bredtmann, T., Diestler, D.J., Li, S.-D., Manz, J., Pérez-Torres, J.F., Tian, W.-J., Wu, Y.-B., Yangaf, Y., Zhaie, H.-J., 2015. Quantum theory of concerted electronic and nuclear fluxes associated with adiabatic intramolecular processes. Phys. Chem. Chem. Phys. 17, 29421–29464.

Bredtmann, T., Manz, J., Zhao, J.-M., 2016. Concerted electronic and nuclear fluxes during coherent tunneling in asymmetric double-well potentials. J. Phys. Chem. A 120, 3142–3154.

Breidung, J., Thiel, W., 2011. Prediction of vibrational spectra from ab initio theory. In: Quack, M., Merkt, F. (Eds.), Handbook of High-Resolution Spectroscopy, Vol. 1. Wiley, Chichester, New York, pp. 389–404 (Chapter 8).

Brickmann, J., Zimmermann, H., 1968. Zur Theorie des Tunneleffekts eines Teilchens im Doppelminimumpotential. Z. Naturforsch. A 23, 11–18.

Brickmann, J., Zimmermann, H., 1969. Lingering time of the proton in the wells of the double-minimum potential of hydrogen bonds. J. Chem. Phys. 50, 1608–1618.

Brillouin, M.L., 1926. Remarques sur la mécanique ondulatoire. J. Phys. Radium 7, 353–368.

Bromley, S.T., Goumans, T.P.M., Herbst, E., Jonese, A.P., Slater, B., 2014. Challenges in modelling the reaction chemistry of interstellar dust. Phys. Chem. Chem. Phys. 16, 18623–18643.

Brorsen, K.R., Yang, Y., Hammes-Schiffer, S., 2017. Multicomponent density functional theory: impact of nuclear quantum effects on proton affinities and geometries. J. Phys. Chem. Lett. 8, 3488–3493.

Buchenau, H., Toennies, J.P., Arnold, J., Wolfrum, J., 1990. H + H_2: the current status. Ber. Bunsenges. Phys. Chem. 94, 1231–1248.

Buchowiecki, M., Vaníček, J., 2010. Direct evaluation of the temperature dependence of the rate constant based on the quantum instanton approximation. J. Mol. Model. 16, 1779–1787.

Buekenhoudt, A., Vandemaele, G., Van Gerven, L., 1990. Spin conversion of tunneling CH_3 rotors in copper acetate. Phys. Rev. B 41, 9038–9044.

Bunker, P.R., Jensen, P., 1998. Molecular Symmetry and Spectroscopy. NRC Press, Ottawa.

Caminati, W., 2011. Microwave spectroscopy of large molecules and molecular complexes. In: Quack, M., Merkt, F. (Eds.), Handbook of High-Resolution Spectroscopy, Vol. 2. Wiley, Chichester, New York, pp. 829–852 (Chapter 21).

Car, R., Parrinello, M., 1985. Unified approach for molecular dynamics and density-functional theory. Phys. Rev. Lett. 55, 2471–2474.

Carrington Jr., T., 2011. Using iterative methods to compute vibrational spectra. In: Quack, M., Merkt, F. (Eds.), Handbook of High-Resolution Spectroscopy, Vol. 1. Wiley, Chichester, New York, pp. 573–586 (Chapter 14).

Carrington, A., Buttenshaw, J., Kennedy, R.A., 1982. Observation of the infrared spectrum of H_3^+ at its dissociation limit. Mol. Phys. 45, 735–758.

Carrington, A., Kennedy, R.A., 1984. Infrared predissociation spectrum of the H_3^+ ion. J. Chem. Phys. 81, 91–112.

Carter, S., Sharma, A.R., Bowman, J.M., 2013. Multimode calculations of rovibrational energies and dipole transition intensities for polyatomic molecules with torsional motion: application to H_2O_2. J. Chem. Phys. 135, 014308.

Cattaneo, L., Vos, J., Lucchini, M., Gallmann, L., Cirelli, C., Keller, U., 2016. Comparison of attosecond streaking and RABBITT. Opt. Express 24, 29060–29076.

Cavalli, S., Aquilanti, V., Mundim, K.C., De Fazio, D., 2014. Theoretical reaction kinetics astride the transition between moderate and deep tunneling regimes: the F + HD case. J. Phys. Chem. A 118, 6632–6641.

Cendagorta, J.R., Powers, A., Hele, T.J.H., Marsalek, O., Bačić, Z., Tuckerman, M.E., 2016. Competing quantum effects in the free energy profiles and diffusion rates of hydrogen and deuterium molecules through clathrate hydrates. Phys. Chem. Chem. Phys. 18, 32169–32176.

Ceriotti, M., Manolopoulos, D.E., Parrinello, M., 2011. Accelerating the convergence of path integral dynamics with a generalized Langevin equation. J. Chem. Phys. 134, 084104.

Chan-Huot, M., Dos, A., Zander, R., Sharif, S., Tolstoy, P.M., Compton, S., Fogle, E., Toney, M.E., Shenderovich, I., Gleb, S., Denisov, G.S., Limbach, H.-H., 2013. NMR studies of protonation and hydrogen bond states of internal aldimines of pyridoxal 5′-phosphate acidbase in alanine racemase, aspartate aminotransferase, and poly-L-lysine. J. Am. Chem. Soc. 135, 18160–18175.

Chapman, S., Garrett, B.C., Miller, W.H., 1975. Semiclassical transition state theory for nonseparable systems: application to the collinear H + H_2 reaction. J. Chem. Phys. 63, 2710–2716.

Chuang, C.C., Tsang, S.N., Klemperer, W., Chang, H.C., 1997. Reassignment of the 11537 cm^{-1} band of hydrogen fluoride dimer and observation of the intermolecular combination mode $3\nu_1 + \nu_4$. J. Phys. Chem. A 101, 6702–6708.

Chesnavich, W.J., Bowers, M.T., 1979. Statistical methods in reaction dynamics. In: Bowers, M.T. (Ed.), Gas Phase Ion Chemistry, Vol. 1. Elsevier Inc., Amsterdam, pp. 119–151.

Ciaćka, P., Fita, P., Listkowski, A., Kijak, M., Nonell, S., Kuzuhara, D., Yamada, H., Radzewicz, C., Waluk, J., 2015. Tautomerism in porphycenes: analysis of rate-affecting factors. J. Phys. Chem. B 119, 2292–2301.

Ciaćka, P., Fita, P., Listkowski, A., Radzewicz, C., Waluk, J., 2016. Evidence for dominant role of tunneling in condensed phases and at high temperatures: double hydrogen transfer in porphycenes. J. Phys. Chem. Lett. 7, 283–288.

Clary, D., 2018. Spiers memorial lecture: quantum dynamics of chemical reactions. Faraday Discuss. Chem. Soc. 212, 9–32.

Cohen, E.R., Cvitaš, T., Frey, J.G., Holmström, B., Kuchitsu, K., Marquardt, R., Mills, I., Pavese, F., Quack, M., Stohner, J., Strauss, H.L., Takami, M., Thor, A., 2011. Quantities, Units and Symbols in Physical Chemistry, third reviewed printing 3rd ed. Royal Soc. Chem., London.

Cole, W.T.S., Farrell, J.D., Sheikh, A.A., Yönder, O., Fellers, R.S., Viant, M.R., Wales, D.J., Saykally, R.J., 2018. Terahertz VRT spectroscopy of the water hexamer-D_{12} prism: dramatic enhancement of bifurcation tunneling upon librational excitation. J. Chem. Phys. 148, 094301.

Cole, W.T.S., Fellers, R.S., Viant, M.R., Leforestier, C., Saykally, R.J., 2015. Far-infrared VRT spectroscopy of the water dimer: characterization of the 20 μm out-of-plane librational vibration. J. Chem. Phys. 143, 154306.

Cole, W.T.S., Fellers, R.S., Viant, M.R., Saykally, R.J., 2017. Hydrogen bond breaking dynamics in the water pentamer: terahertz VRT spectroscopy of a 20 μm libration. J. Chem. Phys. 146, 014306.

Congiu, E., Minissale, M., Baouche, S., Chaabouni, H., Moudens, A., Cazaux, S., Manicó, G., Pirronello, V., Dulieu, F., 2014. Efficient diffusive mechanisms of O atoms at very low temperatures on surfaces of astrophysical interest. Faraday Discuss. Chem. Soc. 168, 151–166.

Coon, J.B., Naugle, N.W., McKenzie, R.D., 1966. The investigation of double-minimum potentials in molecules. J. Mol. Spectrosc. 20, 107–129.

Costain, C.C., Sutherland, G.B.B.M., 1952. A method of determining the potential barriers restricting inversion in ammonia, phosphine and arsine from vibrational force constants. J. Phys. Chem. 56, 321–324.

Cremer, E., Polanyi, M., 1932. Eine Prüfung der "Tunneltheorie" der heterogenenen Katalyse am Beispiel der Hydrierung von Styrol. Z. Phys. Chem. B 19, 443–450.

Császár, A.G., Allen, W.D., Schaefer III, H.F., 1998. In pursuit of the ab initio limit for conformational energy prototypes. J. Chem. Phys. 108, 9751–9764.

Császár, A.G., Fábri, C., Szidarovszky, T., Mátyus, E., Furtenbacher, T., Czakó, G., 2012. The fourth age of quantum chemistry: molecules in motion. Phys. Chem. Chem. Phys. 14, 1085–1114.

Császár, A.G., Furtenbacher, T., 2016. Promoting and inhibiting tunneling via nuclear motions. Phys. Chem. Chem. Phys. 18, 1092–1104.

Császár, A.G., Fábri, C., Szidarovszky, T., 2020. Exact numerical methods for stationary-state-based quantum dynamics of complex polyatomic molecules. In: Marquardt, R., Quack, M. (Eds.), Molecular Spectroscopy and Quantum Dynamics. Elsevier, Amsterdam. Chapter 2 (this book).

Cvitaš, M., Richardson, J., 2020. Quantum dynamics of water clusters. In: Marquardt, R., Quack, M. (Eds.), Molecular Spectroscopy and Quantum Dynamics. Elsevier, Amsterdam. Chapter 9 (this book).

Dahlström, J.M., L'Huillier, A., Maquet, A., 2012. Introduction to attosecond delays in photoionization. J. Phys. B: At. Mol. Phys. 45, 183001.

Dashevskaya, E.I., Litvin, I., Nikitin, E.E., Troe, J., 2010. Locking of the intrinsic angular momentum in the capture of quadrupole diatoms by ions. Mol. Phys. 108, 873–882.

Dashevskaya, E.I., Litvin, I., Nikitin, E.E., Troe, J., 2011. Electron capture by polarizable dipolar targets: numerical and analytically approximated capture probabilities. J. Phys. Chem. A 115, 6825–6830.

Dashevskaya, E.I., Maergoiz, A.I., Troe, J., Litvin, I., Nikitin, E.E., 2003. Low-temperature behavior of capture rate constants for inverse power potentials. J. Chem. Phys. 118, 7313–7320.

Dawadi, M.H., Perry, D.S., 2014. Conical intersections between vibrationally adiabatic surfaces in methanol. J. Chem. Phys. 140, 161101.

de Haag, P.U., Spooren, R., Ebben, M., Meerts, L., Hougen, J.T., 1990. Internal rotation in 1, 4-dimethylnaphthalene studied by high resolution laser spectroscopy. Mol. Phys. 69, 265–280.

de Tudela, R.P., Suleimanov, Y.V., Richardson, J.O., Sáez Rábanos, V., Green, W.H., Aoiz, F.J., 2013. Stress test for quantum dynamics approximations: deep tunneling in the muonium exchange reaction D + HMu → DMu + H. J. Phys. Chem. Lett. 5, 4219–4224.

Dennison, D.M., Uhlenbeck, G.E., 1932. The two-minima problem and the ammonia molecule. Phys. Rev. 41, 313–321.

Devault, D., 1984. Quantum-Mechanical Tunnelling in Biological Systems. Cambridge Univ. Press, Cambridge.

Dietiker, P., Miloglyadov, E., Quack, M., Schneider, A., Seyfang, G., 2015. Infrared laser induced population transfer and parity selection in $^{14}NH_3$: a proof of principle experiment towards detecting parity violation in chiral molecules. J. Chem. Phys. 143, 244305.

Domcke, W., Yarkony, D.R., 2012. Role of conical intersections in molecular spectroscopy and photoinduced chemical dynamics. Annu. Rev. Phys. Chem. 63, 325–352.

Dumez, J.N., Hakansson, P., Mamone, S., Meier, B., Stevanato, G., Hill-Cousins, J.T., Roy, S.S., Brown, R.C.D., Pileio, G., Levitt, M.H., 2015. Theory of long-lived nuclear spin states in methyl groups and quantum-rotor induced polarisation. J. Chem. Phys. 142, 044506.

Dyke, T.R., Howard, B.J., Klemperer, W., 1972. Radio frequency and microwave spectrum of the hydrogen fluoride dimer, a nonrigid molecule. J. Chem. Phys. 56, 2442–2454.

Eckart, C., 1930. The penetration of a potential barrier by an electron. Phys. Rev. 35, 1303–1309.

Eckert, J., Kohen, A., McMahon, R., 2017. In: Proceedings 18th International Workshop on Quantum Atomic and Molecular Tunneling in Solids and Other Phases (QAMTS). Madison/Wisconsin, 20–24 May, 2017.

Eckert, J., Meier, B.H., Merkt, F., Quack, M., 2015. In: Proceedings 17th International Workshop on Quantum Atomic and Molecular Tunneling in Solids and Other Phases (QAMTS). Beatenberg/Interlaken Switzerland, 31 May – 3 June, 2015.

Eckert, J., Georgiev, P., Quack, M., Doslic, N., Stare, J., 2019. In: Proc. 19th Int. Workshop on Quantum Atomic and Molecular Tunneling Systems (QAMTS). Borovets, Bulgaria, 16–20 June, 2019.

Einstein, A., 1922. Grundzüge der Relativitätstheorie. Vieweg, Braunschweig.

Endo, Y., Sumiyoshi, Y., 2011. Rotational spectroscopy of complexes containing free radicals. In: Quack, M., Merkt, F. (Eds.), Handbook of High-Resolution Spectroscopy, Vol. 2. Wiley, Chichester, New York, pp. 897–916 (Chapter 23).

Ernst, R.R., Carrington, T., Seyfang, G., Merkt, F., 2013. Editorial Special Issue. Mol. Phys. 111, 1939–1963.

Ernst, R.R., Bodenhausen, G., Wokaun, A., 1987. Principles of Nuclear Magnetic Resonance in One and Two Dimensions. Clarendon Press, Oxford.

Ertelt, M., Henkel, S., Zhang, W.T.X., Hrovat, D.A., Borden, W.T., 2015. Tunneling in the degenerate rearrengment of semibullvalene at cryogenic temperatures L28. In: Eckert, J., Merkt, F., Meier, B.H., Quack, M. (Eds.), Proc. 17th QAMTS. Beatenberg, Switzerland.

Ertelt, M., Hrovat, D.A., Borden, W.T., Sander, W., 2014. Heavy-atom tunneling in the ring opening of a strained cyclopropene at very low temperatures. Chem. Eur. J. 20, 4713–4720.

Ertl, G., 2008. Reactions at surfaces: from atoms to complexity (Nobel lecture). Angew. Chem., Int. Ed. 47, 3524–3535.

Eyring, H., 1938a. The calculation of activation energies. Trans. Faraday Soc. 34, 3–11.

Eyring, H., 1938b. The theory of absolute reaction rates. Trans. Faraday Soc. 34, 41–48.

Eyring, H., Walter, J., Kimball, G., 1944. Quantum Chemistry. John Wiley, New York.

Fábri, C., Albert, S., Chen, Z., Prentner, R., Quack, M., 2018. A molecular quantum switch based on tunneling in meta-d-phenol C_6H_4DOH. Phys. Chem. Chem. Phys. 20, 7387–7394.

Fábri, C., Horný, L., Quack, M., 2015. Tunneling and parity violation in trisulfane (HSSSH): an almost ideal molecule for detecting parity violation in chiral molecules. ChemPhysChem 16, 3584–3589.

Fábri, C., Marquardt, R., Császár, A., Quack, M., 2019. Controlling tunneling in ammonia isotopomers. J. Chem. Phys. 150, 014102 (and to be published).

Fábri, C., Marquardt, R., Quack, M., 2015. Full-dimensional dynamics and spectroscopy of ammonia isotopomers. In: Eckert, J., Merkt, F., Meier, B.H., Quack, M. (Eds.), Proceedings of XVIIth International Workshop on Quantum Atomic and Molecular Tunneling in Solids and Other Phases. Beatenberg, Switzerland. pp. L-42 (and to be published 2020).

Fábri, C., Mátyus, E., Császár, A.G., 2011. Rotating full- and reduced-dimensional quantum chemical models of molecules. J. Chem. Phys. 130, 074105.

Fábri, C., Mátyus, E., Császár, A.G., 2014. Numerically constructed internal-coordinate Hamiltonian with Eckart embedding and its application for the inversion tunneling of ammonia. Spectrochim. Acta A 119, 84–89.

Fábri, C., Quack, M., Császár, A.G., 2017. On the use of nonrigid-molecular symmetry in nuclear motion computations employing a discrete variable representation: a case study of the bending energy levels of CH_5^+. J. Chem. Phys. 147 (13), 134101.

Farrar, J.M., Lee, Y.T., 1974. Chemical dynamics. Annu. Rev. Phys. Chem. 25, 357–386.

Fehrensen, B., Luckhaus, D., Quack, M., 1999a. Inversion tunneling in aniline from high resolution infrared spectroscopy and an adiabatic reaction path Hamiltonian approach. Z. Phys. Chem. 209, 1–19.

Fehrensen, B., Luckhaus, D., Quack, M., 1999b. Mode selective stereomutation tunnelling in hydrogen peroxide isotopomers. Chem. Phys. Lett. 300 (3–4), 312–320.

Fehrensen, B., Luckhaus, D., Quack, M., 2007. Stereomutation dynamics in hydrogen peroxide. Chem. Phys. 338 (2–3), 90–105.

Fehrensen, B., Luckhaus, D., Quack, M., Willeke, M., Rizzo, T.R., 2003. Ab initio calculations of mode selective tunneling dynamics in $^{12}CH_3OH$ and $^{13}CH_3OH$. J. Chem. Phys. 119 (11), 5534–5544.

Felix, D., Eschenmoser, A., 1968. Slow inversion at pyramidal nitrogen: isolation of diastereomeric 4-chloro-7-azabicyclo-(4.1.0) heptanes at room temperature. Angew. Chem., Int. Ed. 7, 224–225.

Felker, P.M., Bačić, Z., 2016. Communication: quantum six-dimensional calculations of the coupled translation rotation eigenstates of $H_2O@C_{60}$. J. Chem. Phys. 144, 201101.

Felker, P.M., Bačić, Z., 2019. Weakly bound molecular dimers: intramolecular vibrational fundamentals, overtones, and tunneling splittings from full-dimensional quantum calculations using compact contracted bases of intramolecular and low-energy rigid-monomer intermolecular eigenstates. J. Chem. Phys. 153, 024305.

Felker, P., Lauvergnat, D., Scribano, Y., Benoit, D.M., Bačić, Z., 2019. Intramolecular stretching vibrational states and frequency shifts of $(H_2)_2$ confined inside the large cage of clathrate hydrate from an eight-dimensional quantum treatment using small basis sets. J. Chem. Phys. 151, 124311.

Fermi, E., 1932. Sulle bande di oscillazione et rotazione dell'ammoniaca. Nuovo Cimento 9, 277–283.

Firmino, T., Marquardt, R., Gatti, F., Dong, W., 2014. Diffusion rates for hydrogen on Pd(111) from molecular quantum dynamics calculations. J. Phys. Chem. Lett. 5, 4270–4274.

Fleming, D.G., Arseneau, D.J., Sukhorukov, O., Brewer, J.H., Mielke, S.L., Schatz, G.C., Garrett, B.C., Peterson, K.A., Truhlar, D.G., 2011a. Kinetic isotope effects for the reactions of muonic helium and muonium with H_2. Science 331, 448–451.

Fleming, D.G., Arseneau, D.J., Sukhorukov, O., Brewer, J.H., Mielke, S.L., Truhlar, D.G., Schatz, G.C., Garrett, B.C., Peterson, K.A., 2011b. Kinetics of the reaction of the heaviest hydrogen atom with H_2, the $^4He\mu + H_2 \rightarrow {}^4He\mu H + H$ reaction: Experiments, accurate quantal calculations, and variational transition state theory, including kinetic isotope effects for a factor of 36.1 in isotopic mass. J. Chem. Phys. 135, 184310.

Fowler, R.H., Nordheim, L., 1928. Electron emission in intense electric fields. Proc. Roy. Soc. A 119, 173–181.

Francis, K., Sapienza, P.J., Lee, A.L., Kohen, A., 2016. The effect of protein mass modulation on human dihydrofolate reductase. Biochemistry 55, 1100–1106.

Frey, H.M., Kummli, D., Lobsiger, S., Leutwyler, S., 2011. High-resolution rotational Raman coherence spectroscopy with femtosecond pulses. In: Quack, M., Merkt, F. (Eds.), Handbook of High-Resolution Spectroscopy, Vol. 2. Wiley, Chichester, New York, pp. 1237–1266 (Chapter 32).

Friedman, J.I., Telegdi, V., 1957. Nuclear emulsion evidence for parity nonconservation in the decay chain π^+ - π^- - ε^+. Phys. Rev. 105, 1681–1682.

Gallmann, L., Cirelli, C., Keller, U., 2012. Attosecond science: recent highlights and future trends. Annu. Rev. Phys. Chem. 63, 447–469.

Gallmann, L., Keller, U., 2011. Femtosecond and attosecond light sources and techniques for spectroscopy. In: Quack, M., Merkt, F. (Eds.), Handbook of High-Resolution Spectroscopy, Vol. 3. Wiley, Chichester, New York, pp. 1805–1836 (Chapter 50).

Gamov, G., 1928a. The quantum theory of nuclear disintegration. Nature 122, 805–806.

Gamov, G., 1928b. Zur Quantentheorie des Atomkernes. Z. Phys. 51, 204–212.

Garwin, R.L., Lederman, L.M., Weinrich, M., 1957. Observation of the failure of conservation of parity an charge conjugation in meson decays - magnetic moment of the free muon. Phys. Rev. 105, 1415–1417.

Gerbig, D., Schreiner, P.R., 2015. Hydrogen-tunneling in biologically relevant small molecules: the rotamerizations of α-ketocarboxylic acids. J. Phys. Chem. B 119, 693–703.

Ghosh, S., Thomas, J., Huang, W., Xu, Y., Jäger, W., 2015. Rotational spectra of two hydrogen-bonded methyl salicylate monohydrates: relative stability and tunneling motions. J. Phys. Chem. Lett. 6, 3126–3131.

Gioumousis, G., Stevenson, D.P., 1958. Reactions of gaseous molecule ions with gaseous molecules. V. Theory. J. Chem. Phys. 29, 294–299.

Glänzer, K., Quack, M., Troe, J., 1976. A spectroscopic determination of the methyl radical recombination rate constant in shock waves. Chem. Phys. Lett. 39 (2), 304–309.

Glänzer, K., Quack, M., Troe, J., 1977. High temperature UV absorption and recombination of methyl radicals in shock waves. In: 16th International Symposium on Combustion. The Combustion Institute, Pittsburgh, pp. 949–960.

Goldanskii, V.I., 1976. Chemical reactions at very low temperatures. Annu. Rev. Phys. Chem. 27, 85–126.

Gordon, J.P., Zeiger, H.J., Townes, C.H., 1955. The maser – new type of microwave amplifier, frequency standard, and spectrometer. Phys. Rev. 99, 1264–1274.

Gorin, E., 1938. Loose complex model. Acta Phys. Chim. URSS 9, 691.

Grabow, J.-U., 2011. Fourier transform microwave spectroscopy measurement and instrumentation. In: Quack, M., Merkt, F. (Eds.), Handbook of High-Resolution Spectroscopy, Vol. 2. Wiley, Chichester, New York, pp. 723–799 (Chapter 19).

Grohmann, T., Manz, J., Schild, A., 2013. Effects of molecular symmetry on the directions of nuclear flux densities during tunnelling in double well potentials. Mol. Phys. 111, 2251–2262.

Guennoun, Z., Maier, J.P., 2011. Electronic spectroscopy of transient molecules. In: Quack, M., Merkt, F. (Eds.), Handbook of High-Resolution Spectroscopy, Vol. 2. Wiley, Chichester, New York, pp. 1321–1344 (Chapter 33).

Gurney, R.W., Condon, E.U., 1928. Wave mechanics and radioactive disintegration. Nature 122, 439.

Gurney, R.W., Condon, E.U., 1929. Wave mechanics and radioactive disintegration. Phys. Rev. 33, 127–140.

Hall, F.H.J., Eberle, P., Hegi, G., Raoult, M., Aymar, M., Dulieu, O., Willitsch, S., 2013. Ion-neutral chemistry at ultralow energies: dynamics of reactive collisions between lasercooled Ca^+ ions and Rb atoms in an ion-atom hybrid trap. Mol. Phys. 111, 2020–2032.

Hamm, P., Stock, G., 2013. Vibrational conical intersections in the water dimer. Mol. Phys. 111, 2046–2056.

Hänninen, V., Horn, M., Halonen, L., 1999. Torsional motion and vibrational overtone spectroscopy of methanol. J. Chem. Phys. 111, 3018–3026.

Hargis, J.C., Evangelista, F.A., Ingels, J.B., Schaefer III, H.F., 2008. Short intramolecular hydrogen bonds: derivatives of malonaldehyde with symmetrical substituents. J. Am. Chem. Soc. 130, 17471–17478.

Hartmann, C., Joyeux, M., Trommsdorff, H.P., Vial, J.C., von Borczyskowski, J., 1992. Optical measurements of methyl group tunneling in molecular crystals: temperature dependence of the nuclear spin conversion rate. J. Chem. Phys. 96, 6335–6343.

Hauge, E.H., Stoevneng, J.A., 1989. Tunneling times: a critical review. Rev. Modern Phys. 61, 917–936.

Haupt, J., 1972. A new effect of dynamic polarization in a solid obtained by rapid change of temperature. Phys. Lett. A 38, 389–390.

Hauser, A.W., Pototschnig, J.V., Ernst, W.E., 2015. A classic case of Jahn–Teller effect theory revisited: ab initio simulation of hyperfine coupling and pseudorotational tunneling in the $1\,^2E'$ state of Na_3. Chem. Phys. 460, 2–13.

Häusler, W., 1990. Theory of spin-conversion in XH_3-systems. Z. Phys. B, Condens. Matter 81, 265–272.

Havenith, M., Birer, O., 2011. High-resolution IR-laser jet spectroscopy of the formic acid dimer. In: Quack, M., Merkt, F. (Eds.), Handbook of High-Resolution Spectroscopy, Vol. 2. Wiley, Chichester, New York, pp. 1119–1128 (Chapter 29).

Hawbaker, N.A., Blackmond, D.G., 2019. Energy threshold for chiral symmetry breaking in molecular self-replication. Nat. Chem. 11, 957–962.

Hawking, S.W., 1975. Particle creation by black holes. Comm. Math. Phys. 43, 199–220.

He, Y., Müller, H.B., Quack, M., Suhm, M.A., 2007. High resolution FTIR and diode laser supersonic jet spectroscopy of the $N = 2$ HF-stretching polyad in $(HF)_2$ and (HFDF): hydrogen bond switching and predissociation dynamics. Z. Phys. Chem. 221, 1581–1645.

Heidemann, A., Magerl, A., Prager, M., Richter, D., Springer, T., 1987. Quantum Aspects of Molecular Motions in Solids. Springer Proceedings in Physics, vol. 17. Springer, Berlin, Heidelberg.

Heilbronner, E., Günthard, H.H., Gerdil, R., 1956. Linearkombination Hermite'scher Orthogonalfunktionen, ein Verfahren zur Behandlung eindimensionaler Molekel-Modellen der Quanten-Chemie. Helv. Chim. Acta 39, 1171–1181.

Helmchen, G., 2016. 50 Jahre Spezifikation der molekularen Chiralität durch Cahn, Ingold und Prelog. Angew. Chem. 128, 6910–6911.

Henkel, S., Sander, W., 2015. Activation of molecular hydrogen by a singlet carbene through quantum mechanical tunneling. Angew. Chem., Int. Ed. 54, 4603–4607.

Hermann, G., Liu, C.H., Manz, J., Paulus, B., Peérez-Torres, J.F., Pohl, V., Tremblay, J.C., 2016. Multidirectional angular electronic flux during adiabatic attosecond charge migration in excited benzene. J. Phys. Chem. A 120, 5360–5369.

Hernandez, R., Miller, W.H., 1993. Semiclassical transition state theory. A new perspective. Chem. Phys. Lett. 214, 129–136.

Herschbach, D.R., 1973. Reactive scattering. Faraday Discuss. Chem. Soc. 55, 233–251.

Herschbach, D.R., 1987. Molecular dynamics of elementary chemical reactions (Nobel lecture). Angew. Chem., Int. Ed. 26, 1221–1243.

Herzberg, G., 1939. Molekülspektren und Molekülstruktur. Steinkopff, Dresden.

Herzberg, G., 1945. Molecular Spectra and Molecular Structure: II. Infrared and Raman Spectra. Van Nostrand, New York.

Herzberg, G., 1966. Molecular Spectra and Molecular Structure: III. Electronic Spectra and Electronic Structure of Polyatomic Molecules. Van Nostrand, New York.

Hippler, M., Miloglyadov, E., Quack, M., Seyfang, G., 2011. Mass and isotope selective infrared spectroscopy. In: Quack, M., Merkt, F. (Eds.), Handbook of High Resolution Spectroscopy, Vol. 2. Wiley, Chichester, New York, pp. 1069–1118 (Chapter 28).

Hippler, M., Oeltjen, L., Quack, M., 2007. High-resolution continuous-wave-diode laser cavity ring-down spectroscopy of the hydrogen fluoride dimer in a pulsed slit jet expansion: two components of the $n = 2$ triad near 1.3 micrometer. J. Phys. Chem. A 111 (49), 12659–12668.

Hofacker, L., 1963. Quantentheorie chemischer Reaktionen. Z. Naturforsch. A 18, 607–619.

Hoffmann, R., 1998. Qualitative thinking in the age of modern computational chemistry – or what Lionel Salem knows. J. Mol. Struct. 424, 1–6.

Hofmann, C., Landsman, A.S., Keller, U., 2019. Attoclock revisited on electron tunnelling time. J. Modern Opt. 66, 1052–1070.

Hollenstein, U., Merkt, F., Meyer, L., Seiler, R., Softley, T.P., Willitsch, S., 2007. Rovibronic photoionization dynamics of ammonia isotopomers. Mol. Phys. 105, 1711–1722.

Hölsch, N., Beyer, M., Salumbides, E.J., Eikema, K.S.E., Ubachs, W., Jungen, C., Merkt, F., 2019. Benchmarking theory with an improved measurement of the ionization and dissociation energies of H_2. Phys. Rev. Lett. 122, 103002.

Homayoon, Z., Bowman, J.M., Evangelista, F.A., 2014. Calculations of mode-specific tunneling of double-hydrogen transfer in porphycene agree with and illuminate experiment. J. Phys. Chem. Lett. 5, 2723–2727.

Horká-Zelenková, V., Seyfang, G., Dietiker, P., Quack, M., 2019. Nuclear spin symmetry conservation studied for symmetric top molecules (CH_3D, CHD_3, CH_3F, and CH_3Cl) in supersonic jet expansions. J. Phys. Chem. A 123, 6160–6174.

Horný, L., Quack, M., 2015. Computation of molecular parity violation using the coupled-cluster linear response approach. Mol. Phys. 113 (13–14), 1768–1779.

Horný, L., Quack, M., Schaefer III, H.F., Willeke, M., 2016. Chlorine peroxide (Cl_2O_2) and its isomers: structures, spectroscopy, formation and thermochemistry. Mol. Phys. 114, 1135–1147.

Horsewill, A.J., 1999. Quantum tunnelling aspects of methyl group rotation studied by NMR. Prog. Nucl. Magn. Reson. Spectrosc. 35, 359–389.

Hoshina, H., Fushitani, M., Momose, T., Shida, T., 2004. Tunneling chemical reactions in solid parahydrogen: direct measurement of the rate constants of $R + H_2 \rightarrow RH + H$ ($R = CD_3$, CD_2H, CDH_2, CH_3, at 5 K. J. Chem. Phys. 120, 3706–3715.

Hougen, J.T., Bunker, P.R., Johns, J.W.C., 1970. Vibration-rotation problem in triatomic molecules allowing for a large-amplitude bending vibration. J. Mol. Spectrosc. 34, 136–172.

Hund, F., 1927a. Symmetriecharaktere von Termen bei Systemen mit gleichen Partikeln in der Quantenmechanik. Z. Phys. 43, 788–804.

Hund, F., 1927b. Zur Deutung der Molekelspektren. III. -Bemerkungen über das Schwingungs- und Rotationsspektrum bei Molekeln mit mehr als zwei Kernen. Z. Phys. 43, 805–826.

Hund, F., 1927c. Zur Deutung der Molekülspektren I. Z. Phys. 40, 742–764.

Icker, M., Berger, S., 2012. Unexpected multiplet patterns induced by the Haupt-effect. J. Magn. Reson. 219, 1–3.

Inui, H., Sawada, K., Oishi, S., Ushida, K., McMahon, R.J., 2013. Aryl nitrene rearrangements: spectroscopic observation of a benzazirine and its ring expansion to a ketenimine by heavy-atom tunneling. J. Am. Chem. Soc. 135, 10246–10249.

Ivanov, S.D., Grant, I.M., Marx, D., 2015. Quantum free energy landscapes from ab initio path integral metadynamics: double proton transfer in the formic acid dimer is concerted but not correlated. J. Chem. Phys. 143, 124304.

Jambrina, P.G., Aldegunde, J., Aoiz, F.J., Sneha, M., Zare, R.N., 2016. Effects of reagent rotation on interferences in the product angular distributions of chemical reactions. Chem. Sci. 7, 642–649.

Jankunas, J., Sneha, M., Zare, R.N., Bouakline, F., Althorpe, S.C., 2013a. Disagreement between theory and experiment grows with increasing rotational excitation of HD(v′,J′) product for the $H + D_2$ reaction. J. Chem. Phys. 138, 094310.

Jankunas, J., Sneha, M., Zare, R.N., Bouakline, F., Althorpe, S.C., 2013b. Hunt for geometric phase effects in $H + HD \rightarrow$ HD(v′,J′) + H. J. Chem. Phys. 139, 144316.

Jankunas, J., Zare, R.N., Bouakline, F., Althorpe, S.C., Diego Herráez-Aguilar, D., Aoiz, F.J., 2012. Seemingly anomalous angular distributions in $H + D_2$ reactive scattering. Science 336, 1687–1691.

Janoschek, R., 1991. Theories on the origin of biomolecular homochirality. In: Janoschek, R. (Ed.), Chirality - From Weak Bosons to the α-Helix. Springer, Berlin.

Jortner, J., 2006. Conditions for the emergence of life on the early Earth: summary and reflections. Philos. Trans. R. Soc. Lond. B, Biol. Sci. 361, 1877–1891.

Karplus, M., 2014. Development of multiscale models for complex chemical systems: from $H+H_2$ to biomolecules (Nobel lecture). Angew. Chem., Int. Ed. 53, 9992–10005.

Kästner, J., 2013. Path length determines the tunneling decay of substituted carbenes. Chem. Eur. J. 13, 8207–8212.

Kästner, J., 2014a. Der Tunneleffekt in chemischen Reaktionen. Bunsen-Magazin 1, 15–19.

Kästner, J., 2014b. Theory and simulation of atom tunneling in chemical reactions. WIREs Comput. Mol. Sci. 4, 158–168.

Kawasaki, T., Matsumura, Y., Tsutsumi, T., Suzuki, K., Ito, M., Soai, K., 2009. Asymmetric autocatalysis triggered by carbon isotope ($^{13}C/^{12}C$) chirality. Science 324, 492–496.

Keutsch, F.N., Fellers, R.S., Brown, M.G., Viant, M.R., Petersen, P.B., Saykally, R.J., 2001. Hydrogen bond breaking dynamics of the water trimer in the translational and librational band region of liquid water. J. Am. Chem. Soc. 123, 5938–5941.

Kilaj, A., Gao, H., Rösch, D., Rivero, U., Küpper, J., Willitsch, S., 2018. Observation of different reactivities of para and ortho-water towards trapped diazenylium ions. Nat. Commun. 9, 2096.

Klopper, W., Quack, M., Suhm, M.A., 1998a. Explicitly correlated coupled cluster calculations of the dissociation energies and barriers to concerted hydrogen exchange of $(HF)_n$ oligomers ($n = 2, 3, 4, 5$). Mol. Phys. 94 (1), 105–119.

Klopper, W., Quack, M., Suhm, M.A., 1998b. HF dimer: empirically refined analytical potential energy and dipole hypersurfaces from ab initio calculations. J. Chem. Phys. 108 (24), 10096–10115 (88 pages supplementary material published as AIP Document No PAPS JCPS 16-108-303 820-88 by American Institute of Physics, Physics Auxiliary Publication Service).

Klopper, W., Samson, C.C.M., Tarczay, G., Császár, A., 2001. Equilibrium inversion barrier of NH_3 from extrapolated coupled-cluster pair energies. J. Comput. Chem. 22, 1306–1314.

Kneba, M., Wolfrum, J., 1980. Bimolecular reaction of vibrationally excited molecules. Annu. Rev. Phys. Chem. 31, 47–79.

Koeppe, B., Guo, J., Tolstoy, P.M., Denisov, G.S., Limbach, H.-H., 2013. Solvent and H/D isotope effects on the proton transfer pathways in heteroconjugated hydrogen-bonded phenol-carboxylic acid anions observed by combined UV-vis and NMR spectroscopy. J. Am. Chem. Soc. 135, 7553–7568.

Kolesnikov, A.I., Reiter, G.F., Choudhury, N., Prisk, T.R., Mamontov, E., Podlesnyak, A., Ehlers, G., Seel, A.G., Wesolowski, D.J., Anovitz, L.H., 2016. Quantum tunneling of water in beryl: a new state of the water molecule. Phys. Rev. Lett. 116, 167802.

Koszinowski, K., Goldberg, N.T., Pomerantz, A.E., Zare, R.N., Juanes-Marcos, J.C., Althorpe, S.C., 2005. Collision-energy dependence of HD($v' = 1$, J') product rotational distributions for the H + D_2 reaction. J. Chem. Phys. 123, 054306.

Kozuch, S., 2014a. Heavy atom tunneling in the automerization of pentalene and other antiaromatic systems. RSC Adv. 4, 21650–21656.

Kozuch, S., 2014b. The reactivity game: theoretical predictions for heavy atom tunneling in adamantyl and related carbenes. Phys. Chem. Chem. Phys. 16, 7718–7727.

Kozuch, S., Zhang, X., Hrovat, D.A., Borden, W.T., 2013. Calculations on tunneling in the reactions of noradamantyl carbenes. J. Am. Chem. Soc. 135, 17274–117277.

Kramers, H.A., 1926. Wellenmechanik und halbzahlige Quantisierung. Z. Phys. 39, 828–840.

Kuhn, B., Rizzo, T.R., Luckhaus, D., Quack, M., Suhm, M.A., 1998. A new six-dimensional analytical potential up to chemically significant energies for the electronic ground state of hydrogen peroxide. J. Chem. Phys. 111, 2565–2587.

Kundu, A., Piccini, G.M., Sillar, K., Sauer, J., 2016. Ab initio prediction of adsorption isotherms for small molecules in metal-organic frameworks. J. Am. Chem. Soc. 138, 14047–14056.

Kushnarenko, A., Miloglyadov, E., Quack, M., Seyfang, G., 2018. Intramolecular vibrational energy redistribution in $HCCCH_2X$ (X = Cl, Br, I) measured by femtosecond pump-probe experiments in a hollow waveguide. Phys. Chem. Chem. Phys. 20, 10949–10959.

Laane, J., 1999. Spectroscopic determination of ground and excited state vibrational potential energy surfaces. Int. Rev. Phys. Chem. 18, 301–341.

Ladenthin, J.N., Frederiksen, T., Persson, M., Sharp, J.C., Gawinkowski, S., Waluk, J., Kumagai, T., 2016. Force-induced tautomerization in a single molecule. Nat. Chem. 8, 935–940.

Lalowicz, Z.T., Werner, U., Müller-Warmuth, W., 1988. Rotational tunneling of CD_3 groups in molecular crystals as studied by NMR spectra. Z. Naturforsch. A 43, 219–227.

Lamberts, T., Samanta, P.K., Köhn, A., Kästner, J., 2016. Quantum tunneling during interstellar surface-catalyzed formation of water: the reaction $H + H_2O_2 \rightarrow H_2O + OH$. Phys. Chem. Chem. Phys. 18, 33021–33030.

Landauer, R., Martin, T., 1994. Barrier interaction time in tunneling. Rev. Modern Phys. 66, 217–228.

Landsman, A.S., Keller, U., 2014. Tunnelling time in strong field ionisation. J. Phys. B: At. Mol. Phys. 47, 204024.

Landsman, A.S., Keller, U., 2015. Attosecond science and the tunnelling time problem. Phys. Rep. 547, 1–24.

Landsman, A.S., Weger, M., Maurer, J., Boge, R., Ludwig, A., Heuser, S., Cirelli, C., Gallmann, L., Keller, U., 2014. Ultrafast resolution of tunneling delay time. Optica 1, 343–349.

Lauvergnat, D., Felker, P., Scribano, Y., Benoit, D.M., Bačić, Z., 2019. H_2, HD, and D_2 in the small cage of structure II. Clathrate hydrate: vibrational frequency shifts from fully coupled quantum six-dimensional calculations of the vibration-translation-rotation eigenstates. J. Chem. Phys. 150, 154303.

Lee, Y.T., 1987. Molecular beam studies of elementary chemical processes (Nobel lecture). Angew. Chem., Int. Ed. 26, 939–958.

Lee, T.D., Yang, C.N., 1956. Question of parity conservation in weak interactions. Phys. Rev. 104, 254–258.

Leforestier, C., 2012. Infrared shifts of the water dimer from the fully flexible ab initio HBB2 potential. Philos. Trans. R. Soc. Lond. Ser. A 370, 2675–2690.

Lehn, J.M., 1970. Nitrogen inversion – experiment and theory. In: Davison, A., Dewar, M.J.S., Hafner, K., Heilbronner, E., Hofmann, K., Niedenzu, U., Schäfer, K., Wittig, C. (Eds.), Fortschritte der Chemischen Forschung – Topics in Current Chemistry, Vol. 15. Springer, Berlin, Heidelberg, New York, pp. 311–377.

Lehn, J.M., 2002. Toward complex matter: supramolecular chemistry and self-organization. Proc. Nat. Acad. Sci. 99, 4763–4768.

Lépine, F., Ivanov, M.Y., Vrakking, M.J.J., 2014. Attosecond molecular dynamics: fact or fiction? Nat. Photonics 8, 195–204.

Levine, R.D., 2005. Molecular Reaction Dynamics, first ed. Cambridge Univ. Press, Cambridge.

Levine, R.D., Bernstein, R.B., 1989. Molecular Reaction Dynamics and Chemical Reactivity. Oxford Univ. Press, Oxford.

Levitt, M., 2014. Birth and future of multiscale modeling for macromolecular systems (Nobel lecture). Angew. Chem., Int. Ed. 53, 10006–10018.

Li, L.Y., Seifert, N.A., Xie, F., Heger, M., Xu, Y., Jäger, W., 2018. A spectroscopic and ab initio study of the hydrogen peroxide-formic acid complex: hindering the internal motion of H_2O_2. Phys. Chem. Chem. Phys. 20, 21345–21351.

Light, J.C., Carrington Jr., T., 2001. Discrete-variable representations and their utilization. Adv. Chem. Phys. 114, 263–310.

Limbach, H.H., 2007. Single and multiple hydrogen/deuterium transfer reactions in liquids and solids. In: Hynes, J.T., Klinman, J.P., Limbach, H., Schowen, R.L. (Eds.), Hydrogen-Transfer Reactions. Wiley, Weinheim, pp. 135–221.

Liu, C.M., Manz, J., Yang, Y., 2015. Nuclear fluxes during coherent tunneling in asymmetric double well potentials. J. Phys. B: At. Mol. Phys. 48, 164001.

Longuet-Higgins, H.C., 1963. The symmetry group of non-rigid molecules. Mol. Phys. 6, 445–461.

Low, F.E., 1998. Comments on apparent superluminal propagation. Ann. Phys. (Leipzig) 7, 660–661.

Löwdin, P.O., 1963. Proton tunneling in DNA and its biological implications. Rev. Modern Phys. 35, 724–732.

Lozada-Garcia, R.R., Ceponkus, J., Chevalier, M., Chin, W., Mestdagh, J.M., Crépin, C., 2012. Nuclear spin conversion to probe the methyl rotation effect on hydrogen-bond and vibrational dynamics. Angew. Chem., Int. Ed. 51, 6947–6950.

Lu, C.M., Manz, J., Yang, Y., 2016. Staircase patterns of nuclear fluxes during coherent tunneling in excited doublets of symmetric double well potentials. Phys. Chem. Chem. Phys. 18, 5048–5055.

Luckhaus, D., 2006. Concerted hydrogen exchange tunneling in formic acid dimer. J. Phys. Chem. A 110, 3151–3158.

Luckhaus, D., Quack, M., 1992. Spectrum and dynamics of the CH chromophore in CD_2HF. 1. Vibrational Hamiltonian and analysis of rovibrational spectra. Chem. Phys. Lett. 190 (6), 581–589.

Luckhaus, D., Quack, M., Schmitt, U., Suhm, M.A., 1995. On FTIR spectroscopy in asynchronously pulsed supersonic free jet expansions and on the interpretation of stretching spectra of HF clusters. Ber. Bunsenges. Phys. Chem. 99 (3), 457–468. Also: Abstracts of the Conference on Molecular Spectroscopy and Molecular Dynamics, Theory and Experiment, Grainau 1994, p. 73.

Ludwig, C., Saunders, M., Marin-Montesinos, I., Günther, U.L., 2010. Quantum rotor induced hyperpolarization. Proc. Nat. Acad. Sci. 107, 10799–10803.

Luisi, P.L., 2006. The Emergence of Life – From Chemical Origins to Synthetic Biology. Cambridge University Press, Cambridge.

Luk, L.Y.P., Ruiz-Pernía, J.J., Dawson, W.M., Roca, M., Loveridge, E.J., Glowacki, D.R., Harvey, J.N., Mulholland, A.J., Tunón, I., Moliner, V., Allemann, R.K., 2013. Unraveling the role of protein dynamics in dihydrofolate reductase catalysis. Proc. Nat. Acad. Sci. 110, 16344–16349.

Lüthy, J., Rétey, J., Arigoni, D., 1969. Preparation and detection of chiral methyl groups. Nature 221, 1213–1215.

Lüttschwager, N.O.B., Wassermann, T.N., Coussan, S., Suhm, M.A., 2013. Vibrational tuning of the hydrogen transfer in malonaldehyde – a combined FTIR and Raman jet study. Mol. Phys. 111 (14–15), 2211–2227.

Luzar, A., 2000. Resolving the hydrogen bond dynamics conundrum. J. Chem. Phys. 113, 10663–10675.

Luzar, A., Chandler, D., 1996. Hydrogen-bond kinetics in liquid water. Nature 379, 55–57.

MacColl, L.A., 1932. Note on the transmission and reflection of wave packets by potential barriers. Phys. Rev. 40, 621–626.

Maerker, C., Schleyer, P.V., Liedl, K.R., Ha, T.K., Quack, M., Suhm, M.A., 1997. A critical analysis of electronic density functionals for structural, energetic, dynamic, and magnetic properties of hydrogen fluoride clusters. J. Comput. Chem. 18 (14), 1695–1719. Supp. Material ftp://wiley.com/public/journals/jcc/suppmat/18/1695 or http://journals.wiley.com/jcc.

Mamone, S., Concistrè, M., Carignani, E., Meier, B., Krachmalnicoff, A., Johannessen, O.G., Lei, X., Li, Y., Denning, M., Carravetta, M., Goh, K., Horsewill, A.J., Whitby, R.J., Levitt, M.H., 2014. Nuclear spin conversion of water inside fullerene cages detected by low-temperature nuclear magnetic resonance. J. Chem. Phys. 140, 194306.

Manca Tanner, C., Quack, M., Schmidiger, D., 2013. Nuclear spin symmetry conservation and relaxation in water ($^1H_2\,^{16}O$) studied by cavity ring-down (CRD) spectroscopy of supersonic jets. J. Phys. Chem. A 117 (39), 10105–10118.

Manca Tanner, C., Quack, M., Schmidiger, D., 2018. Nuclear spin symmetry conservation and relaxation of water (H_2O) seeded in supersonic jets of argon and oxygen: measurements by cavity ring-down laser spectroscopy. Mol. Phys. 116, 3718–3730.

Manca, C., Quack, M., Willeke, M., 2008. Vibrational predissociation in hydrogen bonded dimers: the case of $(HF)_2$ and its isotopomers. Chimia 62 (4), 235–239.

Manning, M.F., 1935. Energy levels of a symmetrical double minima problem with applications to the NH_3 and ND_3 molecules. J. Chem. Phys. 3, 136–138.

Manthe, U., 2015. The multi-configurational time-dependent Hartree approach revisited. J. Chem. Phys. 142, 244109.

Marcus, R.A., 1964. Generalization of the activated-complex theory of reaction rates. I. Quantum mechanical treatment. J. Chem. Phys. 41, 2614–2623.

Marcus, R.A., 1965. Generalization of activated-complex theory. III. Vibrational adiabaticity, separation of variables, and a connection with analytical mechanics. J. Chem. Phys. 43, 1598–1605.

Marcus, R.A., Coltrin, M.E., 1977. A new tunneling path for reactions such as $H + H_2 \rightarrow H_2 + H$. J. Chem. Phys. 67, 2609–2613.

Mardyukov, A., Quanz, H., Schreiner, P.R., 2016. Conformer-specific hydrogen atom tunnelling in trifluoromethylhydroxycarbene. Nat. Chem. 9, 71–76.

Marquardt, R., Quack, M., 1989. Infrared-multiphoton excitation and wave packet motion of the harmonic and anharmonic-oscillators – exact-solutions and quasiresonant approximation. J. Chem. Phys. 90 (11), 6320–6327.

Marquardt, R., Quack, M., 2011. Global analytical potential energy surfaces for high resolution molecular spectroscopy and reaction dynamics. In: Quack, M., Merkt, F. (Eds.), Handbook of High-Resolution Spectroscopy, Vol. 1. Wiley, Chichester, New York, pp. 511–549 (Chapter 12).

Marquardt, R., Quack, M., 2020. Molecular Spectroscopy and Quantum Dynamics. Elsevier, Amsterdam. Chapter 1 (this book).

Marquardt, R., Quack, M., Thanopulos, I., Luckhaus, D., 2003a. A global electric dipole function of ammonia and isotopomers in the electronic ground state. J. Chem. Phys. 119 (20), 10724–10732.

Marquardt, R., Quack, M., Thanopulos, I., Luckhaus, D., 2003b. Tunneling dynamics of the NH chromophore in NHD_2 during and after coherent infrared excitation. J. Chem. Phys. 118 (2), 643–658.

Marquardt, R., Sagui, K., Klopper, W., Quack, M., 2005. Global analytical potential energy surface for large amplitude nuclear motions in ammonia. J. Phys. Chem. B 109 (17), 8439–8451.

Marquardt, R., Sagui, K., Zheng, J., Thiel, W., Luckhaus, D., Yurchenko, S., Mariotti, F., Quack, M., 2013. A global analytical potential energy surface for the electronic ground state of NH_3 from high level ab initio calculations. J. Phys. Chem. A 117 (32), 7502–7522.

Marx, D., 2006. Advanced Car-Parrinello techniques: path integrals and nonadiabaticity in condensed matter simulations. Lecture Notes in Phys. 704, 508–539.

Matsumoto, A., Ozaki, H., Harada, S., Tada, K., Ayugase, T., Ozawa, H., Kawasaki, T., Soai, K., 2016. Asymmetric induction by a nitrogen $^{14}N/^{15}N$ isotopomer in conjunction with asymmetric autocatalysis. Angew. Chem., Int. Ed. 55, 15472–15475.

Mátyus, E., Wales, D.J., Althorpe, S.C., 2016. Quantum tunneling splittings from path-integral molecular dynamics. J. Chem. Phys. 144, 114108.

Meier, K., Choutko, A., Dolenc, J., Eichenberger, A.P., Riniker, S., van Gunsteren, W.F., 2013b. Multi-resolution simulation of biomolecular systems: a review of methodological issues. Angew. Chem., Int. Ed. 52, 2820–2834.

Meier, B., Dumez, J.N., Stevanato, G., Hill-Cousins, J.T., Roy, S.S., Hakansson, P., Mamone, S., Brown, R.C.D., Pileio, G., Levitt, M.H., 2013a. Long-lived nuclear spin states in methyl groups and quantum-rotor-induced polarization. J. Am. Chem. Soc. 135, 18746–18749.

Meierhenrich, U., 2008. Aminoacids and Asymmetry of Life. Springer, Berlin.

Meisner, J., Kästner, J., 2016. Reaction rates and kinetic isotope effects of $H_2 + OH \rightarrow H_2O + H$. J. Chem. Phys. 144, 174303.

Meisner, J., Kästner, J., 2018. Dual-level approach to instanton theory. J. Chem. Theory Comput. 14, 1865–1872.

Mengesha, E.T., Zehnacker-Rentien, A., Sepio, J., Kijak, M., Waluk, J., 2015. Spectroscopic study of jet-cooled deuterated porphycenes: unusual isotopic effects on proton tunneling. J. Phys. Chem. B 119, 2193–2203.

Merkel, A., Zülicke, L., 1987. Nonempirical parameter estimate for the statistical adiabatic theory of unimolecular fragmentation. Mol. Phys. 60 (6), 1379–1393.

Merkt, F., Quack, M., 2011. Molecular quantum mechanics and molecular spectra, molecular symmetry, and interaction of matter with radiation. In: Quack, M., Merkt, F. (Eds.), Handbook of High-Resolution Spectroscopy, Vol. 1. Wiley, Chichester, New York, pp. 1–55 (Chapter 1). See also preface to this Handbook.

Merkt, F., Willitsch, S., Hollenstein, U., 2011. High-resolution photoelectron spectroscopy. In: Quack, M., Merkt, F. (Eds.), Handbook of High-Resolution Spectroscopy, Vol. 3. Wiley, Chichester, New York, pp. 1617–1654 (Chapter 44).

Merzbacher, E., 2002. The early history of quantum tunneling. Phys. Today 59, 44–49.

Meyer, R., 1970. Trigonometric interpolation method for one dimensional quantum mechanical problems. J. Chem. Phys. 52 (4), 2053–2059.

Meyer, H.-D., Gatti, F., Worth, G.A., 2009. Multidimensional Quantum Dynamics: MCTDH Theory and Applications. John Wiley-VCH, Weinheim.

Meyer, H.-D., Manthe, U., Cederbaum, L.S., 1990. The multi-configurational time-dependent Hartree approach. Chem. Phys. Lett. 165 (1), 73–78.

Mielke, S.L., Garrett, B.C., Fleming, D.G., Truhlar, D.G., 2015. Zero-point energy, tunnelling, and vibrational adiabaticity in the $Mu + H_2$ reaction. Mol. Phys. 113, 160–175.

Miller, W.H., 1974. Classical-limit quantum mechanics and the theory of molecular collisions. Adv. Chem. Phys. 25, 69–177.

Miller, W.H., 1975a. The classical S-matrix in molecular collisions. Adv. Chem. Phys. 30, 77–136.

Miller, W.H., 1975b. Semiclassical limit of quantum mechanical transition state theory for nonseparable systems. J. Chem. Phys. 62 (1), 1899–1906.

Miller, W.H., 2014. A journey through chemical dynamics. Annu. Rev. Phys. Chem. 65, 1–19.

Miller, W.H., Handy, N.C., Adams, J.E., 1980. Reaction path Hamiltonian for polyatomic molecules. J. Chem. Phys. 72 (1), 99–112.

Mills, I., Quack, M., 2002. The symmetry groups of non-rigid molecules – introductory comment. Mol. Phys. 100 (1), 9–10.

Minissale, M., Congiu, E., Baouche, S., Chaabouni, H., Moudens, A., Dulieu, F., Accolla, M., Cazaux, S., Manicó, G., Pirronello, V., 2013. Quantum tunneling of oxygen atoms on very cold surfaces. Phys. Rev. Lett. 111, 053201.

Minissale, M., Congiu, E., Dulieu, F., 2014. Oxygen diffusion and reactivity at low temperature on bare amorphous olivine-type silicate. J. Chem. Phys. 140, 074705.

Miyamoto, Y., Tsubouchi, M., Momose, T., 2013. Infrared spectroscopy of chloromethyl radical in solid parahydrogen and its nuclear spin conversion. J. Phys. Chem. A 117, 9510–9517.

Morse, P.M., Stückelberg, E.C.G., 1931. Lösung des Eigenwertproblems eines Potentialfeldes mit zwei Minima. Helv. Phys. Acta 4, 337–354.

Moruzzi, G., Winnewisser, B.P., Winnewisser, M., Mukhopadhyay, I., Strumia, F., 2018. Microwave, Infrared and Laser Transitions in Methanol – Atlas of Assigned Lines from 0 to 1258 cm^{-1}, second ed., reprint after first (1995) ed. CRC Press, Boca Raton, FL.

Mukherjee, N., Perreault, W.E., Zare, R.N., 2017. Stark-induced adiabatic Raman ladder for preparing highly vibrationally excited quantum states of molecular hydrogen. J. Phys. B: At. Mol. Phys. 50, 144005.

Mukhopadhyay, A., Xantheas, S.S., Saykally, R.J., 2018. The water dimer II: theoretical investigations. Chem. Phys. Lett. 700, 163–175.

Müller-Dethlefs, K., Riese, M., 2011. Molecular clusters and noncovalent bonds probed by photoionization and photoelectron spectroscopy. In: Quack, M., Merkt, F. (Eds.), Handbook of High-Resolution Spectroscopy, Vol. 3. Wiley, Chichester, New York, pp. 1713–1740 (Chapter 56).

Nesbitt, D.J., 1994. High-resolution, direct laser absorption spectroscopy in slit supersonic jets: intermolecular forces and unimolecular vibrational dynamics in clusters. Annu. Rev. Phys. Chem. 45, 367–399.

Newton, R.R., Thomas, L.H., 1948. Internal molecular motions of large amplitude illustrated by the symmetrical vibration of ammonia. J. Chem. Phys. 16, 310–323.

Nguyen, N.T.T., Furukawa, H., Gándara, F., Nguyen, H.T., Cordova, K.E., Yaghi, O.M., 2014. Selective capture of carbon dioxide under humid conditions by hydrophobic chabazite-type zeolitic imidazolate frameworks. Angew. Chem., Int. Ed. 53, 10645–10648.

Nikitin, E.E., 1965. Statistical theory of endothermic reactions – Part 2. Monomolecular reactions. Teor. Eksp. Khim. 1, 90–94.

Nikitin, E.E., Troe, J., 2010a. Dynamics of ion-molecule complex formation at very low energies and temperatures. Phys. Chem. Chem. Phys. 7, 1540–1551.

Nikitin, E.E., Troe, J., 2010b. Mutual capture of dipolar molecules at low and very low energies. I. Approximate analytical treatment. J. Phys. Chem. A 114, 9762–9767.

Nimtz, G., 2011. Tunneling confronts special relativity. Found. Phys. 41, 1193–1199.

Nimtz, G., Haibel, A., 2004. Tunneleffekt – Räume ohne Zeit: Vom Urknall zum Wurmloch. John Wiley-VCH, Weinheim.

Nunes, C.M., Knezz, S.N., Reva, I., Fausto, R., McMahon, R.J., 2016. Evidence of a nitrene tunneling reaction: spontaneous rearrangement of 2-formyl phenylnitrene to an imino ketene in low-temperature matrixes. J. Am. Chem. Soc. 138, 15287–15290.

Oka, T., 2011. Orders of magnitude and symmetry in molecular spectroscopy. In: Quack, M., Merkt, F. (Eds.), Handbook of High-Resolution Spectroscopy, Vol. 1. Wiley, Chichester, New York, pp. 633–658 (Chapter 17).

Oppenheimer, J.R., 1928. Three notes on the quantum theory of aperiodic effects. Phys. Rev. 13, 66–81.

Ortlieb, M., Havenith, M., 2007. Proton transfer in $(HCOOH)_2$: an IR high-resolution spectroscopic study of the antisymmetric C-O stretch. J. Phys. Chem. A 111, 7355–7363.

Ospelkaus, S., Ni, K.K., Wang, D., de Miranda, M.H.G., Neyenhuis, B., Quéméner, G., Julienne, P.S., Bohn, J.L., Jin, D.S., Ye, J., 2010. Quantum-state controlled chemical reactions of ultracold potassium-rubidium molecules. Science 327, 853–857.

Ottinger, P., Leutwyler, S., 2012. Excitonic splitting and coherent electronic energy transfer in the gas-phase benzoic acid dimer. J. Chem. Phys. 137, 204303.

Ottinger, P., Leutwyler, S., Köppel, H., 2012. Vibrational quenching of excitonic splittings in H-bonded molecular dimers: the electronic Davydov splittings cannot match experiment. J. Chem. Phys. 136, 174308.

Paesani, F., 2016. Getting the right answers for the right reasons: toward predictive molecular simulations of water with many-body potential energy functions. Acc. Chem. Res. 49, 1844–1851.

Pechukas, P., Light, J.C., 1965. On detailed balancing and statistical theories of chemical kinetics. J. Chem. Phys. 42, 3281–3291.

Perreault, W.E., Mukherjee, N., Zare, R.N., 2017. Quantum control of molecular collisions at 1 Kelvin. Science 358, 356–359.

Perry, D.S., 2008. Torsion-vibration coupling in methanol: diabatic behavior in the CH overtone region. J. Phys. Chem. A 112, 215–223.

Perry, D.S., 2009. The adiabatic approximation as a diagnostic tool for torsion-vibration dynamics. J. Mol. Spectrosc. 257, 1–10.

Petersen, J., Pollak, E., 2017. Tunneling flight time, chemistry, and special relativity. J. Phys. Chem. Lett. 8, 4017–4022.

Petersen, J., Pollak, E., 2018. Quantum coherence in the reflection of above barrier wavepackets. J. Chem. Phys. 148, 074111.

Pfeiffer, P., 1983. Molecular structure derived from first-principles quantum mechanics: two examples. In: Hinze, J. (Ed.), Energy Storage and Redistribution in Molecules, Proc. of Two Workshops. Bielefeld, June 1980. Plenum Press, New York, pp. 315–326.

Pham, T., Forrest, K.A., Falcao, E.H.L., Eckert, J., Space, B., 2016a. Exceptional H_2 sorption characteristics in a Mg^{2+}-based metal-organic framework with small pores: insights from experimental and theoretical studies. Phys. Chem. Chem. Phys. 18, 1786–1796.

Pham, T., Forrest, K.A., Space, B., Eckert, J., 2016b. Dynamics of H_2 adsorbed in porous materials as revealed by computational analysis of inelastic neutron scattering spectra. Phys. Chem. Chem. Phys. 18, 17141–17158.

Piccini, G.M., Alessio, M., Sauer, J., 2018. Ab initio study of methanol and ethanol adsorption on Brœnsted sites in zeolite H-MFI. Phys. Chem. Chem. Phys. 20, 19964–19970.

Piecha-Bisiorek, A., Bator, G., Sawka-Dobrowolska, W., Sobczyk, L., Rok, M., Medycki, W., Schneider, G.J., 2014. Structure and tunneling splitting spectra of methyl groups of tetramethylpyrazine in complexes with chloranilic and bromanilic acids. J. Phys. Chem. A 118, 7159–7166.

Pine, A.S., Howard, B.J., 1986. Hydrogen bond energies of the HF and HCl dimers from absolute infrared intensities. J. Chem. Phys. 84, 590–596.

Pine, A.S., Lafferty, W.J., 1983. Rotational structure and vibrational predissociation in the HF stretching bands of the HF dimer. J. Chem. Phys. 78, 2154–2162.

Pine, A.S., Lafferty, W.J., Howard, B.J., 1984. Vibrational predissociation, tunneling, and rotational saturation in the HF and DF dimers. J. Chem. Phys. 81, 2939–2950.

Pittico, L., Schäfer, M., Merkt, F., 2012. Structure and dynamics of the electronically excited C_1 and D_0^+ states of ArXe from high-resolution vacuum ultraviolet spectra. J. Chem. Phys. 136, 074304.

Plazanet, M., Neumann, M.A., Trommsdorff, H.P., 2000. Methyl group rotational tunneling in vibrational spectra of crystals at low temperatures. Chem. Phys. Lett. 320, 651–657.

Polanyi, J.C., 1987. Some concepts in reaction dynamics (Nobel lecture). Angew. Chem., Int. Ed. 26, 952–971.

Pollak, E., 2017. Transition path time distribution, tunneling times, friction, and uncertainty. Phys. Rev. Lett. 116, 070401.

Pomerantz, A.E., Ausfelder, F., Zare, R.N., Althorpe, S.C., Aoiz, F.J., Banares, L., Castillo, J.F., 2004. Disagreement between theory and experiment in the simplest chemical reaction: collision energy dependent rotational distributions for $H + D_2 \rightarrow HD(v' = 3, J') + D$. J. Chem. Phys. 120, 3244–3255.

Poms, J., Hauser, A.W., Ernst, W.E., 2012. Helium nanodroplets doped with xenon and rubidium atoms: a case study of Van der Waals interactions between heliophilic and heliophobic dopants. Phys. Chem. Chem. Phys. 14, 15158–15165.

Potapov, A., Asselin, P., 2014. High-resolution jet spectroscopy of weakly bound binary complexes involving water. Int. Rev. Phys. Chem. 33, 275–300.

Powers, A., Marsalek, O., Xu, M., Ulivi, L., Colognesi, D., Tuckerman, M.E., Bačić, Z., 2016. Impact of the condensed-phase environment on the translation-rotation eigenstates and spectra of a hydrogen molecule in clathrate hydrates. J. Phys. Chem. Lett. 7, 308–313.

Prager, M., Heidemann, A., 1997. Rotational tunneling and neutron spectroscopy: a compilation. Chem. Rev. 97, 2933–2966.

Pratt, D.W., 2011. Electronic spectroscopy in the gas phase. In: Quack, M., Merkt, F. (Eds.), Handbook of High-Resolution Spectroscopy, Vol. 2. Wiley, Chichester, New York, pp. 1291–1320 (Chapter 34).

Prelog, V., Wieland, P., 1944. Über die Spaltung der Trögerschen Base in optisehe Antipoden, ein Beitrag zur Stereochemie des dreiwertigen Stickstoffs. Helv. Chim. Acta 27, 1127–1134.

Prentner, R., Quack, M., Stohner, J., Willeke, M., 2015. Wavepacket dynamics of the axially chiral molecule Cl-O-O-Cl under coherent radiative excitation and including electroweak parity violation. J. Phys. Chem. A 119 (51), 12805–12822.

Press, W., 1981. Single-Particle Rotations in Molecular Crystals. Springer Tracts in Modern Physics, vol. 92. Springer, Berlin, Heidelberg.

Primas, H., 1981. Chemistry, Quantum Mechanics and Reductionism: Perspectives in Theoretical Chemistry. Springer, Berlin.

Pylaeva, S., Allolio, S., Koeppe, B., Gleb, S., Denisov, G.S., Limbach, H.-H., Sebastianid, D., Tolstoy, P.M., 2015. Proton transfer in a short hydrogen bond caused by solvation shell fluctuations: an ab initio MD and NMR/UV study of an $(OHO)^-$ bonded system. Mol. Phys. 114, 2251–2262.

Qu, C., Bowman, J.M., 2013. Full-dimensional, ab initio potential energy surface for $CH_3OH \rightarrow CH_3+OH$. Mol. Phys. 111, 1964–1971.

Qu, C., Bowman, J.M., 2016. An ab initio potential energy surface for the formic acid dimer: zero-point energy, selected anharmonic fundamental energies, and ground-state tunneling splitting calculated in relaxed 1-4-mode subspaces. Phys. Chem. Chem. Phys. 18, 24835–24840.

Quack, M., 1977. Detailed symmetry selection-rules for reactive collisions. Mol. Phys. 34 (2), 477–504.

Quack, M., 1978. Theory of unimolecular reactions induced by monochromatic infrared radiation. J. Chem. Phys. 69 (3), 1282–1307.

Quack, M., 1979. Quantitative comparison between detailed (state selected) relative rate data and averaged (thermal) absolute rate data for complex-forming reactions. J. Phys. Chem. 83 (1), 150–158.

Quack, M., 1980. Statistical models for product energy distributions in bimolecular reactions with metastable intermediates. Chem. Phys. 51, 353–367.

Quack, M., 1981. Statistical-mechanics and dynamics of molecular fragmentation. Nuovo Cimento Soc. Ital. Fis. B 63 (1), 358–377.

Quack, M., 1985. On the densities and numbers of rovibronic states of a given symmetry species: rigid and nonrigid molecules, transition states, and scattering channels. J. Chem. Phys. 82 (7), 3277–3283.

Quack, M., 1986. On the measurement of the parity violating energy difference between enantiomers. Chem. Phys. Lett. 132, 147–153.

Quack, M., 1989. Structure and dynamics of chiral molecules. Angew. Chem., Int. Ed. 28 (5), 571–586. Angew. Chem. 101, 1989 588–604.

Quack, M., 1990. The role of quantum intramolecular dynamics in unimolecular reactions. Philos. Trans. R. Soc. Lond. A 332 (1625), 203–220.

Quack, M., 1999. Intramolekulare Dynamik: Irreversibilität, Zeitumkehrsymmetrie und eine absolute Moleküluhr. Nova Acta Leopoldina 81 (Neue Folge 314), 137–173.

Quack, M., 2001. Molecules in motion. Chimia 55, 753–758.

Quack, M., 2002. How important is parity violation for molecular and biomolecular chirality? Angew. Chem., Int. Ed. 41, 4618–4630. Angew. Chem. 114, 2002 4812–4825.

Quack, M., 2003. Molecular spectra, reaction dynamics, symmetries and life. Chimia 57 (4), 147–160.

Quack, M., 2004. Time and time reversal symmetry in quantum chemical kinetics. In: Brändas, E.J., Kryachko, E.S. (Eds.), Fundamental World of Quantum Chemistry. A Tribute to the Memory of Per-Olov Löwdin, Vol. 3. Kluwer Academic Publishers, Dordrecht, pp. 423–474.

Quack, M., 2006. Electroweak quantum chemistry and the dynamics of parity violation. In: Naidoo, K.J., Brady, J., Field, M.J., Gao, J., Hann, M. (Eds.), Modelling Molecular Structure and Reactivity. Royal Society of Chemistry, Cambridge.

Quack, M., 2011a. Frontiers in spectroscopy (concluding paper to Faraday discussion 150. Faraday Discuss. Chem. Soc. 150, 533–565.

Quack, M., 2011b. Fundamental symmetries and symmetry violations from high resolution spectroscopy. In: Quack, M., Merkt, F. (Eds.), Handbook of High Resolution Spectroscopy, Vol. 1. Wiley, Chichester, New York, pp. 659–722 (Chapter 18).

Quack, M., 2014. The concept of law and models in chemistry. Eur. Rev. 22, S50–S86.

Quack, M., 2015a. Die Spiegelsymmetrie des Raumes und die Chiralität in Chemie, Physik, und in der biologischen Evolution. Nova Acta Leopoldina NF 127 (412), 19 and pp. 119–166 (English summary p. 33).

Quack, M., 2015b. On biomolecular homochirality as a quasi-fossil of the evolution of life. Adv. Chem. Phys. 157, 249–290 (Chapter 18).

Quack, M., Jans-Bürli, S., 1986. Molekulare Thermodynamik und Kinetik, Vol. 1. Vd F Publishers, Zürich, pp. A61–A71.

Quack, M., Merkt, F. (Eds.), 2011. Handbook of High-Resolution Spectroscopy. John Wiley, New York (3 volumes).

Quack, M., Schmitt, U., Suhm, M.A., 1993a. Evidence for the $(HF)_5$ complex in the HF stretching FTIR absorption-spectra of pulsed and continuous supersonic jet expansions of hydrogen-fluoride. Chem. Phys. Lett. 208 (5–6), 446–452.

Quack, M., Schmitt, U., Suhm, M.A., 1997. FTIR spectroscopy of hydrogen fluoride clusters in synchronously pulsed supersonic jets. Isotopic isolation, substitution and 3-d condensation. Chem. Phys. Lett. 269 (1–2), 29–38. Also: J. Aerosol Sci. 28 (Supplement) (1997) 363–364.

Quack, M., Stockburger, M., 1972. Resonance fluorescence of aniline vapor. J. Mol. Spectrosc. 43 (1), 87–116.

Quack, M., Stohner, J., 2000. Influence of parity violating weak nuclear potentials on vibrational and rotational frequencies in chiral molecules. Phys. Rev. Lett. 84 (17), 3807–3810.

Quack, M., Stohner, J., Suhm, M.A., 1993b. Vibrational dynamics of $(HF)_n$ aggregates from an ab initio based analytical $(1+2+3)$-body potential. J. Mol. Struct. 294, 33–36.

Quack, M., Stohner, J., Suhm, M.A., 2001. Analytical three-body interaction potentials and hydrogen bond dynamics of hydrogen fluoride aggregates, $(HF)_n$, $n \geq 3$. J. Mol. Struct. 599 (1–3), 381–425.

Quack, M., Stohner, J., Willeke, M., 2008. High-resolution spectroscopic studies and theory of parity violation in chiral molecules. Annu. Rev. Phys. Chem. 59, 741–769.

Quack, M., Suhm, M.A., 1991a. Potential energy surfaces, quasi-adiabatic channels, rovibrational spectra, and intramolecular dynamics of $(HF)_2$ and its isotopomers from quantum Monte Carlo calculations. J. Chem. Phys. 95 (1), 28–59.

Quack, M., Suhm, M.A., 1991b. Quasi-adiabatic channels and effective transition-state barriers for the disrotatory in-plane hydrogen-bond exchange motion in $(HF)_2$. Chem. Phys. Lett. 183 (3–4), 187–194.

Quack, M., Suhm, M.A., 1995. Accurate quantum Monte-Carlo calculations of the tunneling splitting in $(HF)_2$ on a 6-dimensional potential hypersurface. Chem. Phys. Lett. 234 (1–3), 71–76.

Quack, M., Suhm, M.A., 1997. Potential energy hypersurfaces for hydrogen bonded clusters $(HF)_n$. In: Kryachko, E.S., Calais, J.L. (Eds.), Conceptual Perspectives in Quantum Chemistry. Kluwer, Dordrecht, pp. 415–463.

Quack, M., Suhm, M.A., 1998. Spectroscopy and quantum dynamics of hydrogen fluoride clusters. In: Bačić, Z., Bowman, J. (Eds.), Advances in Molecular Vibrations and Collision Dynamics, Vol. 3, Molecular Clusters. JAI Press Inc., Stamford, Conn. and London, England, pp. 205–248.

Quack, M., Sutcliffe, E., 1985. On the validity of the quasiresonant approximation for molecular infrared-multiphoton excitation. J. Chem. Phys. 83 (8), 3805–3812.

Quack, M., Sutcliffe, B.T., 1986. URIMIR: programs for the calculation of the quantum dynamics of IR multiphoton excitation and dissociation (unimolecular reactions induced by monochromatic infrared radiation), quantum chemistry program exchange, program 515. QCPE Bull. 6, 98. Extended and updated in: Marquardt, R., Quack, M., Stohner, J., Thanopulos, I., Quantum dynamics of the iodine atom in a strong laser field as calculated with the URIMIR package, Mol. Phys. 117 (2019) 3132–3147.

Quack, M., Troe, J., 1974. Specific rate constants of unimolecular processes: statistical adiabatic channel model. Ber. Bunsenges. Phys. Chem. 78, 240–252.

Quack, M., Troe, J., 1975a. Complex formation in reactive and inelastic scattering. Ber. Bunsenges. Phys. Chem. 79, 170–183.

Quack, M., Troe, J., 1975b. Product state distributions after dissociation. Ber. Bunsenges. Phys. Chem. 79, 469–475.

Quack, M., Troe, J., 1976. Information, memory and statistical theories of elementary chemical reactions. Ber. Bunsenges. Phys. Chem. 80, 1140–1149.

Quack, M., Troe, J., 1977a. Maximum free energy criterion for the high pressure limit of dissociation reactions. Ber. Bunsenges. Phys. Chem. 81, 329–337.

Quack, M., Troe, J., 1977b. Unimolecular Reactions and Energy Transfer of Highly Excited Molecules, Vol. 2. The Chemical Society, London, pp. 175–238.

Quack, M., Troe, J., 1977c. Vibrational-relaxation of diatomic-molecules in complex-forming collisions with reactive atoms. Ber. Bunsenges. Phys. Chem. 81 (2), 160–162.

Quack, M., Troe, J., 1981. Statistical methods in scattering. In: Henderson, D. (Ed.), Theoretical Chemistry: Advances and Perspectives (Theory of Scattering, Papers in Honor of Henry Eyring), Vol. 6B. Academic Press, New York, pp. 199–276.

Quack, M., Troe, J., 1998. Statistical adiabatic channel model. In: von Ragué Schleyer, P., Allinger, N., Clark, T., Gasteiger, J., Kollman, P.A., Schaefer III, H.F., Schreiner, P.R. (Eds.), Encyclopedia of Computational Chemistry, Vol. 4. John Wiley and Sons, pp. 2708–2726.

Quack, M., Willeke, M., 1999. Ab initio calculations for the anharmonic vibrational resonance dynamics in the overtone spectra of the coupled OH and CH chromophores in CD_2H-OH. J. Chem. Phys. 110 (24), 11958–11970.

Quack, M., Willeke, M., 2006. Stereomutation tunneling switching dynamics and parity violation in chlorineperoxide Cl–O–O–Cl. J. Phys. Chem. A 110 (9), 3338–3348.

Reiher, M., Wolf, A., 2009. Relativistic Quantum Chemistry: The Fundamental Theory of Molecular Science, first ed. Wiley-VCH, Weinheim.

Reiss, H.R., 2014. The tunnelling model of laser-induced ionization and its failure at low frequencies. J. Phys. B: At. Mol. Phys. 47, 204006.

Reva, I., Nunes, C.M., Biczysko, M., Fausto, R., 2015. Conformational switching in pyruvic acid isolated in Ar and N_2 matrixes: spectroscopic analysis, anharmonic simulation, and tunneling. J. Phys. Chem. A 119, 2614–2627.

Rice, S.A., 1975. Some comments on the dynamics of primary photochemical processes. In: Lim, E.C. (Ed.), Excited States, Vol. 2. Elsevier Inc., Amsterdam, pp. 111–320.

Richard, R.M., Lao, K.U., Herbert, J.M., 2014. Aiming for benchmark accuracy with the many-body expansion. Acc. Chem. Res. 47, 2828–2836.

Richardson, J.O., 2016. Derivation of instanton rate theory from first principles. J. Chem. Phys. 144, 114106.

Richardson, J.O., 2018. Ring-polymer instanton theory. Int. Rev. Phys. Chem. 37, 171–216.

Richardson, J.O., Althorpe, S.C., 2009. Ring-polymer molecular dynamics rate-theory in the deep-tunneling regime: connection with semiclassical instanton theory. J. Chem. Phys. 131, 214106.

Richardson, J.O., Althorpe, S.C., 2011. Ring-polymer instanton method for calculating tunneling splittings. J. Chem. Phys. 134, 054109.

Richardson, J.O., Althorpe, S.C., Wales, D.J., 2011. Instanton calculations of tunneling splittings for water dimer and trimer. J. Chem. Phys. 135, 124109.

Richardson, J.O., Bauer, R., Thoss, T., 2015. Semiclassical Green's functions and an instanton formulation of electron-transfer rates in the nonadiabatic limit. J. Chem. Phys. 143, 134115.

Richardson, J.O., Pérez, C., Lobsinger, S., Reid, A.A., Temelso, B., Shields, G.C., Kisiel, Z., Wales, D.J., Pate, B.H., Althorpe, S.C., 2016. Concerted hydrogen-bond breaking by quantum tunneling in the water hexamer prism. Science 351, 1310–1314.

Richardson, J.O., Wales, D.J., Althorpe, S.C., McLaughlin, R.P., Viant, M.R., Shih, O., Saykally, R.J., 2013. Investigation of terahertz vibration-rotation tunneling spectra for the water octamer. J. Phys. Chem. A 117, 6960–6966.

Riniker, S., Allison, J.A., van Gunsteren, W.F., 2012. On developing coarse-grained models for biomolecular simulation: a review. Phys. Chem. Chem. Phys. 14, 12423–12430.

Rommel, J.P., Goumans, T.M., Kästner, J., 2011. Locating instantons in many degrees of freedom. J. Chem. Theory Comput. 7, 690–698.

Roston, D., Islam, Z., Kohen, A., 2014. Kinetic isotope effects as a probe of hydrogen transfers to and from common enzymatic cofactors. Arch. Biochem. Biophys. 544, 96–104.

Roston, D., Kohen, A., 2013. A critical test of the "tunneling and coupled motion" concept in enzymatic alcohol oxidation. J. Am. Chem. Soc. 135, 13624–13627.

Ruiz-Pernía, J.J., Luk, L.Y.P., Garcia-Meseguer, R., Marti, S., Loveridge, E.J., Tunón, I., Moliner, V., Allemann, R.K., 2013. Increased dynamic effects in a catalytically compromised variant of escherichia coli dihydrofolate reductase. J. Am. Chem. Soc. 135, 18689–18696.

Rungtaweevoranit, B., Baek, J., Araujo, J.R., Archanjo, B.S., Choi, K.M., Yaghi, O.M., Somorjai, G.A., 2016. Asymmetric autocatalysis triggered by carbon isotope ($^{13}C/^{12}C$) chirality. Nano Lett. 16, 7645–7649.

Rutherford, E., 1900. Radioactive substance emitted from thorium compounds. Philos. Mag. 49 (1).

Sala, M., Guérin, S., Gatti, F., Marquardt, R., Meyer, H.-D., 2012. Laser-induced enhancement of tunneling in NHD_2. J. Chem. Phys. 113, 194308.

Sarka, J., Császár, A.G., 2016. Interpretation of the vibrational energy level structure of the astructural molecular ion H_5^+ and all of its deuterated isotopomers. J. Chem. Phys. 18, 154309.

Sarka, J., Császár, A.G., Althorpe, S.A., Wales, D.J., Mátyus, E., 2016a. Rovibrational transitions of the methane–water dimer from intermolecular quantum dynamical computations. Phys. Chem. Chem. Phys. 18, 22816–22826.

Sarka, J., Lauvergnat, D., Brites, V., Császár, A.G., Léonard, C., 2016b. Rovibrational energy levels of the $F^-(H_2O)$ and $F^-(D_2O)$ complexes. Phys. Chem. Chem. Phys. 18, 17678–17690.

Sato, I., Urabe, H., Saori, I., Ishiguro, S., Shibata, T., Soai, K., 2003. Amplification of chirality from extremely low to greater than 99.5% ee by asymmetric autocatalysis. Angew. Chem., Int. Ed. 42, 315–317.

Sauer, J., 2016. Broensted activity of two-dimensional zeolites compared to bulk materials. Faraday Discuss. Chem. Soc. 188, 227–234.

Sauer, J., Freund, H.J., 2015. Models in catalysis. Catal. Lett. 145, 109–125.

Schmitt, M., Meerts, W.L., 2011. Rotationally resolved electronic spectroscopy and automatic assignment techniques using evolutionary algorithms. In: Quack, M., Merkt, F. (Eds.), Handbook of High-Resolution Spectroscopy, Vol. 2. Wiley, Chichester, New York, pp. 1345–1372 (Chapter 36).

Schnell, M., 2011. Group theory for high-resolution spectroscopy of nonrigid molecules. In: Quack, M., Merkt, F. (Eds.), Handbook of High-Resolution Spectroscopy, Vol. 1. Wiley, Chichester, New York, pp. 607–623 (Chapter 16).

Schnitzler, E.G., Jäger, W., 2014. The benzoic acid–water complex: a potential atmospheric nucleation precursor studied using microwave spectroscopy and ab initio calculations. Phys. Chem. Chem. Phys. 16, 2305–2314.

Schoedel, A., Ji, Z., Yaghi, O.M., 2016. The role of metal-organic frameworks in a carbon-neutral energy cycle. Nat. Energy 1, 16034–16046.

Schopper, H., 1957. Circular polarization of gamma rays – further proof for parity-failure in beta decay. Philos. Mag. 2, 710–713.

Schottky, W., 1931. Leitungs- und Photoeffekte an Sperrschichten. Z. Phys. 32, 833–842.

Schreiner, P.R., 2017. Tunneling control of chemical reactions: the third reactivity paradigm. J. Am. Chem. Soc. 139, 15276–15283.

Schreiner, P.R., Reisenauer, H.P., Ley, D., Gerbig, D., Wu, C.H., Allen, W.D., 2011. Methylhydroxycarbene: tunneling control of a chemical reaction. Science 332, 1300–1303.

Schreiner, P.R., Wagner, J.P., Reisenauer, H.P., Gerbig, D., Ley, D., Sarka, J., Császár, A.G., Vaughn, A., Allen, W.D., 2015. Domino tunneling. J. Am. Chem. Soc. 137, 7828–7834.

Schulenburg, A.M., Merkt, F., 2010. Internal rotation in Jahn–Teller coupled systems: the ethene and allene cations. Chem. Phys. 377, 66–77.

Schultz, A., Cruse, H.W., Zare, R.N., 1972. Laser-induced fluorescence: a method to measure the internal state distribution of reaction products. J. Chem. Phys. 57, 1354–1355.

Seyfang, G., Quack, M., 2018. Atomare und molekulare Tunnelprozesse. Nachr. Chem. 66, 307–315.

Shan, X., Burd, T.A.H., Clary, D.C., 2019. New developments in semiclassical transition-state theory. J. Phys. Chem. A 123, 4639–4657.

Shenderovich, I.G., Lesnichin, S.B., Tu, C., Silverman, D.N., Tolstoy, P.M., Denisov, G.S., Limbach, H.-H., 2015. NMR studies of active-site properties of human carbonic anhydrase II by using ^{15}N-labeled 4-methylimidazole as a local probe and histidine hydrogen-bond correlations. Chem. Eur. J. 21, 2915–2929.

Shipman, S.T., Pate, B.H., 2011. New techniques in microwave spectroscopy. In: Quack, M., Merkt, F. (Eds.), Handbook of High-Resolution Spectroscopy, Vol. 2. Wiley, Chichester, New York, pp. 801–827 (Chapter 20).

Shubert, V.A., Schmitz, D., Schnell, M., 2013. Communication through the phenyl ring: internal rotation and nuclear quadrupole splitting in p-halotoluenes. Mol. Phys. 111, 2189–2197.

Signorell, R., Merkt, F., 2000. PFI-ZEKE photoelectron spectra of the methane cation and the dynamic Jahn-Teller effect. Faraday Discuss. Chem. Soc. 115, 205–228.

Signorell, R., Sommavilla, M., Merkt, F., 1999. Jahn–Teller distortion in $CD_2H_2^+$ from a rotationally resolved photoelectron spectrum. Chem. Phys. Lett. 312, 139–148.

Simmen, B., Mátyus, E., Reiher, M., 2013. Elimination of the translational kinetic energy contamination in pre-Born–Oppenheimer calculations. Mol. Phys. 111, 2086–2092.

Smith, I.W.M., 2006. Reactions at very low temperatures: gas kinetics at a new frontier. Angew. Chem., Int. Ed. 45, 2842–2861.

Snels, M., Fusina, L., Hollenstein, H., Quack, M., 2000. The ν_1 and ν_3 bands of ND_3. Mol. Phys. 98 (13), 837–854.

Snels, M., Hollenstein, H., Quack, M., 2003. The NH and ND stretching fundamentals of $^{14}ND_2H$. J. Chem. Phys. 119 (15), 7893–7902.

Snels, M., Hollenstein, H., Quack, M., 2006a. Mode selective tunneling dynamics observed by high resolution spectroscopy of the bending fundamentals of $^{14}NH_2D$ and $^{14}ND_2H$. J. Chem. Phys. 125, 194319.

Snels, M., Hollenstein, H., Quack, M., 2006b. The NH and ND stretching fundamentals of $^{14}NH_2D$. J. Mol. Spectros. 237 (2), 143–148.

Snels, M., Horká-Zelenková, V., Hollenstein, H., Quack, M., 2011. High resolution FTIR and diode laser spectroscopy of supersonic jets. In: Quack, M., Merkt, F. (Eds.), Handbook of High Resolution Spectroscopy; High Resolution FTIR and Diode Laser Spectroscopy of Supersonic Jets, Vol. 2. Wiley, Chichester, New York, pp. 1021–1067 (Chapter 27).

Soai, K., Shibata, T., Morioka, H., Choji, K., 1995. Asymmetric autocatalysis and amplification of enantiomeric excess of a chiral molecule. Nature 373, 767–768.

Song, L., Kästner, J., 2016. Formation of the prebiotic molecule NH_2CHO on astronomical amorphous solid water surfaces: accurate tunneling rate calculations. Phys. Chem. Chem. Phys. 18, 29278–29285.

Sprecher, D., Beyer, M., Merkt, F., 2013. Precision measurement of the ionisation energy of the 3dσ GK state of H_2. Mol. Phys. 111, 2100–2107.

Stocker, T.F., Johnson, E.R., 1991. The trapping and scattering of topographic waves by estuaries and headlands. J. Fluid Mech. 222, 501–524.

Stohner, J., Quack, M., 2011. Conventions, symbols, quantities, units and constants for high resolution molecular spectroscopy. In: Quack, M., Merkt, F. (Eds.), Handbook of High-Resolution Spectroscopy, Vol. 1. Wiley, Chichester, New York, pp. 263–324 (Chapter 5).

Stratt, R.M., Handy, N.C., Miller, W.H., 1979. On the quantum mechanical implications of classical ergodicity. J. Chem. Phys. 71, 3311–3322.

Swalen, J.D., Ibers, J.A., 1962. Potential function for the inversion of ammonia. J. Chem. Phys. 36, 1914–1918.

Tan, X.Q., Majewski, W.A., Plusquellic, D.F., Pratt, D.W., 1989. Methyl torsional barriers in different electronic states. Simultaneous determination from the rotationally resolved fluorescence excitation spectrum of a large molecule. J. Chem. Phys. 90, 2521–2522.

Tan, X.Q., Majewski, W.A., Plusquellic, D.F., Pratt, D.W., 1991. Methyl group torsional dynamics from rotationally resolved electronic spectra. 1- and 2-methylnaphthalene. J. Chem. Phys. 94, 7721–7733.

Tanaka, K., Harada, K., Endo, Y., 2019. FTMW Spectroscopy of ^{13}C-Tropolone. In: Proc. HRMS. Dijon.

Tanaka, K., Harada, K., Yamada, K.M.T., 2011. THz and submillimeter-wave spectroscopy of molecular complexes. In: Quack, M., Merkt, F. (Eds.), Handbook of High-Resolution Spectroscopy, Vol. 2. Wiley, Chichester, New York, pp. 853–896 (Chapter 22).

Tanaka, K., Honjo, H., Tanaka, T., Kohguchi, H., Ohshima, Y., Endo, Y., 1999. Determination of the proton tunneling splitting of tropolone in the ground state by microwave spectroscopy. J. Chem. Phys. 110, 1969–1978.

Tautermann, C.S., Voegele, A.F., Loerting, T., Liedl, K.R., 2002. The optimal tunneling path for the proton transfer in malonaldehyde. J. Chem. Phys. 117, 1962–1966.

Telle, H.R., Steinmeyer, G., Dunlop, A.E., Stenger, J., Sutter, D.H., Keller, U., 1999. Carrier-envelope offset phase control: a novel concept for absolute optical frequency measurement and ultrashort pulse generation. Appl. Phys. B 69, 327–332.

Tennyson, J., 2011. High accuracy rotation-vibration calculations on small molecules. In: Quack, M., Merkt, F. (Eds.), Handbook of High-Resolution Spectroscopy, Vol. 1. Wiley, Chichester, New York, pp. 551–572 (Chapter 13).

Thomas, J., Liu, X., Jäger, W., Xu, Y., 2015. Unusual H-bond topology and bifurcated H-bonds in the 2-fluoroethanol trimer. Angew. Chem., Int. Ed. 54, 11711–11715.

Toh, S.Y., Djuricanin, P., Momose, T., Miyazaki, J., 2015. UV photochemistry of benzene and cyclohexadienyl radical in solid parahydrogen. J. Phys. Chem. A 119, 2683–2691.

Tomaselli, M., Degen, C., Meier, B.H., 2003. Haupt magnetic double resonance. J. Chem. Phys. 118, 8559–8562.

Tomaselli, M., Meier, U., Meier, B.H., 2004. Tunneling-induced spin alignment at low and zero field. J. Chem. Phys. 120, 4051–4054.

Townes, C.H., 1964. Production of coherent radiation by atoms and molecules. The Noble Foundation, Stockholm.

Townes, C.H., 1965. Production of coherent radiation by atoms and molecules. Science 149, 831–841.

Townes, C.H., Schawlow, A.L., 1975. Microwave Spectroscopy. Dover Books, New York. Corrected republication of the first edition, McGraw-Hill, 1955.

Troe, J., 1987. Statistical adiabatic channel model for ion-molecule capture processes. J. Chem. Phys. 87, 2773–2780.

Troe, J., 1992. Statistical aspects of ion-molecule reactions. Adv. Chem. Phys. 82, 485–529.

Troe, J., 1996. Statistical adiabatic channel model for ion-molecule capture processes II. Analytical treatment of ion-dipole capture. J. Chem. Phys. 105, 6249–6262.

Truhlar, D.G., Garrett, B.C., 1984. Variational transition-state theory. Annu. Rev. Phys. Chem. 35, 159–189.

Truhlar, D.G., Garrett, B.C., Klippenstein, S.J., 1996. Current status of transition-state theory. J. Phys. Chem. 100, 12771–12800.

Tuma, C., Sauer, J., 2015. Quantum chemical ab initio prediction of proton exchange barriers between CH_4 and different H-zeolites. J. Chem. Phys. 143, 102810.

Uy, D., Cordonnier, M., Oka, T., 1997. Observation of ortho-para H_3^+ selection rules in plasma chemistry. Phys. Rev. Lett. 78, 3844–3847.

Vaillant, C.L., Wales, D.J., Althorpe, S.C., 2018. Tunneling splittings from path-integral molecular dynamics using a Langevin thermostat. J. Chem. Phys. 148, 234102.

van Gunsteren, W.F., Bakowies, D., Baron, R., Chandrasekhar, I., Christen, M., Daura, X., Gee, P., Geerke, D.P., Glättli, A., Hünenberger, P.H., Kastenholz, M.A., Oostenbrink, C., Schenk, M., Trzesniak, D., an der Vegt, N.F.A., Yu, H.B., 2006. Biomolecular modeling: goals, problems, perspectives. Angew. Chem., Int. Ed. 45, 4064–4092.

Vandemaele, G., Coppens, P., Van Gerven, L., 1986. New method to study spin conversion of a nuclear-spin rotor with low tunnei splitting. Phys. Rev. Lett. 56, 1202–1205.

Vasilatou, K., Schäfer, M., Merkt, F., 2010. The rotational structure of the origin band of the pulsed-field-ionization, zero-kinetic-energy photoelectron spectra of propene-H_6 and propene-D_6. J. Phys. Chem. A 114, 11085–11090.

Veillard, A., Lehn, J.M., Munsch, B., 1968. An ab initio SCF-LCAO-MO study of the nitrogen inversion barriers in ammonia and in ethylenimine. Theor. Chim. Acta 9, 275–277.

Vilenkin, A., 1988. Quantum cosmology and the initial state of the Universe. Phys. Rev. D 37, 888–897.

Vissers, G.W.M., Groenenboom, G.C., van der Avoird, A., 2003. Spectrum and vibrational predissociation of the HF dimer. I. Bound and quasibound states. J. Chem. Phys. 119, 277–285.

von Puttkamer, K., Quack, M., 1985. High-resolution FTIR overtone spectra and hydrogen-bond dissociation dynamics in $(HF)_2$. Chimia 39 (11), 358–360.

von Puttkamer, K., Quack, M., 1987. High-resolution interferometric FTIR spectroscopy of $(HF)_2$ - analysis of a low-frequency fundamental near 400 cm^{-1}. Mol. Phys. 62 (5), 1047–1064.

von Puttkamer, K., Quack, M., 1989. Vibrational-spectra of $(HF)_2$, $(HF)_n$ and their D-isotopomers – mode selective rearrangements and nonstatistical unimolecular decay. Chem. Phys. 139 (1), 31–53.

von Puttkamer, K., Quack, M., Suhm, M.A., 1988. Observation and assignment of tunnelling rotational transitions in the far infrared-spectrum of $(HF)_2$. Mol. Phys. 65 (5), 1025–1045.

Wagner, J.P., Reisenauer, H.P., Hirvonen, V., Wu, C.-H., Tyberg, J.L., Allen, W.D., Schreiner, P.R., 2016. Tunnelling in carbonic acid. Chem. Commun. 52, 7858–7861.

Wall, F.T., Glockler, G., 1937. The double minimum problem applied to the ammonia molecules. J. Chem. Phys. 5, 314–315.

Waluk, J., 2017. Spectroscopy and tautomerization studies of porphycenes. Chem. Rev. 117, 2447–2480.

Wang, Y., Bowman, J.M., 2013. Mode-specific tunneling using the Q_{im} path: theory and an application to full-dimensional malonaldehyde. J. Chem. Phys. 139, 154303.

Wang, X., Perry, D.S., 1998. An internal coordinate model of coupling between the torsion and C-H vibrations in methanol. J. Chem. Phys. 109, 10795–10805.

Warshel, A., 2014. Multiscale modeling of biological functions: from enzymes to molecular machines (Nobel lecture). Angew. Chem., Int. Ed. 53, 10020–10031.

Weber, B., Nagata, Y., Ketzetzi, S., Tang, F., Smit, W.J., Bakker, H.J., Backus, E.H.G., Bonn, M., Bonn, D., 2018. Molecular insight into the slipperiness of ice. J. Phys. Chem. Lett. 9, 2838–2842.

Wentzel, G., 1926. Eine Verallgemeinerung der Quantenbedingungen für die Zwecke der Wellenmechanik. Z. Phys. 38, 518–529.

Wichmann, G., Miloglyadov, E., Seyfang, G., Quack, M., 2020. Nuclear spin symmetry conservation studied by cavity ring-down spectroscopy of ammonia in a seeded supersonic jet from a pulsed slit nozzle. Mol. Phys., e1752946. https://doi.org/10.1080/00268976.2020.1752946.

Wigner, E., 1932. Über das Überschreiten der Potentialschwelle bei chemischen Reaktionen. Z. Phys. Chem. B 19, 203–216.

Wigner, E., 1938. The transition state method. Trans. Faraday Soc. 34, 29–41.

Willitsch, S., 2011. Experimental methods in cation spectroscopy. In: Quack, M., Merkt, F. (Eds.), Handbook of High-Resolution Spectroscopy, Vol. 3. Wiley, Chichester, New York, pp. 1691–1712 (Chapter 46).

Wolfrum, J., 2001. Advanced laser spectroscopy in combustion chemistry: from elementary steps to practical devices. Faraday Discuss. Chem. Soc. 119, 1–26.

Wong, Y.T.A., Toh, S.Y., Djuricanin, P., Momose, T., 2015. Conformational composition and population analysis of β-alanine isolated in solid parahydrogen. J. Mol. Spectrosc. 310, 23–31.

Wörner, H.J., Corkum, P.B., 2011. Attosecond spectroscopy. In: Quack, M., Merkt, F. (Eds.), Handbook of High-Resolution Spectroscopy, Vol. 3. Wiley, Chichester, New York, pp. 1781–1804 (Chapter 49).

Wörner, H.J., Merkt, F., 2009. Jahn–Teller effects in molecular cations studied by photoelectron spectroscopy and group theory. Angew. Chem., Int. Ed. 48, 6404–6424.

Wörner, H.J., Merkt, F., 2011. Fundamentals of electronic spectroscopy. In: Quack, M., Merkt, F. (Eds.), Handbook of High-Resolution Spectroscopy, Vol. 1. Wiley, Chichester, New York, pp. 175–262 (Chapter 4).

Wörner, H.J., Qian, X., Merkt, F., 2007. Jahn-Teller effect in tetrahedral symmetry: large-amplitude tunneling motion and rovibronic structure of CH_4^+ and CD_4^+. J. Chem. Phys. 126, 144305.

Wu, C.S., Ambler, E., Hayward, R.W., Hoppes, D.D., Hudson, R.P., 1957. Experimental test of parity conservation in beta decay. Phys. Rev. 105, 1413–1415.

Wu, Q., Zhang, D.H., Zhang, J.Z.H., 1995. 6D quantum calculation of energy levels for HF stretching excited $(HF)_2$. J. Chem. Phys. 103, 2548–2554.

Würger, A., 1990. Nuclear spin conversion of methyl groups. Z. Phys. B, Condens. Matter 81, 273–279.

Xantheas, S.S., 1996. Significance of higher-order many-body interaction energy terms in water clusters and bulk water. Philos. Mag. B 76, 107–115.

Xu, M., Felker, P.M., Mamone, S., Horsewill, A.J., Rols, S., Whitby, R.J., Bačić, Z., 2019. The endofullerene $HF@C_{60}$: inelastic neutron scattering spectra from quantum simulations and experiment, validity of the selection rule, and symmetry breaking. J. Phys. Chem. Lett. 10, 5365–5371.

Xu, L.H., Hougen, J.T., Fisher, J.M., Lees, R.M., 2010. Symmetry and Fourier analysis of the ab initio-determined torsional variation of structural and Hessian-related quantities for application to vibration-torsion-rotation interactions in CH_3OH. J. Mol. Spectrosc. 260, 88–104.

Xu, L.H., Hougen, J.T., Lees, R.M., 2011. Comparison of independently calculated ab initio normal-mode displacements for the three C-H stretching vibrations of methanol along the internal rotation path. J. Mol. Spectrosc. 293, 38–59.

Xu, L.H., Hougen, J.T., Lees, R.M., Hougen, J.T., Bowman, J.M., Huang, X., Carter, S., 2014. On the physical interpretation of ab initio normal-mode coordinates for the three C–H stretching vibrations of methanol along the internal-rotation path. J. Mol. Spectrosc. 299, 11–16.

Xu, L.H., Wang, X., Cronin, T.J., Perry, D.S., Fraser, G.T., Pine, A.S., 1997. Sub-Doppler infrared spectra and torsion–rotation energy manifold of methanol in the CH-stretch fundamental region. J. Mol. Spectrosc. 185, 158–172.

Xu, M., Ye, S., Powers, A., Lawler, R., Turro, N.J., Bačić, Z., 2013. Inelastic neutron scattering spectrum of $H_2@C_{60}$ and its temperature dependence decoded using rigorous quantum calculations and a new selection rule. J. Chem. Phys. 139, 064309.

Xue, Z., Suhm, M.A., 2009. Probing the stiffness of the simplest double hydrogen bond: the symmetric hydrogen bond modes of jet-cooled formic acid dimer. J. Chem. Phys. 131 (5), 054301.

Yaghi, O.M., 2016. Reticular chemistry - construction, properties, and precision reactions of frameworks. J. Am. Chem. Soc. 138, 15507–15509.

Yaghi, O.M., 2019. Reticular chemistry in all dimensions. ACS Central Sci. 5, 1295–1300.

Yaghi, O.M., Kalmutzki, M.J., Diercks, C.S., 2019. Introduction to Reticular Chemistry Metal - Organic Frameworks and Covalent Organic Frameworks, first ed. Wiley-VCH, Weinheim.

Yamaguchi, Y., Schaefer III, H.F., 2011. Analytic derivative methods in molecular electronic structure theory: a new dimension to quantum chemistry and its applications to spectroscopy. In: Quack, M., Merkt, F. (Eds.), Handbook of High-Resolution Spectroscopy, Vol. 1. Wiley, Chichester, New York, pp. 325–362 (Chapter 6).

Yang, Y., Brorsen, K.R., Culpitt, T., Pak, M.V., Hammes-Schiffer, S., 2017. Development of a practical multicomponent density functional for electron–proton correlation to produce accurate proton densities. J. Chem. Phys. 147, 114113.

Yarkony, D.R., 1996. Diabolical conical intersections. Rev. Modern Phys. 68, 985–1013.

Yoder, B.L., Bravaya, K.B., Bodi, A., West, A.H.C., Sztáray, B., Signorell, R., 2015. Barrierless proton transfer across weak CH···O hydrogen bonds in dimethyl ether dimer. J. Chem. Phys. 142, 114303.

Yu, Z., Hammam, E., Klemperer, W., 2005. The (4, 0) mode of HF dimer at 14700 cm^{-1}. J. Chem. Phys. 122, 194318.

Zare, R.N., 1988. Angular Momentum, first ed. John Wiley, New York.

Zeller, S., Kunitski, M., Voigtsberger, J., Kalinin, A., Schottelius, A., Schober, C., Waitz, M., Sann, H., Hartung, A., Bauer, T., Pitzer, M., Trinter, F., Goihl, C., Janke, C., Richter, M., Kastirke, G., Weller, M., Czasch, A., Kitzler, M., Braune, M., Grisenti, R.E., Schöllkopf, W., Schmidt, L.P.H., Schöffler, M.S., Williams, J.B., Jahnke, T., Dörner, R., 2016. Imaging the He_2 quantum halo state using a free electron laser. Proc. Nat. Acad. Sci. 113, 14651–14655.

Zhang, X., Hrovat, D.A., Borden, W.T., 2010. Calculations predict that carbon tunneling allows the degenerate Cope rearrangement of semibullvalene to occur rapidly at cryogenic temperatures. Org. Lett. 12, 2798–2801.

Zhou, H.C., Long, J.R., Yaghi, O.M., 2016. Introduction to metal–organic frameworks. Chem. Rev. 112, 673–674.

Zielke, P., Suhm, M.A., 2007. Raman jet spectroscopy of formic acid dimers: low frequency vibrational dynamics and beyond. Phys. Chem. Chem. Phys. 9, 4528–4534.

Zimmermann, H., 1964. Protonic states in chemistry. Angew. Chem., Int. Ed. 3, 157–164.

Zimmermann, T., Vaníček, J., 2010. Three applications of path integrals: equilibrium and kinetic isotope effects, and the temperature dependence of the rate constant of the [1, 5] sigmatropic hydrogen shift in (z)-1, 3-Pentadiene. J. Mol. Mod. 16, 1779–1787.

CHAPTER 8

Ultrafast Femtosecond Dynamics and High-Resolution Spectroscopy of Molecular Cations

TOSHIAKI ANDO • ATSUSHI IWASAKI • KAORU YAMANOUCHI
Department of Chemistry, School of Science, the University of Tokyo, Tokyo, Japan

Abstract

Our recent studies on ultrafast nuclear dynamics and high-resolution spectroscopy of isolated molecular ions in which intense few-cycle near-IR laser pulses are used are introduced. By the pump–probe measurements of the fragmentation processes of methanol cations, we investigate ultrafast hydrogen migration in methanol cation in the time domain and determine vibrational frequencies of methanol cations in the frequency domain by Fourier transform of the yields of parent and fragment ions. We further demonstrate that the rovibrational level energy separations of D_2 and D_2^+ can be determined with ultrahigh resolution by strong-field Fourier transform spectroscopy.

8.1 INTRODUCTION

It has been revealed during the past two decades that, when molecules are exposed to an intense near-IR laser field whose intensity is in the range of 10^{12}–10^{15} W/cm^2, they exhibit a variety of dynamical phenomena such as tunnel ionization, chemical bond rearrangement, and Coulomb explosion (Yamanouchi, 2002). It was shown from momentum imaging of the fragment ions produced after Coulomb explosion of simple triatomic molecules that their skeletal structure changes within an ultrashort period of the laser field from linear to bent (Hishikawa et al., 1999a) or from bent to linear (Hishikawa et al., 1999b). It was also shown for ethanol that chemical bond breaking pathways can be changed sensitively by the characteristics of the laser field such as the pulse duration, pulse energy and carrier wavelength (Itakura et al., 2003; Yazawa et al., 2007, 2006).

Among the characteristic responses of polyatomic molecules to an intense laser field, ultrafast hydrogen migration occurring in hydrocarbon molecules is noteworthy. In 2004, it was revealed for acetonitrile, CH_3CN, that a hydrogen atom in the methyl group migrates to the nitrile group prior to the Coulomb explosion from the dication produced in a femtosecond intense laser field (Hishikawa et al., 2004). From the series of studies on methanol, CH_3OH, it was shown that more than one hydrogen atom migrates from methyl group to hydroxyl group prior to the C–O bond breaking in a femtosecond intense laser field (Okino et al., 2006a). Furthermore, from the study on methyl acetylene, CH_3C_2H, and its isotopologues, we learned that hydrogen atom migration proceeds simultaneously from the methyl to the ethynyl group and vice versa, which was referred to as hydrogen scrambling (Okino et al., 2012).

In the experimental identification of the hydrogen migration processes, a method called coincidence momentum imaging (CMI) played a crucial role, in which fragment ions produced from Coulomb explosion of multiply charged cations are detected in coincidence by a position sensitive detector (Hasegawa et al., 2001). In the course of the CMI measurements of hydrocarbon molecules ionized by an intense laser field, it was found that a triatomic hydrogen molecular ion, H_3^+, is formed from a variety of hydrocarbon dication species. For example, methanol dication produces $H_3^+ + CHO^+$ (Okino et al., 2006b,c), and from the fact that the ejection directions of these fragment ions are isotropic, the precursor dication, CH_3OH^{2+}, from which H_3^+ and CHO^+ are generated, was found to be long lived, showing that the lifetime is comparable with the period of the rotational motion of the dication or longer, that is, the lifetime is of the order of 1 ps or longer (Okino et al., 2006b). From the investigation of partially deuterated methanol, CD_3OH, it was shown that H_3^+ can be produced from the three hydrogen atoms in methyl group as well as from two hydrogen atoms in the methyl group and one hydrogen atom in the hydroxyl group (Okino et al., 2006c). The fact that one hydrogen atom in the hydroxyl group is involved in the formation of

Molecular Spectroscopy and Quantum Dynamics. https://doi.org/10.1016/B978-0-12-817234-6.00013-1

H_3^+ ejected from the dication shows clearly that hydrogen migration proceeds within the long lived dication. By the theoretical simulation of the H_3^+ ejection processes of methanol dications (Nakai et al., 2013), it was suggested that a neutral hydrogen molecule, H_2, is produced and can migrate within the dication. It should also be mentioned that the production of H_3^+ from the allene dication, $CH_2=C=CH_2$, shows that H_3^+ can be generated only after the hydrogen migration (Hoshina et al., 2008). In order to explore the hydrogen migration processes revealed in the series of our studies, we started investigating these processes in time domain by pump–probe measurements using near-IR few-cycle laser pulses whose pulse duration is in the range of 4–7 fs. Pump-probe spectroscopic techniques are also described in Chapter 3 of this book (Braun et al., 2020).

In this chapter, we show that there are two different time scales in the dynamical processes in which hydrogen atoms are involved by referring to our recent studies on time-resolved CMI measurements of the Coulomb explosion of methanol, i.e., (i) an ultrafast hydrogen migration from the methyl group to the hydroxyl group occurring in the period shorter than 25 fs, and (ii) a much slower process resulting in the formation of H_3^+ (Ando et al., 2018b, 2015). For the ultrafast hydrogen migration, we show detailed dynamics of the motion of a wave packet created in the migrated species, $CH_2\text{-}OH_2^{2+}$, in real time (Ando et al., 2015), and for the slow hydrogen migration, we show that H_3^+ is generated periodically with a period of 38 fs, which was found to be the period of the vibrational motion of the C–O stretching mode of the methanol cation, CH_3OH^+ (Ando et al., 2018b). We then show that this periodical ejection of H_3^+ and other fragmentation processes of CH_3OH^+ can be used to determine vibrational frequencies of the methanol cation via Fourier transform of the time-domain data and that this method called strong-field Fourier transform spectroscopy is very promising in determining spectroscopic data of molecular cations with high-resolution (Ando et al., 2019).

In the last section, we show by introducing our recent study (Ando et al., 2018a) that the rotational and vibrational level energies of D_2^+ can be determined with extremely high precision of as high as 0.0002 cm^{-1}, which is one-order of magnitude smaller than the radiative corrections (~0.005 cm^{-1}), by strong-field Fourier transform spectroscopy, and conclude that, by pump–probe measurements using intense few-cycle near-IR laser pulses, we can extract not only the ultrafast nuclear dynamics of molecules and molecular cations in the time domain, but also their rovibrational level energy separations with extremely high accuracy in the frequency domain.

8.2 ULTRAFAST HYDROGEN MIGRATION IN METHANOL CATION

We investigate the ultrafast dynamics of the hydrogen migration process in methanol cations by pump–probe coincidence momentum imaging with few-cycle laser pulses (Ando et al., 2015) by detecting the fragment ions produced via the following two Coulomb explosion pathways:

non-migration pathway

$$CH_3OH^{2+} \rightarrow CH_3^+ + OH^+, \tag{8.1}$$

migration pathway

$$CH_3OH^{2+} \rightarrow CH_2^+ + OH_2^+, \tag{8.2}$$

and recorded the distribution of the released kinetic energies, E_{kin}, as a function of the pump–probe time delay Δt ranging between −20 and 500 fs.

For the non-migration pathway, we found that the C–O bond breaking,

$$CH_3OH^+ \rightarrow CH_3^+ + OH, \tag{8.3}$$

proceeds via the energized states in the first electronically excited $\tilde{A}$ state. For the migration pathway, we have found that a nuclear wave packet flows from the electronic states whose energies are higher than the $\tilde{A}$ state in the non-migrated geometrical configuration into the bound well of the migrated geometrical configuration, where it first oscillates along the C–O bond, and then bifurcates at ~150 fs into the bound component and the dissociative component that leads to the C–O bond breaking,

$$CH_3OH^+ \rightarrow CH_2^+ + H_2O. \tag{8.4}$$

In the E_{kin} distribution in Fig. 8.1A, we can identify only one distinct component exhibiting a monotonic decrease as a function of the pump-probe time delay. On the other hand, in Fig. 8.1B, we can identify two distinct components: (i) the high kinetic energy component at E_{kin} ~ 4.5 eV exhibiting a broad distribution with a characteristic periodical structure peaked at 35, 120, and 200 fs, and (ii) the low kinetic energy component which starts appearing at 120 fs and decreases monotonically as the pump–probe delay time increases.

When the C–O distance in CH_3OH^+ increases, the kinetic energy released by the Coulomb explosion,

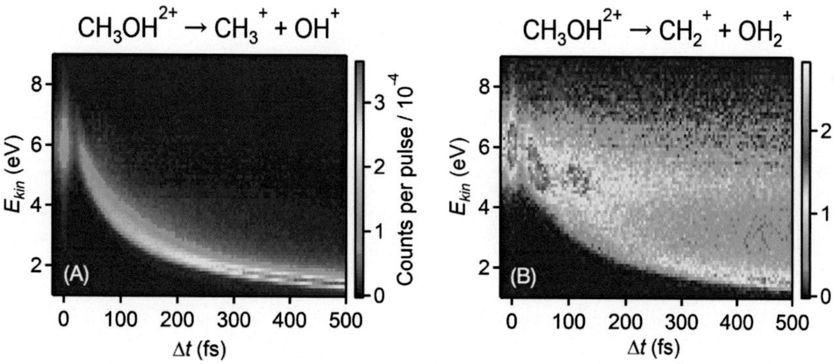

FIG. 8.1 The E_{kin} distributions in the non-migration pathway (A) and the migration pathway (B) as a function of Δt obtained by the pump–probe measurements using few-cycle near-infrared intense laser pulses (6.0(5) fs, 770 nm, 2.1(2) $\times$ 10^{14} W/cm^2). The signals generated by the pump laser pulses only and those by the probe laser pulses only are subtracted. In both (A) and (B), the intense and narrow distribution appearing at $\Delta t = 0$ stretching in the kinetic energy range between 5.3 and 6.7 eV is ascribed to the non-linearly increased ion signals that were not subtracted by the subtraction correction.

$CH_3OH^{2+} \rightarrow CH_3^+ + OH^+$, proceeding immediately after the ionization into CH_3OH^{2+} from CH_3OH^+ by a probe laser pulse, decreases. Therefore, the monotonically decreasing component in Fig. 8.1A can be assigned to the C–O bond breaking pathway (8.3). It has been known by photoelectron–photoion coincidence (PEPICO) spectroscopy using tunable monochromatic vacuum ultraviolet radiation (Borkar et al., 2011) that CH_3OH^+ prepared in the electronically excited $\tilde{B}$ or $\tilde{C}$ states decomposes into the fragments via the pathway (8.3) after the internal conversion to the dissociative manifold of the $\tilde{A}$ state, which forms a degenerate pair with the ground $\tilde{X}$ state when the C–O internuclear distance becomes infinitely large.

In Fig. 8.1A, the low E_{kin} component starts appearing at $\Delta t \sim 20$ fs in the non-migration pathway. The low E_{kin} component may reflect the temporal duration required for the dissociative component of the wave packet prepared first on the $\tilde{B}$ and/or $\tilde{C}$ states to reach the internuclear distance at which the potential energy surface (PES, see also Marquardt and Quack, 2020, Chapter 1 of this book) of CH_3OH^{2+} becomes repulsive, leading to the decomposition into $CH_3^+ + OH^+$ after the second ionization by the pump pulse. The two distinct components in Fig. 8.1B are considered to reflect two different types of nuclear dynamics in the migrated geometry, $CH_2OH_2^+$. The low E_{kin} component can be assigned to the dissociative motion of a wave packet along the C–O bond into $[CH_2\text{–}OH_2]^+$, which dissociates into $CH_2^+ + H_2O$ (8.4). On the other hand, the high E_{kin} component exhibiting the characteristic oscillatory structures can be assigned to the motion of a vibrational wave packet trapped in the bound potential at the migrated geometry, $CH_2OH_2^+$ (Xu et al., 2010). The delay-time dependences of these two components show that the wave packet prepared by a pump laser pulse vibrates until ~120 fs, and then bifurcates into two components: one component continuing the vibration in the bound well and the other dissociating along the C–O coordinate.

Although the electronic states of the C–O dissociation pathway (8.4) were not understood well in the previous study (Borkar et al., 2011), the vibrational wave packet needs to be prepared on the $\tilde{B}$ or higher-lying electronically excited states so that the total energy becomes larger than the threshold energy of the decomposition into $CH_2^+ + H_2O$. It is probable that the wave packet initially prepared in the $\tilde{B}$ and/or $\tilde{C}$ states of CH_3OH^+ is transferred to the electronic ground state of migrated $CH_2OH_2^+$ via internal conversion in the course of the migration.

As shown in Fig. 8.2 obtained by our quantum chemical calculation (UB3LYP/aug-cc-pVTZ) (Xu et al., 2010), there is a saddle point on the PES between the non-migrated (CH_3OH^+) and migrated ($CH_2OH_2^+$) geometrical configurations, which can be regarded as a transition state of the hydrogen migration reaction, separating the well bound along the C–O distance in the non-migrated geometrical configuration and the well bound along the C–O distance in the migrated configuration. If a wave packet migrating from the non-migrated configuration flows into the bound well in the electronic ground state in the migrated geometrical configuration, the motion of the wave packet in the

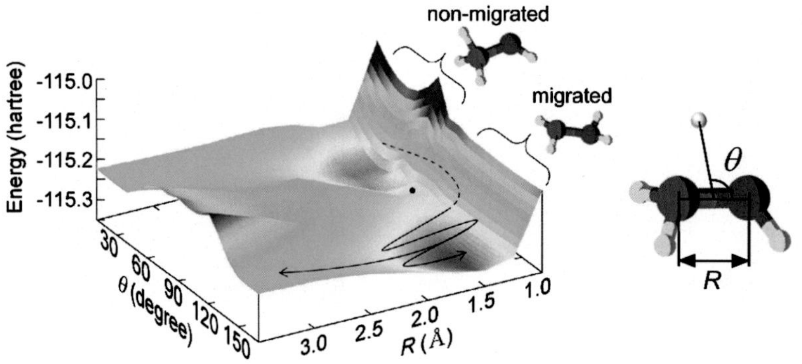

FIG. 8.2 The trajectory of the wave packet migrating into the bound well of the migrated geometrical configuration of $CH_2OH_2^+$ estimated on the basis of the recorded E_{kin} distributions of the migration pathway shown in Fig. 8.1B. The dashed-line part of the trajectory line represents that the wave packet migrates from the or higher-lying electronically excited states in the non-migrated geometrical configuration. The two-dimensional PES $V(R, \theta)$ is that of the electronic ground $\tilde{X}$ state of CH_3OH^+ calculated by Gaussian09 (UB3LYP/aug-cc-pVTZ) obtained by Xu et al. (2010). All the structural parameters other than R and θ are optimized in the calculation of the PES. The definitions of R and θ are shown in the inset. The black dot represents the position of the transition state between CH_3OH^+ and $CH_2OH_2^+$.

migrated geometrical configuration may be interpreted using the PES shown in Fig. 8.2.

We have probed ultrafast hydrogen migration in CH_3OH^+ in real time by pump–probe measurements using few-cycle near-IR laser pulses. From the resultant temporal evolution of the distributions of the released kinetic energies of the fragment ions, we have reached the following conclusions: (i) The C–O bond breaking, $CH_3OH^+ \rightarrow CH_3^+ + OH$, proceeds on the PES of the first electronically excited $\tilde{A}$ state in the non-migrated geometrical configuration (CH_3OH^+). (ii) The hydrogen migration, $CH_3OH^+ \rightarrow CH_2OH_2^+$, starts from CH_3OH^+ prepared on the PES of the $\tilde{B}$ or the higher lying electronically excited states and the nuclear wave packet is transferred through the internal conversion into the bound part of the PES of the electronic ground state in the migrated geometrical configuration in 25 fs. (iii) The wave packet flown into the migrated geometrical configuration vibrates at 1.5 cycles in the bound well and bifurcates at ~150 fs into the bound component kept in the bound well to vibrate further and the dissociative component leading the C–O bond breaking, $CH_2OH_2^+ \rightarrow CH_2^+ + H_2O$. On the basis of our experimental data, we have constructed rough landscape images of the two-dimensional PES of the different electronic states on which the migrating nuclear wave packet evolves in the femtosecond time domain.

8.3 PERIODICAL EMISSION OF H_3^+ FROM METHANOL

As introduced in the preceding section, by the pump–probe measurements of the kinetic energies released from the Coulomb explosion pathways of the methanol dication occurring after the hydrogen migration from methyl group to hydroxyl group, we were able to extract the ultrafast evolution of the vibrational wave packet proceeding in the electronically excited state of methanol cation prepared by an ultrashort pump laser pulse. As will be introduced below in this section, by the pump–probe measurements of the kinetic energy released from the H_3^+ ejection pathway,

$$CH_3OH^{2+} \rightarrow CHO^+ + H_3^+, \tag{8.5}$$

and those of the H^+ ejection pathway,

$$CH_3OH^{2+} \rightarrow CH_3O^+ + H^+, \tag{8.6}$$

we are able to monitor in real time the C–O stretching vibration of methanol cation in the electronic ground state (Ando et al., 2018b).

Through a series of our studies on the formation of H_3^+ from hydrocarbon molecules by the irradiation of ultrashort intense laser pulses (Furukawa et al., 2005; Kanya et al., 2012; Okino et al., 2006c,b; Xu et al., 2009), it has been revealed that the hydrocarbon dications from which H_3^+ is ejected are all long lived, and that their lifetimes are comparable to or longer than a

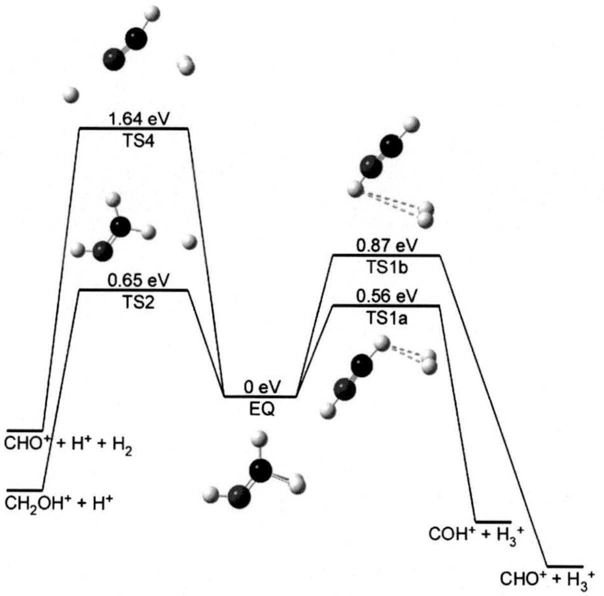

FIG. 8.3 Theoretical energy diagram of CH_3OH^{2+} referred from Ando et al. (2018b). The zero point energies at the three transition states (TS1a, TS1b, TS2 and TS4) measured from the zero point energy at the equilibrium structure (EQ) are shown.

period of the overall rotation of the dications. We performed first-principle molecular dynamics simulations of methanol dication and showed that the H_3^+ formation from CH_3OH^{2+} is a unimolecular decomposition process in which a long-lived neutral moiety of H_2, formed within a dication molecule, abstracts a proton in the other moiety having the charge of +2 (Nakai et al., 2013).

As plotted in Fig. 8.3, there are two transition states, TS1a (0.56 eV) and TS1b (0.85 eV), in the H_3^+ ejection pathway (8.5) of the methanol dication. At TS1a and TS1b, CH_3OH^{2+} takes two different characteristic geometrical structures, both of which show that neutral H_2 is attached to the $CHOH^{2+}$ moiety. These two transition state structures at TS1a and TS1b, representing a loosely bound complex in which the charge induced dipole moment of the H_2 moiety is bound by the positive charge of the $CHOH^{2+}$ moiety. When the unimolecular reaction proceeds through the lowest energy transition state, TS1a, H_3^+ is expected to be ejected from the methyl group, and, when it proceeds through TS1b, H_3^+ is expected to be ejected from the hydroxyl group after the migration of neutral H_2 from the methyl group as discussed by Nakai et al. (2013). The energy of the transition state of the H^+ ejection pathway (8.6) (TS2), 0.65 eV, is located between the energies of TS1a and TS1b.

When the hydrocarbon dication is prepared by sequential ionization from the neutral species by pump and probe ultrashort laser pulses, the vibrational energy of the dication is varied depending on the delay time of the second ionization pulse with respect to the first one. During the time delay, a vibrational wave packet prepared on the potential energy surface of the monocation evolves, and this temporal evolution of the wave packet can be reflected to the yield of the H_3^+ ejection of from dication.

Fig. 8.4A shows the distribution of the released kinetic energy, E_{kin}, and the ion yield of the H_3^+ ejection pathway (8.5) as a function of the pump–probe time delay Δt. The ion yield exhibits a periodic increase with the period of ~38 fs starting from $\Delta t \sim 58$ fs, and this periodic structure continues beyond 500 fs. In order to extract the frequency of the ion yield oscillation, we perform the Fourier transform of the yield in the range of $120 < \Delta t < 500$ fs as shown in Fig. 8.4C, in which a distinct peak appears at 26.3 THz, corresponding to the period of 38.0 fs. As shown in Fig. 8.4D, the Fourier transform of the yield of the H^+ ejection pathway (8.6) also exhibits a peak at ~26 THz.

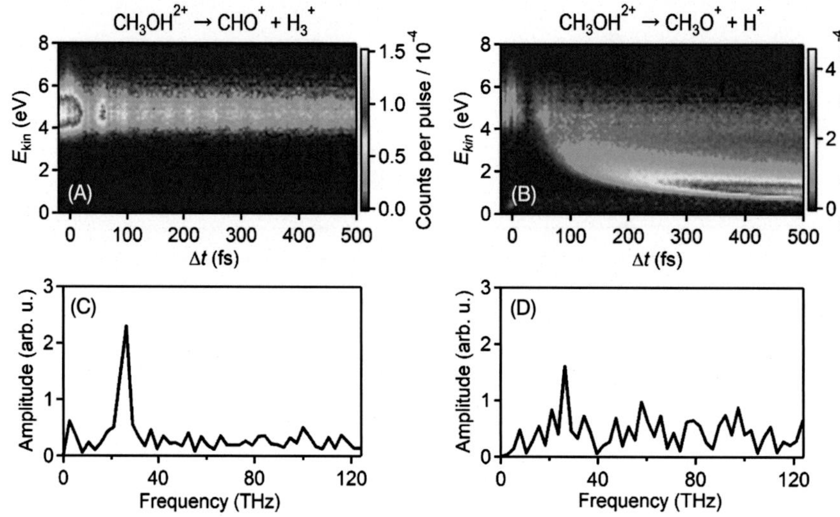

FIG. 8.4 The E_{kin} distribution of (A) the H_3^+ ejection pathway (8.5) and (B) the H^+ ejection pathway (8.6) obtained by the pump–probe measurements using few-cycle near-infrared intense laser pulses (6.0(5) fs, 770 nm, 2.1(2) × 10^{14} W/cm^2). The Fourier-transformed ion yields of (C) the H_3^+ ejection pathway (8.5) and (B) the H^+ ejection pathway (8.6) obtained using the data in the range of 120 < Δt < 500 fs after subtracting DC components.

The long-lasting oscillation of the yield of H_3^+ shown in Fig. 8.4A can only be ascribed to the motion of the vibrational wave packet on the electronic ground state of CH_3OH^+. This is because a wave packet prepared in the electronically excited states is expected to decay into the dissociation continuum and could not oscillate many times. Indeed, even in the lowest-lying electronically excited state of CH_3OH^+, the lowest energy is located 0.5 eV above the dissociation threshold for $CH_2OH^+ + H$ (Borkar et al., 2011; Karlsson et al., 1977). The oscillation frequency of 26.3 THz is closed to the C–O stretching frequency 26.8 THz (895 cm^{-1}) of the electronic ground state of CH_3OH^+ (Karlsson et al., 1977). Therefore, it is highly probable that we monitor the oscillatory motion of the vibrational wave packet, prepared in the electronic ground state of CH_3OH^+, along the C–O stretching vibration.

In order to relate the vibrational frequency of the C–O stretching mode of the electronic ground state of CH_3OH^+ to the periodical increase in the yield of the H_3^+ ejection pathway of CH_3OH^{2+}, we plot the schematic potential energy curves of the electronic ground state of CH_3OH ($\tilde{X}^1A'$), CH_3OH^+ ($(2a'')^{-1}\tilde{X}^2A''$), and CH_3OH^{2+} ($(2a'')^{-2}\tilde{X}^1A'$) along the C–O stretching coordinate in Fig. 8.5. According to the previous experimental and theoretical studies, the equilibrium C–O internuclear distances of CH_3OH, CH_3OH^+, and CH_3OH^{2+} are 1.428(3) Å (Kimura and Kubo, 1959), 1.36 Å (Thapa and Schlegel, 2014), and 1.19 Å (Thapa and Schlegel, 2014), respectively, reflecting the fact that the 2a″ orbital has an antibonding character, and therefore, upon the first ionization, the vibrational wave packet is expected to be prepared at the outer turning point of the potential energy curve. Consequently, as the delay time increases, the energy of the dication decreases and takes the minimum value at $\Delta t = 19$ fs, a half of the vibrational period of the C–O stretch of the monocation, and then, the energy becomes maximum at $\Delta t = 38$ fs. The delay times at which the yield of H_3^+ became maximum, 58, 96, 134 fs, ..., shown in Fig. 8.4A, correspond to the timings when the vibrational wave packet reached the inner turning point of the potential energy curve of the monocation, i.e., the timing when the energy of the dication takes the minimum value. Therefore, it is probable that, at the lowest energy of the dication, only the lowest energy channels (8.5) and (8.6) are opened energetically, and that once the other dissociation channels are opened energetically, they dominate over the lowest energy channels, leading to the decrease in the yield of H_3^+ through the channel (8.5).

The time-resolved pump–probe measurements of the yield of H_3^+ ejected from the methanol dication has given us an opportunity to determine the vibrational

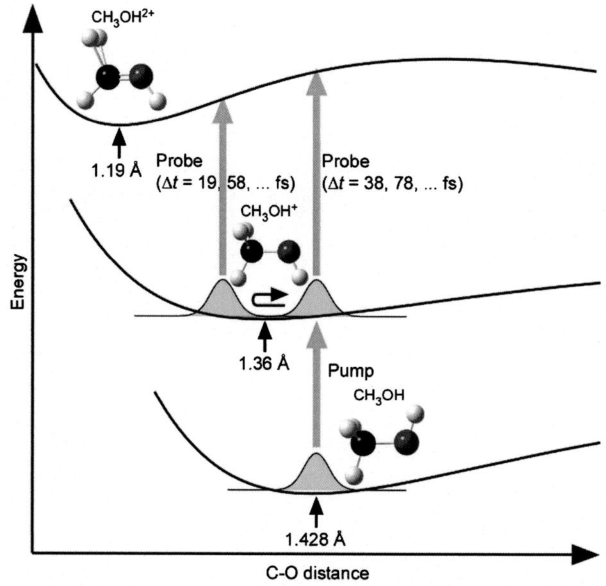

FIG. 8.5 Schematics of the potential energy curves of CH_3OH, CH_3OH^+ and CH_3OH^{2+} along the CO internuclear distance reprinted from Ando et al. (2018b). The vibrational wave packet of CH_3OH^+ prepared by the first laser pulse at the outer turning point starts moving towards the shorter internuclear distance region and the oscillation continues. The wave packet prepared in CH_3OH^{2+} by the second laser pulse has the lowest energy every time when the wave packet of CH_3OH^+ comes back to the inner turning point periodically with the interval of 38 fs starting from at $\Delta t = 19$ fs.

frequency of the C–O stretching mode. The H_3^+ ejection has commonly been found in the Coulomb explosion processes of hydrocarbon molecules having more than two hydrogen atoms (Hoshina et al., 2008). Therefore, the pump-probe measurements of the yield of the very slow H_3^+ ejection pathway followed by the Fourier transform of the delay time dependent yield can be a promising spectroscopic method to determine the vibrational frequencies of hydrocarbon cations that have not been reported before.

8.4 STRONG FIELD VIBRATIONAL SPECTROSCOPY OF METHANOL CATION AND ITS ISOTOPOLOGUES

Frequencies of the vibrational modes of neutral molecules have been determined with high precision by conventional IR and Raman vibrational spectroscopy, and these frequency data have been accumulated as a form of the established database (Shimanouchi, 1972). However, the determination of vibrational mode frequencies of molecular cations is a challenging task even nowadays mainly because of the extremely low densities of the sample gases of ions. The vibration frequencies of molecular ions have been determined by IR spectroscopy using ion beams and trapped ions as well as by photoelectron spectroscopy (Biesheuvel et al., 2017; Bieske and Dopfer, 2000; Carrington and Kennedy, 1984; Dai et al., 2014; Germann et al., 2014; Heine and Asmis, 2015; Mosley et al., 2015; Müller-Dethlefs and Schlag, 1998; Polfer, 2011). These spectroscopic techniques have been developed during the past two decades by the combination of frequency tunable narrowband lasers and sufficiently cooled molecular ions.

The frequency of the C–O stretching vibration of the methanol cation, CH_3OH^+ was obtained by photoelectron spectroscopy (Karlsson et al., 1977; Macneil and Dixon, 1977). Recently, the torsional level energies (corresponding to wave numbers less than 300 cm^{-1}) of CH_3OH^+, CH_3OD^+ and CD_3OD^+ and the vibrational wave number of the C–O stretching mode of CD_3OD^+ (710(4) cm^{-1}) was obtained by zero kinetic energy (ZEKE) spectroscopy (Dai et al., 2014) and the vibrational mode frequencies in the high frequency region (2000–4000 cm^{-1}) were determined by IR laser photodissociation spectroscopy of $CH_3OH^+\cdots Ar$ complex (Mosley et al., 2015). However, the wave numbers of the vibration modes in the intermediate frequency range between 1000 and 2000 cm^{-1} have not been determined experimentally.

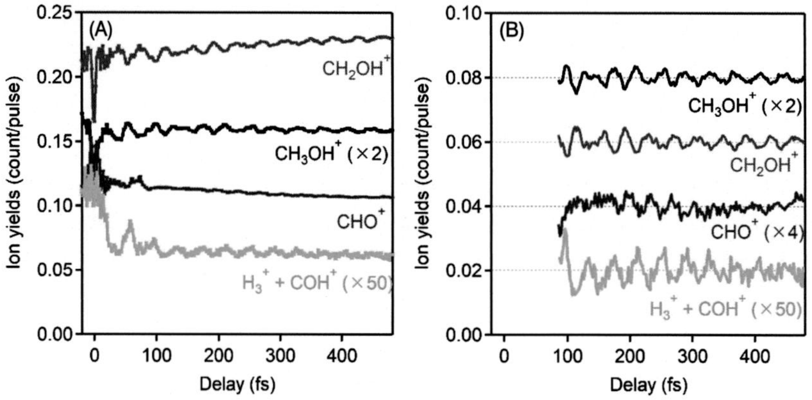

FIG. 8.6 (A) Ion yields of CH_3OH^+, CH_2OH^+, CHO^+, and H_3^+ obtained by the pump–probe measurements using few-cycle near-infrared intense laser pulses (6.0(5) fs, 770 nm, 2.1(2) $\times$ 10^{14} W/cm^2) as a function of the pump–probe time delay. (B) Oscillatory components of the ion yields obtained after the subtraction of the respective DC components.

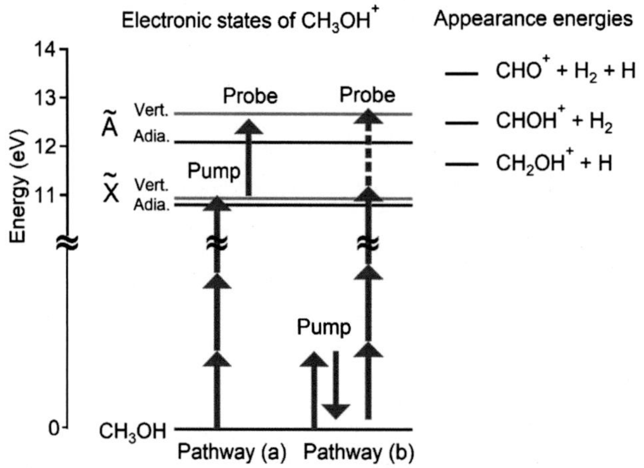

FIG. 8.7 Energy diagram for CH_3OH and CH_3OH^+ and the appearance energies of dissociation pathways from CH_3OH^+.

In this section, we show that the pump–probe measurements of the yield of fragment ions produced from the methanol cation, CH_3OH^+, can also be used to determine vibrational mode frequencies of CH_3OH^+. The fragmentation processes of CH_3OH^+ is more preferable for the strong field Fourier transform spectroscopy than those of CH_3OH^{2+} because the yield of fragment ions from CH_3OH^+ is much larger than the yield of H_3^+ from CH_3OH^{2+}. We also show in this section that the vibrational mode frequencies of the isotopologues of CH_3OH^+ can also be determined by strong field Fourier transform spectroscopy.

Fig. 8.6A shows the ion yields of CH_3OH^+, CH_2OH^+, CHO^+ and H_3^+ as a function of the time delay, Δt. As can be seen in this figure, the ion yields oscillate with the period of ~40 fs. In order to extract the oscillatory structure from the ion yields, we fitted the ion yields in the range of $90 < \Delta t < 480$ fs with exponential curves and subtracted the exponential curves from the respective ion yields as shown in Fig. 8.6B, by which the non-oscillating and smoothly-varying components in the ion yields originating only from the pump laser pulses and those originating only from the probe laser pulses are subtracted.

Fig. 8.7 shows the energy diagram of methanol and the methanol cation. The adiabatic ionization energy of methanol is 10.846(2) eV, which was determined from photoelectron spectroscopy experiments by Macneil and Dixon (1977). It was reported on the basis of the photoelectron–photoion coincidence measurements that the appearance energies of CH_2OH^+, $CHOH^+$ and CHO^+ are 11.646(3), ~12.5, and ~13.5 eV, respectively (Borkar et al., 2011). It was also reported that the production of CH_2OH^+ is a dominant channel when the internal energy of CH_3OH^+ is in the range between 11.6 and 15.2 eV and that the production of CHO^+ is dominant when the internal energy exceeds 15.2 eV (Borkar et al., 2011). The oscillatory structures identified in the ion yields of CH_3OH^+, CH_2OH^+ and CHO^+ shown in Fig. 8.6B can be ascribed to the following two possible processes:

(a) Ionization by the pump laser pulse

First, the pump pulse creates a vibrational wave packet of methanol cation in the $\tilde{X}^2A''$ state by the strong field ionization. After the field free propagation of the prepared vibrational wave packet during the time delay, the probe pulse projects the wave packet onto the potential energy surface of an electronically excited state. The electronically excited CH_3OH^+ is decomposed into fragment ions because the appearance energy of CH_2OH^+ (11.646(3) eV, from Borkar et al., 2011) is lower than the energy of the first electronically excited state, $\tilde{A}^2A'$ (12.1 eV, see Karlsson et al., 1977). Because the probability of the excitation to the $\tilde{A}^2A'$ state depends on the instantaneous geometrical structure of CH_3OH^+ in the $\tilde{X}^2A''$ state, the motion of the vibrational wave packet can be converted into the variations in the yields of the parent and fragment ions.

For example, the energy gap between the $\tilde{X}^2A''$ and $\tilde{A}^2A'$ states of CH_3OH^+ is larger than the one-photon energy (~1.6 eV) of the probe pulse when the wave packet is prepared by the Franck–Condon projection from neutral CH_3OH, and the energy gap is known to decrease as the C–O internuclear distance increases (Borkar et al., 2011). Therefore, the efficiency of the electronic excitation by the probe pulse is expected to be enhanced when the vibrational wave packet moves towards the outer turning point of the electronic ground state of the potential energy curve along the C–O coordinate.

(b) Ionization by the probe laser pulse

The pump pulse can also create a vibrational wave packet in neutral methanol through the impulsive stimulated Raman process and/or through the structure-dependent ionization depletion (Ergler et al., 2006; Wei et al., 2017). In this scheme, the vibrational wave packet is created in neutral methanol in the electronic ground state and the time evolution of the wave packet is detected as the variation in the yield of a parent cation when the wave packet is prepared in a region of the electronic ground state whose energy is lower than the dissociation thresholds or as the variation in the yield of a fragment ion when the wave packet is prepared in the electronically excited states, leading to the dissociation. Because the ionization probability of neutral CH_3OH by a probe pulse depends sensitively on the instantaneous geometrical structure of CH_3OH, not only the yield of the parent ion but also the yields of the fragment ions are expected to oscillate with the periods of vibrational modes of CH_3OH.

These two different processes can be distinguished one from the other on the basis of the relative phase ϕ between the oscillation of the yield of the parent ion and that of the fragment ions. In the process (a), the sum of the yield of the parent ions and the yields of the fragment ions does not oscillate in time because the total number of the ions is kept constant even when a parent ion is decomposed into the fragment ion and the counterpart neutral fragment. Therefore, the oscillations of the yield of the parent ion and the oscillation of the yield of the fragment ions are expected to be in antiphase, that is, the relative phases of the yields of the fragment ions oscillate with the phase shift of π (radian), that is, the relative phase of $\phi = \pi$. On the other hand, in the process (b), the sum of the yield of the parent ions and that of the fragment ions is expected to oscillate in phase, reflecting the fact that the ionization probability is influenced by the oscillation of the wave packet created by the pump pulse in CH_3OH. Therefore, the phase of the oscillation of the yield of the fragment ions with respect to the phase of the parent ions is expected to be $\phi = 0$ or 2π.

The FT spectra of CH_3OH and CH_3OH^+ obtained from the time delay dependent yields of CH_3OH^+ and the fragment ions are shown in Fig. 8.8. By referring to the relative phases of the peak profiles and the frequencies of the vibrational modes derived by the density functional theory (DFT) calculations, we assigned the peak profiles in the FT spectra to CH_3OH and to CH_3OH^+ as summarized in Table 8.1.

The DFT calculations were performed by the Gaussian09 program at B3LYP/aug-cc-pVTZ level and the resultant vibrational frequencies were scaled by the scaling factor of 0.968, according to Johnson (2018). The theoretical wave numbers for the vibrational modes of CH_3OH and CH_3OH^+ are compared with the experimental values in Table 8.1.

In Table 8.1, the potential energy distributions (PEDs) of the vibration modes of CH_3OH^+ calculated using the VEDA 4.0 program (Jamroz, 2004) with the results obtained by the DFT calculations are also com-

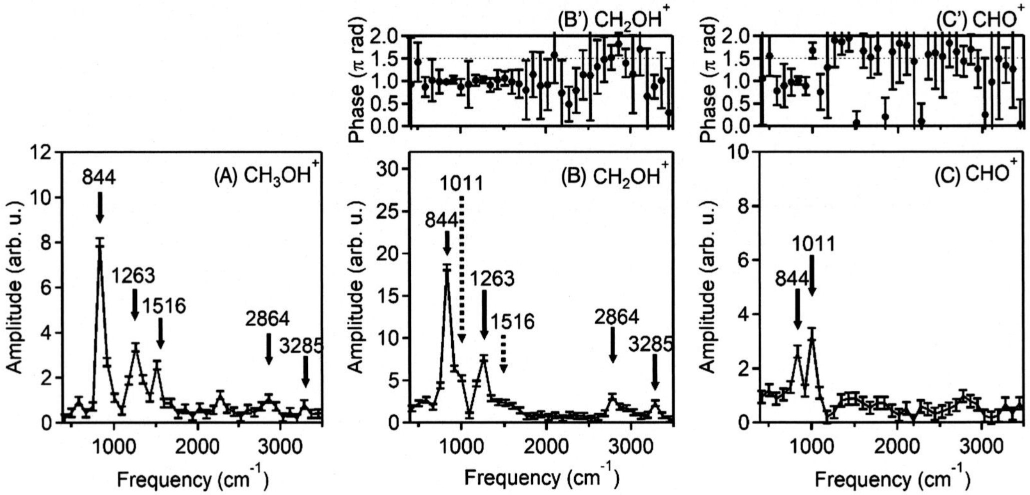

FIG. 8.8 Strong-field Fourier transform spectra of CH_3OH and CH_3OH^+ obtained by the yields of (A) CH_3OH^+, (B) CH_2OH^+, and (C) COH^+. The relative phases of the Fourier transform spectra for the yields of (B) CH_2OH^+ and (C) COH^+ measured from the phases for CH_3OH^+ are shown in (B) and (C), respectively.

TABLE 8.1
Comparison of the vibrational wave numbers (in cm^{-1}) of CH_3OH and CH_3OH^+.

No.	Sym.	CH_3OH			CH_3OH^+			
		Obs.[a]	This work		Obs.	This work		PED (%)
			Obs.[b]	Calc.		Obs.[b]	Calc.	
ν_1	a'	3681		3714	3156[c]	3285	3434	OH str. (100)
ν_2		3000		3009	2902[c]		3077	CH_3 d-str. (81)
ν_3		2844	2864	2898	(2413)[c]		2657	CH_3 s-str. (81)
ν_4		1477		1462	1356(30)[d]	1516	1433	CH_3 s-def. (43), CH_3 d-def. (39)
ν_5		1455		1433		1263	1244	OH bend (57), CH_3 rock (26)
ν_6		1345		1322			1195	CO str. (55), CH_3 s-def. (20)
ν_7		1060		1044			1011	CH_3 rock (46), OH bend (35)
ν_8		1033	1011	1008	895(80)[e] 926(30)[d]	844	907	CH_3 d-def. (41), CO str. (35)
ν_9	a''	2960		2943	(2413)[c]		2549	CH_3 d-str. (99)
ν_{10}		1477		1451			1175	CH_3 d-def. (70), Torsion (20)
ν_{11}		1165		1135			941	CH_3 rock (79), Torsion (20)
ν_{12}		295		278	13(4)[f]		289	Torsion (57), CH_3 d-def. (29)

[a] Infrared spectroscopy (Shimanouchi, 1972).
[b] Uncertainties of 42 cm^{-1}.
[c] Laser dissociation spectroscopy of $[CH_3OH]^+Ar$ (Mosley et al., 2015). The two candidates of the vibrational modes, ν_3 and ν_9, were suggested for the vibrational wave number of 2413 cm^{-1}.
[d] Photoelectron spectroscopy of CH_3OH with the peak deconvolution (Macneil and Dixon, 1977).
[e] Photoelectron spectroscopy of CH_3OH (Karlsson et al., 1977).
[f] PFI-ZEKE spectroscopy of CH_3OH (Dai et al., 2014).

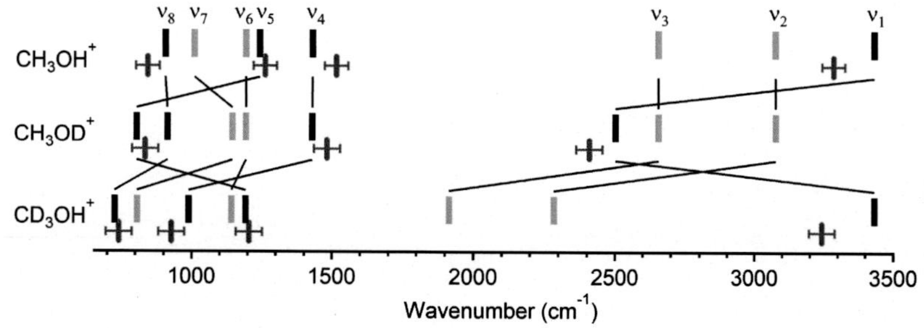

FIG. 8.9 Wave numbers of the vibrational modes with a' symmetry ($\nu_1 \sim \nu_8$) of methanol cation. The observed vibrational mode wave numbers are represented by red vertical bars and the corresponding theoretical vibrational wave numbers are represented by black vertical bars. The vibrational wave numbers of the other modes are represented by gray vertical bars.

pared. As shown in Table 8.1, all the observed peak profiles are assigned to the vibrational modes having the a' symmetry. This symmetry selection suggests that the vibrational symmetry is kept in the course of the vertical ionization through which the vibrational wave packet is prepared from the vibrational ground state of CH_3OH (a' symmetry) by the pump pulse.

The strong field Fourier transform (FT) spectroscopy has also been applied to CH_3OD and CD_3OH. The frequencies we have determined by the analysis of the FT spectra and those derived from the DFT calculations are summarized in Fig. 8.9. The experimentally determined frequency shifts of the vibrational mode frequencies induced by the isotope substitutions are consistent with those obtained by the DFT calculations.

Although the resolution of the spectra is around 90 cm^{-1} in the present measurement and the uncertainties in the vibrational frequencies are around 40 cm^{-1}, the resolution can be improved easily to be 10^{-2} cm^{-1} when the time delay range is lengthened up to ~3 ns as long as the vibrational coherence is maintained as demonstrated in the next section. Indeed, strong-field FT spectroscopy is a promising approach to obtain vibrational mode frequencies of non-polar and polar cations whose high-resolution data have not been determined before experimentally.

8.5 HIGH-RESOLUTION ROVIBRATIONAL SPECTROSCOPY OF D_2 AND D_2^+

In the preceding section, we showed that the fragmentation process of a molecular ion can be used to obtain vibrational spectra of molecular ions. In principle, the spectral resolution of the Fourier transform spectrum can be improved simply by increasing the scan range of the pump–probe time delay as long as the coherence time of the rovibrational states are longer than the time delay. In this section, we present our recent study on the pump–probe measurements of D_2^+ with the time delay up to 527 ps and show that the results of the measurements in the time domain can be used for high-resolution rovibrational spectroscopy in the frequency domain (Ando et al., 2018a).

A hydrogen molecular ion is the simplest molecular species and the high-precision measurements of its rovibrational energy levels have been used for the test of quantum electrodynamics (QED) theory (Balint-Kurti et al., 1990; Karr et al., 2016b). It has been known that the fundamental constants such as the proton–electron mass ratio (Korobov et al., 2014) and the proton charge radius (Karr et al., 2016a) can be determined from the rovibrational level energies of hydrogen molecular ions if they are measured with sufficiently high precision. In the case of HD^+, thanks to the existence of the dipole moment, 0.87 D (Bunker, 1974), vibrational transitions are allowed, and the vibrational transition frequencies were determined with the precision of $\delta\nu/\nu = 10^{-9}$ by the frequency-comb spectroscopy of HD^+ in the cold ion trap (Biesheuvel et al., 2016; Bressel et al., 2012; Koelemeij et al., 2007). On the other hand, in the cases of H_2^+ and D_2^+, because there is no dipole moment, no vibrational transitions have been observed, and the vibrational level energies have been determined by photoelectron spectroscopy (Öhrwall and Baltzer, 1998; Peatman et al., 1983; Pollard et al., 1982) and zero-kinetic energy (ZEKE) photoelectron spectroscopy (Beyer and Merkt, 2017, 2016; Chang et al., 2007; Stimson et al., 1998) whose frequency resolutions have been around 1 cm^{-1} in most cases. Recently, PFI-ZEKE photoelectron spectroscopy of high-lying vibrational states near the dissociation limit on H_2^+ and D_2^+ were performed with the uncertainty of 0.06 cm^{-1} and 0.02 cm^{-1}, respec-

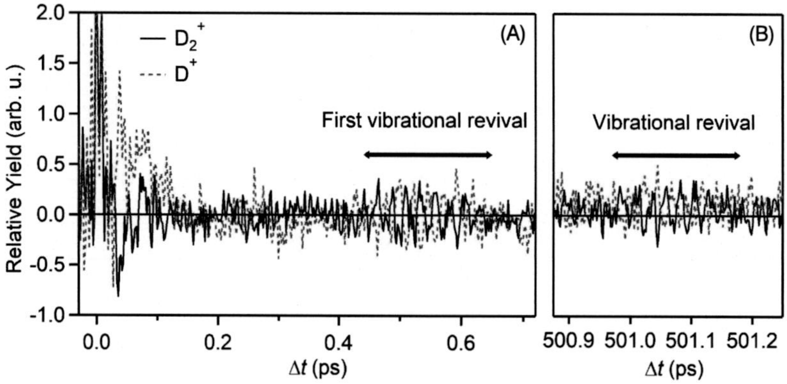

FIG. 8.10 Ion yields of D_2^+ (black solid line) and D^+ (red broken line) in the different time delay ranges of (A) $0 < \Delta t < 0.7$ ps and (B) $500.88 < \Delta t < 501.23$ ps recorded by the pump–probe measurements using few-cycle near-infrared intense laser pulses (5 fs, 780 nm, 3.2×10^{14} W/cm^2).

tively (Beyer and Merkt, 2017, 2016), and the lifetimes of the quasibound levels were determined on the basis of the spectral line broadening.

In theory, the vibrational level energies of hydrogen molecular ions have been obtained with the highest accuracy among those of molecular ion species. By the *ab initio* calculations, the non-adiabatic energy with relativistic and radiative corrections have been obtained (Karr and Hilico, 2006; Korobov, 2008; Korobov et al., 2014; Moss, 1993; Ning and Yan, 2014). Because the accuracies of the rovibrational energy levels obtained by these theoretical calculations are considered to be sufficiently high, the validity of the frequencies and their uncertainties obtained experimentally can be examined by the comparison with these theoretical values.

Fig. 8.10 shows the ion yields of D_2^+ and D^+ as a function of the time delay in the ranges from 0 to 0.7 ps and from 500.9 to 501.2 ps. As shown in Fig. 8.10A, the ion yields of D^+ and D_2^+ oscillates with a period of ~25 fs in the time delay range between 0.45 and 0.65 ps, and the amplitudes of the oscillations in the yields of D^+ and D_2^+ are almost the same, but their phases differ by π.

As shown in Fig. 8.10B, the yields of D^+ and D_2^+ in the delay time range between 500.95 and 501.15 ps exhibit the oscillations whose amplitudes and frequencies are almost the same as those appearing in the range between 0.45 and 0.65 ps, showing that the coherence time of the vibrational wave packet is sufficiently longer than 500 ps.

By taking advantage of the observation that the phases of the ion yields of D_2^+ and D^+ are flipped by π with respect to one another, we performed the Fourier transform (FT) of the difference between the ion yields of D_2^+ and D^+ in the time delay range of 370 fs < Δt < 527 ps. As shown in Fig. 8.11, the peaks appearing in the FT spectra can be assigned to the rovibrational energy separations of D_2^+ and neutral D_2, originating from the quantum beats among the rovibrational levels of D_2^+ and D_2. The peak frequencies correspond to the difference among the rovibrational energy separations of D_2^+, $E^+(v^{+''}, N^{+''}) - E^+(v^{+'}, N^{+'})$, and D_2, $E(v'', N'') - E(v', N')$, where v and N represent the vibrational quantum number and the rotational quantum number, respectively. In the FT spectrum, the peaks appearing in the frequency range from 0 to 700 cm^{-1} are assigned to the rotation level separations ($\Delta v^+ = 0$, $\Delta N^+ = 2, 4$) of D_2^+, the peaks appearing in the frequency range from 800 to 1800 cm^{-1} are assigned to the vibration level separations, ($\Delta v^+ = 1$, $\Delta N^+ = 0, \pm 2, \pm 4$) of D_2^+, and the peaks appearing in the frequency range from 2300 to 2960 cm^{-1} are assigned to the vibrational separations corresponding to the vibrational overtones ($\Delta v^+ = 2$, $\Delta N^+ = 0, \pm 2$) of D_2^+.

In addition, seven rotational level separations ($N' = 0 \sim 6$, $\Delta v = 0$, $\Delta N = 2$) of D_2 are observed in the frequency range between 170 and 870 cm^{-1} and seven vibrational level separations ($N' = 0 \sim 6$, $\Delta v = 1$, $\Delta N = 0$) of D_2 are observed in the frequency range between 2940 and 3000 cm^{-1}. These peaks originating from the quantum beats of neutral D_2 show that a pump pulse also creates a rovibrational wave packet of D_2, and therefore, the yields of D_2^+ and D^+ are influenced by the rotational and vibrational excitations in D_2. Fig. 8.12A shows the potential energy curves of D_2 and D_2^+. The vibrational wave packet of D_2 is prepared via

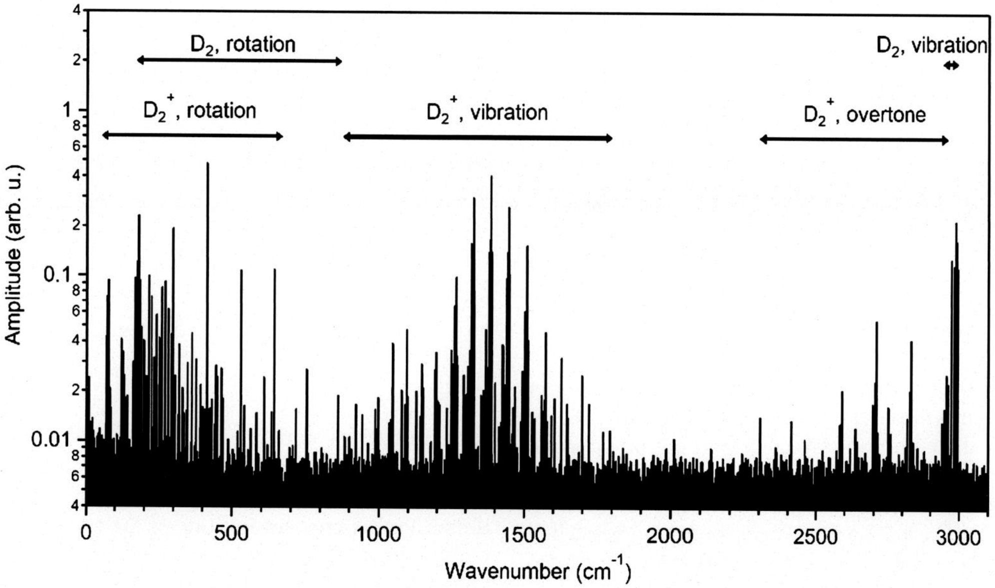

FIG. 8.11 The absolute value of the discrete Fourier transform of the difference between the yield of D_2^+ and D^+.

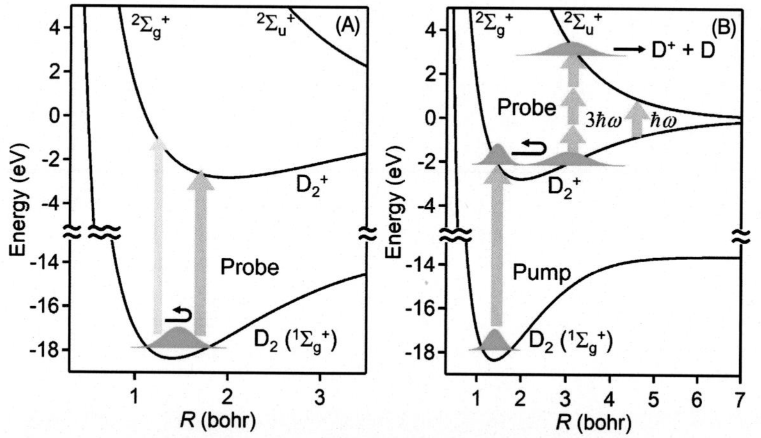

FIG. 8.12 Ionization and dissociation processes of (A) D_2 and (B) D_2^+. (Black curves) Potential energy curves of the electronic ground state $^1\Sigma_g^+$ of D_2 and the electronic ground $^2\Sigma_g^+$ state and the first electronically excited $^2\Sigma_u^+$ state of D_2^+. The energy of the dissociation limit of D_2^+ is set to be zero. The unit bohr stands for the Bohr radius $a_0 \approx 52.9$ pm.

the impulsive stimulated Raman process and/or via the R-dependent ionization depletion process (Ergler et al., 2006; Wei et al., 2017). Because the ionization probability of diatomic molecules in an intense laser field is enhanced when the internuclear distance increases (Seideman et al., 1995), the probability of the ionization by the probe pulse becomes large when the vibrational wave packet of D_2 is in the vicinity of the outer turning point.

The mechanism of the oscillation of the dissociation process, $D_2^+ \rightarrow D^+ + D$, associated with the vibration of D_2^+ can be explained as follows. Through the ionization of D_2 by the pump pulse, the rovibrational wave packets of D_2^+ can be produced in the electronic ground $^2\Sigma_g^+$

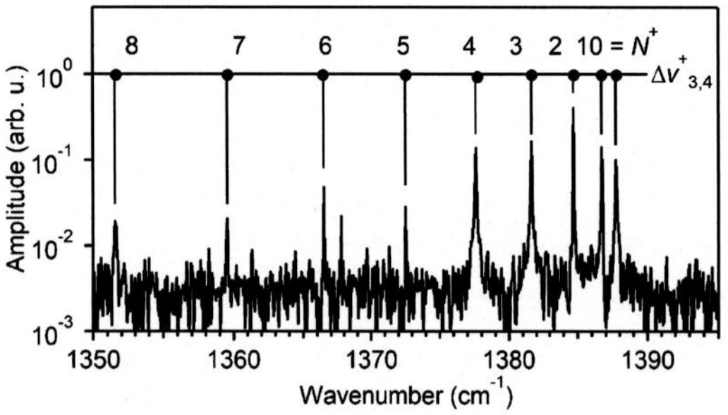

FIG. 8.13 Q-branch transitions of D_2^+ (1350–1390 cm^{-1}).

state, and then, the wave packet starts oscillating under the absence of the laser pulse as shown in Fig. 8.12B. Because the three-photon resonance and the one-photon resonance are located in the longer internuclear distance region than the equilibrium internuclear distance, the wave packet at the outer turning point can be excited by the probe pulse to the electronically excited dissociative $^2\Sigma_u^+$ state with a higher probability than the wave packet at the inner turning point. The wave packet prepared by the probe pulse on the dissociative potential curve, $^2\Sigma_u^+$, decomposes into D^+ and D. Because the excitation probability depends on the internuclear distance and the angle between the molecular axis and the polarization direction of the probe pulse, the variations in the ion yields of D_2^+ and/or D^+ reflect the motion of rotational and vibrational wave packet created in the electronic ground $^2\Sigma_u^+$ state of D_2^+. Therefore, from the Fourier transform of the ion yields, the energy separations among the rotational and vibrational levels of D_2^+ can be obtained.

In the expanded view of the FT spectrum in the frequency range between 1300 and 1400 cm^{-1} shown in Fig. 8.13, rotationally resolved Q-branch transitions are clearly identified, showing that rovibrational level energies of molecular cations can be determined with high precision by strong field high-resolution FT spectroscopy.

In order to determine the vibrational level energy separations, we fitted the peaks appearing in the FT spectra with a sinc-function,

$$S(\omega) = P\,\mathrm{sinc}\left(\frac{\pi(\omega-\omega_0)}{\Delta\omega}\right) \times \exp\left(\mathrm{i}\left(\frac{\pi(\omega-\omega_0)}{\Delta\omega}+\varphi\right)\right), \tag{8.7}$$

in which P, ω_0 and ϕ denote respectively the amplitude, frequency, and phase of each of the peak profiles appearing in the FT spectra. The frequency axis of the FT spectra was calibrated by comparing the rovibrational level energy separations of D_2 ($N = 0, 1, 2$) determined by our measurements with those determined previously with high precision by REMPI measurements (Niu et al., 2014). The systematic uncertainty originating from the calibration is $\Delta\nu_{\mathrm{statistical}}/\nu = 8 \times 10^{-8}$. As shown in Table 8.2, the resultant rovibrational frequencies determined from the Q-branch transitions in Fig. 8.13 are compared with the theoretical calculations (Moss, 1993) whose accuracy is considered to be 1×10^{-4} cm^{-1}. As can be seen in Table 8.2, the rovibrational frequencies we determined by strong field FT spectroscopy are in excellent agreement with those reported before by the theoretical calculations (Moss, 1993). Furthermore, the experimental uncertainties of the frequencies we determined are smaller in magnitude than the relativistic corrections (~0.02 cm^{-1}) and the radiative corrections (~0.005 cm^{-1}) included in the theoretical calculations.

TABLE 8.2
Comparison of the vibrational wave numbers of D_2^+, $E^+(v^{+\prime\prime} = 3, N^+) - E^+(v^{+\prime} = 2, N^+)$ (in cm^{-1}).

N^+	Present Experiment	Previous calculation[a]				Exp. – Calc.
		Non-relativistic	Relativistic	Radiative	Total	
0	1387.7523(6)	1387.7365	0.0209	−0.0056	1387.7518	0.0005
1	1386.7294(6)	1386.7148	0.0209	−0.0057	1386.7301	−0.0007
2	1384.6896(2)	1384.6745	0.0208	−0.0056	1384.6897	−0.0001
3	1381.6368(5)	1381.6216	0.0206	−0.0055	1381.6367	0.0001
4	1377.5794(5)	1377.5653	0.0205	−0.0055	1377.5802	−0.0008
5	1372.5359(33)	1372.5176	0.0203	−0.0055	1372.5324	−0.0035

[a] Uncertainty is 0.0001 cm^{-1} (Moss, 1993).

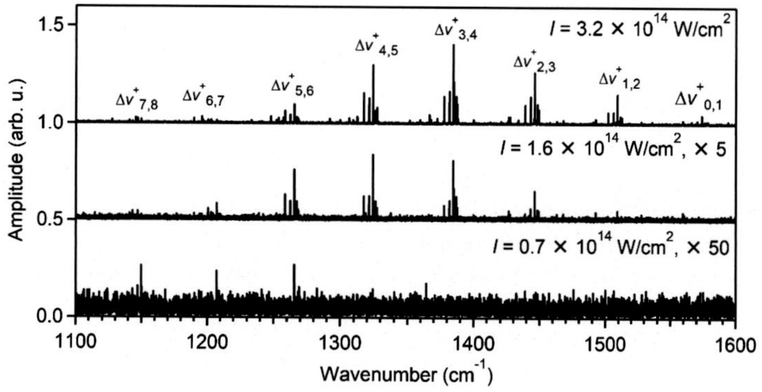

FIG. 8.14 Intensity dependence of the strong field FT spectra of D_2^+.

So far, the frequencies for the rovibrational energy separations have been regarded as the frequencies under the field-free conditions. However, if the picosecond pulse contrast of the laser pulses is poor so that the tail part of the laser pulses persistently exist during the time separation between the pump and probe pulses, the frequencies of the peaks derived from the FT spectrum may be the frequencies of D_2^+ in the light field, which can be different from the frequencies of D_2^+ under the field-free conditions because the AC Stark effect (Kołos and Wolniewicz, 1967; Schiller et al., 2014) can be induced by the picosecond pedestal. However, even when the remaining light field intensity in the tail part is as large as 1%, corresponding to the pulse contrast of 10^{-7}, the frequency shifts of vibrational level energies originating from the AC Stark effect are expected to be at most 1.5×10^{-5} cm^{-1}, which is sufficiently small compared with the uncertainties in the present measurements.

Fig. 8.14 shows the dependence of the FT spectra on the laser field intensity. When the laser field intensity decreases, the intensities of the peaks assigned to the vibrational level separations for the lower vibrational levels decreases, and, at the focal intensity of 0.7×10^{14} W/cm^2, only the peaks for the higher vibrational states ($v^{+\prime} > 4$) are identified. This dependence of the light-field intensity can be explained using the light-dressed potential energy surfaces (Giusti-Suzor et al., 1995; Jiang et al., 2017; Maruyama et al., 2004; Zavriyev et al., 1990). When the intensity of the probe pulse is relatively weak, the dissociation into $D^+ + D$ proceeds only from the higher vibrational states of D_2^+ through the mechanisms called the bond softening and the below-threshold dissociation, and therefore, only the peaks representing the rovibrational energy separations for the higher vibrational levels can appear in the FT spectrum. As the light-field intensity increases, the three-photon crossing point opens and the dissociation starts proceeding through the lower-lying vibrational levels through the mechanism called the above-threshold dissociation, and consequently, the peaks representing the rovibrational energy separations for the lower vibrational levels also appear in the FT spectrum.

In the present strong-field FT spectroscopy, as long as ions created by an ultrashort pump pulse in the bound rovibrational levels in the electronic ground states keep staying in the same spatial domain until they are irradiated with an ultrashort probe pulse, the resolution in the frequency domain can be raised more simply by lengthening the pump–probe time delay range. When we employ an effusive molecular beam whose mean translational velocity is around 300 m/s, ions created by a pump pulse whose focal diameter at the laser–molecule interaction region is 30 μm pass through the interaction region within around 50 ns. Therefore, the frequency resolution could not be raised better than 20 MHz, which is the reciprocal of 50 ns. In other words, by extending the delay line length up to 15 m, we can achieve the frequency resolution as high as 20 MHz. In the present measurements on D_2^+, we were able to determine the rovibrational energy separations with uncertainties whose magnitude is 1/300 of the spectral resolution. Therefore, when the resolution is raised to 20 MHz, uncertainties associated with the rovibrational energy separations will be as small as 60 kHz (20 MHz/300), which means that $\Delta\nu/\nu$ can be as small as 2×10^{-9}. If we introduce only ions species having lower velocities using a velocity selector into the laser–molecule interaction volume, $\Delta\nu/\nu$ can be lowered more to the level of 0.1×10^{-9} so that the ratio of the mass of an electron and the mass of a proton (Korobov et al., 2014) can be determined experimentally.

8.6 SUMMARY

We have shown that the pump–probe measurements of molecules and molecular ions using intense few-cycle laser pulses enable us to investigate not only the ultrafast nuclear dynamics in the time domain but also the rovibrational energy level structure in the frequency domain and that the precision of the rovibrational level energies determined by the strong-field Fourier transform spectroscopy can be raised simply by extending the scan range of the pump-probe time delay. We can straightforwardly apply this new Fourier transform spectroscopy to a variety of molecular ion species including homonuclear diatomic molecular ions for which it has been a difficult task to observe the rovibrational transitions (Germann et al., 2014), and determine their rovibrational level structures with ultrahigh precision of the order of 1×10^{-9}.

ACKNOWLEDGMENT

Our research results introduced in the present article were those obtained from our research projects supported by JSPS KAKENHI Grants No. JP19002006 and JP15H05696.

REFERENCES

Ando, T., Iwasaki, A., Yamanouchi, K., 2018a. Strong-field Fourier transform vibrational spectroscopy of D_2^+ using few-cycle near-infrared laser pulse. Phys. Rev. Lett. 120, 263002.

Ando, T., Iwasaki, A., Yamanouchi, K., 2019. Strong-field Fourier transform vibrational spectroscopy of methanol cation and its isotopologues using few-cycle near-infrared laser pulses. Mol. Phys. 117, 1732–1740.

Ando, T., Shimamoto, A., Miura, S., Iwasaki, A., Nakai, K., Yamanouchi, K., 2018b. Coherent vibrations in methanol cation probed by periodic H_3^+ ejection after double ionization. Commun. Chem. 1, 7.

Ando, T., Shimamoto, A., Miura, S., Nakai, K., Xu, H., Iwasaki, A., Yamanouchi, K., 2015. Wave packet bifurcation in ultrafast hydrogen migration in CH_3OH^+ by pump-probe coincidence momentum imaging with few-cycle laser pulses. Chem. Phys. Lett. 624, 78–82.

Balint-Kurti, G.G., Moss, R.E., Sadler, I.A., Shapiro, M., 1990. Calculations of vibration-rotation energy levels of HD^+. Phys. Rev. A 41, 4913–4921.

Beyer, M., Merkt, F., 2016. Observation and calculation of the quasibound rovibrational levels of the electronic ground state of H_2^+. Phys. Rev. Lett. 116, 093001.

Beyer, M., Merkt, F., 2017. High-resolution photoelectron spectroscopy and calculations of the highest bound levels of D_2^+ below the first dissociation threshold. J. Phys. B, At. Mol. Opt. Phys. 50, 154005.

Biesheuvel, J., Karr, J.P., Hilico, L., Eikema, K.S.E., Ubachs, W., Koelemeij, J.C.J., 2016. Probing QED and fundamental constants through laser spectroscopy of vibrational transitions in HD^+. Nat. Commun. 7, 10385.

Biesheuvel, J., Karr, J.P., Hilico, L., Eikema, K.S.E., Ubachs, W., Koelemeij, J.C.J., 2017. High-precision spectroscopy of the HD^+ molecule at the 1-p.p.b. level. Appl. Phys. B, Lasers Opt. 123, 23.

Bieske, E.J., Dopfer, O., 2000. High-resolution spectroscopy of cluster ions. Chem. Rev. 100, 3963–3998.

Borkar, S., Sztáray, B., Bodi, A., 2011. Dissociative photoionization mechanism of methanol isotopologues (CH_3OH, CD_3OH, CH_3OD and CD_3OD) by iPEPICO: energetics, statistical and non-statistical kinetics and isotope effects. Phys. Chem. Chem. Phys. 13, 13009.

Braun, H., Bayer, T., Wollenhaupt, M., Baumert, T., 2020. 2D strong-field spectroscopy to elucidate impulsive and adiabatic ultrafast electronic control schemes in molecules. In: Marquardt, R., Quack, M. (Eds.), Molecular Spectroscopy and Quantum Dynamics. Elsevier, Amsterdam. Chapter 3 (this book).

Bressel, U., Borodin, A., Shen, J., Hansen, M., Ernsting, I., Schiller, S., 2012. Manipulation of individual hyperfine states in cold trapped molecular ions and application to HD^+ frequency metrology. Phys. Rev. Lett. 108, 183003.

Bunker, P.R., 1974. Allowed transitions and the dipole moment of HD^+. Chem. Phys. Lett. 27, 322–324.

Carrington, A., Kennedy, R.A., 1984. Infrared predissociation spectrum of the H_3^+ ion. J. Chem. Phys. 81, 91.

Chang, C., Ng, C.-Y., Stimson, S., Evans, M., Hsu, C.W., 2007. Rotationally resolved vacuum ultraviolet pulsed field ionization-photoelectron vibrational bands for H_2^+ ($X^2\Sigma_g^+$, $v^+ = 0$–18). Chinese J. Chem. Phys. 20, 352–364.

Dai, Z., Gao, S., Wang, J., Mo, Y., 2014. Torsional energy levels of CH_3OH^+/CH_3OD^+/CD_3OD^+ studied by zero-kinetic energy photoelectron spectroscopy and theoretical calculations. J. Chem. Phys. 141, 144306.

Ergler, T., Feuerstein, B., Rudenko, A., Zrost, K., Schröter, C.D., Moshammer, R., Ullrich, J., 2006. Quantum-phase resolved mapping of ground-state vibrational D_2^+ wave packets via selective depletion in intense laser pulses. Phys. Rev. Lett. 97, 103004.

Furukawa, Y., Hoshina, K., Yamanouchi, K., Nakano, H., 2005. Ejection of triatomic hydrogen molecular ion from methanol in intense laser fields. Chem. Phys. Lett. 414, 117–121.

Germann, M., Tong, X., Willitsch, S., 2014. Observation of electric-dipole-forbidden infrared transitions in cold molecular ions. Nat. Phys. 10, 820–824.

Giusti-Suzor, A., Mies, F.H., DiMauro, L.F., Charron, E., Yang, B., 1995. Dynamics of H_2^+ in intense laser fields. J. Phys. B, At. Mol. Opt. Phys. 28, 309–339.

Hasegawa, H., Hishikawa, A., Yamanouchi, K., 2001. Coincidence imaging of Coulomb explosion of CS_2 in intense laser fields. Chem. Phys. Lett. 349, 57–63.

Heine, N., Asmis, K.R., 2015. Cryogenic ion trap vibrational spectroscopy of hydrogen-bonded clusters relevant to atmospheric chemistry. Int. Rev. Phys. Chem. 34, 1–34.

Hishikawa, A., Hasegawa, H., Yamanouchi, K., 2004. Hydrogen migration in acetonitrile in intense laser fields in competition with two-body Coulomb explosion. J. Electron Spectros. Relat. Phenomena 141, 195–200.

Hishikawa, A., Iwamae, A., Yamanouchi, K., 1999a. Ultrafast deformation of the geometrical structure of CO_2 induced in intense laser fields. Phys. Rev. Lett. 83, 1127–1130.

Hishikawa, A., Iwamae, A., Yamanouchi, K., 1999b. Ultrafast structural deformation of NO_2 in intense laser fields studied by mass-resolved momentum imaging. J. Chem. Phys. 111, 8871–8878.

Hoshina, K., Furukawa, Y., Okino, T., Yamanouchi, K., 2008. Efficient ejection of H_3^+ from hydrocarbon molecules induced by ultrashort intense laser fields. J. Chem. Phys. 129, 104302.

Itakura, R., Yamanouchi, K., Tanabe, T., Okamoto, T., Kannari, F., 2003. Dissociative ionization of ethanol in chirped intense laser fields. J. Chem. Phys. 119, 4179–4186.

Jamroz, M.H., 2004. Vibrational Energy Distribution Analysis VEDA 4. Warsaw.

Jiang, S., Yu, C., Yuan, G., Wu, T., Lu, R., 2017. Dissociation and ionization of quasi-periodically vibrating H_2^+ in intense few-cycle mid-infrared laser fields. Sci. Rep. 7, 42086.

Johnson, R.D., 2018. https://cccbdb.nist.gov/vibscalejust.asp [WWW Document]. Natl. Institutes Stand. Technol. Comput. Chem. Comp. Benchmark Database, NIST Stand. Ref. Database Number 101. III.B.3.a. http://cccbdb.nist.gov/vibscalejust.asp.

Kanya, R., Kudou, T., Schirmel, N., Miura, S., Weitzel, K.-M., Hoshina, K., Yamanouchi, K., 2012. Hydrogen scrambling in ethane induced by intense laser fields: statistical analysis of coincidence events. J. Chem. Phys. 136, 204309.

Karlsson, L., Jadrny, R., Mattsson, L., Chau, F.T., Siegbahn, K., 1977. Vibrational and vibronic structure in the valence electron spectra of CH_3X molecules (X=F, Cl, Br, I, OH). Phys. Scr. 16, 225–234.

Karr, J.P., Hilico, L., 2006. High accuracy results for the energy levels of the molecular ions H_2^+, D_2^+ and HD^+, up to $J = 2$. J. Phys. B, At. Mol. Opt. Phys. 39, 2095–2105.

Karr, J.P., Hilico, L., Koelemeij, J.C.J., Korobov, V.I., 2016a. Hydrogen molecular ions for improved determination of fundamental constants. Phys. Rev. A 94, 050501.

Karr, J.P., Patra, S., Koelemeij, J.C.J., Heinrich, J., Sillitoe, N., Douillet, A., Hilico, L., 2016b. Hydrogen molecular ions: new schemes for metrology and fundamental physics tests. J. Phys., Conf. Ser. 723, 012048.

Kimura, K., Kubo, M., 1959. Structures of dimethyl ether and methyl alcohol. J. Chem. Phys. 30, 151–158.

Koelemeij, J.C.J., Roth, B., Wicht, A., Ernsting, I., Schiller, S., 2007. Vibrational spectroscopy of HD^+ with 2-ppb accuracy. Phys. Rev. Lett. 98, 173002.

Kołos, W., Wolniewicz, L., 1967. Polarizability of the hydrogen molecule. J. Chem. Phys. 46, 1426–1432.

Korobov, V.I., 2008. Relativistic corrections of $m\alpha^6$ order to the rovibrational spectrum of H_2^+ and D_2^+ molecular ions. Phys. Rev. A 77, 022509.

Korobov, V.I., Hilico, L., Karr, J.-P., 2014. $m\alpha^7$-order corrections in the hydrogen molecular ions and antiprotonic helium. Phys. Rev. Lett. 112, 103003.

Macneil, K.A.G.A.G., Dixon, R.N.N., 1977. High-resolution photoelectron spectroscopy of methanol and its deuterated derivatives: internal rotation in the ground ionic state. J. Electron Spectros. Relat. Phenomena 11, 315–331.

Marquardt, R., Quack, M., 2020. Foundations of time dependent quantum dynamics of molecules under isolation and in coherent electromagnetic fields. In: Marquardt, R., Quack, M. (Eds.), Molecular Spectroscopy and Quantum Dynamics. Elsevier, Amsterdam. Chapter 1 (this book).

Maruyama, I., Sako, T., Yamanouchi, K., 2004. Time-dependent nuclear wavepacket dynamics of H_2^+ by quasi-stationary Floquet approach. J. Phys. B, At. Mol. Opt. Phys. 37, 3919–3936.

Mosley, J.D., Young, J.W., Huang, M., McCoy, A.B., Duncan, M.A., 2015. Infrared spectroscopy of the methanol cation and its methylene-oxonium isomer. J. Chem. Phys. 142.

Moss, R.E., 1993. The $2p\sigma_u$–$1s\sigma_g$ electronic spectrum of D_2^+. J. Chem. Soc. Faraday Trans. 89, 3851–3855.

Müller-Dethlefs, K., Schlag, E.W., 1998. Chemical applications of zero kinetic energy (ZEKE) photoelectron spectroscopy. Angew. Chem., Int. Ed. 37, 1346–1374.

Nakai, K., Kato, T., Kono, H., Yamanouchi, K., 2013. Communication: long-lived neutral H_2 in hydrogen migration within methanol dication. J. Chem. Phys. 139, 181103.

Ning, Y., Yan, Z.C., 2014. Variational energy bounds for the hydrogen molecular ion. Phys. Rev. A 90, 032516.

Niu, M.L., Salumbides, E.J., Dickenson, G.D., Eikema, K.S.E., Ubachs, W., 2014. Precision spectroscopy of the $X^1\Sigma_g^+$, $v = 0 \rightarrow 1$ $(J = 0 - 2)$ rovibrational splittings in H_2, HD and D_2. J. Mol. Spectrosc. 300, 44–54.

Öhrwall, G., Baltzer, P., 1998. Angle-resolved photoelectron spectrum of rotational states in H_2^+. Phys. Rev. A 58, 1960–1965.

Okino, T., Furukawa, Y., Liu, P., Ichikawa, T., Itakura, R., Hoshina, K., Yamanouchi, K., Nakano, H., 2006a. Coincidence momentum imaging of ultrafast hydrogen migration in methanol and its isotopomers in intense laser fields. Chem. Phys. Lett. 423, 220–224.

Okino, T., Furukawa, Y., Liu, P., Ichikawa, T., Itakura, R., Hoshina, K., Yamanouchi, K., Nakano, H., 2006b. Coincidence momentum imaging of ejection of hydrogen molecular ions from methanol in intense laser fields. Chem. Phys. Lett. 419, 223–227.

Okino, T., Furukawa, Y., Liu, P., Ichikawa, T., Itakura, R., Hoshina, K., Yamanouchi, K., Nakano, H., 2006c. Ejection dynamics of hydrogen molecular ions from methanol in intense laser fields. J. Phys. B, At. Mol. Opt. Phys. 39, S515–S521.

Okino, T., Watanabe, A., Xu, H., Yamanouchi, K., 2012. Ultrafast hydrogen scrambling in methylacetylene and methyl-d3-acetylene ions induced by intense laser fields. Phys. Chem. Chem. Phys. 14, 10640.

Peatman, W., Wolf, F.-P., Unwin, R., 1983. Threshold electron photoionisation study of D_2: vibration-rotation states of D_2^+ $(X^2\Sigma_g^+)$. Chem. Phys. Lett. 95, 453–457.

Polfer, N.C., 2011. Infrared multiple photon dissociation spectroscopy of trapped ions. Chem. Soc. Rev. 40, 2211–2221.

Pollard, J.E., Trevor, D.J., Reutt, J.E., Lee, Y.T., Shirley, D.A., 1982. Rotationally resolved photoelectron spectroscopy of n-H_2, p-H_2, HD, and D_2. J. Chem. Phys. 77, 34–46.

Schiller, S., Bakalov, D., Bekbaev, A.K., Korobov, V.I., 2014. Static and dynamic polarizability and the Stark and blackbody-radiation frequency shifts of the molecular hydrogen ions H_2^+, HD^+, and D_2^+. Phys. Rev. A 89, 052521.

Seideman, T., Ivanov, M.Y., Corkum, P.B., 1995. Role of electron localization in intense-field molecular ionization. Phys. Rev. Lett. 75, 2819–2822.

Shimanouchi, T., 1972. Tables of Molecular Vibrational Frequencies Consolidated Volume I. National Bureau of Standards, Washington, DC.

Stimson, S., Chen, Y.-J., Evans, M., Liao, C.-L., Ng, C., Hsu, C.-W., Heimann, P., 1998. Rotational-resolved pulsed field ionization photoelectron bands for H_2^+ ($X^2\Sigma_g^+$, $v^+ = 0$, 2, 9 and 11). Chem. Phys. Lett. 289, 507–515.

Thapa, B., Schlegel, H.B., 2014. Molecular dynamics of methanol monocation (CH_3OH^+) in strong laser fields. J. Phys. Chem. A 118, 1769–1776.

Wei, Z., Li, J., Wang, L., See, S.T., Jhon, M.H., Zhang, Y., Shi, F., Yang, M., Loh, Z.H., 2017. Elucidating the origins of multimode vibrational coherences of polyatomic molecules induced by intense laser fields. Nat. Commun. 8, 735.

Xu, H., Marceau, C., Nakai, K., Okino, T., Chin, S.-L., Yamanouchi, K., 2010. Communication: two stages of ultrafast hydrogen migration in methanol driven by intense laser fields. J. Chem. Phys. 133, 071103.

Xu, H., Okino, T., Yamanouchi, K., 2009. Ultrafast hydrogen migration in allene in intense laser fields: evidence of two-body Coulomb explosion. Chem. Phys. Lett. 469, 255–260.

Yamanouchi, K., 2002. Laser chemistry and physics: the next frontier. Science 295, 1659–1660.

Yazawa, H., Shioyama, T., Suda, Y., Kannari, F., Itakura, R., Yamanouchi, K., 2006. Dissociative ionization of ethanol by 400 nm femtosecond laser pulses. J. Chem. Phys. 125, 184311.

Yazawa, H., Shioyama, T., Suda, Y., Yamanaka, M., Kannari, F., Itakura, R., Yamanouchi, K., 2007. Controlling the dissociative ionization of ethanol with 800 and 400 nm two-color femtosecond laser pulses. J. Chem. Phys. 127, 124312.

Zavriyev, A., Bucksbaum, P.H., Muller, H.G., Schumacher, D.W., 1990. Ionization and dissociation of H_2 in intense laser fields at 1.064 μm, 532 nm, and 355 nm. Phys. Rev. A 42, 5500–5513.

CHAPTER 9

Quantum Dynamics in Water Clusters

MARKO T. CVITAŠ* • JEREMY O. RICHARDSON†
*Department of Physical Chemistry, Ruđer Bošković Institute, Zagreb, Croatia
†Laboratory of Physical Chemistry, ETH Zurich, Zurich, Switzerland

Abstract

The ring-polymer instanton method is used to calculate tunneling pathways and hence tunneling splitting patterns of the water dimer, trimer, pentamer, and hexamer prism. We discuss recent developments to improve the efficiency of this method and present results utilizing the MB-pol potential-energy surface. We explain how the symmetry analysis can be performed in an automatic manner in order to obtain the splitting pattern and to assign the levels.

9.1 INTRODUCTION

Water clusters exhibit interesting spectroscopic features which are a consequence of their structure and dynamics (Keutsch et al., 2003; Liu et al., 1996d). By comparison of theoretical and experimental results, we can probe the accuracy of our description of intermolecular forces. Because the simulation of liquid water is limited by the accuracy of the intermolecular forces (Ceriotti et al., 2013, 2016), it is expected that through the study of clusters one will gain a better understanding also of the structure and dynamics of the liquid phase (Keutsch and Saykally, 2001).

The minimum-energy structures of the water clusters have been predicted using geometry optimization and *ab initio* electronic structure (Fowler and Schaefer, 1995; Lee et al., 1994; Smith et al., 1990; Temelso et al., 2011). Typically, the water molecules arrange themselves such that each molecule can make hydrogen bonds to its neighbors in the cluster by donating up to two and accepting up to two hydrogen bonds. These structures are in agreement with experimental observations of the vibrational (Keutsch and Saykally, 2001) and rotational spectra (Pérez et al., 2013, 2012, 2014). A particularly difficult problem was predicting the lowest-energy state of the hexamer. Due to zero-point energy effects, it was found that the cage and prism forms are roughly identical (Babin and Paesani, 2013; Wang et al., 2012).

However, it is not just the structure of the water clusters which can be probed by spectroscopic techniques. Of particular interest is the tunneling splitting pattern, which provides information on the quantum dynamics of the clusters (Saykally and Blake, 1993). Tunneling is an essentially non-classical process in which a particle can pass under a barrier even if its energy is lower than that of the barrier height (Hund, 1927; see also Quack and Seyfang, 2020, Chapter 7 of this book). The tunneling splitting pattern depends not just on the depth of the potential wells, but on the shape of the barrier to hydrogen-bond rearrangement. It therefore encodes information which will be of use in understanding the intermolecular forces which hold water molecules together (Stone, 2013).

In order to perform theoretical calculations of the quantum dynamics, one typically defines a potential-energy surface (PES, see also Marquardt and Quack, 2020, Chapter 1 of this book) as a function of geometry. This function should in principle return a solution of the electronic Schrödinger equation, but in practice one has to rely on approximations. There are a number of approaches which can be used to obtain the PES at various levels of accuracy.

One approach is to employ a semiempirical function for the PES with parameters chosen to reproduce experimental data. This approach has been employed to obtain the PES of the water dimer by inversion of spectroscopic data (Groenenboom et al., 2000; Leforestier et al., 2002). Potentials for larger water clusters (or the condensed phase) have been defined by fitting parameters to reproduce a number of known properties of various clusters such as structures and energetics (Fanourgakis and Xantheas, 2008).

An alternative approach is to describe the intermolecular potential based only on first-principles calculations. This approach has been used with some success for many years in describing the intermolecular potential between two rigid water molecules based on fits to electronic-structure calculations over a range of geometries (Matsuoka et al., 1976) or a multipole analysis (Millot et al., 1998). With modern improvements in computational power, the accuracy and applicability

Molecular Spectroscopy and Quantum Dynamics. https://doi.org/10.1016/B978-0-12-817234-6.00014-3

of these methods have dramatically increased, making them probably the most accurate approach now available (Bukowski et al., 2007; Shank et al., 2009; Szalewicz et al., 2009).

Larger water cluster potentials are commonly defined in terms of a multibody expansion (Paesani, 2016). The one-body term describes the covalent bonds in a single water molecule, the two-body term the intermolecular interaction between two molecules, and so on. It is expected that the series converges quickly and is often truncated after the three-body term. This has led to the development of a number of different PESs, such as WHBB (Wang and Bowman, 2010; Wang et al., 2011, 2009), MB-pol (Babin et al., 2013, 2014; Reddy et al., 2016), and CC-pol (Góra et al., 2014). It has been suggested that similar levels of accuracy are expected from all of these approaches (Nguyen et al., 2018).

The quantum tunneling effect in water clusters has been the focus of many spectroscopic experiments. The earliest was a microwave spectrum of the water dimer (Dyke et al., 1977) which showed a splitting pattern that could be explained qualitatively by a group-theory analysis (Dyke, 1977). Later experiments employed far-infrared spectroscopy to observe the tunneling splitting patterns in the trimer (Liu et al., 1996c, 1994; Pugliano and Saykally, 1992), tetramer (Cruzan et al., 1996a,b, 1997), pentamer (Harker et al., 2005; Liu et al., 1996b, 1997a), and hexamer cage (Liu et al., 1996a, 1997b). No tunneling was observed in the octamer (Richardson et al., 2013). This approach is now being used to study the spectrum of the clusters in excited vibrational states (Cole et al., 2017; Cole and Saykally, 2017).

The tunneling splittings of larger clusters have been observed by the new experimental technique of chirped-pulse cavity ring-down rotational spectroscopy (Pérez et al., 2012). This has led to the observation and assignment of the spectra of hexamer, heptamer, octamer, nonamer, and decamer (Pérez et al., 2013, 2014).

A number of theoretical techniques have been employed to predict and explain the tunneling splitting patterns observed in these experiments. Quantum dynamics calculations on the water dimer are possible by numerically solving the nuclear Schrödinger equation (Althorpe and Clary, 1994, 1995; Coudert and Hougen, 1988; Hougen, 1985; Leforestier et al., 1997; Wang and Carrington, 2018). These approaches are only applicable to the larger clusters if reduced-dimensionality models are used (Sabo et al., 1995; van der Avoird et al., 1996), and in order to treat them in full dimensionality, more efficient approaches are required. One option is to use Diffusion Quantum Monte Carlo (DQMC) (Gregory and Clary, 1995a,b; Suhm and Watts, 1991). However, to use this method to obtain accurate tunneling splittings, a nodal surface must be defined *a priori*, which is in general not known exactly.

An alternative approach is to use semiclassical Wentzel-Kramers-Brillouin (WKB) theory (Wales, 1993a,b). This is an efficient computational method and can therefore be easily combined with accurate potential-energy surfaces. However, again an *a priori* choice must be made to define the tunneling pathway and the prediction can be highly dependent on this choice.

Ring-polymer instanton theory (Richardson, 2018a) overcomes these problems and can be used to locate the uniquely defined dominant tunneling pathway and hence give reliable tunneling splitting predictions. In this chapter, we will focus on how this method has been used to compute the tunneling splitting pattern of various water clusters and give information on the mechanisms involved.

9.2 RING-POLYMER INSTANTON APPROACH

The instanton approach (Miller, 1975) was first used to obtain tunneling splittings by the particle physics community (Coleman, 1977). It was later applied to simple one- and two-dimensional models of molecular rearrangements (Benderski et al., 1994), but it was not until the development of more efficient algorithms (Mil'nikov and Nakamura, 2001), such as the ring-polymer instanton method (Richardson and Althorpe, 2011), that it could be applied to molecules in full dimensionality.

In this section, we describe the theoretical basis and computational methods used. An open-source implementation of the method is available in the i-PI package (Kapil et al., 2019). Note that a more in-depth review of the first-principles derivation is given in Richardson (2018b) and here we concentrate on the applications of instanton theory to water clusters.

9.2.1 Semiclassical Theory

The semiclassical limit of quantum mechanics can be defined as an asymptotic approximation in the limit that the parameter $\hbar$ tends to 0. When applied to approximate the path-integral formulation of quantum mechanics, the integral over all paths is replaced by a sum over classical trajectories (Gutzwiller, 1990). According to Hamilton's principle, these are paths for which the Euclidean action,

$$S[x(\tau)] = \int_0^{\beta\hbar} \left[\frac{m}{2}||\dot{x}||^2 + V(x)\right] d\tau, \tag{9.1}$$

is stationary. In this way, one can show how classical mechanics is a limiting case of quantum mechanics and the approximation is used to drastically simplify the problem.

Although no classical trajectory exists which tunnels, the quantum dynamics of the particle can nonetheless be approximated using trajectories. Using the equation, $E = p^2/2m + V(x)$, one can see that if the particle energy, E, is lower than the potential energy, $V(x)$, then the kinetic energy term, $p^2/2m$, must be negative. Because the mass, m, is positive, this implies that the momentum, p, is imaginary and that therefore particles tunneling through barriers can be considered to be moving in imaginary time (Miller, 1975).

Between a pair of degenerate wells, there exists a dominant tunneling pathway, defined as the imaginary-time classical trajectory traveling from the bottom of one well to the other. The semiclassical approximation for the coupling between these wells is proportional to $e^{-S/\hbar}$ and has a prefactor which depends on the fluctuations around the path. Clearly, the paths with the minimal value of S will dominate this result.

In this way the instanton approach predicts the mechanism leading to the tunneling splitting. In certain cases there is a competition between different mechanisms and the instanton approach is able to justify why one is preferred—the one with the smallest S. This may not always be the mechanism with the lowest barrier as tunneling through a tall, narrow barrier can be easier than that through a short, wide barrier (Carpenter, 2011).

Instanton theory gives fairly accurate results for small tunneling splittings in systems with a high barrier between the wells. However, if the barrier height is so small that it becomes similar to the harmonic zero-point energy of the wells then the semiclassical approximation breaks down and the tunneling splitting can be overestimated by about a factor of two (Richardson and Althorpe, 2011). Instanton theory thus complements the diffusion Monte Carlo method, which is accurate for computing large tunneling splittings but can be dominated by statistical errors for calculations of a small splitting.

Performing a full-dimensional calculation including all the perpendicular modes has been shown to be necessary for obtaining quantitative results (Richardson, 2017; Yagi et al., 2004). Although in the instanton method the modes perpendicular to the pathway are treated within a harmonic approximation, they are not ignored and do at least contribute with a zero-point energy effect. This makes the instanton method the state-of-the-art approach for computing tunneling effects rates in polyatomic systems.

9.2.2 Optimization Algorithm

In order to apply the method to predict the tunneling splitting, it is necessary to obtain the tunneling pathway joining two degenerate wells in multidimensional space. Historically the pathways were located by numerically solving the imaginary-time classical equations of motion and searching for boundary conditions leading to the required solution connecting the two wells (Benderski et al., 1994). This method, known as "shooting", is a simple procedure for one-dimensional systems but quickly becomes practically impossible to perform as the number of degrees of freedom increases beyond two. This is because when searching for an unstable trajectory in a multidimensional potential, the final position at the end of the trajectory changes exponentially strongly as the initial condition is varied.

The disadvantages of the shooting method for solving general double-ended boundary problems are clearly outlined in *Numerical Recipes* (Press et al., 2007) and the alternative method of "relaxation" is recommended instead. To apply this approach to the problem of locating instantons, the path is represented by a discretized set of points and the action function defined in terms of these. According to the variational principle, the first variation of S is zero for a classical trajectory, and thus discretized trajectories can be obtained by searching for minima of the action.

A number of schemes for discretizing the instanton path have been suggested based on a Fourier or polynomial expansion (Doll et al., 1989; Kryvohuz, 2012; Mil'nikov and Nakamura, 2001; Tautermann et al., 2002). However, the most common discretization scheme is that of the ring polymer (Barker, 1979; Chandler and Wolynes, 1981). Here the path is represented by N "beads", x_i, each representing an image of the full multidimensional system and specifying the Cartesian coordinates of all atoms. The use of Cartesian coordinates makes the method practical, as it can be applied to any molecular system without modification, whereas the exact quantum methods usually work, for efficiency reasons, in internal curvilinear coordinates and require the derivation of specialized system-dependent Hamiltonians.

In the original formulation of the ring-polymer instanton method (Richardson and Althorpe, 2011), the optimal tunneling pathway, the instanton, is located in many dimensions using a discretization of the path in terms of equal imaginary time steps. The potential energy and its gradients along the pathway are evaluated explicitly at beads and not fit to any predetermined shape. Minimum-action paths are then determined using standard computational approaches in the extended multidimensional space of a ring polymer. Once the minimum action is found, the ring-polymer Hessian is evaluated in order to evaluate the contributions of other paths in the harmonic vicinity of the minimum-action path. It has been possible to obtain instantons in many complex systems using this approach, but even more efficient methods have recently been developed.

The ring-polymer potential requires the specification of the imaginary-time duration $\beta\hbar$, related to the system temperature T through $\beta = 1/kT$. Tunneling splittings are obtained in the zero-temperature limit of the quantum partition functions in path integral formulation. Thus the trajectories are of very long time. For a two-well system, this means that the trajectory asymptotically approaches the potential energy minima, defined to be at zero energy, at both of its ends. The discretization in terms of equal time steps results in most beads settling close to the bottom of the wells, while only a minority describe the pathway over the barrier. This is, in fact, the origin of the name "instanton"; the trajectory is such that it is almost stationary at the beginning and end, but somewhere in the middle the tunneling event happens very quickly, almost in an instant.

A number of suggestions have been made to redistribute the beads closer to the part of the trajectory which is moving and to avoid them clustering in the wells. One idea is to redefine the imaginary time discretization in a non-uniform way (Rommel and Kästner, 2011). However, this requires *a priori* knowledge of the pathway in order to choose the discretization in an optimal way.

A more promising approach is to use the Jacobi form of the variational principle in which the time dependence of the action integral is factored out by a Legendre transform,

$$S(x', x'', \beta\hbar) = W(x', x'', E) + E\beta\hbar, \tag{9.2}$$

where W, the abbreviated or Jacobi action, is given by an integral along the path

$$W(x', x'', E) = \int_{x'}^{x''} p(x, E)\,\mathrm{d}x, \tag{9.3}$$

which does not involve time (Faccioli et al., 2006). In Eq. (9.3), the magnitude of the system momentum $|p|$ at energy E is obtained from the potential using $|p(x, E)| = \sqrt{2m[V(x) - E]}$, and the imaginary time duration $\beta\hbar$ is no longer a convergence parameter. The object of the minimization above was to find the time-dependence of the trajectory with a preset duration, $\beta\hbar$. The object now is to determine the shape of the trajectory at zero energy, which corresponds to a trajectory of infinite duration, without reference to time. We are free to choose any discretization that is best suited for defining the shape. Equal interbead distances in coordinate space are commonly used. Once the shape of the zero-energy trajectory is found, the positions of beads that specify the ring polymer need to be determined in order to calculate ring-polymer hessian, which provides an estimate of the contribution of the neighboring paths to the tunneling matrix element. For this purpose, bead positions along the minimum-action path are determined by another optimization in one dimension along the instanton, where full action S is minimized on the interpolated potential.

The above time-independent approach, based on the minimization of Jacobi action W has also been used for rate calculations (Ásgeirsson et al., 2018) but becomes particularly efficient for tunneling splitting calculations (Cvitaš, 2018; Cvitaš and Althorpe, 2016). The gradient-based search for a minimum of action is performed, whereby beads are kept at equal distances in course of a minimization using either a spring force, in a nudged elastic band method (Einarsdóttir et al., 2012), or by reinterpolation to equally-spaced beads in terms of Euclidean distances, in the string method (E et al., 2002). This is equivalent to the original method with variable time steps, where the time steps are set at every iteration in a way that is consistent with the classical motion discretized over a uniform space grid. This approach reduces the number of beads N needed to represent the minimum-action path and the lower number of degrees of freedom in the minimization algorithm leads to the smaller number of iterations needed to obtain a converged path. The efficiency savings in tunneling splitting calculations are such that about two orders of magnitude fewer potential evaluations are needed to obtain the instanton in examples below.

9.2.3 Symmetry Analysis

Unlike simpler examples of molecules exhibiting tunneling splittings such as NH_3, water clusters are not described by a double well but have a larger number of degenerate wells which may be connected together by various tunneling pathways. Instanton theory is, however, easily extended to treat such problems using a graph-theoretical analysis to enumerate all paths (Richardson et al., 2011). The implementation of the ring-polymer instanton method is exactly equivalent to the double-well case and one simply needs to compute a tunneling matrix element for each of the possible rearrangement pathways. These matrix elements are combined into an adjacency matrix to describe the connectivity of the wells in a similar way to Hückel theory. The eigenvalues of this matrix give the energy levels and hence the tunneling splitting pattern.

A symmetry analysis can also be performed which allows these energy levels to be labeled such that nuclear spin statistics can be computed and allowed spectroscopic transitions worked out (Bunker and Jensen, 2005; Longuet-Higgins, 1963; Wales, 1993a). Most of the symmetry analysis can be done in an automatic way, which is particularly important for water clusters such as the pentamer which has a group of 320 elements. Here we will outline the method. The reader is also referred to Chapters 1 (Marquardt and Quack, 2020) and 7 (Quack and Seyfang, 2020) of this book.

First one should obtain an optimized minimum-energy geometry of the cluster. Each atom is then labeled by a distinct letter or number. This defines the reference "version".

In principle, any permutation of like atoms or inversion operation will generate a new version which is also an optimized geometry with the same energy. An instanton connecting the wells can be obtained using an appropriate optimization algorithm. In many cases, the action of the instanton will be very large and it will thus contribute negligibly to the tunneling splitting pattern. These pathways are then termed unfeasible and neglected in the following analysis. Some tunneling pathways found in this way will pass through other versions, as is easily noticed if the instantaneous potential energy drops back down to the initial value before the end of the path. These composite pathways are formed of multiple instantons and will be taken care of by the graph theory analysis and thus are not counted in the list of instantons. The remaining direct pathways are the instantons corresponding to feasible rearrangements in the cluster. In practice, one does not need to search for an instanton for every possible permutation as it is clear that many permutations will be unfeasible if they involve breaking covalent bonds or rearranging heavy atoms.

The permutation-inversion operations corresponding to the instanton pathways are identified. These are used to generate the molecular symmetry (MS) group, $\mathcal{G}$, by considering multiplications and inverses of the operations and adding new elements until the group is complete. The elements of this group are the corresponding nuclear permutations which link the set of versions which mix via the tunneling dynamics. It is this mixing which gives rise to the observed tunneling splitting pattern. The number of elements in the group is equal to the number of local minima which are coupled together by the tunneling pathways, n, multiplied by the number of point-group symmetry operations of the minimum-energy geometry. The elements can be placed in classes according to similarity. The irreducible representations (irreps) and character table can be found by using isomorphisms between the group and standard point groups.

The symmetry labels of the set of tunneling states, Γ_{tun}, can be found by reducing the representation of the set of non-tunneling states. This has character n for the identity operation, but character 0 for all other permutation-inversion operations. The standard reduction formula can be used to perform this task automatically.

A set of nuclear-spin states can be constructed by considering all combinations of spins in the cluster. The symmetry of this set is found by calculating the character corresponding to each element in the group. The character is equal to the number of spin states which are unaffected by the permutation operator. The irreps, Γ_{ns}, are again found by the reduction formula.

A shortcut to obtaining the character, $\chi^{\mathcal{P}}$, of the spin states is given by (Bunker and Jensen, 2006):

$$\chi^{\mathcal{P}} = \prod_{\alpha}(2I_\alpha + 1), \tag{9.4}$$

where the product contains one factor for each set of indistinguishable nuclei permuted by $\mathcal{P}$ and I_α is the spin of each of the nuclei in the set.

The total wave function is the product of the tunneling state with a nuclear-spin state. This total wave function must obey the Pauli principle and it must therefore have the symmetry of the irrep with character -1 for operations which exchange pairs of fermions and character $+1$ otherwise. This irrep is called Γ_+. For permutations including the inversion operation, a second possibility exists, called Γ_-, which has the characters of Γ_+ for pure permutation operations, and the opposite sign for operations involving inversion. Then by testing all combinations of the tunneling states with nuclear-spin states to find those which include $\Gamma_\pm$, one obtains the statistical weights of the tunneling states.

To build the tunneling matrix, one creates a list of the n versions by operating on the reference by each permutation-inversion operation in the MS group and considers only those which are not related by point-group symmetry. Then by considering the permutations corresponding to instanton pathways on each version in turn, the version index of the final state is obtained.

The matrix element corresponding to this instanton is inserted in the matrix at this point.

The eigenvalues of the tunneling matrix are the energy levels of the tunneling splitting pattern, and the eigenvectors will be used to determine their symmetry labels. Using the projection operator on a trial vector in the basis of the *n* versions, one obtains a vector corresponding to each irrep. This vector will be orthogonal to all eigenvectors except those which are of the same symmetry and thus one can determine the symmetry label and hence statistical weight of each level.

9.3 TUNNELING IN VARIOUS WATER CLUSTERS

In this section, we present results from the application of the instanton approach to a number of water clusters. The MB-pol potential (Babin et al., 2014) will be used here and can be compared to previous work (Richardson et al., 2011) using the WHBB potential (Wang et al., 2009). It is seen that, in all cases, no large deviations are found, implying that both PESs have similar accuracy. We do, however, find that MB-pol is computationally faster, at least for our examples.

9.3.1 Water Dimer

The water dimer, $(H_2O)_2$, and its isotopomer, $(D_2O)_2$, provide ideal test cases for the instanton method as the splitting pattern is non-trivial but well understood (Coudert and Hougen, 1988; Dyke, 1977), and the various tunneling splittings span four orders of magnitude. The dimer has been studied extensively by spectroscopy to quantify the experimental splitting pattern (Fraser et al., 1989; Zwart et al., 1991) as well as its deuterated form (Karyakin et al., 1993; Paul et al., 1998; Suenram et al., 1989). Many theoretical studies have also computed the tunneling splittings for the dimer by the direct solution of the Schrödinger equation, originally using fixed monomers (Althorpe and Clary, 1994, 1995; Chen et al., 1999; Fellers et al., 1999; Groenenboom et al., 2000; Huang et al., 2008), or using adiabatic separation of the intermolecular and intramolecular modes (Leforestier et al., 2002), but more recently it has been possible to solve the full problem numerically exactly (Wang and Carrington, 2018). There have also been approaches based on model Hamiltonians (Coudert and Hougen, 1988), diffusion Monte Carlo (Gregory and Clary, 1995a), and the WKB approximation (Watanabe et al., 2004).

It is well-known that the water dimer's minimum geometry has a "trans-linear" geometry with C_s symmetry (Dyke et al., 1977) as depicted in Fig. 9.1. The water molecule whose hydrogen atom forms a hydrogen bond is called the *donor* monomer; the other is the *acceptor*. There are 8 degenerate isomers connected by feasible pathways, i.e., those which do not break covalent bonds (Dyke, 1977), and only five of these pathways are unrelated by symmetry. Here we ignore exchange of hydrogens as this breaks covalent bonds and has a large barrier (Wang et al., 2010). It is therefore expected to lead to unfeasible pathways and can be neglected. The molecular symmetry group of the water dimer has 16 elements and is isomorphic to the point group D_{4h}. The character table of the molecular symmetry group is given in Table 9.1.

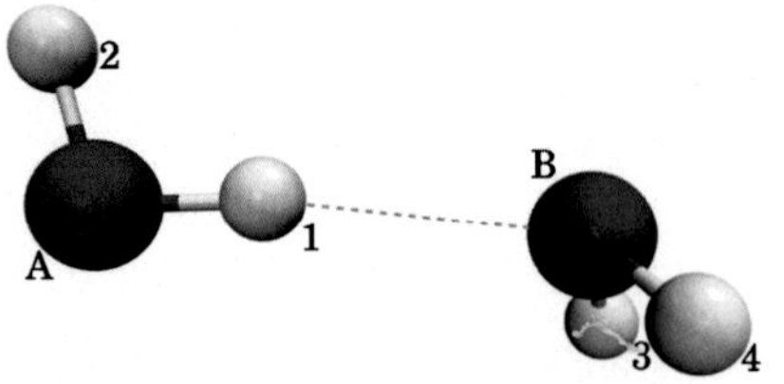

FIG. 9.1 The minimum-energy geometry of the water dimer labeled so as to represent the reference version. Reprinted from Richardson et al. (2011), with the permission of AIP Publishing.

TABLE 9.1 Character table appropriate for the water dimer. There are two notations in use in the literature for labeling the irreps which are both given here. The size of each class is shown underneath a representative permutation-inversion operation.

		E	(12)	(AB)(13)(24)	(AB)(1324)	(12)(34)	E^*	(12)*	(AB)(13)(24)*	(AB)(1324)*	(12)(34)*
LH[a]	MW[b]	1	2	2	2	1	1	2	2	2	1
A_1^+	A_1^+	1	1	1	1	1	1	1	1	1	1
A_2^+	A_2^+	1	−1	−1	1	1	1	−1	−1	1	1
B_1^+	B_2^+	1	−1	1	−1	1	1	−1	1	−1	1
B_2^+	B_1^+	1	1	−1	−1	1	1	1	−1	−1	1
E^+	E^+	2	0	0	0	−2	2	0	0	0	−2
A_1^-	B_1^-	1	1	1	1	1	−1	−1	−1	−1	−1
A_2^-	B_2^-	1	−1	−1	1	1	−1	1	1	−1	−1
B_1^-	A_2^-	1	1	−1	−1	1	−1	−1	1	1	−1
B_2^-	A_1^-	1	−1	1	−1	1	−1	1	−1	1	−1
E^-	E^-	2	0	0	0	−2	−2	0	0	0	2

[a] Standard notation from Longuet-Higgins (1963).
[b] Alternative notation from Merer and Watson (1973).

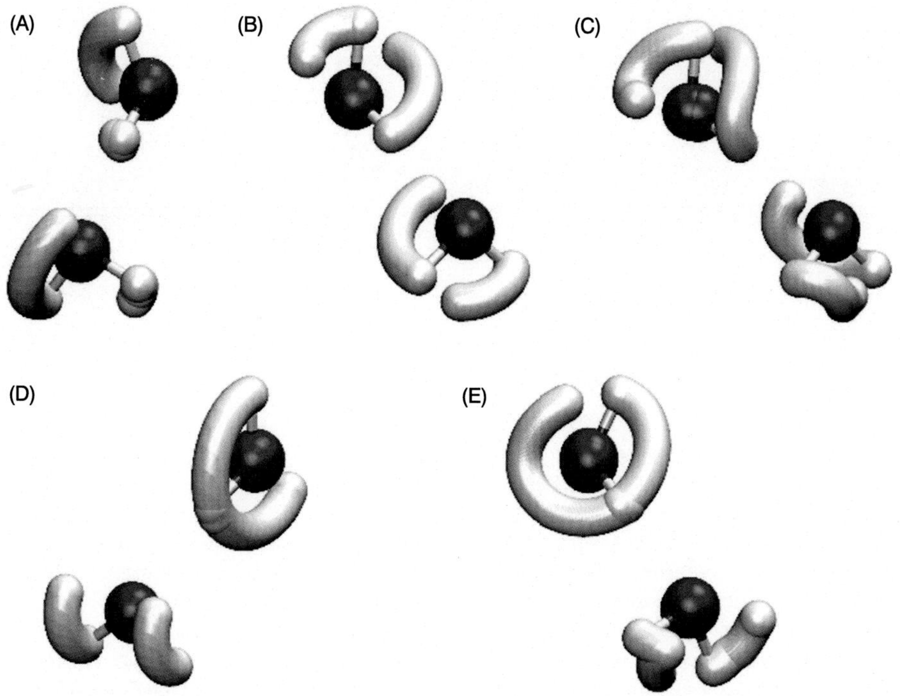

FIG. 9.2 Conflated snapshots of instanton tunneling paths for $(H_2O)_2$. The paths shown are (A) acceptor tunneling, (B) geared interchange, (C) anti-geared interchange, (D) bifurcation tunneling, and (E) donor exchange.

Only five elements of the tunneling matrix are independent and need to be calculated by the instanton method (Richardson et al., 2011). The three most important low-lying transition states of the water dimer were first studied by Smith et al. (1990) using *ab initio* potential-energy surfaces, which were later refined with explicitly-correlated computations (Tschumper et al., 2002). Unconstrained minimum-energy pathways for these rearrangements were calculated by Wales (1999). The instanton approach, however, is able to locate the minimum-action pathways for all five rearrangements (see Fig. 9.2), some of which are similar but not equivalent to the minimum-energy pathways.

The pathway corresponding to the permutation (34) is referred to as *acceptor tunneling*. Previous work (Wales, 1999) found that this motion does not correspond to a simple internal rotation of the acceptor water molecule but rather to a concerted twist of the donor monomer and wag of the acceptor. The computed instanton is entirely consistent with this type of motion. The instanton differs, however, from the minimum-energy path used in the WKB calculations of Watanabe et al. (2004) in that it passes through a point with a plane of symmetry close to the second-order C_s saddle point instead of going through the first-order C_1 saddle point. This occurs because the action is smaller for the instanton pathway which, although it passes through regions of higher potential energy, is shorter than the minimum-energy path. One effect of this difference which must be considered is that there are two symmetrically-equivalent instantons, but four symmetrically-equivalent minimum-energy paths. As a result, the correct adjacency matrix element is 2, whereas Taketsugu and Wales (2002) and Watanabe et al. (2004) used 4, which is the appropriate number for their calculation.

The instanton predictions for the (AB)(1324) and (12)(34) tunneling paths agree with previous work (Taketsugu and Wales, 2002), both in terms of the adjacency matrix element of 1, and in terms of the tunneling dynamics along the paths. The most important of these paths is the (AB)(1324) path, which corresponds to a *geared interchange* of the acceptor and donor monomers. The *anti-geared interchange*, (AB)(14)(23), behaves similarly, but also performs an extra hydrogen permutation during the pathway. Its existence has been predicted

TABLE 9.2
Instanton tunneling pathways located on the MB-pol potential surface for water dimer. The actions, S, and tunneling matrix elements, h, were obtained from the instanton calculations.

		$(H_2{}^{16}O)_2$		$(D_2{}^{16}O)_2$		$(H_2{}^{18}O)_2$	
Pathway	$\mathcal{P}$	$S/\hbar$	$-h/\text{cm}^{-1}$	$S/\hbar$	$-h/\text{cm}^{-1}$	$S/\hbar$	$-h/\text{cm}^{-1}$
Acceptor tunneling (AT)	(34)	5.39	6.2	7.28	7.7(−1)	5.42	6.0
Geared interchange (GI)	(AB)(1324)	7.25	1.9(−1)	10.10	9.8(−3)	7.26	1.8(−1)
Antigeared interchange (AI)	(AB)(14)(23)	10.56	3.3(−2)	14.67	5.0(−4)	10.59	3.2(−2)
Bifurcation tunneling (BT)	(12)(34)	11.81	2.2(−2)	16.52	1.8(−4)	11.83	2.1(−2)
Donor exchange (DE)	(12)	15.61	1.7(−3)	21.77	3.3(−6)	15.64	1.7(−3)

from experimental measurements (Fraser et al., 1989), but is frequently neglected in theoretical studies because although its inclusion shifts the energy levels, it does not cause an extra splitting. The (12)(34) pathway is called either *donor* or *bifurcation tunneling* and exchanges the hydrogen atoms on the donor molecule accompanied by a wag of the acceptor. Unlike the minimum-energy paths that were obtained in previous work, the instanton paths for (AB)(1324), (AB)(14)(23) and (12)(34) do not pass directly through the saddle points, but do pass close to them through points of C_i, C_2, and C_{2v} symmetry, respectively.

The remaining (12) path has been little studied in the literature, and is thought to contribute only a small amount to the tunneling pattern (Coudert and Hougen, 1988). No minimum-energy pathway can be found which connects these wells directly (Taketsugu and Wales, 2002). Nevertheless a minimum-action path does exist and we were able to find an instanton describing this tunneling pathway, which we call *donor exchange*. Unlike the other four paths, the (12) instanton does not pass through a point with any symmetry, and for this reason (and also because the donor could rotate either clockwise or anticlockwise) there are other degenerate pathways such that the adjacency matrix element is 4.

The predicted matrix elements, h, for each pathway are presented in Table 9.2. These elements are seen to span four orders of magnitude and to be significantly smaller for the deuterated cluster, but to have only a small dependence on isotopic substitution of the oxygens. This is because it is the hydrogens which move along the tunneling pathways and the oxygens are almost stationary.

When arranged from largest to smallest, the tunneling matrix elements retain the same order as was found by the WKB analysis (Watanabe et al., 2004). This is because the action of a pathway is still loosely connected to the height of the barrier, and the pathways did not deviate too strongly from the minimum-energy pathways. The instanton method will not therefore predict a qualitatively different splitting pattern for the water dimer from those obtained by previous studies. However, as we will show in later sections, the instanton predictions of the splitting patterns of larger clusters can be quite different from what would have been predicted from the minimum-energy pathways.

The adjacency matrix obtained from the instantons gives rise to the following tunneling matrix:

$$\mathsf{W} = \begin{pmatrix} 0 & h_{BT} & 4h_{DE} & 2h_{AT} & h_{GI} & h_{GI} & h_{AI} & h_{AI} \\ h_{BT} & 0 & 2h_{AT} & 4h_{DE} & h_{GI} & h_{GI} & h_{AI} & h_{AI} \\ 4h_{DE} & 2h_{AT} & 0 & h_{BT} & h_{AI} & h_{AI} & h_{GI} & h_{GI} \\ 2h_{AT} & 4h_{DE} & h_{BT} & 0 & h_{AI} & h_{AI} & h_{GI} & h_{GI} \\ h_{GI} & h_{GI} & h_{AI} & h_{AI} & 0 & h_{BT} & 4h_{DE} & 2h_{AT} \\ h_{GI} & h_{GI} & h_{AI} & h_{AI} & h_{BT} & 0 & 2h_{AT} & 4h_{DE} \\ h_{AI} & h_{AI} & h_{GI} & h_{GI} & 4h_{DE} & 2h_{AT} & 0 & h_{BT} \\ h_{AI} & h_{AI} & h_{GI} & h_{GI} & 2h_{AT} & 4h_{DE} & h_{BT} & 0 \end{pmatrix}. \tag{9.5}$$

The splitting pattern, obtained by diagonalizing W, is shown in Fig. 9.3 with the states labeled by the irreps of the molecular symmetry group given in Table 9.1. We use Longuet-Higgins' (LH) notation, which is recommended by a more systematic labeling approach in which the letter and subscript are determined uniquely by the permutation and the ± superscript by the inversion operator (Quack, 1977).

We do not expect the ring-polymer instanton method to give quantitative agreement with experiment because, as mentioned above, the method neglects anharmonicity perpendicular to the tunneling path and assumes that the splittings depend only weakly on the rotational

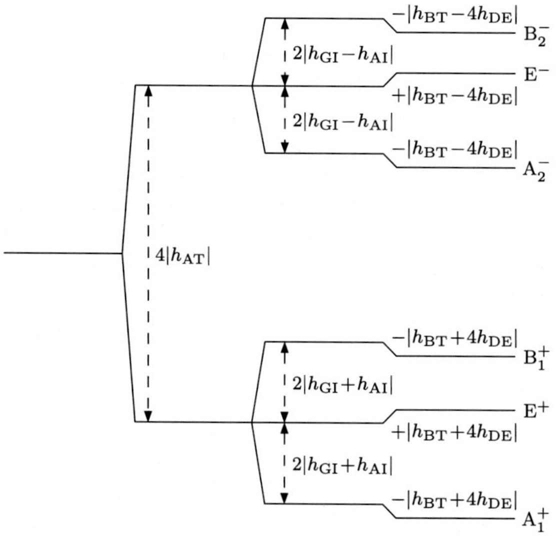

FIG. 9.3 Water dimer tunneling-splitting pattern, obtained by diagonalizing the instanton tunneling matrix W of Eq. (9.5). The values of the tunneling elements, h, and of the splittings are given in Tables 9.2 and 9.3. Adapted from Richardson et al. (2011).

quantum numbers. Nevertheless, the agreement of the instanton results with experiment is often much better than a factor of two.

Two pieces of evidence suggest that most of the quantitative errors in the instanton splittings result from the neglect of anharmonicity. First, the errors are smaller for $(D_2O)_2$ than for $(H_2O)_2$, which is consistent with the expectation that deuteration reduces the fluctuations of the polymers around the wells. Second, the interchange splittings (those dependent mostly on h_{GI}) agree much better with experiment than the acceptor splittings do. This is because the barrier height along the acceptor-tunneling path is close to the change in harmonic zero-point energy, and calculations on a model double-well system show that there are large errors for similarly low barriers (Richardson and Althorpe, 2011). This effect is almost certainly why the instanton acceptor splitting is a factor of two too large in $(H_2O)_2$. This error is, of course, much smaller in $(D_2O)_2$, as deuteration decreases the fluctuations.

The values of h are known to depend only weakly on rotational quantum numbers, except for h_{AT} (Coudert and Hougen, 1988; Huang et al., 2008). The acceptor tunneling pathway involves an effective internal twist of the two monomers, which causes h_{AT} to depend strongly on the rotational state. In Table 9.2, we have quoted the ground-rotational state experimental values for h_{AT}. The instanton results therefore include errors resulting from "contamination" from excited rotational states. The coupling of rotational motion to the acceptor tunneling motion in water dimer can also be quantified using instanton theory (Vaillant and Cvitaš, 2018). The tunneling splitting of a particular rotational symmetry can be obtained by a symmetry-adapted projection (Mátyus and Althorpe, 2016) from instantons with different end-to-end overall orientations. The approach applies the instanton approximation to path integrals in the density matrix formulation of tunneling splittings (Mátyus et al., 2016), where the energy of the state of a particular symmetry is obtained in an analogous fashion to the multiwell theory of Richardson et al. (2011).

9.3.2 Water Trimer

The water trimer, $(H_2O)_3$, and its deuterated form, $(D_2O)_3$, exhibit a rich tunneling splitting pattern (Brown et al., 1999; Liu et al., 1994; Pugliano and

TABLE 9.3
Tunneling splittings (in cm^{-1}) for the water dimer obtained from the instanton calculations compared experiment. See Fig. 9.3 for a diagram of the overall splitting pattern.

		$(H_2O)_2$		$(D_2O)_2$	
	Splitting	**inst.**	**expt**[a]	**inst.**	**expt**[b]
Acceptor splitting	$4\|h_{AT}\|$	25	9.4	3.1	1.8
Interchange (lower)	$4\|h_{GI}+h_{AI}\|$	0.89	0.75	0.041	0.039
Interchange (upper)	$4\|h_{GI}-h_{AI}\|$	0.63	0.65	0.037	0.036
Bifurcation (lower)	$\|h_{BT}+4h_{DE}\|$	0.029	0.02	1.9(−4)	2.3(−4)
Bifurcation (upper)	$\|h_{BT}-4h_{DE}\|$	0.015	0.02	1.7(−4)	2.2(−4)

[a] Zwart et al. (1991).
[b] Karyakin et al. (1993); Paul et al. (1998); Suenram et al. (1989).

Saykally, 1992). Comparisons between theory and experiment provide information on the important three-body terms (Gregory and Clary, 1995b) which are expected to also have a large influence on the structure of liquid water. A good review on the water trimer can be found in Keutsch et al. (2003).

It is not possible to apply exact DVR-type (discrete variable representation; see also Marquardt and Quack, 2020, Chapter 1 of this book) approaches to systems with this many degrees of freedom (Carrington, 2017) even if rigid monomers were used, as the size of the basis set needed would be too large; see also Chapter 1 and 2. However, the trimer has been studied theoretically using DQMC (Gregory and Clary, 1995a) and WKB (Takahashi et al., 2005). These theoretical approaches were not however able to explain the existence of "anomalous" quartets in the spectrum, which we show is a consequence of a number of competing bifurcation pathways. In the DQMC study, the nodal planes were defined such that there was no differentiation between the six possible excited states corresponding to the different bifurcation rearrangements. In this way, the information necessary for explaining the appearance of the spectrum was lost. In the WKB study, depending on the potential-energy surface used, different minimum-energy rearrangement pathways were found such that two types of splitting pattern were suggested (Walsh and Wales, 1996), neither of which agreed completely with experiment. The method gives no way of calculating which of the many mixed patterns suggested was correct (Keutsch et al., 2002). Seeing as the instanton method does not rely on predefined nodal planes or on the minimum-energy pathways between two isomers, it is able to resolve some of these problems as is shown below.

The stationary points of the water trimer have been located using *ab initio* potentials (Fowler and Schaefer, 1995; Schütz et al., 1993). The equilibrium geometry is found to be a cyclic arrangement with C_1 symmetry where each monomer acts as a single hydrogen-bond donor and single acceptor as shown in Fig. 9.4. The dangling hydrogen atoms on two of the monomers point up (u) relative to the plane: these are called the *majority* monomers, and are referred to individually as the *acceptor* and the *donor*, depending on their relation to the remaining *minority* monomer, in which the loose hydrogen points down (d).

There are 96 permutation-inversion isomers for the water trimer equilibrium geometry if covalent bonds are conserved (Wales, 1993b). The permutations include combinations of flipping a majority u to d and *vice versa*, exchanging the two hydrogen atoms in one monomer

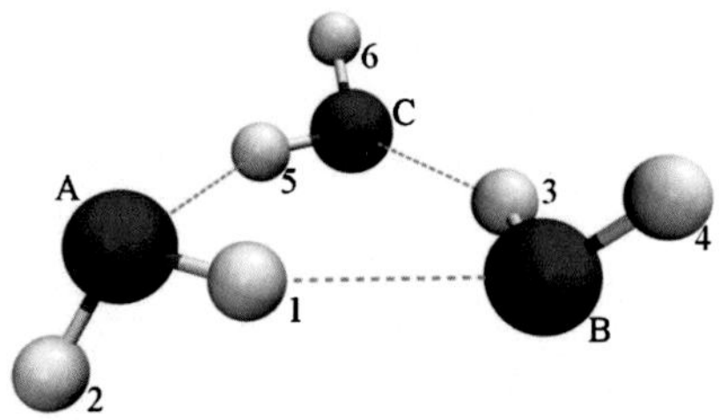

FIG. 9.4 The minimum-energy geometry of the water trimer labeled in its reference version. Reprinted from Richardson et al. (2011), with the permission of AIP Publishing.

and changing the direction of the cyclic formation of hydrogen bonds. There are also higher-energy crown-shaped wells with uuu or ddd geometries (Fowler and Schaefer, 1995), but these have only a minor effect on the tunneling-splitting pattern (Takahashi et al., 2005), which is dominated by tunneling between the degenerate uud wells (and the equivalent ddu, udu, etc., wells).

Instantons were optimized for each possible permutation-inversion operation (Richardson et al., 2011). Many of these were found to be formed of two or more combinations of the others and were thus not considered further as combinations are implicitly included in the diagonalization of the adjacency matrix. We could also neglect the unfeasible motions which involve the clockwise to counterclockwise (cwccw) transition. The remaining pathways are shown in Fig. 9.5 alongside one cwccw path for comparison. These paths replicate qualitatively the major tunneling pathways as those previously identified in WKB calculations (Takahashi et al., 2005; Walsh and Wales, 1996), from where we borrow the notation. However, we emphasize that no prior assumptions were made about the nature of the instantons, which were found by minimizing the instanton action as described above. The resulting tunneling matrix elements are given in Table 9.4. Again it is seen that isotopic substitution of the oxygen makes very little effect, but that deuteration significantly reduces the tunneling.

The simplest instanton pathway is the flip which breaks no hydrogen bonds and is thus very facile. Since the (harmonic) zero-point energy in the wells is slightly greater than the barrier height, the flip is not a tunneling processes. We thus expect the instanton result to overpredict the flip matrix element by about a factor of two as in the acceptor tunneling of the water dimer. Pathways also exist which describe double flips where two dangling hydrogen atoms flip at the same time. However, it was found that the actions were, as expected, about equal to twice that of the single-flip instanton

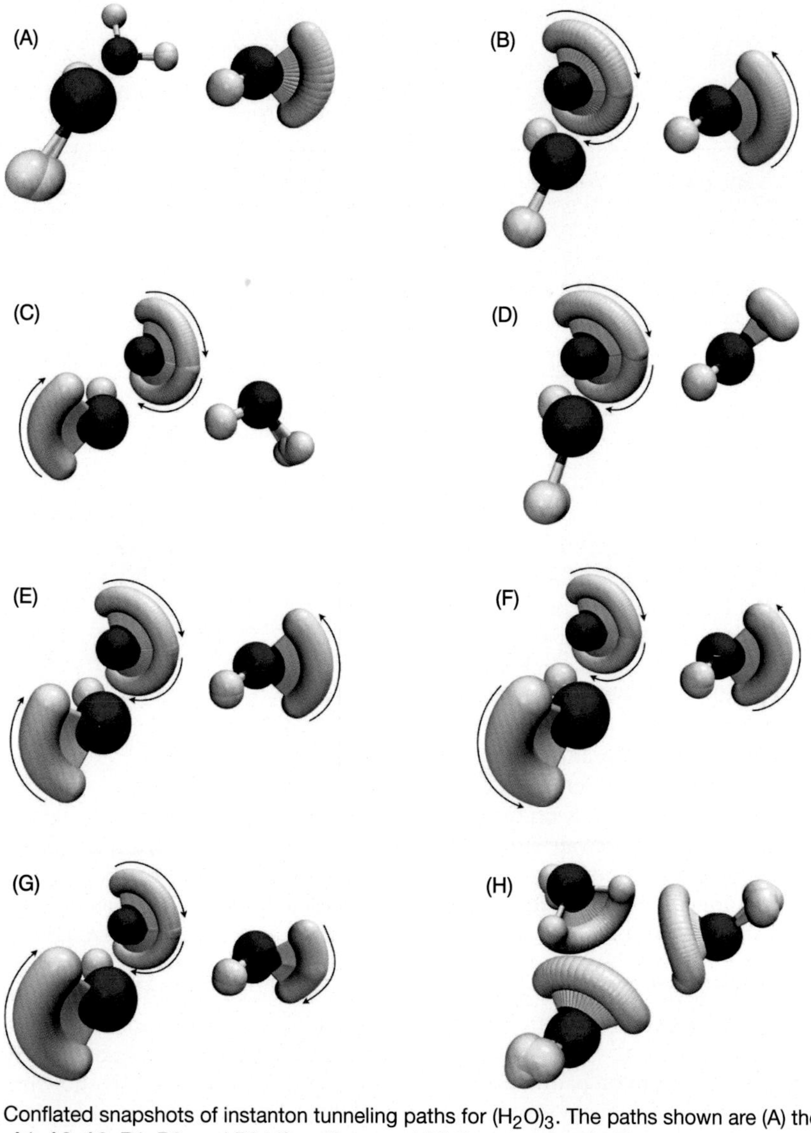

FIG. 9.5 Conflated snapshots of instanton tunneling paths for $(H_2O)_3$. The paths shown are (A) the flip, (B)–(G) the A1, A2, A3, B1, B2, and B3 bifurcations, and (H) the cwccw path. The arrows show the direction of motion of the hydrogen atoms. Note that the B2 path (F) does not contribute to the tunneling because it is a saddle point (see text). Reprinted from Richardson et al. (2011), with the permission of AIP Publishing.

and would not therefore make a significant effect to the overall splitting pattern.

The remaining six paths in Table 9.4 describe *bifurcation tunneling* (previously known as *donor tunneling*). In each case, the dangling hydrogen on one of the monomers swaps roles with the bonding hydrogen, such that the geometry of the cluster passes through a bifurcated structure rather than performing a C_2 rotation (Wales, 1993b). Here the barriers are much larger than the zero-point energy of the well and so the predictions are expected to be more accurate than for the case of the flip. These instantons give the same qualitative description of the tunneling as one of the paths identified in previous WKB calculations (Takahashi et al., 2005; Walsh and Wales, 1996), from where we have taken the notation. Three of these pathways are self-reversible, i.e., the monomers have the same roles within the cluster at the start and finish of the path,

TABLE 9.4
Instanton tunneling pathways located on the MB-pol potential surface for water trimer. The actions, S, and tunneling matrix elements, h, were obtained from the instanton calculations.

		$(H_2{}^{16}O)_3$		$(D_2{}^{16}O)_3$		$(H_2{}^{18}O)_3$	
Pathway	$\mathcal{P}$	$S/\hbar$	$-h/\text{cm}^{-1}$	$S/\hbar$	$-h/\text{cm}^{-1}$	$S/\hbar$	$-h/\text{cm}^{-1}$
flip	(ABC)(135)(246)*	2.04	49	2.74	21	2.06	49
min+acc/don+min	(ABC)(135246)	13.28	3.5(−3)	18.60	1.6(−5)	13.29	3.3(−3)
min+don/acc+min	(ACB)(153264)	14.69	9.6(−3)	20.31	2.9(−5)	14.73	9.2(−3)
don/acc	(ABC)(136245)*	12.90	4.4(−3)	17.98	2.5(−5)	12.93	4.2(−3)
min+double	(12)*	14.80	1.2(−2)	20.60	3.3(−5)	14.83	1.1(−2)
acc+double	(34)*	16.26	8.0(−3)	22.23	1.5(−5)	16.33	7.6(−3)

and are given the label B (as opposed to A for the others).

Previous studies (Walsh and Wales, 1996) identified a possible bifurcation path, denoted B2. There is in fact a stationary point of the instanton action corresponding to B2 (see Fig. 9.5F), but it is a first order saddle; initial guesses set to this geometry relax into combinations of the flip and the A1 path. As a result, semi-classical tunneling between the wells connected by the B2 path proceeds via combinations of the flip and A1, *not* via B2. We therefore set the corresponding elements of W to zero. No changes needed to be made to incorporate the combined flip and A1 paths, since these are already included individually W, and the path-integral theory ensures that all possible combinations of individual paths are included. We note that other potential-energy surfaces tested also returned first-order saddles for B2, so we can be quite confident that the B2 path does not contribute significantly to the splitting pattern in water trimer.

No combination of the six instanton paths described above is able to convert a minimum-energy geometry of water trimer into the permutation-inversion isomer corresponding to reversal of the hydrogen-bonding pattern in the ring in Fig. 9.4. The generator for this operation is (35)(46)(BC)*.

Nevertheless, the possibility of a tunneling path corresponding to this rearrangement in water trimer has been discussed in the literature (Pugliano and Saykally, 1992), and we applied the instanton method to locate the instanton. The mechanism is a concerted breaking and reforming of all three of the hydrogen bonds in a motion described as *clockwise–counterclockwise* (cwccw) tunneling. It was straightforward to locate the cwccw instanton, which is shown in Fig. 9.5H. As expected, the action of this instanton is very large, and the tunneling matrix element negligibly small. For the purposes of calculating the tunneling-splitting pattern, this pathway is considered unfeasible. Other pathways similar to this but accompanied by flips or even bifurcations will have similarly large actions and are also expected to have a negligible effect on the tunneling-splitting pattern.

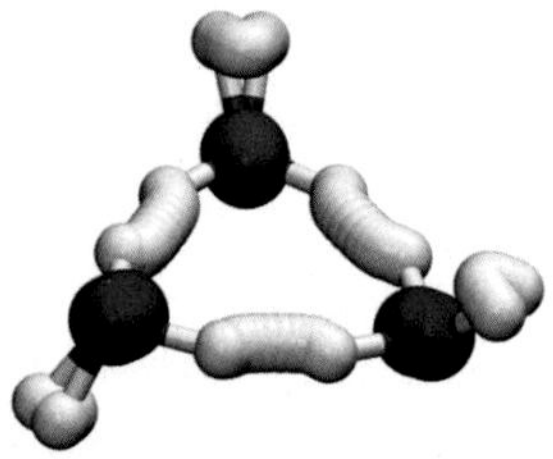

FIG. 9.6 The unfeasible concerted proton transfer pathway in water trimer.

Another degenerate rearrangement pathway which should be considered is that of concerted proton transfer. The instanton pathway is shown in Fig. 9.6, which was obtained using a variant of the WHBB PES allowing for hydrogen exchange. However, as the barrier is so high, instanton theory predicts its matrix element to be about 3×10^{-12} cm^{-1}. Like the cwccw pathway, this is not feasible and makes no discernible difference to the splitting pattern on the resolution available to experiment.

As a result, the tunneling can be described adequately using the $\mathcal{G}_{48}$ molecular symmetry group instead of $\mathcal{G}_{96}$ (Wales, 1993b), and the wells split into two sets, which are not coupled by W. This finding is consistent with all recent studies of the water trimer (Keutsch et al., 2003), in which tunneling between these subsets is thought not to contribute to the splitting patterns. The

TABLE 9.5
Calculated tunneling-splitting pattern in cm^{-1} for both isotopomers of the water trimer using MB-pol.

$(H_2O)_3$		$(D_2O)_3$	
Term	Irrep	Term	Irrep
98.003	$^{1}A_1^-$	42.000 01	$^{76}A_1^-$
98.001	$^{3}T_1^-$	42.000 00	$^{54}T_2^- \oplus {}^{108}T_1^-$
97.999	$^{9}T_2^-$	41.999 99	$^{11}A_2^-$
97.997	$^{11}A_2^-$	21.000 03	$^{108}T_1^+$
49.004	$^{9}T_2^+$	21.000 02	$^{70}E_2^+$
49.003	$^{3}T_1^+$	21.000 01	$^{54}T_2^+$
49.002	$^{8}E_1^+$	20.999 99	$^{108}T_1^+$
48.998	$^{0}E_2^+$	20.999 98	$^{8}E_1^+$
48.997	$^{9}T_2^+$	20.999 97	$^{54}T_2^+$
48.996	$^{3}T_1^+$	−20.999 89	$^{108}T_1^-$
−48.967	$^{3}T_1^-$	−20.999 93	$^{70}E_2^-$
−48.971	$^{0}E_2^-$	−20.999 94	$^{54}T_2^-$
−48.986	$^{9}T_2^-$	−21.000 06	$^{108}T_1^-$
−49.014	$^{3}T_1^-$	−21.000 07	$^{8}E_1^-$
−49.029	$^{8}E_1^-$	−21.000 11	$^{54}T_2^-$
−49.033	$^{9}T_2^-$	−41.999 81	$^{11}A_2^+$
−97.945	$^{11}A_2^+$	−41.999 94	$^{54}T_2^+$
−97.982	$^{9}T_2^+$	−42.000 06	$^{108}T_1^+$
−98.018	$^{3}T_1^+$	−42.000 19	$^{76}A_1^+$
−98.055	$^{1}A_1^+$		

character tables for $\mathcal{G}_{48}$ are tabulated in Wales (1993b). It is important to take account of the fact that the E irreps are each formed of two accidentally degenerate irreps with complex characters.

By following the symmetry analysis described above, one finds that $\Gamma_+ = A_2^+$, $\Gamma_- = A_2^-$, and $\Gamma_{dip} = A_1^-$. For $(H_2O)_3$, the nuclear spin statistics are such that $\Gamma_{ns} = 11A_1^+ \oplus A_2^+ \oplus 8E_1^+ \oplus 9T_1^+ \oplus 3T_2^+$, but for $(D_2O)_3$, $\Gamma_{ns} = 76A_1^+ \oplus 11A_2^+ \oplus 8E_1^+ \oplus 70E_2^+ \oplus 108T_1^+ \oplus 54T_2^+$.

The splitting patterns are obtained by diagonalization of W and results are given in Table 9.5. The instanton splitting pattern has an overall quartet structure. This result agrees with previous experimental and theoretical studies of the trimer, in which this pattern has also been shown to result from the flipping of one of the dangling hydrogen atoms. When this operation is performed six times in succession, two of the principal moments of inertia complete a full rotation. For this reason the flipping process is dubbed a pseudorotation (Liu et al., 1994) and assigned pseudorotational quantum numbers $k = 0, \pm 1, \pm 2, 3$.

The more detailed structure in the instanton splitting patterns is produced by the instantons involving bifurcation. The contributions of the various bifurcation instantons to the splitting pattern are difficult to disentangle. The five bifurcation paths make contributions to the splitting pattern of roughly equal importance. This leads to a far more complex splitting pattern than would appear if tunneling only occurred along one bifurcation pathway (Keutsch et al., 2002), and explains the appearance of the "anomalous" quartets with varying line-widths.

Note that in Richardson et al. (2011), an error was made in building the W matrix such that the matrix elements of B2 and B3 were exchanged. For this reason, the eigenstates appear in a slightly different order, but the findings are not significantly affected. Here, not only is this result corrected, but we also use the MB-pol potential, which gives results similar but not identical to those of WHBB. It should be noted that the same order results from either PES when treated correctly.

The instanton splitting patterns are in good overall agreement with the results of experiment, (Keutsch et al., 2003; Liu et al., 1996c, 1994; Olthof et al., 1996; Pugliano and Saykally, 1992; Suzuki and Blake, 1994; Viant et al., 1997) in which far-infrared spectra for $(H_2O)_3$ and $(D_2O)_3$ and other isotopomers have been interpreted in terms of an overall quartet structure caused by the flip pseudorotation, each branch of which is further split by bifurcation tunneling. The latter gives rise to quartets in the spectrum, which are either regular (equally spaced with equal line-widths) or "anomalous" (similar but irregular spacing with varying line-widths) (Keutsch et al., 2002). The former are associated with the outer $k = 0, 3$ branches; the latter with the $k = \pm 1, \pm 2$ inner branches (for the ground rotational state).

Previous studies based on the WKB approximation, which define the tunneling pathways to lie along minimum-energy paths (Takahashi et al., 2005), assumed that only one bifurcation pathway would contribute to the spectrum. If this were true, and the pathway was one of A1, A2, or A3, one would obtain a qualitatively different spectrum from that if it were one of B1, B2, or B3. By analyzing the known experimental spectrum, it was shown that the A-type pattern was incorrect (Keutsch et al., 2002) because it leads to unevenly spaced $k = \pm 1, \pm 2$ quartets, whereas the experimental "anomalous" quartets are roughly evenly spaced. Although the B-type pattern does reproduce the evenly spaced quartets, it is not necessarily the only possible explanation. Our results in Table 9.5, calculated from a combination of the five tunneling pathways (of both A and B type), also show approximately evenly spaced patterns, provided one assumes that the inner sextets give rise to what appear to be quartets in the experimental spectrum. This case would also explain the varying line-widths of the "anomalous" quartets (Keutsch et al., 2002).

A more detailed comparison with experiment is difficult for the trimer, owing to complexities in the rovibrational spectrum caused by, amongst other things, a dependence of the $k = \pm 1, \pm 2$ bifurcation splitting pattern on Coriolis coupling (Keutsch et al., 2002). As mentioned above, the ring-polymer instanton method assumes that dependence of the splitting on overall rotation can be neglected, but the effects of rotational coupling are expected to be much weaker here than in the water dimer acceptor splitting.

Details of the $k = \pm 1, \pm 2$ bifurcation splitting pattern are sensitive to small changes in the matrix elements, since the pattern results from the interplay of several competing paths of roughly equal weight. It might be that neglect of anharmonicity in the instanton method introduces some small errors into the contributions made by some of the bifurcation paths. An attempt to replicate this pattern in detail would go therefore beyond the capabilities of the method.

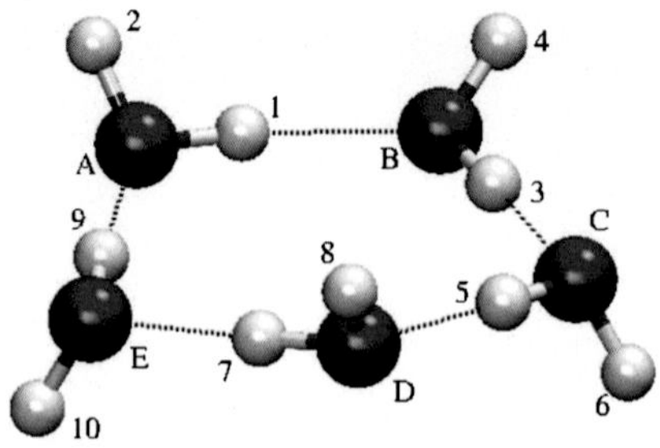

FIG. 9.7 The minimum energy geometry of water pentamer, labeled so as to represent the reference version. Adapted from Cvitaš and Richardson (2020).

One property, however, which the instanton approach should be able to describe well is the overall dependence of the bifurcation splitting on k. The instanton calculations predict that this splitting decreases markedly with increase in k. This is contrary to the results which would be gained if only one B-type pathway dominated the spectrum (Takahashi et al., 2005). The disagreement arises because the instanton calculations predict that five of the bifurcation tunneling paths contribute roughly equally to the splitting. Further work will be needed to decide if an analysis of the experimental data is consistent with this prediction.

9.3.3 Water Pentamer

The water pentamer, $(H_2O)_5$, has a similar cyclic structure to the water trimer with external hydrogens pointing up or down out of the plane (see Fig. 9.7). The tunneling dynamics are therefore similar in that they consist of flips and bifurcations. It has been suggested that the intermolecular forces are at a maximum in this water cluster, making it a particularly useful system for study (Cruzan et al., 1998).

The flip motion leads to pseudorotation dynamics for the cluster similarly to trimer (Liu et al., 1996b, 1997a). Although the effect of the bifurcations in the per-deuterated version, $(D_2O)_5$, could not be resolved (Cruzan et al., 1998) it has been observed in $(H_2O)_5$ (Brown et al., 1998; Harker et al., 2005).

Our recent study of the tunneling dynamics in the pentamer (Cvitaš and Richardson, 2020) employs instanton theory to predict the tunneling splitting pattern

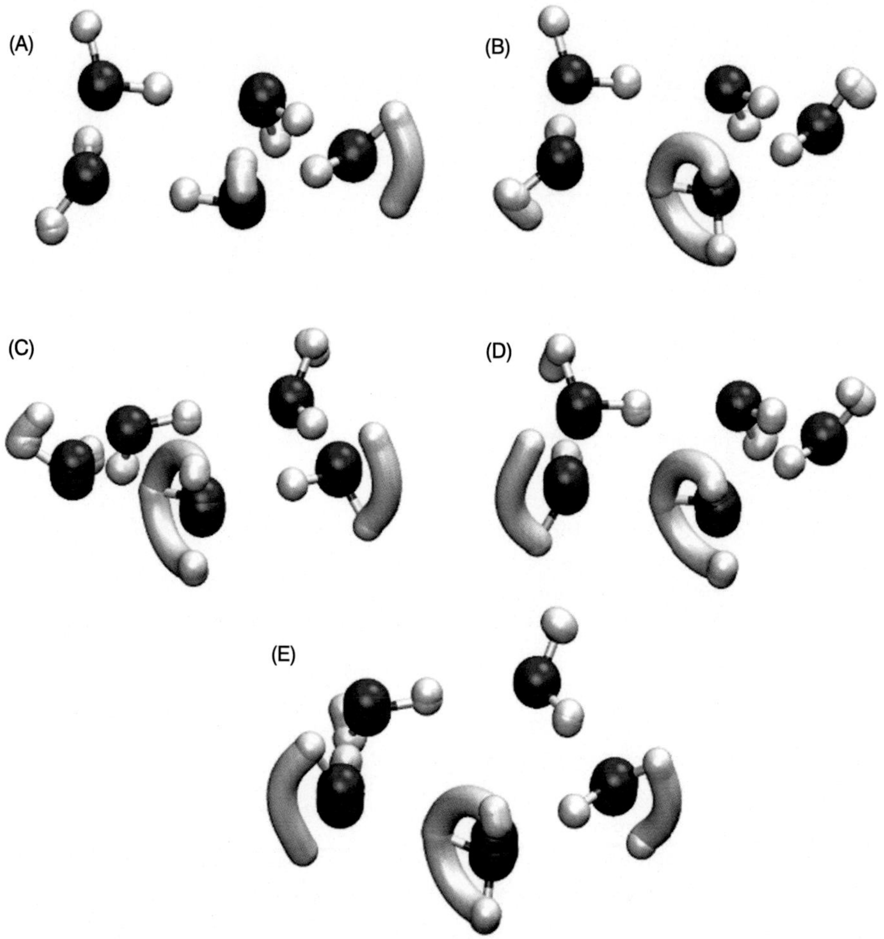

FIG. 9.8 Conflated snapshots of instanton tunneling paths for $(H_2O)_5$. The paths shown are: (A) A flip, (B) A bifurcation, (C) A+E, (D) B+C, and (E) C+BD. Adapted from Cvitaš and Richardson (2020).

in both $(H_2O)_5$ and $(D_2O)_5$. As well as the flip motion, we found four distinct bifurcation pathways as shown in Fig. 9.8. As for the trimer, the pathways discovered by instanton theory are not the same as the minimum-energy pathways (Wales and Walsh, 1996). The tunneling matrix elements corresponding to each instanton pathway are presented in Table 9.6.

One major difference between the trimer and pentamer is that the five oxygen atoms of pentamer do not lie on a perfect plane, but are slightly puckered. Of course, for trimer with only three oxygens, no puckering is possible. Due to the puckering in pentamer, any flip rearrangement pathway requires at least a small movement of the oxygen atoms. This motion happens slowly as the oxygen is heavier than hydrogens and would require a much longer imaginary-time pathway when using the standard ring-polymer instanton approach. However, the time-independent optimization algorithm described in Section 9.2.2 is able to successfully converge the calculations in an efficient manner. It is due to this slight oxygen motion that the tunneling is slightly reduced on isotopic substitution of the oxygen atoms. The effect is more pronounced than in water trimer, although still quite weak.

Due to the effect of the puckering, one would expect the flip tunneling to be suppressed relative to the equivalent motion in the trimer. However, this effect is balanced by the reduced strain experienced in the pentamer which lowers the barrier resulting in a splitting of

TABLE 9.6
Instanton tunneling pathways located on the MB-pol potential surface for water pentamer. The actions, S, and tunneling matrix elements, h, were obtained from the instanton calculations.

		$(H_2{}^{16}O)_5$		$(D_2{}^{16}O)_5$		$(H_2{}^{18}O)_5$	
Pathway	$\mathcal{P}$	$S/\hbar$	$-h/\mathrm{cm}^{-1}$	$S/\hbar$	$-h/\mathrm{cm}^{-1}$	$S/\hbar$	$-h/\mathrm{cm}^{-1}$
flip A/flip B	(AEDCB)(19753)(2 10 864)*	1.64	50	2.17	24	1.65	49
bif. A/bif. B	(AEDCB)(197532 10 864)*	14.76	4.7(−4)	20.58	1.3(−6)	14.80	4.4(−4)
A+E/C+B	(ADBEC)(17395284 10 6)	16.30	5.0(−4)	22.57	1.0(−6)	16.36	4.5(−4)
B+C/E+A	(ACEBD)(1594826 10 37)	15.65	2.2(−4)	21.83	4.9(−7)	15.69	2.0(−4)
C+BD/E+AD	(ADBEC)(17396284 10 5)*	17.27	1.7(−4)	23.91	2.8(−7)	17.34	1.4(−4)

a similar magnitude in both pentamer and trimer (Liu et al., 1997a).

The MS group of the water pentamer is called $\mathcal{G}_{320}$. Its character table is tabulated by Wales and Walsh (1996). It is important to note that the E irreps are each formed of two accidentally degenerate irreps with complex characters. The symmetry analysis gives $\Gamma_+ = A_2^+$, $\Gamma_- = A_2^-$, and $\Gamma_{\mathrm{dip}} = A_1^-$. For $(H_2O)_5$, $\Gamma_{\mathrm{ns}} = 51A_1^+ \oplus A_2^+ \oplus 48E_1^+ \oplus 48E_2^+ \oplus 27H_1^+ \oplus 27H_2^+ \oplus 9H_3^+ \oplus 9H_4^+ \oplus 3H_5^+ \oplus 81H_6^+$, but for $(D_2O)_5$, $\Gamma_{\mathrm{ns}} = 1560A_1^+ \oplus 51A_2^+ \oplus 1554E_1^+ \oplus 1554E_2^+ \oplus 48E_3^+ \oplus 48E_4^+ \oplus 1944H_1^+ \oplus 1944H_2^+ \oplus 972H_3^+ \oplus 972H_4^+ \oplus 486H_5^+ \oplus 3888H_6^+$.

By building and diagonalizing the tunneling matrix, we obtain our prediction for the splitting patterns in both $(H_2O)_5$ and $(D_2O)_5$. The energy levels are given in Table 9.7.

In both cases, the overall pattern is a sextet due to the 10-fold pseudorotation of the flip motions. Each of these is further split by the bifurcations into a complex pattern as in the trimer, in some cases even causing accidental degeneracies for the number of significant figures given.

Our prediction for the bifurcation splittings in $(H_2O)_5$ is in the same order of magnitude as the 4.8 MHz (1.6×10^{-4} cm^{-1}) splitting observed in experiment (Brown et al., 1998). The bifurcation matrix elements are smaller in pentamer than in trimer due to stronger cooperative interactions between the five water molecules. For this reason, the bifurcation splitting has not been observed in $(D_2O)_5$ and an upper limit of 450 kHz (1.5×10^{-5} cm^{-1}) has been determined. Our calculations predict a splitting of about one order of magnitude smaller than this, which suggests that they may be observed with further improvements to the experimental resolution.

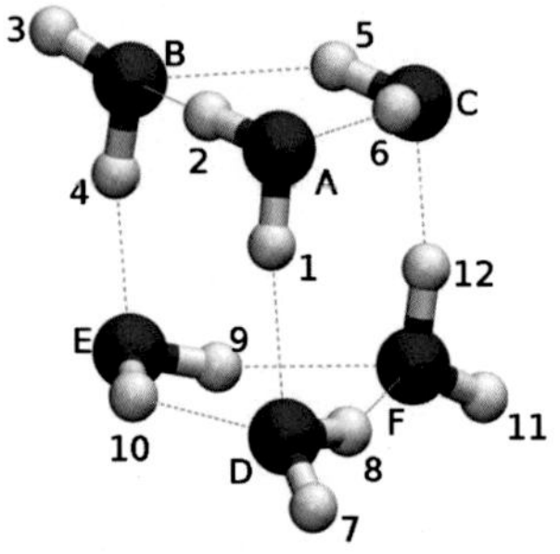

FIG. 9.9 Minimum-energy structure of hexamer prism. From Richardson et al. (2016). Reprinted with the permission from AAAS.

We conclude therefore that the dynamics of the pentamer is similar to that of the trimer. There are again many possible bifurcation pathways involving different combinations of flips, although due to higher barriers, the bifurcation tunneling splittings of pentamer are smaller.

9.3.4 Water Hexamer Prism

The water hexamer, $(H_2O)_6$, is the first water cluster which has low-lying geometries not based on a ring shape. Instead it can form a number of structures including a cage, prism or book (Tsai and Jordan, 1993). A number of studies have been made of the spectroscopic lines of the cage (Gregory and Clary, 1997; Liu et al., 1996a, 1997b; Pérez et al., 2012; Wales, 1999). However, the rearrangement pathways in this cluster are non-degenerate and not therefore directly related to the tunneling splitting pattern. More recently tunneling was observed in the hexamer prism (see Fig. 9.9) and the mechanisms were found using ring-polymer instanton theory (Clary, 2016; Richardson et al., 2016).

The water hexamer prism has two different tunneling mechanisms shown in Fig. 9.10, one which breaks one

TABLE 9.7
Calculated tunneling-splitting pattern in cm^{-1} for both isotopomers of the water pentamer using MB-pol.

$(H_2O)_5$		$(D_2O)_5$	
Term	**Irrep**	**Term**	**Irrep**
100.000 16	$^{51}A_2^-$	48.000 0002	$^{1560}A_1^-$
100.000 10	$^{81}H_5^-$	48.000 0001	$^{3888}H_6^-$
100.000 03	$^{27}H_3^- \oplus {}^{27}H_4^-$	48.000 0000	$^{972}H_3^- \oplus {}^{972}H_4^- \oplus {}^{1944}H_1^- \oplus {}^{1944}H_2^-$
99.999 97	$^{9}H_1^- \oplus {}^{9}H_2^-$	47.999 9999	$^{486}H_5^-$
99.999 90	$^{3}H_6^-$	47.999 9998	$^{51}A_2^-$
99.999 84	$^{1}A_1^-$	38.832 8167	$^{3888}H_6^+ \oplus {}^{1554}E_2^+$
80.901 97	$^{3}H_6^+ \oplus {}^{9}H_1^+$	38.832 8166	$^{1944}H_1^+$
80.901 91	$^{0}E_2^+$	38.832 8162	$^{1944}H_2^+$
80.901 88	$^{27}H_3^+$	38.832 8161	$^{972}H_3^+$
80.901 83	$^{9}H_2^+$	38.832 8160	$^{3888}H_6^+$
80.901 74	$^{27}H_4^+$	38.832 8158	$^{972}H_4^+$
80.901 71	$^{81}H_5^+$	38.832 8157	$^{1944}H_2^+$
80.901 69	$^{3}H_6^+$	38.832 8155	$^{486}H_5^+$
80.901 66	$^{9}H_2^+$	38.832 8153	$^{972}H_4^+ \oplus {}^{1944}H_1^+$
80.901 57	$^{27}H_4^+$	38.832 8149	$^{972}H_3^+$
80.901 52	$^{9}H_1^+$	38.832 8147	$^{48}E_4^+ \oplus {}^{486}H_5^+$
80.901 49	$^{48}E_4^+$	14.832 8185	$^{1554}E_1^-$
80.901 43	$^{27}H_3^+ \oplus {}^{81}H_5^+$	14.832 8184	$^{3888}H_6^-$
30.902 88	$^{0}E_1^-$	14.832 8179	$^{1944}H_2^-$
30.902 80	$^{3}H_6^-$	14.832 8169	$^{1944}H_1^-$
30.902 58	$^{9}H_2^-$	14.832 8168	$^{972}H_4^-$
30.902 18	$^{9}H_1^-$	14.832 8164	$^{3888}H_6^-$
30.902 10	$^{27}H_4^-$	14.832 8158	$^{972}H_3^-$
30.902 01	$^{3}H_6^-$	14.832 8157	$^{1944}H_1^-$
30.901 71	$^{27}H_3^-$	14.832 8150	$^{486}H_5^-$
30.901 69	$^{9}H_1^-$	14.832 8147	$^{1944}H_2^-$
30.901 39	$^{81}H_5^-$	14.832 8146	$^{972}H_3^-$
30.901 29	$^{9}H_2^-$	14.832 8136	$^{972}H_4^-$
30.901 22	$^{27}H_3^-$	14.832 8131	$^{486}H_5^-$
30.900 82	$^{27}H_4^-$	14.832 8130	$^{48}E_3^-$
30.900 60	$^{81}H_5^-$	−14.832 8126	$^{1944}H_2^+$
30.900 52	$^{48}E_3^-$	−14.832 8128	$^{3888}H_6^+$
−30.900 29	$^{3}H_6^+$	−14.832 8134	$^{972}H_4^+$
−30.900 30	$^{9}H_2^+$	−14.832 8137	$^{1554}E_1^+$
−30.900 55	$^{0}E_1^+$	−14.832 8143	$^{1944}H_1^+$
−30.900 76	$^{27}H_4^+$	−14.832 8151	$^{972}H_3^+$
−30.901 02	$^{9}H_1^+$	−14.832 8153	$^{486}H_5^+$
−30.901 48	$^{27}H_3^+$	−14.832 8162	$^{3888}H_6^+$

continued on next page

TABLE 9.7 *(continued)*

$(H_2O)_5$ Term	Irrep	$(D_2O)_5$ Term	Irrep
−30.901 67	$^{81}H_5^+$	−14.832 8164	$^{1944}H_1^+$
−30.901 73	$^{3}H_6^+$	−14.832 8172	$^{972}H_3^+$
−30.901 92	$^{9}H_1^+$	−14.832 8178	$^{48}E_3^+$
−30.902 38	$^{27}H_3^+$	−14.832 8180	$^{1944}H_2^+$
−30.902 64	$^{9}H_2^+$	−14.832 8186	$^{486}H_5^+$
−30.902 85	$^{48}E_3^+$	−14.832 8189	$^{972}H_4^+$
−30.903 10	$^{27}H_4^+$	−38.832 8115	$^{972}H_3^-$
−30.903 11	$^{81}H_5^+$	−38.832 8118	$^{486}H_5^-$
−80.899 88	$^{27}H_3^-$	−38.832 8127	$^{1944}H_1^-$
−80.900 05	$^{81}H_5^-$	−38.832 8129	$^{48}E_4^-$
−80.900 32	$^{9}H_1^-$	−38.832 8138	$^{972}H_4^-$
−80.900 60	$^{48}E_4^-$	−38.832 8149	$^{1944}H_2^-$
−80.900 87	$^{27}H_4^-$	−38.832 8152	$^{3888}H_6^-$
−80.901 31	$^{9}H_2^-$	−38.832 8163	$^{486}H_5^-$
−80.901 37	$^{3}H_6^-$	−38.832 8165	$^{972}H_4^-$
−80.902 03	$^{81}H_5^-$	−38.832 8177	$^{1944}H_2^-$
−80.902 09	$^{27}H_4^-$	−38.832 8186	$^{1554}E_2^-$
−80.902 53	$^{9}H_2^-$	−38.832 8188	$^{972}H_3^-$
−80.902 80	$^{0}E_2^-$	−38.832 8197	$^{3888}H_6^-$
−80.903 08	$^{27}H_3^-$	−38.832 8199	$^{1944}H_1^-$
−80.903 35	$^{3}H_6^-$	−47.999 9939	$^{51}A_2^+$
−80.903 52	$^{9}H_1^-$	−47.999 9963	$^{486}H_5^+$
−99.997 28	$^{51}A_2^+$	−47.999 9988	$^{972}H_3^+ \oplus {}^{972}H_4^+$
−99.998 37	$^{81}H_5^+$	−48.000 0012	$^{1944}H_1^+ \oplus {}^{1944}H_2^+$
−99.999 46	$^{27}H_3^+ \oplus {}^{27}H_4^+$	−48.000 0037	$^{3888}H_6^+$
−100.000 54	$^{9}H_1^+ \oplus {}^{9}H_2^+$	−48.000 0061	$^{1560}A_1^+$
−100.001 63	$^{3}H_6^+$		
−100.002 72	$^{1}A_1^+$		

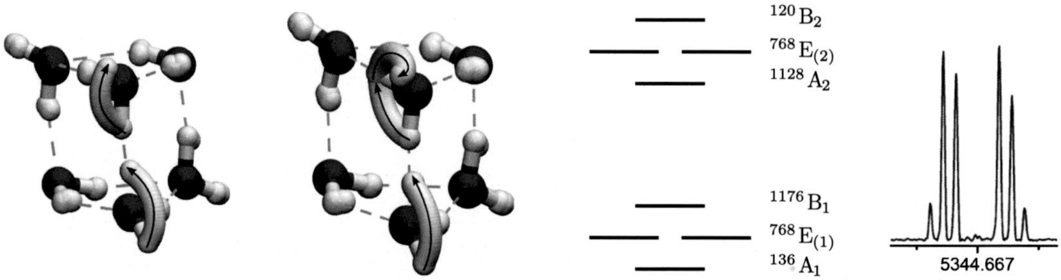

FIG. 9.10 The tunneling pathways calculated for the water hexamer prism (left) showing concerted hydrogen-bond breaking. The instanton method predicts the correct splitting pattern (a doublet of triplets, middle) as observed in experiment (right). Adapted from Richardson et al. (2016).

hydrogen bond, and one which simultaneously breaks two. Surprisingly, it turns out that both mechanisms have similar tunneling probabilities. Note that it would not have been possible to know the tunneling pathways *a priori* as they lie far from the minimum-energy paths. The two pathways are named antigeared and geared and the tunneling matrix elements predicted from instanton theory are given in Table 9.8.

The permutations corresponding to the two feasible tunneling pathways are identified as

$$P_a = (AD)(BF)(CE)(17)(28)(3\,11)(4\,12)(59)(6\,10) \text{ and}$$
$$P_g = (AD)(BF)(CE)(1827)(3\,11)(4\,12)(59)(6\,10).$$

We identity five more permutations (as well as the identity) generated by products of P_a and P_g which are necessary to complete the MS group. These are the inverse of the geared pathway, P_g^3, the single bifurcations (12) and (78), the double bifurcation (12)(78) and a "doubly bifurcating exchange" corresponding to the permutation $P_g P_a P_g^3 = (AD)(BF)(CE)(18)(27)(3\,11)(4\,12)(59)(6\,10)$, which performs an antigeared exchange of the roles of monomers A and D while simultaneously rotating the monomers to bifurcate their hydrogen bonds to O_B and O_F. The geared pathway forms a class with its inverse, the "doubly bifurcating exchange" with the anti-geared pathway, the single bifurcations with each other and the double bifurcation forms a class on its own. Instantons were calculated for the five generated permutations and found to have tunneling matrix elements at least two orders of magnitude smaller than the two considered above and can thus be set to zero.

The complete group is $D_{2d}(M)$, so called because it is isomorphic to the point group D_{2d}. The character table is tabulated in the supplementary material of Richardson et al. (2016). The MS group connects eight versions which can be identified according to the location of their labeled nuclei. The versions form a basis of a vector space which spans the representations of the group. They transform according to Γ_{vib} which reduces to give the irreps corresponding to the vibrational states, $A_1 \oplus A_2 \oplus B_1 \oplus B_2 \oplus 2E$. Note that the cluster is chiral in the equilibrium configuration, and that no operations involving inversion are feasible. All states are therefore degenerate with those of the other enantiomer.

We now take account of the Pauli principle which, in this case, requires that the total internal wave function have B_1 symmetry such that it changes sign under permutations which swap an odd number of hydrogen atoms. For $J = 0$, the total internal wave function is the product of the vibrational states with nuclear spin functions. The symmetry of the nuclear spin functions was

TABLE 9.8
Tunneling matrix elements, h_a and h_g in cm^{-1}, for the antigeared and geared pathways in the water hexamer prism. The results were computed by instanton theory using MB-pol.

Matrix element	$(H_2{}^{16}O)_6$	$(H_2{}^{18}O)_6$
$-h_a$	2.5(−5)	2.1(−5)
$-h_g$	3.7(−6)	3.0(−6)

computed as follows. All combinations of the spin orientation $+\frac{1}{2}$ or $-\frac{1}{2}$ for each of the 12 hydrogen atoms were generated and the symmetry under the MS group operations explored to give a representation of Γ_{ns}. Reducing to the irreps gives a nuclear spin symmetry of $\Gamma_{ns} = 1176A_1 \oplus 120A_2 \oplus 136B_1 \oplus 1128B_2 \oplus 768E$. Vibrational levels must combine with these nuclear spin functions to give an overall B_1 symmetry. Using products of irreps (e.g., A_2 must combine with $^{1128}B_2$), we can therefore deduce that the vibrational levels with symmetry $A_1 : A_2 : B_1 : B_2 : E$ have relative weights $17 : 141 : 147 : 15 : 96$.

The energies and states of the tunneling system correspond to the eigenvalues and eigenvectors of the tunneling matrix

$$W = \begin{pmatrix} 0 & h_g & 0 & h_g & h_a & 0 & 0 & 0 \\ h_g & 0 & h_g & 0 & 0 & h_a & 0 & 0 \\ 0 & h_g & 0 & h_g & 0 & 0 & h_a & 0 \\ h_g & 0 & h_g & 0 & 0 & 0 & 0 & h_a \\ h_a & 0 & 0 & 0 & 0 & h_g & 0 & h_g \\ 0 & h_a & 0 & 0 & h_g & 0 & h_g & 0 \\ 0 & 0 & h_a & 0 & 0 & h_g & 0 & h_g \\ 0 & 0 & 0 & h_a & h_g & 0 & h_g & 0 \end{pmatrix}. \quad (9.6)$$

This leads to a splitting of the ground-state energy level of

$$\begin{array}{lll} h_a + 2h_g\,(A_1), & h_a\,(E_{(1)}), & h_a - 2h_g\,(B_1), \\ -h_a + 2h_g\,(A_2), & -h_a\,(E_{(2)}), & -h_a - 2h_g\,(B_2), \end{array}$$

where symmetry labels of the vibrational states are assigned according to the characters of the eigenvectors when operated upon by the permutations. We have distinguished between the two pairs of E states using the labels $E_{(1)}$ and $E_{(2)}$, although both have E symmetry.

The numerical values of h_a and h_g, calculated using the ring-polymer instanton method, are presented in Table 9.8. These results are combined with the analysis above to provide a full quantitative prediction of the splitting of the ground-state energy level in the hexamer prism. An analysis of the experimental spectrum gives the values of $-h_a = 1.27 \times 10^{-5}$ cm^{-1} and $-h_g = 2.43 \times 10^{-6}$ cm^{-1} for the ground state of $(H_2O)_6$, which can be compared directly to the theoretical values given above. The small decrease of the matrix elements on ^{18}O-substitution, is a result of the slight motion of O-atoms along the tunneling pathway. This predicted kinetic isotope effects of about 82% closely matches the experimental spectra, which has a kinetic isotope effect of 85%. The splittings in $(D_2O)_6$ are expected to be far too small to be observed.

The rotational symmetry is complicated by the torsion of individual water molecules in the cluster and cannot be completely separated from the vibration-tunneling motion (Longuet-Higgins, 1963). We cannot therefore rigorously assign symmetry labels to the *rovibrational* states for $J > 0$. However, because the heavy oxygen atom structure is not significantly rearranged by the tunneling processes, it will be a good approximation to assume that they are decoupled. Excited rotational states are then expected to show similar splitting patterns with symmetry labels obtained as the direct product of the vibrational and rotational symmetry label.

The symmetry of the internal dipole moment, which allows transitions between *vibrational* states, can be found by considering the equivalent rotations. The equivalent rotations which return the heavy oxygen atoms to their initial configuration after the permutation are defined as R_0^π for the classes including P_a or P_g but the identity operator E for the other classes. The R_0^π rotation is around an axis perpendicular to a plane approximately passing through oxygens B, C, E and F. This internal dipole therefore has $\Gamma_{dip} = A_1 \oplus 2B_2$ symmetry and is used to compute the observed transitions.

The hexamer prism is an asymmetric top and its rotational states $J_{K_a K_c}$ have symmetries A_1 for ee/oe or B_2 for eo/oo. It is a near prolate rotor and thus has rotational transitions (Bunker and Jensen, 2006)

$$\begin{aligned} &\Delta K_a = 0, && \Delta K_c = \text{odd}, && \text{a-type}, \\ &\Delta K_a = \pm 1, && \Delta K_c = \text{odd}, && \text{b-type}, \\ &\Delta K_a = \pm 1, && \Delta K_c = \text{even}, && \text{c-type}. \end{aligned}$$

Therefore, a- and b-type transitions are coupled to a B_2 dipole whereas c-type are coupled to A_1. Although the c-type transitions were not observed due to a very weak dipole, we can predict that they would allow only pure rotational transitions and would not show tunneling splittings.

The strongest dipole is μ_a, which points approximately perpendicular to the two triangular faces. It is easy to see that this has B_2 symmetry under the MS group because it is reversed whenever monomers A and D exchange roles. This dipole couples the vibrational states $A_1 \leftrightarrow B_2$, $A_2 \leftrightarrow B_1$, and $E_{(1)} \leftrightarrow E_{(2)}$. Note that in principle, all $E \leftrightarrow E$ transitions are allowed, but, as we shall explain, the pure rotational transitions are weaker and may not be observable.

In order to discover which $E \leftrightarrow E$ transitions are strong, we consider the allowed transitions in the case that some of the tunneling pathways were unfeasible. For instance, if the P_g pathway were not feasible, the MS group would be $C_2(M)$ and the splitting pattern would be a simple doublet with A and B levels, where transitions are only allowed between vibrational states $A \leftrightarrow B$. Because $E_{(1)}$ correlates with A and $E_{(2)}$ with B, we expect the pure rotational transitions $E_{(1)} \leftrightarrow E_{(1)}$ and $E_{(2)} \leftrightarrow E_{(2)}$ to be very weak, although they are not forbidden by symmetry. This effect also occurs in the water dimer (Fraser et al., 1989) and is a consequence of the fact that these transitions do not reverse the dipole moment.

We therefore predict six observed lines with the pattern:

$$\begin{aligned} &2h_a + 4h_g\,(17), && 2h_a\,(96), && 2h_a - 4h_g\,(147), \\ &-2h_a + 4h_g\,(141), && -2h_a\,(96), && -2h_a - 4h_g\,(15). \end{aligned}$$

This thus explains the doublet of triplet splitting pattern observed in experiment where the outermost lines are considerably smaller. Small central peaks may be visible due to the weak $E \leftrightarrow E$ transitions as well as other pure rotational transitions originating from deviations from the rigid rotor approximation (Bunker and Jensen, 2005).

This example has again shown that instanton theory can explain the tunneling mechanisms of water clusters and, at least, qualitatively predict the splitting. It is likely that larger water clusters with prism structures, such as the decamer, will also exhibit similar tunneling mechanisms.

9.4 DISCUSSION

We have discussed the quantum dynamics of water clusters, focusing in particular on the resulting tunneling splittings as described by ring-polymer instanton theory. Here we have only considered the ground-state tunneling splitting, but instanton theory can be extended to study tunneling from vibrationally-excited (Mil'nikov and Nakamura, 2005) and rotationally-excited states (Vaillant and Cvitaš, 2018). More work is needed to apply these approaches to the larger water clusters to help explain recent experimental work (Cole and Saykally, 2017).

In some cases the semiclassical approximation at the heart of instanton theory is not applicable, and an exact but computationally more expensive imaginary-time path-integral simulation can be employed instead (Mátyus and Althorpe, 2016; Mátyus et al., 2016). So far, this approach has only been applied to the water dimer (Vaillant et al., 2018) and it greatly improves the agreement with experiment, particularly for the acceptor splitting. In principle, this approach could also be applied to the larger clusters, which will be particularly important for studying the flip motion and for accurate benchmarking of PESs.

We have discussed applications of the instanton method to the water dimer, trimer, pentamer, and hexamer prism. However, other water clusters could, of course, also be studied using this approach as it is not limited by the size of the cluster. Instantons were, for instance, applied to the water octamer, but it was found that the splittings are much smaller than experimental resolution (Richardson et al., 2013). The tetramer forms a four-membered ring similar to the trimer's, but with point group symmetry S_4 (Cruzan et al., 1996a). Because there are an even number of dangling hydrogen atoms and therefore equal numbers of u and d, single flips are not degenerate rearrangements. Minimum-energy pathways describing bifurcations have not been found (Wales and Walsh, 1997), and the expectation is that tunneling motions occur in a concerted fashion, passing through second-order saddle points. However, minimum-action paths may exist where minimum-energy paths do not, and it has already been shown that the ring-polymer instanton method is able to pass through second-order saddles (Richardson et al., 2011). We expect that a detailed study of this system will clarify the tunneling dynamics.

An ortho–para transition has been observed in molecular beams of water clusters (Manca Tanner et al., 2013, 2018). Assuming that the transition is forbidden for isolated molecules (Miani and Tennyson, 2004), it implies that a fast intermolecular proton transfer mechanisms exist in water clusters. These are yet to be identified, and instanton theory may be the tool which can provide the answer. The thermal rates of proton transfer in water dimer and trimer have been found to be too small, so one may have to look to larger clusters for the answer or to non-equilibrium effects.

The family of protonated water clusters, $H^+(H_2O)_n$, has also been studied by spectroscopy (Douberly et al., 2010; Headrick et al., 2005), although typically with a lower resolution than the neutral clusters. Work is in progress to build accurate and efficient potential-energy surfaces for these systems (Huang et al., 2005; Natarajan et al., 2015; Pinski and Csányi, 2013; Yu and Bowman, 2016). The quantum dynamics of $H_5O_2^+$ have been studied by wave function approaches (Vendrell et al., 2007) and other hydrated-ion systems (Bajaj et al., 2019) with instanton theory. However, there is still much work to be done to understand the quantum dynamical effects in larger protonated water clusters. In particular, the vibrational spectra require assignment and even the minimum-energy geometries are disputed (Chang et al., 2005; Yu and Bowman, 2017). Because the resolution of the experiments is currently quite low, tunneling splittings have not yet been observed, except in the simplest hydronium molecule (Cvitaš, 2018; Liu et al., 1985). However, instanton theory nonetheless has the capabilities to compute them, which will drive the design of high-resolution experiments and help with the interpretation of their results.

ACKNOWLEDGMENTS

The authors gratefully acknowledge the support and guidance of Stuart Althorpe throughout the formative years of their own research on the topic of instanton theory. M.T.C. has been supported by Croatian Science Foundation under projects IP-2016-06-1142 and IP-2014-09-7540, and in part by the QuantiXLie Centre of Excellence, a project cofinanced by the Croatian government and European Union through the European Regional Development Fund–the Competitiveness and Cohesion Operational Programme (Grant No. KK.01.1.1.01.0004). J.O.R.'s research has been financially supported by the Swiss National Science Foundation (Project No. 175696). The authors also received support from COST action CM1405, MOLIM: Molecules in Motion.

REFERENCES

Althorpe, S.C., Clary, D.C., 1994. Calculation of the intermolecular bound states for water dimer. J. Chem. Phys. 101 (5), 3603.

Althorpe, S.C., Clary, D.C., 1995. A new method for calculating the rovibrational states of polyatomics with application to water dimer. J. Chem. Phys. 102 (11), 4390.

Ásgeirsson, V., Arnaldsson, A., Jónsson, H., 2018. Efficient evaluation of atom tunneling combined with electronic structure calculations. J. Chem. Phys. 148 (10), 102334.

Babin, V., Leforestier, C., Paesani, F., 2013. Development of a "first principles" water potential with flexible monomers: dimer potential energy surface, VRT spectrum, and second virial coefficient. J. Chem. Theory Comput. 9 (12), 5395–5403.

Babin, V., Medders, G.R., Paesani, F., 2014. Development of a "first principles" water potential with flexible monomers. II: trimer potential energy surface, third virial coefficient, and small clusters. J. Chem. Theory Comput. 10 (4), 1599–1607.

Babin, V., Paesani, F., 2013. The curious case of the water hexamer: cage vs. prism. Chem. Phys. Lett. 580, 1–8.

Bajaj, P., Richardson, J.O., Paesani, F., 2019. Ion-mediated hydrogen bond rearrangement through tunneling in the iodide-dihydrate complex. Nat. Chem. 11, 367–374.

Barker, J.A., 1979. A quantum-statistical Monte Carlo method; path integrals with boundary conditions. J. Chem. Phys. 70, 2914.

Benderski, V.A., Makarov, D.E., Wight, C.A., 1994. Chemical Dynamics at Low Temperatures. Adv. Chem. Phys., vol. 88. Wiley, New York.

Brown, M.G., Keutsch, F.N., Saykally, R.J., 1998. The bifurcation rearrangement in cyclic water clusters: breaking and making hydrogen bonds. J. Chem. Phys. 109 (22).

Brown, M.G., Viant, M.R., McLaughlin, R.P., Keoshian, C.J., Michael, E., Cruzan, J.D., Saykally, R.J., van der Avoird, A., et al., 1999. Quantitative characterization of the water trimer torsional manifold by terahertz laser spectroscopy and theoretical analysis. II. $(H_2O)_3$. J. Chem. Phys. 111 (17).

Bukowski, R., Szalewicz, K., Groenenboom, G.C., van der Avoird, A., 2007. Predictions of the properties of water from first principles. Science 315 (5816), 1249–1252.

Bunker, P.R., Jensen, P., 2005. Fundamentals of Molecular Symmetry. Institute of Physics Publishing, Bristol.

Bunker, P.R., Jensen, P., 2006. Molecular Symmetry and Spectroscopy, 2nd ed. NRC Research Press, Ottawa.

Carpenter, B.K., 2011. Taking the high road and getting there before you. Science 332, 1269.

Carrington Jr, T., 2017. Perspective: computing (ro-)vibrational spectra of molecules with more than four atoms. J. Chem. Phys. 146 (12), 120902.

Ceriotti, M., Cuny, J., Parrinello, M., Manolopoulos, D.E., 2013. Nuclear quantum effects and hydrogen bond fluctuations in water. Proc. Natl. Acad. Sci. USA 110 (39), 15591–15596.

Ceriotti, M., Fang, W., Kusalik, P.G., McKenzie, R.H., Michaelides, A., Morales, M.A., Markland, T.E., 2016. Nuclear quantum effects in water and aqueous systems: experiment, theory, and current challenges. Chem. Rev. 116 (13), 7529–7550.

Chandler, D., Wolynes, P.G., 1981. Exploiting the isomorphism between quantum theory and classical statistical mechanics of polyatomic fluids. J. Chem. Phys. 74 (7), 4078–4095.

Chang, H.C., Wu, C.C., Kuo, J.L., 2005. Recent advances in understanding the structures of medium-sized protonated water clusters. Int. Rev. Phys. Chem. 24 (3–4), 553–578.

Chen, H., Liu, S., Light, J.C., 1999. Six-dimensional quantum calculation of the intermolecular bound states for water dimer. J. Chem. Phys. 110 (1), 168.

Clary, D.C., 2016. Quantum dynamics in the smallest water droplet. Science 351 (6279), 1267–1268.

Cole, W.T.S., Fellers, R.S., Viant, M.R., Saykally, R.J., 2017. Hydrogen bond breaking dynamics in the water pentamer: terahertz VRT spectroscopy of a 20 μm libration. J. Chem. Phys. 146 (1), 014306.

Cole, W.T.S., Saykally, R.J., 2017. Hydrogen bond network rearrangement dynamics in water clusters: effects of intermolecular vibrational excitation on tunneling rates. J. Chem. Phys. 147 (6), 064301.

Coleman, S., 1977. The uses of instantons. In: Proc. Int. School of Subnuclear Physics. Erice. Also in Coleman, S., 1985. Aspects of Symmetry. Cambridge University Press, pp. 265–350 (Chapter 7).

Coudert, L.H., Hougen, J.T., 1988. Tunneling splittings in the water dimer: further development of the theory. J. Mol. Spectrosc. 130, 86–119.

Cruzan, J.D., Braly, L.B., Liu, K., Brown, M.G., Loeser, J.G., Saykally, R.J., 1996a. Quantifying hydrogen bond cooperativity in water: VRT spectroscopy of the water tetramer. Science 271 (5245), 59–62.

Cruzan, J.D., Brown, M.G., Liu, K., Braly, L.B., Saykally, R.J., 1996b. The far-infrared vibration–rotation–tunneling spectrum of the water tetramer-*d*8. J. Chem. Phys. 105 (16), 6634–6644.

Cruzan, J.D., Viant, M.R., Brown, M.G., Lucas, D.D., Liu, K., Saykally, R.J., 1998. Terahertz laser vibration–rotation–tunneling spectrum of the water pentamer–d_{10}. Constraints on the bifurcation tunneling dynamics. Chem. Phys. Lett. 292 (4), 667–676.

Cruzan, J.D., Viant, M.R., Brown, M.G., Saykally, R.J., 1997. Terahertz laser vibration-rotation tunneling spectroscopy of the water tetramer. J. Phys. Chem. A 101 (48), 9022–9031.

Cvitaš, M.T., 2018. Quadratic string method for locating instantons in tunneling splitting calculations. J. Chem. Theory Comput. 14, 1487–1500.

Cvitaš, M.T., Althorpe, S.C., 2016. Locating instantons in calculations of tunneling splittings: the test case of malonaldehyde. J. Chem. Theory Comput. 12 (2), 787–803.

Cvitaš, M.T., Richardson, J.O., 2020. Quantum tunnelling pathways of the water pentamer. Phys. Chem. Chem. Phys. 22 (3), 1035–1044.

Doll, J.D., Beck, T.L., Freeman, D.L., 1989. Classical Monte Carlo dynamics: a simulated annealing approach to the construction of double-ended classical trajectories. Int. J. Quant. Chem. 36 (S23), 73–78.

Douberly, G.E., Walters, R.S., Cui, J., Jordan, K.D., Duncan, M.A., 2010. Infrared spectroscopy of small protonated water clusters, $H^+(H_2O)_n$ ($n = 2$–5): isomers, argon tagging, and deuteration. J. Phys. Chem. A 114 (13), 4570–4579.

Dyke, T.R., 1977. Group theoretical classification of the tunneling-rotational energy levels of water dimer. J. Chem. Phys. 66 (2), 492.

Dyke, T.R., Mack, K.M., Muenter, J.S., 1977. The structure of water dimer from molecular beam electric resonance spectroscopy. J. Chem. Phys. 66 (2), 498.

E, W., Ren, W., Vanden-Eijnden, E., 2002. String method for the study of rare events. Phys. Rev. B 66, 052301.

Einarsdóttir, D.M., Arnaldsson, A., Óskarsson, F., Jónsson, H., 2012. Path optimization with application to tunneling. In: Jónasson, K. (Ed.), Applied Parallel and Scientific Computing, 10th International Conference. PARA 2010. In: Lecture Notes in Computer Science, vol. 7134. Springer-Verlag, Berlin, pp. 45–55.

Faccioli, P., Sega, M., Pederiva, F., Orland, H., 2006. Dominant pathways in protein folding. Phys. Rev. Lett. 97, 108101.

Fanourgakis, G.S., Xantheas, S.S., 2008. Development of transferable interaction potentials for water. V. Extension of the flexible, polarizable, Thole-type model potential (TTM3-F, v. 3.0) to describe the vibrational spectra of water clusters and liquid water. J. Chem. Phys. 128, 074506.

Fellers, R.S., Braly, L.B., Saykally, R.J., Leforestier, C., 1999. Fully coupled six-dimensional calculations of the water dimer vibration-rotation-tunneling states with split Wigner pseudospectral approach. II. Improvements and tests of additional potentials. J. Chem. Phys. 110 (13), 6306.

Fowler, J.E., Schaefer III, H.F., 1995. Detailed study of the water trimer potential energy surface. J. Am. Chem. Soc. 117 (1), 446–452.

Fraser, G.T., Suenram, R.D., Coudert, L.H., 1989. Microwave electric-resonance optothermal spectroscopy of $(H_2O)_2$. J. Chem. Phys. 90 (11), 6077.

Góra, U., Cencek, W., Podeszwa, R., van der Avoird, A., Szalewicz, K., 2014. Predictions for water clusters from a first-principles two-and three-body force field. J. Chem. Phys. 140 (19), 194101.

Gregory, J.K., Clary, D.C., 1995a. Calculations of the tunneling splittings in water dimer and trimer using diffusion Monte Carlo. J. Chem. Phys. 102 (20), 7817.

Gregory, J.K., Clary, D.C., 1995b. Three-body effects on molecular properties in the water trimer. J. Chem. Phys. 103 (20), 8924.

Gregory, J.K., Clary, D.C., 1997. Theoretical study of the cage water hexamer structure. J. Phys. Chem. A 101 (36), 6813–6819.

Groenenboom, G.C., Wormer, P.E.S., van der Avoird, A., Mas, E.M., Bukowski, R., Szalewicz, K., 2000. Water pair potential of near spectroscopic accuracy. II. Vibration-rotation-tunneling levels of the water dimer. J. Chem. Phys. 113, 6702.

Gutzwiller, M.C., 1990. Chaos in Classical and Quantum Mechanics. Springer-Verlag, New York.

Harker, H.A., Viant, M.R., Keutsch, F.N., Michael, E.A., McLaughlin, R.P., Saykally, R.J., 2005. Water pentamer: characterization of the torsional-puckering manifold by terahertz VRT spectroscopy. J. Phys. Chem. A 109 (29), 6483–6497.

Headrick, J.M., Diken, E.G., Walters, R.S., Hammer, N.I., Christie, R.A., Cui, J., Myshakin, E.M., Duncan, M.A., Johnson, M.A., Jordan, K.D., 2005. Spectral signatures of hydrated proton vibrations in water clusters. Science 308 (5729), 1765–1769.

Hougen, J.T., 1985. A generalized internal axis method for high barrier tunneling problems, as applied to the water dimer. J. Mol. Spectrosc. 114 (2), 395–426.

Huang, X., Braams, B.J., Bowman, J.M., 2005. Ab initio potential energy and dipole moment surfaces for $H_5O_2^+$. J. Chem. Phys. 122 (4), 044308.

Huang, X., Braams, B.J., Bowman, J.M., Kelly, R.E.A., Tennyson, J., Groenenboom, G.C., van der Avoird, A., 2008. New *ab initio* potential energy surface and the vibration-rotation-tunneling levels of $(H_2O)_2$ and $(D_2O)_2$. J. Chem. Phys. 128, 034312.

Hund, F., 1927. Zur Deutung der Molekelspektren. III. Z. Phys. 43 (11–12), 805–826.

Kapil, V., Rossi, M., Marsalek, O., Petraglia, R., Litman, Y., Spura, T., Cheng, B., Cuzzocrea, A., Meißner, R.H., Wilkins, D.M., Juda, P., Bienvenue, S.P., Fang, W., Kessler, J., Poltavsky, I., Vandenbrande, S., Wieme, J., Corminboeuf, C., Kühne, T.D., Manolopoulos, D.E., Markland, T.E., Richardson, J.O., Tkatchenko, A., Tribello, G.A., Van Speybroeck, V., Ceriotti, M., 2019. i-PI 2.0: a universal force engine for advanced molecular simulations. Comput. Phys. Commun. 236, 214–223.

Karyakin, E.N., Fraser, G.T., Suenram, R.D., 1993. Microwave spectrum of the $K_a = 1 \leftarrow 0$ rotation-tunneling band of $(D_2O)_2$. Mol. Phys. 78 (5), 1179–1189.

Keutsch, F.N., Cruzan, J.D., Saykally, R.J., 2003. The water trimer. Chem. Rev. 103 (7), 2533–2578.

Keutsch, F.N., Saykally, R.J., 2001. Water clusters: untangling the mysteries of the liquid, one molecule at a time. Proc. Natl. Acad. Sci. USA 98 (19), 10533–10540.

Keutsch, F.N., Saykally, R.J., Wales, D.J., 2002. Bifurcation tunneling dynamics in the water trimer. J. Chem. Phys. 117 (19), 8823–8835.

Kryvohuz, M., 2012. Calculation of chemical reaction rate constants using on-the-fly high level electronic structure computations with account of multidimensional tunneling. J. Chem. Phys. 137 (23), 234304.

Lee, C., Chen, H., Fitzgerald, G., 1994. Structures of the water hexamer using density functional methods. J. Chem. Phys. 101 (5), 4472–4473.

Leforestier, C., Braly, L.B., Liu, K., Elrod, M.J., Saykally, R.J., 1997. Fully coupled six-dimensional calculations of the water dimer vibration-rotation-tunneling states with a split Wigner pseudospectral approach. J. Chem. Phys. 106 (20), 8527.

Leforestier, C., Gatti, F., Fellers, R.S., Saykally, R.J., 2002. Determination of a flexible (12D) water dimer potential via direct inversion of spectroscopic data. J. Chem. Phys. 117 (19), 8710.

Liu, D.J., Haese, N.N., Oka, T., 1985. Infrared spectrum of the ν_2 vibration-inversion band of H_3O^+. J. Chem. Phys. 82 (12), 5368–5372.

Liu, K., Brown, M.G., Carter, C., Saykally, R.J., Gregory, J.K., Clary, D.C., 1996a. Characterization of a cage form of the water hexamer. Nature 381 (6582), 501–503.

Liu, K., Brown, M.G., Cruzan, J.D., Saykally, R.J., 1996b. Vibration-rotation tunneling spectra of the water pentamer: structure and dynamics. Science 271 (5245), 62.

Liu, K., Brown, M.G., Cruzan, J.D., Saykally, R.J., 1997a. Terahertz laser spectroscopy of the water pentamer: structure and hydrogen bond rearrangement dynamics. J. Phys. Chem. A 101 (48), 9011–9021.

Liu, K., Brown, M.G., Saykally, R.J., 1997b. Terahertz laser vibration-rotation tunneling spectroscopy and dipole moment of a cage form of the water hexamer. J. Phys. Chem. A 101 (48), 8995–9010.

Liu, K., Brown, M.G., Viant, M.R., Cruzan, J.D., Saykally, R.J., 1996c. Far infrared VRT spectroscopy of two water trimer isotopomers: vibrationally averaged structures and rearrangement dynamics. Mol. Phys. 89 (5), 1373–1396.

Liu, K., Cruzan, J.D., Saykally, R.J., 1996d. Water clusters. Science 271, 929–932.

Liu, K., Loeser, J.G., Elrod, M.J., Host, B.C., Rzepiela, J.A., Pugliano, N., Saykally, R.J., 1994. Dynamics of structural rearrangements in the water trimer. J. Am. Chem. Soc. 116 (8), 3507–3512.

Longuet-Higgins, H.C., 1963. The symmetry groups of non-rigid molecules. Mol. Phys. 6 (5), 445–460.

Manca Tanner, C., Quack, M., Schmidiger, D., 2013. Nuclear spin symmetry conservation and relaxation in water ($^1H_2{}^{16}O$) studied by cavity ring-down (CRD) spectroscopy of supersonic jets. J. Phys. Chem. A 117 (39), 10105–10118.

Manca Tanner, C., Quack, M., Schmidiger, D., 2018. Nuclear spin symmetry conservation and relaxation of water ($H_2{}^{16}O$) seeded in supersonic jets of argon and oxygen: measurements by cavity ring-down laser spectroscopy. Mol. Phys. 116 (23–24), 3718–3730.

Marquardt, R., Quack, M., 2020. Foundations of time dependent quantum dynamics of molecules under isolation and in coherent electromagnetic fields. In: Marquardt, R., Quack, M. (Eds.), Molecular Spectroscopy and Quantum Dynamics. Elsevier, Amsterdam. Chapter 1 (this book).

Matsuoka, O., Clementi, E., Yoshimine, M., 1976. CI study of the water dimer potential surface. J. Chem. Phys. 64 (4), 1351.

Mátyus, E., Althorpe, S.C., 2016. Calculating splittings between energy levels of different symmetry using path-integral methods. J. Chem. Phys. 144, 114109.

Mátyus, E., Wales, D.J., Althorpe, S.C., 2016. Quantum tunneling splittings from path-integral molecular dynamics. J. Chem. Phys. 144, 114108.

Merer, A.J., Watson, J.K.G., 1973. Symmetry considerations for internal rotation in ethylene-like molecules. J. Mol. Spectrosc. 47 (3), 499–514.

Miani, A., Tennyson, J., 2004. Can ortho–para transitions for water be observed? J. Chem. Phys. 120 (6), 2732–2739.

Miller, W.H., 1975. Semiclassical limit of quantum mechanical transition state theory for nonseparable systems. J. Chem. Phys. 62 (5), 1899–1906.

Millot, C., Soetens, J.C., Martins Costa, M.T.C., Hodges, M.P., Stone, A.J., 1998. Revised anisotropic site potentials for the water dimer and calculated properties. J. Phys. Chem. A 102 (4), 754–770.

Mil'nikov, G.V., Nakamura, H., 2001. Practical implementation of the instanton theory for the ground-state tunneling splitting. J. Chem. Phys. 115 (15), 6881–6897.

Mil'nikov, G.V., Nakamura, H., 2005. Instanton theory for the tunneling splitting of low vibrationally excited states. J. Chem. Phys. 122 (12), 124311.

Natarajan, S.K., Morawietz, T., Behler, J., 2015. Representing the potential-energy surface of protonated water clusters by high-dimensional neural network potentials. Phys. Chem. Chem. Phys. 17 (13), 8356–8371.

Nguyen, T.T., Székely, E., Imbalzano, G., Behler, J., Csányi, G., Ceriotti, M., Götz, A.W., Paesani, F., 2018. Comparison of permutationally invariant polynomials, neural networks, and Gaussian approximation potentials in representing water interactions through many-body expansions. J. Chem. Phys. 148 (24), 241725.

Olthof, E.H.T., van der Avoird, A., Wormer, P.E.S., Liu, K., Saykally, R.J., 1996. Tunneling dynamics, symmetry, and far-infrared spectrum of the rotating water trimer. II. Calculations and experiments. J. Chem. Phys. 105 (18), 8051.

Paesani, F., 2016. Getting the right answers for the right reasons: toward predictive molecular simulations of water with many-body potential energy functions. Acc. Chem. Res. 49 (9), 1844–1851.

Paul, J.B., Provencal, R.A., Saykally, R.J., 1998. Characterization of the $(D_2O)_2$ hydrogen-bond-acceptor antisymmetric stretch by IR cavity ringdown laser absorption spectroscopy. J. Phys. Chem. A 102 (19), 3279–3283.

Pérez, C., Lobsiger, S., Seifert, N.A., Zaleski, D.P., Temelso, B., Shields, G.C., Kisiel, Z., Pate, B.H., 2013. Broadband Fourier transform rotational spectroscopy for structure determination: the water heptamer. Chem. Phys. Lett. 571, 1–15.

Pérez, C., Muckle, M.T., Zaleski, D.P., Seifert, N.A., Temelso, B., Shields, G.C., Kisiel, Z., Pate, B.H., 2012. Structures of cage, prism, and book isomers of water hexamer from broadband rotational spectroscopy. Science 336 (6083), 897–901.

Pérez, C., Zaleski, D.P., Seifert, N.A., Temelso, B., Shields, G.C., Kisiel, Z., Pate, B.H., 2014. Hydrogen bond cooperativity and the three-dimensional structures of water nonamers and decamers. Angew. Chem., Int. Ed. 53 (52), 14368–14372.

Pinski, P., Csányi, G., 2013. Reactive many-body expansion for a protonated water cluster. J. Chem. Theory Comput. 10 (1), 68–75.

Press, W.H., Teukolsky, S.A., Vetterling, W.T., Flannery, B.P., 2007. Numerical Recipes: The Art of Scientific Computing, 3rd ed. Cambridge University Press, Cambridge.

Pugliano, N., Saykally, R.J., 1992. Measurement of quantum tunneling between chiral isomers of the cyclic water trimer. Science 257, 1937.

Quack, M., 1977. Detailed symmetry selection rules for reactive collisions. Mol. Phys. 34 (2), 477–504.

Quack, M., Seyfang, G., 2020. Atomic and molecular tunneling processes in chemistry. In: Marquardt, R., Quack, M. (Eds.), Molecular Spectroscopy and Quantum Dynamics. Elsevier, Amsterdam. Chapter 7 (this book).

Reddy, S.K., Straight, S.C., Bajaj, P., Huy Pham, C., Riera, M., Moberg, D.R., Morales, M.A., Knight, C., Götz, A.W., Paesani, F., 2016. On the accuracy of the MB-pol many-body potential for water: interaction energies, vibrational frequencies, and classical thermodynamic and dynamical properties from clusters to liquid water and ice. J. Chem. Phys. 145 (19), 194504.

Richardson, J.O., 2017. Full- and reduced-dimensionality instanton calculations of the tunneling splitting in the formic acid dimer. Phys. Chem. Chem. Phys. 19, 966–970.

Richardson, J.O., 2018a. Perspective: ring-polymer instanton theory. J. Chem. Phys. 148, 200901.

Richardson, J.O., 2018b. Ring-polymer instanton theory. Int. Rev. Phys. Chem. 37, 171.

Richardson, J.O., Althorpe, S.C., 2011. Ring-polymer instanton method for calculating tunneling splittings. J. Chem. Phys. 134, 054109.

Richardson, J.O., Althorpe, S.C., Wales, D.J., 2011. Instanton calculations of tunneling splittings for water dimer and trimer. J. Chem. Phys. 135, 124109.

Richardson, J.O., Pérez, C., Lobsiger, S., Reid, A.A., Temelso, B., Shields, G.C., Kisiel, Z., Wales, D.J., Pate, B.H., Althorpe, S.C., 2016. Concerted hydrogen-bond breaking by quantum tunneling in the water hexamer prism. Science 351, 1310–1313.

Richardson, J.O., Wales, D.J., Althorpe, S.C., McLaughlin, R.P., Viant, M.R., Shih, O., Saykally, R.J., 2013. Investigation of terahertz vibration-rotation tunneling spectra for the water octamer. J. Phys. Chem. A 117, 6960–6966.

Rommel, J.B., Kästner, J., 2011. Adaptive integration grids in instanton theory improve the numerical accuracy at low temperature. J. Chem. Phys. 134, 184107.

Sabo, D., Bačić, Z., Bürgi, T., Leutwyler, S., 1995. 3-dimensional model calculation of torsional levels of $(H_2O)_3$ and $(D_2O)_3$. Chem. Phys. Lett. 244, 283–294.

Saykally, R.J., Blake, G.A., 1993. Molecular interactions and hydrogen bond tunneling dynamics: some new perspectives. Science 259, 12.

Schütz, M., Bürgi, T., Leutwyler, S., Bürgi, H.B., 1993. Fluxionality and low-lying transition structures of the water trimer. J. Chem. Phys. 99 (7), 5228–5238.

Shank, A., Wang, Y., Kaledin, A., Braams, B.J., Bowman, J.M., 2009. Accurate *ab initio* and "hybrid" potential energy surfaces, intramolecular vibrational energies, and classical ir spectrum of the water dimer. J. Chem. Phys. 130, 144314.

Smith, B.J., Swanton, D.J., Pople, J.A., Schaefer III, H.F., Radom, L., 1990. Transition structures for the interchange of hydrogen atoms within the water dimer. J. Chem. Phys. 92 (2), 1240.

Stone, A., 2013. The Theory of Intermolecular Forces, 2nd ed. Oxford.

Suenram, R.D., Fraser, G.T., Lovas, F.J., 1989. Microwave spectrum of $(D_2O)_2$. J. Mol. Spectrosc. 138, 440.

Suhm, M.A., Watts, R.O., 1991. Quantum Monte Carlo studies of vibrational states in molecules and clusters. Phys. Rep. 204 (4), 293–329.

Suzuki, S., Blake, G.A., 1994. Pseudorotation in the D_2O trimer. Chem. Phys. Lett. 229, 499–505.

Szalewicz, K., Leforestier, C., van der Avoird, A., 2009. Towards the complete understanding of water by a first-principles computational approach. Chem. Phys. Lett. 482, 1–14.

Takahashi, M., Watanabe, Y., Taketsugu, T., Wales, D.J., 2005. An *ab initio* study of tunneling splittings in the water trimer. J. Chem. Phys. 123, 044302.

Taketsugu, T., Wales, D.J., 2002. Theoretical study of rearrangements in water dimer and trimer. Mol. Phys. 100 (17), 2793–2806.

Tautermann, C.S., Voegele, A.F., Loerting, T., Liedl, K.R., 2002. The optimal tunneling path for the proton transfer in malonaldehyde. J. Chem. Phys. 117 (5), 1962.

Temelso, B., Archer, K.A., Shields, G.C., 2011. Benchmark structures and binding energies of small water clusters with anharmonicity corrections. J. Phys. Chem. A 115 (43), 12034–12046.

Tsai, C.J., Jordan, K.D., 1993. Theoretical study of the $(H_2O)_6$ cluster. Chem. Phys. Lett. 213 (1), 181–188.

Tschumper, G.S., Leininger, M.L., Hoffman, B.C., Valeev, E.F., Schaefer III, H.F., Quack, M., 2002. Anchoring the water dimer potential energy surface with explicitly correlated computations and focal point analyses. J. Chem. Phys. 116 (2), 690–701.

Vaillant, C.L., Cvitaš, M.T., 2018. Rotation-tunneling spectrum of the water dimer from instanton theory. Phys. Chem. Chem. Phys. 20 (42), 26809–26813.

Vaillant, C.L., Wales, D.J., Althorpe, S.C., 2018. Tunneling splittings from path-integral molecular dynamics using a Langevin thermostat. J. Chem. Phys. 148 (23), 234102.

van der Avoird, A., Olthof, E.H.T., Wormer, P.E.S., 1996. Tunneling dynamics, symmetry, and far-infrared spectrum of the rotating water trimer. I. Hamiltonian and qualitative model. J. Chem. Phys. 105 (18), 8034–8050.

Vendrell, O., Gatti, F., Meyer, H.D., 2007. Dynamics and infrared spectroscopy of the protonated water dimer. Angew. Chem., Int. Ed. 46 (36), 6918–6921.

Viant, M.R., Cruzan, J.D., Lucas, D.D., Brown, M.G., Liu, K., Saykally, R.J., 1997. Pseudorotation in water trimer isotopomers using terahertz laser spectroscopy. J. Phys. Chem. A 101 (48), 9032–9041.

Wales, D.J., 1993a. Some estimates of tunneling splittings in small clusters. J. Am. Chem. Soc. 115 (24), 11191–11201.

Wales, D.J., 1993b. Theoretical study of water trimer. J. Am. Chem. Soc. 115 (24), 11180–11190.

Wales, D.J., 1999. Rearrangements of water dimer and hexamer. In: Jellinek, J. (Ed.), Theory of Atomic and Molecular Clusters: With a Glimpse at Experiments. Springer-Verlag, Berlin, pp. 86–110.

Wales, D.J., Walsh, T.R., 1996. Theoretical study of the water pentamer. J. Chem. Phys. 105 (16), 6957.

Wales, D.J., Walsh, T.R., 1997. Theoretical study of the water tetramer. J. Chem. Phys. 106, 7193–7207.

Walsh, T.R., Wales, D.J., 1996. Rearrangements of the water trimer. J. Chem. Soc. Faraday Trans. 92 (14), 2505–2517.

Wang, X.G., Carrington Jr, T., 2018. Using monomer vibrational wavefunctions to compute numerically exact (12D) rovibrational levels of water dimer. J. Chem. Phys. 148 (7), 074108.

Wang, Y., Babin, V., Bowman, J.M., Paesani, F., 2012. The water hexamer: cage, prism, or both. Full dimensional quantum simulations say both. J. Am. Chem. Soc. 134 (27), 11116–11119.

Wang, Y., Bowman, J.M., 2010. Towards an *ab initio* flexible potential for water, and post-harmonic quantum vibrational analysis of water clusters. Chem. Phys. Lett. 491, 1–10.

Wang, Y., Bowman, J.M., Huang, X., 2010. Communication: prediction of the rate constant of bimolecular hydrogen exchange in the water dimer using an *ab initio* potential energy surface. J. Chem. Phys. 133, 111103.

Wang, Y., Huang, X., Shepler, B.C., Braams, B.J., Bowman, J.M., 2011. Flexible, *ab initio* potential, and dipole moment surfaces for water. I. Tests and applications for clusters up to the 22-mer. J. Chem. Phys. 134, 094509.

Wang, Y., Shepler, B.C., Braams, B.J., Bowman, J.M., 2009. Full-dimensional, *ab initio* potential energy and dipole moment surfaces for water. J. Chem. Phys. 131, 054511.

Watanabe, Y., Taketsugu, T., Wales, D.J., 2004. An *ab initio* study of tunneling splittings in the water dimer. J. Chem. Phys. 120 (13), 5993.

Yagi, K., Mil'nikov, G.V., Taketsugu, T., Hirao, K., Nakamura, H., 2004. Effect of out-of-plane vibration on the hydrogen atom transfer reaction in malonaldehyde. Chem. Phys. Lett. 397 (4), 435–440.

Yu, Q., Bowman, J.M., 2016. How the Zundel ($H_5O_2^+$) potential can be used to predict the proton stretch and bend frequencies of larger protonated water clusters. J. Phys. Chem. Lett. 7 (24), 5259–5265.

Yu, Q., Bowman, J.M., 2017. High-level quantum calculations of the IR spectra of the Eigen, Zundel, and ring isomers of $H^+(H_2O)_4$ find a single match to experiment. J. Am. Chem. Soc. 139 (32), 10984–10987.

Zwart, E., ter Meulen, J.J., Meerts, W.L., Coudert, L.H., 1991. The submillimeter rotation tunneling spectrum of the water dimer. J. Mol. Spectrosc. 147, 27–39.

Author Index

Note: Page numbers followed by "*f*" indicate figures, and "*t*" indicate tables.

D

E

H

M

Q

Index

Printed in the United States
By Bookmasters